Glencoe
Mathematics
Applications and Connections

Course 3

Glencoe
McGraw-Hill

New York, New York   Columbus, Ohio   Woodland Hills, California   Peoria, Illinois

Visit the Glencoe Mathematics Internet Site for
Mathematics: Applications and Connections at

www.glencoe.com/sec/math/mac/mathnet

You'll find:

Chapter Review

Test Practice

Data Collection

Games

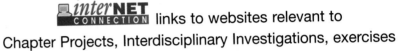 links to websites relevant to
Chapter Projects, Interdisciplinary Investigations, exercises

and much more!

About the Space Shuttle Hologram
The optimum viewing angle for the space shuttle hologram on the cover of this textbook is a 45° angle. For best results, view the hologram at this angle under a direct light source, such as sunlight or incandescent lighting.

Glencoe/McGraw-Hill
A Division of The McGraw-Hill Companies

Send all inquiries to:
Glencoe/McGraw-Hill
8787 Orion Place
Columbus, OH 43240-4027

ISBN: 0-07-822852-2

2 3 4 5 6 7 8 9 10 071/043 08 07 06 05 04 03 02 01 00

Dear Students, Teachers, and Parents,

Mathematics students are very special to us! That's why we wrote **Mathematics: Applications and Connections,** a math program designed specifically for you. The exciting, relevant content and up-to-date design will hold your interest and answer the question "When am I ever going to use this?"

As you page through your text, you'll notice the variety of ways math is presented for you. You'll see real-world applications as well as connections to other subjects like science, history, language arts, and music. You'll have opportunities to use technology tools such as the Internet, CD-ROM, graphing calculators, and computer applications like spreadsheets.

You'll appreciate the easy-to-follow lesson format. Each new concept is introduced with an interesting application or connection followed by clear explanations and examples. As you complete the exercises and solve interesting problems, you'll learn a great deal of useful math. You'll also have the opportunity to complete relevant Chapter Projects, Hands-On Labs, and Interdisciplinary Investigations. Test Practice, Test-Taking Tips, and Reading Math Study Hints will help you improve your test-taking skills.

Each day, as you use **Mathematics: Applications and Connections,** you'll see the practical value of math. You'll quickly grow to appreciate how often math is used in ways that relate directly to your life. If you don't already realize the importance of math in your life, you soon will!

Sincerely, The Authors

Kay M° Clain

Patricia S. Wilson

Patricia Frey

Linda Dritsas

Barbara Smith

Arthur Coulard

Jack Price

Jack M. Ott

Ron Pelfrey

Beatrice Moore-Harris

David Molina

Authors

William Collins
Director of The Sisyphus
Mathematics Learning Center
W. C. Overfelt High School
San Jose, CA

Linda Dritsas
District Coordinator
Fresno Unified School District
Fresno, CA

Patricia Frey
Mathematics Department
   Chairperson
Buffalo Academy for Visual
   And Performing Arts
Buffalo, NY

*"**Mathematics: Applications and Connections** helps students make the connection between mathematics and the real world. Applications lead students through the classroom door into the world of art, geography, science, and beyond."*—**Linda Dritsas**

Arthur C. Howard
Program Director for Secondary
   Mathematics
Aldine Independent School District
Houston, TX

Kay McClain
Lecturer
George Peabody College
Vanderbilt University
Nashville, TN

David Molina
Adjunct Professor of Mathematics
   Education
The University of Texas at Austin
Austin, TX

Beatrice Moore-Harris
Staff Development Specialist
Bureau of Education and
 Research
Houston, TX

Jack M. Ott
Distinguished Professor of
 Mathematics Education
University of South Carolina
Columbia, SC

Ronald Pelfrey
Mathematics Consultant
Lexington, KY

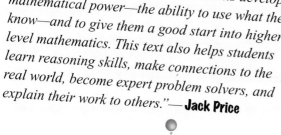

"*Mathematics: Applications and Connections* is designed to help middle school students develop mathematical power—the ability to use what they know—and to give them a good start into higher level mathematics. This text also helps students learn reasoning skills, make connections to the real world, become expert problem solvers, and explain their work to others."—**Jack Price**

"*The strongest focus of any middle school mathematics program should be on problem solving. **Mathematics: Applications and Connections** not only provides such a focus, but it makes the problem solving alive for students through its applications and connections."—**Ronald Pelfrey**

Jack Price
Professor, Mathematics
 Education
California State Polytechnic
 University
Pomona, CA

Barbara Smith
Mathematics Supervisor
Unionville-Chadds Ford
 School District
Kennett Square, PA

Patricia S. Wilson
Associate Professor of
 Mathematics Education
University of Georgia
Athens, GA

Academic Consultants and Teacher Reviewers

Each of the Academic Consultants read all 39 chapters in Courses 1, 2, and 3, while each Teacher Reviewer read two chapters. The Consultants and Reviewers gave suggestions for improving the Student Editions and the Teacher's Wraparound Editions.

ACADEMIC CONSULTANTS

Richie Berman, Ph.D.
Mathematics Lecturer and Supervisor
University of California, Santa Barbara
Santa Barbara, California

Robbie Bonneville
Mathematics Coordinator
La Joya Unified School District
Alamo, Texas

Cindy J. Boyd
Mathematics Teacher
Abilene High School
Abilene, Texas

Gail Burrill
Mathematics Teacher
Whitnall High School
Hales Corners, Wisconsin

Georgia Cobbs
Assistant Professor
The University of Montana
Missoula, Montana

Gilbert Cuevas
Professor of Mathematics Education
University of Miami
Coral Gables, Florida

David Foster
Mathematics Director
Robert Noyce Foundation
Palo Alto, California

Eva Gates
Independent Mathematics
    Consultant
Pearland, Texas

Berchie Gordon-Holliday
Mathematics/Science Coordinator
Northwest Local School District
Cincinnati, Ohio

Deborah Grabosky
Mathematics Teacher
Hillview Middle School
Whittier, California

Deborah Ann Haver
Principal
Great Bridge Middle School
Virginia Beach, Virginia

Carol E. Malloy
Assistant Professor, Math Education
The University of North Carolina,
    Chapel Hill
Chapel Hill, North Carolina

Daniel Marks, Ed.D.
Associate Professor of Mathematics
Auburn University at Montgomery
Montgomery, Alabama

Melissa McClure
Mathematics Consultant
Teaching for Tomorrow
Fort Worth, Texas

TEACHER REVIEWERS

Course 1

Carleen Alford
Math Department Head
Onslow W. Minnis, Sr. Middle School
Richmond, Virginia

Margaret L. Bangerter
Mathematics Coordinator K-6
St. Joseph School District
St. Joseph, Missouri

Diana F. Brock
Sixth and Seventh Grade Math Teacher
Memorial Parkway Junior High
Katy, Texas

Mary Burkholder
Mathematics Department Chair
Chambersburg Area Senior High
Chambersburg, Pennsylvania

Eileen M. Egan
Sixth Grade Teacher
Howard M. Phifer Middle School
Pennsauken, New Jersey

Melisa R. Grove
Sixth Grade Math Teacher
King Philip Middle School
West Hartford, Connecticut

David J. Hall
Teacher
Ben Franklin Middle School
Baltimore, Maryland

Ms. Karen T. Jamieson, B.A., M.Ed.
Teacher
Thurman White Middle School
Henderson, Nevada

David Lancaster
Teacher/Mathematics Coordinator
North Cumberland Middle School
Cumberland, Rhode Island

Jane A. Mahan
Sixth Grade Math Teacher
Helfrich Park Middle School
Evansville, Indiana

Margaret E. Martin
Mathematics Teacher
Powell Middle School
Powell, Tennessee

Diane Duggento Sawyer
Mathematics Department Chair
Exeter Area Junior High
Exeter, New Hampshire

Susan Uhrig
Teacher
Monroe Middle School
Columbus, Ohio

Cindy Webb
Title 1 Math Demonstration Teacher
Federal Programs LISD
Lubbock, Texas

Katherine A. Yule
Teacher
Los Alisos Intermediate School
Mission Viejo, California

Course 2

Sybil Y. Brown
Math Teacher Support Team-USI
Columbus Public Schools
Columbus, Ohio

Ruth Ann Bruny
Mathematics Teacher
Preston Junior High School
Fort Collins, Colorado

BonnieLee Gin
Junior High Teacher
St. Mary of the Woods
Chicago, Illinois

Larry J. Gonzales
Math Department Chair
Desert Ridge Middle School
Albuquerque, New Mexico

Susan Hertz
Mathematics Teacher
Revere Middle School
Houston, Texas

Rosalin McMullan
Mathematics Teacher
Honea Path Middle School
Honea Path, South Carolina

Mrs. Susan W. Palmer
Teacher
Fort Mill Middle School
Fort Mill, South Carolina

Donna J. Parish
Teacher
Zia Middle School
Las Cruces, New Mexico

Ronald J. Pischke
Mathematics Coordinator
St. Mary of the Woods
Chicago, Illinois

Sister Edward William Quinn I.H.M.
Chairperson Elementary Mathematics
Curriculum
Archdiocese of Philadelphia
Philadelphia, Pennsylvania

Marlyn G. Slater
Title 1 Math Specialist
Paradise Valley USD
Paradise Valley, Arizona

Sister Margaret Smith O.S.F.
Seventh and Eighth Grade Math Teacher
St. Mary's Elementary School
Lancaster, New York

Pamela Ann Summers
Coordinator, Secondary Math/Science
Lubbock ISD
Lubbock, Texas

Dora Swart
Teacher/Math Department Chair
W. F. West High School
Chehalis, Washington

Rosemary O'Brien Wisniewski
Middle School Math Chairperson
Arthur Slade Regional School
Glen Burnie, Maryland

Laura J. Young, Ed.D.
Eighth Grade Mathematics Teacher
Edwards Middle School
Conyers, Georgia

Susan Luckie Youngblood
Teacher/Math Department Chair
Weaver Middle School
Macon, Georgia

Course 3

Beth Murphy Anderson
Mathematics Department Chair
Brownell Talbot School
Omaha, Nebraska

David S. Bradley
Mathematics Teacher
Thomas Jefferson Junior High School
Salt Lake City, Utah

Sandy Brownell
Math Teacher/Team Leader
Los Alamos Middle School
Los Alamos, New Mexico

Eduardo Cancino
Mathematics Specialist
Education Service Center, Region One
Edinburg, Texas

Sharon Cichocki
Secondary Math Coordinator
Hamburg High School
Hamburg, New York

Nancy W. Crowther
Teacher, retired
Sandy Springs Middle School
Atlanta, Georgia

Charlene Mitchell DeRidder, Ph.D.
Mathematics Supervisor K-12
Knox County Schools
Knoxville, Tennessee

Ruth S. Garrard
Mathematics Teacher
Davidson Fine Arts School
Augusta, Georgia

Lolita Gerardo
Secondary Math Teacher
Pharr San Juan Alamo High School
San Juan, Texas

Donna Jorgensen
Teacher of Mathematics/Science
Toms River Intermediate East
Toms River, New Jersey

Statha Kline-Cherry, Ed.D.
Director of Elementary Education
University of Houston – Downtown
Houston, Texas

Charlotte Laverne Sykes Marvel
Mathematics Instructor
Bryant Junior High School
Bryant, Arkansas

Albert H. Mauthe, Ed.D.
Supervisor of Mathematics
Norristown Area School District
Norristown, Pennsylvania

Barbara Gluskin McCune
Teacher
East Middle School
Farmington, Michigan

Laurie D. Newton
Teacher
Crossler Middle School
Salem, Oregon

Indercio Abel Reyes
Mathematics Teacher
PSJA Memorial High School
Alamo, Texas

Fernando Rosa
Mathematics Department Chair
Edinburg High School
Edinburg, Texas

Mary Ambriz Soto
Mathematics Coordinator
PSJA I.S.D.
Pharr, Texas

Judy L. Thompson
Eighth Grade Mathematics Teacher
Adams Middle School
North Platte, Nebraska

Karen A. Universal
Eighth Grade Mathematics Teacher
Cassadaga Valley Central School
Sinclairville, New York

Tommie L. Walsh
Teacher
S. Wilson Junior High School
Lubbock, Texas

Marcia K. Ziegler
Mathematics Teacher
Pharr-San Juan- Alamo North High School
Pharr, Texas

Student Advisory Board

The Student Advisory Board gave the editorial staff and design team feedback on the design, content, and covers of the Student Editions. We thank these students from Crestview Middle School in Columbus, Ohio, and McCord Middle School in Worthington, Ohio, for their hard work and creative suggestions in making *Mathematics: Applications and Connections* more student friendly.

Mara Flood

Anna Hoza

Neil Gardner

Lareva Summerall

Mike Leonelli

Chris Franklin

Kelly McLoughlin

Jennifer Shoemaker

Maghee Disch

CHAPTER 1

Problem Solving and Algebra

Applications, Connections, and
Integration Index, pages xxii–1.

Let the Games Begin
- You're the Greatest, **47**

SCHOOL to CAREER
- Technology, **26**

MATH IN THE MEDIA
- Fill It Up!, **20**

interNET CONNECTION
- Chapter Project, **3**
- Data Update, **15**
- School to Career, **26**
- Let the Games Begin, **47**
- Chapter Review, **48**
- Test Practice, **53**

Standardized Test Practice
10, 15, 20, 25, 29, 31, 36, 42, 47,
52–53

Algebra: Using Integers

Let the Games Begin

- Absolutely!, **61**
- Matrix Madness, **76**

interNET CONNECTION

Standardized Test Practice

MATH IN THE MEDIA

Using Proportion and Percent

Applications, Connections, and
Integration Index, pages xxii–1.

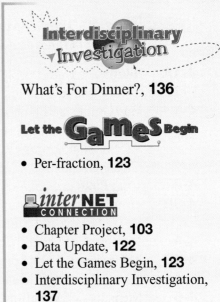

Statistics: Analyzing Data

- You're the Winner...Bar None!, **146**

- History, **152**

- Computer Comparisons, **151**

inter NET
CONNECTION

Standardized Test Practice

Geometry: Investigating Patterns

Let the Games Begin

• Shape Relations, **218**

SCHOOL to CAREER

• Advertising, **219**

interNET CONNECTION

Standardized Test Practice

Applications, Connections, and
Integration Index, pages xxii–1.

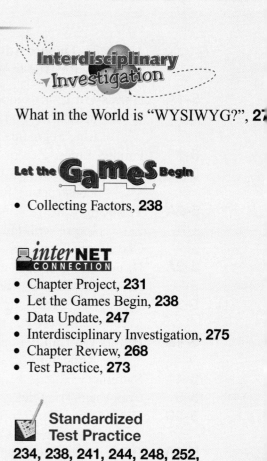

Interdisciplinary Investigation

What in the World is "WYSIWYG?", 2?

Let the Games Begin

interNET CONNECTION

Standardized Test Practice

MATH IN THE MEDIA

Algebra: Using Rational Numbers

Applications, Connections, and
Integration Index, pages xxii–1.

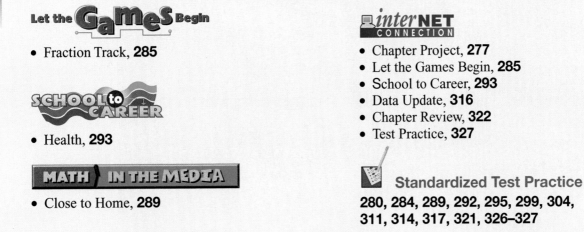

Let the Games Begin

• Fraction Track, **285**

SCHOOL to CAREER

• Health, **293**

MATH IN THE MEDIA

• Close to Home, **289**

interNET CONNECTION

• Chapter Project, **277**
• Let the Games Begin, **285**
• School to Career, **293**
• Data Update, **316**
• Chapter Review, **322**
• Test Practice, **327**

Standardized Test Practice
280, 284, 289, 292, 295, 299, 304,
311, 314, 317, 321, 326–327

CHAPTER 8
Applying Proportional Reasoning

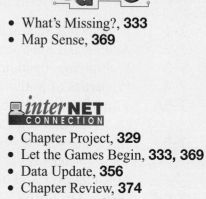

Let the **Games** Begin

- What's Missing?, **333**
- Map Sense, **369**

interNET
C O N N E C T I O N

**Standardized
Test Practice**

- B.C., **364**

Algebra: Exploring Real Numbers

Applications, Connections, and
Integration Index, pages xxii–1.

Interdisciplinary Investigation

Math at the Mall, **424**

Let the Games Begin

• Estimate and Eliminate, **389**

SCHOOL to CAREER

• Law Enforcement, **395**

interNET CONNECTION

• Chapter Project, **381**
• Let the Games Begin, **389**
• Data Update, **394**
• School to Career, **395**
• Interdisciplinary Investigation, **425**
• Chapter Review, **418**
• Test Practice, **423**

Standardized Test Practice
384, 389, 394, 401, 403, 407, 413, 417, 422–423

Algebra: Graphing Functions

- Guess My Rule, **431**
- Parabola Hit or Miss, **455**

- Computer Programming, **436**

- Herman, **440**

interNET
CONNECTION

- Chapter Project, **427**
- Data Update, **430**
- Let the Games Begin, **431, 455**
- School to Career, **436**
- Chapter Review, **468**
- Test Practice, **473**

Standardized Test Practice
431, 435, 440, 444, 449, 451, 455,
459, 463, 467, 472–473

CHAPTER 11

Geometry: Using Area and Volume

Let the **GAMES** **Begin**

- Architest, **498**

interNET CONNECTION

Standardized Test Practice

MATH IN THE MEDIA

CHAPTER 12
Investigating Discrete Math and Probability

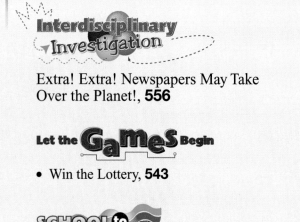

Interdisciplinary Investigation

Extra! Extra! Newspapers May Take
Over the Planet!, **556**

Let the Games Begin

• Win the Lottery, **543**

• Health, **549**

13 Algebra: Exploring Polynomials

Let the **GaMeS** Begin

- Factor Challenge, **591**

SCHOOL to CAREER

- Sports Management, **582**

interNET
CONNECTION

- Chapter Project, **559**
- Data Update, **561**
- School to Career, **582**
- Let the Games Begin, **591**
- Chapter Review, **592**
- Test Practice, **597**

Standardized Test Practice
564, 569, 572, 576, 581, 585, 587, 591, 596–597

Applications, Connections, and Integration Index

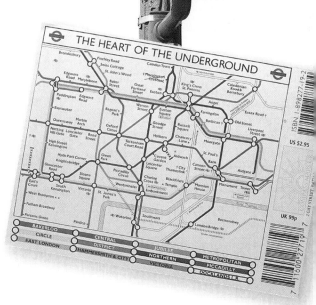

Snow on December 25?

Boston, MA		23%
Cheyenne, WY		35%
Denver, CO		42%
Detroit, MI		50%
Minneapolis, MN		73%
Seattle, WA		5%

Source: Northeast Regional Climate Center at Cornell University

Problem Solving and Algebra

**What** you'll
learn in Chapter 1

- to solve problems by using the four-step problem solving plan,
- to evaluate numerical and algebraic expressions,
- to write and solve equations,
- to solve problems by working backward,
- to find the areas and perimeters of rectangles, squares, and parallelograms, and
- to identify and solve inequalities.

CHAPTER Project

HOME PAGE BOUND

The Internet has changed the way we communicate. One way we can communicate to others about ourselves is through a home page on the World Wide Web. Software is available that makes designing a web page a fairly simple task. In this project, you will design your own home page for the World Wide Web and organize your results into a report.

Getting Started

- Find out about software that can be used to design a home page that is available for your computer. Decide on the software you will use.

- Learn how to use the software. Be sure to follow any instructions carefully. Experiment to learn how to best use the capabilities of your software. Work towards designing a page that is attractive and one to which people will be attracted.

- Write a report about what you learned about designing a home page and how you designed your page.

Technology Tips

- Use **graphing software, paint programs,** or **photo editing software** to create graphics and to design your home page.

- Use a **word processor** to write your report.

- Surf the **Internet** to study other home pages.

interNET CONNECTION

Research **For more information on creating a home page, visit:**

www.glencoe.com/sec/math/mac/mathnet

Working on the Project

You can use what you'll learn in Chapter 1 to help you design your own home page.

Page	Exercise
7	7
25	34
41	21
51	Alternative Assessment

What you'll learn

You will learn to solve problems by using the four-step plan.

When am I ever going to use this?

You will graph data in science class and look for a pattern in the graph to help analyze the data.

The London Underground in England is the oldest subway system in the world. The first lines in the system opened in 1863. Now the network contains 272 stations and 267 miles of track!

Part of the map displayed in the Underground stations is shown. Each interchange is shown as a circle or connected circles, but the map is not drawn to scale. How can you get from Bond Street to Tower Hill with the fewest train changes?

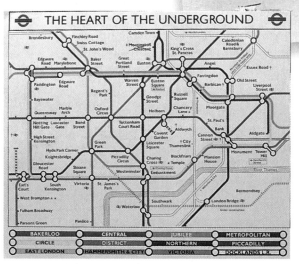

THE HEART OF THE UNDERGROUND

Where would you begin to solve this problem? In mathematics, we have a four-step plan to solve problems like this.

1. *Explore* Determine what information is given in the problem and what you need to find. Do you have all the information you need to solve the problem? Is there too much information?

2. *Plan* After you understand the problem, select a strategy for solving it. There may be several strategies that you can use. It is usually helpful to make an estimate of what you think the answer should be.

3. *Solve* Solve the problem by carrying out your plan. If your plan doesn't work, try another, and maybe even another.

4. *Examine* Finally, examine your answer carefully. See if it fits the facts given in the problem. Compare it to your estimate. You may also want to check your answer by solving the problem again in a different way. If the answer is not reasonable, make a new plan and start again.

Refer to the beginning of the lesson. Find the route from Bond Street to Tower Hill with the fewest train changes.

Explore   *What do you know?*
There are several lines that run through the city in the area of Bond Street and Tower Hill.

What are you trying to find?
A route from Bond Street to Tower Hill with the fewest train changes.

Plan   You could find every route between Bond Street and Tower Hill. However, since you could change trains in many interchanges to get on and off of different lines, this method would take a great deal of time. Another alternative is to find a route and then look for one with fewer train changes until no more changes can be eliminated.

Solve   There are two lines that run through the Bond Street station – Jubilee and Central. Look for the route with the fewest train changes using Jubilee. Then look for the route with the fewest train changes using Central and compare.

Jubilee

Using the Jubilee line, one possible route is Bond Street – Green Park – Victoria – Tower Hill. This involves two train changes.

Another option on the Jubilee line is Bond Street – Charring Cross – Embankment – Tower Hill, which also has two train changes.

There are no routes that use fewer than two train changes using the Jubilee line.

Central

On the Central line, a possible route is Bond Street – Bank – Tower Hill. This route uses only one train change.

There are no routes on the Central line that use fewer than one train change.

So, the route from Bond Street to Tower Hill with the fewest train changes is Bond Street – Bank – Tower Hill.

Examine   Is your answer reasonable? Only a route using one line has fewer train changes than the one found. Since there is no line that stops at both Bond Street and Tower Hill, the best possible route must have at least one train change.

Throughout this textbook, you will be solving many kinds of problems. Some can be solved easily by adding, subtracting, multiplying, or dividing. Others can be solved by using a strategy like finding a pattern, solving a simpler problem, making a model, drawing a graph, and so on. No matter which strategy you use, you can always use the four-step plan to solve a problem.

② **Medicine** The table shows how the amount of a medicine in the bloodstream is related to the time since it was taken. If the doctor wants the concentration to be at least 0.30 mg/L all of the time, how often should the medicine be taken?

Time (h)	Concentration (mg/L)
0	0
1	0.60
2	0.75
3	0.69
4	0.60
5	0.52
6	0.45

Explore   You know the concentrations for 0 to 6 hours after the medication is taken. You need to predict when the concentration reaches 0.30 mg/L.

Plan   One good way to find a pattern in a set of data is to show the data on a graph. In this case, make a graph that shows time on the horizontal axis and the concentration on the vertical axis. Then look for a pattern in the graph.

Solve

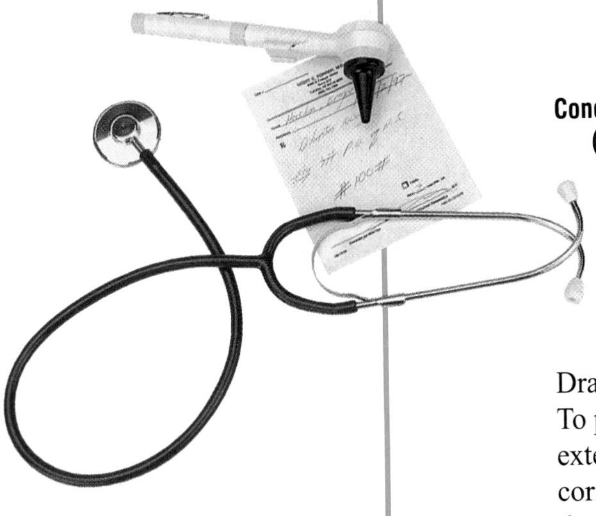

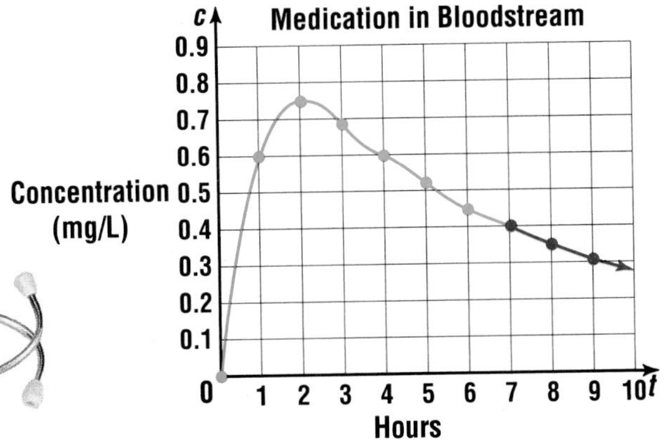

Medication in Bloodstream

Draw a smooth curve through the points on the graph. To predict when the concentration will reach 0.30 mg/L, extend the curve. The extended line is shown in red. This corresponds to about 9 on the horizontal axis. Therefore, the medicine should be taken about every 9 hours.

Examine   Look for another pattern in the data. The time 9 hours seems reasonable.

Time (h)	Concentration (mg/L)	
4	0.60	⟩ 0.08
5	0.52	⟩ 0.07
6	0.45	⟩ 0.06
7	0.39	⟩ 0.05
8	0.34	⟩ 0.04
9	0.30	

CHECK FOR UNDERSTANDING

Communicating Mathematics

Read and study the lesson to answer each question.

1. *Tell* what each step in the four-step plan means.

2. *Explain* what to do when your plan to solve the problem doesn't work.

Math

Journal

3. *Write* two or three sentences in your journal that describe what you expect to learn in this course.

Guided Practice

Use the four-step plan to solve each problem.

4. *Physical Science* A chemist pours sodium chloride, or table salt, into a beaker. If the beaker plus the sodium chloride have a mass of 84.8 grams, and the beaker itself has a mass of 63.3 grams, what is the mass of the sodium chloride that was poured into the beaker?

5. *Transportation* Refer to the beginning of the lesson. Find a route from Oxford Circus to Old Street using the fewest train changes.

EXERCISES

Applications and Problem Solving

Use the four-step plan to solve each problem.

6. *Look for a Pattern* The table shows the cost of mailing a first-class letter.
   a. Make a graph of the data.
   b. Predict the cost of mailing a letter that weighs 6 ounces.

Weight (ounces)	Cost
1	$0.33
2	$0.55
3	$0.77
4	$0.99
5	$1.21

7. *Working on the* **CHAPTER Project**
   Designing a home page is like solving a problem. Use the plan for problem solving to help you devise a plan for your home page. Record your plan and follow it as you work on the Chapter Project.

8. *Television* The graph shows the number of prime-time television viewers in millions for each night of the week.
   a. List the nights of the week from most viewers to least viewers.
   b. Estimate the total number of viewers for the entire week.
   c. About how many more viewers watch television on Thursday than on Friday?

Prime-time Viewers (millions)

Monday	91.9
Tuesday	89.8
Wednesday	90.6
Thursday	93.9
Friday	78.0
Saturday	77.1
Sunday	87.7

Source: Nielsen Media Research

9. *Use a Model* Suppose you had 100 sugar cubes. What is the largest cube you could build with the sugar cubes?

10. *Technology* Armando used his calculator to find $7,129 + 6,859 + 7,523 + 6,792 + 6,928$. How can he tell if the answer shown is reasonable?

    7129 [+] 6859 [+] 7523 [+] 6792 [+] 6928 [=] *35231*

11. *Critical Thinking* Replace each ■ in the figure at the right so that the total of the numbers along each side is the same.

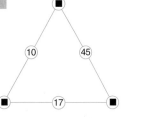

For **Extra Practice,** see page 605.

1-2 Powers and Exponents

What you'll learn

You'll learn to use powers and exponents in expressions.

When am I ever going to use this?

You can use exponents to determine how much money will be in your savings account after time.

Word Wise

exponent
factor
power
base
associative property
commutative property
evaluate

In a recent movie, Walter Matthau portrays the famous physicist Albert Einstein. He tries to set up his niece, mathematician Catherine Boyd, played by Meg Ryan, with Ed Walters, an auto mechanic played by Tim Robbins. When Ed asks Catherine "How many stars do you think there are in the sky?" she replies "10 to the twelfth plus 1."

Catherine's answer uses an **exponent**. An exponent is a way of writing a repeated multiplication in a simple way.

When two or more numbers are multiplied, these numbers are called **factors**. When the same factor is repeated, you may use an exponent.

$$32 = 2 \cdot 2 \cdot 2 \cdot 2 \cdot 2 \rightarrow 2^5 \quad \textit{5 is the exponent.}$$

An expression like 2^5 is called a **power** of 2. The 2 in this expression is called the **base**. The base names the factor being repeated.

Example

1 Write the expression $3 \cdot 3 \cdot 3 \cdot 3$ using exponents.

There are four factors of 3.

$$3 \cdot 3 \cdot 3 \cdot 3 = 3^4$$

> **Study Hint**
>
> Reading Math To read a power, say the base and then the exponent. For example, 3^4 is read as 3 to the fourth power.

You can use these properties to help write expressions involving exponents.

Property	Arithmetic	Algebra
Associative	$3 + (4 + 10) = (3 + 4) + 10$	$a + (b + c) = (a + b) + c$
	$5 \cdot (2 \cdot 12) = (5 \cdot 2) \cdot 12$	$a \cdot (b \cdot c) = (a \cdot b) \cdot c$
Commutative	$1 + 2 = 2 + 1$	$a + b = b + a$
	$3 \cdot 2 = 2 \cdot 3$	$a \cdot b = b \cdot a$

Examples

Write each expression using exponents.

2 $2 \cdot 2 \cdot 5 \cdot 5 \cdot 7 \cdot 7 \cdot 7$

Use the associative property to group the factors that are the same.

$2 \cdot 2 \cdot 5 \cdot 5 \cdot 7 \cdot 7 \cdot 7$

$= (2 \cdot 2) \cdot (5 \cdot 5) \cdot (7 \cdot 7 \cdot 7)$

$= 2^2 \cdot 5^2 \cdot 7^3$

3 $3 \cdot 4 \cdot 3 \cdot 3 \cdot 4 \cdot 6 \cdot 6$

Use the commutative property to rearrange the factors. Then use the associative property to group them.

$3 \cdot 4 \cdot 3 \cdot 3 \cdot 4 \cdot 6 \cdot 6$

$= 3 \cdot 3 \cdot 3 \cdot 4 \cdot 4 \cdot 6 \cdot 6$

$= (3 \cdot 3 \cdot 3) \cdot (4 \cdot 4) \cdot (6 \cdot 6)$

$= 3^3 \cdot 4^2 \cdot 6^2$

When you find the value of a power, you are **evaluating** the power. The $\boxed{y^x}$ key on your calculator allows you to evaluate one or more powers.

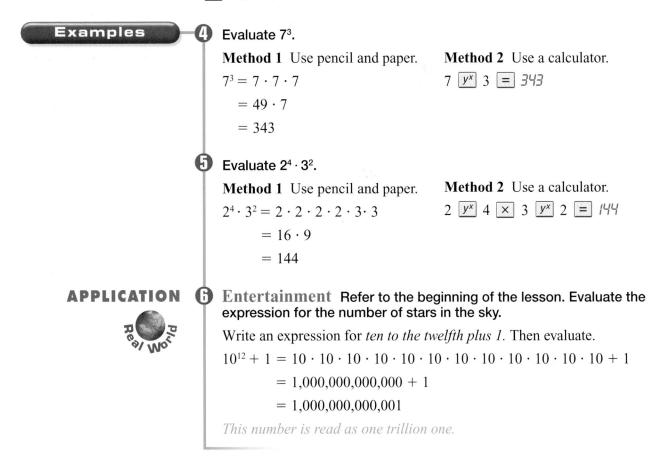

Examples

4 Evaluate 7^3.

Method 1 Use pencil and paper.

$7^3 = 7 \cdot 7 \cdot 7$

$\quad = 49 \cdot 7$

$\quad = 343$

Method 2 Use a calculator.

7 $\boxed{y^x}$ 3 $\boxed{=}$ *343*

5 Evaluate $2^4 \cdot 3^2$.

Method 1 Use pencil and paper.

$2^4 \cdot 3^2 = 2 \cdot 2 \cdot 2 \cdot 2 \cdot 3 \cdot 3$

$\quad = 16 \cdot 9$

$\quad = 144$

Method 2 Use a calculator.

2 $\boxed{y^x}$ 4 $\boxed{\times}$ 3 $\boxed{y^x}$ 2 $\boxed{=}$ *144*

APPLICATION

6 **Entertainment** Refer to the beginning of the lesson. Evaluate the expression for the number of stars in the sky.

Write an expression for *ten to the twelfth plus 1*. Then evaluate.

$10^{12} + 1 = 10 \cdot 10 \cdot 10 \cdot 10 \cdot 10 \cdot 10 \cdot 10 \cdot 10 \cdot 10 \cdot 10 \cdot 10 \cdot 10 + 1$

$\qquad = 1{,}000{,}000{,}000{,}000 + 1$

$\qquad = 1{,}000{,}000{,}000{,}001$

This number is read as one trillion one.

CHECK FOR UNDERSTANDING

Communicating Mathematics

Read and study the lesson to answer each question.

1. *Tell* what the 3 represents in 2^3.

2. *Express* *four to the 5th power* using exponents.

3. *Write* a few sentences about why scientists working with very large numbers often use exponents.

Guided Practice

Write each expression using exponents.

4. $7 \cdot 7 \cdot 7$ **5.** $16 \cdot 16 \cdot 20 \cdot 20$ **6.** $2 \cdot 3 \cdot 3 \cdot 2 \cdot 2 \cdot 7$

Evaluate each expression.

7. 4^3 **8.** 9^2 **9.** $6^2 \cdot 2^3$ **10.** $7 \cdot 6^3 \cdot 10^3$

11. *Astronomy* The average distance from Jupiter to the Sun is 4.84×10^8 miles.

a. Write an expression for 10^8.

b. Evaluate 4.84×10^8.

Practice

Write each expression using exponents.

12. $6 \cdot 6 \cdot 6 \cdot 6$

13. $10 \cdot 10$

14. $8 \cdot 8 \cdot 8 \cdot 8 \cdot 8 \cdot 8$

15. $4 \cdot 4 \cdot 4 \cdot 8 \cdot 8$

16. $2 \cdot 2 \cdot 2 \cdot 5 \cdot 5$

17. $12 \cdot 14 \cdot 14 \cdot 5 \cdot 5 \cdot 5$

18. $5 \cdot 5 \cdot 4 \cdot 2 \cdot 5 \cdot 4$

19. $18 \cdot 5 \cdot 5 \cdot 5 \cdot 18$

20. $a \cdot b \cdot b \cdot a \cdot a \cdot b$

Evaluate each expression.

21. 6^4

22. 8^3

23. 1^{10}

24. 10^1

25. $4^2 \cdot 5^3$

26. $5^2 \cdot 8^2 \cdot 3^3$

27. $1,000^2$

28. 16^4

29. $7^2 - 3^3$

30. $2 \cdot 2^2 \cdot 2^3$

31. $3 \cdot 2^2 \cdot 4^3$

32. $9 \cdot 6^2 \cdot 2 \cdot 3^3$

33. Write the product $10 \cdot 10 \cdot 10 \cdot 24 \cdot 24 \cdot 50$ using exponents.

34. Evaluate $2 \cdot 2^3 \cdot 3^2$.

Applications and Problem Solving

35. *Geometry* To find the volume of a cube, multiply its length, its width, and its depth.

   a. Write an expression for the volume of a cube whose edges are each 12 centimeters long.

   b. If each edge of a cube measures 2.3 meters, what is the volume of the cube?

36. *Shipping* The busiest port in the United States is the Port of South Louisiana. Recently, it handled 1.93×10^8 tons of cargo. Express this number without exponents.

37. *Life Science* Every person has 2 biological parents. Each of their parents has 2 parents, so every person has $2 \cdot 2$ or 4 grandparents, and so on. How many great-great grandparents does every person have?

38. *Critical Thinking* Write each of the following as a power of 10 or the product of a whole number and a power of 10.

   a. 100,000

   b. fifty million

   c. 3,000,000,000

   d. sixty thousand

39. *Money Matters* At the school bookstore, a pen costs $0.28, and a small writing tablet costs $0.23. What combination of pens and tablets could you buy for exactly $0.74? *(Lesson 1-1)*

Mixed Review

40. **Standardized Test Practice** Shelby selected 6 items at the craft store that ranged in price from $3.97 to $5.31. Which is a reasonable total for her purchases? *(Lesson 1-1)*

   A less than $20

   B between $20 and $25

   C between $25 and $30

   D between $30 and $35

   E more than $35

For **Extra Practice**, see page 605.

1-3 Variables, Expressions, and Equations

What you'll learn

You'll learn to evaluate expressions and to find the solutions of equations.

When am I ever going to use this?

You can use equations to enter information into a spreadsheet.

Word Wise

numerical expression
substitute
order of operations
variable
algebraic expression
equation
open sentence
solution
replacement set

Study Hint

Reading Math
Grouping symbols include:
- parentheses (),
- brackets [], and
- fraction bars, as in $\frac{5+3}{2}$, which means $(5+3) \div 2$.

The Chicago Bulls became the first team in the history of the National Basketball Association to win 72 games. Michael Jordan led the Bulls in their record-breaking season, but did you know that Toni Kukoc was the NBA's sixth man award winner? The sixth man award is given to the league's best substitute player.

In mathematics, the concept of substitution is just as important as it is in sports. Consider the **numerical expression** $8 + 4$. It has a value of 12. However, the expression $x + 4$ does not have a value until a value for x is given. Suppose you let $x = 13$. To find the value of the expression, you must **substitute** 13 for x, or put 13 in place of x, in the expression. It becomes the numerical expression $13 + 4$, which has a value of 17. Therefore, if $x = 13$, then $x + 4 = 17$.

When you evaluate a numerical expression, you need to follow an order of operations. That is, you need to know which operations to do first when there is more than one operation in the expression. The following rules are used when evaluating numerical expressions.

Order of Operations	1. Do all operations within grouping symbols first; start with the innermost grouping symbols. 2. Do all powers before other operations. 3. Multiply and divide in order from left to right. 4. Add and subtract in order from left to right.

Example

1 Evaluate $(9 + 6) \div 5 \times 4 + (2^3 - 3)$.

Follow the order of operations. Do operations inside the parentheses first.

$(9 + 6) \div 5 \times 4 + (2^3 - 3)$

$= 15 \div 5 \times 4 + (8 - 3)$ *$9 + 6 = 15$ and $2^3 = 8$*

$= 15 \div 5 \times 4 + 5$ *$8 - 3 = 5$*

$= 3 \times 4 + 5$ *$15 \div 5 = 3$*

$= 12 + 5$ *$3 \times 4 = 12$*

$= 17$

Algebra is a language of symbols. In algebra, letters, called **variables**, are used to represent unknown quantities. In the expression $x + 6$, x is a variable. Expressions that contain variables are called **algebraic expressions**. In order to evaluate numerical and algebraic expressions, you must know how to read algebraic expressions.

$3(2)$ $\boxed{\text{means}}$ 3×2

$5x$ $\boxed{\text{means}}$ $5 \times x$

xy $\boxed{\text{means}}$ $x \times y$

$4 \cdot 7a$ $\boxed{\text{means}}$ $4 \times 7 \times a$

$8ab^2$ $\boxed{\text{means}}$ $8 \times a \times b \times b$

$a[b(cd)]$ $\boxed{\text{means}}$ $a \times [b \times (c \times d)]$

$\dfrac{x}{10z}$ $\boxed{\text{means}}$ $x \div (10 \times z)$

$a\left(\dfrac{b}{3}\right)$ $\boxed{\text{means}}$ $a \times (b \div 3)$

Examples

2 Evaluate $a - b + 7$ if $a = 15$ and $b = 9$.

Begin by using substitution. Replace each variable in the expression with its value. Then use the order of operations.

$$\begin{aligned}
a - b + 7 &= 15 - 9 + 7 & \textit{Replace a with 15 and b with 9.} \\
&= 6 + 7 & \textit{Follow the order of operations.} \\
&= 13 & \textit{Check your answer mentally.}
\end{aligned}$$

3 Evaluate $3a + 4b$ if $a = 5$ and $b = 17$.

$$\begin{aligned}
3a + 4b &= 3(5) + 4(17) & \textit{Replace a with 5 and b with 17.} \\
&= 15 + 68 & \textit{Do multiplication before addition.} \\
&= 83 & \textit{Add.}
\end{aligned}$$

4 Evaluate $\dfrac{m^3}{2n}$ if $m = 6$ and $n = 9$.

The fraction bar, which means division, is also a grouping symbol. Evaluate the expressions in the numerator and denominator separately before dividing.

$$\begin{aligned}
\frac{m^3}{2n} &= \frac{6^3}{2 \cdot 9} & \textit{Replace m with 6 and n with 9.} \\
&= \frac{216}{18} & \textit{Evaluate the numerator and the denominator separately.} \\
&= 216 \div 18 & \textit{Then divide.} \\
&= 12
\end{aligned}$$

A mathematical sentence that contains an "=" is called an **equation**. Some examples of equations are $5 + 9 = 14$, $12(3) = 36$, and $8 - (4 \div 2) = 6$. An equation that contains a variable is an **open sentence**. When a number is substituted for the variable in an open sentence, the sentence may be true or false.

Equation: $34 + 16 = c$

Replace c with 40. → $34 + 16 = 40$
This equation is false.

Replace c with 50. → $34 + 16 = 50$
This equation is true.

The values of the variable that make the equation true are called the **solutions** of the equation. In the equation $34 + 16 = c$, the solution is 50. The process of finding a solution is called *solving the equation*.

 **Example 5**

Find the solution of $82 + a = 96$ if the value of a can be selected from the set {10, 12, 14}.

Try each value from the solution set to see which is a solution.

Try 10.	Try 12.	Try 14.
$82 + a = 96$	$82 + a = 96$	$82 + a = 96$
$82 + 10 \stackrel{?}{=} 96$	$82 + 12 \stackrel{?}{=} 96$	$82 + 14 \stackrel{?}{=} 96$
$92 = 96$ *false*	$94 = 96$ *false*	$96 = 96$ *true*

14 is the solution of $82 + a = 96$.

When you are given a set of numbers from which to choose the value of the variable, the set is called the **replacement set** for the equation.

Example 6
Real World APPLICATION

Technology E-mail is the most widely used and rapidly growing online activity among Internet subscribers. Suppose an e-mail message contains 72,000 bytes, and 24,000 of this is used to include a picture. How many bytes of the message were used to write the text? Use the equation $24,000 + t = 72,000$ and the replacement set {48,000, 52,000, 56,000}.

$24,000 + t = 72,000$

$24,000 + 48,000 \stackrel{?}{=} 72,000$ *Replace t with 48,000.*

$72,000 = 72,000$ *true*

The solution is 48,000. The text uses 48,000 bytes.

Communicating Mathematics

Read and study the lesson to answer each question.

1. *Tell* the difference between a numerical expression and an algebraic expression.

2. *Write* a definition for a replacement set. Explain how the replacement set is related to the solution.

Guided Practice

Evaluate each expression.

3. $17 + 2 \cdot 8$

4. $3(7) - 4 \div 2$

5. $\frac{36}{3^2 - 3}$

Evaluate each expression if $a = 5$, $b = 6$, $c = 3$, and $d = 4$.

6. $2a + 3b - 4c$

7. $bdc \div 12$

8. $(3a + 2c)d$

Find the solution for each equation from the given replacement set.

9. $x = 16 + 28$, $\{42, 44, 46\}$

10. $2.30 - y = 0.50$, $\{1.40, 0.90, 1.80\}$

11. $4m = 60$, $\{20, 15, 10\}$

12. $15 = \frac{480}{p}$, $\{32, 30, 28\}$

13. *Transportation* There were 7,000 miles of concrete roads in the United States in 1917. Just ten years later in 1927, a network of 50,000 miles of roads had been built. Find the number of miles of concrete roads built between 1917 and 1927. Use the equation $7,000 + b = 50,000$ and the replacement set $\{41,000, 43,000, 45,000\}$.

Practice

Evaluate each expression.

14. $(3 + 6)^2$

15. $5 \cdot (4 + 2^2)$

16. $(8 - 3)^2 + \frac{8 \cdot 3}{12}$

17. $27 \div 9 + 5$

18. $(5^2 + 3) \div 7$

19. $[(6 + 5)2] \div 11$

20. $4^2 - 2 \cdot 5 + (8 - 2)$

21. $\frac{28}{4^2 - 2}$

22. $3 \cdot (18 - 7) + \frac{25 - 9}{8}$

Evaluate each expression if $w = 4$, $x = 7$, $y = 6$, and $z = 3$.

23. $3x + 4y - 2w$

24. $xyz \div 21$

25. zw^2

26. $(zw)^2$

27. $(3y + 2z) \cdot w$

28. $(2z \cdot w) + 3y$

29. $\frac{3yz}{w}$

30. $\frac{x^2}{2z + 1}$

31. $\frac{(wz)^2}{y + 10}$

Find the solution for each equation from the given replacement set.

32. $r + 35 = 80$, $\{35, 40, 45, 50\}$

33. $216 = 127 + n$, $\{81, 89, 91, 99\}$

34. $w - 18 = 62$, $\{80, 85, 90, 95\}$

35. $4x = 124$, $\{29, 31, 33, 35\}$

36. $108 = 6h$, $\{20, 18, 16, 14\}$

37. $\frac{576}{w} = 72$, $\{4, 6, 8, 12\}$

38. $2x - 6 = 18$, $\{8, 10, 12, 14\}$

39. $9 = 25 - 4a$, $\{0, 1, 3, 4\}$

40. Solve $d = \$7.00 - \$2.33$ if the replacement set is $\{\$5.67, \$4.67, \$4.33\}$.

41. Find the solution for $6m = 42$. Use the replacement set $\{6, 7, 8, 9\}$.

Applications and Problem Solving

For the latest information on drive-in theaters, visit:

www.glencoe.com/sec/ math/mac/mathnet

42. *Entertainment* Today there are 2,170 fewer drive-in theaters in the United States than there were in the 1980s. Find the number of drive-in theaters now if there were approximately 3,000 drive-in theaters in the 1980s. Use the equation $3{,}000 - 2{,}170 = x$ to represent the situation.

　a. What does the variable x represent?

　b. Use the replacement set $\{830, 850, 880, 930\}$ with the equation to find the approximate number of drive-in theaters that remain.

Carol Moseley Braun

43. *Politics* In 1992, Carol Moseley Braun became the first African-American woman elected to the U.S. Senate. Shirley Chisholm had become the first African-American woman to enter the U.S. House of Representatives in 1968. The equation $1968 + y = 1992$ can be used to represent the situation.

　a. What does the variable y represent?

　b. Use the replacement set $\{19, 24, 29, 34\}$ with the equation to find the number of years between Ms. Chisholm's and Ms. Braun's elections.

44. *Critical Thinking* Write two different open sentences whose solution is 4.

Shirley Chisholm

Mixed Review

45. Find the value of the expression 2^6. *(Lesson 1-2)*

46. **Standardized Test Practice** How can $5 \cdot 5 \cdot 7 \cdot 7 \cdot 7 \cdot q \cdot q$ be written in exponential notation? *(Lesson 1-2)*

　A $5 \cdot 12^2 \cdot q^2$

　B $5^2 \cdot 7^3 \cdot q^2$

　C $35^2 \cdot q^2$

　D $70q^2$

47. *Hobbies* Jerome put 4 pounds of sunflower seeds in his bird feeder on Sunday. On Friday, the bird feeder was empty, so Jerome put 4 more pounds of seed in it. The following Sunday, the seeds were half gone. How many pounds of sunflower seeds were consumed by the birds that week? *(Lesson 1-1)*

48. *Food* On a typical day, 2 million gallons of ice cream are produced in the United States. About how many gallons of ice cream are produced each year? *(Lesson 1-1)*

For **Extra Practice**, see page 605.

Lesson 1-3 Variables, Expressions, and Equations **15**

1-3B Evaluating Expressions

A Follow-Up of Lesson 1-3

graphing
calculator

Graphing calculators follow the order of operations. So there is no need to perform each operation separately. To evaluate an expression, enter it just as it is written. If an expression contains parentheses, enter them in the calculator just as they are written, and the expression will be evaluated correctly.

On many calculators, the multiplication and division signs do not appear on the screen as they do on the keys. Instead, the calculator displays the symbols used in computer language. That is, * means multiplication and / means division.

TRY THIS

Work with a partner.

Evaluate $3(x - 6) \div 2 + (x^2 - 15)$ for $x = 8$ and for $x = 12$.

You can replace each x with 8 as you enter the expression. However, if you will evaluate an expression for more than one value of x, it is helpful to enter the variables so that you can reevaluate quickly.

Step 1 Use the following keystrokes to evaluate for $x = 8$.

8 [STO→] [X,T,θ,n] [ALPHA] [:] *Stores 8 as the value of x.*

3 [(] [X,T,θ,n] [−] 6 [)] [÷] 2 [+] [(] [X,T,θ,n] [x²] [−] 15 [)] [ENTER] 52

Step 2 You do not need to reenter the whole expression to reevaluate for $x = 12$. Just change the value for x.

[2nd] [ENTRY] *Redisplays the last entry.*

Use the arrow keys to scroll to the beginning of the display. Press [DEL] to delete the 8. Then press [2nd] [INS] 12 to insert the new value for x. Reevaluate the expression by pressing [ENTER]. When $x = 12$, the value of the expression is 138.

ON YOUR OWN

Use a graphing calculator to evaluate each expression for *x* = 3, *x* = 6, and *x* = 15.

1. $x^2 - 9$

2. $2x^2 + 10$

3. $\frac{16x^2}{3}$

4. $x(20 - x)$

5. How would you evaluate xy^2 for $x = 4$ and $y = 7$ on a graphing calculator? (*Hint:* Enter Y by pressing [ALPHA] [Y].)

1-4

Solving Subtraction and Addition Equations

What you'll learn

You'll learn to solve equations by using the subtraction and addition properties of equality.

When am I ever going to use this?

You can use addition and subtraction equations to find increases and decreases in prices.

Word Wise

addition property
subtraction property
inverse operation

In World War II, the Navajo language was used to devise a secret code. It is one of the few unbroken codes in history. There were 29 original "code talkers," Navajo Marines who used the code. By the end of the war, 420 Navajos served as code talkers. How many recruits were added to the original Navajo code talkers?

Suppose we let r represent the number of recruits added to the code talkers. Then we could use the equation $29 + r = 420$ to find the number of recruits. *This problem will be solved in Example 3.*

In Lesson 1-3, you learned that the value of a variable that makes an equation true is different for different equations. You used the values in a given replacement set to find the solutions.

Often when you need to find the solution to an equation, a replacement set is not given or is too large to try all of the numbers. You can then use the properties of algebra to find the solution of the equation.

HANDS-ON MINI-LAB

Work with a partner.

Solve $x + 3 = 7$ using cups and counters.

⬚: cups and counters

▱ equation mat

Try This

- Let a cup represent x. Put a cup and 3 counters on one side of the mat and 7 counters on the other side. These two quantities are equal.

- The goal is to get the cup all by itself on one side of the mat. Take 3 counters away from each side. What you have left is the value of the cup, which is the value of x.

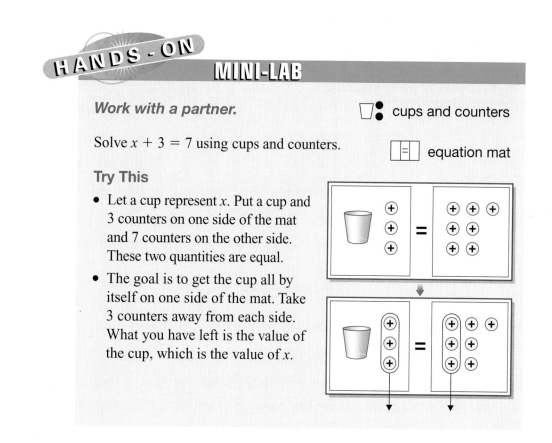

Lesson 1-4 Solving Subtraction and Addition Equations **17**

1. What is the value of x?
2. If there were 5 counters on the side with the cup and 8 on the other side of the mat, how many would you have taken from each side to solve for x? Explain how you know.
3. In the equation $x + 3 = 7$, 3 is added to x. To solve it, you subtract 3. Suppose you were solving the equation $y - 4 = 8$. What operation would you use with 4 to find the value of y?

The Mini-Lab suggests that the following two properties can help us solve equations.

Addition Property of Equality	**Words:** If you add the same number to each side of an equation, then the two sides remain equal.
	
	 Symbols: Arithmetic Algebra
	          $3 = 3$ $a = b$
	    $3 + 5 = 3 + 5$ $a + c = b + c$
	          $8 = 8$

Subtraction Property of Equality	**Words:** If you subtract the same number from each side of an equation, then the two sides remain equal.
	
	 Symbols: Arithmetic Algebra
	          $3 = 3$ $a = b$
	    $3 - 2 = 3 - 2$ $a - c = b - c$
	          $1 = 1$

To solve an equation in which a number is added to or subtracted from the variable, you can use the opposite, or **inverse**, operation. Remember, it is always wise to check your solution.

Examples

1 Solve $y - 27 = 18$. Check your solution.

27 is subtracted from y. To solve, add 27 to each side.

Solve the equation.	Check the solution.
$y - 27 = 18$	Replace y with 45.
$y - 27 + 27 = 18 + 27$	$y - 27 = 18$
$y = 45$	$45 - 27 \stackrel{?}{=} 18$
	$18 = 18$ ✓

The solution is 45.

2 Solve $9.84 = 5.75 + m$.

5.75 is added to m. To solve, subtract 5.75 from each side.

$$9.84 = 5.75 + m \qquad\qquad \textbf{Check:} \quad 9.84 = 5.75 + m$$
$$9.84 - 5.75 = 5.75 + m - 5.75 \qquad\qquad 9.84 \stackrel{?}{=} 5.75 + 4.09$$
$$4.09 = m \qquad\qquad\qquad 9.84 = 9.84 \text{ ✓}$$

The solution is 4.09.

History Refer to the beginning of the lesson. Use the equation $29 + r = 420$ to find the number of recruits added to the code talkers after their formation.

$$29 + r = 420$$
$$29 + r - 29 = 420 - 29$$
$$r = 391$$

There were 391 recruits added.

Check: $29 + r = 420$
$$29 + 391 \stackrel{?}{=} 420$$
$$420 = 420 \checkmark$$

Did you know

According to William McCabe, one of the code designers, there were no code books for the Navajo code. The code talkers used the 411-word code from memory.

CHECK FOR UNDERSTANDING

Communicating Mathematics

Read and study the lesson to answer each question.

1. *Tell* how you determine whether to add or subtract to solve an equation.

2. *Explain* how the addition and subtraction properties of equality are used to solve equations.

3. *You Decide* Taina says that you need to use an inverse operation to solve $y = 8 + 12$. Tess disagrees. Who is correct and why?

HANDS-ON MATH

4. *Model* and solve the equation $x + 5 = 8$ by using cups and counters.

Guided Practice

Solve each equation. Check your solution.

5. $2 + x = 9$
6. $30 + n = 55$
7. $p - 1 = 8$
8. $y = 3.4 + 5.9$
9. $m - 24 = 19$
10. $8.3 - 4.6 = z$

11. *Sports* The Women's National Basketball Association (WNBA) began play in 1997. Lisa Leslie and Rebecca Lobo were among the league stars who earned as much as $250,000 that year. That salary was $2,500 more than the minimum salary in the men's National Basketball Association (NBA). Use the equation $s + 2,500 = \$250,000$ to find the minimum salary in the NBA for 1997.

EXERCISES

Practice

Solve each equation. Check your solution.

12. $y = 29 - 7$
13. $q + 3 = 7$
14. $21 = 17 + s$
15. $m + 40 = 75$
16. $p = 17 - 8$
17. $z - 1 = 0$
18. $16 + t = 23$
19. $s - 9.6 = 17.8$
20. $q - 62 = 153$
21. $115 - 32 = r$
22. $0.7 + x = 2.62$
23. $36 = s - 16$
24. $c + 2.4 = 11.3$
25. $25 + t = 52$
26. $6.01 - 5.28 = b$
27. $w = 8.26 + 6.2$
28. $143 = h - 98$
29. $183 = y + 97$

Lesson 1-4 Solving Subtraction and Addition Equations **19**

30. *Medicine* The first kidney transplant was performed in 1954. In 1968, the first transplant of a pancreas was performed. Use the equation $1954 + y = 1968$ to find the number of years between these two medical breakthroughs.

31. *Life Science* An African elephant calf weighs approximately 250 pounds at birth. Male adult elephants can weigh up to 12,000 pounds and adult females up to 8,000 pounds.
 a. Use the equation $250 + w = 12,000$ to find the amount of weight gained by a male elephant between birth and adulthood.
 b. Find the difference in weight between an adult female elephant and an adult male using the equation $12,000 - d = 8,000$.

32. *Geometry* Suppose the measure of angle $X = x$ and the measure of angle $Y = y$. If angle X and angle Y are complementary, then $x + y = 90°$. Angle X has a measure of $30°$. Use the equation to find the measure of angle Y.

33. *Critical Thinking* Solve $2x + 3 = x + 7$.

Mixed Review

34. Evaluate the expression $6 \cdot (18 - 14) + \frac{22 - 4}{9}$. *(Lesson 1-3)*

35. Solve $456 = 76y$ if the replacement set is $\{4, 6, 7, 9\}$. *(Lesson 1-3)*

36. Standardized Test Practice A painting company uses the formula $A = \ell w$ where A is the area of the wall, ℓ is the wall length, and w is the wall width. What is the area of a wall 13 feet long and 9 feet wide? *(Lesson 1-3)*
 A 22 sq ft   **B** 52 sq ft   **C** 117 sq ft   **D** 127 sq ft

For **Extra Practice**, see page 606.

37. Write $5 \cdot 5 \cdot 8 \cdot 8 \cdot 8$ using exponents. *(Lesson 1-2)*

MATH IN THE MEDIA

Fill It Up!

The portion of the article below appeared in *USA TODAY* on August 13, 1997.

After a long, steady decline, gasoline prices are suddenly soaring at the peak of the summer driving season.

Nationally, the average price of a gallon of unleaded self-serve regular spurted 5.4 cents in one week to $1.24 a gallon in the latest check by Lundberg Survey, an industry price-taking service.

Pumped-up Prices
Average retail prices for one-gallon of regular unleaded self-serve gasoline:

City	Aug 9, '96	July 25	Aug 8
Atlanta	$1.09	$1.02	$1.07
Chicago	$1.32	$1.31	$1.36
Long Island, N.Y.	$1.31	$1.28	$1.32
Los Angeles	$1.33	$1.24	$1.32
U. S. Average	**$1.24**	**$1.19**	**$1.24**

Source: Lundberg Survey

"To see things pop up this quickly is very unusual," says Mike Morrissey, spokesman for the American Automobile Association. "It caught people off guard."

1. The equation $p + 0.054 = 1.24$ could be used to find the previous week's average price per gallon. Solve for the price.

2. Does the information in the table match the information in the article? Explain.

Solving Division and Multiplication Equations

On April 21, 1997, Lameck Aguta of Kenya won the 26.2-mile Boston Marathon with a time of 2 hours 10 minutes 33 seconds. On average, how fast would you have had to run each mile if your goal was to complete the race in 3 hours?

The formula $d = rt$ can be used to solve this problem. In the formula, d represents the distance traveled, r represents the rate, and t represents the time. Substitute 26.2 for d and 3 for t. The formula becomes $26.2 = 3r$. *You will solve this equation in Example 3.*

In Lesson 1-4, you learned that equations could be solved by using the inverse operation. This is also true for multiplication and division equations. The Mini-Lab below shows how this works.

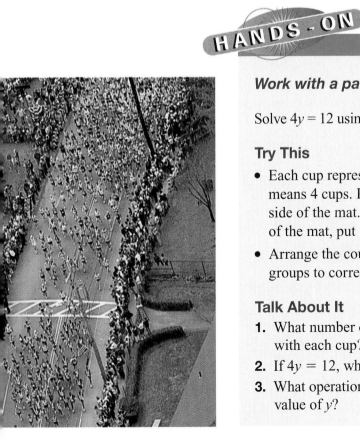

HANDS-ON MINI-LAB

Work with a partner. cups and counters equation mat

Solve $4y = 12$ using cups and counters.

Try This

- Each cup represents y or $1y$. So $4y$ means 4 cups. Put 4 cups on one side of the mat. On the other side of the mat, put 12 counters.

- Arrange the counters into 4 equal groups to correspond to the cups.

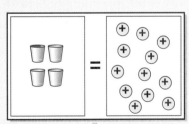

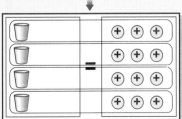

Talk About It

1. What number of counters matches with each cup?
2. If $4y = 12$, what is the value of y?
3. What operation does the lab suggest you could use to find the value of y?

In the Mini-Lab, you solved the multiplication equation $4y = 12$ by separating, or dividing, the counters into 4 equal groups. You would get the same result if you simply divided each side of the equation by 4. When you divide each side of an equation by the same number, you are using the **division property of equality** to solve the equation. There is also a **multiplication property of equality**.

Division Property of Equality	**Words:**	If you divide each side of an equation by the same number, then the two sides remain equal.
	Symbols:	**Arithmetic** $\qquad\qquad$ **Algebra**
		$8 = 8$ $\qquad\qquad\qquad$ $a = b$
		$8 \div 2 = 8 \div 2$ $\qquad$ $\frac{a}{c} = \frac{b}{c}, c \neq 0$
		$4 = 4$
Multiplication Property of Equality	**Words:**	If you multiply each side of an equation by the same number, then the two sides remain equal.
	Symbols:	**Arithmetic** $\qquad\qquad$ **Algebra**
		$8 = 8$ $\qquad\qquad\qquad$ $a = b$
		$8 \cdot 2 = 8 \cdot 2$ $\qquad\qquad$ $ac = bc$
		$16 = 16$

To solve an equation in which the variable is multiplied or divided by a number, use the opposite, or inverse operation. Multiplication and division are inverse operations.

Examples

1 Solve $183 = 3m$. Check your solution.

m is multiplied by 3. To solve, divide each side by 3.

Solve the equation. $\qquad\qquad\qquad\qquad$ Check the solution.

$183 = 3m$ $\qquad\qquad\qquad\qquad\qquad$ **Check:** $\quad 183 = 3m$

$\dfrac{183}{3} = \dfrac{3m}{3}$ $\quad$ *Divide each side by 3* $\qquad\qquad 183 \stackrel{?}{=} 3 \cdot 61$
$\qquad\qquad\quad$ *to undo multiplication by 3.*

$61 = m$ $\qquad\qquad\qquad\qquad\qquad\qquad\qquad 183 = 183$ ✓

The solution is 61.

2 Solve $\dfrac{y}{5} = 3$. Check your solution.

y is divided by 5. To solve, multiply each side by 5.

Solve the equation. $\qquad\qquad\qquad\qquad$ Check the solution.

$\dfrac{y}{5} = 3$ $\qquad\qquad\qquad\qquad\qquad$ **Check:** $\quad \dfrac{y}{5} = 3$

$\dfrac{y}{5} \cdot 5 = 3 \cdot 5$ $\quad$ *Multiply each side by 5* $\qquad\qquad \dfrac{15}{5} \stackrel{?}{=} 3$
$\qquad\qquad\qquad$ *to undo division by 5.*

$y = 15$ $\qquad\qquad\qquad\qquad\qquad\qquad\qquad 3 = 3$ ✓

The solution is 15.

Example ③ **Sports** Refer to the beginning of the lesson. Use the equation $26.2 = r \cdot 3$ to find the average rate needed to run a marathon in 3 hours.

$$26.2 = r \cdot 3$$

$$\frac{26.2}{3} = \frac{r \cdot 3}{3} \quad \textit{Divide each side by 3 to undo the multiplication.}$$

$$\frac{26.2}{3} = r$$

26.2 ÷ 3 = *8.733333333*

In order to run a marathon in 3 hours, you would have to run about 8.7 miles per hour.

CHECK FOR UNDERSTANDING

Communicating Mathematics

Read and study the lesson to answer each question.

1. ***Tell*** whether you would first multiply or divide to solve the equation $\frac{x}{4} = 20$. Explain how you know.

2. ***Show*** how to use a calculator to solve $138{,}336 = 7{,}833r$ to determine Lameck Aguta's average rate in feet per second. Round to the nearest tenth.

HANDS-ON MATH

3. ***Write*** the equation shown by the model at the right. Then solve.

Guided Practice

Solve each equation. Check your solution.

4. $16x = 64$ 5. $40 = 8p$ 6. $n = \frac{280}{70}$

7. $\frac{n}{6} = 13$ 8. $3.7 \cdot 4 = g$ 9. $23.8 \div 7 = s$

10. ***Medicine*** In 1997, the American Medical Association (AMA) elected its first female president, Texas physician Dr. Nancy Dickey. About 32,120 or 0.11 of the doctors in the AMA are women. Use the equation $0.11w = 32{,}120$ to find the number of doctors in the AMA. Round to the nearest whole number.

Dr. Nancy Dickey

Practice   **Solve each equation. Check your solution.**

11. $9 \cdot 25 = a$ **12.** $13m = 182$ **13.** $78 = 13b$

14. $6 = \dfrac{k}{50}$ **15.** $56 \times 8 = n$ **16.** $\dfrac{z}{0.7} = 4.2$

17. $\dfrac{q}{9} = 6$ **18.** $245 \div 5 = f$ **19.** $3.18 \div 6 = q$

20. $6 = \dfrac{p}{35}$ **21.** $8 = \dfrac{x}{25}$ **22.** $322 = 14h$

23. $43.4 = 6.2t$ **24.** $\dfrac{n}{0.19} = 8$ **25.** $16.8 = 0.4r$

26. $350 = 14d$ **27.** $\dfrac{t}{3} = 61$ **28.** $\$7.56 = \$0.27b$

29. Solve $\dfrac{d}{0.11} = 5$.

30. Find the value of x for $\$5.44 = \$0.34x$.

Applications and Problem Solving

31. *Recreation* The Boundary Waters of Minnesota and Canada is a popular area for canoeing and camping. The outfitters in nearby Ely will arrange for your permits and rent all of the equipment for a trip.

  a. Cliff Wold's offers a three-day package for two people for $322. Use the equation $2c = 322$ to find the per-person cost.

  b. Wilderness Outfitters rents canoes on a daily basis. If the daily charge for a rental is $32, use the equation $\dfrac{t}{32} = 7$ to find the total charge for a seven-day rental.

Dr. Pedro José Greer Jr.

32. *Social Work* Dr. Pedro José Greer Jr. has founded four free clinics to provide health care for homeless people. One of those, the Camillus Health Concern, in Miami, Florida, gets about 30,000 visits per year. Use the equation $365v = 30,000$ to find the average number of visits per day. Round to the nearest tenth.

33. *Critical Thinking* In order for a bill that has been vetoed by the president to become law, it must be passed by two-thirds of both the House of Representatives and the Senate. That means that 290 members of the House of Representatives would have to pass a bill to override a veto. Use the equation $\dfrac{2}{3}x = 290$ to find the number of members of the House of Representatives.

34. Working on the CHAPTER Project Your home page should include information about your interests. For example, if you are interested in music, you might want to include some information about your favorite artists. Make a list of your interests and choose a few to incorporate into that plan you have made for your home page.

Mixed Review

35. Archaeology A magazine reports that stone tools found in Ethiopia are estimated to be 2.5 million years old. That is about 700,000 years older than similar tools found in Tanzania. Use the equation $2,500,000 = a + 700,000$ to find the age of the tools found in Tanzania. *(Lesson 1-4)* **Source:** *Nature* magazine

36. Algebra Solve $t - 9 = 23$. Check your solution. *(Lesson 1-4)*

37. Write $4 \cdot 4 \cdot 8 \cdot 8 \cdot 4$ as an expression using exponents. *(Lesson 1-2)*

38. Standardized Test Practice One centimeter is about 0.3937 inch. If Dale is 68 inches tall, then his height in centimeters must be — *(Lesson 1-1)*
  A less than 140 centimeters.
  B between 140 and 150 centimeters.
  C between 150 and 160 centimeters.
  D between 160 and 170 centimeters.
  E more than 170 centimeters.

For **Extra Practice**, see page 606.

CHAPTER 1

Mid-Chapter Self Test

1. **Transportation** A jet carries 342 passengers with 36 in first-class and the rest in coach. Suppose a first-class ticket to fly from Los Angeles to Chicago costs $750 and a coach ticket costs $450. Use the four-step problem solving plan to find the greatest possible ticket sales on one flight. *(Lesson 1-1)*

2. Write the product $5 \cdot 5 \cdot 5$ using exponents. *(Lesson 1-2)*

3. Evaluate the expression $3^3 \cdot 4^4$. *(Lesson 1-2)*

4. **Algebra** Find the value of $a^2 + b^2 - c^2$ if $a = 10$, $b = 5$, and $c = 4$. *(Lesson 1-3)*

5. **Algebra** Solve $y + 45 = 60$ if the replacement set is $\{10, 15, 20, 25\}$. *(Lesson 1-3)*

Solve each equation. Check your solution. *(Lessons 1-4 and 1-5)*

6. $m + 30 = 100$

7. $s - 5.8 = 14.3$

8. $16x = 48$

9. $18 = \dfrac{y}{7}$

10. **Technology** In 1996, shares of a company that created a popular world wide web search engine, were offered for sale. At $33 each, the shares owned by a company co-founder were valued at $132 million. Use the equation $33p = 132,000,000$ to find how many shares the co-founder owned. *(Lesson 1-5)*

TECHNOLOGY

Tim Berners-Lee
WORLD WIDE WEB INVENTOR

Tim Berners-Lee had a problem he needed to solve. In 1989, he was a physicist in Switzerland and wanted to exchange information with colleagues all over the world. So, he created the World Wide Web to link him with this information. His invention has opened up a wealth of information for people all over the world. He now heads the W3 Consortium at MIT's Laboratory for Computer Science in Boston, Massachusetts.

Most companies prefer that Internet developers have college degrees. Courses in computer science, graphics design, and mathematics are helpful in preparing for a career as an Internet developer.

For more information:
American Electronics Association
5201 Great American Pkwy.
Suite 520
Santa Clara, CA 95054

*inter*NET
CONNECTION
www.glencoe.com/sec/
math/mac/mathnet

Someday, I'd like to develop a computer product to help people solve problems.

Your Turn

Use the World Wide Web to gather information about the future of careers in Internet or software development. Will more developers be needed in the next decade? What trends are expected in the industry over the next 10 to 20 years?

Writing Expressions and Equations

What you'll learn

You'll learn to write algebraic expressions and equations from verbal phrases and sentences.

When am I ever going to use this?

You can use an equation to change Fahrenheit temperatures to Celsius.

Digital television will become common as technology improves. There will be better sound and video quality and Internet access through your television. Have you ever wondered how digital television works? Signals are sent through the air in a computer-based language of 0s and 1s. Then a digital converter box in the television translates the signals back into the images you see.

You can act as a translator in mathematics, interpreting words and ideas and translating them into mathematical expressions and equations. There are many words and phrases that suggest arithmetic operations. Any variable can be used to represent a number.

Verbal Phrase	Algebraic Expression
five less than a number	$x - 5$
a number increased by 12	$y + 12$
twice a number decreased by 3	$2c - 3$
the quotient of a number and 4	$\frac{n}{4}$

The chart shows common phrases that usually indicate the four operations.

Addition or Subtraction		Multiplication or Division	
plus	minus	times	divided
sum	difference	product	quotient
more than	less than	multiplied	separated
increased by	subtract	each	
total	decreased by	of	
in all			

Examples

Write each sentence or phrase as an algebraic expression.

1 The number of e-mail users on a server increased by 300.

Let u represent the number of e-mail users before the increase. The word *increased* means there are more e-mail users.

The algebraic expression is $u + 300$.

2 three times the number of points required for an A

Let a = the number of points required for an A.

Three times means multiply by 3.   → $3 \cdot a$ or $3a$

The algebraic expression is $3a$.

You can also translate a verbal sentence into an equation. Then the equation can be used to solve a problem.

Verbal Sentence	**Algebraic Equation**
24 is 6 more than a number.	$24 = n + 6$
Five times a number is 60.	$5n = 60$

 **Example 3 APPLICATION**

Retail Sales In a sports store in New York City, shoppers are surrounded by sports monuments and video screens showing their favorite sports heroes. There are 6,600 more customers each day than there are full– or part–time employees. If there are 7,000 customers each day, write an equation to find the average number of employees.

Let n = the number of employees.

6,600 more than the number of employees → $n + 6{,}600$

The number of customers, 7,000, equals this. → $7{,}000 = n + 6{,}600$

CHECK FOR UNDERSTANDING

Communicating Mathematics

Read and study the lesson to answer each equation.

1. *Write* two different verbal phrases that could be represented by the algebraic expression $y + 3$.

2. *State* one or more words that tell you that a verbal sentence can be written as an equation.

3. *You Decide* Kathy says that the phrases *7 less than y* and *7 less y* are translated into the same algebraic expression. Chapa disagrees. Who is correct and why?

Guided Practice

Write each phrase or sentence as an algebraic expression or equation.

4. 15 more than x

5. the product of 7 and n

6. the quotient of m and 8

7. 9 less than w is 15.

8. The difference between 29 and t is 8.

9. The desks in the classroom can be separated into 7 groups of 4.

10. *Life Science* Did you know that a mosquito has 15 more teeth than an adult human? If t represents the number of teeth a human has, write an expression for the number of teeth in a mosquito.

Practice

Write each phrase or sentence as an algebraic expression or equation.

11. the sum of k and 9

12. 28 less n is 25.

13. 5 more than a number

14. 7 less than y is 25.

15. The product of 16 and r is 80.

16. thirteen fewer than the total

17. the sum of 6 and b

18. Four times w is 72.

19. 10 more than Juanita scored

20. half of Tyrone's sales

21. 48 less v is 15.

22. 5 points more than the class average

23. two times as many touchdowns as the Cowboys

24. Five more than m is 7.

25. The difference between a number and 8 is 25.

26. The product of a number and 7 is 56.

27. The number of desks divided into 3 classrooms is 25.

28. $100 more than the amount made yesterday

Family Activity

Have family members tell you phrases to write as expressions and equations. For example, your brother may say "If you are 3 years older than I am and you are 14, how old am I?"

Applications and Problem Solving

Real World

29. *Geography* Niagara Falls is one of the most visited waterfalls in North America, but it is not the tallest. Yosemite Falls is 2,249 feet taller!

a. If n is the height of Niagara Falls, write an expression for the height of Yosemite Falls.

b. Yosemite Falls is 2,425 feet high. Write an equation to find the height of Niagara Falls.

30. *Space Exploration* In 1983, Sally Ride became the first woman in space, traveling aboard the *Challenger*. Three years later, *Challenger* tragically exploded in the first seconds of a flight. Write an equation to find the year in which the *Challenger* accident occurred.

31. *Nutrition* Use the information in the table to answer each question.

a. Write an expression for the amount of vitamin C in a serving of orange if there are s milligrams in 5.5 ounces of strawberries.

b. There is 2.3 times the recommended daily allowance of vitamin C in 5.5 ounces of kiwifruit. Write an equation for the amount of vitamin C recommended for each day.

Sally Ride

Fruit	Vitamin C (mg in 5.5 oz)
Orange	52
Strawberries	63
Kiwifruit	103.5

Source: Food and Drug Administration

32. *Critical Thinking* Write an equation to represent *5 times the sum of twice x and 4 is 40.*

Mixed Review

33. *Algebra* Solve $7b = 105$. *(Lesson 1-5)*

34. *Standardized Test Practice* Sydney's packages weighed 12.5 pounds, 9 pounds, and 11.25 pounds. What was the total weight of the packages? *(Lesson 1-4)*

    A 35.25 lb   **B** 22.75 lb   **C** 16.25 lb   **D** 12.50 lb   **E** Not Here

35. *Algebra* Evaluate $6a^2 + b$ if $a = 4$ and $b = 1$. *(Lesson 1-3)*

36. Find the value of the expression 2^6. *(Lesson 1-2)*

For **Extra Practice,** see page 606.

PROBLEM SOLVING

1-7A Work Backward

A Preview of Lesson 1-7

Alonso and Ron have just finished buying supplies for their camping trip. They think they may have gotten the wrong change back at one of the stores. Let's listen in!

I think we should have more money left. We started with $60 and I only have about $2 left.

Ron

Wow, it does seem like we should have more than that left!

Well, we just left the grocery. The receipt says we spent $15.89. With the $2 or so we have left, that makes about $18.

Alonso

Before that we were at the sporting goods store. We got all that equipment. The total there was $21.91. Adding that on, that's about $40.

And the first thing we did was use a third of the money as a deposit on the campsite.

So do we have the right change or not?

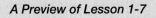

THINK ABOUT IT

Work with a partner.

1. *Tell* how Alonso and Ron were working backward to solve the problem.

2. *Work backward* to finish the problem and determine whether Alonso and Ron have the correct change.

3. *Explain* when is it easier to work backward to solve a problem than it is to work forward.

4. *Show* the order that you would need to use to solve this problem by **working backward.**

   Start with a number. Multiply it by 4, divide by 10, and add 14. Then double. If the result is 124, what was the starting number?

For **Extra Practice,** see page 607.

ON YOUR OWN

5. The last step of the 4-step plan for problem solving is to *examine* the solution. ***Explain*** how you can examine the solution when you solve a problem by working backward.

6. ***Write a Problem*** that could be solved by working backward.

7. ***Look Ahead*** If following the order of operations is working forward, how is the solution to Example 1 on page 33 working backward?

MIXED PROBLEM SOLVING

STRATEGIES

Look for a pattern.
Solve a simpler problem.
Act it out.
Guess and check.
Draw a diagram.
Make a chart.
Work backward.

Solve. Use any strategy.

8. *Tourism* An amusement park in Texas features giant statues of comic strip characters. If you multiply one character's height by 4 and add 1 foot, you will find its length. If the statue of the character is 65 feet long, how tall is it?

9. *Money Matters* Mrs. Navarro's class bought food for the Adopt-A-Family program. They spent $127.68. The 19 students divided the cost equally. How much did each student contribute?

10. *Earth Science* According to a magazine, the average American produces 110,000 pounds of garbage by the time he or she is 75 years old. How many pounds of garbage does the average American produce each year?
Source: *Health* magazine

11. *Money Matters* Shelley Green just bought her first new car. She will pay $400 a month for five years. The actual amount of the loan is $15,900. How much extra will Ms. Green pay in order to spread the payment out over five years?

12. **Standardized Test Practice** In one month, Bryant's father spent $51.90, $22.78, $33.11, and $45.08 on groceries. A good estimate of the total amount spent on groceries during the month is —

A $100. **B** $120. **C** $140.
D $150. **E** $200.

13. *Games* A popular board game began selling in 1935. The rules haven't changed since then. The chart shows how the game would have changed if it had kept up with inflation.

IT'S ONLY MONEY

	1935	1995
Pool of Money	$15,140	$184,794
Dollars per Player	$1,500	$18,308
Park Place Rent	$35	$428
Passing GO	$200	$2,441

a. How much greater would the rent be on Park Place in 1995 than in 1935?

b. It costs $400 to buy Boardwalk. How much would it be if it was adjusted for inflation?

14. *Money Matters* A store tripled the price it paid for a pair of jeans. After a month, the jeans were marked down $5. Two weeks later, the price was divided in half. Finally, the price was reduced by $3 and the jeans sold for $14.99.

a. How much did the store pay for the pair of jeans?

b. Did the store make or lose money on the jeans?

Solving Two-Step Equations

What you'll learn
You'll learn to solve two-step equations.

When am I ever going to use this?
Doctors use two-step equations to determine how many doses of medication a patient needs.

Word Wise
distributive property

Have you ever been to a drive-in theater? Richard Hollingshead, Jr. opened the first drive-in theater in Camden, New Jersey, in 1933. He first tested his idea for a drive-in theater by setting up a projector and screen in his driveway. He even turned on the sprinkler to see if he could still watch the movie in the rain!

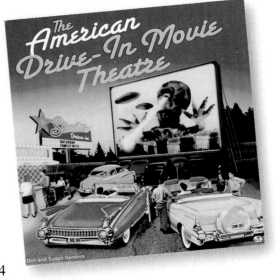

Don and Susan Sanders

Admission to Mr. Hollingshead's theater was $1 per carload or $0.25 per person. If there were 214 carload admissions one night and the total sales was $285.50, how many individual tickets were sold? *This problem will be solved in Example 3.*

In order to find the number of individual tickets sold, you will need to solve a two-step equation. You can investigate solving two-step equations using cups and counters.

HANDS-ON

MINI-LAB

Work with a partner.

⎵• cups and counters

Solve $4 + 2x = 10$ using cups and counters.

⊡ equation mat

Try This

- Place 4 counters and 2 cups on one side of the mat to represent $4 + 2x$. On the other side of the mat, place 10 counters.

- Remove 4 counters from each side of the mat.

- Separate the remaining counters into two equal groups to correspond to the two cups.

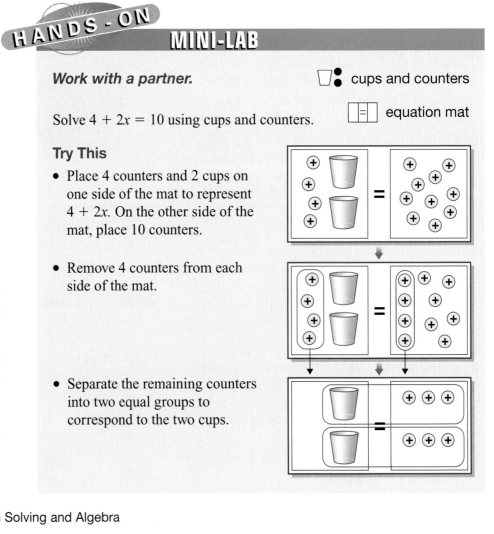

In the Mini-Lab, you solved the equation $4 + 2x = 10$. Notice that in the equation, the x is first multiplied by 2, and then 4 is added. When you solved the equation, you first "undid" the addition by subtracting 4. Then you "undid" the multiplication by dividing by 2. Notice that this is the opposite order than the order of operations. Many equations can be solved algebraically in a similar manner.

Examples

Study Hint

Problem Solving You can use the work backward strategy to solve two-step equations.

1 Solve $6m - 4 = 38$. Check your solution.

$$6m - 4 = 38$$

$$6m - 4 + 4 = 38 + 4 \quad \textit{Add 4 to each side to undo the subtraction of 4.}$$

$$6m = 42$$

$$\frac{6m}{6} = \frac{42}{6} \quad \textit{Divide each side by 6 to undo the multiplication by 6.}$$

$$m = 7$$

The solution is 7.

Check: $\quad 6m - 4 = 38$

$$6(7) - 4 \stackrel{?}{=} 38 \quad \textit{Replace m with 7.}$$

$$42 - 4 \stackrel{?}{=} 38 \quad \textit{Multiply before subtracting.}$$

$$38 = 38 \quad \checkmark \quad \textit{It checks.}$$

2 Solve $16.45 = \frac{c}{3.6} - 5.2$. Check your solution.

$$16.45 = \frac{c}{3.6} - 5.2$$

$$16.45 + 5.2 = \frac{c}{3.6} - 5.2 + 5.2 \quad \textit{Add 5.2 to each side to undo the subtraction of 5.2.}$$

$$21.65 = \frac{c}{3.6}$$

$$21.65 \cdot 3.6 = \frac{c}{3.6} \cdot 3.6 \quad \textit{Multiply each side by 3.6 to undo the division by 3.6.}$$

$$77.94 = c$$

The solution is 77.94.

Check: $\quad \frac{c}{3.6} - 5.2 = 16.45$

$$\frac{77.94}{3.6} - 5.2 \stackrel{?}{=} 16.45$$

$$21.65 - 5.2 \stackrel{?}{=} 16.45$$

$$16.45 = 16.45 \quad \checkmark \quad \textit{The solution of 77.94 checks.}$$

Example

APPLICATION

3 Entertainment Refer to the beginning of the lesson. Find the number of individual tickets sold.

Explore   You know that admission was $1 per carload or $0.25 per person. There were 214 carload admissions and the total sales was $285.50. You need to find how many individual tickets were sold.

Plan   Write and solve an equation.

Solve   Let a represent the number of individual tickets.

$$\underbrace{\text{sales from}}_{} \underbrace{}_{} \underbrace{\text{sales from}}_{} \underbrace{}_{} \underbrace{}_{}$$

carload admissions and individual admissions is total sales

$$\$1 \cdot 214 \quad + \quad \$0.25 \cdot a \quad = \quad \$285.50$$

$$\$214 + \$0.25a = \$285.50$$

$$\$214 + \$0.25a - \$214 = \$285.50 - \$214 \qquad \textit{Subtract 214 from each side.}$$

$$\$0.25a = \$71.50$$

$$\frac{\$0.25a}{\$0.25} = \frac{\$71.50}{\$0.25} \qquad \textit{Divide each side by 0.25.}$$

$$a = 286$$

There were 286 individual tickets sold.

Examine   The sales from 214 carload admissions and 286 individual tickets is $214 \cdot \$1 + 286 \cdot \$0.25$ which is $\$214 + 71.50$ or $\$285.50$. The solution checks.

You can use the distributive property to help solve two-step equations.

Distributive Property	**Words:**	The sum of two addends multiplied by a number is the sum of the products of each addend and the number.
	Symbols: **Arithmetic**	$6 \cdot (5 + 7) = (6 \cdot 5) + (6 \cdot 7)$
	Algebra	$a \cdot (b + c) = (a \cdot b) + (a \cdot c)$

Example ——④ Solve $3(x - 5) = 27$. Check your solution.

$$3(x - 5) = 27$$
$$3 \cdot x - 3 \cdot 5 = 27 \qquad \textit{Apply the distributive property.}$$
$$3x - 15 = 27$$
$$3x - 15 + 15 = 27 + 15 \qquad \textit{Add 15 to each side.}$$
$$3x = 42$$
$$\frac{3x}{3} = \frac{42}{3} \qquad \textit{Divide each side by 3.}$$
$$x = 14 \qquad \text{The solution is 14.}$$

Check: $3(x - 5) = 27$
$$3(14 - 5) \stackrel{?}{=} 27$$
$$3(9) \stackrel{?}{=} 27$$
$$27 = 27 \quad \checkmark \quad \textit{The solution checks.}$$

CHECK FOR UNDERSTANDING

Communicating Mathematics

Read and study the lesson to answer each question.

1. *Tell* what step you would do first to solve $3x + 5 = 17$.

2. *Explain* how the order of operations is used to solve two-step equations.

HANDS-ON MATH

3. *Show* how to use the model to solve the equation $4x + 3 = 15$. How can you check your solution?

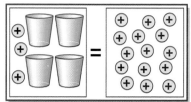

Guided Practice

Solve each equation. Check your solution.

4. $8c - 3 = 13$

5. $9 + 3d = 45$

6. $7 + \frac{m}{3} = 9$

7. $\frac{m}{8} - 6 = 2$

8. $\frac{y}{7} + 0.8 = 2.3$

9. $5(x + 1) = 25$

10. Write *four more than twice a number is 38* as an equation and solve.

11. *Puzzles* The first crossword puzzle was published in 1913. Sixty-six years later, the world's largest crossword puzzle was created with 20 more than 384 times the number of clues that the first puzzle had! If the world's largest puzzle has 12,308 clues, how many clues were in the first puzzle?

EXERCISES

Practice

Solve each equation. Check your solution.

12. $4c - 3 = 25$

13. $3r + 7 = 16$

14. $\frac{m}{16} - 2 = 10$

15. $9 + 11s = 53$

16. $\frac{r}{4} - 1 = 12$

17. $\frac{g}{8} + 2 = 21$

18. $4(x + 2) = 20$

19. $\frac{r}{8} + 3 = 27$

20. $6f - 15 = 3$

21. $5g + 1.8 = 4.3$

22. $2.7 + 0.8n = 4$

23. $0.68 = 0.17 + 1.7k$

24. $2.3 + 6t = 10.1$

25. $2b + 7.5 = 8.0$

26. $11y - 5.3 = 5.7$

27. $4(w - 3) = 36$

28. $18 = 2(t + 1)$

29. $1.3 = \frac{n}{9} - 0.8$

30. Find the solution of $3x - 4 = 7$.

31. What is the value of g for $7 = 4g - 9$?

Write each sentence as an equation. Then solve.

32. Eight more than three times a number is 59.

33. The quotient of a number and four decreased by one is 7.

34. Ruben scored 12 points less than half Jean's total in Laser Tag. Ruben scored 103 points. What was Jean's score?

Applications and Problem Solving

35. *Business* A rental agency charges $52 per day for a car. The first 100 miles are free, but any miles after that cost $0.32 each. Ms. Misel rented a car for three days, and the rental fee was $237.60. How many miles did Ms. Misel drive?

36. *Fund-raising* The Live Aid Concert on July 13, 1985, was the longest television concert ever, with more than 60 bands performing in Philadelphia and London. It raised $65 million for food for African countries experiencing a drought. Ticket sales for the Philadelphia concert totaled $3,375,000, with 15,000 of these tickets sold at $50 each.

   a. If general admission tickets were $35 each, write an equation to find how many general admission tickets were sold.

   b. Find the number of general admission tickets sold.

37. *Write a Problem* that can be solved using the equation $2x + 3 = 15$.

38. *Critical Thinking* In diving, seven judges score each dive, and the sum of the three middle scores is multiplied by the degree of difficulty for the dive. After eleven dives, the scores for the dives are added to find the diver's final score.

   a. Beatriz's score is 366.5 going into the last dive. The degree of difficulty is 2.7, and the leader's score is 439.4. Write an equation to find the total of the judges' scores for her to tie for first.

   b. What must Beatriz score to tie for first?

Mixed Review

39. *Algebra* Write an equation to represent *three more than the number of cars is fifteen.* *(Lesson 1-6)*

40. **Standardized Test Practice** A day-care center's thermostat used to be set on 70° for 24 hours each day. To save money, the owner decided to turn down the furnace for 9 hours during the night between the hours of 8 P.M. and 5 A.M. If she saves $1.70 per day on her heating bill, how much will she save during a 30-day month? *(Lesson 1-5)*

   A $15.30   **B** $33.50   **C** $51.00   **D** $72.80   **E** Not Here

41. *Algebra* Solve $b - 19 = 73$. *(Lesson 1-4)*

42. Evaluate $\dfrac{45 - 9}{3^2 + 3}$. *(Lesson 1-3)*

43. Evaluate $2^2 \cdot 7^2$. *(Lesson 1-2)*

For **Extra Practice,** see page 607.

COOPERATIVE LEARNING

1-7B Function Machines

A Follow-Up of Lesson 1-7

Input: Cacao beans

Do you like chocolate? It takes seven steps to turn the input of cacao beans in chocolate-making machines to the output of chocolate.

In mathematics, we have *functions* that work like machines. In a function, you *input* a number. Since the *output* depends on the input, it is *a function of* the input number. If the input number is represented by *x,* then the function of the input number can be represented by *f(x),* which is read "*f* of *x.*"

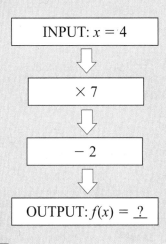

Output: Chocolate

TRY THIS

Work in pairs.

Put a number in the top of the "function machine" at the right. At each stage, an operation is performed. Then the result moves to the next stage.

- What number is being used as the input?
- What is the first operation performed? What is the result after the first operation?
- What is the second operation? What is the result after the second operation?
- What is the final output of the function?

INPUT: $x = 4$

⬇

× 7

⬇

− 2

⬇

OUTPUT: $f(x) = $?

ON YOUR OWN

1. Write a sentence or two to describe the operations performed on the input by the function machine.

2. Each stage of the function machine represents a mathematical operation. If the input is represented by *x*, write an expression for the value of the output.

3. Suppose you were given the output. How could you find the number that was the input?

Find the missing input or output for each function machine.

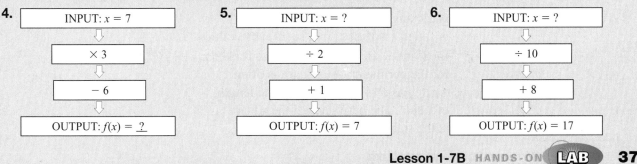

4.
INPUT: $x = 7$
⬇
× 3
⬇
− 6
⬇
OUTPUT: $f(x) = $?

5.
INPUT: $x = $?
⬇
÷ 2
⬇
+ 1
⬇
OUTPUT: $f(x) = 7$

6.
INPUT: $x = $?
⬇
÷ 10
⬇
+ 8
⬇
OUTPUT: $f(x) = 17$

Integration: Geometry
Perimeter and Area

What you'll learn

You'll learn to find the perimeters and areas of rectangles, squares, and parallelograms.

When am I ever going to use this?

You can estimate the amount of paint needed for your room by finding the areas of the walls.

Word Wise

rectangle
perimeter
square
parallelogram
area
base
altitude
height

Stop the presses! The big news is that the Newseum in Arlington, Virginia, is open. The Newseum is the first museum dedicated to news. Interactive displays allow visitors to the Newseum to imagine themselves as reporters, editors, and news anchors. One display includes a wall that is a **rectangle** 126 feet long and 10 feet high with video screens that carry news broadcasts. How much material was needed to build the frame around the video screens? *This problem will be solved in Example 1.*

The **perimeter** of a geometric figure is the sum of the measures of all of its sides. The formula for the perimeter of any rectangle is $P = 2\ell + 2w$, where ℓ represents the length and w represents the width.

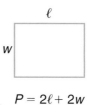

$$P = 2\ell + 2w$$

Example **1**

Real World APPLICATION

Construction Refer to the beginning of the lesson. Find the amount of material needed to build the frame around the video screens.

The length of the wall is 126 feet, and the width is 10 feet.

$P = 2\ell + 2w$

$P = 2(126) + 2(10)$ *Replace ℓ with 126 and w with 10.*

$P = 252 + 20$

$P = 272$

The frame around the video screens required 272 feet of material.

A rectangle whose sides are all equal is called a **square**. In a square, the values for ℓ and w are the same. So, the formula for the perimeter of a square is often written as $P = 4s$, where s is the length of a side. The perimeter of the square at the right is 24 inches.

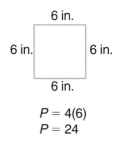

6 in.
6 in.   6 in.
6 in.

$P = 4(6)$
$P = 24$

Squares and rectangles are special types of **parallelograms**. In a parallelogram, each pair of opposite sides are parallel and have the same length. To find the perimeter of a parallelogram, add the lengths of the sides. So, the formula for the perimeter of a parallelogram can be written as $P = 2a + 2b$, where a and b are the lengths of adjacent sides.
The parallelogram at the right has a perimeter of 26 meters.

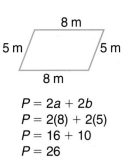

$P = 2a + 2b$
$P = 2(8) + 2(5)$
$P = 16 + 10$
$P = 26$

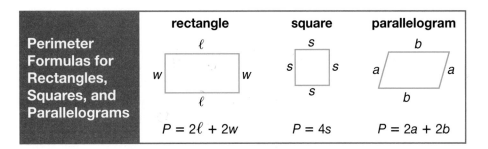

	rectangle	**square**	**parallelogram**
Perimeter Formulas for Rectangles, Squares, and Parallelograms	$P = 2\ell + 2w$	$P = 4s$	$P = 2a + 2b$

In addition to the perimeter, we often solve problems by using the **area** of a geometric figure. The area is the measure of the surface enclosed by the figure. The area of any rectangle can be found by multiplying the width and the length.

Examples Find the area of each rectangle.

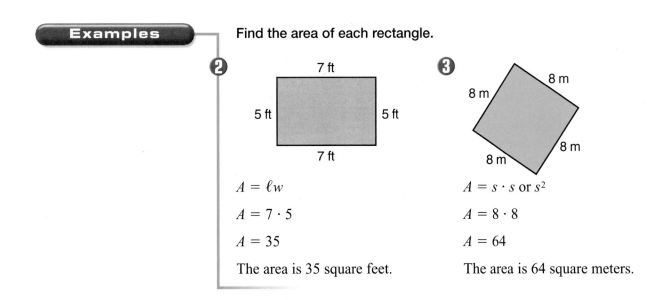

2

$A = \ell w$

$A = 7 \cdot 5$

$A = 35$

The area is 35 square feet.

3

$A = s \cdot s \text{ or } s^2$

$A = 8 \cdot 8$

$A = 64$

The area is 64 square meters.

The formula for the area of a parallelogram is *not* the product of the sides. However, it is related to the formula for the area of a rectangle.

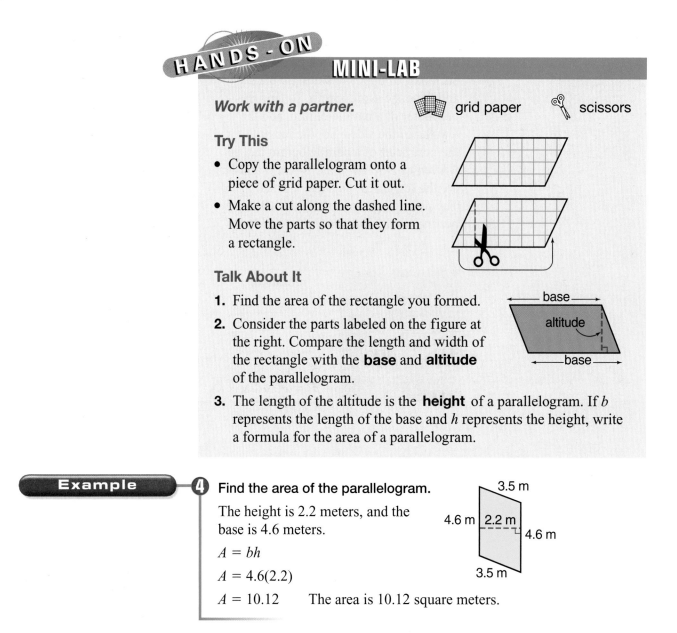

Work with a partner. grid paper scissors

Try This

- Copy the parallelogram onto a piece of grid paper. Cut it out.
- Make a cut along the dashed line. Move the parts so that they form a rectangle.

Talk About It

1. Find the area of the rectangle you formed.

2. Consider the parts labeled on the figure at the right. Compare the length and width of the rectangle with the **base** and **altitude** of the parallelogram.

3. The length of the altitude is the **height** of a parallelogram. If b represents the length of the base and h represents the height, write a formula for the area of a parallelogram.

Example

④ **Find the area of the parallelogram.**

The height is 2.2 meters, and the base is 4.6 meters.

$A = bh$

$A = 4.6(2.2)$

$A = 10.12$ The area is 10.12 square meters.

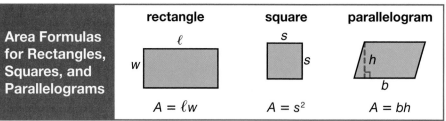

Area Formulas for Rectangles, Squares, and Parallelograms

	rectangle	square	parallelogram
	$A = \ell w$	$A = s^2$	$A = bh$

CHECK FOR UNDERSTANDING

Communicating Mathematics

Read and study the lesson to answer each question.

1. **Draw** and label a square that has an area of 25 square inches. What is its perimeter?

2. **Tell** how parallelograms, rectangles, and squares are related.

3. **Write** in your own words how the area of a figure differs from its perimeter.

Math Journal

Guided Practice

Find the perimeter and area of each figure.

4.
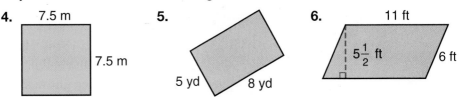
7.5 m
7.5 m

5.
5 yd
8 yd

6.
11 ft
$5\frac{1}{2}$ ft
6 ft

7. A rectangle has a length that is twice its width. Find the area and perimeter of the rectangle if the width is 8 centimeters.

8. *Celebrations* In celebration of Flag Day in 1997, a food company made a giant flag cake at the base of the Statue of Liberty. The cake was $62\frac{1}{2}$ feet by 88 feet. Find the perimeter and area of the cake.

EXERCISES

Practice

Find the perimeter and area of each figure.

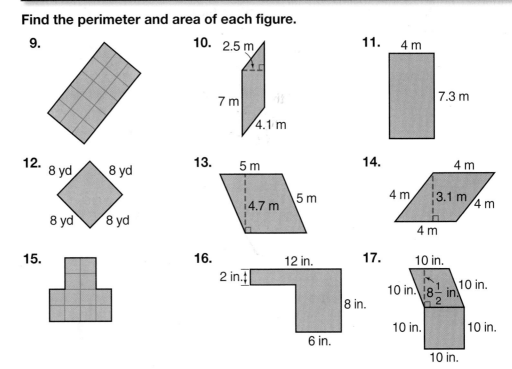

9.

10. 2.5 m
7 m
4.1 m

11. 4 m
7.3 m

12. 8 yd   8 yd
8 yd   8 yd

13. 5 m
4.7 m   5 m

14. 4 m
4 m   3.1 m   4 m
4 m

15.

16. 12 in.
2 in.
8 in.
6 in.

17. 10 in.
10 in. $8\frac{1}{2}$ in.   10 in.
10 in.   10 in.
10 in.

18. Write an expression for the perimeter of a rectangle whose length is three times its width.

19. Use an equation to find the base of a parallelogram whose height is 8 yards and whose area is the same as a square with sides each 12 yards long.

20. The area of a parallelogram is 91 square meters. Find the height of the parallelogram if its base is 13 meters long.

Applications and Problem Solving

21. *Working on the* **CHAPTER Project** Part of designing a home page is planning the layout. The different parts of the page, such as text, art, and photographs, should be placed so that the page is attractive and easy to read. Use what you have learned about area to help you lay out your home page.

22. **History** Did you know that the poem that became our national anthem, "The Star Spangled Banner," was inspired by the flag flying over Fort McHenry in Maryland in 1814? The flag now hangs in the Smithsonian Institution.

   a. Mary Pickersgill was paid $405.60 to make the Fort McHenry flag in 1813. When commissioned, it was 42 feet wide and 30 feet high. What was the area and perimeter of the flag?

   b. In 1997, a project to clean and restore the flag was started. Experts found that the area of the flag had been reduced by 270 square feet because of battle, souvenir hunters, and deterioration. What is the remaining area to be restored?

23. **Marketing** If you had bought a box of a certain cereal in 1955, you would own a tiny piece of land in Canada's Yukon Territory. The cereal company gave deeds to 21 million one-inch by one-inch plots of land in boxes of cereal!

   a. What is the area of one plot of the land given away?

   b. One man claimed he had collected the deeds to 75 square feet of land. How many boxes of cereal did he have to buy? (*Hint:* He bought more than 900 boxes.)

24. **Patterns** Copy and complete the table of perimeters and areas of rectangles.

Length	Width	Perimeter	Area
1	2		
2	4		
4	8		

   a. Notice that the dimensions of each rectangle are doubled. How do the perimeters of the rectangles compare?

   b. How are the areas of the rectangles changed when the length and width are doubled?

25. **Critical Thinking** The figure is an ancient Chinese tangram puzzle. The square is cut into seven pieces: five triangles, one square, and one parallelogram. Suppose the area of the entire tangram is 1 square unit. Find the area of each piece.

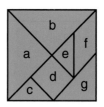

Mixed Review

26. **Algebra** Write an equation to represent *three times a number less five is 16.* Then solve the equation. *(Lesson 1-7)*

27. **Standardized Test Practice** Lenora has saved $2.26, Karen has saved $3.47, and Kevin has saved $2.93 toward the purchase of a new video game that costs $25.95. What is the total the friends have saved? *(Lesson 1-4)*

   A $8.66   **B** $8.56   **C** $7.66   **D** $7.56   **E** Not Here

For **Extra Practice,** see page 607.

28. **Algebra** Solve $x - 32 = 20$ if the replacement set is $\{47, 52, 57, 63\}$. *(Lesson 1-3)*

Solving Inequalities

What you'll learn

You'll learn to write and solve inequalities.

When am I ever going to use this?

You can find the minimum score you need on a test to get a certain grade in the course.

Word Wise

inequality

Have you wondered what it feels like to jump from an airplane? At an amusement park in California, you can get the feel of skydiving in the world's tallest free-fall ride "The DROP ZONE Stunt Tower." On this ride, up to four people at a time plummet from a height of 224 feet reaching speeds in excess of 60 miles per hour! While there is no minimum age requirement to go on this ride, you must be greater than 54 inches tall.

If h represents your height, then $h > 54$ represents the height you must be to go on the ride. The sentence $h > 54$ is called an **inequality**. Inequalities are sentences that compare quantities. They contain symbols like $>$ and $<$.

Study Hint

Reading Math
Recall that $>$ is read as *is greater than* and $<$ is read as *is less than*. The symbol $\geq$ is read as *is greater than or equal to* and $\leq$ is read as *is less than or equal to*.

	Words	**Symbols**
Arithmetic	6 is greater than 4.	$6 > 4$
	2 is less than 15.	$2 < 15$
Algebra	$2x + 9$ is greater than 11.	$2x + 9 > 11$
	$\frac{m}{3} - 5$ is less than 8.	$\frac{m}{3} - 5 < 8$

Some inequalities use the symbols $\geq$ or $\leq$. They are combinations of the equals sign and the inequality symbols.

	Words	**Symbols**
Arithmetic	8 is greater than or equal to 7.	$8 \geq 7$
	16 is less than or equal to 16.	$16 \leq 16$
Algebra	11 is greater than or equal to $5x$.	$11 \geq 5x$
	$\frac{t}{3} + 4$ is less than or equal to 12.	$\frac{t}{3} + 4 \leq 12$

Equations with one variable often have one solution. Unlike an equation, an inequality may have many solutions. The solution can be written as a set of numbers.

MINI-LAB

Work with a partner.

Try This

- Copy the number line.

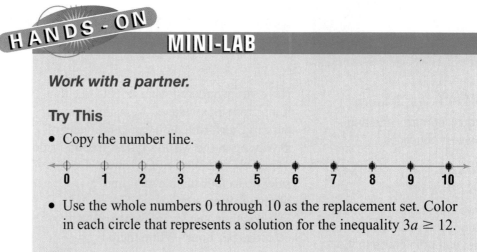

- Use the whole numbers 0 through 10 as the replacement set. Color in each circle that represents a solution for the inequality $3a \geq 12$.

Talk About It

1. What does the number line suggest about the solution set?
2. If all real numbers are included in the replacement set, are there solutions that are not marked on the number line? If so, name one.
3. Solve the equation $3a = 12$. How is the solution of this equation related to the solution of the inequality?

There are many situations in real life that can be described using an inequality. The table below shows some common phrases and corresponding inequalities.

<	>	≤	≥
• less than • fewer than • up to	• greater than • more than • exceeds • in excess of	• less than or equal to • no more than • at most	• greater than or equal to • no less than • at least

Example

CONNECTION

1 **Literature** *Alice In Wonderland* is one of the most popular children's books ever. But its author, Lewis Carroll, said he would be surprised if it sold more than 2,000 copies. Before Mr. Carroll died in 1898, *Alice In Wonderland* had sold more than 180,000 copies.

a. Write an inequality to show the number of copies Mr. Carroll expected to sell. Then graph the inequality.

copies expected to sell	*no more than*	*2,000*
c	$\leq$	2,000

The copies Mr. Carroll expected to sell can be represented by $c \leq 2,000$.

When $\leq$ or $\geq$ is used, the circle on the graph is solid to show that the number is included in the solution.

b. Write an inequality to show how many books were actually sold before Mr. Carroll died. Then graph the inequality.

copies sold	*were more than*	*180,000*
s	$>$	180,000

The copies sold before Mr. Carroll died can be represented by $s > 180{,}000$.

$$\xleftarrow{\hspace{2cm}} \underset{\substack{150{,}000 \;\; 160{,}000 \;\; 170{,}000 \;\; 180{,}000 \;\; 190{,}000 \;\; 200{,}000}}{} \xrightarrow{\hspace{2cm}}$$

When $<$ or $>$ is used, the circle on the graph is open to show that the number is not included in the solution.

As we learned in solving equations, the guess-and-check method is not always the quickest way to solve an equation. Likewise, it is usually not the best way to solve an inequality. You can use your knowledge of solving equations to solve an inequality.

Example

2 Solve $5x - 15 > 17$. Graph the solution on a number line.

Solve the related equation, $5x - 15 = 17$. Remember to add to undo the subtraction. Then divide to undo the multiplication.

$$5x - 15 = 17$$
$$5x - 15 + 15 = 17 + 15 \quad \textit{Add 15 to each side.}$$
$$5x = 32$$
$$\frac{5x}{5} = \frac{32}{5} \quad\quad \textit{Divide each side by 5.}$$
$$x = 6.4$$

The solution will either be numbers greater than 6.4 or numbers less than 6.4. Let's test a number to see which is correct.

numbers greater than 6.4	*numbers less than 6.4*
Try 9.	Try 5.
$5(9) - 15 > 17$	$5(5) - 15 > 17$
$45 - 15 \overset{?}{>} 17$	$25 - 15 \overset{?}{>} 17$
$30 > 17 \quad \textit{true}$	$10 > 17 \quad \textit{false}$

The numbers greater than 6.4 make up the solution set. So, $x > 6.4$.

To graph the solution, draw an empty circle at 6.4. Then draw a large arrow to indicate the numbers that are solutions.

$$\xleftarrow{\hspace{1cm}} \underset{\substack{0 \;\; 1 \;\; 2 \;\; 3 \;\; 4 \;\; 5 \;\; 6 \;\; 7 \;\; 8 \;\; 9 \;\; 10}}{} \xrightarrow{\hspace{1cm}}$$

Try other numbers in the set to check your solution.

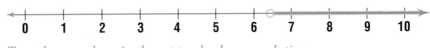

Communicating Mathematics

Read and study the lesson to answer each question.

1. *Tell* what is meant by the phrase "in excess of" in describing the speed of the DROP ZONE Stunt Tower in the beginning of the lesson.

2. *Describe* the three different ways two numbers can compare.

HANDS-ON MATH

3. *Draw* a number line for a solution set that is all numbers less than 4.

Guided Practice

Solve each inequality. Graph the solution on a number line.

4. $a - 4 > 3$

5. $21 > 12 + d$

6. $x - 3 \leq 14$

7. $3x < 27$

8. $\frac{x}{8} \geq 5$

9. $7 < 2c - 3$

10. Write an inequality for *five times a number is greater than sixty*. Then solve the inequality.

11. *Recreation* A magazine recommends that when you go hiking, three times the weight of your backpack and its contents should be no more than your body weight.

   a. If you weigh 90 pounds and your empty backpack weighs 12 pounds, write an inequality to find how much the contents of your backpack should weigh.

   b. Solve the inequality.

   Source: *Women's Sports and Fitness* magazine

Practice

Solve each inequality. Graph the solution on a number line.

12. $x - 5 < 18$

13. $y + 9 \geq 34$

14. $8 < k - 7$

15. $3d \leq 12$

16. $28 \leq 4x$

17. $\frac{m}{5} > 2$

18. $30 \geq 6a$

19. $18 > 2h$

20. $5x \geq 25$

21. $5r \leq 35$

22. $8 < \frac{b}{3}$

23. $\frac{k}{2} \leq 8$

24. $7 < 2x - 3$

25. $\frac{y}{3} + 8 \leq 11$

26. $7 < 5x - 8$

27. $2 > \frac{r}{5} - 8$

28. $9q + 4 \leq 22$

29. $\frac{m}{6} + 1 \leq 4$

Write an inequality for each sentence. Then solve the inequality.

30. Three times a number is less than sixty.

31. Four more than a number is greater than fifteen.

32. Six less than four times a number is greater than thirty.

Applications and Problem Solving

33. *Law* In many states, a person has to be greater than 16 years old to obtain a driver's license. Write an inequality that tells the ages at which a person can obtain a driver's license in those states.

34. *Space* More than 50,000 UFO sightings have been reported in the last 20 years. Between 1947 and 1969, the U.S. Air Force collected and studied reports of UFO sightings. In more than 10,800 cases, more than 90% of those studied, they were able to conclude that these were meteors, stars, satellites, aircraft, clouds, or jokes!

    a. Write an inequality for the UFO sightings in the last 20 years.

    b. Write and solve an inequality for the number of cases the Air Force studied.

35. *Critical Thinking* Find the possible values of x for the inequality $10 < 4x + 2 \le 30$.

Mixed Review

36. *Home Improvement* Mr. Delgado decided to extend his living room to add a sunroom to his home. If the length of the room is increased by 10 feet and the width of the room is 16 feet, find the area of the new sunroom. *(Lesson 1-8)*

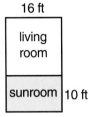

37. *Algebra* Solve $\frac{n}{2} + 31 = 45$. Check your solution. *(Lesson 1-7)*

38. *Standardized Test Practice* Last week Carlos ran 3.75 kilometers on Monday, 5 kilometers on Tuesday, and 4.5 kilometers on Wednesday. How many kilometers did Carlos run on those three days? *(Lesson 1-4)*

    A 5.75 km   **B** 13.25 km   **C** 17.25 km   **D** 20.75 km   **E** Not Here

39. Evaluate $3[14 - (8 - 5)^2] + 20$. *(Lesson 1-3)*

For **Extra Practice**, see page 608.

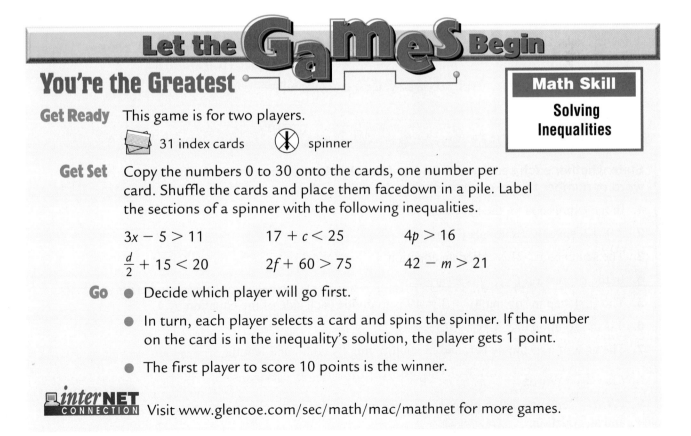

Let the Games Begin

You're the Greatest

Math Skill

Solving Inequalities

Get Ready This game is for two players.

    31 index cards   spinner

Get Set Copy the numbers 0 to 30 onto the cards, one number per card. Shuffle the cards and place them facedown in a pile. Label the sections of a spinner with the following inequalities.

    $3x - 5 > 11$ $17 + c < 25$ $4p > 16$

    $\frac{d}{2} + 15 < 20$ $2f + 60 > 75$ $42 - m > 21$

Go ● Decide which player will go first.

    ● In turn, each player selects a card and spins the spinner. If the number on the card is in the inequality's solution, the player gets 1 point.

    ● The first player to score 10 points is the winner.

interNET **CONNECTION** Visit www.glencoe.com/sec/math/mac/mathnet for more games.

Study Guide and Assessment

inter NET CONNECTION Chapter Review **For more review, visit:** www.glencoe.com/sec/math/mac/mathnet

Vocabulary

After completing this chapter, you should be able to define each term, concept, or phrase and give an example or two of each.

Numbers and Operations
base (p. 8)
equation (p. 13)
evaluate (p. 9)
exponent (p. 8)
factor (p. 8)
numerical expression (p. 11)
order of operations (p. 11)
power (p. 8)

Geometry
altitude (p. 40)
area (p. 39)
base (p. 40)
height (p. 40)
parallelogram (p. 39)
perimeter (p. 38)
rectangle (p. 38)
square (p. 38)

Algebra
addition property of equality (p. 18)
algebraic expression (p. 12)
associative property (p. 8)
commutative property (p. 8)
distributive property (p. 34)
division property of equality (p. 22)
inequality (p. 43)
inverse operation (p. 18)
multiplication property of equality (p. 22)
open sentence (p. 13)
replacement set (p. 13)
solution (p. 13)
substitute (p. 11)
subtraction property of equality (p. 18)
variable (p. 12)

Problem Solving
four-step plan (p. 4)
work backward (p. 30)

Understanding and Using the Vocabulary

State whether each sentence is *true* or *false*. If false, replace the underlined word or number to make a true sentence.

1. In the expression 6^3, the <u>base</u> is 3.

2. $9 + 12$ is an <u>algebraic</u> expression.

3. The sentence $x > 4$ is called an <u>equation</u>.

4. In the expression $t - 8$, t is a <u>variable</u>.

5. The <u>first</u> step in solving $4y + 3 = 19$ is to divide each side of the equation by 4.

6. 4 is the <u>solution</u> of $x + 6 = 10$.

7. The <u>area</u> of a rectangle is found by multiplying its width and length.

In Your Own Words

8. *Write* a sentence that explains the difference between an algebraic expression and an equation.

Objectives & Examples

Upon completing this chapter, you should be able to:

● solve problems using the four-step plan
(Lesson 1-1)

Explore	What do you know? What are you trying to find?
Plan	How will I go about solving this?
Solve	Carry out your plan. Does it work? Do you need another plan?
Examine	Does your answer seem reasonable? If not, check your solution.

● use powers and exponents in expressions
(Lesson 1-2)

Find $4^2 \cdot 3^3$.

$4^2 \cdot 3^3 = 4 \cdot 4 \cdot 3 \cdot 3 \cdot 3$
$= 16 \cdot 27$ or 432

● evaluate expressions *(Lesson 1-3)*

Evaluate $5x + y$ if $x = 3$ and $y = 2$.

$5x + y = 5(3) + 2$
$= 15 + 2$ or 17

● find the solutions of equations *(Lesson 1-3)*

Find the solution for $9 + m = 15$. Use the replacement set $\{6, 7, 8, 9\}$.

Let $m = 6$. $9 + 6 = 15$
6 is the solution. $15 = 15$ ✓

● solve equations by using the subtraction and addition properties of equality *(Lesson 1-4)*

Solve $d - 14 = 20$.

$d - 14 + 14 = 20 + 14$ *Add to undo*
$d = 34$ *subtraction.*

Review Exercises

Use these exercises to review and prepare for the chapter test.

Use the four-step plan to solve each problem.

9. **Time** The distance between Alicia's house and Paquita's house is 1,600 feet. If it takes Alicia 3 seconds to walk 10 feet, how long will it take her to walk to Paquita's house?

10. **Construction** The concrete slab for a floor is 40 feet long, 30 feet wide, and 2 feet deep. If the volume of the concrete is measured in cubic feet and is found by multiplying the length, width, and height, find the volume of the concrete used.

Write each expression using exponents.

11. $5 \cdot 5 \cdot 6 \cdot 6 \cdot 6$
12. $2 \cdot 4 \cdot 2 \cdot 2 \cdot 4 \cdot 7$

Evaluate each expression.

13. 5^4 14. $4^2 \cdot 2^3$

Evaluate each expression if $a = 1$, $b = 2$, $c = 5$, and $d = 8$.

15. $cd^2 + (3b - 4)$ 16. $abd + 3a$
17. $(4c) \div (ab)$ 18. $7a + 2c$

Find the solution for each equation from the given replacement set.

19. $3z = 480, \{160, 140, 110\}$
20. $x + 10 = 58, \{68, 58, 48\}$
21. $87 - b = 37, \{124, 50, 40\}$

Solve each equation. Check your solution.

22. $n + 40 = 90$ 23. $s - 13 = 62$
24. $143 = a + 35$ 25. $7.6 = 4.7 + t$
26. $r - 12 = 27$ 27. $1.6 = p - 2.5$

solve equations by using the division and multiplication properties of equality *(Lesson 1-5)*

Solve $5f = 45$.

$$\frac{5f}{5} = \frac{45}{5} \quad \textit{Divide to undo multiplication.}$$

$$f = 9$$

Solve each equation. Check your solution.

28. $2.7m = 8.1$

29. $\frac{s}{7} = 42$

30. $60 = 15t$

31. $3 = \frac{d}{24}$

32. $\frac{r}{0.2} = 6.6$

33. $6.72 = 0.56k$

---

write algebraic expressions and equations from verbal phrases and sentences *(Lesson 1-6)*

Write an algebraic expression to represent *the sum of a number and 7.*

The algebraic expression is $n + 7$.

Write each phrase or sentence as an algebraic expression or equation.

34. the sum of 9 and z

35. the product of 6 and x

36. Ten less a number is 25.

37. Four times a number is 48.

---

solve two-step equations *(Lesson 1-7)*

Solve $3d + 3 = 30$.

$$3d + 3 - 3 = 30 - 3$$

$$3d = 27$$

$$\frac{3d}{3} = \frac{27}{3} \quad \textit{Divide each side by 3.}$$

$$d = 9$$

Solve each equation. Check your solution.

38. $5f - 12 = 18$

39. $\frac{h}{6} + 30 = 41$

40. $16 = \frac{t}{2} - 8$

41. $3(m + 1) = 27$

---

find the perimeters and areas of rectangles, squares, and parallelograms *(Lesson 1-8)*

Perimeter and Area Formulas		
Figure	Perimeter	Area
rectangle	$P = 2\ell + 2w$	$A = \ell w$
square	$P = 4s$	$A = s^2$
parallelogram	$P = 2a + 2b$	$A = bh$

Find the perimeter and area of each figure.

42.

43.

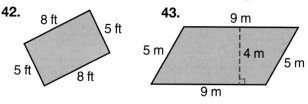

---

write and solve inequalities *(Lesson 1-9)*

Solve $4x - 5 > 15$.

Solve $4x - 5 = 15$. $\rightarrow$ $x = 5$

Test a number greater than 5.

Try 6. $4(6) - 5 \overset{?}{>} 10$

$$19 \overset{?}{>} 15 \quad \textit{true}$$

The solution to $4x - 5 > 15$ is all numbers greater than 5.

Solve each inequality. Graph the solution on a number line.

44. $r - 4 < 1$

45. $9 < \frac{a}{4} + 6$

Write an inequality for each sentence. Then solve the inequality.

46. Three times a number is less than ninety.

47. Five more than a number is greater than sixteen.

48. *Medicine* The doctor gave Mitsu 30 capsules for strep throat. She was to take two capsules with every meal for the first two days and then one capsule with every meal after that until the capsules are gone. If Mitsu starts the medication with lunch on Monday and eats three meals a day, when will she take the last capsule? *(Lesson 1-1)*

49. *Money Matters* The Village Produce Market charges $1.79 a pound for red grapes. Mr. Franklin paid $7.16 for a bag of red grapes. How many pounds of grapes did he buy? *(Lesson 1-5)*

50. *Meteorology* A degree day is the difference between 64°F and the day's average temperature. Write an equation to find the degree day, if you are given the average temperature. Then find the degree day for a day when the average temperature is 83°F. *(Lesson 1-6)*

51. *Work Backward* A cup, two saucers, and three bowls cost $44. One bowl costs as much as one saucer and one cup. If a saucer costs $4, how much is each bowl and cup? *(Lesson 1-7A)*

52. *Health* Neal lives on the block shown in the diagram below. If Neal jogs around the block four times, how many feet has he jogged? *(Lesson 1-8)*

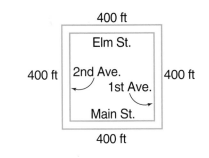

● *Open Ended*

Your family plans to rent a car on your next vacation. Company A charges a flat rate of $39 a day plus $0.10 per mile. Company B charges a daily fee of $45 plus $0.05 for each mile. Suppose you plan to rent the car for 4 days. How many miles would you need to drive for Company A to be less expensive?

Create your own car rental charges other than the ones given. Choose a flat rate to start with and then a per day or per mile charge.

● *Completing the* CHAPTER Project

Use the following checklist to make sure your home page design is complete.

☑ The information included on the page is accurate and easy to read.

☑ The graphics on the page are attractive and clear.

☑ The layout of the page is well-organized.

Add any finishing touches that will make your home page more attractive.

PORTFOLIO Select a problem you solved in this chapter and place it in your portfolio. Attach a note explaining how the problem illustrates an important concept covered in this chapter.

A practice test for Chapter 1 is provided on page 647.

Section One: Multiple Choice

There are twelve multiple-choice questions in this section. Choose the best answer. If a correct answer is *not here,* choose the letter for Not Here.

1. The scale weighs a shoe in pounds. What is the weight of the shoe to the nearest pound?

A 3 pounds

B $3\frac{1}{2}$ pounds

C 4 pounds

D $4\frac{1}{2}$ pounds

2. $(2 \times 8) + (3 \times 8)$ is equivalent to —

F 6×8.

G 6×64.

H $8(2 + 5)$.

J $8(2 + 3)$.

3. Evaluate $(3 + 4)^2 + (6 - 2)^2$.

A 11

B 121

C 49

D 65

4. How is the product $4 \times 4 \times 4$ expressed using exponents?

F 4×3

G 4^3

H 3^4

J 4^4

Please note that Questions 5–12 have five answer choices.

5. Mrs. Numkena and 4 of her students planned to leave a 15% tip when they had lunch. Two of the students had hamburgers, two of the students had salads, and Mrs. Numkena bought a salad and soup. What other information is needed to determine how much to leave for a tip?

A the cost of a salad

B the cost of hamburgers

C where they had lunch

D the cost of the meals

E the time the waiter was at the table

6. During a 3-day time period, Lindsay spent $3.00, $3.25, and $2.98 for her lunches. A good estimate for the total amount Lindsay spent on lunches during the 3 days is —

F $7   **G** $8

H $9   **J** $10

K $11

7. The graph shows how often families eat together at breakfast.

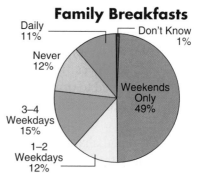

Family Breakfasts

Daily 11%
Don't Know 1%
Never 12%
Weekends Only 49%
3–4 Weekdays 15%
1–2 Weekdays 12%

What percent of the families did *not* say that they eat together 1–4 weekdays?

A 27%   **B** 73%

C 60%   **D** 13%

E 64%

8. Chet had $20 to spend at the fair. His admission was $3.50, and he gave the cashier a $5 bill. How much change should he receive?

   F $1.50   **G** $2.50

   H $3.50   **J** $16.50

   K Not Here

9. Robbie has saved $11.42, Beth has saved $12.00, and Monsa has saved $6.78. They want to purchase a collectible baseball card that costs $40. What total have they saved?

   A $23.42   **B** $30.20

   C $33.68   **D** $40.20

   E Not Here

10. If 33% was deducted from a paycheck for income taxes and 5% was deducted for a charitable contribution, what is the total percent deducted from the paycheck?

    F 28%   **G** 33%

    H 38%   **J** 43%

    K Not Here

11. When three bunches of bananas are weighed at the check-out, they weighed 2.5 pounds, 3.04 pounds, and 1 pound. What was the total weight of the bananas?

    A 3.5 lb   **B** 5.54 lb

    C 4.04 lb   **D** 6.04 lb

    E Not Here

12. Wesley's movie ticket cost $5.75, and his snacks cost $4.35. If he had $23, how much does Wesley have now?

    F $8.55   **G** $12.90

    H $14.50   **J** $17.25

    K Not Here

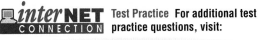

Test Practice For additional test practice questions, visit:

www.glencoe.com/sec/math/mac/mathnet

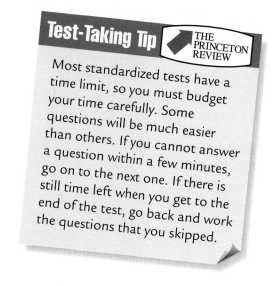

Section Two: Free Response

This section contains seven questions for which you will provide short answers. Write your answers on your paper.

13. Bonnie is reading a 186-page book. She needs to read twice as many pages as she has already read. How many pages has she read?

14. A certain number is divided by 4 and then 5 is subtracted from the result. The final answer is 25. What is the number?

15. What is the solution to the inequality $12 > t + 8$?

16. Write an equation to represent *four more than twice a number is 12*.

17. If $c = 10$ and $p = 2$, what is the value of cp?

18. If you add 30 centimeters to Rosa's height and take a third of the sum, you have half of her younger brother Hugo's height. Hugo is 140 centimeters tall. How tall is Rosa?

19. Find the area and perimeter of a rectangle that is 2 inches wide and 5 inches long.

CHAPTER 2

Algebra: Using Integers

What you'll learn in Chapter 2

- to compare, order, and compute with integers,
- to organize and use data in matrices,
- to solve problems by eliminating possibilities,
- to solve problems involving integers, and
- to graph points on the coordinate plane.

CHAPTER Project

WHERE THE WILD THINGS ARE

In this project, you will work in a group to investigate a threatened or endangered species in the United States or elsewhere in the world and report on changes in its population. You will present your findings as a script for a 1- or 2-minute television public service announcement and illustrate your points using algebra, graphs, and statistics.

Getting Started

- Research an endangered species of your choice.
- Make a list of information that you will include in the script for your television public service announcement. The tables show population information on four endangered or threatened species.

Bald Eagle

Year	Number in Lower 48 States
1782	25,000
1963	900
1994	9,000

Source: U.S. Fish & Wildlife Service

California Condor

Year	Number in World
1979	35
1987	27
1997	119

Source: California Dept. of Fish & Game

Grizzly Bear

Year	Number in Lower 48 States
1800	50,000
1975	1,000
1995	665

Source: U.S. Fish & Wildlife Service

Whooping Crane

Year	Number in North America
1870	1,400
1939	18
1997	180

Source: Texas Parks & Wildlife Dept.

Technology Tips

- Use an **electronic encyclopedia** to research your selected species.
- Use a **word processor** to write the script for your television public service announcement.
- Use a **spreadsheet** to organize any data that you collect.

inter NET CONNECTION Data Update For up-to-date information on endangered species, visit:

www.glencoe.com/sec/math/mac/mathnet

Working on the Project

You can use what you'll learn in Chapter 2 to help you make a television public service announcement script about your endangered species.

Page	Exercise
65	38
83	44
95	48
99	Alternative Assessment

What you'll learn

You'll learn to graph integers on a number line and find absolute value.

When am I ever going to use this?

Integers are used to describe scores in sports and to measure temperature.

Word Wise

integer
graph
coordinate
absolute value

Study Hint

Reading Math
The periods (...) in the set of integers are called an *ellipsis*. They indicate that the set continues without end, following the same pattern.

What is your favorite way to beat the heat? People in the Daloi Danakil Depression in Ethiopia, Africa, really have to learn to tolerate hot days. They have the hottest average annual temperature in the world: 95°F! On the other end of the spectrum, Plateau Station, Antarctica, has the coldest average annual temperature, −71°F.

Negative numbers like −71, as well as positive whole numbers like 95, are part of the set of **integers**. The set of integers can be written as {..., −4, −3, −2, −1, 0, +1, +2, +3, +4, ...}.

Positive integers can be written with or without the + sign. So, +95 and 95 are the same.

You can graph integers on a number line. 0 is considered the starting point on the number line, with positive numbers to the right and negative numbers to the left. Zero is neither negative nor positive.

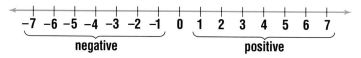

The arrows show that the numbers continue without end.

To **graph** an integer, locate the number and draw a dot at that point on the line. Sometimes letters are used to name points on a number line. The integer that corresponds to a letter is called the **coordinate** of the point.

Examples

1 Name the coordinate of each point graphed on the number line.

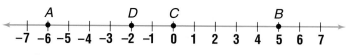

The coordinate of *A* is −6.   The coordinate of *B* is 5.
The coordinate of *C* is 0.   The coordinate of *D* is −2.

2 Graph the following points on a number line.
E has the coordinate −3.   *F* has the coordinate 4.
G has the coordinate −7.   *H* has the coordinate 1.

Find each number. Draw a dot there. Write the letter above the dot.

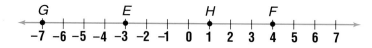

3 **Earth Science** The table at the right shows the average January temperatures in Celsius for selected cities. Graph each temperature on a number line.

City	Temperature (°C)
Boston	2
Chicago	−2
Detroit	−1
Milwaukee	−3
New York	3
Pittsburgh	1

You can use the first letter of each city name to label the points.

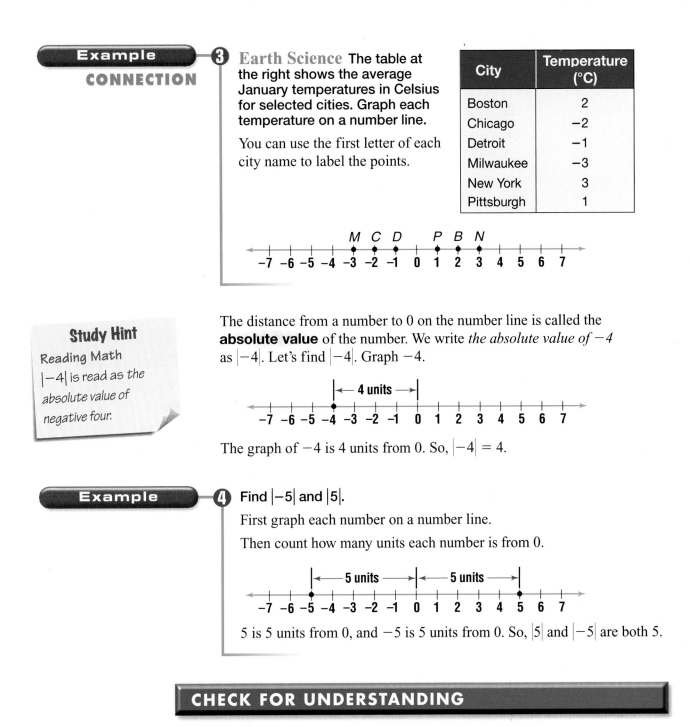

Study Hint

Reading Math

$|-4|$ is read as the absolute value of negative four.

The distance from a number to 0 on the number line is called the **absolute value** of the number. We write *the absolute value of* −4 as $|-4|$. Let's find $|-4|$. Graph −4.

The graph of −4 is 4 units from 0. So, $|-4| = 4$.

Example

4 Find $|-5|$ and $|5|$.

First graph each number on a number line.

Then count how many units each number is from 0.

5 is 5 units from 0, and −5 is 5 units from 0. So, $|5|$ and $|-5|$ are both 5.

CHECK FOR UNDERSTANDING

Communicating Mathematics

Read and study the lesson to answer each question.

1. *Write* how to graph an integer on a number line.

2. *Draw* a number line from −8 to 8. Graph two points that are the same distance from 0. Write a number sentence about these two points.

3. *Write a Problem* about a real-life situation involving a negative integer.

Guided Practice

4. Name the coordinate of each point graphed on the number line.

Graph each set of points on a number line.

5. $\{-3, 1, 4\}$

6. $\{5, 0, -2\}$

7. $\{6, 4, -4\}$

Find each absolute value.

8. $|-7|$ **9.** $|23|$ **10.** $|-44|$ **11.** $|0|$

12. Write an integer to describe a low temperature that is 10 degrees below zero.

13. *Earth Science* The Caspian Sea between Europe and Asia is the only major natural lake in the world below sea level. It is 92 feet below sea level. Write an integer to describe the level of the Caspian Sea.

EXERCISES

Practice

14. Name the coordinate of each point graphed on the number line.

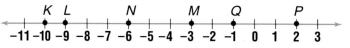

Graph each set of points on a number line.

15. $\{4, 5, 7\}$ **16.** $\{-1, -3, 5\}$ **17.** $\{-3, 5, 6, 0\}$

18. $\{-6, -8, 0, 1\}$ **19.** $\{9, -5, -7, 12\}$ **20.** $\{3, -6, -8, 1\}$

21. $\{9, 5, 8, -2\}$ **22.** $\{2, 5, -6, -1\}$ **23.** $\{3, -2, 2, -3, 0\}$

Find each absolute value.

24. $|-4|$ **25.** $|5|$ **26.** $|-16|$ **27.** $|12|$

28. $|-15|$ **29.** $|45|$ **30.** $|-65|$ **31.** $|88|$

32. $|4 + 8|$ **33.** $|13 - 9|$ **34.** $|5| + |-9|$ **35.** $|-17| - |12|$

Write an integer to describe each situation.

36. Amparo finished the race 3 seconds ahead of the 2nd place finisher.

37. Myron ended his round of golf 3 under par. (Par is a score of 0.)

38. Denver, the mile high city, is 5,280 feet above sea level.

Jeff Gordon

Applications and Problem Solving

39. *Sports* Jeff Gordon won the 1997 Daytona 500. But Mark Martin led the pack for 12 laps more than Gordon. Express 12 laps more as an integer.

40. *Algebra* Consider two numbers A and B on a number line. Determine which is greater, $|A + B|$ or $|A| + |B|$. Always? Sometimes? Never?

41. *Critical Thinking* a and b are integers. Is it *true* or *false* that if $|a| > |b|$, then $a > b$? Use examples to prove your answer.

Mixed Review

42. *Algebra* Solve $4s < 36$. Graph the solution on a number line. *(Lesson 1-9)*

43. **Standardized Test Practice** Shanté wants to paint a wall that is 8 feet by 12 feet. A door that measures 7 feet by 3 feet is in the middle of the wall and will not be painted. How much surface area will be painted? *(Lesson 1-8)*

    A 20 sq ft   **B** 21 sq ft   **C** 75 sq ft   **D** 96 sq ft   **E** Not Here

44. *Algebra* Solve $4y = 196$. Check your solution. *(Lesson 1-5)*

45. *Algebra* Solve $30 = k - 141$. Check your solution. *(Lesson 1-4)*

46. Write $5 \cdot 5 \cdot 8 \cdot 8 \cdot 8$ using exponents. *(Lesson 1-2)*

For **Extra Practice**, see page 608.

2-2

Comparing and Ordering Integers

What you'll learn

You'll learn to compare and order integers.

When am I ever going to use this?

You can compare and order integers to determine the winner in a miniature golf game.

Did you know The original snowboards were called "snurfers" because the sport was compared to surfing on snow.

Have you ever gone snowboarding? Snowboarding is a rather new sport, which began in the 1960s. It is one of several sports that is increasing in popularity. The table shows the percent of increase or decrease of people participating in different sports from 1992 to 1993. Which sport has had the greatest percent of increase? Which had the greatest percent of decrease?

SPORTING TRENDS	
Sport	**Percent Increase or Decrease**
In-line Skating	27
Racquetball	−18
Snowboarding	50
Step Aerobics	15
Windsurfing	−22

Source: National Sporting Goods Association

Let's sketch a number line for the percents and graph each number.

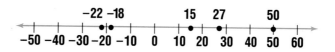

Use the number line to list the numbers from greatest to least. They would be in order from *right to left* on the number line.

$$50 \qquad 27 \qquad 15 \qquad -18 \qquad -22$$

You can use one of three symbols to compare any two numbers.

= *is equal to*
> *is greater than*
< *is less than*

Examples

Use the number line to compare integers. Replace each ● with >, <, or = to make a true sentence.

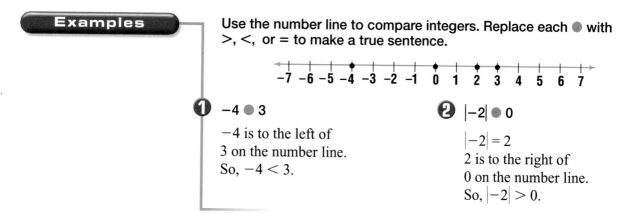

1 −4 ● 3

−4 is to the left of 3 on the number line.
So, −4 < 3.

2 |−2| ● 0

|−2| = 2
2 is to the right of 0 on the number line.
So, |−2| > 0.

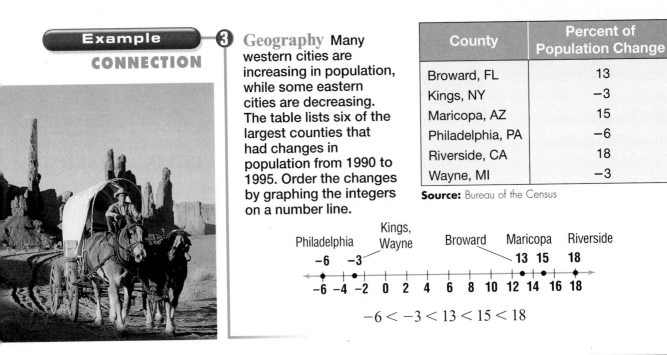

Example 3

CONNECTION

③ Geography Many western cities are increasing in population, while some eastern cities are decreasing. The table lists six of the largest counties that had changes in population from 1990 to 1995. Order the changes by graphing the integers on a number line.

County	Percent of Population Change
Broward, FL	13
Kings, NY	−3
Maricopa, AZ	15
Philadelphia, PA	−6
Riverside, CA	18
Wayne, MI	−3

Source: Bureau of the Census

$$-6 < -3 < 13 < 15 < 18$$

CHECK FOR UNDERSTANDING

Communicating Mathematics

Read and study the lesson to answer each question.

1. **Write** a comparison of −3 and 4 using an inequality sign. Write a second comparison using a different inequality sign.

2. **Tell** which of the following are true and which are false.
   a. $-5 < -3$ **b.** $-5 < |-3|$ **c.** $|-5| < -3$ **d.** $|-5| < |-3|$

3. **You Decide** Mariano says that the inequality symbol always points to the lesser number. Judy says it points to the greater number. Who is correct? Explain.

Guided Practice

Replace each ● with >, <, or = to make a true sentence.

4. $-3 ● 2$ 5. $2 ● -3$ 6. $|-23| ● 23$ 7. $34 ● 16$

8. Order the integers $\{45, -23, 55, 0, -12, -37\}$ from least to greatest.

9. **Space Exploration** In 1997, NASA established websites to allow the public to follow the Pathfinder Mission to Mars. The internal temperature of the Rover fluctuated between 40C° and −22C°. Write two inequalities to show the relationship between these temperatures.

EXERCISES

Practice

Replace each ● with >, <, or = to make a true sentence.

10. $4 ● -4$ 11. $5 ● |-5|$ 12. $0 ● |3|$ 13. $-6 ● -12$

14. $-35 ● -16$ 15. $0 ● |-12|$ 16. $|-8| ● 0$ 17. $|3| ● |-3|$

18. $-19 ● -22$ 19. $90 ● 21$ 20. $|-34| ● |-9|$ 21. $|-821| ● |821|$

22. Order the integers $\{-12, 12, -34, 56, -22, 34\}$ from least to greatest.

23. Order the integers $\{-450, 564, -356, -100, 254\}$ from greatest to least.

24. Order the integers $\{-467, -237, 1,276, -3,456, -943\}$ from greatest to least.

Applications and Problem Solving

For **Extra Practice,** see page 608.

25. *Recreation* For Nate's birthday, his family played miniature golf. On the first hole, Nate scored a 1, and his sister Jamila scored −1. Write two inequalities to compare their scores.

26. *Earth Science* The table shows the record low temperatures in the U.S. Graph the temperatures on a number line. Order them from least to greatest.

City	Temp. (°F)	Date
Island Park Dam, ID	−60	1/18/43
McIntosh, SD	−58	2/17/36
Moran, WY	−63	2/9/33
Prospect Creek Camp, AK	−80	1/23/71
Rogers Pass, MT	−70	1/20/54
Seneca, OR	−54	1/18/43

Source: National Climatic Data Center

27. *Critical Thinking* Consider the inequality $|x| < 3$.

a. Is 3 a solution to this inequality? Explain.

b. Graph all integer solutions for the equation on a number line.

Mixed Review

28. *Algebra* Evaluate $2 + |1 - 0|$. *(Lesson 2-1)*

29. *Algebra* Solve $102 = 17p$. Check your solution. *(Lesson 1-5)*

30. *Standardized Test Practice* Mrs. Acosta buys 2 flag pins for each of the 168 band members. Pins cost $0.09 each. Estimate the cost of the pins. *(Lesson 1-1)*

A $8   **B** $20   **C** $30   **D** $50   **E** $70

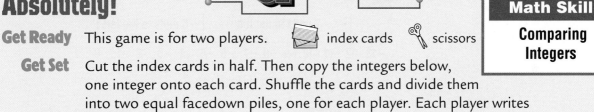

Let the Games Begin

Absolutely!

Math Skill

Comparing Integers

Get Ready This game is for two players. 📇 index cards ✂ scissors

Get Set Cut the index cards in half. Then copy the integers below, one integer onto each card. Shuffle the cards and divide them into two equal facedown piles, one for each player. Each player writes "absolute value" on two other cards and places these cards aside.

$$-17 \quad -3 \quad 0 \quad 19 \quad 16 \quad 5 \quad 25 \quad -10 \quad 3 \quad -2 \quad -8 \quad -7$$
$$7 \quad 6 \quad 9 \quad 22 \quad 11 \quad 12 \quad 1 \quad 14 \quad -20 \quad -13 \quad -16 \quad -18$$

Go ● Each player places the top card from his or her pile faceup. The partner with the greater card takes both cards and puts them facedown in a separate pile. When there are no more cards in your original pile, shuffle the cards in the second pile and use them.

● Twice during the game, each player can use an "absolute value" card after the two other cards have been played. When an absolute value card is played, players compare the absolute values of the integers on the cards. When there is a tie, continue play; the first player with a greater integer takes all of the upturned cards.

● The winner is the player who takes all of the cards.

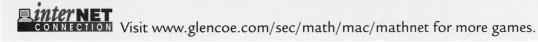

inter NET CONNECTION Visit www.glencoe.com/sec/math/mac/mathnet for more games.

2-3 Adding Integers

What you'll learn

You'll learn to add integers.

When am I ever going to use this?

You can add integers to find your profit at a bake sale.

Word Wise

zero pair

"Are you a spender or a saver?" A girl's magazine asked its website visitors this question in its weekly poll. They found that chances are you have a full piggy bank at home. Of those who answered the poll, 53 percent called themselves savers. About 27 percent like to save half and spend half when they get some money. And for 20 percent, money gets spent right away.

Source: *American Girl* magazine

You can use integers to keep track of your savings and spending.

- Receiving money can be modeled as a positive number. For example, earning $10 from mowing the lawn can be written as $+10$.

- When you spend money or give it to someone else, that can be modeled as a negative number. So if you spent $3 on a magazine, that is written as -3.

Suppose Amy borrowed some money from her friend Yoko. She borrowed $5 for lunch and later borrowed $7 for a movie. How much does Amy owe Yoko? Let a = the total amount Amy owes.

$$
\underbrace{lunch\ \$}\quad \underbrace{plus}\quad \underbrace{movie\ \$}\quad \underbrace{is}\quad \underbrace{total\ owed}
$$
$$
-5 \quad + \quad (-7) \quad = \quad a
$$

Addition can be modeled on a number line. First graph -5. Since -7 is negative, you will move 7 units to the left of -5.

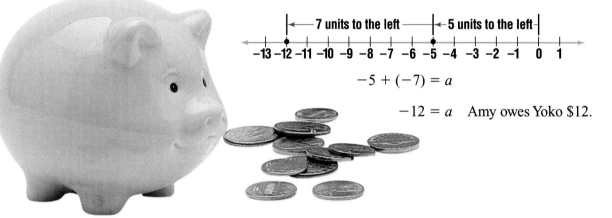

$$-5 + (-7) = a$$

$$-12 = a \quad \text{Amy owes Yoko \$12.}$$

You can also use counters to add integers.

Example

1 Solve $-9 + (-6) = s$.

Put 9 negative counters on a mat. Add 6 more negative counters.

There are 15 negative counters. Therefore, $s = -15$. The solution is -15.

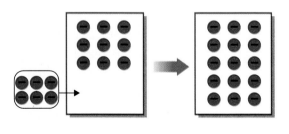

You already know how to add two positive integers. For example, to solve $r = 23 + 34$, you simply add 23 and 34. The solution is 57. In Example 1, you found that the sum $-9 + (-6)$ is -15. Notice that the value of $-(9 + 6)$ is the same. In each case, you add the absolute values of the addends. The sum has the same sign as the integers.

Adding Integers with Same Sign	To add integers with the same sign, add their absolute values. Give the result the same sign as the integers.

If you add a negative integer and a positive integer, what do you suppose happens? Let's use counters to find a rule.

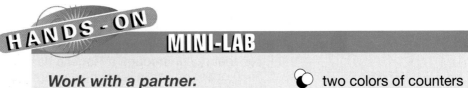

MINI-LAB

Work with a partner.

two colors of counters

Find the sum $5 + (-4)$ using counters.

integer mat

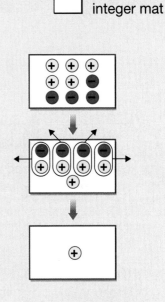

Try This

- Let one color counter represent positive and the other color represent negative. Here we used yellow counters for positive and red for negative.
- Place 5 positive and 4 negative counters on the mat.
- When a positive counter is paired with a negative counter, the result is called a **zero pair**. Since a zero pair has a value of zero, you can add or remove zero pairs without changing the value of the set. Remove all of the zero pairs from the mat.
- The counters you have left represent the solution.

Talk About It

1. Is the sum of $5 + (-4)$ positive or negative?
2. Which number, 5 or -4, has the greater absolute value?
3. Use counters to find the sum $-5 + 4 = t$. Compare the sign of the sum to the sign of the number with the greater absolute value.
4. Make a conjecture about the sign of the sum $-9 + 8 = m$.

The results of the Mini-Lab suggest the following rule for adding two integers with different signs.

Adding Integers with Different Signs	To add integers with different signs, subtract their absolute values. Give the result the same sign as the integer with the greater absolute value.

Examples

2 Solve $f = (-17) + 20$.

$|20| > |-17|$, so the sum is positive.

The difference of 20 and 17 is 3, so $f = 3$.

3 Solve $52 + (-60) = m$.

$|-60| > |52|$, so the sum is negative.

The difference of 60 and 52 is 8, so $m = -8$.

APPLICATION

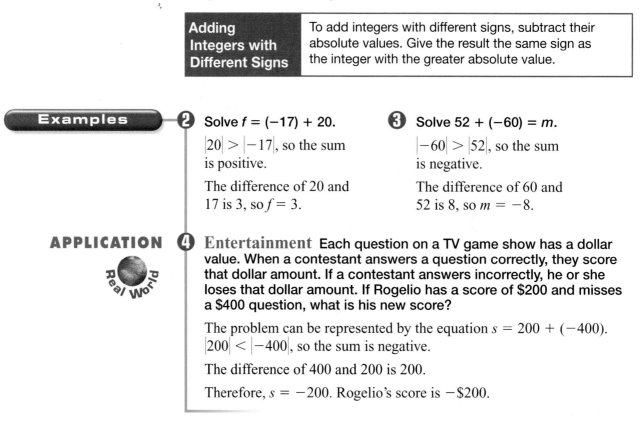

4 **Entertainment** Each question on a TV game show has a dollar value. When a contestant answers a question correctly, they score that dollar amount. If a contestant answers incorrectly, he or she loses that dollar amount. If Rogelio has a score of $200 and misses a $400 question, what is his new score?

The problem can be represented by the equation $s = 200 + (-400)$. $|200| < |-400|$, so the sum is negative.

The difference of 400 and 200 is 200.

Therefore, $s = -200$. Rogelio's score is $-\$200$.

CHECK FOR UNDERSTANDING

Communicating Mathematics

Read and study the lesson to answer each question.

1. *Demonstrate* how to find the sum of two negative numbers on a number line.

2. *Explain* how the absolute values of numbers are used when adding two integers.

HANDS-ON MATH

3. *Find* the sums $(-3) + 2$ and $2 + (-3)$ using counters. Does the order of the addends make a difference? Explain.

Guided Practice

Solve each equation.

4. $t = 4 + 6$
5. $-4 + (-3) = q$
6. $6 + (-3) = s$
7. $g = 3 + (-8)$
8. $w = (-32) + 44$
9. $-4 + 4 = y$

10. Find the sum $45 + (-67)$.

11. *Algebra* Evaluate $w + (-7)$ if $w = -3$.

12. *Entertainment* Did you know that a recent movie with Steve Martin was a remake of a movie made 41 years earlier? The remake was made in 1991.

 a. Let y = the year the original movie was made. Write an addition equation for y involving integers.

 b. Solve the equation to find when the original movie was made.

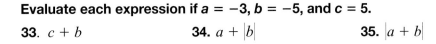
Practice

Solve each equation.

13. $-45 + (-5) = p$ **14.** $5 + 17 = r$ **15.** $y = -13 + 13$

16. $s = (-31) + (-28)$ **17.** $v = 71 + (-60)$ **18.** $k = 5 + (-12)$

19. $a = 34 + (-60)$ **20.** $d = -34 + 75$ **21.** $(-9) + 29 = c$

22. $-18 + (-41) = u$ **23.** $h = 45 + 63$ **24.** $m = 35 + (-32)$

25. $56 + (-34) = m$ **26.** $-19 + (-37) = z$ **27.** $b = -98 + (-32)$

28. $t = 56 + (-2)$ **29.** $-60 + 30 = r$ **30.** $-23 + (-456) = n$

31. Find the sum $76 + (-45)$.

32. What is the value of $-319 + (-100)$?

Evaluate each expression if $a = -3$, $b = -5$, and $c = 5$.

33. $c + b$ **34.** $a + |b|$ **35.** $|a + b|$

Applications and Problem Solving

36. *Earth Science* The highest waterfall in the world is Angel on the Carrao River in Venezuela. The longest single drop is 2,648 feet! The water falls another 268 feet in the rest of the drops.

    a. Let f = the change in the water level. Write an addition equation for f involving integers.

    b. Solve the equation to find the length that the water falls.

37. *Crime* Recently, thefts in the United States totaled $14,608 million. Police were able to recover goods worth $5,202 million.

    a. Let v = the value of the unrecovered goods. Write an addition equation for v involving integers.

    b. Solve the equation to find the value of the unrecovered goods.

38. *Working on the* **CHAPTER Project** Refer to the data on bald eagles on page 55. There were 3,000 fewer bald eagles in 1990 than in 1994.

    a. Use $p + 3,000 = 9,000$ to find the number of bald eagles in 1990.

    b. Write an equation to compare the populations of your species in two different years.

39. *Critical Thinking* Are values of $|-3| + |5|$, $|-3 + 5|$, and $-3 + 5$ the same? Explain.

Mixed Review

40. Order the integers $\{226, -3, 18, -157, 2, -28\}$ from least to greatest. *(Lesson 2-2)*

41. *Standardized Test Practice* Jen had $15 and wanted to spend no more than that to buy 4 notebooks for school. Which inequality could she use to find the price, p, of each notebook? *(Lesson 1-9)*

  A $p + 4 \leq 15$ **B** $4p \geq 15$ **C** $15 > 4p$

  D $4p \leq 15$ **E** $15 - p \leq 4$

42. *Algebra* Solve $6x = 42$. Check your solution. *(Lesson 1-5)*

43. Evaluate $[3(18 - 2)] - 4^2$. *(Lesson 1-3)*

For **Extra Practice**, see page 609.

44. Write $4 \cdot 4 \cdot 6 \cdot 6 \cdot 6$ using exponents. *(Lesson 1-2)*

More About Adding Integers

What you'll learn

You'll learn to add three or more integers.

When am I ever going to use this?

You can add integers to find the total yardage gained by a football team in a series of plays.

You have probably been in or seen a tug-of-war. But did you know that the history of the sport goes back to the ancient Egyptians? Teams of three men would hold each other by the waist and try to pull the other team across a mark. No rope was used then. From there the sport spread to ancient Greece, China, Scandinavia, and western Europe. Now the Tug-of-War International Federation organizes championship tournaments for men's and women's teams.

The progress of a tug-of-war team can be modeled by adding integers. Pulling the opponent toward the mark is a positive number, and being pulled is a negative number.

Suppose a United States team is competing with a Canadian team and the U.S. moves at the end of each of the first five minutes of competition are $+1, -1, +1, +2, -1$. If each integer represents a number of feet, what is the position of the mark at the end of the first five minutes?

LOOK BACK
You can refer to Lesson 1-2 to review the commutative and associative properties.

Let p = the position of the mark.

$p = 1 + (-1) + 1 + 2 + (-1)$

$p = [1 + (-1)] + (1 + 2) + (-1)$ *Use the associative property to group the addends.*

$p = 0 + 3 + (-1)$

$p = (0 + 3) + (-1)$

$p = 3 + (-1)$

$p = 2$ The United States team has pulled the Canadian team 2 feet.

Sometimes using both the associative and commutative properties allows you to find the result in fewer steps. One way is to group all of the positive numbers and all of the negative numbers.

$$p = 1 + (-1) + 1 + 2 + (-1) \quad \rightarrow \quad p = (1 + 1 + 2) + (-1 + (-1))$$
$$p = 4 + (-2)$$
$$p = 2$$

Example

1 Solve $d = 7 + 20 + (-5)$.

Use the associative property to group the first two addends.

$d = (7 + 20) + (-5)$

$d = 27 + (-5)$

$d = 22$

Check: Use the associative property to group the last two addends.

$d = 7 + (20 + (-5))$

$d = 7 + 15$

$d = 22$ ✓

2 Solve $w = -4 + 3 + (-3) + 4$.

Look for groupings that make the addition simpler. Adding 0 is easy, and since $-4 + 4 = 0$ and $(-3) + 3 = 0$, group these.

$w = -4 + 3 + (-3) + 4$ **Check:** $w = -4 + 3 + (-3) + 4$

$w = (-4 + 4) + (-3 + 3)$ $w = (-4 + (-3)) + (3 + 4)$

$w = 0 + 0$ $w = -7 + 7$

$w = 0$ $w = 0$ ✓

APPLICATION

Real World

3 **Music** They say that rock and roll is here to stay! But the sales of rock music are not the same from one year to the next. In 1991, 35% of the music sold was rock. The table shows the change in sales in the successive years. What percent of the music sold in 1995 was rock?

Years	Change
1991-1992	-3
1992-1993	-1
1993-1994	+5
1994-1995	-1

Source: Recording Industry Assn. of America

$t = 35 + (-3) + (-1) + 5 + (-1)$

$t = (35 + 5) + (-3 + (-1) + (-1))$

$t = 40 + (-4 + (-1))$

$t = 40 + (-5)$

$t = 35$

Check: $35 - 3 - 1 + 5 - 1 = 35$ ✓

Study Hint

Technology To enter -54 in a scientific calculator, press 54 and then press the [+○−] key. A negative sign will appear.

CHECK FOR UNDERSTANDING

Communicating Mathematics

Read and study the lesson to answer each question.

1. *Write* a sentence to tell how you might use the commutative and associative properties to solve $x = -13 + (-45) + 15 + (-55) + 25 + 13$ mentally.

2. *Show* three different ways you can solve $t = -3 + 4 + (-5) + 6$.

Guided Practice

Solve each equation. Check by solving another way.

3. $h = 8 + (-4) + 9$

4. $n = -3 + 8 + 9$

5. $16 + (-33) + (-14) + 33 = q$

6. $-235 + 613 + (-844) + 361 = y$

7. *Algebra* Evaluate the expression $d + (-8) + 4 + 3$ if $d = 5$.

8. *Sports* During a fourth-quarter possession in Super Bowl XXXI, the Green Bay Packers gained or lost the following yards in 10 plays. What was the net gain on the series?

Play	1	2	3	4	5	6	7	8	9	10
Yards Gained/Lost	0	7	18	7	1	-5	7	3	1	0

Practice

Solve each equation. Check by solving another way.

9. $d = -3 + 5 + (-9)$

10. $x = 3 + 21 + (-6)$

11. $-7 + 12 + 9 = v$

12. $3 + (-2) + (-10) + 6 = b$

13. $m = 4 + (-8) + 12 + (-11)$

14. $g = 23 + 19 + (-8) + 12$

15. $9 + 50 + 3 + (-50) = k$

16. $14 + 7 + (-23) + 10 = p$

17. $8 + (-4) + 10 + (-2) = s$

18. $-6 + 12 + (-11) + 1 = r$

19. $t = 92 + 73 + (-51) + 100$

20. $y = 17 + (-21) + 10 + (-17)$

Evaluate each expression if $x = -4$, $y = -5$, and $z = 4$.

21. $2 + (-6) + x + 10$

22. $-7 + y + z$

23. $z + 4 + (-9)$

24. $-144 + x + z + 2$

25. *Write a Problem* that can be more easily solved by using at least two mental math strategies.

Applications and Problem Solving

26. *Geography* In 1970, the population of Worcester, Massachusetts, was 177 thousand. By 1980, the population had decreased by 12 thousand. From 1980 to 1990, the population increased 8 thousand. What was the population of Worcester in 1990?

27. *Entertainment* The table shows the revenue of the top ten money-losing movies.

a. How much money was lost by these movies?

b. One company produced both Movie A and Movie B. Suppose they are able to make $30,000 on video sales of Movie B and $120,000 on television showings for Movie A. What would their total gain or loss be on these movies?

Movie	Revenue
A	−$48,100,000
B	−$47,300,000
C	−$47,000,000
D	−$44,100,000
E	−$38,100,000
F	−$37,000,000
G	−$34,200,000
H	−$33,000,000
I	−$30,300,000
J	−$30,000,000

28. *Critical Thinking* Beng was trying to evaluate $6 + (-9) + 14$ using a calculator. He accidentally pressed the [+○−] key before 9 instead of after. Is the result still correct? Explain.

Mixed Review

29. **Standardized Test Practice** A stock on the New York Stock Exchange opened at $52 on Monday morning. During the week, the stock lost $2, gained $1, gained $3, lost $1, and lost $4. What was the stock worth at the close of business on Friday? *(Lesson 2-3)*

   A $41   **B** $49   **C** $57   **D** $63   **E** Not Here

30. Find $|4^2|$. *(Lesson 2-1)*

31. *Geometry* Find the area of a square with sides of 8 meters. *(Lesson 1-8)*

32. *Algebra* Write an expression for *17 more than p*. *(Lesson 1-6)*

For **Extra Practice**,
see page 609.

Subtracting Integers

How do you spend your leisure time? The graph shows how the way people spend leisure time has changed over a five-year period. What is the difference between the change in percent watching videos and the change of percent watching TV? *This problem will be solved in Example 3.*

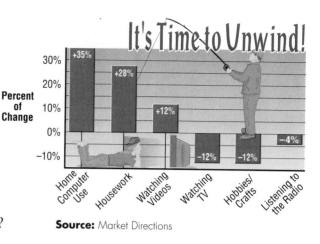

Source: Market Directions

In order to solve this problem, you will need to subtract integers. You can use counters to show subtraction.

HANDS-ON MINI-LAB

Work with a partner.

⚫⚪ two colors of counters

☐ integer mat

Try This

Solve $x = -3 - (-1)$.

- Put 3 negative counters on the mat.
- Remove 1 negative counter.
- What is the value of x?

Solve $z = 4 - (-2)$.

- Put 4 positive counters on the mat.
- You need to remove 2 negative counters, but there are none on the mat. Add 2 zero pairs to the mat. Now remove the 2 negative counters.
- What is the value of z?

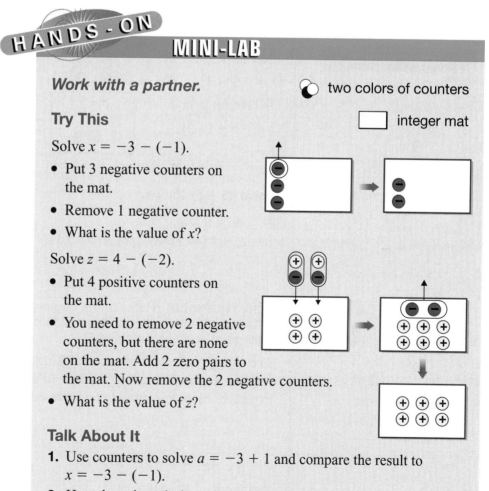

Talk About It

1. Use counters to solve $a = -3 + 1$ and compare the result to $x = -3 - (-1)$.
2. How does the solution of $c = 4 + 2$ compare to $z = 4 - (-2)$?

Each integer has an opposite. The **opposite** is the number that is the same distance from zero but in the opposite direction. For example, the opposite of 3 is −3. The opposite of any number is called its **additive inverse**. Additive inverses can be used to subtract integers.

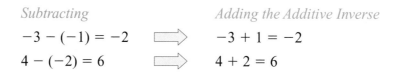

Additive Inverse	**Words:** The sum of an integer and its additive inverse is 0.
	Symbols: **Arithmetic** **Algebra**
	$4 + (-4) = 0$ $a + (-a) = 0$

In the Mini-Lab, you compared the result of subtracting an integer with the result of adding its inverse.

Subtracting   *Adding the Additive Inverse*

$-3 - (-1) = -2$ ⟹ $-3 + 1 = -2$

$4 - (-2) = 6$ ⟹ $4 + 2 = 6$

Notice that adding the additive inverse of an integer produces the same result as subtracting the integer.

Subtracting Integers	To subtract an integer, add its additive inverse.

Examples

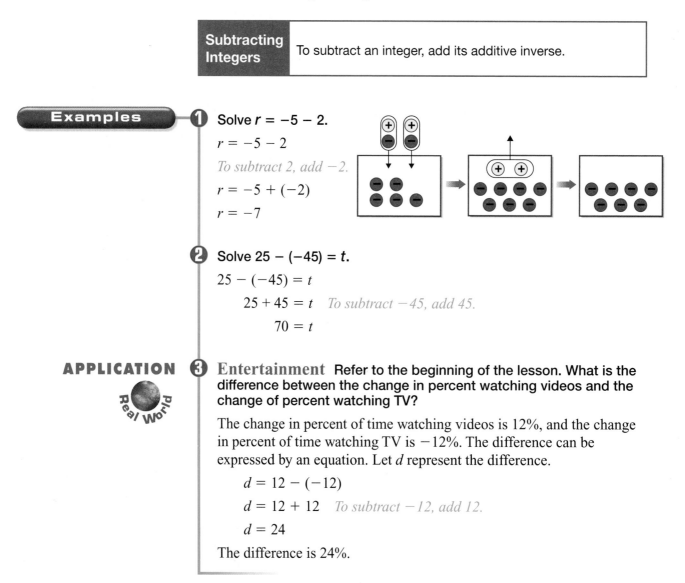

① Solve $r = -5 - 2$.

$r = -5 - 2$

To subtract 2, add −2.

$r = -5 + (-2)$

$r = -7$

② Solve $25 - (-45) = t$.

$25 - (-45) = t$

$25 + 45 = t$ *To subtract −45, add 45.*

$70 = t$

APPLICATION

Real World

③ **Entertainment**   Refer to the beginning of the lesson. What is the difference between the change in percent watching videos and the change of percent watching TV?

The change in percent of time watching videos is 12%, and the change in percent of time watching TV is −12%. The difference can be expressed by an equation. Let d represent the difference.

$d = 12 - (-12)$

$d = 12 + 12$ *To subtract −12, add 12.*

$d = 24$

The difference is 24%.

Communicating Mathematics

Read and study the lesson to answer each question.

1. *Write* how you find the additive inverse of an integer. Use the term *opposite* in your explanation. Give an example of an integer and its additive inverse.

2. *Tell* how you know whether a subtraction problem will have a positive or a negative answer.

HANDS-ON MATH

3. *Draw* a model showing $3 - (-1) = a$. Then solve for a.

4. *Write* the additive inverse of -3.

Guided Practice

Solve each equation.

5. $s = -4 - (-3)$
6. $r = 3 - (-2)$
7. $t = 4 - (-5)$
8. $-23 - (34) = y$
9. $b = 3 - 3$
10. $0 - (-4) = d$

11. *Algebra* Find the value of $68 - a$ if $a = -7$.

12. *Spreadsheets* Millions of people use software to organize their finances. The spreadsheet below is an example printout of such software. The software finds the balance after each transaction is entered. Write an addition or subtraction equation and find the balance for each transaction.

Date	Num	Payee/Category/Memo	Payment		Ck	Deposit		Balance	
1/27/98		Opening Balance				534	00	534	00
		(Checking)							
1/30/98	234	Food King	50	00					
		Groceries							
2/01/98		Deposit				30	00		
		Rebate							
2/03/98	ATM	Cash	20	00					
		Entertainment							

Practice

Write the additive inverse of each integer.

13. 7
14. -37
15. 0

Solve each equation.

16. $h = 4 - (-7)$
17. $-23 - (-2) = g$
18. $14 - 14 = d$
19. $x = 53 - 78$
20. $y = 17 - (-26)$
21. $w = -43 - 88$
22. $-78 - (-98) = z$
23. $k = 44 - (-11)$
24. $u = 4 - (-89)$
25. $n = 56 - (-22)$
26. $435 - (-878) = b$
27. $w = 63 - 92$
28. $r = -9 - (-4)$
29. $-345 - 67 = t$
30. $s = -34 - (-25)$
31. $v = 823 - (-19)$
32. $-13 - (-12) = x$
33. $d = 0 - (-6)$

34. Find the value of s for $s = 56 - (-78)$.

35. What value of f makes $-18 - 0 = f$ true?

Evaluate each expression if $a = 9$, $b = -6$, and $c = -2$.

36. $78 - b$
37. $c - a$
38. $12 - a - b$

39. *Write a Problem* involving subtraction of integers for which the answer is −3.

40. *Geography* The table shows the elevations above sea level of the surface and the deepest point of each of the Great Lakes.

Lake	Elevation of Deepest Point (m)	Elevation of Surface (m)
Erie	−64	174
Huron	−229	176
Michigan	−281	176
Ontario	−244	75
Superior	−406	183

Source: National Ocean Service

a. How far below the surface is the deepest part of each lake?

b. How does the deepest part of Lake Ontario compare with the deepest part of Lake Superior?

c. Find the difference between the deepest part of Lake Erie and the deepest part of Lake Superior.

41. *History* The calendar that astronomers use began on Jan 1, 4713 B.C. On that day, the Julian calendar, the lunar calendar, and the Roman tax system calendar all coincided. This won't happen again until A.D. 3267!

a. How many years ago was the astronomer's calendar started? (*Hint:* There was no year 0.)

b. Find the number of years between times that the three calendars coincide.

42. *Critical Thinking* Do an integer and its additive inverse always have different signs? Is there an integer that is its own inverse?

Mixed Review

43. Solve $44 + 8 + (−20) + 15 = s$. Check your solution. *(Lesson 2-4)*

44. Replace ● with >, <, or = to make $10 ● −10$ a true sentence. *(Lesson 2-2)*

45. **Standardized Test Practice** If $12 + 5d = 72$, what is the value of d? *(Lesson 1-7)*

    A 10   **B** 12   **C** 14   **D** 16

46. *Language* There are 999 million people who speak Mandarin. This is 512 million more than speak English. Write an equation to find the number of people who speak English. *(Lesson 1-6)*

For **Extra Practice**, see page 609.

47. *Algebra* Solve $c + 9 = 27$ if the replacement set is {12, 17, 18, 22}. *(Lesson 1-3)*

CHAPTER 2

Mid-Chapter Self Test

1. Graph {8, 4, −3, 2, 0, −8, −1} on a number line. *(Lesson 2-1)*

2. Write {−7, 8, 0, −3, −2, 5, 6} in order from least to greatest. *(Lesson 2-2)*

Solve each equation. *(Lessons 2-3, 2-4, and 2-5)*

3. $x = −5 + 3$ **4.** $514 − 600 = s$ **5.** $y = 2 + (−4) + (−6) + 8$

2-6

Integration: Statistics
Matrices

What you'll learn

You'll learn to use matrices to organize data.

When am I ever going to use this?

You can use a matrix to record sales of different items in a bookstore.

Word Wise

matrix
row
column
element

Keeping up with the fashion fads is hard work. And for the corporations that make the fashions, being behind the times can mean going out of business. The table lists the sales of the top three clothing makers in the United States for 1995 and 1996.

Corporation	1995 Sales (millions)	1996 Sales (millions)
Company A	$4,761	$6,471
Company B	$5,061	$5,137
Company C	$3,518	$3,483

Another way to organize information is by using a **matrix**. A matrix is a rectangular arrangement of numbers in **rows** and **columns**. Each number in a matrix is called an **element** of the matrix.

The matrix at the right has 2 rows and 3 columns. Elements are named by telling the row and the column in which they appear.

$$\begin{array}{c} \textit{column 3} \\ \downarrow \\ \begin{bmatrix} 3 & 8 & 0 \\ \boxed{5} & -2 & -3 \end{bmatrix} \leftarrow \textit{row 2} \\ \uparrow \\ \textit{element (2, 1)} \end{array}$$

Example
APPLICATION
Real World

1 **Business** Refer to the beginning of the lesson. Write a matrix for the apparel sales data.

The matrix will have a row for each corporation and a column for each year. Only the numbers are part of the matrix. The labels are written outside of the matrix.

$$\begin{array}{ccc} & \textbf{1995} & \textbf{1996} \\ & \textbf{Sales} & \textbf{Sales} \\ \textbf{Company A} & \begin{bmatrix} 4,761 & 6,471 \\ \textbf{Company B} & 5,061 & 5,137 \\ \textbf{Company C} & 3,518 & 3,483 \end{bmatrix} \end{array}$$

You can add or subtract matrices that have the same number of rows and columns. You add and subtract matrices by adding or subtracting the corresponding elements.

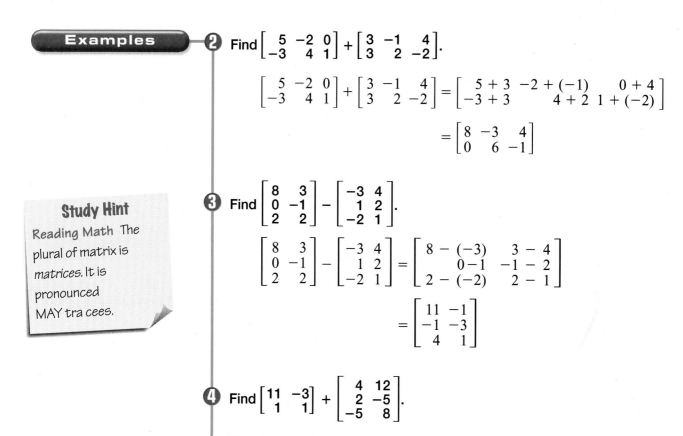

Examples

2 Find $\begin{bmatrix} 5 & -2 & 0 \\ -3 & 4 & 1 \end{bmatrix} + \begin{bmatrix} 3 & -1 & 4 \\ 3 & 2 & -2 \end{bmatrix}$.

$$\begin{bmatrix} 5 & -2 & 0 \\ -3 & 4 & 1 \end{bmatrix} + \begin{bmatrix} 3 & -1 & 4 \\ 3 & 2 & -2 \end{bmatrix} = \begin{bmatrix} 5+3 & -2+(-1) & 0+4 \\ -3+3 & 4+2 & 1+(-2) \end{bmatrix}$$

$$= \begin{bmatrix} 8 & -3 & 4 \\ 0 & 6 & -1 \end{bmatrix}$$

Study Hint

Reading Math The plural of matrix is matrices. It is pronounced MAY tra cees.

3 Find $\begin{bmatrix} 8 & 3 \\ 0 & -1 \\ 2 & 2 \end{bmatrix} - \begin{bmatrix} -3 & 4 \\ 1 & 2 \\ -2 & 1 \end{bmatrix}$.

$$\begin{bmatrix} 8 & 3 \\ 0 & -1 \\ 2 & 2 \end{bmatrix} - \begin{bmatrix} -3 & 4 \\ 1 & 2 \\ -2 & 1 \end{bmatrix} = \begin{bmatrix} 8-(-3) & 3-4 \\ 0-1 & -1-2 \\ 2-(-2) & 2-1 \end{bmatrix}$$

$$= \begin{bmatrix} 11 & -1 \\ -1 & -3 \\ 4 & 1 \end{bmatrix}$$

4 Find $\begin{bmatrix} 11 & -3 \\ 1 & 1 \end{bmatrix} + \begin{bmatrix} 4 & 12 \\ 2 & -5 \\ -5 & 8 \end{bmatrix}$.

The first matrix has 2 rows and 2 columns, and the second matrix has 3 rows and 2 columns, so these matrices cannot be added or subtracted.

CHECK FOR UNDERSTANDING

Communicating Mathematics

Read and study the lesson to answer each question.

1. *Tell* how many rows and columns there are in the matrix at the right. $\begin{bmatrix} 4 & -1 \\ 1 & 3 \\ 2 & -4 \end{bmatrix}$

2. *Explain* when you can add or subtract two matrices.

Guided Practice

Find each sum or difference. If there is no sum or difference, write *impossible*.

3. $\begin{bmatrix} 1 & 4 \\ 3 & -2 \end{bmatrix} + \begin{bmatrix} -2 & 3 \\ 2 & 0 \end{bmatrix}$

4. $\begin{bmatrix} 0 & 6 & 3 \\ -12 & -7 & 9 \end{bmatrix} - \begin{bmatrix} -3 & 5 & 5 \\ -8 & -7 & -1 \end{bmatrix}$

5. $\begin{bmatrix} -1 & 2 & 0 \\ 2 & 1 & 0 \\ 4 & 1 & 3 \end{bmatrix} - \begin{bmatrix} -1 & -2 & -1 \\ 4 & 4 & 2 \\ 1 & 1 & 0 \end{bmatrix}$

6. $\begin{bmatrix} 3 & 7 \\ -2 & -4 \\ 6 & -6 \end{bmatrix} + \begin{bmatrix} 4 & 16 & 8 \\ -9 & 14 & 5 \end{bmatrix}$

7. *Life Science* Manatees have been living off of the coast of Florida for centuries. A decline in population has led to preservation and protection programs. The table shows the number of manatees that died from January to April of 1996. Write a matrix of the data.

Cause of Death	J	F	M	A
Watercraft	8	4	7	6
Weather	13	2	0	0
Natural Causes	7	5	47	24
Undetermined	13	12	47	36

Source: Florida Department of Environmental Protection

EXERCISES

Practice

Find each sum or difference. If there is no sum or difference, write *impossible*.

8. $\begin{bmatrix} 3 & -2 \\ 4 & -1 \end{bmatrix} + \begin{bmatrix} -3 & 8 \\ 0 & 6 \end{bmatrix}$

9. $\begin{bmatrix} 7 & 20 & 6 \\ -1 & 15 & 1 \end{bmatrix} - \begin{bmatrix} 4 & 3 \\ -5 & -3 \\ -1 & 0 \end{bmatrix}$

10. $\begin{bmatrix} 12 & -7 \\ -5 & 9 \\ 7 & 11 \end{bmatrix} - \begin{bmatrix} 13 & -7 \\ -7 & 3 \\ 2 & 13 \end{bmatrix}$

11. $\begin{bmatrix} 6 & -9 \\ -11 & 17 \\ 8 & 10 \end{bmatrix} + \begin{bmatrix} 5 & -3 \\ 12 & 9 \\ -5 & 18 \end{bmatrix}$

12. $\begin{bmatrix} 1 & 8 \\ -2 & 16 \end{bmatrix} - \begin{bmatrix} 8 & -5 & 0 \\ -6 & 1 & 1 \\ 1 & -3 & 0 \end{bmatrix}$

13. $\begin{bmatrix} 14 & 23 & 9 \\ -5 & -7 & 4 \end{bmatrix} - \begin{bmatrix} 9 & 18 & 6 \\ -6 & -4 & -6 \end{bmatrix}$

14. $\begin{bmatrix} 4 & 3 & -10 \\ 0 & -1 & 4 \\ 0 & 0 & -6 \end{bmatrix} - \begin{bmatrix} -4 & 5 & 2 \\ -4 & 9 & 6 \\ 1 & 0 & 0 \end{bmatrix}$

15. $\begin{bmatrix} 7 & -3 & -1 \\ 5 & -4 & 14 \\ 7 & 22 & -6 \end{bmatrix} - \begin{bmatrix} -9 & -7 & 6 \\ 0 & -3 & 9 \\ 6 & 13 & 6 \end{bmatrix}$

16. $\begin{bmatrix} 1 \\ 2 \\ 9 \end{bmatrix} - \begin{bmatrix} -5 & -2 & -7 \end{bmatrix}$

17. $\begin{bmatrix} 1 & 23 & -11 \\ 0 & -8 & 6 \\ 0 & 2 & -9 \end{bmatrix} + \begin{bmatrix} -1 & -15 & 7 \\ 0 & -1 & 4 \\ 1 & 19 & 6 \end{bmatrix}$

18. $\begin{bmatrix} 18 & -3 & 10 \\ 14 & 28 & 7 \\ -9 & -6 & -7 \end{bmatrix} + \begin{bmatrix} 5 & 17 & 1 \\ -6 & -15 & 0 \\ -8 & 14 & 4 \end{bmatrix}$

19. $\begin{bmatrix} 3 & 12 & 22 \\ 4 & 8 & 21 \\ 9 & -7 & -6 \\ -5 & 18 & -9 \end{bmatrix} + \begin{bmatrix} -3 & -2 & -2 \\ 5 & -9 & 27 \\ -6 & -4 & -2 \\ -1 & 16 & 18 \end{bmatrix}$

Applications and Problem Solving

20. *Agriculture* The table shows the number of thousands of livestock on farms in the United States in 1994 to 1996. Write a matrix for the data.

	Livestock on U.S. Farms (thousands)		
Year	Beef Cattle/ Milk Cows	Sheep	Hogs
1994	110,516	9,742	57,904
1995	112,252	8,886	59,900
1996	113,231	8,457	58,700

Source: The World Almanac

21. Business The table shows the expenses for the top three apparel corporations for 1995 and 1996.

Corporation	1995 Expenses (millions)	1996 Expenses (millions)
Company A	$4,362	$5,917
Company B	$4,904	$4,838
Company C	$3,353	$3,344

   a. Write a matrix for the expenses data.

   b. Find the profit matrix by subtracting the matrix in part a from the sales matrix in Example 1.

22. Critical Thinking In which row and column of the matrix will the number 100 occur?

$$\begin{bmatrix} 1 & 3 & 6 & 10 & 15 & \dots \\ 2 & 5 & 9 & 14 & 20 & \dots \\ 4 & 8 & 13 & 19 & 26 & \dots \\ 7 & 12 & 18 & 25 & 33 & \dots \\ 11 & 17 & 24 & 32 & 41 & \dots \\ 16 & 23 & 31 & 40 & 50 & \\ \vdots & \vdots & \vdots & \vdots & \vdots & \end{bmatrix}$$

Mixed Review

23. Standardized Test Practice Solve $b = 65 - (-87)$. *(Lesson 2-5)*

   A 152   **B** 22   **C** -22 **D** -152 **E** Not Here

24. Solve $h = -8 + 4 - (-3)$. *(Lesson 2-5)*

25. Education Do you or someone you know attend school at home? According to the U.S. Department of Education, more than fifty-six times as many students are being taught at home now as there were in the 1970s. If 12,500 students were home-schooled in the 1970s, how many are home-schooled now? *(Lesson 1-9)*

For **Extra Practice**, see page 610.

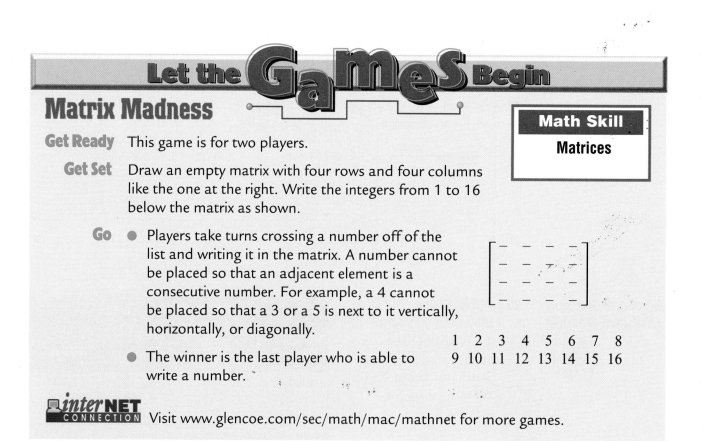

Let the GameS Begin

Matrix Madness

Math Skill
Matrices

Get Ready This game is for two players.

Get Set Draw an empty matrix with four rows and four columns like the one at the right. Write the integers from 1 to 16 below the matrix as shown.

Go ● Players take turns crossing a number off of the list and writing it in the matrix. A number cannot be placed so that an adjacent element is a consecutive number. For example, a 4 cannot be placed so that a 3 or a 5 is next to it vertically, horizontally, or diagonally.

$$\begin{bmatrix} - & - & - & - \\ - & - & - & - \\ - & - & - & - \\ - & - & - & - \end{bmatrix}$$

1 2 3 4 5 6 7 8
9 10 11 12 13 14 15 16

   ● The winner is the last player who is able to write a number.

interNET CONNECTION Visit www.glencoe.com/sec/math/mac/mathnet for more games.

GRAPHING CALCULATORS

2-6B Matrices

A Follow-Up of Lesson 2-6

graphing calculator

You can use most graphing calculators to perform operations with matrices. On some calculators, the [MATRX] key accesses the matrix operations.

TRY THIS

Work with a partner.

Enter matrix $A = \begin{bmatrix} -5 & 1 \\ 2 & -1 \\ 0 & 4 \end{bmatrix}$ and $B = \begin{bmatrix} 8 & -1 \\ -4 & 3 \\ 2 & -2 \end{bmatrix}$. Find $A + B$.

Step 1 Begin by entering matrix A into the calculator's memory.

Enter: [MATRX] [◄] [ENTER] *Choose the edit option and matrix A.*

3 [ENTER] 2 [ENTER] *Enter number of rows and columns.*

[(−)] 5 [ENTER] 1 [ENTER] *Enter each matrix element.*

2 [ENTER] [(−)] 1 [ENTER]

0 [ENTER] 4 [ENTER]

Step 2 Next enter matrix B.

Enter: [MATRX] [◄] 2 *Choose the edit option and matrix B.*

3 [ENTER] 2 [ENTER] *Enter number of rows and columns.*

8 [ENTER] [(−)] 1 [ENTER] *Enter each matrix element.*

[(−)] 4 [ENTER] 3 [ENTER] 2 [ENTER] [(−)] 2 [ENTER]

Step 3 Find the sum of matrices A and B.

Enter: [2nd] [QUIT] *Exit EDIT screen.*

[MATRX] 1 *Choose matrix A.*

[+] [MATRX] *Add matrix B.*

2 [ENTER]

```
[A]+[B]
          [[3   0]
           [-2  2]
           [2   2]]
```

ON YOUR OWN

Enter the matrices into a graphing calculator. Then find each sum or difference.

$$D = \begin{bmatrix} 4 & 0 & 2 \\ 2 & 1 & 1 \\ 3 & -1 & 4 \end{bmatrix} \qquad E = \begin{bmatrix} 1 & 5 & 3 \\ -3 & 10 & 6 \\ 6 & -7 & -2 \end{bmatrix} \qquad F = \begin{bmatrix} 5 & -3 \\ -3 & 1 \\ 11 & 0 \end{bmatrix}$$

1. $E + D$ **2.** $D + F$ **3.** $D - E$ **4.** $D + E$

5. Observe the solutions for Exercises 1 and 4. Do you think that addition of matrices is commutative? Explain.

Multiplying Integers

What you'll learn

You'll learn to multiply integers.

When am I ever going to use this?

You can multiply integers to find temperature changes over time.

The first occupied hot air balloon carried a duck, a rooster, and a sheep on an eight-minute flight on September 19, 1783. Ballooning has come a long way since then. Now, six teams are attempting to be the first to fly a balloon around the world.

One of the most serious challenges the balloonists must overcome is weather. The temperature drops about 2°F for each rise of 530 feet. The teams expect to travel about 40,000 feet high on an around-the-world trip. About how many degrees difference will there be between the ground temperature and the temperature at 40,000 feet? *This problem will be solved in Example 5.*

This problem can be solved by multiplying integers. Let's investigate how to multiply integers using counters.

HANDS-ON MINI-LAB

Work with a partner.

⬤ two colors of counters

Try This

☐ integer mat

Solve $y = 3 \cdot (-2)$.

- Begin with an empty mat.
- The 3 means put 3 sets of counters on the mat. The -2 means that each set contains 2 negative counters.

Solve $z = -2 \cdot (-5)$.

- Begin with an empty mat.
- The -2 means *remove* 2 sets of counters from the mat. The -5 means that each set contains 5 negative counters.
- Since the mat contains no counters, add zero pairs to the mat. Add just enough so that you can remove 2 sets of 5 negative counters.

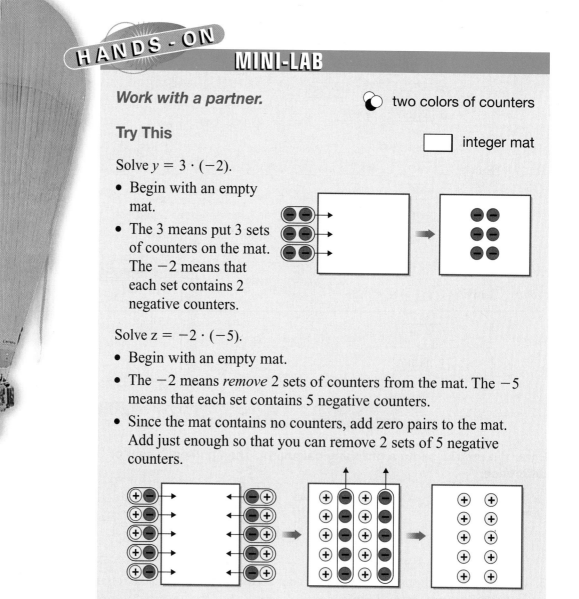

Talk About It

1. What are the values of y and z?
2. Describe the factors when the product was positive.
3. Describe the factors when the product was negative.

You can generalize the results of the Mini-Lab as follows.

Multiplying Integers	The product of two integers with the same sign is positive. The product of two integers with different signs is negative.

Examples

1 Solve $r = (-3)(-2)$.
The two integers have the same sign. The product will be positive.

$r = (-3)(-2)$
$r = 6$

2 Solve $(-4)(7) = t$.
The two integers have different signs. The product will be negative.

$(-4)(7) = t$
$-28 = t$

3 Solve $s = (-3)(-2)(-5)$.
Group the factors by using the associative property.

$s = [(-3)(-2)](-5)$
$s = 6(-5)$
$s = -30$

4 Solve $a = (-5)^2$.
The exponent says there are two factors of -5.

$a = (-5)(-5)$
$a = 25$

APPLICATION

5 **Sports** Refer to the beginning of the lesson. About how many degrees difference will there be between the ground temperature and the temperature at 40,000 feet?

Explore   You know the height and the temperature change per 530 feet. You need to find the change for 40,000 feet.

Plan   Write and solve an equation to find the number of 2° temperature changes. Then solve for the temperature change at 40,000 feet.

Solve   Let n represent the number of 2° temperature changes.

$n = 40,000 \div 530$

40000 ÷ 530 = 75.47169811

The temperature t will drop 2°F about 75 times.

$t \approx 75(-2)$
$t \approx -150$

The difference in temperature is 150°F.

Examine   Check the solution by finding the number of 2° decreases it would take to change the temperature by 150°. Then multiply by 530 feet to verify the distance.

Communicating Mathematics

Read and study the lesson to answer each question.

1. *Show* two different ways to find the product $(-35)(45)(-2)$.

2. *Write a Problem* that requires the multiplication of 2 and -3. Write about another situation that requires the multiplication of -2 and 3. Are the problems the same? Are the products the same? Explain.

HANDS-ON MATH

3. *Draw* a model using counters to illustrate $c = (-2)(-4)$. Then solve for c.

Guided Practice

Solve each equation.

4. $(4)(5) = k$
5. $(-3)(-2) = y$
6. $n = 14(-7)$
7. $p = (-2)(-3)(-9)$
8. $q = (4)(-67)(0)$
9. $(-4)^2 = x$

10. Evaluate $4xy$ if $x = -5$ and $y = 2$.

11. *Life Science* Did you know that most people lose 100 to 200 hairs per day? If you were to lose 120 hairs each day in a week, what is the change in the number of hairs you have?

Practice

Solve each equation.

12. $k = (-35)(245)$
13. $t = 8(6)$
14. $9(-9) = y$
15. $-8(-3) = m$
16. $-9(12) = n$
17. $-6(-13) = r$
18. $55(-11) = s$
19. $-5(-14) = t$
20. $v = (-8)^2$
21. $a = 7(-6)(-12)$
22. $z = 50(-4)(-1)$
23. $c = 5(23)(7)$
24. $(-2)(12)(3) = d$
25. $(-21)^2 = f$
26. $g = (6)(-2)(13)$
27. $h = (9)(-8)(6)^2$
28. $k = 15(-3)^2$
29. $m = (5)^2 \cdot (-4)^2$

Evaluate each expression if $a = -5$, $b = -9$, and $c = 10$.

30. $6cb$
31. $-4ac$
32. $5abc$

Applications and Problem Solving

For **Extra Practice**, see page 610.

33. *Energy* Degree-days help heating companies determine the needs of their customers. The number of degree–days for a day is the temperature minus 65°. If the mean temperature in St. Louis has been 55 degrees for 6 days, what is the total number of degree-days?

34. *Business* A theme park is the number one tourist attraction in France. But in 1993, its revenue was about $-\$930,000,000$! If this continued, what would the revenue at the theme park have been after five years?

35. *Critical Thinking* What is the sign of y if $y = x^2$? Explain.

Mixed Review

36. Write two matrices that have no sum. *(Lesson 2-6)*

37. *Sports* Jeff Sluman won the 1997 Tucson Classic for golf. His scores for the four rounds were 3, -4, -7, and -5. What was his total score? *(Lesson 2-4)*

38. **Standardized Test Practice** Find the area of a parallelogram whose height is 6.5 feet and whose base is 9 feet. *(Lesson 1-8)*

   A 58.5 ft^2 B 31 ft^2 C 42.5 ft^2 D 56.5 ft^2

Dividing Integers

What you'll learn

You'll learn to divide integers.

When am I ever going to use this?

You can divide integers to find the average amount of decrease in prices over time.

Are the tigers among your favorite animals to visit at the zoo? The natural habitat of tigers is the steamy hot jungles or the icy forests of Asia. Experts estimate that there may have been 100,000 tigers living 100 years ago. Now there are only about 6,000. What was the average change in tiger population each of the last 100 years? *This problem will be solved in Example 3.*

You can solve this problem by dividing integers. Since division is related to multiplication, it uses the same rules of signs.

Dividing Integers	The quotient of two integers with the same sign is positive. The quotient of two integers with different signs is negative.

Examples

1 Solve $r = -56 \div 8$.

$r = -56 \div 8$ *The signs are different.*

$r = -7$ *The quotient is negative.*

2 Solve $(-42) \div (-7) = t$.

$(-42) \div (-7) = t$ *The signs are the same.*

$6 = t$ *The quotient is positive.*

CONNECTION

3 **Life Science** **Refer to the beginning of the lesson. Find the average change in tiger population each of the last 100 years.**

The change in tiger population can be expressed as $6{,}000 - 100{,}000$ or $-94{,}000$. The change occurred over 100 years. The average change can be found by dividing $-94{,}000$ by 100. Let c represent the average change.

$c = -94{,}000 \div 100$ *The signs are different.*

$c = -940$ *The quotient is negative.*

The average change in population was -940 tigers per year. This means that each year there were about 940 fewer tigers in the world than the year before.

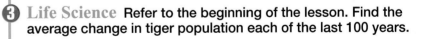

Remember that fractions are also a way of showing division. Another way to show $c = a \div b$ is $c = \frac{a}{b}$.

4 Solve $r = \frac{96}{-12}$.

$r = \frac{96}{-12}$ *The signs are different.*

$r = -8$ *The quotient is negative.*

CHECK FOR UNDERSTANDING

Communicating Mathematics

Read and study the lesson to answer each question.

1. *Tell* whether the dividend is positive or negative if the divisor is positive and the quotient is negative.

2. *Explain* how to check a solution of an equation solved by dividing integers.

3. *You Decide* Lesharo says that the next division sentence in the pattern is $-15 \div 3 = -5$. Rachel says the next division sentence is $-15 \div 5 = -3$. Who is correct and why?

$5 \div 5 = 1$
$0 \div 5 = 0$
$-5 \div 5 = -1$
$-10 \div 5 = -2$

Guided Practice

Solve each equation.

4. $-12 \div (-4) = m$
5. $320 \div (-8) = n$
6. $p = -240 \div 60$
7. $b = \frac{-56}{8}$
8. $s = 90 \div 10$
9. $\frac{365}{-5} = v$

Evaluate each expression if $x = 3$, $y = -10$, and $z = 9$.

10. $\frac{800}{y}$
11. $z \div (-3)$
12. $5yz \div x$

13. *Energy* Many homes built before 1960 were heated by coal. In 1950, about 115 million tons of coal were used for fuel each year. By the year 2000, it is estimated that coal use will be only about 5 million tons each year. What was the average change in coal use per year?

EXERCISES

Practice

Solve each equation.

14. $q = \frac{400}{-10}$
15. $t = -26 \div (-13)$
16. $295 \div 5 = h$

17. $-63 \div (-9) = w$
18. $c = \frac{-88}{44}$
19. $z = 49 \div (-7)$

20. $112 \div (-4) = f$
21. $-56 \div (-2) = r$
22. $x = -930 \div (-30)$

23. $g = \frac{245}{-5}$
24. $62 \div 2 = u$
25. $90 \div (-15) = h$

26. $p = -76 \div (-4)$
27. $s = 216 \div (-18)$
28. $\frac{224}{-32} = b$

29. $-143 \div 11 = d$
30. $f = -195 \div 65$
31. $k = 588 \div (-6)$

Evaluate each expression if $a = -8$, $b = 4$, and $c = -16$.

32. $\dfrac{72}{a}$

33. $\dfrac{-32}{b}$

34. $-48 \div c$

35. $52 \div b$

36. $c \div (-2)$

37. $-424 \div a$

38. $ac \div (-32)$

39. $bc \div a$

40. $(abc)^2 \div 64$

Applications and Problem Solving

41. *Employment* There were 7,404,000 unemployed American workers in 1995. In 1980, there were 7,637,000 Americans unemployed.

   a. What was the average change in unemployment for each of these 15 years?

   b. If this decline continues at the same rate, how many Americans will be unemployed in the year 2055?

42. *Agriculture* Texas is the number one agricultural state in the United States. The graph shows the number of farms and the average size of farms in Texas in 1987 and 1992.

   a. Find the average annual change in number of farms of 50 to 179 acres.

   b. There were 35,610,951 acres of land farmed in Texas in 1987. In 1992, 36,381,847 acres were farmed. Explain how this is related to the graph.

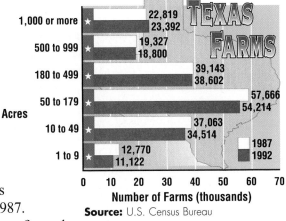

Source: U.S. Census Bureau

43. *Write a Problem* about the photograph at the left that requires division of integers in the solution.

44. *Working on the* **CHAPTER Project**
Research your species. Then use the procedure in Example 3 on page 81 to find an average change in population.

45. *Critical Thinking* Use examples to support your answers.

   a. The commutative property of multiplication is $a \cdot b = b \cdot a$. Is division commutative?

   b. The associative property of multiplication is $a(b \cdot c) = (a \cdot b) \cdot c$. Is division associative?

Mixed Review

46. Solve $t = -4(12)$. *(Lesson 2-7)*

47. Solve $h = -28 + 15 + 6 + (-30)$. Check your solution. *(Lesson 2-4)*

48. Graph $\{-3, -1, 0, 2\}$ on a number line. *(Lesson 2-1)*

49. **Standardized Test Practice** ReadyRent charges $24 to rent a power saw for 4 hours. The rent is $4.50 for each hour after the first 4. Which sentence could be used to find c, the cost for keeping the saw 10 hours? *(Lesson 1-6)*

   A $c = \$24 + 10(\$4.50)$ **B** $c = \$24 + 6(\$4.50)$ **C** $c = 10(\$4.50)$

   D $\$24 + c = 6(\$4.50)$ **E** $\$4.50 + 6(\$24) = c$

50. *Algebra* Evaluate $6a^2 + b$ if $a = 4$ and $b = 1$. *(Lesson 1-3)*

For **Extra Practice**, see page 610.

2-9A Solving Equations

A Preview of Lesson 2-9

You have used counters to model operations with integers. You can also use counters to solve equations that involve integers.

TRY THIS

Work with a partner.

1 Solve $x + (-2) = 7$.

- Start with an empty mat.

- Let a cup represent the unknown x value. Put the cup and 2 negative counters on one side of the mat. Place 7 positive counters on the other side of the mat.

- The goal is to get the cup by itself on one side of the mat. Then the counters on the other side will be the value of the cup, or x.

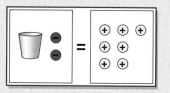

- Add 2 positive counters to each side of the mat to eliminate the 2 negative counters on the side with the cup.

- Group the counters to form zero pairs. Then remove all of the zero pairs.

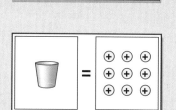

- Now the cup is by itself on one side of the mat. The counters are on the other side.

ON YOUR OWN

1. What is the solution of $x + (-2) = 7$?

2. How do you know what type of counter to add to each side of the mat in order to be able to get the cup by itself?

Work with a partner.

❷ Solve $2x - (-3) = 7$.

Before using the counters, rewrite the expression using the additive inverse. $2x - (-3) = 7$ becomes $2x + 3 = 7$.

- Start with an empty mat.
- Place 2 cups on one side of the mat to represent $2x$. Add 3 positive counters on that side of the mat. Place 7 positive counters on the other side of the mat.

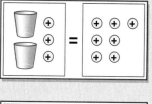

- Add 3 negative counters to each side of the mat.

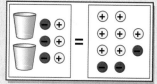

- Remove all of the zero pairs that can be formed.

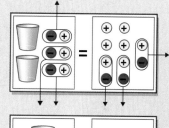

- Arrange the remaining counters on the right side of the mat into 2 equal groups so that they correspond to the 2 cups.

3. How many counters correspond to each cup?

4. What is the solution of $2x - (-3) = 7$?

Write an equation for each model.

5.

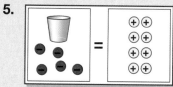

6.

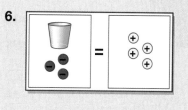

Solve each equation by using models.

7. $x + 2 = 8$
8. $x - 4 = 6$
9. $x + (-3) = 2$
10. $x - (-2) = 4$
11. $x + 6 = -2$
12. $x + (-4) = 6$
13. $4x = 8$
14. $3x = -15$
15. $2x + (-1) = 11$

Solving Equations

You'll learn to solve equations with integer solutions.

When am I ever going to use this?

You can solve an equation to determine how to place the seeds in a garden for best growth.

In the 1800s, forests in the United States were cleared to provide firewood and lumber for housing and to make way for farms. Now that the country is settled, trees are being replanted. Other countries in the growth stage of development are clearing forests. In Brazil, an area just 802 square miles less than the state of Washington was cleared in five years!

You can use an equation to find the average amount of land cleared each year in Brazil. Equations involving integers are solved in the same way that you solved equations involving whole numbers. *This problem will be solved in Example 5.*

Examples

1 Solve $r + 5 = -10$.

$$r + 5 = -10$$
$$r + 5 - 5 = -10 - 5 \quad \text{\textit{Subtract 5 from each side.}}$$
$$r = -15$$

Check: $\quad r + 5 = -10$
$$-15 + 5 \stackrel{?}{=} -10 \quad \text{\textit{Replace r with} } -15.$$
$$-10 = -10 \quad \checkmark$$

2 Solve $p - (-3) = -6$.

$$p - (-3) = -6$$
$$p + 3 = -6 \quad \text{\textit{Rewrite using the additive inverse.}}$$
$$p + 3 - 3 = -6 - 3 \quad \text{\textit{Subtract 3 from each side.}}$$
$$p = -9$$

Check: $\quad p - (-3) = -6$
$$-9 - (-3) \stackrel{?}{=} -6 \quad \text{\textit{Replace p with} } -9.$$
$$-6 = -6 \quad \checkmark$$

3 Solve $2t = -98$.

$$2t = -98$$
$$\frac{2t}{2} = \frac{-98}{2} \quad \text{\textit{Divide to undo multiplication.}}$$
$$t = -49$$

Check: $\quad 2t = -98$
$$2(-49) \stackrel{?}{=} -98 \quad \text{\textit{Replace t with} } -49.$$
$$-98 = -98 \quad \checkmark$$

Some equations involving integers require two steps to solve. You can work backward to solve for the variable.

4 Solve $-5x + 8 = -7$.

$$-5x + 8 = -7$$

$$-5x + 8 - 8 = -7 - 8 \quad \textit{Subtract 8 from each side.}$$

$$-5x = -15$$

$$\frac{-5x}{-5} = \frac{-15}{-5} \quad \textit{Divide each side by } -5.$$

$$x = 3$$

Check: $\quad -5x + 8 = -7$

$$-5(3) + 8 \stackrel{?}{=} -7 \quad \textit{Replace x with 3.}$$

$$-15 + 8 \stackrel{?}{=} -7$$

$$-7 = -7 \quad \checkmark$$

CONNECTION **5** **Geography** Refer to the beginning of the lesson. If the area of the state of Washington is 71,302 square miles, what was the average amount of land cleared in Brazil each year?

Explore You know the area of Washington and that the area cleared in Brazil was 802 square miles less than the area of Washington. You need to find the average area cleared each year.

Plan Let $a =$ the average amount of land cleared each year. Therefore, the amount cleared in five years was $5a$. Write an equation to solve the problem.

$$\overbrace{\text{land cleared in five years}}^{} \; \underbrace{\text{plus}}_{} \; \overbrace{\text{802 sq mi}}^{} \; \underbrace{\text{is}}_{} \; \overbrace{\text{area of Washington}}^{}$$

$$5a \qquad\quad + \quad 802 \quad = \qquad 71{,}302$$

Solve $\qquad 5a + 802 = 71{,}302$

$$5a + 802 - 802 = 71{,}302 - 802 \quad \textit{Subtract 802 from each side.}$$

$$5a = 70{,}500$$

$$\frac{5a}{5} = \frac{70{,}500}{5} \qquad\qquad \textit{Divide each side by 5.}$$

$$a = 14{,}100$$

There was an average of 14,100 square miles of forest cleared each year.

Examine Check the solution against the words of the problem. If 14,100 square miles were cleared each year, then $5 \cdot 14{,}100$ or 70,500 square miles were cleared. This is $71{,}302 - 70{,}500$ or 802 square miles less than the size of Washington.

Communicating Mathematics

Read and study the lesson to answer each question.

1. *Compare* the steps used to solve $3r + 5 = -10$ and $3r - (-5) = -10$.

2. *Write* a two-step equation that has a solution of 0.

3. *Write* a sentence to explain why it is a good idea to check your work. What should you do if your check does not agree?

Guided Practice

Solve each equation. Check your solution.

4. $3b = -36$

5. $v + 35 = 32$

6. $\frac{x}{-14} = 32$

7. $x - (-35) = -240$

8. $8c - 12 = 36$

9. $75 = -3y + 15$

10. The sum of two integers is -24. One of the integers is -13. Write an equation to find the other integer. Then find the integer.

EXERCISES

Practice

Solve each equation. Check your solution.

11. $x - 13 = -22$

12. $2y = -90$

13. $-30 = 42 + k$

14. $\frac{y}{15} = 22$

15. $w - (-350) = 32$

16. $-4,968 = -69n$

17. $y - 13 = 45$

18. $-200 = \frac{r}{3}$

19. $\frac{x}{-7} = -5$

20. $n + 34 = 16$

21. $4c = -60$

22. $v - 16 = -64$

23. $300 = 120 + 5h$

24. $-15 + 4m = 45$

25. $6b - (-3) = 105$

26. $4x - (-7) = -17$

27. $\frac{x}{6} - 5 = -13$

28. $12 + \frac{n}{4} = 0$

Write an equation for each problem and solve.

29. The difference of two integers is 8. The greater integer is -2. What is the other integer?

30. Two integers have a product of 35. One of the integers is -7. Find the other integer.

31. Two times a number plus 7 is -21. What is the number?

Applications and Problem Solving

Real World

32. *Health* Would you like to brush your teeth with hogs hair? Toothbrushes were invented in China in the late 1400s and were made of hogs hair! The first nylon toothbrush was made in 1938, 23 years before the first electric toothbrush. Write an equation for the year that the electric toothbrush was invented. Then find the year.

33. *Aviation* Wilbur Wright made the first flight in Kitty Hawk, North Carolina, on December 17, 1903. His brother Orville made the second flight that same day. Orville flew 4 yards more than 6 times as far as Wilbur. If Orville's flight was 244 yards, how far did Wilbur fly?

For **Extra Practice**, see page 611.

34. *Critical Thinking* Solve $6x + 7 = 8x - 13$.

35. Solve $-325 \div 25 = q$. *(Lesson 2-8)*

36. Order the integers $\{-219, -52, 18, 3, -24, 120, -186\}$ from greatest to least. *(Lesson 2-2)*

37. *Algebra* Solve $5 = \frac{k}{2} + 3$. Check your solution. *(Lesson 1-7)*

38. **Standardized Test Practice** Rey has $12.64. He wants to buy a soccer ball that is $17.95. How much more will Rey need to save? *(Lesson 1-1)*

　　A $30.59　　　　**B** $20.59　　　　**C** $15.31　　　　**D** $5.31　　　　**E** Not Here

Broadway Bound?

The financial comparison shown below appeared in the *New York Times* on November 21, 1994.

Basic Economics: Calculating Against Theatrical Disaster

Neil Simon plans to open his newest play, "London Suite," Off Broadway. His producer Emanuel Azenberg, provided this financial comparison.

MONEY NEEDED TO OPEN		BROADWAY 1,000 seats $55 top ticket price	OFF BROADWAY 500 seats $40 top ticket price
What the producer spends	Sets, costumes, lights	$357,000	$87,000
	Loading in (building set, etc.)	175,000	8,000
	Rehearsal salaries	102,000	63,000
	Director and designer fees	126,000	61,000
	Advertising	300,000	121,000
	Administration	235,000	100,000
		$1,295,000	**$440,000**
THE WEEKLY BUDGET			
Revenues	Weekly receipts at 75 percent capacity	**$250,000**	**$109,000**
Minus expenses	Theater rent, house crew	45,000	12,000
	Salaries	54,000	20,000
	Advertising	30,000	15,000
	Lights and sound rental	6,000	3,500
	Administration	32,000	11,000
	Royalties	23,500	14,000
	Extra rent, salaries based on ticket sales	16,000	6,500
		−206,500	**−82,000**
Equals weekly profit		**= 43,500**	**= 27,000**

1. Use an equation to find the number of performances that it would take to make up the money needed to open on Broadway and Off Broadway.

2. Do you agree with Neil Simon's decision to produce the play Off Broadway? Explain why or why not.

2-9B Eliminate Possibilities

A Follow-Up of Lesson 2-9

Crystal and Jackie are choosing a time to meet their friends Opa and Ellen for doubles tennis. Let's listen in!

Jackie

Crystal

I have to work Monday nights and Saturday mornings, so that's out.

Ok. And I volunteer at the library on Wednesdays and Thursdays after school.

I know that Ellen baby-sits for her neighbors after school on Monday, Wednesday, and Friday. And Sundays are out for her too.

This isn't leaving many options is it? I think Opa said that she works on Saturdays, too.

Let's organize all of this. I'll use a table with the days on the top and each of our names on the side. Then we can mark out the times we can't play and see what's left.

	Mon.	Tues.	Wed.	Thurs.	Fri.	Sat.	Sun.
Crystal	X					X	
Jackie			X	X			
Ellen	X		X		X		X
Opa						X	

Looks like Tuesday is the day! Let's go practice so we're ready!

THINK ABOUT IT

Work with a partner.

1. **Tell** how Crystal and Jackie eliminated possibilities to solve the problem.

2. **Eliminate possibilities** to solve the multiple-choice problem.

   $16,340 \div 19 =$

   A 80   **B** 86

   C 860   **D** 8,600

3. **Describe** a real-life situation in which you could eliminate possibilities to solve a problem.

4. **Tell** what day the girls could play tennis if Ellen's baby-sitting job changed to Tuesdays, Thursdays, and Saturdays.

For **Extra Practice,** see page 611.

ON YOUR OWN

5. The second step of the 4-step plan for problem solving is to *plan* the solution. **Tell** what is involved in planning to solve a problem by eliminating possibilities.

6. *Write a Problem* that could be solved by eliminating possibilities.

7. *Reflect Back* Explain how you could use eliminating possibilities to solve Exercise 38 on page 89.

MIXED PROBLEM SOLVING

STRATEGIES

Look for a pattern.
Solve a simpler problem.
Act it out.
Guess and check.
Draw a diagram.
Make a chart.
Work backward.

Solve. Use any strategy.

8. *Technology* The price of calculators has been decreasing. A 4-function calculator sold for $12.50 in 1980. A similar calculator sold for $8.90 in 1990. If the price decrease continues at the same rate, what would be the price in 2010?

9. Trey and Matias are going to take a bus to Baltimore for the day. The buses run every hour on the hour from 7:00 A.M. to 11:00 P.M. It will take one hour to get to Baltimore. They don't want to leave home before 7:30 A.M. and they want to return before 10:00 P.M. While they are in Baltimore, they want to spend $2\frac{1}{2}$ hours at the Aquarium and at least 2 hours at the Science Center. If they have time, they would like to take the 30-minute submarine tour.

   a. If they take an hour to eat lunch at the Inner Harbor, what times might they take the bus to and from Baltimore?

   b. Will they have time for the tour of the submarine?

10. *Employment* Three after-school jobs are posted on the job board. The first job pays $5.15 per hour for 15 hours of work each week. The second job pays $10.95 per day for 2 hours of work, 5 days each week. The third job pays $82.50 for 15 hours of work each week. If you want to apply for the best-paying job, which job should you choose?

11. **Standardized Test Practice** A human heart beats an average of 72 times in one minute. Estimate the number of times a human heart beats in one year.

   A 37,800,000   **B** 378,000
   C 37,800   **D** 3,780

12. **Standardized Test Practice** The table below shows the results of a survey of popular cookie flavors. How many flavors were chosen by fewer than 10 students?

   A 3
   B 4
   C 5
   D 6

Students	Favorite Cookie
12	Sugar
15	Molasses
32	Oatmeal raisin
45	Peanut butter
8	Vanilla creme
3	Mint
14	Coconut
56	Chocolate chip
6	Pecan

13. *Sports* What non-motor sport do you think is the fastest? It's snow skiing. The record was set at 142.165 mph in 1992. The second fastest record is for speed skating. It was set at 30.68 mph in 1988. How much faster is the skiing record?

14. **Standardized Test Practice** The Meadowlands Arena seats 20,039 people. If each ticket for an event sells for $12.75, how much should the receipts for a sell-out be?

   A $2,554,972   **B** $255,497.25
   C $25,549.72   **D** $255,497.20

Integration: Geometry
The Coordinate System

What you'll learn

You'll learn to graph points on the coordinate plane.

When am I ever going to use this?

You can graph points to look for trends in data collected in science class.

Word Wise

coordinate system
origin
x-axis
y-axis
quadrant
ordered pair
x-coordinate
y-coordinate

Most of us have used a globe to locate places in the world. The first globe was made in about 300 B.C., by the Greek scientist Dicaearchus. Since then, globes have changed quite a bit. New continents have been discovered, and the longitude and latitude system has been perfected.

When reading a globe, you can use the longitude and latitude lines to pinpoint a location on Earth. In mathematics, you can locate a point precisely by using a **coordinate system** similar to the longitude and latitude system used on a globe. The coordinate system is formed by two number lines that intersect at their zero points. This intersection point is called the **origin**. The horizontal number line is called the **x-axis**, and the vertical number line is the **y-axis**. The two axes separate the coordinate plane into four sections called **quadrants**.

Any point on the coordinate plane can be graphed by using an **ordered pair** of numbers. The first number in an ordered pair is called the **x-coordinate**. The second number is the **y-coordinate**. The coordinates are your directions to find the point.

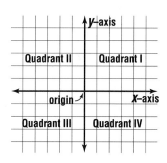

Example

1 Graph the point whose coordinates are (4, −3).

Start at the origin. The *x*-coordinate is 4. This tells you to go 4 units to the right of the origin.

The *y*-coordinate is −3. It tells you to go down 3 units.

Draw a dot. The dot is the graph of the point whose coordinates are (4, −3).

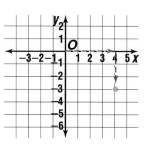

Letters are often used to name points. The symbol $B(-3, 2)$ means point B has an *x*-coordinate of −3 and a *y*-coordinate of 2.

Example 2

Name the ordered pair for point *D*.

Move left on the *x*-axis to find the *x*-coordinate of point *D*. The *x*-coordinate is −1.

Move up along the *y*-axis to find the *y*-coordinate. The *y*-coordinate is 5.

The ordered pair for point *D* is (−1, 5).

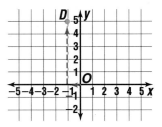

You can use ordered pairs to show how data are related.

Example 3

INTEGRATION

Algebra Evaluate the expression $3x - 2$ for $x = -2, 0, 1,$ and 3. Graph the ordered pairs formed by *x* and the corresponding value of the expression for *x*.

Use a table to evaluate the expression.

x	3x – 2	(x, y)
−2	3(−2) − 2 = −6 − 2 or −8	(−2, −8)
0	3(0) − 2 = 0 − 2 or −2	(0, −2)
1	3(1) − 2 = 3 − 2 or 1	(1, 1)
3	3(3) − 2 = 9 − 2 or 7	(3, 7)

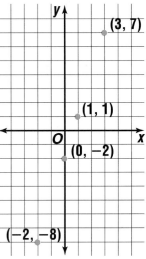

Now graph each ordered pair.

> **Study Hint**
>
> **Reading Math** In this book, when no numbers are given on the *x*- and *y*-axis, you can assume that each grid square is one unit.

CHECK FOR UNDERSTANDING

Communicating Mathematics

Read and study the lesson to answer each question.

1. *Tell* how you know whether to go left or right for the *x*-coordinate and whether to go up or down for the *y*-coordinate when graphing a point.

2. *Describe* how you would graph the points whose coordinates are (0, 3) and (3, 0).

3. *Write* how you can remember whether the first number of the set of ordered pairs is the *x*-coordinate or the *y*-coordinate.

Math Journal

Guided Practice

Name the ordered pair for the coordinates of each point graphed on the coordinate plane.

4. *A*

5. *B*

6. *C*

7. *D*

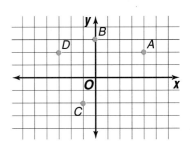

Graph each point on the same coordinate plane.

8. $E(-4, 3)$ **9.** $F(-3, -9)$ **10.** $G(9, -7)$

11. $H(5, 0)$ **12.** $I(0, 0)$ **13.** $J(7, 7)$

14. *History* Since gaining the right to vote in 1920, women have made slow gains in Congress. The table shows the number of females in the U.S. Senate for several years since 1975. Use the years as the x-coordinates and the numbers of female Senators as the y-coordinates to graph the data.

Year	Female Senators
1975	0
1977	2
1979	1
1981	2
1983	2
1985	2
1987	2
1989	2
1991	3
1993	7
1995	9
1997	9

Source: National Women's Political Caucus

EXERCISES

Practice

Name the ordered pair for the coordinates of each point graphed on the coordinate plane.

15. Z **16.** X

17. W **18.** Y

19. T **20.** V

21. U **22.** S

23. Q **24.** R

25. P **26.** M

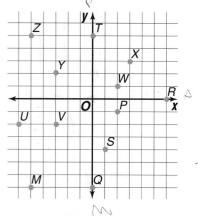

Family Activity

Find a street map that includes your neighborhood. Also find maps that include the homes of two of your relatives or friends who do not live with you. List the coordinates of your home and the other two homes.

Graph each point on the same coordinate plane.

27. $A(4, 7)$ **28.** $C(1, 0)$ **29.** $B(0, 7)$

30. $E(-1, -2)$ **31.** $D(-4, -7)$ **32.** $F(-10, 3)$

33. $G(9, 9)$ **34.** $J(7, -8)$ **35.** $K(-6, 0)$

36. $H(0, -3)$ **37.** $I(4, 0)$ **38.** $M(2, 7)$

39. $N(8, -1)$ **40.** $L(-1, -1)$ **41.** $P(3, 3)$

42. $Q(-2, 0)$ **43.** $S(-3, -3)$ **44.** $R(5, -4)$

Applications and Problem Solving

Real World

45. *Geometry* Graph the points $A(4, 2)$, $B(2, -1)$, $C(-4, -1)$, and $D(-2, 2)$ on the same coordinate plane. Draw the line segments from A to B, from B to C, from C to D, and from D to A. Is the shape formed a rectangle? Explain.

46. Sports And they're off! Post position, where a horse begins a race, may give a horse an edge on the victory. The table shows how many horses at the various post positions have won the Kentucky Derby since 1900. Use the post positions as the *x*-coordinates and the number of winners as the *y*-coordinates to graph the data.

Post Position	Winners	Post Position	Winners
1	12	11	3
2	9	12	3
3	7	13	3
4	10	14	2
5	9	15	1
6	6	16	1
7	7	17	0
8	8	18	1
9	4	19	0
10	9	20	1

47. Life Science At 100 feet long and 150 tons, the blue whale is the largest living mammal today. At full speed, a blue whale can travel 36 feet per second. If *x* represents the number of seconds, the expression 36*x* gives the total distance a whale can travel.

a. Evaluate the expression to find the distances traveled in 0, 1, 3, 4, and 6 seconds.

b. Graph the ordered pairs formed by the times and distances.

48. Working on the **CHAPTER Project** Use the populations that you found for your species. Graph each population on the same coordinate plane, using the year as the *x*-coordinate and the population as the *y*-coordinate. Write a description of the population changes and add the graph and the description to your television public service announcement script.

49. Critical Thinking The points at $(-3, 3)$, $(-3, -3)$, and $(-6, 3)$ are three of the vertices, or corners, of a square. What are the coordinates of the fourth vertex?

Mixed Review

50. Algebra Solve $6d + 18 = 36$. *(Lesson 2-9)*

51. Find $\begin{bmatrix} -5 & -1 & 0 \\ 2 & 3 & -1 \end{bmatrix} + \begin{bmatrix} -1 & -1 & 3 \\ 2 & 0 & 6 \end{bmatrix}$. *(Lesson 2-6)*

52. Standardized Test Practice Solve $g = -56 - 77$. *(Lesson 2-5)*

    A -21 **B** -133 **C** 21 **D** 133 **E** Not Here

53. Replace ● with $>$, $<$, or $=$ to make -114 ● -97 a true sentence. *(Lesson 2-2)*

54. Geometry Find the perimeter and area of the figure. *(Lesson 1-8)*

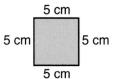

5 cm
5 cm   5 cm
5 cm

55. Geography Do you live in Midway? Midway is the most common place name in the United States. There are 15 more occurrences of the name Midway than of the second most common name, Fairview. If there are 192 Fairviews in the U.S., how many Midway's are there? *(Lesson 1-6)*

For **Extra Practice,** see page 611.

2

inter NET
CONNECTION Chapter Review **For additonal lesson-by-lesson review, visit:**
www.glencoe.com/sec/math/mac/mathnet

Vocabulary

After completing this chapter, you should be able to define each term, concept, or phrase and give an example or two of each.

Numbers and Operations
absolute value (p. 57)
additive inverse (p. 70)
integer (p. 56)
opposite (p. 70)
zero pair (p. 63)

Algebra
column (p. 73)
element (p. 73)
graph (p. 56)
matrix (p. 73)
row (p. 73)

Geometry
coordinate (p. 56)
coordinate system (p. 92)
ordered pair (p. 92)
origin (p. 92)
quadrant (p. 92)
x-axis (p. 92)
x-coordinate (p. 92)
y-axis (p. 92)
y-coordinate (p. 92)

Problem Solving
eliminate possibilities (p. 90)

Understanding and Using the Vocabulary

Choose the letter of the term that best matches each statement or phrase.

1. the distance a number is from zero on a number line
2. what you add to subtract an integer
3. the sign of the product of two integers with the same sign
4. the sign of the quotient of two integers with different signs
5. one of the four parts that the two axes separate a coordinate plane into
6. the first number in an ordered pair
7. one of the numbers in a matrix
8. the point where the x- and y-axes intersect
9. the number that corresponds to a point graphed on a number line

a. positive
b. negative
c. coordinate
d. x-coordinate
e. absolute value
f. additive inverse
g. quadrant
h. origin
i. element

In Your Own Words

10. *Write* a sentence that explains why the additive inverse of zero is zero.

Objectives & Examples

Upon completing this chapter, you should be able to:

● graph integers on a number line and find absolute value *(Lesson 2-1)*

Graph −3 and 2.

Find each number on a number line.

Draw a dot at that point.

$$-4 \;-3\;-2\;-1\quad 0\quad 1\quad 2\quad 3$$

● compare and order integers *(Lesson 2-2)*

Compare −5 and −10.

−5 is to the right of −10 on a number line, so −5 > −10.

● add integers *(Lesson 2-3)*

Solve −14 + (−20) = x.

$|-14| + |-20| = 14 + 20$ or 34

So, x = −34.

● add three or more integers *(Lesson 2-4)*

Solve y = 22 + (−17) + 3.

y = [22 + (−17)] + 3   *Associative*
                        property

y = 5 + 3

y = 8

● subtract integers *(Lesson 2-5)*

Solve z = −30 − 13.

z = −30 − 13

z = −30 + (−13)   *To subtract 13,*
                   add −13.

z = −43

Review Exercises

Use these exercises to review and prepare for the chapter test.

Graph each set of points on a number line.

11. {3, −2, 6, 0}

12. {−5, 4, −1, 2}

Find each absolute value.

13. $|11|$ **14.** $|-88|$

Replace each ● with >, <, or = to make a true sentence.

15. −49 ● −340   **16.** −13 ● 2

17. $|-483|$ ● 483   **18.** −6 ● $|-6|$

Solve each equation.

19. 125 + (−75) = a

20. −66 + (−119) = b

21. c = −46 + 38

22. y = −54 + 21

Solve each equation. Check by solving another way.

23. 20 + 16 + (−5) + 6 = k

24. −14 + 37 + (−20) + 2 = x

25. v = 52 + (−78) + 8

26. 28 + (−50) + 12 + (−15) = p

Solve each equation.

27. 45 − (−63) = b

28. y = −58 − 34

29. m = −76 − (−56)

30. −16 − 24 = n

● use matrices to organize data *(Lesson 2-6)*

Find the sum.

$$\begin{bmatrix} 4 & -2 \\ 3 & 0 \end{bmatrix} + \begin{bmatrix} -3 & 2 \\ 5 & -4 \end{bmatrix}$$

$$= \begin{bmatrix} 4 + (-3) & -2 + 2 \\ 3 + 5 & 0 + (-4) \end{bmatrix}$$

$$= \begin{bmatrix} 1 & 0 \\ 8 & -4 \end{bmatrix}$$

Find each sum or difference. If there is no sum or difference, write *impossible*.

31. $\begin{bmatrix} 2 & -1 \\ -3 & 9 \end{bmatrix} + \begin{bmatrix} 10 & -3 \\ 8 & -3 \end{bmatrix}$

32. $\begin{bmatrix} 3 & -2 \\ 4 & -1 \end{bmatrix} + \begin{bmatrix} -3 & 8 \\ 0 & 6 \\ -4 & 2 \end{bmatrix}$

33. $\begin{bmatrix} -4 & -2 \\ 5 & 6 \\ 7 & -5 \end{bmatrix} - \begin{bmatrix} 2 & 2 \\ 4 & -4 \\ 9 & -0 \end{bmatrix}$

● multiply integers *(Lesson 2-7)*

Solve $a = (-7)(3)(5)$.

$a = [(-7)(3)](5)$ *Associative property*

$a = (-21)(5)$

$a = -105$

Solve each equation.

34. $u = -7(11)$

35. $w = -4(-25)$

36. $(-20)(-2)(3) = q$

37. $(-15)(-4)(-1) = g$

● divide integers *(Lesson 2-8)*

Solve $b = -48 \div (-6)$.

$b = -48 \div (-6)$

$b = 8$

Solve each equation.

38. $-88 \div 8 = c$ 39. $\frac{-210}{7} = h$

40. $z = 170 \div 5$ 41. $b = \frac{180}{-15}$

● solve equations with integer solutions *(Lesson 2-9)*

Solve $c - (-30) = 255$.

$c - (-30) = 255$

$c + 30 = 255$

$c + 30 - 30 = 255 - 30$

$c = 225$

Solve each equation. Check your solution.

42. $-3s = -51$

43. $3p - (-7) = 52$

44. $24 + m = -93$

45. $\frac{r}{4} - 12 = -14$

● graph points on the coordinate plane *(Lesson 2-10)*

Graph the point whose coordinates are $(3, -4)$.

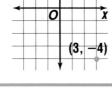

Graph each point on the same coordinate plane.

46. $A(-1, -6)$

47. $B(-3, -2)$

48. $C(3, 6)$

49. $D(5, -4)$

Applications & Problem Solving

50. *Recreation* The Chess Club at Mason Middle School is holding a tournament in which each player earns $+1$ point for a win, -1 point for a loss, and 0 for a draw. The scoreboard below shows the final standings of five players. *(Lessons 2-1 and 2-4)*

Player	Wins	Losses
Neva	6	0
Kaitlin	4	2
Antoine	2	4
Trenna	3	3
Sarah	0	6

  a. Determine each player's total points.

  b. Graph the total scores on a number line. Let the first letter of each player's name be the letter for each coordinate.

51. *Sports* Alberto scored -2 on five miniature golf holes. What was his score for these 5 holes? *(Lesson 2-7)*

52. *Patterns* Third grader Koleka Smith of Chehalis, Washington, discovered a new way of multiplying by 5. To multiply an even number by 5, divide by 2 and then add a 0 to the end. To multiply an odd number by 5, subtract 1, divide by 2, and then add a 5 to the end. *(Lesson 2-8)*

  a. Will Koleka's method work with negative integers?

  b. If not, give another pattern that will work.

53. *Eliminate Possibilities* Yuria and Evan have the same favorite color. It is either blue, red, or green. Yuria does not like green. Evan does not like red. What is each person's favorite color? *(Lesson 2-9B)*

Alternative Assessment

● Open Ended

You just received a statement from your bank showing that your account is overdrawn by $25.60, which is shown as -25.60. Your records show a balance of $15.90, or $+15.90$, in your account. You balance your checkbook and decide that your balance is correct. After discussing the error with the bank, you found that they made just one error. What error could the bank have made?

You discover that you withdrew $10.00, recorded as -10.00, from your account that you forgot to record. How does this change the mistake made by the bank? Is your account overdrawn now? Explain the amount of the mistake now.

A practice test for Chapter 2 is provided on page 648.

● Completing the CHAPTER Project

Use the following checklist to make sure that your television public service announcement script is complete.

☑ Summarize your research on the species that you chose. Why is it endangered or threatened?

☑ Include population figures for various years and discuss how the population has changed.

☑ Add descriptions of video clips to accompany the audio.

 PORTFOLIO   Select one of the problems you solved in this chapter and place it in your portfolio. Attach a note explaining how your problem illustrates one or more of the important concepts covered in this chapter.

Section One: Multiple Choice

There are twelve multiple-choice questions in this section. Choose the best answer. If a correct answer is *not here*, choose the letter for Not Here.

1. How is $2 \cdot 2 \cdot 7 \cdot a \cdot a \cdot a$ written in exponential notation?

  A $2 \cdot 5 \cdot a$

  B $2^7 \cdot 7^2 \cdot a$

  C $2^2 \cdot 7 \cdot a^3$

  D $14^2 \cdot a^3$

2. What is the perimeter of the rectangle?

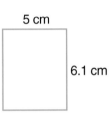

5 cm

6.1 cm

  F 11.1 cm

  G 22.2 cm

  H 33.3 cm

  J 30.5 cm

3. If $b = 5$ and $c = 8$, what is the value of $b(10 - c)$?

  A 10

  B 42

  C 50

  D 52

4. There are approximately 6,479,000 bicycles in North Carolina. What is this number rounded to the nearest million?

  F 6,000,000

  G 6,500,000

  H 6,480,000

  J 7,000,000

Please note that Questions 5–12 have five answer choices.

5. The emergency squad responded to 120 calls for help in the first three months of the year. They responded to 42 calls the fourth month. Which would be a reasonable answer for the number of calls that they responded to?

  A less than 35 calls per month

  B between 35 and 45 calls per month

  C between 45 and 55 calls per month

  D between 55 and 65 calls per month

  E more than 65 calls per month

6. Manuela wanted to spend no more than $20 at the grocery. If she bought 9 bottles of sports drink, which inequality could be used to find the price, p, of each bottle so that the total would be at or below $20 excluding tax?

  F $p + 9 < 20$

  G $p + 9 \leq 20$

  H $9p < 20$

  J $9p \leq 20$

  K $9 - p < 20$

7. On a math test, Lee scored 10 points less than twice the lowest score. If his score was 96, what was the lowest score?

  A 86

  B 53

  C 48

  D 43

  E 33

8. It costs approximately $60 per month to heat a 2,000 square-foot home. At that rate, how much should it cost per month to heat a 3,000 square-foot home?

  F $40

  G $70

  H $90

  J $110

  K $120

9. Victor bought a stereo that cost $532.16. He paid for the stereo in 12 equal payments. The best estimate for the amount of each payment is —
   A less than $10.
   B between $10 and $20.
   C between $25 and $35.
   D between $40 and $50.
   E between $55 and $65.

10. Three friends will equally share the cost of a $28.95 board game. Which is a *not* a reasonable estimate of each person's share?
    F $9.90
    G $9.65
    H $8.80
    J $9.50
    K All are reasonable.

11. Missy bought 3 sweaters for $108. If each sweater cost the same amount, how much did each sweater cost?
    A $30.24
    B $32.40
    C $36.00
    D $54.00
    E Not Here

12. The Lewis Center 4-H Club sold 312 pocket calendars last year. This year they sold 447 pocket calendars. How many more calendars did they sell this year than last year?
    F 135
    G 125
    H 155
    J 189
    K Not Here

inter NET
CONNECTION Test Practice For additional test practice questions, visit:
www.glencoe.com/sec/math/mac/mathnet

Test-Taking Tip THE PRINCETON REVIEW

Most basic formulas that you need to answer the questions on a standardized test are usually given to you in the test booklet. However, it is a good idea to review common formulas before the test. Make sure to familiarize yourself with how to use formulas. A quick review may allow you to answer more questions correctly.

Section Two: Free Response

This section contains seven questions for which you will provide short answers. Write your answers on your paper.

13. Solve $-64 \div 8 = q$.

14. Find the product of 20 and -9.

15. If $\frac{c}{-3} = 6$, what is the value of c?

16. Simplify $(3a + 1) + (2a + 5)$.

17. Find the product $4(3x - 2)$.

18. The product of a number and itself is 196. What is the number?

19. What is the area of the parallelogram?

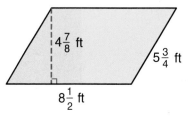

$4\frac{7}{8}$ ft $5\frac{3}{4}$ ft

$8\frac{1}{2}$ ft

Using Proportion and Percent

What you'll learn in Chapter 3

- to express ratios as fractions and determine unit rates,
- to solve problems using proportions,
- to express percents as fractions and decimals,
- to estimate and compute mentally with percents, and
- to determine reasonable answers in real-world problems.

CHAPTER Project

PACK YOUR BAGS

In this project, you will use ratios, proportions, and percents to help you plan a week-long trip to a foreign country of your choice. You will write a report describing your plan and the cost of your trip.

Getting Started

- Choose a foreign country you would like to visit. Then make a list of the cities and attractions you would like to visit.

- Estimate how much money your trip will cost. It will be helpful to organize your expenses into these categories: transportation, housing, food, and souvenirs.

- When you travel to a foreign country, you must exchange U.S. dollars for that country's currency. Find the exchange rate for your chosen country. Some exchange rates are shown in the table.

Country	Currency Name	Dollar Equivalent as of 7/7/97
Brazil	Real	1.0777
Canada	Dollar	1.3767
China	Yuan	8.3214
Egypt	Pound	3.3938
Ireland	Punt	0.6573
Japan	Yen	112.77
Mexico	Peso	7.9210
Pakistan	Rupee	40.01
South Africa	Rand	4.5365
Switzerland	Franc	1.4610

Technology Tips

- Use an **electronic encyclopedia** to do your research.

- Use a **spreadsheet** to organize your expenses.

- Use a **word processor** to write your report.

 interNET CONNECTION **Data Update For up-to-date information on foreign currency rates, visit:**

www.glencoe.com/sec/math/mac/mathnet

Working on the Project

You can use what you'll learn in Chapter 3 to help you make your travel plan.

Page	Exercise
106	28
113	29
129	35
133	Alternative Assessment

Ratios and Rates

What you'll learn

You'll learn to express ratios as fractions in simplest form and determine unit rates.

When am I ever going to use this?

Ratios and rates are frequently used in advertising and commercials.

Word Wise

ratio
rate
unit rate

The characters that appear in many television commercials are entertaining as well as effective. One example is the Dough Boy. In a 1996 Louis Harris Poll, about 38 out of 100 people who were surveyed rated the Dough Boy commercials very effective. The expression *38 out of 100* is an example of a **ratio**.

Ratio	A ratio is a comparison of two numbers by division.

The ratio that compares 38 to 100 can be expressed as follows.

Say: 38 to 100, 38 out of 100

Write: $38:100, \frac{38}{100}$ *A fraction bar is used to indicate division.*

Since a ratio can be written as a fraction, it can be simplified.

$$\frac{38}{100} \overset{\div 2}{\underset{\div 2}{=}} \frac{19}{50}$$ *Divide both the numerator and denominator by 2.*

The equation $\frac{38}{100} = \frac{19}{50}$ indicates that the two ratios are equivalent. This means that 38 out of 100 is equivalent to 19 out of 50.

Examples

Express each ratio in simplest form.

1 number of shaded squares:total number of squares

15 out of 25 squares are shaded.

$\frac{15}{25} = \frac{3}{5}$ *Divide the numerator and denominator by 5.*

The ratio in simplest form is $\frac{3}{5}$ or 3 out of 5.

Study Hint

Reading Math In Example 1, the ratio 3 out of 5 means that for every 5 squares in the grid, 3 are shaded.

2 24 inches to 1 yard

To express these measurements as a ratio in simplest form, they must have the same unit.

$\frac{24 \text{ inches}}{1 \text{ yard}} = \frac{24 \text{ inches}}{36 \text{ inches}}$

24 inches

1 yard = 36 inches

$\frac{24}{36} = \frac{2}{3}$ *Divide the numerator and denominator by 12.*

The ratio in simplest form is $\frac{2}{3}$ or 2:3.

A **rate** is a special kind of ratio. It is a comparison of two measurements with different units, such as miles to gallons or cents to pounds. For example, suppose you spend $9.75 on 2.5 pounds of candy. The rate $\frac{\$9.75}{2.5\text{ pounds}}$ compares the money spent to the number of pounds of candy.

Rate	A rate is a ratio of two measurements with different units.

When a rate is simplified so it has a denominator of 1, it is called a **unit rate**. This type of rate is frequently used when referring to statistics.

Example **3**

CONNECTION

Civics Membership in the U.S. House of Representatives is based on population in the preceding census. In 1990, the population of the United States was about 248,000,000. There are 435 members in the House. On average, how many people are represented by each member of the House?

Write the rate that compares the population to the number of members of the House. Then divide both the numerator and the denominator by 435.

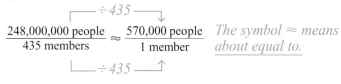

$$\frac{248{,}000{,}000 \text{ people}}{435 \text{ members}} \approx \frac{570{,}000 \text{ people}}{1 \text{ member}}$$

The symbol $\approx$ means about equal to.

Each member of the House of Representatives represents about 570,000 people.

Study Hint

Estimation
You can also estimate the unit rate by dividing the numerator by the denominator.
240,000,000 ÷ 400 = 600,000

CHECK FOR UNDERSTANDING

Communicating Mathematics

Read and study the lesson to answer each question.

1. *Explain* how ratios and rates are alike and how they are different.

2. *Write* a ratio about your class.

3. *Draw* a figure in which the ratio of shaded squares to total number of squares is 3:8.

Guided Practice

Express each ratio or rate in simplest form.

4. 2 cups:18 cups

5. $20 in 25 days

6. shaded squares in the figure at the right to total number of squares

Express each rate as a unit rate.

7. 24 pounds in 8 weeks

8. 3 inches of rain in 2 hours

9. *Environment* In 1996, six California condors were released into the Vermillion Cliffs in Arizona. The rarest birds in North America, they have a wingspan of about 9 feet. The common Barn Owl has a 3-foot wingspan. Use a ratio in simplest form to compare the wingspan of a California condor to the wingspan of a Barn Owl.

Practice

Express each ratio or rate in simplest form.

10. shaded squares in the figure at the right to total number of squares

11. 18 brown-eyed students:12 blue-eyed students

12. 6 absences in 180 school days

13. 1 foot:1 yard   **14.** 99 wins:99 losses

15. 15 out of 45   **16.** 45 minutes per hour

17. 36 to 24   **18.** 150 miles per 6 gallons

Express each rate as a unit rate.

19. $1.75 for 5 minutes   **20.** $25 for 10 disks

21. 300 students to 20 teachers **22.** 100 meters in 10.5 seconds

23. $8.80 for 11 pounds   **24.** $0.96 per dozen

Applications and Problem Solving

25. Jason and Billy have some marbles. The ratio of their marbles is 5:1. When Jason gives Billy 10 marbles, the ratio is 2:1. How many marbles does Jason have?

26. Use ratios to convert the following rates.

   a. 50 ft/min = _____ in./s

   b. 120 gal/h = _____ oz/min

   c. 30 min/h = _____ ft/s

27. *Life Science* A science class is studying a section of weeds that has an area of 12 square meters. They count 46 dandelion plants and 212 grass plants. On average, how many of each kind of plant are in one square meter?

28. *Working on the* **CHAPTER Project**   Refer to the foreign exchange rate you found on page 103. Write a ratio that shows the relationship between the unit of currency used in your foreign country and 1 U.S. dollar.

29. *Critical Thinking* Three people set out at the same time on a trip from town A to town B. Marcel drove the whole trip at a steady speed of 60 miles per hour. Rachel averaged 75 miles per hour for the first half of the distance and 55 miles per hour the second half. Jamal averaged 30 miles per hour for the first third, 45 miles per hour for the second third, and 85 miles per hour for the last third. Who arrived at town B first?

Mixed Review

30. *Geometry* Graph the points $C(2, 5)$, $H(-3, 3)$, $E(4, 1)$, and $G(-1, -2)$ on the same coordinate plane. *(Lesson 2-10)*

For **Extra Practice,** see page 612.

31. *Algebra* Solve $-2y + 15 = 55$. *(Lesson 2-9)*

32. *Algebra* Write an algebraic expression to represent *eight million less than four times the population of Africa*. *(Lesson 1-6)*

33. **Standardized Test Practice** Tonio uses his calculator to divide 685,300 by 86.3. Which is the best estimate of the result? *(Lesson 1-1)*

   A 100   **B** 800   **C** 8,000   **D** 10,000   **E** 80,000

Ratios and Percents

What you'll learn

You'll learn to express ratios as percents and vice versa.

When am I ever going to use this?

Knowing how to express ratios as percents will help you estimate the sale price of your next pair of jeans.

Word Wise

percent

Where is your ideal vacation spot? In 1996, the George Gallup International Institute reported that about 75 out of 100 U.S. teenagers would like to visit the Grand Canyon. The ratio 75 out of 100 can be expressed as a **percent**.

Percent	A percent is a ratio that compares a number to 100.	
	Ratio: 75 out of 100	**Model:**
	Words: seventy-five percent	
	Symbol: 75%	

Example 1

Express each ratio as a percent.

a. **Nine out of 100 U.S. Senators are women.**
   9 out of $100 = 9\%$

b. **He answered 87.5 questions out of 100 correctly.**
   87.5 out of $100 = 87.5\%$

Ratios like 1:2, $\frac{3}{10}$, and 4 out of 5 can also be expressed as percents.

HANDS-ON MINI-LAB

Work with a partner.   grid paper

You can use grid paper to express the ratio 1:2 as a percent.
- Since $1:2 = \frac{1}{2}$, shade $\frac{1}{2}$ of the squares in the decimal model.
- $\frac{1}{2} = \frac{50}{100}$
  $= 50\%$ So, $\frac{1}{2} = 50\%$.

Try This

Use decimal models to express each ratio as a percent.

1. $\frac{3}{10}$ **2.** 4:5   **3.** 1 out of 3

Talk About It

4. Use what you have learned in this lab to express the ratio 7 out of 10 as a percent without using models.

Did you know In the Gallup survey, 58% of teenagers mentioned the White House and 41% mentioned Graceland as desirable vacation spots.

One way to express a fraction or a ratio as a percent is by finding an equivalent fraction with a denominator of 100.

Example ② Express each ratio or fraction as a percent.

a. **In 1995, 2 out of 25 households had a fax machine.**

$$\overset{\times 4}{\underset{\times 4}{\frac{2}{25} = \frac{8}{100}}}$$

So, $\frac{2}{25} = 8\%$.

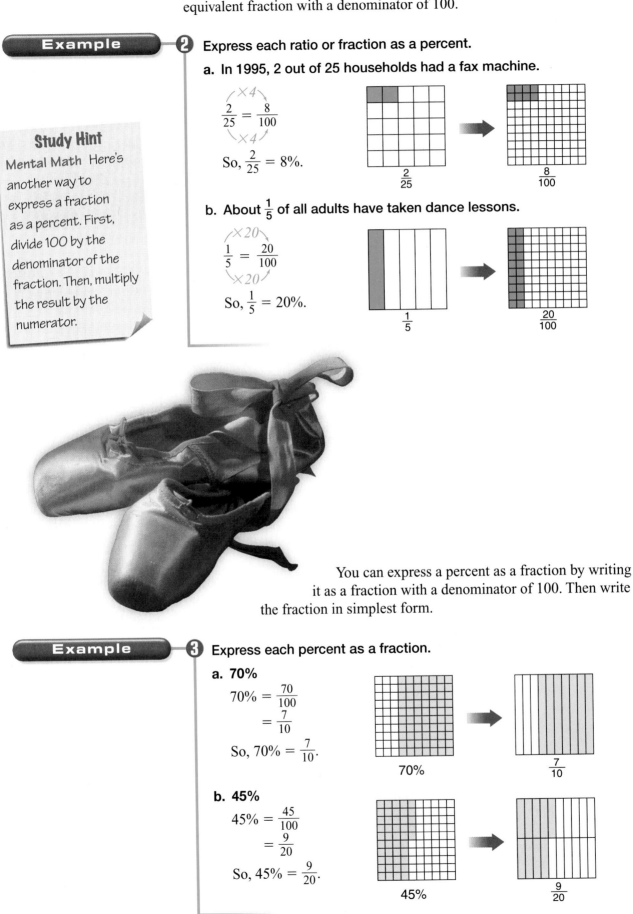

$\frac{2}{25}$ $\frac{8}{100}$

b. **About $\frac{1}{5}$ of all adults have taken dance lessons.**

$$\overset{\times 20}{\underset{\times 20}{\frac{1}{5} = \frac{20}{100}}}$$

So, $\frac{1}{5} = 20\%$.

$\frac{1}{5}$ $\frac{20}{100}$

Study Hint

Mental Math Here's another way to express a fraction as a percent. First, divide 100 by the denominator of the fraction. Then, multiply the result by the numerator.

You can express a percent as a fraction by writing it as a fraction with a denominator of 100. Then write the fraction in simplest form.

Example ③ Express each percent as a fraction.

a. **70%**

$$70\% = \frac{70}{100}$$
$$= \frac{7}{10}$$

So, $70\% = \frac{7}{10}$.

70% $\frac{7}{10}$

b. **45%**

$$45\% = \frac{45}{100}$$
$$= \frac{9}{20}$$

So, $45\% = \frac{9}{20}$.

45% $\frac{9}{20}$

The parts of a circle graph are often labeled with percents. The sum of the percents in a circle graph is 100%.

④ **Statistics** The circle graph shows an estimate of the percent of people who traveled to their vacations by car, plane, bus, or other methods. What fraction of the circle graph is represented by each section?

Car: $65\% = \frac{65}{100}$ or $\frac{13}{20}$

Plane: $30\% = \frac{30}{100}$ or $\frac{3}{10}$

Bus: $3\% = \frac{3}{100}$

Other: $2\% = \frac{2}{100}$ or $\frac{1}{50}$

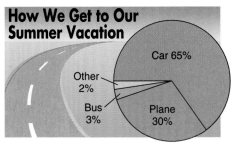

How We Get to Our Summer Vacation

Car 65%

Other 2%

Bus 3%

Plane 30%

Source: Daniel J. Edelman Inc., 1995

CHECK FOR UNDERSTANDING

Communicating Mathematics

Read and study the lesson to answer each question.

1. *Write* the percent and the fraction in simplest form for the model shown at the right.

2. *Explain* how to express the ratio 3 out of 50 as a percent.

HANDS-ON MATH

3. *Use* grid paper to show 17%.

Guided Practice

Express each ratio or fraction as a percent.

4. $\frac{3}{5}$ 5. 1:10

6. Nearly $\frac{90}{100}$ of students ages 9 to 13 play video games.

Express each percent as a fraction in simplest form.

7. 75%   8. 8%   9. 100%

10. *History* In 1860, Abraham Lincoln was elected president with less than 40% of the popular vote. Express 40% as a fraction in simplest form.

EXERCISES

Practice

Express each ratio or fraction as a percent.

11. 2 out of 5   12. $\frac{3}{4}$ 13. $\frac{9}{10}$

14. 19:100   15. 21 out of 25   16. 12.5:100

Express each ratio as a percent.

17. Ninety-nine out of 100 U.S. households have at least one radio.

18. About 1 out of 4 mammals in the world are bats.

19. In 1995, 23 out of 50 girls ages 12 to 17 played basketball.

Express each percent as a fraction in simplest form.

20. 40%   **21.** 5%   **22.** 10%   **23.** 25%

24. 95%   **25.** 1%   **26.** 60%   **27.** 80%

28. Nearly 55% of migrating birds use wetlands as rest stops.

29. Every day, 77% of Americans age 12 or older listen to the radio.

30. There is a 20% chance of rain tomorrow.

31. Which is greater, $\frac{3}{5}$ or 70%?

32. Which is less, $\frac{1}{4}$ or 30%?

Applications and Problem Solving

Take Me Out to the Ballgame

Ages of fans who attended major league baseball games

18-24	14%
25-34	23%
35-44	28%
45-54	18%
55-64	8%
65 and older	9%

Source: Mediamark Research

33. *Baseball* The graph shows the ages of adults who went to baseball games recently. Out of every 100 people who went to baseball games, how many would you expect to be between the ages of 25 and 34?

34. *Technology* According to a cable TV magazine, eleven out of 20 cable TV subscribers decide what program to watch by surfing the channels. Two out of 5 decide by looking in a weekly television guide. Which way of finding a program represents the greater percent?   **Source:** *Cablevision*

35. *Critical Thinking* Suppose you choose a blue marble from a bag of different-colored marbles 7 of every 10 times. What is the probability that you will *not* choose a blue marble?

Mixed Review

36. **Standardized Test Practice** A car traveled 828.8 miles on 28 gallons of gas. How many miles per gallon did the car average? *(Lesson 3-1)*

    A 17.96 mpg   **B** 28.6 mpg   **C** 29.6 mpg

    D 62.42 mpg   **E** Not Here

37. Solve $f = 320 \div (-40)$. *(Lesson 2-8)*

38. Order the set $\{52, -3, 128, 4, -22, 15, 0, -78\}$ from greatest to least. *(Lesson 2-2)*

For **Extra Practice**, see page 612.

39. *Geometry* Find the perimeter and area of the parallelogram. *(Lesson 1-8)*

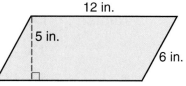

Solving Proportions

What you'll learn

You'll learn to solve proportions.

When am I ever going to use this?

Knowing how to solve proportions can help you halve a recipe when you are cooking.

Word Wise
proportion
cross products

When a marching band goes by, you can easily recognize the brass section because of its brilliant color. Trumpets, trombones, and baritone horns are made of an alloy called brass, which is a mixture of copper and zinc. The copper and zinc are combined in a ratio of 80 parts copper to 20 parts zinc.

The ratio $\frac{80}{20}$ can be simplified to $\frac{4}{1}$. The equation $\frac{80}{20} = \frac{4}{1}$ indicates that the two ratios are equivalent. This is an example of a **proportion**.

Proportion	**Words:**	A proportion is an equation that shows that two ratios are equivalent.
	Symbols:	**Arithmetic** $\frac{80}{20} = \frac{4}{1}$ **Algebra** $\frac{a}{b} = \frac{c}{d}, b \neq 0, d \neq 0$

In a proportion, the two **cross products** are equal. The cross products in the proportion below are 2×12 and 3×8.

$$\frac{2}{3} = \frac{8}{12} \qquad 2 \times 12 = 24, 3 \times 8 = 24$$

Study Hint

Mental Math If both ratios simplify to the same fraction, they form a proportion.

$\frac{8}{12} = \frac{2}{3}, \frac{12}{18} = \frac{2}{3}$

So, $\frac{8}{12} = \frac{12}{18}$.

Property of Proportions	**Words:**	The cross products of a proportion are equal.
	Symbols:	If $\frac{a}{b} = \frac{c}{d}$, then $ad = bc$.

You can use cross products to determine whether a pair of ratios forms a proportion. If the cross products of two ratios are equal, then the ratios form a proportion. If the cross products are *not* equal, the ratios do *not* form a proportion.

Example

1 Determine whether the ratios $\frac{9}{12}$ and $\frac{18}{24}$ form a proportion.

Find the cross products. $\frac{9}{12} = \frac{18}{24}$

$9 \times 24 = 216$ $12 \times 18 = 216$

Since the cross products are equal, the ratios form a proportion.

In a proportion like $\frac{3}{7} = \frac{2.1}{d}$, one of the terms is not known. You can use cross products to find the value of d. This is known as *solving the proportion*.

Examples

INTEGRATION

2 **Algebra** Solve $\frac{3}{7} = \frac{2.1}{d}$.

$$\frac{3}{7} = \frac{2.1}{d}$$

$3 \times d = 7 \times 2.1$ *Find the cross products.*

$3d = 14.7$ *Simplify.*

$\frac{3d}{3} = \frac{14.7}{3}$ *Solve the equation by dividing each side by 3.*

$d = 4.9$ The solution is 4.9.

You can check the solution by substituting the value of d into the original proportion and checking the cross products.

LOOK BACK
You can refer to Lesson 1-5 to review how to solve equations.

APPLICATION

3 **Recycling** When 2,000 pounds of paper are recycled or reused, 17 trees are saved. How many trees would be saved if 5,000 pounds of paper are recycled?

Write a proportion. Let t represent the number of trees.

$$\begin{array}{ccc} paper & \rightarrow & \frac{2,000}{17} = \frac{5,000}{t} & \leftarrow & paper \\ trees & \rightarrow & & \leftarrow & trees \end{array}$$

Find the cross products. Then solve for t.

Method 1 Use paper and pencil.

$2,000 \times t = 17 \times 5,000$

$2,000t = 85,000$

$\frac{2,000t}{2,000} = \frac{85,000}{2,000}$

$t = 42.5$

Method 2 Use a calculator.

17 $\boxed{\times}$ 5000 $\boxed{\div}$ 2000 $\boxed{=}$ *42.5*

If 5,000 pounds of paper are recycled, 42.5 trees are saved.

CHECK FOR UNDERSTANDING

Communicating Mathematics

Math Journal

Read and study the lesson to answer each question.

1. ***Tell*** two methods for determining whether two ratios form a proportion.

2. ***List*** two ratios that do not form a proportion. Explain why they do not.

3. ***Write*** a paragraph explaining how you use algebra when solving a proportion.

Determine whether each pair of ratios forms a proportion.

4. $\dfrac{75}{100}, \dfrac{3}{4}$ **5.** $\dfrac{16}{12}, \dfrac{12}{9}$ **6.** $\dfrac{3}{11}, \dfrac{55}{200}$

Solve each proportion.

7. $\dfrac{2}{34} = \dfrac{5}{x}$ **8.** $\dfrac{3}{5} = \dfrac{c}{10}$ **9.** $\dfrac{a}{0.9} = \dfrac{0.6}{2.7}$

10. *Life Science* When a robin flies, it beats its wings an average of 23 times in 10 seconds. How many times will its wings beat in two minutes?

EXERCISES

Practice

Determine whether each pair of ratios forms a proportion.

11. $\dfrac{7}{6}, \dfrac{6}{7}$ **12.** $\dfrac{8}{12}, \dfrac{10}{15}$ **13.** $\dfrac{10}{12}, \dfrac{5}{6}$ **14.** $\dfrac{6}{9}, \dfrac{8}{12}$

15. $\dfrac{10}{16}, \dfrac{25}{45}$ **16.** $\dfrac{18}{14}, \dfrac{54}{42}$ **17.** $\dfrac{0.2}{0.6}, \dfrac{0.5}{1.5}$ **18.** $\dfrac{0.6}{2.4}, \dfrac{0.4}{0.16}$

Solve each proportion.

19. $\dfrac{4}{100} = \dfrac{12}{n}$ **20.** $\dfrac{9}{1} = \dfrac{n}{14}$ **21.** $\dfrac{5}{4} = \dfrac{y}{12}$ **22.** $\dfrac{120}{b} = \dfrac{24}{60}$

23. $\dfrac{2}{3} = \dfrac{7}{y}$ **24.** $\dfrac{10}{6} = \dfrac{y}{26}$ **25.** $\dfrac{0.35}{3} = \dfrac{c}{18}$ **26.** $\dfrac{n}{2} = \dfrac{0.7}{0.4}$

Applications and Problem Solving

27. *Cooking* A recipe for chocolate chip cookies calls for 2.25 cups of flour and 0.75 cup of granulated sugar. Suppose you only have 1.5 cups of flour. How much granulated sugar should you use when you bake the cookies?

28. *Hobbies* Model railroads are scaled-down replicas of real trains. One of the most popular modeling sizes is HO, where 1 inch on the model represents 87 inches on a real train. An HO model of a modern diesel locomotive is 8 inches long. How long is the real locomotive?

29. *Working on the* **CHAPTER Project** Refer to the foreign exchange rate you found on page 103.
   a. Estimate the amount of U.S. currency you will need to exchange for transportation, housing, food, and souvenirs.
   b. Write and solve a proportion to find the amount of foreign currency you will receive for each category.

For **Extra Practice,** see page 612.

30. *Critical Thinking* In the proportion $\dfrac{10}{x} = \dfrac{y}{5}$, how does the value of y change as the value of x gets larger?

Mixed Review

31. Express 35% as a fraction in simplest form. *(Lesson 3-2)*

32. *Algebra* Solve $-460 = -16a + 52$. *(Lesson 2-9)*

33. Find $-7(-31)$. *(Lesson 2-7)*

34. **Standardized Test Practice** What is the value of the expression $3^2 + 4 \cdot 2 - 6 \div 2$? *(Lesson 1-3)*

   A 14   **B** 10   **C** 5.5   **D** -26

Fractions, Decimals, and Percents

What would you do without your blow dryer? How about the telephone? The table shows that 42% of adults say they can't live without a telephone.

In order to compute with percents, you must first express them as fractions or decimals. In Lesson 3-2, you expressed percents as fractions by writing the percent as a fraction with a denominator of 100. The word percent also means *hundredths* or *per hundred*.

Famous Inventions	
% of adults who say they can't live without them	
automobile	63%
light bulb	54%
telephone	42%
television	22%
aspirin	19%
microwave	13%
blow dryer	8%

Source: *American Demographics*

You can express percents as decimals by dividing by 100 and expressing the answer as a decimal.

TECHNOLOGY

MINI-LAB

Work with a partner.

🖩 calculator

You can use the percent key on your calculator to express 42% as a decimal.

- Enter 42 [2nd] [%].
- The display shows *0.42*. So, 42% = 0.42.

Try This

Use a calculator to express each percent as a decimal.

1. 16%   **2.** 55%   **3.** 93%   **4.** 100%

5. 40%   **6.** 2%   **7.** 12.5%   **8.** 66.7%

Talk About It

9. Explain what happened to the number in the display when you pressed the percent key.

10. Use what you have learned in this lab to express 5% as a decimal without using a calculator.

Example

1 Express 60% as a fraction and as a decimal.

THINK: 60 hundredths

$60\% = \dfrac{60}{100}$ or $\dfrac{3}{5}$

$60\% = 0.60$ or 0.6

To express a decimal as a percent, first express the decimal as a fraction with a denominator of 100. Then express the fraction as a percent.

Express each decimal as a percent.

a. 0.78

$$0.78 = \frac{78}{100}$$
$$= 78\%$$

b. 0.3

$$0.3 = \frac{3}{10} \text{ or } \frac{30}{100}$$
$$= 30\%$$

You can use these shortcuts when working with decimals and percents.

Decimals and Percents	• To write a decimal as a percent, multiply by 100 and add the % symbol. $0.24 = 0.24 = 24\%$ • To write a percent as a decimal, divide by 100 and remove the % symbol. $24\% = 24\% = 0.24$

You have learned to express a fraction as a percent by finding an equivalent fraction with a denominator of 100. This method works well if the denominator of the fraction is a number like 2, 4, 5, 10, or 25. That is, the denominator is a factor of 100. If the denominator is *not* a factor of 100, you can solve a proportion.

Express each fraction as a percent.

a. $\frac{1}{8}$

$$\frac{1}{8} = \frac{n}{100} \qquad \textit{Write a proportion.}$$
$$1 \times 100 = 8 \times n \quad \textit{Find the cross products.}$$
$$100 = 8n$$
$$\frac{100}{8} = \frac{8n}{8} \qquad \textit{Solve the equation by dividing each side by 8.}$$
$$12.5 = n$$

So, $\frac{1}{8} = 12.5\%$.

b. $\frac{2}{3}$

$$\frac{2}{3} = \frac{n}{100} \qquad \textit{Write a proportion.}$$
$$2 \times 100 = 3 \times n \quad \textit{Find the cross products.}$$

2 ⊠ 100 ⊡ 3 ⊟ 66.66666667

So, $\frac{2}{3} \approx 66.67\%$.

Example 4

CONNECTION

Life Science There are 250 known species of sharks, and of that number, only 27 species have been involved in attacks on humans. What percent of known species of sharks have attacked humans?

Express the fraction $\frac{27}{250}$ as a percent by solving a proportion. Let n represent the percent of sharks.

$$\frac{27}{250} = \frac{n}{100}$$

27 × 100 ÷ 250 = 10.8

So, 10.8% of the known species of sharks have attacked humans.

CHECK FOR UNDERSTANDING

Communicating Mathematics

Read and study the lesson to answer each question.

1. *Write* a fraction, decimal, and percent to represent the model at the right.

2. *Explain* two different methods you can use to express fractions as percents. Use $\frac{13}{20}$ and $\frac{13}{19}$ in your explanation.

HANDS-ON MATH

3. *Use models* to show which is greater, 15% or $\frac{1}{8}$.

Guided Practice

Express each percent as a decimal.

4. 58%   **5.** 9%   **6.** 23.8%

Express each decimal as a percent.

7. 0.29   **8.** 0.7   **9.** 0.042

Express each fraction as a percent.

10. $\frac{3}{4}$ **11.** $\frac{5}{8}$ **12.** $\frac{1}{6}$

13. *Earth Science* About $\frac{3}{20}$ of the total land surface of Earth is covered by deserts. What percent is this?

Family Activity

Search magazines or newspapers for examples of ratios that are used in advertisements. Rewrite the advertisement using percents.

EXERCISES

Practice

Express each percent as a decimal.

14. 26%   **15.** 74%   **16.** 6%   **17.** 1%

18. 80%   **19.** 15.5%   **20.** 3.2%   **21.** 3.8%

Express each decimal as a percent.

22. 0.19   **23.** 0.02   **24.** 0.9   **25.** 0.026

26. 0.375   **27.** 0.667   **28.** 0.083   **29.** 0.0625

Express each fraction as a percent.

30. $\frac{23}{50}$ **31.** $\frac{7}{8}$ **32.** $\frac{19}{20}$ **33.** $\frac{5}{8}$

34. $\frac{1}{3}$ **35.** $\frac{5}{16}$ **36.** $\frac{27}{30}$ **37.** $\frac{15}{40}$

Replace each ● with <, >, or = to make a true sentence.

38. 45% ● 4.5 **39.** 9.3 ● 93% **40.** 0.05 ● 50%

41. 57.8% ● 0.578 **42.** 0.3 ● 30% **43.** 1% ● 0.09

44. Express the ratio 1:40 as a percent.

45. Which is greater: $\frac{3}{8}$ or 40%?

Applications and Problem Solving

46. *Geometry* What percent of the area of the square at the right is shaded?

47. *Transportation* According to the Association of American Railroads, about 2 out of every 5 trains in use in 1995 were built in the 1970s. Express 2 out of 5 as a percent.

48. *Critical Thinking* Which is greater: 2% of 98 or 98% of 2?

Mixed Review

49. *Algebra* Solve $\frac{3}{7} = \frac{n}{28}$. *(Lesson 3-3)*

50. What percent of the squares at the right are shaded? *(Lesson 3-2)*

For **Extra Practice,** see page 613.

51. Express the ratio *20 out of 50 students* in simplest form. *(Lesson 3-1)*

52. *Algebra* Solve $4m > 24$. Graph the solution on a number line. *(Lesson 1-9)*

53. Evaluate 3^4. *(Lesson 1-2)*

54. *Standardized Test Practice* Kyung had $17. His dinner cost $5.62, and he gave the cashier a $10 bill. How much change should he receive from the cashier? *(Lesson 1-1)*

 A $4.32   **B** $4.38   **C** $11.38   **D** $15.62   **E** Not Here

CHAPTER 3

Mid-Chapter Self Test

1. Express the rate $420 for 15 tickets as a unit rate. *(Lesson 3-1)*

2. Express the ratio 3 out of 20 as a percent. *(Lesson 3-2)*

3. *Algebra* Solve $\frac{x}{36} = \frac{15}{24}$. *(Lesson 3-3)*

4. *Physical Science* Light travels approximately 1,860,000 miles in 10 seconds. How long will it take light to travel the 93,000,000 miles from the Sun to Earth? *(Lesson 3-3)*

5. Express $\frac{3}{50}$ as a percent using a proportion. *(Lesson 3-4)*

3-4B The Golden Ratio

A Follow-Up of Lesson 3-4

🔲 grid paper

✂️ scissors

🖩 calculator

📏 tape measure

What do a cereal box, a skyscraper, and a famous painting have in common? One thing is their rectangular-shaped faces.

Mathematicians have studied rectangles for centuries. Many ancient Greek mathematicians and philosophers looked for a special rectangle that was the most perfect, visually appealing rectangle. Some rectangles are shown below. Take an informal survey of your classmates to find which one they like best.

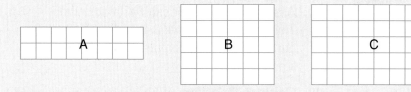

Most people choose rectangle C. It's not too skinny and not too square. This rectangle is an example of a *golden rectangle*.

Work in groups of three.

Step 1   Cut a rectangle out of grid paper that is 34 squares long and 21 squares wide. The ratio of length to width is $\frac{34}{21}$ or about 1.62.

Step 2   Cut the rectangle into two parts, in which one part is the largest possible square and the other part is a rectangle. Measure the rectangle and record its length and width. Write the ratio of length to width and then express it as a decimal. Record your data in a table like the one shown below.

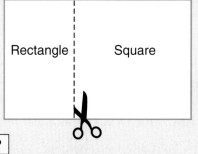

length	34	21	?	?	?	?	?
width	21	13	?	?	?	?	?
ratio	$\frac{34}{21}$	?	?	?	?	?	?
decimal	1.62	?	?	?	?	?	?

Step 3   Use the same method to cut the smaller rectangle into two parts until the remaining rectangle is 1 square by 2 squares. Measure each rectangle to the nearest tenth of a centimeter, and find the ratio of length to width.

1. Look for a pattern in the ratios you recorded. Describe the pattern.

2. If the rectangles you cut out are described as golden rectangles, what do you think the value of the *golden ratio* is?

3. Write a definition of *golden rectangle.* Use the word *ratio* in your definition.

4. The ancient Greeks considered the value of the golden ratio to be 1.618.
   a. Measure the height of your face from the top of your head to your chin. Then measure the width of your face at your cheekbone. Find the ratio for each person in your group.
   b. Find the average ratio for your group. Is it close to the golden ratio?

5. There are many examples of the golden ratio in music, architecture, art, and nature. Some are shown below.

Composition with Grey, Red, Yellow, and Blue by Piet Mondrian

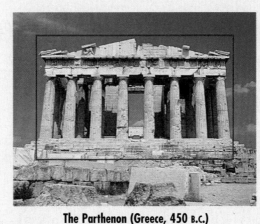

The Parthenon (Greece, 450 B.C.)

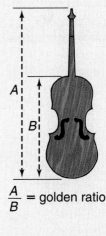

$\frac{A}{B}$ = golden ratio

Do research to find another example of the golden ratio.

6. The spiral of the chambered nautilus follows the pattern of the golden ratio.
   a. Trace the figure at the right. It is a smaller version of the golden rectangle you cut apart at the beginning of this lab.
   b. An arc connects the opposite corners of the large square. Starting at the endpoint of the first arc, sketch an arc that connects the opposite corners of the next largest square. Continue the process.
   c. What do the combined arcs resemble?

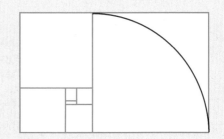

7. The *Fibonacci sequence* is a famous mathematical sequence. It begins 1, 1, 2, 3, 5, 8,... .
   a. Find the next six numbers in the sequence.
   b. What is the relationship between the numbers in the Fibonacci sequence and the golden ratio?

3-5 Finding Percents

What you'll learn

You'll learn to compute mentally with percents.

When am I ever going to use this?

You'll use mental math strategies to help estimate the amount of a tip to leave at a restaurant.

The idea of percent may have originated when the Roman emperor Augustus (63 B.C.–A.D. 14) levied a tax on all goods sold at auction. The tax rate was $\frac{1}{100}$ or 1%.

Look for a pattern in the following problems to find 1% of a number mentally.

Number	Computation			Result
48	1% of 48	→	0.01 × 48	0.48
6	1% of 6	→	0.01 × 6	0.06
125	1% of 125	→	0.01 × 125	1.25

These and other examples suggest that you can find 1% of a number mentally by moving the decimal point two places to the left.

Examples

1 Find 1% of 235.

1% of 235 is 2.35.

2 Find 1% of 12.8.

1% of 12.8 is 0.128.

You can also find 10% of a number mentally.

$$10\% \text{ of } 82 \quad \rightarrow \quad 0.1 \times 82 = 8.2$$

$$10\% \text{ of } 235 \quad \rightarrow \quad 0.1 \times 235 = 23.5$$

These and other examples suggest that you can find 10% of a number mentally by moving the decimal point one place to the left.

Example

APPLICATION

3 **Journalism** A weekly magazine receives 100 to 150 letters each week. They publish about 10% of these letters in each issue. Suppose one week they receive 138 letters. About how many will they publish?

Method 1
Use paper and pencil.

$$\begin{array}{r} 138 \\ \times\ 0.10 \\ \hline 13.80 \end{array}$$

Method 2
Use mental math.

10% of 138 is 13.8.

The magazine will publish about 14 letters.

When you compute with common percents like 25%, 12.5%, or $33\frac{1}{3}\%$, it may be easier to use the fraction form of a percent. This number line shows some fraction-percent equivalents.

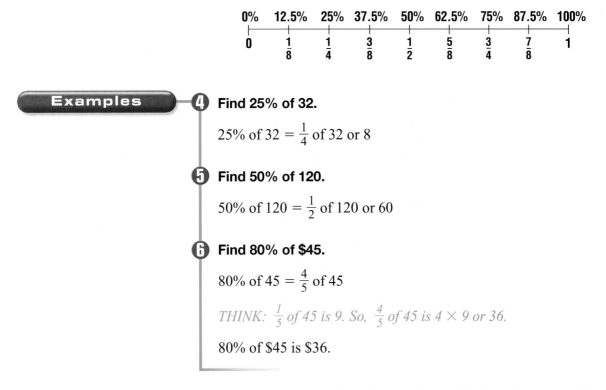

Examples

4 **Find 25% of 32.**

25% of 32 = $\frac{1}{4}$ of 32 or 8

5 **Find 50% of 120.**

50% of 120 = $\frac{1}{2}$ of 120 or 60

6 **Find 80% of $45.**

80% of 45 = $\frac{4}{5}$ of 45

THINK: $\frac{1}{5}$ of 45 is 9. So, $\frac{4}{5}$ of 45 is 4 × 9 or 36.

80% of $45 is $36.

Some percents are used much more frequently than others. For this reason, it's a good idea to memorize the equivalent fractions, decimals, and percents listed in the table below.

Equivalent Fractions, Decimals, and Percents			
$\frac{1}{2}$, 0.5, 50%	$\frac{2}{3}$, $0.66\frac{2}{3}$, $66\frac{2}{3}\%$	$\frac{4}{5}$, 0.8, 80%	$\frac{7}{8}$, 0.875, 87.5%
$\frac{1}{4}$, 0.25, 25%	$\frac{1}{5}$, 0.2, 20%	$\frac{1}{8}$, 0.125, 12.5%	$\frac{3}{10}$, 0.3, 30%
$\frac{3}{4}$, 0.75, 75%	$\frac{2}{5}$, 0.4, 40%	$\frac{3}{8}$, 0.375, 37.5%	$\frac{7}{10}$, 0.7, 70%
$\frac{1}{3}$, $0.33\frac{1}{3}$, $33\frac{1}{3}\%$	$\frac{3}{5}$, 0.6, 60%	$\frac{5}{8}$, 0.625, 62.5%	$\frac{9}{10}$, 0.9, 90%

CHECK FOR UNDERSTANDING

Communicating Mathematics

Read and study the lesson to answer each question.

1. *Explain* how to find 1% of a number mentally.

2. *Name* the fraction you would use to find 62.5% of 16 mentally.

3. *You Decide* Lawana thinks 10% of 95 is 0.95. Elisa thinks 10% of 95 is 9.5. Who is correct? Explain your reasoning.

Guided Practice

Compute mentally.

4. 10% of 25

5. 1% of $1,355

6. 75% of $120

7. 37.5% of 16

8. Which is greater, 25% of 16 or 5?

9. *Health* Many health authorities recommend that a healthy diet contains no more than 30% of its Calories from fat. If you consume 1,500 Calories each day, what is the maximum number of Calories you should consume from fat?

EXERCISES

Practice

Compute mentally.

10. 10% of $105

11. 1% of 17.2

12. 20% of 5

13. 87.5% of 160

14. 50% of $36

15. 12.5% of 48

16. $33\frac{1}{3}$% of 24

17. 90% of 1,000

18. 30% of 50

19. 40% of 75

20. 1% of 15,389

21. 10% of $38.41

Replace each ● with <, >, or = to make a true sentence.

22. 10 ● 10% of 90

23. 75% of 20 ● 16

24. $66\frac{2}{3}$% of 18 ● 60% of 15

25. 1% of 120 ● 10% of 12

26. Find 1% of $36,700 mentally.

27. Find 10% of $12.95 mentally.

Applications and Problem Solving

28. *Statistics* Refer to the graph. Suppose 1,600 women were surveyed.

   a. How many women said they bought more than seven pairs of shoes?

   b. How many women responded that they didn't know how many pairs of shoes they had purchased?

29. *Population* In 1996, there were about 15,167,000 people in the United States who were ages 14 to 17. That section of the population is expected to increase by about 10% by the year 2010. Estimate how many people in the United States will be ages 14 to 17 in 2010.

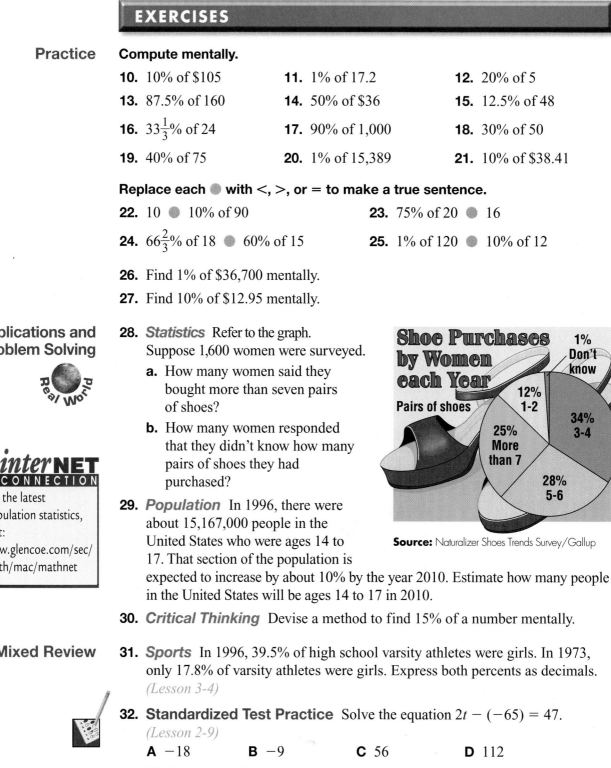

Shoe Purchases by Women each Year

Pairs of shoes

1% Don't know

12% 1-2

34% 3-4

25% More than 7

28% 5-6

Source: Naturalizer Shoes Trends Survey/Gallup

30. *Critical Thinking* Devise a method to find 15% of a number mentally.

Mixed Review

31. *Sports* In 1996, 39.5% of high school varsity athletes were girls. In 1973, only 17.8% of varsity athletes were girls. Express both percents as decimals. *(Lesson 3-4)*

32. *Standardized Test Practice* Solve the equation $2t - (-65) = 47$. *(Lesson 2-9)*

   A -18 **B** -9 **C** 56   **D** 112

33. Find $-15 + (-2)$. *(Lesson 2-3)*

For **Extra Practice,**
see page 613.

34. *Geometry* Find the perimeter and area of a rectangle that is 2 centimeters wide and 5 centimeters long. *(Lesson 1-8)*

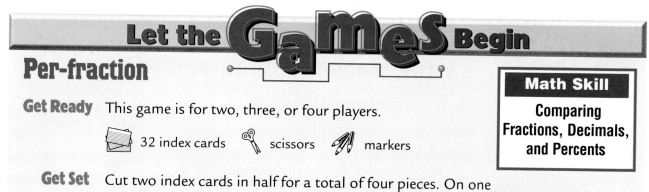

Per-fraction

Get Ready   This game is for two, three, or four players.

32 index cards   scissors   markers

Get Set   Cut two index cards in half for a total of four pieces. On one piece, write a fraction from the table on page 121. On another piece, write any fraction that is equivalent to the first fraction. On the third piece, write the equivalent decimal. On the last piece, write the equivalent percent. Here's a sample.

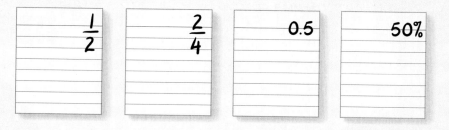

Repeat these steps until you have used all 16 fractions from the table. You will have 64 cards in all.

Go
- Shuffle the cards and deal seven cards to each player. Place the remaining cards facedown in the middle of the table. Take the top card and place it face up next to the deck, forming the discard pile.

Scoring Set

- All players check their cards for scoring sets. A scoring set consists of three equivalent numbers. Scoring sets should be placed face up on the table.

- The player to the left of the dealer draws the top card from the deck or the top card in the discard pile. If the player has a scoring set, he or she should place it faceup on the table. The player may also build onto another player's scoring set by placing a card faceup on the table and announcing the set on which the player is building. The player's turn ends when he or she discards a card.

- Play continues until one player has no cards remaining in his or her hand. That player is the winner.

 Visit www.glencoe.com/sec/math/mac/mathnet for more games.

PROBLEM SOLVING

3-6A Reasonable Answers

A Preview of Lesson 3-6

Tanya wants to buy in-line skates at a sporting goods store. She and her friend Krista are trying to decide whether she has enough money for the skates and the sales tax.

The skates are on sale for $89.99 and the sales tax is 5%. I brought $95, just in case. Do you think that's enough?

Well, the skates cost about $90. I know that 1% of $90 is $0.90. Will that help?

Tanya

Sure! $0.90 times 5 is $4.50. Since $90 and $4.50 is $94.50, I have just enough money!

Krista

THINK ABOUT IT

Work with a partner.

1. *Explain* how Krista knows that 1% of $90 is $0.90.

2. *Think* of another way to help Tanya decide whether she has enough money.

3. *Apply* the strategy of determining **reasonable answers** to solve this problem.

   Lucia knows that it is customary to leave a tip for the server at a restaurant. The tip is usually 15% of the value of the meal. How much should Lucia leave for a meal that costs $24.90?

For **Extra Practice,** see page 613.

ON YOUR OWN

4. *Explain* how you can use mental math skills to help plan the solution to a problem.

5. Do research to find the state sales tax rate for your state. *Write a Problem* using the tax rate.

6. *Look Ahead* Explain whether the statement *11% of 98 is 0.99* is reasonable or not. If not, rewrite the statement so it is reasonable.

MIXED PROBLEM SOLVING

STRATEGIES

Look for a pattern.
Solve a simpler problem.
Act it out.
Guess and check.
Draw a diagram.
Make a chart.
Work backward.

Solve. Use any strategy.

7. *Money Matters* You spend $5.55 plus $0.44 tax at the store for makeup and pay with a $10 bill. Would it be more reasonable to expect about $3.00 or $4.00 in change?

8. *Ecology* In a survey of 1,413 shoppers, 6% said they would be willing to pay more for environmentally safe products. Is 8.4, 84, or 841 a reasonable estimate for the number of shoppers willing to pay more?

9. *Transportation* Yesterday you noted that the mileage on the family car read 60,094.8 miles. Today it reads 60,099.1 miles. Was the car driven about 4 or 40 miles?

10. *Agriculture* An orange grower harvested 1,260 pounds of oranges from one grove, 874 pounds from another, and 602 pounds from a third. What is a reasonable number of crates to have on hand if each crate holds 14 pounds of oranges?

11. *Money Matters* Della has only nickels in her pocket. Ayita has only dimes in hers. Kareem has only quarters in his. Marta approached them for a donation for their school fundraiser. What is the least each person could donate so that each one gives the same amount?

12. *Earth Science* Geothermal energy is heat from inside Earth. Underground temperatures generally increase 9° C for every 300 feet of depth. How deep would you have to dig so that the underground temperature is 90° C greater than the ground temperature?

13. *Sports* The graph shows the number of injuries treated in emergency rooms for the top seven sports. About how many injuries were treated in all?

Sports Injuries

Top sports for number of injuries treated in emergency rooms

Basketball	693,933
Cycling	599,874
Football	390,180
Snow skiing	330,289
Skating (all types)	322,311
Baseball	219,023
Soccer	157,251

Source: Consumer Product Safety Commission, American Academy of Orthopaedic Surgeons

14. *Standardized Test Practice* Diego wants to buy a pair of shoes that cost $59.95. The sales tax is 6%. Which amount is reasonable for the total cost?

 A $60.31

 B $63.55

 C $65.95

 D $89.95

 E $95.95

3-6

Percent and Estimation

What you'll learn
You'll learn to estimate by using equivalent fractions, decimals, and percents.

When am I ever going to use this?
Knowing how to estimate with percents can help you determine prices in a 20%-off sale.

Word Wise
compatible numbers

Fire fighters use geometry and aerial photography to estimate how much of a forest has been damaged by fire. A grid is superimposed on a photograph of the forest.

For example, suppose the tan part of the figure at the right represents the area damaged by a forest fire. About 25 small squares out of 49 squares are shaded tan.

$$\frac{25}{49} \approx \frac{25}{50} \text{ or } \frac{1}{2}$$

$$\frac{1}{2} = 50\%$$

So, about 50% of the area has been damaged by the fire.

You can estimate a percent of a number by using **compatible numbers**. Compatible numbers are two numbers that are easy to divide mentally.

Examples

1 Estimate 19% of 60.

19% is about 20% or $\frac{1}{5}$.   *5 and 60 are compatible numbers.*

$\frac{1}{5}$ of 60 is 12.

So, 19% of 60 is about 12.

2 Estimate 25% of 78.

25% = $\frac{1}{4}$, and 78 is about 80.   *4 and 80 are compatible numbers.*

$\frac{1}{4}$ of 80 is 20.

So, 25% of 78 is about 20.

3 Estimate 65% of 34.

65% is about $66\frac{2}{3}\%$ or $\frac{2}{3}$, and 34 is about 33.   *3 and 33 are compatible numbers.*

$\frac{2}{3}$ of 33 is 22.

So, 65% of 34 is about 22.

Circle graphs are one way to display statistical data. You can use estimation to sketch a circle graph.

Example **4**

CONNECTION

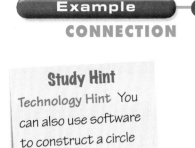

Study Hint

Technology Hint You can also use software to construct a circle graph.

Earth Science Seventy percent of Earth's surface is covered by water. The remaining 30% is land. Sketch a circle graph of this data.

Explore   Circle graphs show how parts are related to a whole. The circle represents the whole quantity. In this case, it represents the surface of Earth.

Plan   Use estimation to find the size of each part.
70% is a little less than 75% or $\frac{3}{4}$. So, the part representing water should be a little less than $\frac{3}{4}$ of the circle. The remaining part will represent land.

Solve   Draw a circle and divide it into fourths using dotted lines to visualize $\frac{3}{4}$ of the circle. Using solid lines, sketch the graph to represent 70% and 30%. Give the graph a title and label the parts.

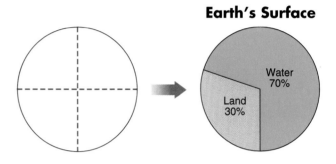

Examine   Since 70% is a little more than twice 30%, the part representing 70% should be more than twice as large as the part representing 30%. The graph is reasonable.

HANDS-ON MINI-LAB

Work with a partner.

calculator   compass

markers

Try This

1. Choose a topic of interest to you and conduct a survey of 20 of your classmates. Here are some suggestions.

    a. number of siblings   **b.** state where born
    c. kinds of pets   **d.** favorite sports team

2. Show your data as percents.
3. Sketch a circle graph of the data.

Talk About It

4. Name an advantage of displaying data in a graph instead of a table.

Communicating Mathematics

Read and study the lesson to answer each question.

1. *Tell* how you can estimate 23% of $98.95 using fractions and compatible numbers.

2. *Draw* a triangle on graph paper. Shade about $\frac{1}{3}$ of the area.

3. *You Decide* Darnell thinks 46% of 80 is less than 40. Elaina thinks it is greater than 40. Who is correct? Explain your reasoning.

Guided Practice

Estimate the percent of the area shaded.

4.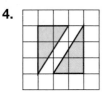

5.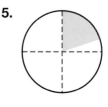

Estimate.

6. 29% of 50

7. 88% of 64

Estimate the percent.

8. 7 out of 16

9. 22 out of 60

10. *Sports* Estimate Shaun's foul shooting percent in basketball if he made 13 foul shots in 22 attempts.

EXERCISES

Practice

Estimate the percent of the area shaded.

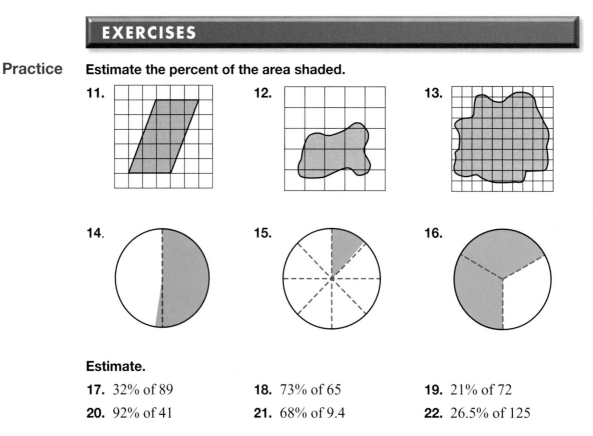

11. 12. 13.

14. 15. 16.

Estimate.

17. 32% of 89

18. 73% of 65

19. 21% of 72

20. 92% of 41

21. 68% of 9.4

22. 26.5% of 125

Estimate the percent.

23. 8 out of 13   **24.** 4 out of 25   **25.** 12 out of 60

26. 7 out of 57   **27.** 9.2 out of 11   **28.** 16 out of 45.8

29. What compatible numbers could you use to estimate 32% of 154?

30. Estimate what percent 8 out of 15 represents.

31. *Write a Problem* in which the solution is about 50%.

Applications and Problem Solving

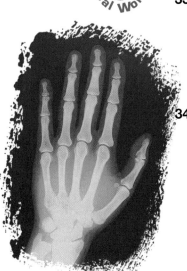

32. *Money Matters* Use the tip table at the right to estimate the tip for a $26.85 bill.

33. *Life Science* The adult skeleton has 206 bones. Sixty of them are in the arms and hands. Estimate the percent of bones that are in the arms and hands.

34. *Statistics* Refer to the data at right.

    a. Sketch this data in a circle graph and a bar graph.

    b. What advantage does the circle graph have over the bar graph?

    c. What advantage does the bar graph have over the circle graph?

Tip Table	
Bill (dollars)	**Tip (dollars)**
5.00	0.75
10.00	1.50
15.00	2.25
20.00	3.00
25.00	3.75
30.00	4.50
35.00	5.25

Portion of books purchased by children ages 13 and younger and adults

Adult Books 59%   Juvenile Books 41%

Source: The NPD Group for Book Industry Study Group

35. *Working on the* **CHAPTER Project** Refer to the budget you planned on page 103.

    a. Of your total expenses, estimate the percent that you'll spend on transportation, housing, food, and souvenirs.

    b. Sketch a circle graph of your expenses.

36. *Critical Thinking* How would you determine whether 8% of 400 is greater than 4% of 180?

Mixed Review

37. Compute 10% of 1,358 mentally. *(Lesson 3-5)*

38. **Standardized Test Practice** Wesley can run 3.5 miles in 40 minutes. How long would it take him to run 8 miles at this same rate? *(Lesson 3-3)*

    A 182 minutes   **B** 91 minutes   **C** 17 minutes   **D** 0.7 minutes

39. Evaluate $(17 - 8) \div 3 + 5^2$. *(Lesson 1-3)*

40. *Transportation* A service station along I-64 in Lexington, Virginia, charges $1.39 a gallon for diesel fuel and $1.10 a quart for oil. How much will 38 gallons of diesel fuel and 2 quarts of oil cost? *(Lesson 1-1)*

For **Extra Practice,** see page 614.

3

inter**NET**
CONNECTION Chapter Review **For additonal lesson-by-lesson review, visit:**
www.glencoe.com/sec/math/mac/mathnet

Vocabulary

After completing this chapter, you should be able to define each term, concept, or phrase and give an example or two of each.

Number and Operations
compatible numbers (p. 126)
cross products (p. 111)
percent (p. 107)
proportion (p. 111)
rate (p. 105)
ratio (p. 104)
unit rate (p. 105)

Problem Solving
reasonable answers (p. 124)

Understanding and Using the Vocabulary

Choose the correct term or number to complete each sentence.

1. A (ratio, proportion) is a comparison of two numbers by division.
2. The expression \$25/5 hours is an example of a (unit rate, rate).
3. An equation that indicates that two ratios are equivalent is a (proportion, percent).
4. A (unit rate, ratio) always has a denominator of 1.
5. The model at the right represents (60%, 6%).
6. The decimal 0.03 written as a percent is (30%, 3%).
7. The fraction $\frac{1}{2}$ written as a percent is $\left(\frac{1}{2}\%, 50\%\right)$.
8. Numbers that are easy to divide mentally are called (compatible, rates).
9. You can estimate 26% of 16 by using the fraction $\left(\frac{1}{4}, \frac{2}{5}\right)$.

In Your Own Words

10. *Explain* the differences between a ratio and a rate.

Upon completing this chapter, you should be able to:

express ratios as fractions in simplest form and determine unit rates *(Lesson 3-1)*

Express the ratio 9 out of 15 in simplest form.

$$\frac{9}{15} = \frac{3}{5}$$

$\div 3$ / $\div 3$

Use these exercises to review and prepare for the chapter test.

Express each ratio or rate in simplest form.

11. 12 peaches:18 pears

12. 40 centimeters per meter

13. 25 to 9

14. 5 girls:40 people

Express each rate as a unit rate.

15. $5 for 2 minutes

16. 150 students to 3 buses

17. 110 miles on 5 gallons

express ratios as percents and vice versa *(Lesson 3-2)*

Express $\frac{4}{5}$ as a percent.

$$\frac{4}{5} = \frac{80}{100}$$

$\times 20$ / $\times 20$

So, $\frac{4}{5} = 80\%$.

Express each ratio or fraction as a percent.

18. 3 out of 5

19. 16.5:100

20. $\frac{1}{4}$

21. $\frac{14}{25}$

Express each percent as a fraction in simplest form.

22. 20%

23. 7%

24. 85%

25. 90%

solve proportions *(Lesson 3-3)*

Solve $\frac{7}{4} = \frac{n}{2}$.

$7 \times 2 = 4 \times n$ *Cross products*

$14 = 4n$

$\frac{14}{4} = \frac{4n}{4}$ *Divide each side by 4.*

$3.5 = n$

Solve each proportion.

26. $\frac{5}{8} = \frac{n}{72}$

27. $\frac{3}{r} = \frac{6}{8}$

28. $\frac{9}{x} = \frac{4}{18}$

29. $\frac{30}{0.5} = \frac{y}{0.25}$

● express percents as fractions and decimals and vice versa *(Lesson 3-4)*

Express 0.18 as a percent.

$0.18 = \dfrac{18}{100}$ or 18%

Express $\dfrac{1}{3}$ as a percent.

$\dfrac{1}{3} = \dfrac{n}{100}$

$3n = 100$

$n \approx 33.3\%$

Express each percent as a decimal.

30. 13%

31. 90%

32. 24%

33. 4.3%

Express each decimal as a percent.

34. 0.35

35. 0.7

36. 0.04

37. 0.655

Express each fraction as a percent.

38. $\dfrac{3}{50}$

39. $\dfrac{7}{20}$

40. $\dfrac{9}{25}$

41. $\dfrac{1}{6}$

● compute mentally with percents *(Lesson 3-5)*

Find 10% of 65.

10% of 65 is 6.5.

Find 25% of 8.

25% of 8 = $\dfrac{1}{4}$ of 8 or 2

Compute mentally.

42. 10% of 18.3

43. 20% of 60

44. 90% of 100

45. 50% of $42

46. $66\frac{2}{3}\%$ of 24

● estimate by using equivalent fractions, decimals, and percents *(Lesson 3-6)*

Estimate 40% of 83.

$40\% = \dfrac{2}{5}$

$\dfrac{2}{5}$ of 80 is 32.   *Round 83 to 80.*

So, 40% of 83 ≈ 32.

Estimate.

47. 8% of 104

48. 62% of 50

49. 99% of 35

Estimate the percent.

50. 11 out of 24

51. 20 out of 52

Applications & Problem Solving

52. *Reasonable Answers* At a restaurant, entrees cost from $9.95 to $18.95, appetizers cost from $4.95 to $6.95, a salad costs $3.25, and soup costs $2.50. A full meal includes soup, salad, appetizer, and an entree. Without tax and tip, would a couple expect to pay more or less than $40 for a full meal at this restaurant? Explain. *(Lesson 3-6A)*

53. *Hobbies* Javier has 130 baseball cards in his collection. Forty-eight of the cards are of Cleveland players. Find the ratio of the Cleveland cards to all cards. *(Lesson 3-1)*

54. *Money Matters* Penelope spends 42% of her monthly income on housing. If her monthly income is $1,900, about how much does she spend on housing? *(Lesson 3-6)*

55. *Writing* Americans spend about $1.7 billion each year on writing instruments. Refer to the graph. Express each percent as a decimal. *(Lesson 3-4)*

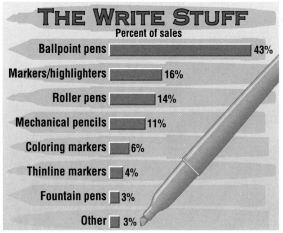

THE WRITE STUFF
Percent of sales

Ballpoint pens	43%
Markers/highlighters	16%
Roller pens	14%
Mechanical pencils	11%
Coloring markers	6%
Thinline markers	4%
Fountain pens	3%
Other	3%

Source: Writing Instrument Manufacturers Association

Alternative Assessment

Open Ended

Sam estimated that Sarah made 75% of the 2-point shots she attempted in a basketball game. If Sarah attempted 18 shots, how many shots could she have made for Sam to arrive at this estimate? Explain your reasoning.

In the next game, Sam estimated that Sarah made 50% of the 2-point shots she attempted. Did Sarah score more or fewer points in the second game? Explain your reasoning.

A practice test for Chapter 3 is provided on page 649.

Completing the CHAPTER Project

Use the following checklist to make sure your report is complete.

☑ The foreign exchange rate is included.

☑ The computation of the number of U.S. dollars you will need to take on your trip is accurate.

☑ Your circle graph is accurate.

Add any finishing touches that you would like to make your report more attractive.

PORTFOLIO Select one of the assignments from this chapter and place it in your portfolio. Attach a note to it explaining why you selected it.

Section One: Multiple Choice

There are ten multiple choice questions in this section. Choose the best answer. If a correct answer is *not here,* choose the letter for Not Here.

1. What is the perimeter of the rectangle?

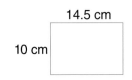

14.5 cm

10 cm

A 24.5 cm

B 34.5 cm

C 49 cm

D 145 cm

2. Which integers are graphed on the number line?

-5   0   5

F $\{-5, -2\}$

G $\{-5, -4, -3, \ldots\}$

H $\{-5, -2, 0, 1\}$

J $\{-5, 0, 5\}$

3. In a random sample of 150 students, 60 ride the bus to school, 54 ride in car pools, and 36 walk. If there will be 800 students next year, about how many will need bus transportation?

A 320

B 160

C 80

D 60

Please note that Questions 4–10 have five answer choices.

4. Including tax, the cost of renting a steam cleaner is $24.99 for the first day and $5.99 for each day after that. Which number sentence could be used to find C, the cost in dollars for keeping the cleaner for 4 days?

F $C = \$24.99 + \$5.99$

G $C = 4(\$24.99) + \$5.99$

H $C = 4(\$5.99) + \$24.99$

J $C = 3(\$5.99) + \$24.99$

K $C = 4(\$5.99) + 4(\$24.99)$

5. Markers ordered for the art room cost $0.04 per marker for the first 100. After that the cost drops to $0.02 per marker. Which is a reasonable total cost for 250 markers?

A The cost is less than $5.

B The cost is between $5 and $10.

C The cost is between $10 and $15.

D The cost is between $15 and $20.

E The cost is greater than $20.

6. Mallory works 4 hours a day for 4 days each week. She earns $5.25 per hour. Which problem *cannot* be solved using the information given?

F How many days must Mallory work to earn $600?

G How much more would Mallory earn each week if she earns $5.75 per hour?

H How many CDs can Mallory buy with the money she earns in 2 weeks?

J How much does Mallory earn each week?

K How much money would Mallory earn in a week in which she was ill and didn't work one day?

7. The lengths of the sides of a five-sided field are 524 feet, 498 feet, 519 feet, 502 feet, and 486 feet. Estimate the amount of fencing needed to enclose the field.

   A 2,000 feet

   B 2,300 feet

   C 2,500 feet

   D 2,800 feet

   E 3,000 feet

8. The recycling center pays 5¢ for every two aluminum cans turned in. If a group of students brought in 1,200 cans, how much money would they get back?

   F $3

   G $6

   H $30

   J $300

   K Not Here

9. Mr. Mitchell drove his car 1,048 miles in 8 months. What is the average distance he drove per month?

   A 131 miles

   B 1,040 miles

   C 1,056 miles

   D 8,384 miles

   E Not Here

10. The Perfect Painting Company determines how much paint they will need for a wall by finding the area of the wall. What is the area of the wall if its length is 13 feet and its width is 9 feet?

    F 22 ft²

    G 52 ft²

    H 117 ft²

    J 127 ft²

    K Not Here

Test-Taking Tip THE PRINCETON REVIEW

You can prepare for taking a standardized test by working through practice tests like this one. The more you work with questions similar in format to the actual test, the better you become at test taking. Do not wait until the night before the test to review. Allow yourself plenty of time to review the basic skills and formulas needed.

Section Two: Free Response

This section contains three questions for which you will provide short answers. Write your answers on your paper.

11. Max is 4 years younger than three times Carl's age. Carl is 14. Write an equation you can use to find Max's age.

12. $(3 \times 8) + (17 \times 8) =$

13. The scale measures the mass of an item in grams. To the nearest gram, what is the mass of the calcium carbonate?

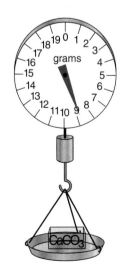

Test Practice For additonal test practice questions, visit:

www.glencoe.com/sec/math/mac/mathnet

Interdisciplinary Investigation

WHAT'S FOR DINNER?

Do you love to go out to eat? Isn't it great to order whatever you want and get it within minutes? Today, more than 500 fast-food chains are operating in the United States.

Some people have claimed that fast food is unhealthy because it contains too much fat and sodium. What do you think? Can such easy meals be good for you?

What You'll Do

In this investigation, you will research data about fast food and plan a healthy menu from your favorite fast-food restaurants.

Materials
- nutritional guides from various fast-food restaurants
- calculator

Procedure

1. Work individually. Suppose you were going to eat at fast-food restaurants for an entire day. List items you would order for breakfast, lunch, and dinner.

2. The number of daily Calories needed by a teenager is about 2,400. Nutrition experts say that a teenager should have about 46 grams of protein per day. In addition, the number of milligrams of sodium should be less than 2,400. For your menu, find the total number of Calories, grams of protein, and milligrams of sodium. How did your menu compare to the recommendations of nutrition experts?

3. Work with a partner. Another recommendation is that the total Calories from fat should not exceed 30% of the total daily Calories. One gram of fat contains 9 Calories. Find the percent of daily Calories from fat in your menus from Step 1.

4. Plan a one-day menu that meets the requirements in Steps 2 and 3.

Technology Tips

- Use a **spreadsheet** to calculate totals and percents.

- Surf the **Internet** to find nutrition information.

Making the Connection

Use the data collected about fast food to help in these investigations.

Language Arts

Do you think that eating fast food is healthy or unhealthy? Write a newspaper article supporting your opinion. Use facts from the investigation.

Health

Choose several other nutritional aspects of food such as calcium, iron, or vitamins. Did your menu meet the daily requirements of these nutrients?

Science

Research the effects of fast-food containers on the environment.

Go Further

- Estimate the cost for your family to eat out every day for one year.

- Survey students in your school or families in your neighborhood. How many times per week do they eat at fast-food restaurants? Which restaurants are their favorites? Prepare a display to present your findings to the class.

interNET **CONNECTION** **Research For current information on fast-food restaurants, visit:**

www.glencoe.com/sec/math/mac/mathnet

You may want to place your work on this investigation in your portfolio.

What you'll learn in Chapter 4

- to solve problems by making a table,
- to make and interpret bar graphs, histograms, circle graphs, line plots, and scatter plots,
- to find measures of central tendency and variation,
- to choose an appropriate display of data, and
- to recognize misleading graphs and statistics.

CHAPTER Project

OLDIES BUT GOODIES!

In this project, you will work in a group to investigate how to run a museum and present your findings in a brochure or poster.

Getting Started

- Choose a museum that houses things of interest to your group. For example, the National Air and Space Museum in Washington, D.C., records the history of flight with items like the Wright Brothers' airplane, and the National Museum of American Art in Washington, D.C., collects art from all regions, cultures, and traditions in the United States.
- Make a list of information that you would like to know if you were considering giving a donation to your museum. Research the museum and its collections and other museums that have similar collections.

Top 5 Art Museums in the United States by Number of Annual Visitors	
Museum	**Annual Visitors**
National Gallery of Art, Washington, D.C.	7,500,000
Metropolitan Museum of Art, New York, NY	3,700,000
Art Institute of Chicago, IL	1,800,000
Museum of Modern Art, New York, NY	1,600,000
Hirshhorn Museum and Sculpture Garden, Washington, D.C.	1,300,000

Source: *The Top Ten of Everything*

Technology Tips

- Use a **spreadsheet** to keep track of the data you gather about your museum and others like it.
- Use a **word processor** to write a brochure or poster about your museum.

 **Research** For up-to-date information on the Smithsonian museums, visit:

www.glencoe.com/sec/math/mac/mathnet

Working on the Project

You can use what you'll learn in Chapter 4 to help you make a brochure or poster about your favorite museum.

Page	Exercise
146	11
161	22
173	21
181	Alternative Assessment

PROBLEM SOLVING

4-1A Make a Table

A Preview of Lesson 4-1

Orlando and Michelle are helping their drama teacher Mrs. Moesley choose the next play the club will perform. Mrs. Moesley made a list of plays to consider. Now Orlando and Michelle are discussing how to poll the club members. Let's listen in!

Orlando

Well, our club meeting is tonight and we have lots of things to talk about. So how can we vote fast and make sure that we keep track of the votes?

Good idea. Let's see. I'll write the play names in a *frequency table* like this. Then they can make tally mark votes. I'll add up the votes after the voting sheet goes around and we'll put on the play that gets the most votes.

Michelle

I wrote down descriptions of the plays, which we can read out loud. Then we can pass around a sheet for people to vote for the play they like best.

Play	Tally	Frequency
The Diary of Anne Frank	卌 卌 II	12
I Know Why the Caged Bird Sings	卌 II	7
The Music Man	III	3
You're a Good Man, Charlie Brown	卌 IIII	9

THINK ABOUT IT

Work with a partner.

1. *Tell* why you think Orlando used tally marks to record the votes.

2. *Brainstorm* types of information you have seen recorded in a table.

3. *Find* survey information that is recorded in a table, in a newspaper, or on the Internet. Which category received the most votes? Which received the fewest votes?

4. *Compare and contrast* the table Orlando made with the table at the right.

Sample Tree Diameters from Cumberland National Forest		
Diameter (in.)	Tally	Frequency
2–4	卌 I	6
4–6	卌 卌 卌 卌 卌 卌	30
6–8	卌 卌 卌 卌 卌 卌 卌 III	38
8–10	卌 卌 卌 卌 卌 卌 卌 III	33
10–12	卌 IIII	9
12–14	IIII	4

For **Extra Practice**, see page 614.

ON YOUR OWN

5. The first step of the 4-step plan for problem solving is to *explore* the problem. **Describe** two things you would need to explore in a set of data before you could **make a table** of the data. Explain your answer.

6. *Write a Problem* that can be answered using a table in this lesson.

7. Refer to the frequency table on page 142. How many states have 501–1,000 malls?

MIXED PROBLEM SOLVING

STRATEGIES

Look for a pattern.
Solve a simpler problem.
Act it out.
Guess and check.
Draw a diagram.
Make a chart.
Work backward.

Solve. Use any strategy.

8. *Technology* The average Internet user spends $6\frac{1}{2}$ hours on-line each week. What percent of the week does the average user spend on-line?

9. *Money Matters* The list shows weekly allowances for a group of 13- and 14-year-olds.

$2.50	$3.00	$3.75	$4.25	$4.25
$4.50	$4.75	$4.75	$5.00	$5.00
$5.00	$5.00	$5.50	$5.50	$5.75
$5.80	$6.00	$6.00	$6.00	$6.50
$6.75	$7.00	$8.50	$10.00	$10.00
$12.00	$15.00			

a. Make a frequency table of the allowances using $1.00 intervals.

b. What is the most common interval of allowance amounts?

10. *Sports* About 17% of American high-school athletes are injured each year. If there are 133 student athletes at Richmond High School, how many would you expect to be injured this year?

11. *Patterns* I am thinking of a number. If I divide it by 2, then multiply the result by itself, I have 5,184. What was my original number?

12. *Money Matters* Kwan has $12 to spend at the movies. After she pays the $4.75 admission, she estimates that she can buy a tub of popcorn that costs $4.25 and a medium drink that is $2.25. Is this reasonable?

13. *Sleep* Use the frequency table. Which age group has the greatest percent of people who talk in their sleep?

GOOD NIGHT!

Age	Percent who talk in their sleep
18-24	29
25-34	23
35-49	15
50+	9

Source: The Better Sleep Council

14. **Standardized Test Practice** If $|a| = 4$, choose the possible values of a.

A -4

B 0 and 4

C -4 and 4

D -16 and 16

15. **Standardized Test Practice** A number divided by 0.4 equals 20. What is the number?

A 4   **B** 8

C 10   **D** 12

Bar Graphs and Histograms

Do you like to shop till you drop? Thousands of malls across America are filled with shoppers. The **frequency table** shows the number of states with different numbers of malls.

Statistics involves collecting, organizing, and analyzing data. Statisticians sometimes use **bar graphs** and special bar graphs, called **histograms** to display data like that in the table. A bar graph compares different categories of data by showing each as a bar whose length is related to the frequency.

Number of Malls	Number of States*
1–500	24
501–1,000	15
1,001–1,500	6
1,501–2,000	3
2,001–2,500	0
2,501–3,000	1
3,001–3,500	1
3,501–4,000	0
4,001–4,500	0
4,501–5,000	0
5,001–5,500	1

* Includes Washington D.C.
Source: National Research Bureau

Example 1
Real World APPLICATION

Technology The table shows the percent of people in the United States who own different consumer electronic products. Make a bar graph of the data.

First draw a horizontal axis and a vertical axis. Label the axes as shown. Be sure to include the title, Electronics at Home.

Each category has a bar to represent it. The vertical scale is the percent of the population.

Product	Percent Who Own
Television	98
Color, stereo television	52
CD player	48
Rack or component audio system	34
Camcorder	23
Car CD player	15
Laserdisc player	2

Source: Electronic Industries Association

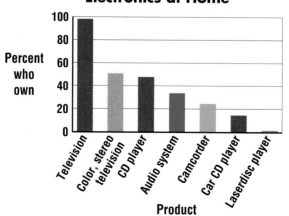

A histogram uses bars to display numerical data that have been organized into equal intervals. The intervals cover the entire range of the data with no overlapping intervals.

Example
Real World APPLICATION

② **Shopping** Refer to the beginning of the lesson.

a. Construct a histogram of the data.

First draw a horizontal axis and a vertical axis. Label the axes as shown. Be sure to include the title, America's Malls.

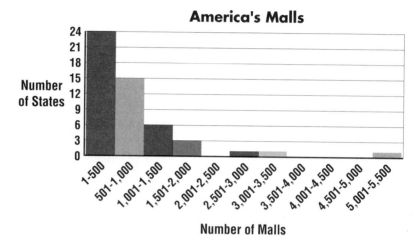

America's Malls

Number of States — vertical axis with values 0, 3, 6, 9, 12, 15, 18, 21, 24

Number of Malls — horizontal axis with intervals 1-500, 501-1,000, 1,001-1,500, 1,501-2,000, 2,001-2,500, 2,501-3,000, 3,001-3,500, 3,501-4,000, 4,001-4,500, 4,501-5,000, 5,001-5,500

- Equal intervals of 500 are shown on the horizontal axis.

- Because the intervals are equal, all of the bars have the same width.

- As in a bar graph, the frequency of data in each interval is shown by the bar height.

- The vertical scale on bar graphs and histograms is often a factor of the greatest frequency. In this case, the greatest frequency is 24, and the vertical scale is 3.

- Intervals with a frequency of 0, like 2,001–2,500, have no bar.

b. How does the number of states with 1–500 malls compare to the number of states with 1,001–1,500 malls?

Compare the heights of the bars. The bar for 1–500 malls is about four times as tall as the bar for 1,001–1,500 malls. There are about four times as many states with 1–500 malls as there are states with 1,001–1,500 malls.

Sometimes when the data in a graph do not start at zero, a jagged line is shown on one of the axes of the histogram.

3 **Earth Science** The table below shows the record high temperatures for each state. Make a histogram of this data.

Record High Temperatures						
Temperature	100–104	105–109	110–114	115–119	120–124	125–129	130–134
States*	3	9	17	12	7	2	1

Source: National Climatic Data Center   *Includes Washington, D.C.

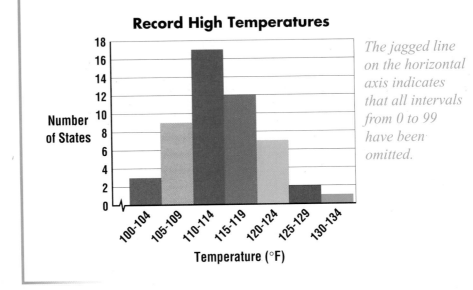

The jagged line on the horizontal axis indicates that all intervals from 0 to 99 have been omitted.

CHECK FOR UNDERSTANDING

Communicating Mathematics

Read and study the lesson to answer each question.

1. *Describe* when a histogram is more useful than a table. When is a table more useful than a histogram?

2. *You Decide* Kayla says that a histogram has no spaces between the bars because the intervals overlap. Koko says there are no spaces because the intervals cover every possible data value. Who is correct? Explain your reasoning.

3. *Write* a few sentences about when to use a bar graph and when to use a histogram.

Guided Practice

4. Use the histogram at the right to answer each question.

   a. Describe the data shown in the graph.

   b. How large is each interval?

   c. Which interval has the greatest number of states?

   d. Why is there a jagged line in the horizontal axis?

   e. Make a frequency table of the data.

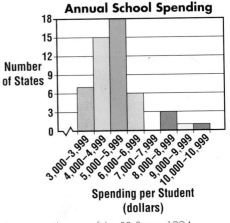

Source: Almanac of the 50 States, 1994

Practice

5. Use the table of incubation times to answer each question.

Egg Incubation Time					
Bird	Chicken	Duck	Goose	Pigeon	Turkey
Time (days)	21	30	30	18	26

a. Describe the data in the table.

b. Construct a bar graph of the data with or without technology.

6. Use the histogram at the right to answer each question.

a. Describe the data in the histogram.

b. How large is each interval?

c. Which interval of heights has the most presidents?

d. Construct a frequency table of the data of presidents' heights.

e. Can you tell from the histogram or the table the exact height of the tallest president? Explain.

f. What percent of the presidents were less than 6 feet tall?

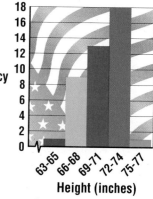

Heights of U.S. Presidents

Frequency / Height (inches): 63-65, 66-68, 69-71, 72-74, 75-77

7. The suggested retail prices of several different bicycle helmets tested by a consumer magazine are shown below. Make a frequency table and a histogram of the prices.

$25	$30	$30	$30	$30	$30	$40	$40
$40	$40	$40	$40	$40	$43	$50	$50

Applications and Problem Solving

Real World

8. *Life Science* Use the table of U.S. threatened species to answer each question.

a. Describe the data in the table.

b. Construct a bar graph of the data.

c. How many species are recorded in the bar graph?

d. *Write a Problem* using the data in the table.

9. *Civics* The United Nations was formed at the end of World War II to maintain international peace and security. The year that each of the 184 member nations entered the U.N. are shown in the table below. Make a histogram of the data.

U.S. Threatened Species	
Species	**Number**
Amphibians	6
Birds	16
Clams	6
Crustaceans	3
Fishes	40
Insects	9
Mammals	9
Reptiles	19
Snails	7

Source: Fish and Wildlife Service

United Nations Entry				
1945–1954	1955–1964	1965–1974	1975–1984	1985–1994
59	55	21	21	28

Source: *The World Almanac*

10. **Natural Resources** This frequency table shows the numbers of states with different amounts of National Forest Service land.

National Forest Service Land	
Acres (thousands)	States
0–2,999	38
3,000–5,999	1
6,000–8,999	1
9,000–11,999	4
12,000–14,999	1
15,000–17,999	2
18,000–20,999	2
21,000–23,999	1

Source: U.S. Forest Service

  a. How much National Forest Service land do most states have?

  b. What fraction of the states have less than 9,000,000 acres of National Forest Service land?

  c. Make a histogram of the data.

11. **Working on the CHAPTER Project** If you chose an art museum, construct a bar graph of the data on museum visitors on page 139. Be sure to include your chosen museum. If you chose another type of museum, find data about visitors to those types of museum and construct a bar graph.

12. **Critical Thinking** Study the information in the table for Exercise 9.

  a. How many nations entered the U.N. in 1994? Explain your answer.

  b. Research world events in the times when there were large numbers of nations that joined the U.N. Explain why the growth occurred.

For **Extra Practice**, see page 614.

Mixed Review

13. Estimate the discount if a $58 jacket is marked down 24%. *(Lesson 3-6)*

14. **Standardized Test Practice** Evaluate $xy + |-4|$ if $x = 2$ and $y = 12$. *(Lesson 2-1)*

  A 24   **B** 27   **C** 28   **D** 30

Let the Games Begin

You're The Winner . . . Bar None!

Math Skill

Making Bar Graphs

Get Ready This game is for two players.   two number cubes

Get Set Make scales for a bar graph as shown at the right.

Go
● Player A chooses a sum. Then player B chooses one of the remaining sums.

● The players take turns rolling the number cubes. With each roll, the player finds the sum on the cubes and creates or lengthens the bar for that sum on the graph.

● The winner is the player whose bar reaches a frequency of 5 first.

● In the next round, player B chooses his or her bar first.

interNET CONNECTION Visit www.glencoe.com/sec/math/mac/mathnet for more games.

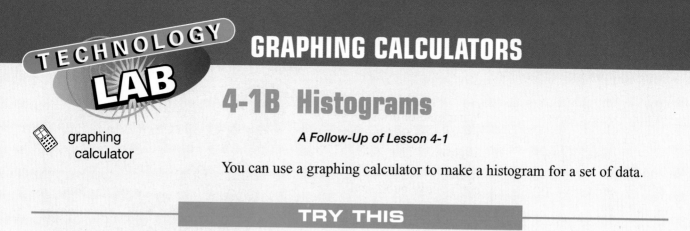

TECHNOLOGY LAB

GRAPHING CALCULATORS

graphing calculator

4-1B Histograms

A Follow-Up of Lesson 4-1

You can use a graphing calculator to make a histogram for a set of data.

TRY THIS

Work with a partner.

The list is the record low temperatures for each state and Washington, D.C.

−27	−80	−40	−29	−45	−61	−32	−17	−15	−2	−17	12	−60	−35
−36	−47	−40	−37	−16	−48	−40	−35	−51	−59	−19	−40	−70	−47
−50	−46	−34	−50	−52	−34	−60	−39	−27	−54	−42	−23	−19	−58
−32	−23	−69	−50	−30	−48	−37	−54	−66					

Step 1 Begin by entering the data into the calculator's memory.
Press STAT ENTER to see the lists. If there are numbers in the first list, press ▲ CLEAR ENTER to clear them. Then enter the data by entering each number and pressing ENTER .

Step 2 Next choose the type of graph. Press 2nd [STAT PLOT] to display the menu. Choose the first plot by pressing ENTER . Use the arrow and ENTER keys to highlight "on", the histogram, L1 for the Xlist, and 1 as the frequency.

Step 3 Choose the display window. Access the menu by pressing WINDOW . Choose appropriate range settings. For this data, use −80 to 20 with a scale of 10 on the *x*-axis and 0 to 15 with a scale of 5 on the *y*-axis.

Step 4 Display the graph by pressing GRAPH .

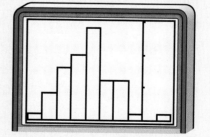

ON YOUR OWN

1. Press the TRACE key. The right and left arrow keys allow you to find the frequency of each interval. How many temperatures are between −60 and −50?
2. Use the data on iced tea prices on page 153 to make a histogram.
3. How does the graphing calculator determine the size of the intervals?

Circle Graphs

What you'll learn

You'll learn to construct and interpret circle graphs.

When am I ever going to use this?

Circle graphs are often used in advertising to compare parts to a whole.

Word Wise

circle graph

What kind of book do you choose for a quiet afternoon? The table shows the number of adult books published recently in different subjects.

A **circle graph** compares parts of a set of data to the whole set. For example, a circle graph of the book subject data could be used to compare the number of fiction books to the total number of books published.

Subject	Books Published
Art, literature, music	4,864
Business, law	9,862
Fiction, general	7,843
History	6,130
Philosophy, psychology	4,346
Science, technology	8,680
Sports, recreation	2,407

Source: *Publishers Weekly*

Example

CONNECTION

Language Arts Construct a circle graph using the data above.

Step 1 Find the total number of books published.

4864 $+$ 9862 $+$ 7843 $+$ 6130 $+$ 4346 $+$ 8680 $+$ 2407 $=$ *44132*

Step 2 Find the ratio that compares the number of books in each subject to the total number of books published. Round each ratio to the nearest hundredth.

Since you are comparing the number of books for each subject to the total, each ratio represents the percent of the whole circle.

Art, literature, music: $\frac{4,864}{44,132}$

4864 $\div$ 44132 $=$ *0.11021481*
$0.11021481 \approx 0.11$

Business, law: $\frac{9,862}{44,132}$

9862 $\div$ 44132 $=$ *0.223465966*
$0.223465966 \approx 0.22$

Fiction, general: $\frac{7,843}{44,132}$

7843 $\div$ 44132 $=$ *0.177716849*
$0.177716849 \approx 0.18$

History: $\frac{6,130}{44,132}$

6130 $\div$ 44132 $=$ *0.138901477*
$0.138901477 \approx 0.14$

Philosophy, psychology: $\frac{4,346}{44,132}$

4346 $\div$ 44132 $=$ *0.098477295*
$0.098477295 \approx 0.10$

Science, technology: $\frac{8,680}{44,132}$

8680 $\div$ 44132 $=$ *0.196682679*
$0.196682679 \approx 0.20$

Sports, recreation: $\frac{2,407}{44,132}$

2407 $\div$ 44132 $=$ *0.054540923*
$0.054540923 \approx 0.05$

Study Hint

Estimation You can check your solution by adding the measures of the angles. If the sum is close to 360, you are probably correct.

Step 3 Since there are 360° in a circle, multiply each ratio by 360 to find the number of degrees for each section of the graph. When necessary, round to the nearest degree.

Art, literature, music: $0.11 \times 360 = 39.6$ *39.6 ≈ 40°*

Business, law: $0.22 \times 360 = 79.2$ *79.2 ≈ 79°*

Fiction, general: $0.18 \times 360 = 64.8$ *64.8 ≈ 65°*

History: $0.14 \times 360 = 50.4$ *50.4 ≈ 50°*

Philosophy, psychology: $0.10 \times 360 = 36$

Science, technology: $0.20 \times 360 = 72$

Sports, recreation: $0.05 \times 360 = 18$

Step 4 Use a compass to draw a circle and a radius. Then use a protractor to draw a 40° angle. *You can start with any of the angles.*

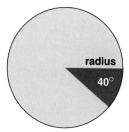

Step 5 From the new radius, draw the next angle. Repeat for each of the remaining angles. Label each section and write each ratio as a percent. Then give the graph a title.

Study Hint

Technology Many spreadsheets and word processing programs will construct circle graphs from data you enter.

Books Published

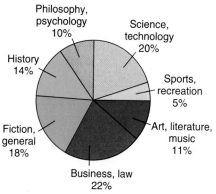

Communicating Mathematics

Read and study the lesson to answer each question.

1. **Identify** the theme park that about one-fourth of the visitors choose.

2. **Describe** when a circle graph is used to show data.

Theme Park Visitors (millions)

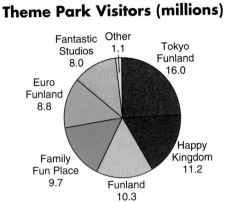

Guided Practice

3. The results of a survey of high school students are shown in the table.

   a. What should the total in the ratio column be?

   b. What total would you expect in the degree column?

   c. Copy and complete the table.

   d. Make a circle graph of the data.

Who Pays When You Date?			
Category	Number	Ratio	Degrees in graph
Boy	1,200		
Split costs	800		
Girl	75		
Girl's parents	25		
Boy's parents	25		
Don't date	375		
Total			

Source: BKG Youth poll

Practice

4. **Music** The table shows the sales of each type of recorded music for a recent year.

   a. Make a circle graph of the data.

   b. Which format accounts for about one-fifth of the total music sales?

5. **History** The Medal of Honor is the highest military award for bravery in the United States. There were 124 medals given to veterans of World War I, 433 for World War II, 131 for the Korean War, 239 for the Vietnam War, 2 for Somalia, and 18 in peacetime. Make a circle graph of the medals of honor.

Format	Sales ($ millions)
CD	9,401.7
CD Single	88.6
Cassette	2,303.6
Cassette Single	236.3
LP/EP	25.1
Vinyl Single	46.7
Video	220.3

Source: Recording Industry Assn. of America

6. **Geography** The table shows the areas of the five counties in the state of Hawaii. Make a circle graph for the areas of the Hawaiian counties.

County	Hawaii	Honolulu	Kalawao	Kauai	Maui
Area (sq mi)	4,028	600	13	623	1,159

Source: U.S. Department of Commerce

7. *Charities* According to the American Association of Fund-Raising Counsel, Americans gave $143.84 billion to charity in 1995. Corporations gave $7.4 billion, foundations gave $10.44 billion, $9.77 billion was left in wills, and individuals gave $116.23 billion. Make a circle graph of the contributions.

8. *History* The table shows the birthplaces of the signers of the Declaration of Independence. Make a circle graph of the data with or without technology.

Location	Signers	Location	Signers	Location	Signers
Connecticut	5	Massachusetts	9	Rhode Island	2
Delaware	2	New York	3	South Carolina	4
Maine	1	New Jersey	3	United Kingdom	8
Maryland	5	Pennsylvania	5	Virginia	9

Source: *World Almanac*

For **Extra Practice,** see page 615.

9. *Critical Thinking* According to a media research company, 98% of U.S. households owned at least one television in 1995. 38% had 2 TVs, 28% had 3 or more TVs, 81% had a VCR, and 65% received cable. Can this information be displayed in a circle graph? Explain. **Source:** Nielsen Media Research

Mixed Review

10. *Entertainment* One hundred students were asked how many hours they watch television each week. The results are given in the table. Make a histogram of the data. *(Lesson 4-1)*

Hours	Frequency
0–2	4
3–5	8
6–8	22
9–11	32
12–14	30
15–17	4

11. **Standardized Test Practice** Solve $h = 18 - (-4)$. *(Lesson 2-5)*

A 22   **B** 14   **C** -22 **D** -14

MATH IN THE MEDIA

Computer Comparisons

The report described below appeared on a nightly news show on May 25, 1995.

Tom Brokaw read a report saying that 14 percent of African-Americans, 13 percent of Hispanics, and 27 percent of Caucasians use computers. He went on to say that computer use by Caucasians was equal to that of African-Americans and Hispanics combined.

The circle graph shows the composition of the United States population by race.

1. Find the percent of the population that each ethnic group represents.

2. Use the figures from the news report with the data from the circle graph to find the number of African-Americans, Hispanics, and Caucasians who use computers.

3. Is it true that "computer use by Caucasians was equal to that of African-Americans and Hispanics combined"? Explain.

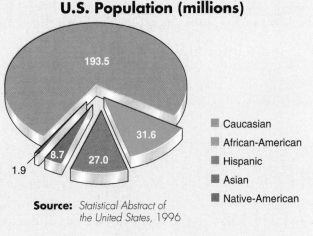

U.S. Population (millions)

193.5
31.6
8.7
27.0
1.9

■ Caucasian
■ African-American
■ Hispanic
■ Asian
■ Native-American

Source: *Statistical Abstract of the United States, 1996*

HISTORY

Spencer R. Crew
MUSEUM DIRECTOR

The National Museum of American History is one of the 14 museums of the Smithsonian Institution. Located in Washington, D.C., it contains over 17 million artifacts including the gowns of the U.S. First Ladies and the original Star-Spangled Banner that inspired the writing of the national anthem. Spencer Crew, director since 1992, leads museum staff in making decisions about research, collecting, and developing new exhibitions and public programs. He is also responsible for overseeing the resources of the museum: its funds, its people, and its spaces.

Museum staffs are skilled in many areas and enjoy sharing their expertise with the public. Curators, historians, museum specialists, museum technicians, and archivists collect, manage, and catalogue artifacts and documents. All museum staff need a good education and hands-on experience. Curators and directors usually have masters or doctoral degrees. To work in one of these museum jobs, you should have a good background in a wide variety of subjects including history, art, science, business, and math. You should also have good team skills, be well-organized, and enjoy working with the public.

I think working in a museum like Spencer Crew does would be fun.

For more information:
American Association of Museums
1225 I St., NW, Suite 200
Washington, DC 20005

interNET
CONNECTION
www.glencoe.com/sec/
math/mac/mathnet

Your Turn
Compile a list of museums in your city or state. At which museums do you think it would be fun to work? Explain why, and describe the types of courses and experience you would need to work there.

One of the ways that you can display data is in a **line plot**. In a line plot, data is organized using a number line.

MINI-LAB

Work with a partner. tape measure   tape   markers

You will make a human line plot of the heights of your class members.

Try This

- Measure and record your partner's height in inches.

- As a class, use masking tape to place a number line on the floor. The distances between consecutive numbers should be the same.

- Have each person stand above his or her height on the number line. People of the same height should form a column above the height.

Talk About It

1. Which height has the most people? How can you tell?
2. Which heights had more boys or more girls?

It is usually not practical to arrange people or objects to form a line plot. You can draw a line plot using a symbol to represent each data point.

A consumer magazine compared the taste and price of 17 lemon-flavored iced teas. The teas were ranked very good (VG), good (G), or fair (F). The results are shown in the table.

Rating	Price per 8-oz serving
G	37¢
G	35¢
VG	8¢
F	21¢
G	8¢
G	34¢
G	42¢
G	37¢
F	21¢
G	7¢
G	30¢
G	23¢
G	36¢
F	21¢
VG	36¢
G	40¢
G	20¢

To make a line plot of the tea prices, use a number line that includes all of the data values. In this case the least value is 7 and the greatest is 42, so a number line from 0 to 45 is a good choice. Mark an × for each price above the number line to complete the line plot.

You can see that there are two groups of prices in which most of the iced teas fall. One group is 20–23 cents per serving, and the other is 34–37 cents per serving. The lowest and highest prices are 7 and 42 cents.

Studying data and making observations is called **data analysis**. Altering the line plot can show more information about the iced tea prices.

Money Matters Make a new line plot for the iced tea prices by replacing each × with a letter rating the quality of the iced tea. Use V for very good, G for good, and F for fair. How much is the least expensive tea with a good rating?

Refer to the original list of iced tea ratings.

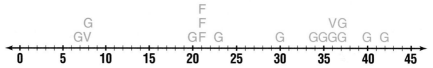

The least expensive tea that was rated good costs 7¢ per serving.

Which tea would you choose and why? Justify your answer.

CHECK FOR UNDERSTANDING

Communicating Mathematics

Read and study the lesson to answer each question.

1. *Describe* how to make a line plot of a set of data.

2. The line plot for a consumer magazine's orange juice test is shown below. *Tell* what you observe about the types and costs per serving of the juices.

   f = frozen concentrate, c = chilled

   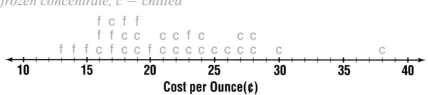

HANDS-ON MATH

3. *Make* a line plot of your classmates' heights found in the Mini Lab.

Guided Practice

4. The table lists the Calories per serving of several brands of yogurt.

   a. Make a line plot of the data.

   b. Analyze the line plot. Write all of the conclusions that you can.

   c. How many yogurts can you choose that have less than 200 Calories per serving?

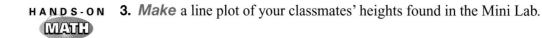

Yogurt Calories		
90	90	90
100	110	110
140	140	140
140	140	140
150	160	170
170	180	190
220	240	250

Practice

5. Make a line plot of the concert ticket prices listed below.

$20	$24	$31	$31	$35	$21	$33	$43	$35	$30
$37	$48	$42	$28	$30	$27	$21	$30	$25	$29

6. Lina and Paige are playing backgammon. The results of their first twenty-six rolls of two dice are shown below.

8	6	5	10	6	9	7	4	6	11	9	2	10
7	3	10	8	6	2	9	7	5	8	3	7	5

   a. Make a line plot of the data.

   b. What are the greatest and least sums?

   c. Which sum occurred most often?

7. In a variation of backgammon, a player gets another turn if they roll 1 and 2. In these 26 rolls, how many times did a player get another turn?

Applications and Problem Solving

8. *Nutrition* The numbers of Calories per serving in selected chocolate-chip and chocolate sandwich cookies are listed at the right. *c = chocolate-chip, s = sandwich*

Cookie Calories		
70-c	80-c	100-s
100-s	130-c	130-s
130-c	140-s	140-c
150-c	150-s	150-s
160-c	160-s	

   a. Make a line plot of the data. Use *c* to represent chocolate chip cookies and *s* to represent chocolate sandwich cookies.

   b. Analyze your graph.

9. *Sports* The National Football League began choosing its champion in the Super Bowl in 1967. The list below shows the margin of victory and the winning league for the first 31 Super Bowl games. *A = AFC, N = NFC*

25-N	19-N	9-A	16-A	3-A	21-N	7-A	17-A
10-A	4-A	18-A	17-N	4-A	12-A	17-A	5-N
10-N	29-A	22-N	36-N	19-N	32-N	4-N	45-N
1-N	13-N	35-N	17-N	10-N	23-N	14-N	

   a. Make a line plot of the data.

   b. What do you observe about the winning margins?

10. *Critical Thinking* The list of Super Bowl margins in Exercise 9 is given in order of years: first 25-N, then 19-N, and so on. Describe any patterns you see in the margins or in the winning league over the years of the Super Bowl.

Mixed Review

11. *Civics* Make a circle graph of the 1996 presidential votes. *(Lesson 4-2)*

12. **Standardized Test Practice** Choose the fraction that is less than 35%. *(Lesson 3-4)*

   A $\frac{2}{5}$ **B** $\frac{3}{8}$ **C** $\frac{1}{6}$ **D** $\frac{5}{12}$

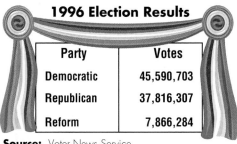

1996 Election Results	
Party	**Votes**
Democratic	45,590,703
Republican	37,816,307
Reform	7,866,284

Source: Voter News Service

For **Extra Practice,** see page 615.

13. *Home Economics* If you were to add lace to the edge of a bedspread that is 66 by 96 inches, how much lace would you need? *(Lesson 1-8)*

COOPERATIVE LEARNING

4-3B Maps and Statistics

A Follow-Up of Lesson 4-3

colored pencils

outline map of the United States

Have you ever seen a map of the United States with statistics shown by shading or coloring individual states? Maps like these are used often in newspapers and magazines. The map below shows how many businesses are owned and operated by African-Americans in each state. Why do you think newspapers use maps instead of lists?

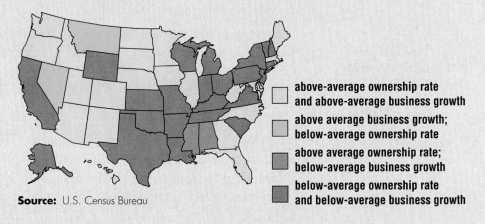

	above-average ownership rate and above-average business growth
	above average business growth; below-average ownership rate
	above average ownership rate; below-average business growth
	below-average ownership rate and below-average business growth

Source: U.S. Census Bureau

TRY THIS

Work with a partner.

Sales of Bird Watching and Feeding Items					
State	Sales (millions)	State	Sales (millions)	State	Sales (millions)
AL	53.6	LA	51.3	OH	123.1
AK	121.3	ME	64.8	OK	55.3
AZ	128.4	MD	83.0	OR	94.3
AR	54.4	MA	124.4	PA	256.4
CA	662.6	MI	267.6	RI	19.1
CO	179.6	MN	97.5	SC	51.6
CT	55.6	MS	34.9	SD	20.7
DE	11.5	MO	165.0	TN	76.2
FL	477.0	MT	76.3	TX	155.3
GA	49.7	NE	23.1	UT	57.0
HI	66.5	NV	56.5	VT	22.7
ID	33.3	NH	57.0	VA	108.3
IL	131.7	NJ	87.5	WA	136.3
IN	64.6	NM	80.9	WV	26.6
IA	30.4	NY	219.0	WI	224.8
KS	23.5	NC	92.4	WY	62.2
KY	57.5	ND	6.6		

Source: Southwick Associates

- The table lists each state's sales of items for watching, feeding, and photographing wild birds. Make a line plot of the data using the state abbreviations instead of ✕s.

- Mapmakers usually organize data into fewer than 7 categories. Separate the data using $0-50 million, $51-100 million, $101-200 million, and more than $200 million as your categories. *Because the intervals are chosen to support a point of view, they are often not equal as they are in a histogram.*

- Choose four colored pencils. Use colors ranging from light to dark to correspond with the ranges from least to greatest. Color each state on a United States map according to its category.

ON YOUR OWN

Refer to your map of birdwatching and feeding item sales.

1. The intervals used to shade the states are not all equal. Why do you think the intervals were chosen as they were?

2. Explain why a mapmaker may not want to use more than seven colors to code a map that shows statistical information.

3. What areas of the country show the highest sales? Why do you think this is true?

4. Compare the map to the line plot that you made. Which would you use if you wanted to show the areas of the country in which bird watching is most popular? Explain.

5. Why do you think newspapers and magazines often use maps to show information instead of using tables?

6. *Earth Science* The table shows the average number of tornadoes that occur in each state each year.

Average Number of Tornadoes Each Year									
State	Average per Year	State	Average per Year	State	Average per Year	State	Average per Year		
AL	22	IN	20	NE	37	SC	10
AK	0	IA	36	NV	1	SD	29
AZ	4	KS	40	NH	2	TN	12
AR	20	KY	10	NJ	3	TX	139
CA	5	LA	28	NM	9	UT	2
CO	26	ME	2	NY	6	VT	1
CT	1	MD	3	NC	15	VA	6
DE	1	MA	3	ND	21	WA	2
FL	53	MI	19	OH	15	WV	2
GA	21	MN	20	OK	47	WI	21
HI	1	MS	26	OR	1	WY	12
ID	3	MO	26	PA	10		
IL	27	MT	6	RI	0		

Source: National Severe Storm Forecast Center

a. Color a map to illustrate this data.

b. How would you change your map if you were a meteorologist trying to get a new National Weather Service tornado research center located in your state?

c. How would you change the map if you were an opponent of having a National Weather Service center located in your state?

7. Compare and contrast a histogram with a map display of data. Which type of graph is easier to adjust to change the impression that is given? Explain your reasoning.

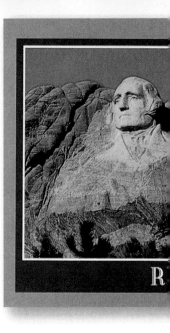

4-4

Measures of Central Tendency

What you'll learn

You'll learn to find the mean, median, and mode of a set of data.

When am I ever going to use this?

You'll use mean, median, and mode to analyze the scores on your tests.

Word Wise

measures of central tendency
mean
median
mode

If you could visit any place in the world, where would you go? Many people from other countries choose to come to the United States! The table shows how much a foreign tourist spends on a trip to the United States.

Tourist Dollars	
Country	Spending (hundreds of dollars)
Canada	4
France	20
Germany	22
Italy	24
Japan	24
Mexico	4
United Kingdom	22
Venezuela	28

Source: *World Almanac*

Suppose one tourist from each of these countries visits your community. How much would each one spend on average? There are three common ways to describe a set of data. These ways are called the **measures of central tendency**. They are the mean, the mode, and the median.

Study Hint

Reading Math When you read a table or graph, be sure to note the units of the data given. For example, in the table, the units are hundreds of dollars. So a tourist from Canada spends 4 hundreds, or $400.

Although the word *average* can be used for any of the measures of central tendency, most people use it to refer to the **mean**. For the dollar amounts above, you can find the mean as shown below.

$$\frac{4 + 20 + 22 + 24 + 24 + 4 + 22 + 28}{8} = \frac{148}{8} \text{ or } 18.5$$

The mean amount spent is 18.5 hundreds or $1,850.

Notice that the mean of a set of data may or may not be a member of the set.

Mean	The mean of a set of data is the sum of the data divided by the number of pieces of data.

A second measure of central tendency is the **mode**. The mode is the piece of data that appears most often.

4　　4　　20　　22　　22　　24　　24　　28

There are two 4s, two 22s, and two 24s in this set of data. So there are three modes, 4, 22, and 24. If there was another 4 in the data set, 4 would be the only mode. A set of data in which no numbers appear more than once has no mode. *A mode is always a member of the data set.*

Mode	The mode of a set of data is the number(s) or item(s) that appear most often.

The final measure of central tendency is the **median**. The median is the middle number when the data are written in order from least to greatest.

ORE

Consider the data set {1, 4, 7, 12, 16, 21, 21, 29, 33}. There are nine numbers in the set, and they are in order from least to greatest. So the median is the fifth number, or 16.

If the number of data is even, as in the set of tourist spending, there are two middle numbers. In that case, the median is the mean of the numbers.

$$4 \quad 4 \quad 20 \quad \underbrace{22 \quad 22}_{\frac{22 + 22}{2} = 22} \quad 24 \quad 24 \quad 28$$

The median of the tourist spending is 22 hundreds or $2,200. *The median is not necessarily a member of the set of data. In this case, it is in the set.*

Median	The median of a set of data is the number in the middle when the data are arranged in order. When there are two middle numbers, the median is their mean.

You can learn more about different sets of data by comparing the mean, median, and mode.

Example

Real World APPLICATION

Sports The table shows the number of home runs hit by each major league baseball team in the 1996 season.

a. Find the mean, mode, and median number of home runs for the National League teams.

Begin by writing the National League numbers in order.

National League		American League	
Team	Home Runs	Team	Home Runs
Colorado	221	Baltimore	257
Atlanta	197	Seattle	245
Cincinnati	191	Oakland	243
Chicago Cubs	175	Kansas City	243
San Francisco	153	Texas	221
Florida	150	Cleveland	218
Los Angeles	150	Boston	209
Montreal	148	Detroit	204
San Diego	147	Chi. White Sox	195
New York Mets	147	California	192
St. Louis	142	Milwaukee	178
Pittsburgh	138	Toronto	177
Philadelphia	132	NY Yankees	162
Houston	129	Minnesota	118

Source: *World Almanac, 1997*

129   132   138   142   147   147   148
150   150   153   175   191   197   221

Mean   Find their mean by dividing their sum by 14.
2220 ÷ 14 = *158.5714286*
The mean is about 159 home runs.

Mode   The mode is the most frequent number. In this data set, 147 and 150 each appear twice. So the modes are 147 and 150.

Median   The median is the middle number. There are an even number of data, so the median is the mean of the 7th and 8th numbers.

$$\frac{148 + 150}{2} = 149$$

The median is 149 home runs. *(continued on the next page)*

b. Find the mean, mode, and median number of home runs for the American League teams.

The numbers for the American League in order are:

118	162	177	178	192	195	204
209	218	221	243	243	245	257

The mean is $\frac{2,862}{14}$ or about 204.

The mode is 243 because it occurs twice.

The median is $\frac{204 + 209}{2}$ or 206.5.

c. Which league has the higher average number of home runs?

The American League has the higher average number of home runs. All of the measures of central tendency for the American League are greater than the corresponding measures for the National League.

CHECK FOR UNDERSTANDING

Communicating Mathematics

Read and study the lesson to answer each question.

1. **Explain** how to find the mean, median, and mode of the data shown in the line plot. Then find the mean, median, and mode.

2. **Analyze** the measures of central tendency for the National League home run data in the Example. Which measure best represents the data? Explain your choice.

3. **You Decide** Dolores says that all measures of central tendency must be members of the data set. Carol disagrees. Who is correct and why?

Guided Practice

Find the mean, median, and mode for each set of data. When necessary, round to the nearest tenth.

4. 9, 8, 15, 8, 20, 23, 16, 5, 6, 14, 12, 25, 18, 22, 24

5. 36, 38, 33, 34, 32, 30, 34, 35

6. 85¢, 87¢, 69¢, 74¢, 70¢, 98¢, 54¢, 89¢, 65¢, 82¢, 81¢, 94¢

7. **Money Matters** The list shows the suggested retail prices of several different personal stereos.

$40	$45	$50	$59	$60	$69	$85	$111	$120

a. Find the mean, median, and mode prices.

b. Which measure of central tendency best represents the prices? Explain.

c. If a price of $17 is added to the list, which measure of central tendency is affected most? Explain.

Practice

Find the mean, median, and mode for each set of data. When necessary, round to the nearest tenth.

8. 20, 17, 20, 14, 19

9. 3, 9, 14, 3, 0, 2, 6, 11

10. 93, 90, 94, 99, 92, 93, 100

11. 78, 80, 75, 73, 84, 81, 84, 79

12. $93, $71, $83, $100, $55, $87, $79, $100, $58, $95, $87

13. 8.8, 10.0, 9.3, 8.3, 9.4, 8.9, 9.5, 8.3, 9.7, 8.1, 9.7, 8.8, 8.3

14. 1.2, 1.78, 1.73, 1.9, 1.19, 1.8, 1.24, 1.92, 1.54, 1.7, 1.42, 1.0

15. 27, 19, 22, 41, 24, 28, 40, 33, 35, 33, 35, 48, 32

16.

17. Give an example of a set of data that has no mode.

18. Is it possible for a set of numerical data to have no mean? Explain.

19. Construct a set of data with at least five different numbers for which the mean, median, and mode are all the same number.

Applications and Problem Solving

20. *Highway Safety* In 1995, President Clinton signed a bill that allows states to set speed limits on their highways. Montana posted its limit as "reasonable and prudent." The speed limits in miles per hour for rural highways in the other states and the District of Columbia are given.

70	65	75	65	70	75	55	65	55	70	70	55	75
65	65	65	70	65	65	65	65	65	65	65	70	70
75	75	65	55	65	65	65	70	65	75	65	65	65
65	75	65	70	75	65	65	70	65	65	75		

a. Find the mean, median, and mode speed limits.

b. What would you say is the "average" speed limit? Explain.

21. *Entertainment* After release of a special edition of a space movie, the average American had seen it seven times. How was that number found?

22. *Working on the* **CHAPTER Project** Research to find the number of items your museum has acquired in each of the last ten years.

a. Find the mean, median, and mode of items acquired.

b. If it takes 15 minutes to catalog each item, about how much employee time would be required for cataloging each year?

c. Estimate the annual budget for catalog staff in your museum. Support your answer with facts and figures.

For **Extra Practice**, see page 615.

23. *Critical Thinking* Of the mean, median, and mode, which is most affected by adding a very large or very small data value to a set of data? Explain.

Mixed Review

24. *Highway Safety* Make a line plot of the data in Exercise 20. *(Lesson 4-3)*

25. *Algebra* Solve $\frac{5}{8} = \frac{m}{24}$. *(Lesson 3-3)*

26. *Standardized Test Practice* Choose the solution of $c = -22 + 5$. *(Lesson 2-3)*

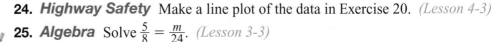

A -27 **B** -17 **C** 17   **D** 27

SPREADSHEETS

4-4B Finding a Mean

A Follow-Up of Lesson 4-4

💻 computer

💿 spreadsheet
software

Would you like to know your grades before you get your report card? One way that you can do that is to use a spreadsheet program.

Many teachers give several tests and determine the final grades by finding the mean of the scores. A portion of a spreadsheet program that Mr. Gutierrez uses to find the mean of his students' test scores is shown.

TRY THIS

Work with a partner.

Use the spreadsheet to determine the mean test score for each student.

	A	B	C	D	E	F
1	Student	Test 1	Test 2	Test 3	Test 4	Mean
2	Alejandra	78	76	81	83	=(B2+C2+D2+E2)/4
3	Jessica	72	83	85	83	=(B3+C3+D3+E3)/4
4	Kelli	84	82	85	88	=(B4+C4+D4+E4)/4
5	Raheem	88	92	90	91	=(B5+C5+D5+E5)/4
6	Steven	90	88	87	92	=(B6+C6+D6+E6)/4

The formula $(B2+C2+D2+E2)/4$ adds the values in cells B2, C2, D2, and E2 and then divides the sum by 4.

Each box in the spreadsheet is called a *cell*. The formulas in the cells in column F find the mean of the scores that are entered in the cells in columns B, C, D, and E. The formula first finds the sum of the scores and then divides the sum by 4 to find the average. The printout below shows the results when the calculations are complete.

	A	B	C	D	E	F
1	Student	Test 1	Test 2	Test 3	Test 4	Mean
2	Alejandra	78	76	81	83	79.5
3	Jessica	72	83	85	83	80.75
4	Kelli	84	82	85	88	84.75
5	Raheem	88	92	90	91	90.25
6	Steven	90	88	87	92	89.25

ON YOUR OWN

1. Use the spreadsheet to determine Editon's average test score if his grades were 92, 84, 89, and 95.

2. If Kelli had scored 2 points higher on each test, how would her average have been affected?

3. What would Raheem have to score on a fifth test to average 90%?

4. How could you change the spreadsheet to find the mean of seven quiz scores?

4-5 Measures of Variation

What you'll learn

You'll learn to find the range and quartiles of a set of data.

When am I ever going to use this?

Doctors use measures of variation to determine whether new medicines are effective.

Word Wise

variation
range
quartile
interquartile range
upper quartile
lower quartile
outlier

**Frances McDormand
Best Actress 1996**

"And the award goes to..." Do you dream of someday accepting an Academy Award? Moviemakers have been honoring their best with Academy Awards since 1927. The ages of the best actress winners for the last 25 years are listed below.

25	26	28	30	30	30	31	32
32	32	33	35	35	36	37	40
40	41	41	44	48	48	60	73	80

The list shows that the data extends from a low of 25 to a high of 80. The *spread* of data is called the **variation**.

One way to measure the variation of a set of data is to find the **range**.

Range	The range of a set of data is the difference between the greatest and the least numbers in the set.

The range of ages of the award winners is 80 − 25 or 55.

When you are considering large sets of data, such as thousands of college entrance exam scores, it is often helpful to separate the data into four equal parts called **quartiles**. The quartiles are used to find another measure of variation called the **interquartile range**.

Interquartile Range	The interquartile range is the range of the middle half of the data.

Find the interquartile range of the actresses' ages.

Step 1 Find the median. The median is the 13th number when the data are in order from least to greatest, 35. It separates the data into two halves, which are shown in brackets.

[25 26 28 30 30 30 31 32 32 32 33 35] 35
[36 37 40 40 41 41 44 48 48 60 73 80]

Step 2 Next find the median of the upper half of the data. This number is the **upper quartile**. There are 12 numbers in the upper half of the data. The median is halfway between the 6th and 7th numbers above the median.

[25 26 28 30 30 30 31 32 32 32 33 35] 35
[36 37 40 40 41 | 41 44 | 48 48 60 73 80]

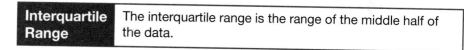

upper quartile = $\frac{41 + 44}{2}$ or 42.5

Step 3 Find the **lower quartile**. The lower quartile is the median of the lower half of the data. In this data set, the lower quartile is $\frac{30 + 31}{2}$ or 30.5.

[25 26 28 30 30 | 30 31 | 32 32 32 33 35] 35
[36 37 40 40 41 41 44 48 48 60 73 80]

Step 4 Find the interquartile range. This is the middle half of the data, which is between the lower quartile and the upper quartile. To find the interquartile range, subtract the lower quartile from the upper quartile. In this data set, the interquartile range is 42.5–30.5 or 12. This means that 50% of the actresses' ages are between 30.5 years and 42.5 years inclusive, a span of 12 years.

Example

Real World APPLICATION

1 **Sports** Thousands of kids have strapped on in-line skates. The frequency table shows the prices of different models of in-line skates reviewed by a consumer magazine.

a. Find the range of this data.

b. What is the interquartile range?

Price	Tally	Frequency
279	I	1
275	IIII	4
239	II	2
229	I	1
220	I	1
200	II	2
185	I	1
159	I	1
139	II	2
116	I	1
115	II	2
109	I	1
100	I	1
79	I	1
69	I	1
59	I	1
50	I	1
49	I	1
39	I	1

a. The range of the data is the difference between the greatest and least numbers. The range of the data is 279–39 or 240.

b. Notice that the data are written in order in the table. The first step to finding the interquartile range is to find the median.

The median is $\frac{139 + 159}{2}$ or 149.

Each half of the data has 13 numbers. The upper quartile, UQ, is the 7th number in the upper half of the data.

UQ = 239

The lower quartile, LQ, is the 7th number in the lower half of the data.

LQ = 100

The interquartile range is 239 − 100 or 139.

In some sets of data, one or both of the extreme values are much greater or less than the other data. Data that are more than 1.5 times the interquartile range from the quartiles are called **outliers**.

Example
Real World APPLICATION

② Entertainment Refer to the beginning of the lesson. Determine whether there are any outliers in the actresses' ages.

We found that the upper quartile is 42.5 and the lower quartile is 30.5. The interquartile range is 12. So data more than 1.5 · 12, or 18, above the upper quartile or below the lower quartile are outliers.

Find the limits for the outliers.

Subtract 18 from the lower quartile. $30.5 - 18 = 12.5$

Add 18 to the upper quartile. $42.5 + 18 = 60.5$

So, 12.5 and 60.5 are the limits for the outliers. There are two outliers in the data, 73 and 80.

CHECK FOR UNDERSTANDING

Communicating Mathematics

Read and study the lesson to answer each question.

1. *Summarize* the differences between the measures of variation and the measures of central tendency.

2. *Show* how to divide the data into quartiles.

{135, 170, 125, 174, 136, 145, 180, 156, 188}

Guided Practice

Find the range, median, upper and lower quartiles, interquartile range, and any outliers for each set of data.

3. 12, 13, 14, 16, 17, 17, 19   4. 1.4, 1.8, 1.0, 1.1, 1.6, 1.2, 3.0, 1.9

5. 43, 55, 49, 49, 53, 48, 57, 60, 57, 60, 47, 51, 59, 22

6. *Nutrition* The list below shows the numbers of Calories per tablespoon of different brands of margarine.

100, 70, 70, 90, 50, 90, 50, 90, 100, 50, 90, 100, 90, 50, 25, 81

Find the range, median, quartiles, interquartile range, and any outliers.

EXERCISES

Practice

Find the range, median, upper and lower quartiles, interquartile range, and any outliers for each set of data.

7. 54, 54, 58, 58, 58, 59, 60, 62, 63   8. 9, 0, 2, 8, 19, 5, 3, 2

9. 38, 43, 36, 37, 32, 37, 29, 51   10. 117, 118, 120, 109, 117, 117, 100

11. 2.3, 2.3, 3.8, 2.6, 3.7, 2.9, 6.1, 2.3, 2.9, 2.5, 3.5

12. 198, 166, 190, 155, 146, 184, 135, 180, 145

13. 55, 76, 104, 65, 62, 79, 63, 57, 52, 72, 57, 73, 55, 60, 80, 53

14. 349, 341, 351, 357, 353, 346, 342, 315, 336, 339, 313, 342, 347, 351, 353, 304, 356, 342, 357

15.

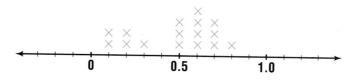

16. Geography The table shows the
lengths of the 16 longest rivers in
North America.

   a. Find the range, the median,
the upper and lower quartiles,
the interquartile range, and any
outliers of the river lengths.

   b. Which river is closest to the median
length?

17. Critical Thinking Create two
different sets of data that meet
the following conditions.

   a. the same range, different
interquartile ranges

   b. the same median and quartiles,
but different ranges

River	Length (mi)
Arkansas	1,459
Churchill, Man.	1,000
Colorado (AZ)	1,450
Columbia	1,243
Mackenzie	1,025
Mississippi	2,340
Upper Mississippi	1,171
Mississippi-Missouri-Red Rock	3,710
Missouri	2,315
Missouri-Red Rock	2,540
Ohio-Allegheny	1,310
Peace	1,210
Red	1,290
Rio Grande (OK-TX-LA)	1,900
Snake	1,038
Yukon	1,979

Source: Geological Survey

18. Civics Are you looking forward to getting a driver's license? The list below
shows the earliest ages at which a person can get a driver's license in each
state and the District of Columbia.

16	16	16	16	14	16	16	16	18	16	16	15	16
15	16	16	16	14	16	15	16	16	16.5	14	15	14
16	16	13	14	14	16	16	15	16	16	14	14	14
16	14	16	16	15	14	14	15	16	16	16	16	

Find the range, median, upper and lower quartiles, the interquartile range, and
any outliers of the data.

For **Extra Practice**,
see page 616.

Mixed Review

19. Life Science Find the mean, median, and mode of the plant heights 22, 4, 1,
12, 5, 22, 5, 25, 25, 19, 23, 24, 11, 16, 3, and 22. *(Lesson 4-4)*

20. Standardized Test Practice Choose the value of the expression 6^3.
(Lesson 1-2)

   A 18   **B** 63   **C** 216   **D** 729

CHAPTER 4 — Mid-Chapter Self Test

**The frequency table shows the grams of sugar per serving
in 28 cereals made for adults.**

1. Use the intervals 0-2, 3-5, 6-8, and 9-11 to make a histogram
of the data. *(Lesson 4-1)*

2. Make a circle graph to show the percent of cereals with 0-2,
3-5, 6-8, and 9-11 grams of sugar. *(Lesson 4-2)*

3. Make a line plot of the data. *(Lesson 4-3)*

4. Find the mean, median, and mode amounts of sugar.
(Lesson 4-4)

5. Find the range, upper and lower quartiles, the interquartile
range, and any outliers of the data. *(Lesson 4-5)*

Grams	Tally	Frequency
0	‖‖‖	5
1		0
2	‖‖‖	3
3	‖‖‖ ‖	6
4	‖	1
5	‖‖‖	5
6	‖‖‖‖	4
7	‖	1
8		0
9	‖	1
10	‖	1
11	‖	1

GRAPHING CALCULATORS

4-5B Box-and-Whisker Plots

A Follow-Up of Lesson 4-5

graphing calculator

A *box-and-whisker plot* uses a number line with the median, the quartiles, and the extreme values to represent a set of data. You can make a box-and-whisker plot with a graphing calculator.

TRY THIS

Work with a partner.

The list shows the ages of the winners of the Academy Award for Best Actor for the last 25 years. Use a graphing calculator to make a box-and-whisker plot of the data.

29	31	31	35	36	36	37
37	38	39	41	41	42	44
47	48	50	51	52	53	53
55	60	61	76			

Step 1 Begin by entering the data into the calculator's memory.

Press STAT ENTER to see the lists. If there are numbers in the first list, press ▲ CLEAR ENTER to clear them. Then enter the data by entering each number and pressing ENTER.

Step 2 Next choose the type of graph. Press 2nd [STAT PLOT] to display the menu. Choose the first plot by pressing ENTER. Use the arrow and ENTER keys to highlight "on", the modified box-and-whisker plot, L1 for the Xlist, and 1 as the frequency.

Step 3 Choose the display window. Access the menu by pressing WINDOW. Choose appropriate range settings for the *x* values. The window 25 to 80 with a scale of 5 includes all of this data.

Step 4 Display the graph by pressing GRAPH.

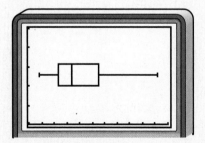

ON YOUR OWN

1. Press the TRACE key. The right and left arrow keys allow you to find the minimum, the maximum, the quartiles, and the median of the data. What are these values for the Academy Award winners' ages?

2. Use the data on the ages of the winners of the Academy Award for Best Actress on page 163 to make a box-and-whisker plot. How does the plot compare to the one above?

3. Does the graphing calculator show outliers? Explain.

4-6

Integration: Algebra
Scatter Plots

What you'll learn

You'll learn to construct and interpret scatter plots.

When am I ever going to use this?

You can use scatter plots to observe trends in data and make predictions.

Word Wise

scatter plot

When you graph two sets of data as ordered pairs, you form a **scatter plot**. You can use scatter plots to look for trends in data.

MINI-LAB

Work in a small group. 🔲 tape measure ▦ grid paper

Try This 📏 ruler

1. Measure each group member's height to the nearest centimeter. Then measure each person's arm from shoulder to wrist.

2. Graph the data on a coordinate graph similar to the one at the right. The point on the graph represents a person whose height is 155 centimeters and whose arm length is 50 centimeters.

Talk About It

3. Use your graph to predict the height of a person who has an arm length of 52 centimeters. Does your graph indicate that taller people tend to have longer or shorter arms?

Scatter plots can suggest whether two sets of data are related. Imagine a line drawn so that half of the points are above it and half are below it. If the line slopes upward to the right, there is a *positive* relationship. A line that slopes downward to the right indicates a *negative* relationship.

Positive Relationship   **Negative Relationship**

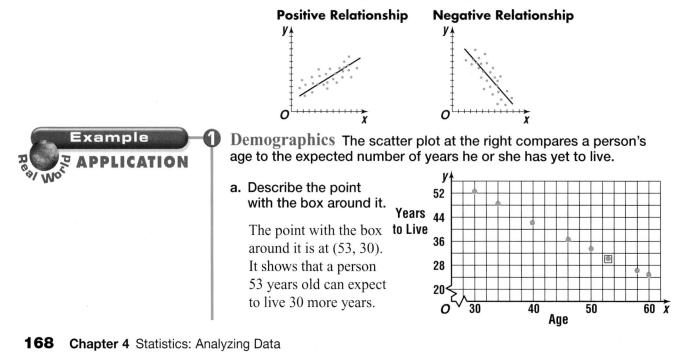

Example

Real World APPLICATION

1 **Demographics** The scatter plot at the right compares a person's age to the expected number of years he or she has yet to live.

a. Describe the point with the box around it.

The point with the box around it is at (53, 30). It shows that a person 53 years old can expect to live 30 more years.

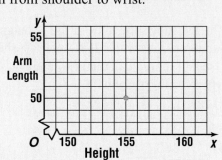

b. What type of relationship is shown by the scatter plot?

The points suggest a line that slopes downward to the right. So the scatter plot shows a negative relationship. This means that the older a person is, the fewer years he or she has yet to live.

Some scatter plots show that there is no relationship between sets of data.

Examples

Determine whether each scatter plot shows a positive, negative, or no relationship.

❷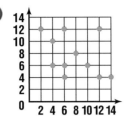

The points in the scatter plot are very spread out. There appears to be no relationship between these data.

❸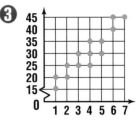

This scatter plot suggests a line that slopes upward to the right. These data appear to have a positive relationship.

CHECK FOR UNDERSTANDING

Communicating Mathematics

Read and study the lesson to answer each question.

1. *Tell* how to draw a scatter plot for two sets of data.

2. *Draw* a scatter plot that shows a negative relationship.

HANDS-ON MATH

3. *Measure* the distance around the closed fist and the length of the forearm of each person in your group. Create a scatter plot of the ordered pairs. Do you see a relationship? Explain.

Guided Practice

Determine whether a scatter plot of the data below would show a *positive, negative,* or *no* relationship.

4. age and height

5. age of a used car and its value

6. hours worked and earnings

7. grades and distance to school

8. *Health* The scatter plot shows the number of Calories in different fruits compared to the number of milligrams of calcium they offer. Does there appear to be a positive, a negative, or no relationship between Calories and calcium in fruit?

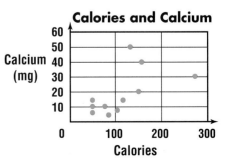

Practice

Determine whether a scatter plot of the data below would show a *positive*, *negative*, or *no* relationship.

9. height and test scores

10. years of education and income

11. speed and distance covered

12. height and month of birth

13. temperature and cooking time

14. bank balance and interest earned

15. miles per gallon and weight of a car

16. playing time and points scored in a basketball game

17. a U.S. president's time in office and age when elected

18. outside temperature and amount of heating fuel used

19. number of people in a household and weekly grocery bill

20. number of pages in a book and number of copies sold

Applications and Problem Solving

21. *Aviation* The table shows the number of seats and the average airborne speed of the most commonly used aircraft.

   a. Make a scatter plot of the seats and speeds.

   b. Does the scatter plot show a positive, a negative, or no relationship?

Aircraft	Seats	Speed (mph)
B747-400	396	538
B747-100	395	521
B747-200/300	342	535
B777	291	512
DC-10-40	288	506
L-1011-100/200	285	490
DC-10-10	283	503
DC-10-30	277	523
A300-600	267	468
MD-11	261	526
L-1011-500	222	524
B757-300ER	219	496
B757-200	186	465
B767-200ER	180	487
A310-300	172	498

Source: Air Transport Assoc. of America

inter NET CONNECTION

For the latest information on the states, visit:

www.glencoe.com/sec/math/mac/mathnet

22. *History* Use a reference book or Internet search to find the area and year of entrance of each state in the United States.

   a. Make a scatter plot of the data.

   b. Does the scatter plot show a positive, a negative, or no relationship?

23. *Critical Thinking* A scatter plot of skateboard sales and swim suit sales shows a positive relationship.

   a. Why might this be true?

   b. Does this mean that one factor caused the other? Explain.

Mixed Review

24. *Statistics* Find the range, median, upper and lower quartiles, interquartile range, and any outliers of {115, 117, 111, 121, 110, 127, 116, 126, 105, 115, 100, 103, 122, 130, 101, 100, 108, 130}. *(Lesson 4-5)*

25. **Standardized Test Practice** Which is in order from least to greatest? *(Lesson 2-2)*

   A $-4, 2, 8$ **B** $4, -1, -6$ **C** $-1, 2, -4, 8$ **D** $0, -1, -4$

26. *Life Science* The United States lists 55 mammals as endangered. That is 8 less than one-fourth of those on international endangered species lists. How many species are on international lists? *(Lesson 1-7)*

For **Extra Practice**, see page 616.

Choosing an Appropriate Display

What you'll learn

You'll learn to choose an appropriate display for a set of data.

When am I ever going to use this?

Knowing how to choose the correct type of graph can help you present information in an understandable way.

The table shows the length of an average fetus at different times of development. How would you display this information in a statistical graph? *This will be solved in Example 1.*

Time (weeks)	Length (mm)
4	7
8	30
12	75
16	180
20	250
24	300
28	350
32	410
36	450
38	500

So far in this chapter you have learned how to construct different types of statistical graphs. But when you are given a set of data to display, you must choose which type of graph would be appropriate for the data.

As you decide what type of graph to use, it helps to ask yourself these two questions:

• What type of information is this?
• What do I want my graph to show?

Consider each type of graph when choosing a display for your data. The displays you have studied are as follows.

• table   • line plot   • circle graph
• histogram   • bar graph   • scatter plot

Example ① CONNECTION

Life Science Refer to the beginning of the lesson. What type of display is most appropriate for displaying the data on fetus size if you are looking for a relationship between time and size?

There are two sets of numerical data to be compared. So we can eliminate displays that are used for one set of data — bar graph, histogram, and circle graph.

A table allows us to see each size at each time. However, because it does not allow a visual interpretation, it is not the best choice.

A line plot is not appropriate. The two sets of data are not similar, so comparing them with a line plot using different symbols would not be appropriate.

A scatter plot is the most appropriate choice. It allows us to plot the data as ordered pairs and determine whether there is a positive, negative, or no relationship between time and size.

2 **Earth Science** You are given data on the amount of space in landfills used for different types of waste. What is the best way to display the data to show that a certain type of waste occupies a much larger amount of space than most others?

A table is a good choice if you want to show the individual data. However, it may be difficult to see quickly which piece of data is much larger.

The line plot is appropriate. The reader would be able to see a number that is much larger than the rest. However, this plot may not be very visually appealing and the reader would not be able to tell which type of waste takes up the greatest amount of space.

A bar graph is a good choice. It would show which type has the greatest amount of space visually. However, a circle graph is most appropriate. A category with a greater amount of space would be shown with a larger section of the graph. The distribution of the other data would be shown for comparison.

CHECK FOR UNDERSTANDING

Communicating Mathematics

Read and study the lesson to answer each question.

1. *Give an example* of a set of data that is best displayed using a histogram.

2. *Make a list* of the advantages and disadvantages of each type of display you have studied.

3. *Find* a display of data in a newspaper or on the Internet. Do you think the most appropriate type of display was used? Explain.

Guided Practice

Choose the most appropriate type of display for each data set and situation. Explain your choice.

4. ages and average heart rates in a television news report

5. test scores in a science class

6. points scored by individual members of a basketball team compared to the team total

7. *Employment* You are given the annual salaries of fifty men and fifty women with similar work experience and education. How would you display the data to show how the salaries of men and women compare?

Practice

Choose the most appropriate type of display for each data set and situation. Explain your choice.

8. ages of state fair attendees in marketing information for the fair

9. the amounts of sales by different divisions of a company in information for budget decisions

10. number of televisions in surveyed homes in a newspaper article

11. plant height measurements made every 2 days in a science fair report

12. math test scores for two classes for comparing teaching techniques

13. home prices in a neighborhood to show that a home is underpriced

14. percent of people who own a certain type of computer compared to all computer owners in advertising

15. annual crime index in an advertisement for a self defense class

16. percent of Americans who are computer owners in each year in a computer advertisement

17. numbers of Americans whose first language is Spanish, Mandarin, or Hindi

18. prices of ice creams rated good or fair in a consumer magazine

Applications and Problem Solving

19. *Money Matters* The table shows the first-class postal rates for different years. Would a histogram be appropriate for this data? Explain.

Year	1974	1975	1978	1981	1985	1988	1991	1995	1999
Rate	10¢	13¢	15¢	18¢	22¢	25¢	29¢	32¢	33¢

Source: USA Today research

20. *Sports* The table below shows the number of gold medals won by each team that won medals in the Atlanta Summer Olympic Games. The United States won 44 gold medals. What type of display would you use if you were writing an article on how well the American athletes performed?

Medals	0	1	2	3	4	5	7	9	13	15	16	20	26	44
Countries	26	18	7	8	7	1	3	3	1	1	1	1	1	1

Source: World Almanac

21. *Working on the* **CHAPTER Project** Study the information you found about your museum. Choose an appropriate display for one of the data sets. Then construct the display and add it to your brochure or poster.

22. *Critical Thinking* Describe a situation in which someone might choose a display other than the one that is most appropriate for the data.

Mixed Review

23. *Life Science* Make a scatter plot of the data on fetus size on page 171. *(Lesson 4-6)*

24. **Standardized Test Practice** What is the area of a rectangle whose length is 14 meters and whose perimeter is 64 meters? *(Lesson 1-8)*

    A 896 square meters   **B** 700 square meters

    C 252 square meters   **D** 50 square meters

25. Write the product $5 \cdot 5 \cdot 8 \cdot 8 \cdot 8$ using exponents. *(Lesson 1-2)*

For **Extra Practice,** see page 616.

Misleading Graphs and Statistics

What you'll learn

You'll learn to recognize when graphs and statistics are misleading.

When am I ever going to use this?

Knowing how to recognize misleading graphs and statistics can help you read and interpret advertisements.

Word Wise

sample

Have you studied president Alf Landon in Social Studies? Probably not. Landon lost the 1936 presidential election to Franklin Roosevelt. But Landon and the readers of *Literary Digest* were confident of a victory.

Literary Digest mailed more than 10 million surveys to households listed in telephone books and car registration books. That method would work well today, but in 1936, only the rich could afford telephones and cars. The people surveyed, called the **sample**, were not representative of the general population. For this reason, the prediction was wrong, and *Literary Digest* was soon out of business.

Statisticians often use samples to represent larger groups. They must make sure that their samples are representative of the larger group.

Example
Real World APPLICATION

1 **Marketing** When a soft-drink company hired Michael Jordan as a spokesperson, they were looking for someone who would appeal to teenagers. Before hiring him, they considered celebrities from sports and entertainment. Would they have gotten good results from each survey?
  a. 200 teens at a professional basketball game in Chicago
  b. 25 teens at a shopping mall
  c. 500 students at a number of middle and high schools

  a. A survey at the Chicago basketball game would probably favor Chicago players. This would not be a good choice unless they were trying to choose between several Chicago players.
  b. The teens at a mall would represent the segment of the population that spend time at the mall. However, they may not represent all of the teens in the United States.
  c. This sample is large, and if the schools are chosen in a number of different areas, this survey would probably give very good results.

Once a survey is taken, the results can be reported in a misleading way. In Lesson 4-4, you learned about the mean, median, and mode of a set of data. In the data set {$76, $76, $78, $78, $78, $80, $92, $95, $97, $97, $99, $100, $100}, the mean is $88, the median is $92, and the mode is $78.

A company whose product sells for $91 could say that the "average price" is $92 to demonstrate that their product is inexpensive. But that is misleading. It does not give a complete picture of the data.

Graphs can also be misleading. Here are some ways a graph may be misleading.

- Numbers are omitted on an axis, but no break is shown.
- The tick marks on the axes are not the same distance apart or do not have the same-sized intervals.

Example
APPLICATION

Entertainment A media research company studies the popularity of television programs. The graphs below show the results of a study of the top five programs for the 1995-1996 season.

a. **Which graph is not misleading? Explain your reasoning.**
b. **How could you change the misleading graph to make it accurate?**

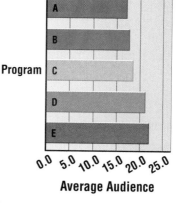

Favorite Programs

Program: A, B, C, D, E

0.0 5.0 10.0 15.0 20.0 25.0
Average Audience

Favorite Programs

Program: A, B, C, D, E

15.0 17.0 19.0 21.0 23.0
Average Audience

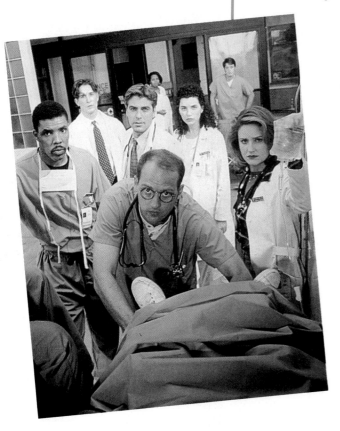

a. Both graphs display the same information. However, the graph on the right has omitted the numbers from 0 to 15 on the horizontal axis. Since the bar representing program E is about twice as long as the bar for program C, it appears that about twice as many people watch program E. Since this is not true, the graph on the right is misleading.

b. Adding a break in the horizontal axis and each of the bars would point out the gap between 0 and 15.

CHECK FOR UNDERSTANDING

Communicating Mathematics

Read and study the lesson to answer each question.

1. *List* two good places and two bad places to conduct a survey on the number of teenagers who have access to a computer.

2. *Transform* one of the graphs in this chapter into a misleading graph.

Guided Practice

Decide whether each location is a good place to find a representative sample for the selected survey. Justify your answer.

3. favorite singer at a Gloria Estefan concert

4. number of books read in a month at a shopping mall

5. passage of a school levy in a television phone-in poll

6. *Fashion* Of the 95% of schools who allow students to wear jeans, many have restrictions such as no jeans with holes. Consider the graph at the right.

DENIM DRESS CODES
72.3%
No restrictions
27.7%
Restrictions

Source: *Seventeen Magazine* Market Research Council

   a. How much greater is the percent of schools who allow jeans with no restrictions than those that allow jeans with restrictions?

   b. How does the difference in the percents compare with the difference in area of the two pairs of jeans?

   c. Is this a misleading graph? Explain.

EXERCISES

Practice

Decide whether each location is a good place to find a representative sample for the selected survey. Justify your answer.

7. favorite kind of entertainment at a movie theater

8. whether families own pets in an apartment complex

9. choice of mayoral candidate in a telephone poll

10. taste test of a snack food at a grocery store

11. Americans who are bilingual in a mall in San Antonio, Texas

12. favorite teacher in a school cafeteria

13. teenagers' favorite magazine at five different high schools

14. favorite drink at a coffee house

15. career choice at a community college

Applications and Problem Solving

Real World

16. *Marketing* The two graphs below show the results of a taste test of Mr. Bill's cookies.

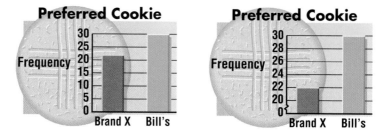
Preferred Cookie — Frequency: 30, 25, 20, 15, 10, 5, 0 — Brand X, Bill's

Preferred Cookie — Frequency: 30, 28, 26, 24, 22, 20, 0 — Brand X, Bill's

   a. Do both graphs contain the same information? Explain.

   b. One of the graphs is misleading. Explain why it is misleading.

   c. Which graph would you choose if you were creating advertising for Mr. Bill's cookies? Explain your reasoning.

17. Math History The timeline shows some of the major events in the history of mathematics.

The Elements published.	*Eratosthenes creates sieve for finding primes.*	*Circle divided into 360 degrees.*	*Hypatia publishes in geometry.*	*Tsu Ch'ung-chih approximates pi.*	*Al-Khowârizmî writes about Algebra.*
300	230	180 B.C.	A.D. 410	480	820

   a. In what way is the timeline misleading? Explain your answer.

   b. Draw a timeline. Include the events in the timeline above and the publication of the Chinese book *Arithmetic in Nine Sections* in 100 B.C., introduction of the Greek number system in 450 B.C., and the destruction of the University of Alexandria in A.D. 641.

   c. Choose one of the events listed on your timeline for further research. Describe the event and how you can see that part of mathematics used today.

18. Careers If you could trade places with an adult, who would you choose? A survey of 5,000 students asked that question. The results are shown in the graph.

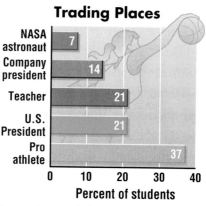

Trading Places

NASA astronaut	7
Company president	14
Teacher	21
U.S. President	21
Pro athlete	37

Percent of students

Source: BKG Youth/Nintendo survey

   a. How does the number of students who chose an astronaut compare to the number who chose the U.S. President?

   b. Are the sizes of the bars consistent with the differences in percents?

   c. Is this graph misleading? Explain.

19. Critical Thinking A toothpaste commercial says "Recommended by 4 out of 5 dentists."

   a. Would this claim be reasonable if 10 dentists were surveyed? Explain.

   b. Dr. Glover was a part of the survey. She also recommended several other toothpastes. Explain how the commercial was written to misrepresent Dr. Glover's recommendation.

Mixed Review

20. Standardized Test Practice What type of graph is most appropriate for displaying the change in house prices over several years? *(Lesson 4-7)*

   A scatter plot

   B circle graph

   C bar graph

   D line plot

21. Express 4.8% as a decimal. *(Lesson 3-4)*

22. Games Hector borrowed $100 from the bank to buy Park Place. If he passes go, collects $200, and repays the bank, how much money will Hector have? *(Lesson 2-3)*

For **Extra Practice,** see page 617.

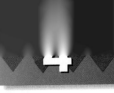

inter**NET**
CONNECTION Chapter Review **For additonal lesson-by-lesson review, visit:**
www.glencoe.com/sec/math/mac/mathnet

Vocabulary

After completing this chapter, you should be able to define each term, concept, or phrase and give an example or two of each.

Statistics
bar graph (p. 142)
box-and-whisker plot (p. 167)
circle graph (p. 148)
data analysis (p. 154)
frequency table (p. 142)
histogram (p. 142)
interquartile range (p. 163)
line plot (p. 153)
lower quartile (p. 164)
mean (p. 158)
measures of central tendency (p. 158)
median (pp. 158, 159)
mode (p. 158)

outlier (p. 164)
quartiles (p. 163)
range (p. 163)
sample (p. 174)
statistics (p. 142)
upper quartile (p. 163)
variation (p. 163)

Algebra
scatter plot (p. 168)

Problem Solving
make a table (p. 140)

Understanding and Using the Vocabulary

Choose the correct term or number to complete each sentence.

1. A (scatter plot, line plot) is a graph whose ordered pairs consist of two sets of data.
2. A (frequency table, histogram) is a bar graph that shows the frequency of data in intervals.
3. The (interquartile range, range) of a set of numbers is the difference between the least and greatest number in the set.
4. (An outlier, A variation) is a piece of data that is more than 1.5 times the interquartile range from one of the quartiles.
5. A small group that is representative of a larger population is called a (sample, mode).
6. The (mean, median) is the sum of the data divided by the number of pieces of data.
7. If you want to show how parts compare to a whole, the best display to use is a (line plot, circle graph).
8. The range is one of the (measures of central tendency, measures of variation).

In Your Own Words

9. *Explain* the difference between the mean and the median of a set of data.

178 **Chapter 4** Statistics: Analyzing Data

Objectives & Examples

Upon completing this chapter, you should be able to:

● construct and interpret bar graphs and histograms *(Lesson 4-1)*

Construct a histogram for {1, 1, 2, 3, 4, 4, 4, 5, 5, 5, 6, 7, 7, 7, 8, 8, 8, 8, 9, 9, 9, 10, 10, 10, 10}.

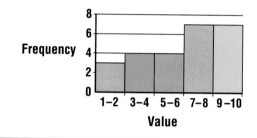

● construct and interpret circle graphs *(Lesson 4-2)*

Each day, an average of 2,749 Americans enroll in a language class. Of these, 46 choose Chinese; 64 Japanese; 93 Russian; 754 French; and 1,127 Spanish. Make a circle graph of the data.

Foreign Language Classes

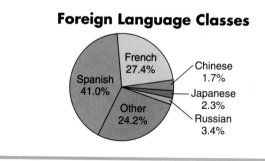

● construct and interpret line plots *(Lesson 4-3)*
Construct a line plot for the data
{1, 1, 2, 3, 4, 4, 4, 4, 5, 5, 5, 6, 7, 7, 7, 8, 8, 8, 8, 9, 9, 9, 10, 10, 10, 10}.

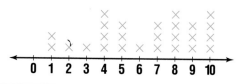

Review Exercises

Use these exercises to review and prepare for the chapter test.

Use the histogram at the left to answer each question.

10. How large is each interval?

11. Which interval has the most data?

12. What percent of the data is below 7?

13. Construct a histogram for the following history test scores.

    56, 87, 67, 74, 77, 84, 94, 80, 72, 58, 87, 90, 68, 90, 60, 73, 74, 82, 68, 64

14. *Geography* Make a circle graph of the data on the area of the Great Lakes.

Lake	Area (sq mi)
Erie	9,910
Huron	23,010
Michigan	22,300
Ontario	7,540
Superior	31,700

Source: Carnegie Library of Pittsburgh

15. Make a circle graph of the time that you spend each week on studying, sleeping, eating, watching television, and other activities.

The list shows ages of the first 24 people to enter an art museum one day.

30, 22, 23, 25, 22, 30, 28, 29, 26, 29, 27, 27, 24, 24, 30, 27, 24, 26, 27, 23, 24, 27, 23, 27

16. Construct a line plot for the data.

17. What is the most common age?

18. What are the ages of the oldest and youngest people?

Objectives & Examples

find the mean, median, and mode of a set of data
(Lesson 4-4)

Find the mean, median, and mode for
{53, 55, 58, 63, 66}.

mean: $\dfrac{53 + 55 + 58 + 63 + 66}{5} = 59$

median: 58   no mode

find the range and quartiles of a set of data
(Lesson 4-5)

Find the range, median, quartiles, and
interquartile range of {1, 2, 2, 3, 3, 3, 4,
4, 5, 6, 6}.

Range: $6 - 1 = 5$ median: 3

LQ: 2   UQ: 5

Interquartile range: $5 - 2 = 3$

construct and interpret scatter plots
(Lesson 4-6)

A scatter plot is a graph of points whose ordered
pairs consist of two sets of data.

choose an appropriate display for a set of data
(Lesson 4-7)

Choose the most appropriate type of display for
a set of data that compares a person's age and
their height. The data can be written in ordered
pairs, so a scatter plot is most appropriate.

recognize when graphs and statistics are
misleading *(Lesson 4-8)*

A sample must be representative of the larger
group. A graph must have an appropriate scale.

Review Exercises

**Find the mean, median, and mode for each
set of data. When necessary, round to the
nearest tenth.**

19. 5.6, 6.5, 6.8, 9.6, 10.1

20. 0, 2, 1, 5, 3, 7, 8, 5, 9

21. $16, $20, $21, $25, $18, $19, $20, $21

**Find the range, median, upper and lower
quartiles, interquartile range, and any
outliers for each set of data.**

22. 8, 9, 5, 10, 7, 6, 2, 4

23. 195, 121, 135, 123, 138, 150, 122, 136, 149,
124, 149, 151, 152

24. 80, 91, 82, 83, 77, 79, 78, 75, 75, 88, 84, 82,
61, 93, 88, 85, 84, 89, 62, 79

**Determine whether the scatter plot of the
data below would show a *positive, negative,*
or *no* relationship.**

25. age and income

26. temperature and day of the week

**Choose the most appropriate type of display
for each data set and situation. Explain your
choice.**

27. number of computers in surveyed homes in
magazine article

28. prices of different types of sandwiches at
a deli

**Decide whether each location is a good
place to find a representative sample for the
selected survey. Justify your answer.**

29. favorite movie at a movie theater

30. favorite dessert at an ice cream shop

Applications & Problem Solving

31. Make a Table Survey your classmates about their favorite dessert. Record the results in a frequency table. *(Lesson 4-1A)*

32. Agriculture The table shows the thousands of tons of several principal crops grown in the United States in 1996. Make a bar graph of the data. *(Lesson 4-1)*

Crop	Tons (thousands)
Almonds	304.3
Hazelnuts	39.0
Pecans	134.0
Walnuts	234.0

33. Wildlife Management The list shows lengths in inches of catfish captured and released in Grand Lake, Oklahoma. 8, 6, 4, 10, 9, 14, 4, 5, 12, 13, 10, 8 Make a line plot of the data. *(Lesson 4-3)*

34. Earth Science The list shows the amount of rainfall in inches in north central Texas for each month of 1994. 1.27, 2.40, 1.78, 2.79, 7.01, 1.68, 3.69, 1.88, 3.14, 6.53, 4.88, 3.37

   a. What is the mean, median, and mode rainfall amount for 1994? *(Lesson 4-4)*

   b. Find the range, upper and lower quartiles, interquartile range, and any outliers of the data. *(Lesson 4-5)*

35. Sales The graphs show Guillermo's sales for January through July. Which graph might be misleading and why? *(Lesson 4-8)*

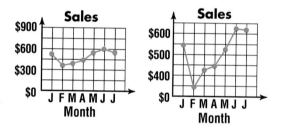

Alternative Assessment

Open Ended

Jacqui Franklin is trying to find the least expensive store in town from which to rent a video. The data she has collected about the costs of renting a video shows that some stores have a one-day rate and others have a two-day rate. Do you think it would help her to display the costs in a statistical graph? Why or why not? If so, what type of display should Jacqui use?

What types of displays would not be appropriate for Jacqui to use? Explain.

Completing the CHAPTER Project
Use the following checklist to make sure that your brochure or poster is complete.

☑ You have included a clear, well-organized description of the museum, its purpose, and its history.

☑ The bar graph of the data visitor information is clear and easy to read.

☑ The measures of central tendency are accurate, and the estimates of the budget requirements are included.

PORTFOLIO   Select a graph you drew in this chapter and place it in your portfolio. Attach a note explaining how it illustrates one of the important concepts in this chapter.

A practice test for Chapter 4 is provided on page 650.

Section One: Multiple Choice

There are twelve multiple-choice questions in this section. Choose the best answer. If a correct answer is *not here,* choose the letter for Not Here.

1. What is 33.67 rounded to the nearest tenth?

   A 33

   B 34

   C 33.6

   D 33.7

2. $36.5\% =$

   F 36.5

   G 3.65

   H 0.365

   J 0.0365

3. If $3m + 8 = 32$, what is the value of m?

   A 12

   B 10

   C 8

   D 6

4. What is $5 \cdot 5 \cdot 5 \cdot 8 \cdot x \cdot x$ written in exponential notation?

   F $5^3 \cdot 8 \cdot x^2$

   G $40x^2$

   H $3^5 \cdot 4 \cdot x^2$

   J $3^5 \cdot 4 \cdot 2x$

5. The line plot shows the heights in inches of the Washington Middle School girls' basketball team.

   Which sentence best describes the data?

   A The range is 6 inches.

   B Most of the players are 63 inches tall.

   C There are 8 players.

   D Most of the players are between 62 and 64 inches tall.

Please note that Questions 6-12 have five answer choices.

6. A bolt of fabric is 36 inches long. It is cut into two pieces so that one piece is 14 inches longer than the other. Which equation could be used to find the shorter piece, z?

   F $z + (z + 14) = 36$

   G $z^2 + 14 = 36$

   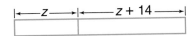

   H $2(z + 14) = 36$

   J $2z - 14 = 36$

   K $2z + (z - 14) = 36$

7. Bryanna invited 8 friends to a soccer game. The cost of tickets is $9.50 per person. Which is the best estimate for the cost of the game?

   A $70

   B $85

   C $90

   D $100

   E $125

8. Stacy and Ebony went to the store to buy 3 bags of chips for a party. They had $8 and wanted to spend no more than that on the chips. Which inequality could they use to find the price, p, of each bag of chips so that the total would be at or below $8?

   F $p + 3 \le 8$

   G $3p \ge 8$

   H $8 > 3p$

   J $3p \le 8$

   K $8 - p \le 3$

9. A tire filled with air has a leaking valve. The change in the amount of air is –16 cubic centimeters per hour. At this rate, in how many hours will the total change be –100 cubic centimeters?

    A 16 hours

    B 12 hours

    C 10.5 hours

    D 6.25 hours

    E Not Here

10. Scoma's Restaurant bought 58 pounds of shrimp at the market for $190.82 and 42 pounds of lobster for $186.90. What was the cost per pound of the shrimp?

    F $2.39

    G $2.84

    H $3.19

    J $4.45

    K Not Here

11. The total price for a jacket and pants was $184.62. This price included $9.97 in sales tax. What was the price of the jacket and pants before tax?

    A $174.65

    B $176.85

    C $177.15

    D $182.65

    E $194.59

12. A truck carrying lumber travels 899.1 miles on 74 gallons of diesel fuel. How many miles per gallon did the truck average?

    F 10.85 mpg

    G 11.21 mpg

    H 11.86 mpg

    J 12.15 mpg

    K 12.85 mpg

Test-Taking Tip   THE PRINCETON REVIEW

You can guess on standardized tests. If you know that one or more answers for a particular problem are definitely wrong, then it can be to your advantage to guess from the choices that remain. Random guessing will not increase your score, but educated guessing can make a difference.

Section Two: Free Response

This section contains four questions for which you will provide short answers. Write your answers on your paper.

13. $-3 + 8 + (-3) + 1 =$

14. Mrs. Campas teaches a class of 23 students. If she wants to buy 9 stickers for each student and the stickers cost $0.07 each, what is the best estimate of the cost of the stickers? Round to the nearest dollar.

15. Write the equation to represent *three more than twice a number equals 14.*

16. George, Justin, and Benito wanted to collect winter hats for a local shelter. Their goal is 120 hats. The table shows their progress so far.

Name	Week 1	Week 2	Week 3	Total
George	4	14	8	26
Justin	12	30	7	49
Benito	5	5	13	23

How many more hats do they need to collect to reach their goal?

 interNET **CONNECTION** Test Practice For additional test practice questions, visit:

www.glencoe.com/sec/math/mac/mathnet

Geometry: Investigating Patterns

What you'll learn in Chapter 5

- to identify lines that are parallel and types of angles formed by parallel lines and transversals,
- to solve problems by using Venn diagrams,
- to classify triangles and quadrilaterals,
- to identify line symmetry and rotational symmetry,
- to identify congruent triangles and similar triangles, and
- to create Escher-like drawings using translations and rotations.

CHAPTER Project

BE TRUE TO YOUR SCHOOL

In this project, you will design a school logo using a tessellation. You will begin by choosing a basic shape that will tessellate. After you determine whether your basic shape has symmetry, you will use one or more transformations to create a new shape. You will use this new shape to cover a poster board to form your school logo.

Getting Started

- A tessellation is a tile-like pattern formed by repeating shapes to fill a plane without gaps or overlaps. Some tessellations require just one shape, while others use more than one shape.
- Study various logos used by companies, sports teams, schools, and so on. List the ideas that you would like to incorporate in your logo.
- Find one or more shapes that will tessellate.
- Pick the colors that you want to use for your school logo.

Technology Tips

- Surf the **Internet** to search for ideas for your school logo.
- Use a **word processor** to write about the symmetry of your basic shape.
- Use **computer software** to design your school logo.

interNET
CONNECTION Research For up-to-date information on tessellations, visit:
www.glencoe.com/sec/math/mac/mathnet

Working on the Project

You can use what you'll learn in Chapter 5 to help you make your school logo.

Page	Exercise
209	20
223	16
227	Alternative Assessment

compass

protractor

ruler

5-1A Measuring and Constructing Line Segments and Angles

A Preview of Lesson 5-1

You know two systems of measurement for determining length: customary and metric. In geometry, both systems may be used to measure line segments.

TRY THIS

Work with a partner.

1 Use a ruler to measure line segments.

- Line segments are named by their endpoints. The segment below may be named $\overline{PF}$ or $\overline{FP}$.

- The sides of a polygon are formed by line segments. *ABCD* is a polygon with four sides. $\overline{AC}$ is a diagonal of polygon *ABCD*. Measure each line segment to the nearest tenth of a centimeter. Record the measurements.

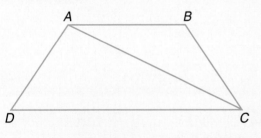

2 Construct a line segment congruent to $\overline{PF}$ by following these steps.

- Draw a long line with a ruler or straightedge. Put a point on the line. Label it *Z*.

- Open your compass to the same width as the length of $\overline{PF}$.

- Put the point of the compass at *Z* and draw an arc that intersects the line. Label this intersection *Y*. Segments with the same length are said to be *congruent*. Segment *ZY* is congruent to segment *PF*, or $\overline{ZY} \cong \overline{PF}$.

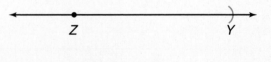

ON YOUR OWN

Trace each line segment. Then construct a line segment congruent to it.

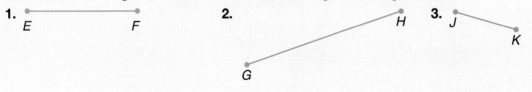

1. E F

2. G H

3. J K

4. Measure each line segment in Exercises 1–3 to the nearest tenth of a centimeter. Then measure each line segment that you constructed. How do these measures compare?

5. *Look Ahead* Suppose the line segments of the polygon above are extended. Which line segments do you think will never intersect?

A circle can be divided into 360 equal sections. Each section is one degree. Degrees are used to measure angles. A protractor can be used to measure an angle in degrees or to draw an angle with a given degree measure.

TRY THIS

Work with a partner.

3 Use a protractor to measure an angle.

- Trace ∠*ADC* from the polygon on page 186. Extend sides $\overrightarrow{DA}$ and $\overrightarrow{DC}$. *(DA means ray DA. When naming a ray, the endpoint must be given first, followed by any other point on the ray. A ray begins at the endpoint and continues in one direction from that point.)*

- Place the protractor over the angle with the center point on vertex *D* and the horizontal line aligned with side $\overrightarrow{DC}$.

- Side $\overrightarrow{DC}$ should intersect the point on the protractor marked 0°.

- Using the scale that begins with 0°, count off the degrees until you reach side $\overrightarrow{DA}$.

- Read the measurement. To say, *the measure of angle ADC is 55 degrees*, we can write *m*∠*ADC* = 55°.

4 Use a protractor to draw an angle with a measure of 115°.

- Draw a ray. Label the endpoint *R* and put a point labeled *S* on the ray.

- Align the protractor so that the center is at *R* and the horizontal line aligns with $\overrightarrow{RS}$.

- Find the scale containing 0° along $\overrightarrow{RS}$. Count along that scale until you reach 115°. Label this point *Q*.

- Draw $\overrightarrow{RQ}$. *m*∠*QRS* = 115°.

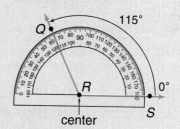

ON YOUR OWN

Use a protractor to draw an angle with the given measure.

6. 70°   **7.** 90°   **8.** 175°

9. Trace polygon *ABCD* on page 186. Measure ∠*BAC* and ∠*ACD*. (*Hint:* Extend any rays that will help you to measure the angles.)

10. *Look Ahead* In polygon *ABCD* on page 186, what is the relationship between ∠*BAC* and ∠*ACD*?

Parallel Lines

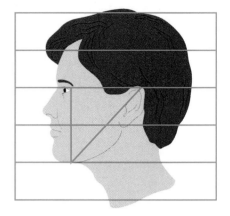

You'll learn to identify lines that are parallel and types of angles formed by parallel lines and transversals.

Knowing how to identify types of angles formed by parallel lines can help you make proportional drawings.

Word Wise

parallel lines
transversal
alternate interior angles
alternate exterior angles
corresponding angles
vertical angles
supplementary angles

Artists must be able to draw people of various proportions. One method that artists use is to draw lines that divide a person's body or face into equal parts. Then they use line segment lengths and angle measures to draw the correct body proportions. The lines that divide the person into equal parts are called **parallel lines**.

Parallel lines are lines in a plane that will never intersect. If line p is parallel to line q, we write $p \parallel q$.

A line that intersects two or more other lines is called a **transversal**. The figure at the right shows a transversal, line ℓ, that intersects two parallel lines, so that eight angles are formed.

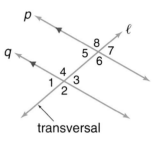

The red arrowheads on the lines p and q mean that $p \parallel q$.

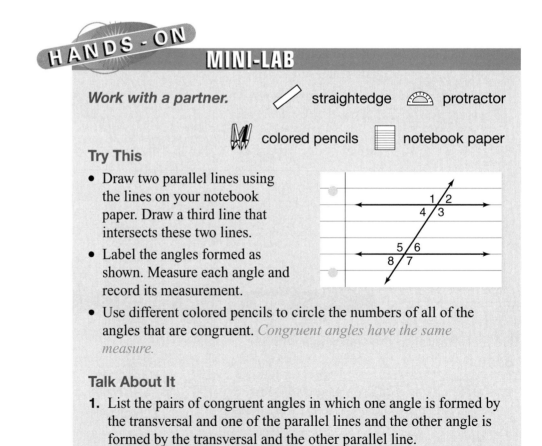

HANDS-ON MINI-LAB

Work with a partner. ⟋ straightedge ⌓ protractor

✏ colored pencils ▤ notebook paper

Try This

- Draw two parallel lines using the lines on your notebook paper. Draw a third line that intersects these two lines.

- Label the angles formed as shown. Measure each angle and record its measurement.

- Use different colored pencils to circle the numbers of all of the angles that are congruent. *Congruent angles have the same measure.*

Talk About It

1. List the pairs of congruent angles in which one angle is formed by the transversal and one of the parallel lines and the other angle is formed by the transversal and the other parallel line.

2. Which pairs of congruent angles are on the same side of the transversal?

3. Which pairs of congruent angles are on opposite sides of the transversal?

4. Using your results from Exercises 1-3, describe the locations of the sets of congruent angles in relationship to the transversal and the parallel lines.

Congruent angles formed by parallel lines and a transversal have special names.

Congruent Angles with Parallel Lines	If a pair of parallel lines is intersected by a transversal, these pairs of angles are congruent.
	alternate interior angles: $\angle 4 \cong \angle 6$, $\angle 3 \cong \angle 5$
	alternate exterior angles: $\angle 1 \cong \angle 7$, $\angle 2 \cong \angle 8$
	corresponding angles: $\angle 1 \cong \angle 5$, $\angle 2 \cong \angle 6$, $\angle 3 \cong \angle 7$, $\angle 4 \cong \angle 8$

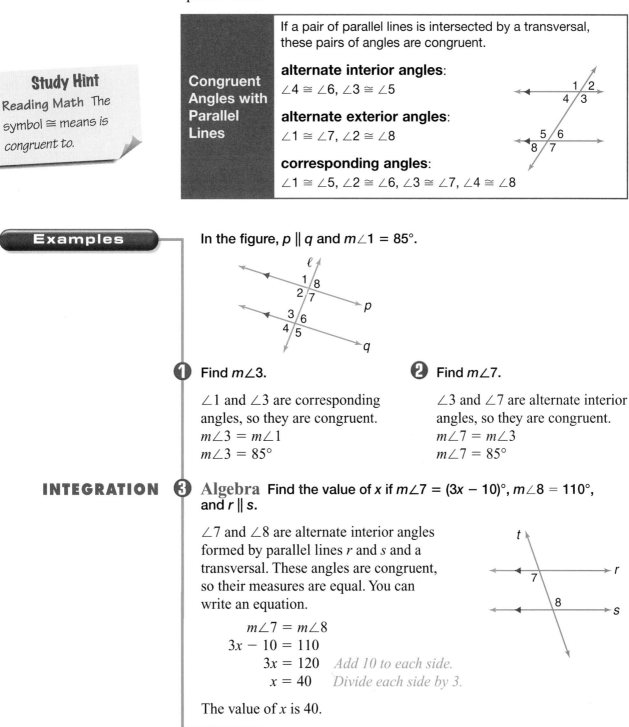

Examples

In the figure, $p \parallel q$ and $m\angle 1 = 85°$.

1 Find $m\angle 3$.

$\angle 1$ and $\angle 3$ are corresponding angles, so they are congruent.
$m\angle 3 = m\angle 1$
$m\angle 3 = 85°$

2 Find $m\angle 7$.

$\angle 3$ and $\angle 7$ are alternate interior angles, so they are congruent.
$m\angle 7 = m\angle 3$
$m\angle 7 = 85°$

INTEGRATION

3 **Algebra** Find the value of x if $m\angle 7 = (3x - 10)°$, $m\angle 8 = 110°$, and $r \parallel s$.

$\angle 7$ and $\angle 8$ are alternate interior angles formed by parallel lines r and s and a transversal. These angles are congruent, so their measures are equal. You can write an equation.

$$m\angle 7 = m\angle 8$$
$$3x - 10 = 110$$
$$3x = 120 \quad \textit{Add 10 to each side.}$$
$$x = 40 \quad \textit{Divide each side by 3.}$$

The value of x is 40.

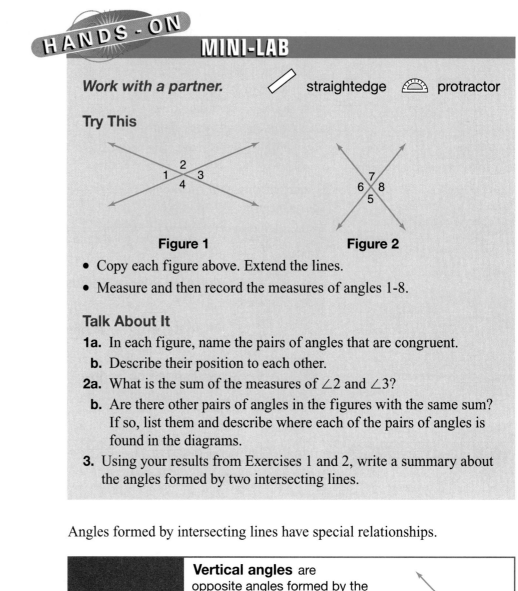

HANDS-ON MINI-LAB

Work with a partner. ✏ straightedge ⬚ protractor

Try This

Figure 1   **Figure 2**

- Copy each figure above. Extend the lines.
- Measure and then record the measures of angles 1-8.

Talk About It

1a. In each figure, name the pairs of angles that are congruent.
  b. Describe their position to each other.
2a. What is the sum of the measures of ∠2 and ∠3?
  b. Are there other pairs of angles in the figures with the same sum? If so, list them and describe where each of the pairs of angles is found in the diagrams.
3. Using your results from Exercises 1 and 2, write a summary about the angles formed by two intersecting lines.

Angles formed by intersecting lines have special relationships.

Straight angles have measures equal to 180°.

Vertical Angles and Supplementary Angles	**Vertical angles** are opposite angles formed by the intersection of two lines. Vertical angles are congruent. ($\angle 1 \cong \angle 3$, $\angle 2 \cong \angle 4$) **Supplementary angles** are two angles whose measures have a sum of 180°. ($\angle 1$ and $\angle 2$ are an example of supplementary angles.)

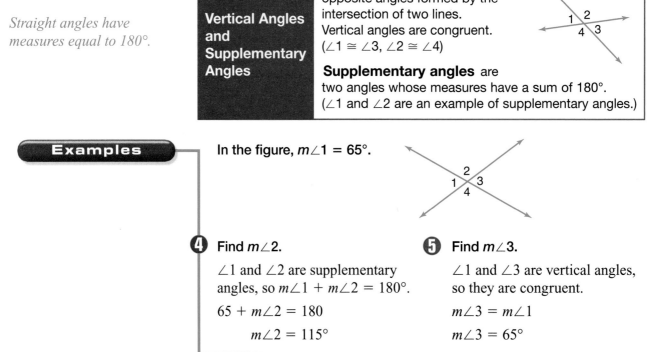

Examples

In the figure, $m\angle 1 = 65°$.

④ Find $m\angle 2$.

∠1 and ∠2 are supplementary angles, so $m\angle 1 + m\angle 2 = 180°$.

$65 + m\angle 2 = 180$

$m\angle 2 = 115°$

⑤ Find $m\angle 3$.

∠1 and ∠3 are vertical angles, so they are congruent.

$m\angle 3 = m\angle 1$

$m\angle 3 = 65°$

Communicating Mathematics

Read and study the lesson to answer each question.

1. *Write* a definition for parallel lines.

2. *Give* two examples of parallel lines in your surroundings.

HANDS-ON MATH

3. *Draw* two parallel lines using the lines on your notebook paper. Draw a third line that intersects these lines.

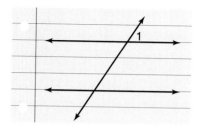

   a. Trace ∠1 on a piece of tracing paper. Use the traced angle to determine which angles are congruent to ∠1. Place a ✓ in each angle that is congruent to ∠1.

   b. Trace one of the angles that is not marked. Use the traced angle to determine which angles are congruent to this angle. Place an X in each of these angles.

Guided Practice

Use the figure at the right for Exercises 4–5.

4. Find $m\angle 5$ if $m\angle 7 = 95°$.

5. Find $m\angle 7$ if $m\angle 4 = 110°$.

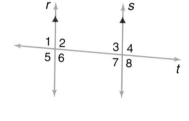

Find the measure of each angle in the figure if $p \parallel q$ and $m\angle 7 = 60°$.

6. $m\angle 1$

7. $m\angle 8$

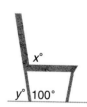

8. *Algebra* Find the value of x in the figure below.

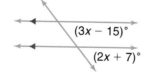

$(3x - 15)°$

$(2x + 7)°$

9. *Furniture* A single piece of wood is used for both the backrest of a chair and its rear legs. If the inside angle that the wood makes with the floor is 100° and the seat is parallel to the floor, what are the values of x and y?

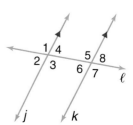

EXERCISES

Practice

Use the figure at the right for Exercises 10–15.

10. Find $m\angle 6$ if $m\angle 2 = 35°$.

11. Find $m\angle 7$ if $m\angle 6 = 45°$.

12. Find $m\angle 3$ if $m\angle 5 = 77°$.

13. Find $m\angle 1$ if $m\angle 3 = 138°$.

14. Find $m\angle 4$ if $m\angle 8 = 122°$.

15. Find $m\angle 7$ if $m\angle 1 = 68°$.

Find the measure of each angle in the figure if
s ∥ t, q ∥ r, and m∠8 = 75°.

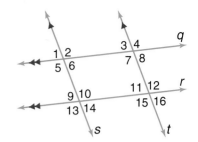

16. $m\angle 1$

17. $m\angle 3$

18. $m\angle 5$

19. $m\angle 7$

20. $m\angle 9$

21. $m\angle 11$

Find the value of x in each figure.

22.

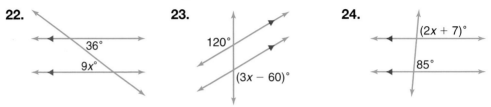

36°

$9x°$

23.

120°

$(3x - 60)°$

24.

$(2x + 7)°$

85°

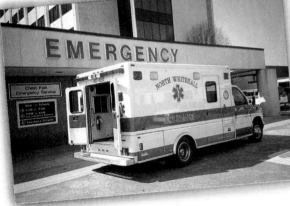

25. The measure of one angle formed by two intersecting lines is 90°. Find the measures of the other angles formed by these lines.

Applications and Problem Solving

26. *Urban Planning* Ambulances can't safely make turns of less than 70°. The angle at the southeast corner of Delavan and Elmwood is 108°. Should the proposed site of the hospital emergency entrance at the northeast corner of Bidwell and Elmwood be approved? Explain your answer.

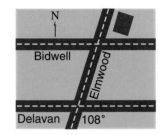

27. *Marching Band* During a performance, a marching band forms two intersecting lines. The measure of ∠1 is 65°. What is the measure of ∠2?

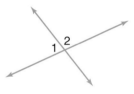

28. *Critical Thinking* The figure at the right is a parallelogram. It has two pairs of parallel sides. If $m\angle R = 110°$, find the measures of the other angles of the parallelogram. (*Hint:* Extend the sides of the parallelogram.)

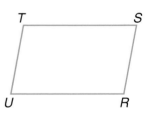

Mixed Review

29. **Standardized Test Practice** Which of the following would be the best place to find a representative sample for a survey of favorite types of music? *(Lesson 4-8)*

A a jazz club

B a symphony performance

C a mall

D a high school dance

E a retirement center

30. *Algebra* Solve $\frac{2}{n} = \frac{7}{98}$. *(Lesson 3-3)*

31. *Geometry* Graph the points $A(-3, 2)$, $B(-3, -1)$, $C(1, -1)$, and $D(1, 2)$ on the same coordinate plane. Draw $\overline{AB}$, $\overline{BC}$, $\overline{CD}$, and $\overline{AD}$. Find the perimeter of the rectangle formed. *(Lesson 2-10)*

For **Extra Practice,** see page 617.

32. Find the value of 5^3. *(Lesson 1-2)*

COOPERATIVE LEARNING

compass

straightedge

5-1B Constructing Parallel Lines

A Follow-Up of Lesson 5-1

Before computers, navigators on ships and airplanes used a compass and a straightedge to plot their course. You can use a compass and a straightedge to construct a line parallel to a given line.

TRY THIS

Work with a partner.

Step 1   Draw a line and label it ℓ.

Step 2   Choose a point P not on line ℓ.

Step 3   Draw a line through point P so that it intersects line ℓ. Label the point of intersection point Q.

Step 4   Place the point of your compass at point Q and draw a large arc. Label the point where the arc crosses line ℓ as Point R, and label where it crosses line PQ as point S.

Step 5   With the same compass opening, place the compass point at P and draw a large arc. Label the point of intersection with line PQ as point T.

Step 6   Use your compass to measure the distance between points S and R.

Step 7   With the compass opened the same amount, place the compass point at point T and draw an arc to intersect the arc already drawn. Label this point of intersection point U.

Step 8   Draw a line through points P and U. Label this line m. You have drawn $\ell \parallel m$.

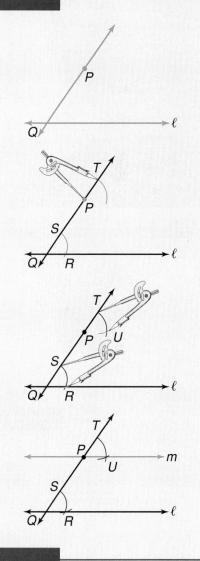

ON YOUR OWN

Trace each line. Then construct a line parallel to it.

1.

2.

3. *Reflect Back*   Examine your constructions. What type of angles did you use to create the parallel lines? Measure the angles. Are the angles congruent?

PROBLEM SOLVING

5-2A Use a Venn Diagram

A Preview of Lesson 5-2

The Spring Carnival committee at Barrington Middle School needs to hire a band. They decided to take a school survey to find the type of band that most of the students like. They are discussing the results. Let's listen in!

Of the students who turned in surveys, 74 like country, 189 like rock, and 43 like rap. 55 students like rock and country, 12 like rock and rap, and 23 like rap and country. Only 8 students like all three types of music.

Will

How many students turned in a survey?

Well, there are 74 + 189 + 43 + 55 + 12 + 23 + 8, or 404 responses in all.

That can't be right because there are only 325 students in our school.

But some of the surveys had more than one response. Let's make a Venn diagram to show the responses.

Dario

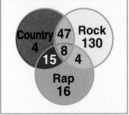

THINK ABOUT IT

Work with a partner.

1. *Explain* what the brown center region of the Venn diagram represents. How many students are represented by this region?

2. *Use* the Venn diagram to find how many students turned in a survey.

3. Use a **Venn diagram** to solve the following problem.

   Napoli's Pizza conducted a survey of 75 customers. The results showed that 35 people like mushroom pizza, 41 like pepperoni, and 11 like both mushroom and pepperoni pizza. How many like neither mushroom nor pepperoni pizza?

4. The Venn diagram below shows the types of books owned by Carter Middle School library. The circle for biographies inside of the nonfiction area indicates that all of the biographies are also nonfiction. Since the nonfiction and fiction areas do not intersect, there are no books that are both fiction and nonfiction. *State* the total number of nonfiction books.

Carter Middle School Library

For **Extra Practice,** see page 617.

ON YOUR OWN

5. The last step of the 4-step plan for problem solving asks you to *examine* the solution. *Explain* how you can use your answer in Exercise 3 to confirm that the number of customers surveyed is accurately represented in your Venn diagram.

6. *Write a Problem* that can be solved by using a Venn diagram.

7. *Look Ahead* Draw a Venn diagram showing the relationship of triangles that have at least two congruent sides and triangles that have all three sides congruent.

MIXED PROBLEM SOLVING

STRATEGIES

Look for a pattern.
Solve a simpler problem.
Act it out.
Guess and check.
Draw a diagram.
Make a chart.
Work backward.

Solve. Use any strategy.

8. *Marketing* The results of a supermarket survey showed that 83 customers chose wheat cereal, 83 chose rice cereal, and 20 chose corn cereal. Of those customers who bought exactly two boxes of cereal, 6 bought corn and wheat, 10 bought rice and corn, and 12 bought rice and wheat. Four customers bought all three. Make a Venn diagram of this information.

9. *Advertising* The graph shows who spent money on radio advertising in a recent year.

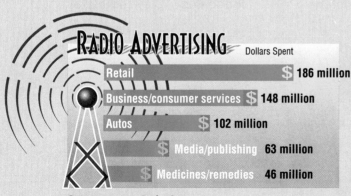

RADIO ADVERTISING   Dollars Spent

Retail	$ 186 million
Business/consumer services	$ 148 million
Autos	$ 102 million
Media/publishing	$ 63 million
Medicines/remedies	$ 46 million

Source: Interep Radio Store

a. About how many times more money was spent by retail companies than by the medicine/remedy companies?

b. A department store chain spent 67 million dollars on radio advertising. How much more money did the chain spend than the entire medicine/remedy industry?

10. *Data Analysis* The school cafeteria surveyed 36 students about their dessert preference. The results are listed below.

Number of Students	Preference of Students
25	cake
20	ice cream
15	pie
2	all three
1	no desserts
15	cake or ice cream
8	pie or cake
3	ice cream only

a. Draw a Venn diagram that will represent the responses.

b. How many students prefer only pie? either pie or ice cream?

c. What two desserts should the cafeteria order? Explain.

11. **Standardized Test Practice** There are about 3,907,000 miles of roads in the United States. Twelve percent of these roads are in Texas and California. About how many miles of roads are located in these two states?

A less than 40,000 miles

B between 50,000 and 100,000 miles

C between 300,000 and 400,000 miles

D between 400,000 and 500,000 miles

E more than 500,000 miles

What you'll learn

You'll learn to classify triangles by their angles and their sides and to find measures of missing angles in triangles.

When am I ever going to use this?

Knowing how to classify triangles can help you describe bicycle frames.

Word Wise

polygon
triangle
acute
right
obtuse
scalene
isosceles
equilateral
perpendicular
complementary
  angles

A **polygon** is a simple closed figure in a plane formed by three or more line segments. A **triangle** is a polygon formed by three line segments. Triangles can be classified by their angles and by their sides.

Examine the Eiffel Tower at the right and locate as many different-shaped triangles as you can. You can see that all of the triangles have at least two angles that are less than 90°. The third angle and its measure determine the classification of a triangle.

Acute angles have measures less than 90°.

Right angles have measures equal to 90°.

Obtuse angles have measures greater than 90°, but less than 180°.

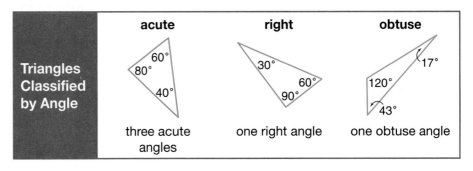

Triangles Classified by Angle

acute	right	obtuse
three acute angles	one right angle	one obtuse angle

Triangles can also be classified by the number of congruent sides. *Congruent sides are often marked with a slash through the sides.*

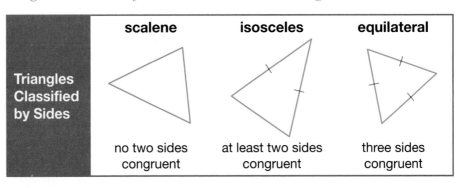

Triangles Classified by Sides

scalene	isosceles	equilateral
no two sides congruent	at least two sides congruent	three sides congruent

These side classifications can be organized in a Venn diagram to show their relationship.

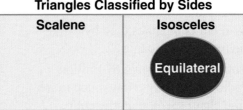

Triangles Classified by Sides

Scalene   Isosceles   Equilateral

Study Hint

Reading Math $\triangle HIJ$ is read as *triangle HIJ*.

Triangles are named by letters at their vertices. The triangle at the right can be named $\triangle HIJ$. *A vertex is the point where two sides meet.*

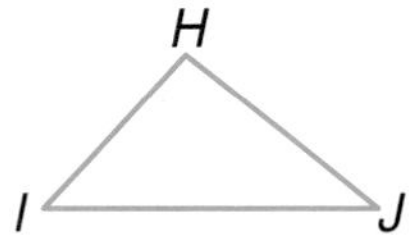

In Example 1, the ∟ symbol in $\triangle EFG$ indicates that $\angle E$ is a right angle. When segments meet to form right angles, they are **perpendicular**.

Examples

Classify each triangle by its angles and by its sides.

1

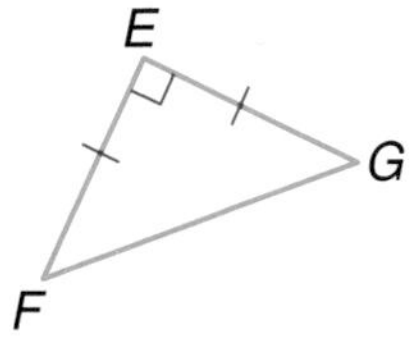

$\triangle EFG$ is a right, isosceles triangle.

2

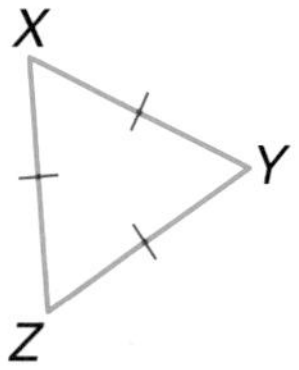

$\triangle XYZ$ is an acute, equilateral triangle.

The sum of the measures of the three angles of a triangle is 180°. You can check this by folding the angles of a triangle as shown in the diagram. Since the three angles together form a straight angle, we know that they add up to 180°.

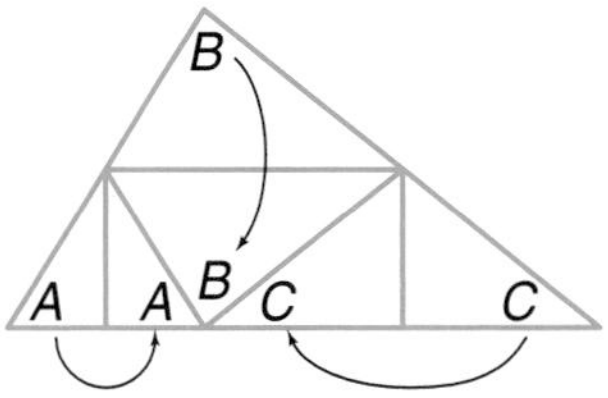

Example INTEGRATION

3 **Algebra** **Find the measure of each angle in $\triangle ABC$ if $\angle C$ is a right angle, $m\angle A = (x + 50)°$, and $m\angle B = (x + 10)°$.**

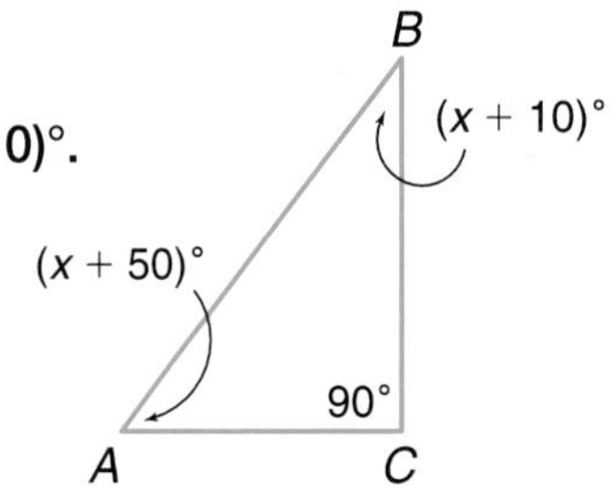

Explore You can draw and label a triangle to represent the problem. You know the expressions for two angles and that the third angle is a right angle. You need to know the measure of each angle.

Plan Find the value of x by using the equation $m\angle A + m\angle B + m\angle C = 180$. Then substitute the value of x into $m\angle A = (x + 50)°$ and $m\angle B = (x + 10)°$, to find $m\angle A$ and $m\angle B$.

Solve

$$m\angle A + m\angle B + m\angle C = 180$$
$$(x + 50) + (x + 10) + 90 = 180$$
$$2x + 150 = 180 \quad \textit{Add like terms.}$$
$$2x = 30 \quad \textit{Subtract 150.}$$
$$x = 15 \quad \textit{Divide by 2.}$$

By substitution, $m\angle A = 15 + 50$ or 65° and $m\angle B = 15 + 10$ or 25°. Therefore, the measures of the three angles of the triangle are 65°, 25° and 90°.

Examine You know that the sum of the measures of the angles of a triangle is 180°. $65 + 25 + 90 = 180$ ✓

The two acute angles in a right triangle have a special relationship. Since the sum of the measures of the angles in the triangle is 180° and the right angle is 90°, the sum of the two acute angles must be 180° − 90° or 90°. Two angles whose sum is 90° are **complementary angles**.

Complementary Angles	**Complementary angles** are two angles whose measures have a sum of 90°. (∠1 and ∠2 are an example of complementary angles.)	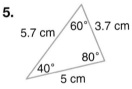

CHECK FOR UNDERSTANDING

Communicating Mathematics

Read and study the lesson to answer each question.

1. *Compare and contrast* isosceles triangles and equilateral triangles.

2. *Describe* the types of angles that are in a right triangle.

3. *You Decide* Jaali and Adia were drawing triangles on paper and then describing them to each other. Adia told Jaali that she drew a triangle that had two obtuse angles. Jaali said that it couldn't be done. Who was correct? Explain.

Guided Practice

Classify each triangle by its angles and by its sides.

4.
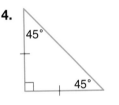

5.
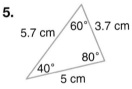

6. *Algebra* Find the measure of each angle in $\triangle RST$ if $m\angle R = x°$, $m\angle S = (x + 20)°$, and $m\angle T = 2x°$.

7. *True* or *false?* A pair of angles can be both complementary and supplementary.

8. In the diagram, $\overleftrightarrow{DE} \parallel \overleftrightarrow{AB}$, $m\angle DEC = 116°$, and $m\angle BAC = 32°$.

   a. Classify $\triangle ABC$ by its angles.

   b. Find the measure of each numbered angle in the drawing.

EXERCISES

Practice

Classify each triangle by its angles and by its sides.

9.   10.   11.
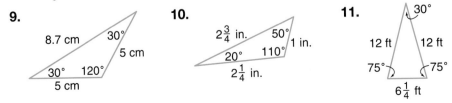

Classify each triangle by its angles and by its sides.

12.
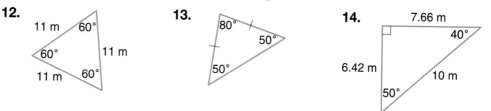
11 m   60°
60°   11 m
11 m   60°

13.
80°
50°
50°

14.
7.66 m
40°
6.42 m
10 m
50°

15. *Algebra* Find the value of x in $\triangle XYZ$ if $m\angle X = 2x°$, $m\angle Y = 64°$, and $m\angle Z = (2x - 16)°$.

16. *Algebra* Find the measure of each angle in $\triangle EFG$ if $\angle E$ is a right angle, $m\angle F = (2x + 5)°$, and $m\angle G = (x + 25)°$.

17. *Algebra* Find the measure of each angle in $\triangle ABC$ if $m\angle A = 65°$, $m\angle B = 3x°$, and $m\angle C = (x + 15)°$.

Tell whether each statement is *true* or *false*. Then draw a figure to justify your answer.

18. A triangle can have three acute angles.

19. An obtuse isosceles triangle has two acute angles.

20. What type of triangle has one pair of perpendicular sides?

21. An equiangular triangle has three congruent angles. What is the measure of each angle of an equiangular triangle?

Applications and Problem Solving

22. *Bicycles* Bicycles with small frames have three tubes in the center that form a triangle. Classify the triangle used in the bike frame by its sides.

23. *Architecture* Triangles are used to stabilize buildings and bridges. Find a photograph in a newspaper or magazine that shows triangles used in architecture. Classify the triangles.

24. *Critical Thinking* Use toothpicks, straws, or pretzel sticks to build the figure below. Remove the stated number of sticks to get the specific number of triangles. Draw your figure.

For **Extra Practice**, see page 618.

	Remove	Number of Triangles
a.	2	3
b.	2	2
c.	3	1
d.	3	2

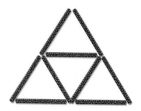

Mixed Review

25. *Geometry* Find the value of x in the figure at the right. *(Lesson 5-1)*

150°
(3x + 30)°

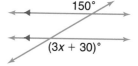

26. Solve $n = -28 + 73$. *(Lesson 2-3)*

27. **Standardized Test Practice** If Rodolfo worked a total of 36 hours over a 4-day period and was paid $5.35 per hour, how much did he earn before deductions? *(Lesson 1-5)*

A $770.40   **B** $192.60   **C** $182.80   **D** $21.40   **E** Not Here

COOPERATIVE LEARNING

HANDS-ON LAB

straightedge

5-3A Polygons as Networks

A Preview of Lesson 5-3

When people talk about networking, they usually mean that they are making contact with other people and exchanging ideas and information. In mathematics, a *network* is a figure consisting of points, called *nodes,* and paths that join various nodes to one another, called *edges*.

TRY THIS

Work with a partner.

Step 1 Make a copy of each polygon. Draw the diagonals from one vertex as shown.

A diagonal is a segment that connects any two nonconsecutive vertices.

rectangle   pentagon   hexagon

Step 2 Label the vertices of each polygon as shown in the rectangle. These figures can be considered networks. The vertices are the nodes, and the line segments are the edges.

Step 3 Vertices or nodes can be called *odd* or *even.* A vertex is odd if it has an odd number of line segments meeting at the vertex. An even vertex has an even number of line segments meeting at the vertex. For each figure, examine the vertices. Which vertices are odd and which are even?

Step 4 A figure is considered *traceable* if you can trace it without covering any line segment twice, and without taking your pencil off the paper. For example, one way the rectangle can be traced is by following *ABCADC*. Therefore, it is traceable. Determine whether the pentagon and hexagon are traceable.

ON YOUR OWN

Draw a heptagon (7-sided polygon) and an octagon (8-sided polygon) and their diagonals from one vertex. Then complete the table.

	Name of Polygon	Number of Odd Vertices	Traceable or Not Traceable
1.	rectangle		
2.	pentagon		
3.	hexagon		
4.	heptagon		
5.	octagon		

6. Examine your answers for Exercises 1–5. Make a conjecture about traceability.

7. **Look Ahead** Will any four-sided figure with one diagonal have the same traceable result as a rectangle? Explain.

Classifying Quadrilaterals

What you'll learn

You'll learn to classify quadrilaterals.

When am I ever going to use this?

Knowing how to identify quadrilaterals can help you design room layouts.

Word Wise

quadrilateral
parallelogram
rectangle
rhombus
square
trapezoid

Have you ever noticed that members of the same family often have the same characteristics? Families of polygons also share certain characteristics. Each member of the **quadrilateral** family has four sides and four angles. Each member of the family has additional characteristics or attributes that distinguish it from other family members.

A **parallelogram** is a quadrilateral with two pairs of opposite sides that are parallel.

A **rectangle** is a parallelogram with four right angles.

A **rhombus** is a parallelogram with all sides congruent.

A **square** is a parallelogram with all sides congruent and four right angles.

A **trapezoid** is a quadrilateral with exactly one pair of opposite sides that are parallel.

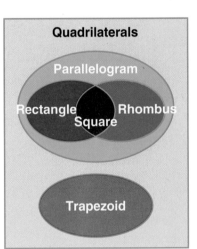

Quadrilaterals

Parallelogram

Rectangle   Rhombus
        Square

Trapezoid

Quadrilaterals are named by the letters at their vertices, in consecutive order. The figure at the right can be called parallelogram *ABCD*.

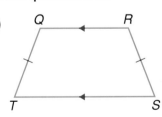

Examples

Identify all of the names that describe each quadrilateral.

1

Quadrilateral *GHJK* is a square. It is also a rhombus, a rectangle, and a parallelogram. When a quadrilateral can fit into several types of definitions, we call it by the name that tells the most about it. In this case, the quadrilateral is called a square.

2

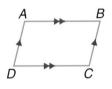

Quadrilateral *QRST* is a trapezoid. A trapezoid in which the two nonparallel sides are congruent is called an *isosceles trapezoid*.

We know about quadrilaterals and their sides. Let's explore quadrilaterals and their angles.

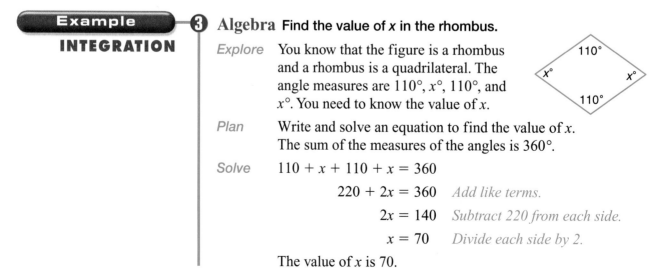

HANDS-ON MINI-LAB

Work in groups of four. 🔲 protractor ✏️ ruler

Try This

- Draw a set of the six quadrilaterals described on page 201.
- Measure the angles of each figure and record them on the figure.
- Measure the sides of each figure and record them on the figure.
- Measure and record the lengths of the diagonals of each figure.

Talk About It

1. Copy and complete the table.

	Quadrilateral	Parallelogram	Trapezoid	Rectangle	Rhombus	Square
Opposite sides parallel?						
Opposite sides congruent?						
All sides congruent?						
Diagonals congruent?						
All right angles?						
Opposite angles congruent?						

2. What is the sum of the angle measures for each figure?

The Mini-Lab suggests that the sum of the measures of the angles of a quadrilateral is 360°.

Example

INTEGRATION

3 **Algebra** **Find the value of x in the rhombus.**

Explore You know that the figure is a rhombus and a rhombus is a quadrilateral. The angle measures are 110°, x°, 110°, and x°. You need to know the value of x.

Plan Write and solve an equation to find the value of x. The sum of the measures of the angles is 360°.

Solve
$$110 + x + 110 + x = 360$$
$$220 + 2x = 360 \quad \textit{Add like terms.}$$
$$2x = 140 \quad \textit{Subtract 220 from each side.}$$
$$x = 70 \quad \textit{Divide each side by 2.}$$

The value of x is 70.

> *Examine* You know the sum of the measures of the angles of
> a quadrilateral is 360°.
>
> $$110 + 70 + 110 + 70 = 360 \quad \checkmark$$

CHECK FOR UNDERSTANDING

Communicating Mathematics

Read and study the lesson to answer each question.

1. *Tell* what characteristics all of the members of the quadrilateral family share.

2. *Explain* why a parallelogram is not considered a trapezoid.

HANDS-ON MATH

3. *Draw* an isosceles trapezoid. Measure the angles, lengths of the sides, and lengths of the diagonals. Record them on the figure.
   a. Are the diagonals congruent?
   b. Which angles are congruent?

Guided Practice

Sketch each figure. Let Q = quadrilateral, P = parallelogram, R = rectangle, S = square, RH = rhombus, and T = trapezoid. Write all of the letters that describe it inside the figure.

4.

5.

Tell whether each statement is *true* or *false*. Then draw a figure to justify your answer.

6. Every square is a parallelogram.

7. The diagonals of a rectangle are perpendicular.

8. *Algebra* In trapezoid $WXYZ$, $m\angle W = 2a°$, $m\angle X = 40°$, $m\angle Y = 110°$, and $m\angle Z = 70°$. Find the value of a.

EXERCISES

Practice

Sketch each figure. Let Q = quadrilateral, P = parallelogram, R = rectangle, S = square, RH = rhombus, and T = trapezoid. Write all of the letters that describe it inside the figure.

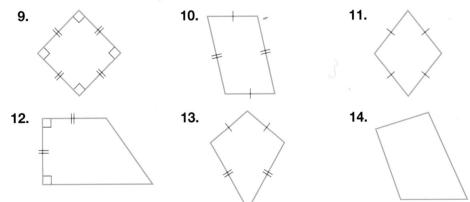

9.

10.

11.

12.

13.

14.

Tell whether each statement is *true* or *false*. Then draw a figure to justify your answer.

15. A rhombus is a square.

16. A trapezoid can have only one right angle.

17. The diagonals of a square cut each other in half.

18. The diagonals of a square cut opposite angles in half.

19. Name all of the quadrilaterals that have congruent diagonals.

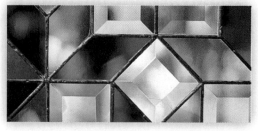

Applications and Problem Solving

20. *Design* The stained glass window at the right uses quadrilaterals in the design. Identify the quadrilaterals.

21. *Algebra* In parallelogram $ABCD$, $m\angle B = (x + 30)°$ and $m\angle D = (2x - 50)°$.
  a. Solve for x.
  b. Find the measure of each angle in parallelogram $ABCD$.

22. *Critical Thinking* I am a quadrilateral. I have one pair of parallel sides. My longest side is twice as long as each of the other sides. My perimeter is 20 centimeters.
  a. What quadrilateral am I?
  b. Draw me and label the lengths of my sides.

For **Extra Practice**, see page 618.

Mixed Review

23. *Algebra* Find the value of x in $\triangle RST$ if $m\angle R = 54°$, $m\angle S = 90°$, and $m\angle T = 3x°$. *(Lesson 5-2)*

24. Solve $d = 28(-12)$. *(Lesson 2-7)*

25. **Standardized Test Practice** Jamilah paid $42.83 for a sweater, which includes $2.78 sales tax. What was the price of the sweater before tax? *(Lesson 1-4)*

  A $40.05   **B** $42.55   **C** $42.61   **D** $45.61   **E** Not Here

---

CHAPTER 5

Mid-Chapter Self Test

In the diagram $n \parallel p$.

1. Find the value of x. *(Lesson 5-1)*

2. Find $m\angle ABC$. *(Lesson 5-1)*

3. Classify $\triangle BCD$ by its angles and by its sides. *(Lesson 5-2)*

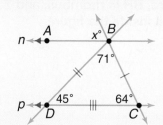

Tell whether each statement is *true* or *false*. Then draw a figure to justify your answer.

4. If the sum of the measures of two angles of a triangle is 90°, it must be an isosceles right triangle. *(Lesson 5-2)*

5. All sides of a parallelogram are congruent. *(Lesson 5-3)*

HANDS-ON LAB

COOPERATIVE LEARNING

5-4A Reflections

A Preview of Lesson 5-4

⬚ geoboards

𝄈 geobands

If you look into a mirror, you see yourself reflected in the mirror. It appears that both you and your reflected image are the same distance from the mirror. When an image is reflected, it is "flipped" over the reflection line. The mirror is the line of reflection.

TRY THIS

Work with a partner.

Step 1 Place two geoboards next to each other. Use a geoband to create the figure shown.

Step 2 The edge where the two boards meet will be the line of reflection. Vertex *A* is 2 pegs left of the line of reflection. Find a peg on the same row that is 2 pegs right of the line of reflection. Place a geoband around this peg.

Step 3 Vertex *B* is 5 pegs left of the line. Find a peg on the same row that is 5 pegs right of the line. Stretch the geoband to fit around this peg.

Step 4 Find the corresponding pegs for vertices *C* and *D*. Complete the figure with the geoband. The figure on the geoboard on the right is the reflection of the image on the geoboard on the left.

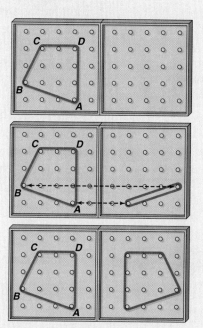

ON YOUR OWN

Use side-by-side geoboards to reflect each figure.

1.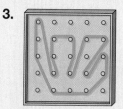

2.

3.

4. Place a third geoboard to the right of the second geoboard and reflect your reflected image from Exercise 1. How does this new image compare to the original figure?

5. **Look Ahead** Compare the original figures and your reflected figures from Exercises 1–3. Are their corresponding sides congruent? Explain the relationship between these two figures.

Symmetry

You'll learn to identify line symmetry and rotational symmetry.

Knowing how to describe and define lines of symmetry can help you design and create logos.

Word Wise

line symmetry
reflection
rotation
rotational symmetry

Many companies use a logo so people can easily identify their products. They often design their logo to have line or rotational symmetry.

A figure has **line symmetry** if it can be folded so that one half of the figure coincides with the other half.

The greeting card company logo, at the right, has line symmetry. The right half is a **reflection** of the left half, and the center line is the *line of reflection*. The line of reflection is also called the *line of symmetry*.

Some figures have more than one line of symmetry. The insurance company logo shown at the right has one vertical, one horizontal, and two diagonal lines of symmetry.

Examples **1** Trace each letter. Draw all lines of symmetry.

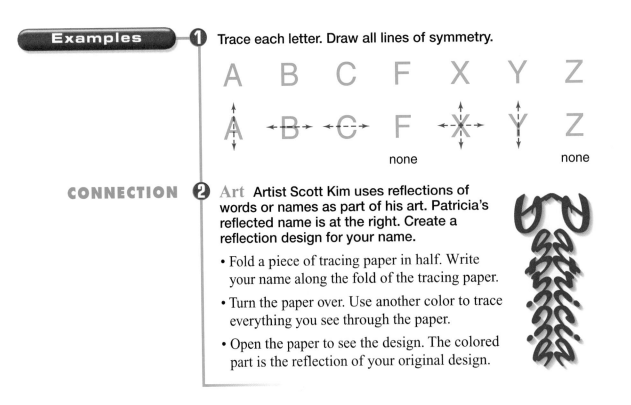

A B C F X Y Z

A B C F X Y Z

none   none

CONNECTION **2** **Art** Artist Scott Kim uses reflections of words or names as part of his art. Patricia's reflected name is at the right. Create a reflection design for your name.

• Fold a piece of tracing paper in half. Write your name along the fold of the tracing paper.

• Turn the paper over. Use another color to trace everything you see through the paper.

• Open the paper to see the design. The colored part is the reflection of your original design.

Another type of symmetry uses **rotations**, or spins. When a compass is turned completely around, it forms a circle. A circle contains 360°. If a figure can be turned less than 360° about its center and still look like the original, then the figure has **rotational symmetry**.

3 **Advertising** Refer to the beginning of the lesson. Trace the insurance company symbol and determine whether it has rotational symmetry.

| original | 90° turn | 180° turn | 270° turn | 360° turn |

When you turn the symbol about its center, it looks like the original. Therefore, it has rotational symmetry.

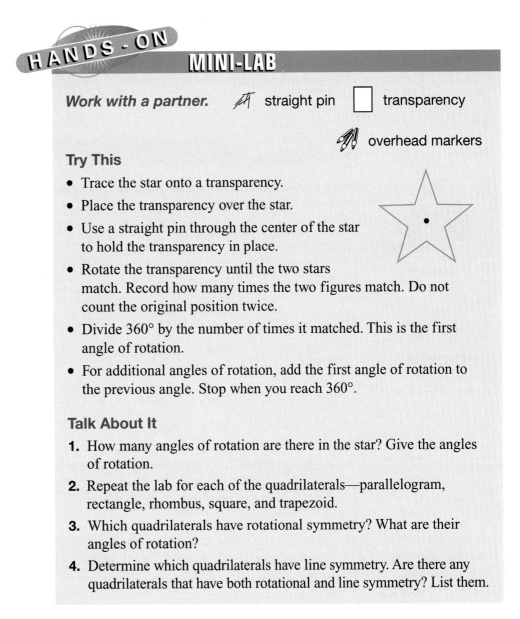

HANDS-ON MINI-LAB

Work with a partner. ✐ straight pin ☐ transparency

✐ overhead markers

Try This

- Trace the star onto a transparency.
- Place the transparency over the star.
- Use a straight pin through the center of the star to hold the transparency in place.
- Rotate the transparency until the two stars match. Record how many times the two figures match. Do not count the original position twice.
- Divide 360° by the number of times it matched. This is the first angle of rotation.
- For additional angles of rotation, add the first angle of rotation to the previous angle. Stop when you reach 360°.

Talk About It

1. How many angles of rotation are there in the star? Give the angles of rotation.
2. Repeat the lab for each of the quadrilaterals—parallelogram, rectangle, rhombus, square, and trapezoid.
3. Which quadrilaterals have rotational symmetry? What are their angles of rotation?
4. Determine which quadrilaterals have line symmetry. Are there any quadrilaterals that have both rotational and line symmetry? List them.

CHECK FOR UNDERSTANDING

Communicating Mathematics

Read and study the lesson to answer each question.

1. *Explain* how a reflection and line symmetry are related.

2. *Draw* a letter of the alphabet that has both line symmetry and rotational symmetry.

HANDS-ON MATH

3. *Draw* several geometric shapes on paper and then draw the same shapes on a transparency. Determine whether the shapes have rotational symmetry. Draw a geometric shape that has an infinite number of angles of rotation.

Guided Practice

Trace each figure. Determine whether the figure has line symmetry. If so, draw the line(s) of reflection.

4. [rectangle]   5. [ellipse]

6. Which of the figures in Exercises 4-5 have rotational symmetry?

7. *Geography* The "Union Jack", a common name for the flag of the United Kingdom, has multiple symmetries. Trace the flag. Draw the lines of symmetry. Name the angles of rotation.

EXERCISES

Practice

Family Activity

Go on a symmetry hunt with family members. First choose a location, such as rooms in your home or an area outside your home. Then, see how many examples of line and rotational symmetry you can find in 15 minutes.

Trace each figure. Determine whether the figure has line symmetry. If so, draw the lines of reflection.

8. [triangle]   9. [hexagon]   10. [parallelogram]

11. [pentagon]   12. [parallelogram]   13. [quadrilateral]

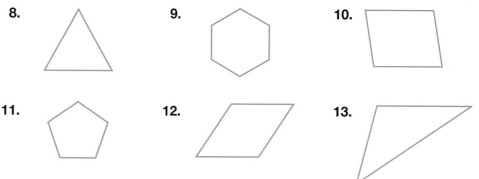

14. Which of the figures in Exercises 8-13 have rotational symmetry?

15. Which types of triangles—*scalene, isosceles, equilateral*—have line symmetry? Which have rotational symmetry?

16. What capital letters of the alphabet produce the same letter after being rotated 180°?

Applications and Problem Solving

17. *Life Science* The diatom is a single-celled marine animal that has multiple symmetries. Trace the diatom.

   a. Draw any lines of symmetry.

   b. Give any angles of rotation.

18. *Sports* Determine whether the emblem for the American Football Conference and the emblem for the National Football Conference display symmetry. If so, what type of symmetry?

19. *Art* Handcrafted tiles were used as decoration in the homes of wealthy merchants and nobles during the Renaissance period in Europe. Usually the artist did not know in advance how the purchaser would use the tile, so the artist made them symmetric. Find all of the lines of symmetry in the tile pattern. Does it also have rotational symmetry? If so, what are the angles of rotation?

20. *Working on the* CHAPTER Project Refer to the shape you found on page 185.

a. Draw all lines of symmetry for your shape.

b. Does your shape have rotational symmetry? If so, what are the angles of rotation?

21. *Critical Thinking* A printer puts four pages on each side of a large sheet of paper when printing books. Decide how and where to print pages 1 through 8 so that the pages will be in the correct order after assembling.

Cut

Fold

Mixed Review

22. *Algebra* In trapezoid *LMNO*, $m\angle L = 120°$, $m\angle M = 120°$, $m\angle N = 2x°$, and $m\angle O = 60°$. Find the value of x. *(Lesson 5-3)*

23. Standardized Test Practice
The bar graph shows the results of a survey on popular bagel flavors. How many flavors were chosen by fewer than 5 students? *(Lesson 4-1)*

A 2 flavors   **B** 3 flavors
C 4 flavors   **D** 5 flavors
E 6 flavors

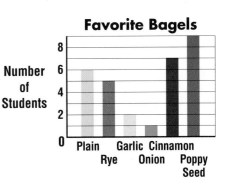

Favorite Bagels

Number of Students

Plain Garlic Cinnamon
Rye Onion Poppy Seed

For **Extra Practice,** see page 618.

24. *Algebra* The product of 9 and *r* is 54. Find the value of *r*. *(Lesson 1-6)*

Congruent Triangles

What you'll learn

You'll learn to verify congruent triangles by using SSS, ASA, and SAS.

When am I ever going to use this?

Knowing how to verify congruence in triangles can help in the construction of buildings.

Word Wise

congruent triangles
corresponding parts

Nature is composed of many shapes and patterns. If we look closely at birds, reptiles, and plants, we can identify specific shapes. The skin pattern on the head of a snake and the wings on a hummingbird are shaped like triangles.

Triangles that have the same size and shape are called **congruent triangles**.

The following triangles are congruent.

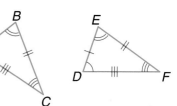

$$\triangle ABC \cong \triangle DEF$$

Congruent Angles	Congruent Sides
$\angle A \cong \angle D$	$\overline{AB} \cong \overline{DE}$
$\angle B \cong \angle E$	$\overline{BC} \cong \overline{EF}$
$\angle C \cong \angle F$	$\overline{AC} \cong \overline{DF}$

Study Hint

Reading Math
Matching arcs on angles show that the angles are congruent.

Parts of congruent triangles that "match" are called **corresponding parts**. For example, in the triangles above, $\angle B$ corresponds to $\angle E$, and $\overline{AC}$ corresponds to $\overline{DF}$.

Two triangles are congruent when all of the corresponding parts are congruent. However, you do not need to know that all six corresponding parts are congruent to know that the triangles are congruent. Sometimes you only need to know that three of the six corresponding parts are congruent.

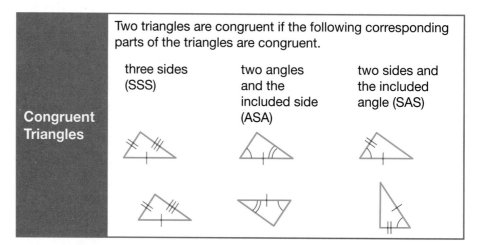

Congruent Triangles — Two triangles are congruent if the following corresponding parts of the triangles are congruent.

three sides (SSS)

two angles and the included side (ASA)

two sides and the included angle (SAS)

Determine whether each pair of triangles is congruent. If so, write a congruence statement and tell why the triangles are congruent.

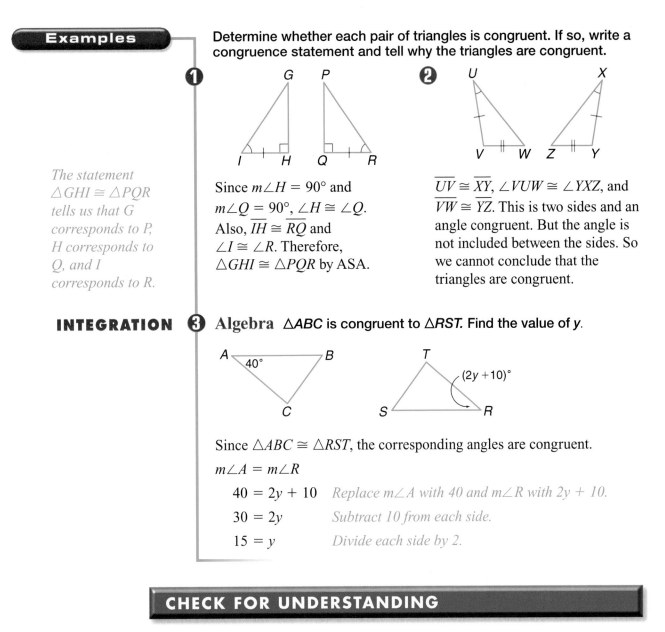

❶

❷

*The statement
△GHI ≅ △PQR
tells us that G
corresponds to P,
H corresponds to
Q, and I
corresponds to R.*

Since $m\angle H = 90°$ and $m\angle Q = 90°$, $\angle H \cong \angle Q$. Also, $\overline{IH} \cong \overline{RQ}$ and $\angle I \cong \angle R$. Therefore, $\triangle GHI \cong \triangle PQR$ by ASA.

$\overline{UV} \cong \overline{XY}$, $\angle VUW \cong \angle YXZ$, and $\overline{VW} \cong \overline{YZ}$. This is two sides and an angle congruent. But the angle is not included between the sides. So we cannot conclude that the triangles are congruent.

INTEGRATION ❸ **Algebra** △ABC is congruent to △RST. Find the value of *y*.

Since $\triangle ABC \cong \triangle RST$, the corresponding angles are congruent.

$m\angle A = m\angle R$

$40 = 2y + 10$ *Replace m∠A with 40 and m∠R with 2y + 10.*

$30 = 2y$ *Subtract 10 from each side.*

$15 = y$ *Divide each side by 2.*

CHECK FOR UNDERSTANDING

Communicating Mathematics

Read and study the lesson to answer each question.

1. *Explain* what must be true for two figures to be congruent.

2. *Describe* an *included* angle. Draw an example.

3. *Write* a paragraph describing three ways to show that two triangles are congruent. Draw a diagram to illustrate each method.

Guided Practice

Determine whether each pair of triangles is congruent. If so, write a congruence statement and tell why the triangles are congruent.

4.

5.

6. Find the value of *x* in the two congruent triangles.

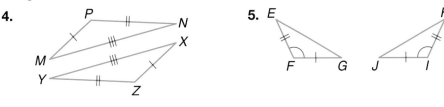

7. *Statistics* The circle graph shows the percent breakdown of all music sales in 1996. If the intersections of each consecutive radius with the circle are connected with a line segment, are any of the triangles formed congruent? If so, tell why the triangles are congruent.

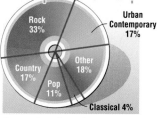

Music Sales

Rock 33%
Urban Contemporary 17%
Country 17%
Pop 11%
Other 18%
Classical 4%

Source: Recording Industry Association of America

EXERCISES

Practice

Determine whether each pair of triangles is congruent. If so, write a congruence statement and tell why the triangles are congruent.

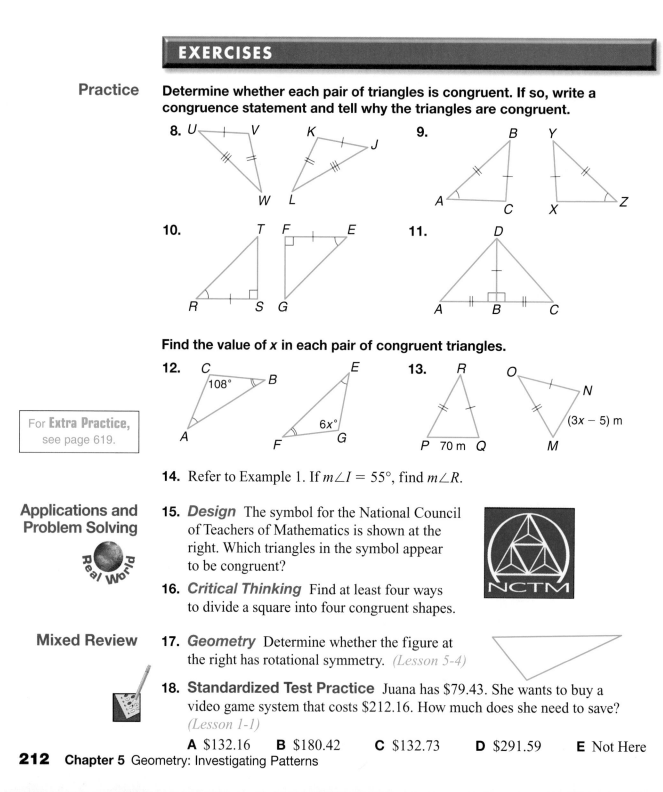

8. U V K J W L

9. B Y A C X Z

10. T F E R S G

11. D A B C

Find the value of *x* in each pair of congruent triangles.

12. C 108° B A E F 6x° G

13. R O N P 70 m Q M (3x − 5) m

For **Extra Practice**, see page 619.

14. Refer to Example 1. If $m\angle I = 55°$, find $m\angle R$.

Applications and Problem Solving

Real World

15. *Design* The symbol for the National Council of Teachers of Mathematics is shown at the right. Which triangles in the symbol appear to be congruent?

NCTM

16. *Critical Thinking* Find at least four ways to divide a square into four congruent shapes.

Mixed Review

17. *Geometry* Determine whether the figure at the right has rotational symmetry. *(Lesson 5-4)*

18. Standardized Test Practice Juana has $79.43. She wants to buy a video game system that costs $212.16. How much does she need to save? *(Lesson 1-1)*

  A $132.16   **B** $180.42   **C** $132.73   **D** $291.59   **E** Not Here

COOPERATIVE LEARNING

5-5B Constructing Congruent Triangles

A Follow-Up of Lesson 5-5

✏ compass

📐 straightedge

Construction engineers and architects often need to copy figures from one part of their drawing to other places. They can use a compass and a straightedge to do this.

TRY THIS

Work with a partner.

Construct a triangle congruent to △ABC.

Step 1 Use a straightedge to draw a line. Put a point on it labeled *G*.

Step 2 Open your compass to the same width as the length of $\overline{AB}$. Put the compass point at *G*. Draw an arc that intersects the line. Label this point of intersection *H*.

Step 3 Open your compass to the same width as the length of $\overline{AC}$. Place your compass point at *G* and draw an arc above the line.

Step 4 Open your compass to the same width as the length of $\overline{BC}$. Place the compass point at *H* and draw an arc above the line so that it intersects the arc that you drew in Step 3. Label this point of intersection *K*.

Step 5 Draw $\overline{HK}$ and $\overline{GK}$.
△ABC ≅ △GHK

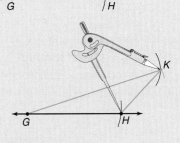

ON YOUR OWN

Trace each triangle. Then construct a triangle congruent to it.

1.

2.

3.

4. Reflect Back How do the corresponding angles of two congruent triangles compare?

HANDS-ON LAB

COOPERATIVE LEARNING

5-6A Dilations

A Preview of Lesson 5-6

3 cardboard strips

3 brass fasteners

map

ruler

Before copy machines were available to enlarge and reduce maps, cartographers, or mapmakers, used pantographs to enlarge or reduce maps and charts. In mathematics, an enlargement or a reduction of an image is called a *dilation*.

TRY THIS

Work with a partner.

Draw an enlargement of a map.

Step 1 Cut one of the strips of cardboard into two congruent pieces.

Step 2 Assemble the pantograph as shown in the diagram using brass fasteners at locations *A, B,* and *C.* Make sure that $\overline{BC} \cong \overline{AP}$ and $\overline{AB} \cong \overline{PC}$.

Step 3 Punch holes in the cardboard strips at locations *P* and *Q* and insert two pencils into the holes.

Step 4 To make an enlargement of your map, hold the pantograph stationary at location *H*. Trace the original map with the pencil at location *P*. The pencil at location *Q* will draw an enlargement of your map.

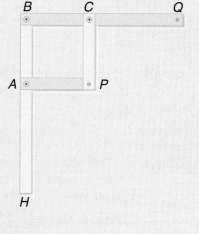

ON YOUR OWN

Trace each drawing. Then use the pantograph to draw an enlargement of it.

1.

2.

3.

4. *Look Ahead* Measure the angles in the drawing in Exercise 1. Then measure the corresponding angles in your enlargement. How do these measures compare?

Similar Triangles

What you'll learn

You'll learn to identify similar triangles.

When am I ever going to use this?

Knowing how to identify similar triangles can help you recognize triangular relationships in architecture.

Word Wise

similar triangles

The Alcoa Office Building in San Francisco, California, uses triangular braces to help secure the building in the event of an earthquake. The triangles marked in the photograph have the same shape, but are different sizes. Triangles that have the same shape, but may differ in size, are called **similar triangles**.

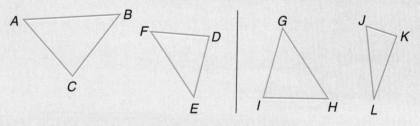

HANDS-ON MINI-LAB

Work with a partner. tracing paper scissors

Try This

Trace each pair of triangles and cut them out.

- Compare the angles in each pair of triangles.
- For each pair of triangles, see if you can place one figure inside the other with equal amount of space separating the corresponding sides in every case.

Talk About It

1. What did you find when you compared the angles?
2. If you can fit one triangle inside another with equal space separating all corresponding sides, the triangles are similar. Name the similar figures.
3. Make a conjecture about triangles with congruent corresponding angles.

Study Hint

Reading Math The symbol ~ means *is similar to.*

The Mini-Lab suggests that if corresponding angles of two triangles are congruent, the triangles are similar. In the figure below, $\triangle ABC$ is similar to $\triangle PQR$. This is written $\triangle ABC \sim \triangle PQR$.

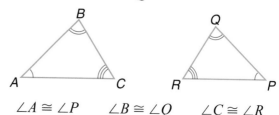

$$\angle A \cong \angle P \qquad \angle B \cong \angle Q \qquad \angle C \cong \angle R$$

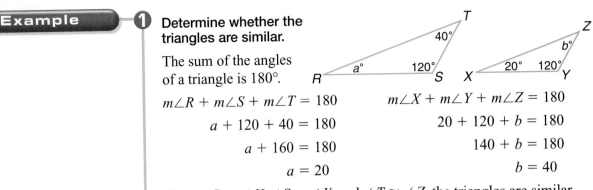

Example ① Determine whether the triangles are similar.

The sum of the angles of a triangle is 180°.

$m\angle R + m\angle S + m\angle T = 180$

$a + 120 + 40 = 180$

$a + 160 = 180$

$a = 20$

$m\angle X + m\angle Y + m\angle Z = 180$

$20 + 120 + b = 180$

$140 + b = 180$

$b = 40$

Since $\angle R \cong \angle X$, $\angle S \cong \angle Y$, and $\angle T \cong \angle Z$, the triangles are similar.

You can use the relationships of similar figures to find missing measures.

Example ②

INTEGRATION

Algebra If $\triangle ABC \sim \triangle UVW$, find the value of *x*.

Since $\triangle ABC \sim \triangle UVW$, the corresponding angles are congruent.

$m\angle A = m\angle U$

$2x - 13 = 85$

$2x = 98$ *Add 13 to each side.*

$x = 49$ *Divide each side by 2.*

CHECK FOR UNDERSTANDING

Communicating Mathematics

Read and study the lesson to answer each question.

1. *Draw* a pair of similar triangles.

2. *Compare and contrast* a pair of triangles that are similar and a pair that are not similar.

3. *You Decide* Helki says that the triangles are similar. Edith disagrees. She says that the unmarked angles may not be congruent. Who is correct? Explain.

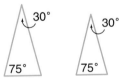

HANDS-ON
MATH

4. *Draw* a triangle. Trace the triangle onto tracing paper. Place the corresponding angles on top of each other.

   a. Are the angles congruent?   **b.** Are the sides congruent?

   c. Are the two triangles congruent?   **d.** Are the two triangles similar?

   e. Explain why two congruent triangles are also similar.

Guided Practice

5. Tell whether the triangles are *congruent*, *similar*, or *neither*. Justify your answer.

6. Find the value of *x* in the pair of similar triangles.

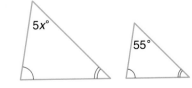

7. The figure at the right is known as the Sierpinski Triangle. How does this figure relate to the congruent and similar triangles?

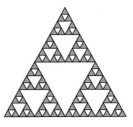

EXERCISES

Practice

Tell whether each pair of triangles is *congruent*, *similar*, or *neither*. Justify your answer.

8.

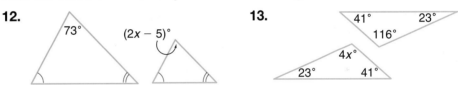

6.5 cm

13 cm

6.5 cm   13 cm

9.

50°

55°

75° 55°

10.

25°   35° 115°

115°

11.

45°

45°

Find the value of *x* in each pair of similar triangles.

12.

73° $(2x - 5)°$

13.

41°   23°
116°

$4x°$

23°   41°

Applications and Problem Solving

Real World

14. *Architecture* The bell tower at Old City Hall in Toronto, Canada, is shown at the left. Describe the four triangles that form the peak of the tower as *congruent, similar,* or *neither.*

15. *Earth Science* Two trees are standing near each other in a park. At any time of day, the rays of sun will hit the ground in the park at the same angle. If both of the trees are perpendicular to the ground, what can you say about the triangles formed by the trees and their shadows as shown below? Explain.

16. **Critical Thinking** Refer to $\triangle RST$. If $\overline{XY} \parallel \overline{ST}$, what is true about $\triangle RST$ and $\triangle RXY$? Explain.

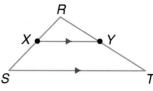

Mixed Review

17. **Geometry** Determine whether the pair of triangles is congruent. If so, write a congruence statement and tell why the triangles are congruent. *(Lesson 5-5)*

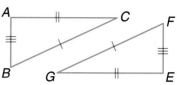

18. **Standardized Test Practice** Vanessa, Seth, Sue, and Jordan are collecting aluminum cans to recycle. Their goal is 100 pounds of cans. The table shows how their collection has progressed so far. *(Lesson 4-1A)*

Name	M	T	W	T	F	Total
Vanessa	1	2	1	2	3	9
Seth	1	3	4	2	0	10
Sue	3	1	3	3	0	10
Jordan	5	3	2	1	4	15

According to the information in the table, how many more pounds of cans do they need to collect to reach their goal?

A 44 lb   **B** 54 lb   **C** 56 lb   **D** 65 lb   **E** 72 lb

For **Extra Practice**, see page 619.

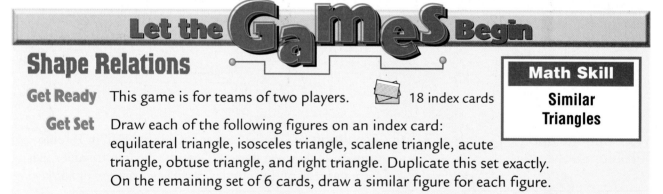

Let the Games Begin

Shape Relations

Math Skill
Similar Triangles

Get Ready   This game is for teams of two players.   18 index cards

Get Set   Draw each of the following figures on an index card: equilateral triangle, isosceles triangle, scalene triangle, acute triangle, obtuse triangle, and right triangle. Duplicate this set exactly. On the remaining set of 6 cards, draw a similar figure for each figure.

Go
- Mix the original set of 6 cards. Then mix the duplicate set and the similar set of cards together.

- Place each stack facedown. One member from each team turns over the top card on each pile. All four players write down all relationships they know between those two figures.

- Then, each team makes one list of the relationships without duplication from their two lists. Each team reads their list. Each team that has a true relationship earns one point for their team. Any team that has listed a true relationship that the other team did not list gets two points instead of one point for that particular relationship.

- Continue these steps until the larger pile has all been used once and the smaller pile has been used twice. The team with the most points wins.

inter NET CONNECTION   Visit www.glencoe.com/sec/math/mac/mathnet for more games.

ADVERTISING

I love art class! I would like to work on advertising for big companies someday!

Elena Baca
GRAPHIC DESIGNER

Elena Baca was born in Peru and raised in Egypt. She has been a graphic designer for more than ten years. Her list of clients includes many major national corporations. She has won many awards in her field, including awards for T-shirt designs and for web page designs.

A person who is interested in becoming a graphic designer needs artistic talent, creativity, and imagination. Math will help a graphic designer see patterns, make scale models, and create designs that are pleasing to the eye. Two out of three people entering the field today have a college degree or some college coursework. Whether a person goes to college or not, a good portfolio containing samples is a necessity.

For more information:
Design International
3748 22nd Street
San Francisco, CA 94114

interNET CONNECTION
www.glencoe.com/sec/
math/mac/mathnet

Your Turn
Design your own wrapping paper. Write a paragraph summarizing how you used math to create your design.

Transformations and M. C. Escher

What you'll learn

You'll learn to create Escher-like drawings using translations and rotations.

When am I ever going to use this?

Knowing how to use translations and rotations can help you analyze how to make products without wasting materials.

Word Wise

tessellation
transformation
translation

A **tessellation** is a tiling made up of copies of the same shape or shapes that fit together without gaps and without overlapping. Brick walls and some quilts are tessellations. Maurits Cornelis Escher (1898–1972) was a Dutch artist whose work used tessellations.

Unique, interlocking, puzzle-like patterns can be made by using **transformations** to change tessellations. Transformations are movements of geometric figures. One transformation commonly used is a slide, or **translation**, of a figure.

HANDS-ON MINI-LAB

Work with a partner.   paper clips   scissors   tape

Try This

- Stack two sheets of paper together. Fold the paper in half twice to make eight layers and then paperclip them together.
- Draw a square on the top layer. Cut out the square so that you have 8 identical squares. Paper clip them together.
- On the left side of the squares, cut out a section. Tape that section onto the right side of all eight squares in the same position. Moving the part from one side to the opposite side is a translation.
- On the top of the squares, cut out a section. Tape that section onto the bottom of all eight squares in the same position.

- Arrange the eight pieces in two rows of four pieces so that they cover a flat surface without overlapping and without leaving gaps.

Talk About It

1. Create another tessellation by repeating the steps above, but use a parallelogram instead of a square. Describe your design.
2. What are the differences between using a square and using a parallelogram to create a tessellation?

1 Make an Escher-like drawing using the modification shown at the right.

Draw a square. Copy the pattern shown. Translate the change shown on the right side to the left side.

Now translate the change shown on the bottom side to the top side to complete the pattern unit.

Repeat this pattern on a tessellation of squares. *It is sometimes helpful to complete one pattern unit, cut it out, and trace it for the other units.*

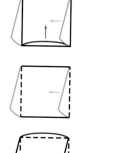

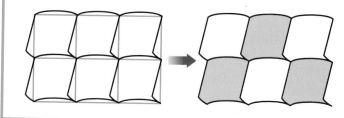

Other tessellations can be modified by using changes with a rotation. Rotations are transformations that involve a turn around a point.

2 Create a tile for tessellating by using a rotation around a vertex of a parallelogram base figure.

base figure   change   change rotated around right vertex   final tile

Now repeat this pattern unit on a tessellation of parallelograms.

CONNECTION

3 **Art** The Escher lithograph entitled *Study of Regular Division of the Plane with Reptiles* was created in 1939. Study the figures carefully.

a. Trace the outside edge of one pattern unit.

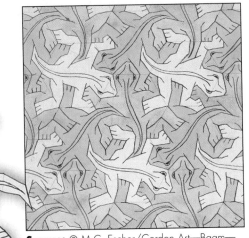

Source: © M.C. Escher/Cordon Art—Baam—Holland Collection Haags Gemeentemuseum—The Hague

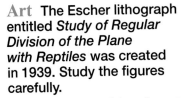

b. What geometric shape do you think was used for this tessellation?

By looking at what was deleted and then added to make the reptiles, you can see that the shape that was used was a six-sided polygon, or a hexagon.

CHECK FOR UNDERSTANDING

Communicating Mathematics

Read and study the lesson to answer each question.

1. *Describe* a tessellation.

2. *Name* five shapes that will tessellate.

HANDS-ON MATH

3. *Trace* the equilateral triangle.
   a. Measure each angle of the triangle.
   b. There are 360° around a point. Divide 360° by the measure of each angle of the triangle.
   c. Use the triangle to create a tessellation using rotation about one of the vertices. How many figures are needed to fill all the space around the vertex?
   d. What is the relationship between the measure of a rotated angle and the number of figures needed to fill the space around the vertex?

Guided Practice

4. Make an Escher-like drawing for the pattern described at the right. Use a tessellation of two rows of three squares as the base.

5. Tell whether the pattern in Exercise 4 involves *translations* or *rotations*.

6. *Hobbies* Name the polygon and the transformation used to produce a sheet of the Pacific 97 stamps at the left.

Practice

Make an Escher-like drawing for each pattern described. For squares, use a tessellation of two rows of three squares as the base. For the triangle, use a tessellation of two rows of five equilateral triangles as the base.

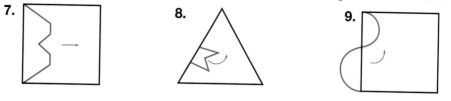

7. 8. 9.

Tell whether each pattern involves translations or rotations.

10. Exercise 7 **11.** Exercise 8 **12.** Exercise 9

13. Create an Escher-like drawing from a pattern of your own.

Applications and Problem Solving

14. *Food Design* Crackers are often baked into shapes that tessellate. Explain why a company would want to design tessellating crackers.

15. *Construction* A floor installer lays a two-by-two tile at the right. Examine each design and determine whether *translations* or *rotations* were used to create each pattern.

a.  **b.**  **c.**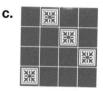

16. *Working on the* **CHAPTER Project** Refer to the shape you found on page 185. Tessellate your shape as described in this lesson. Use your shape to cover a poster board. Color your design.

17. *Critical Thinking* Refer to the tessellation.

a. What polygon was transformed to produce the pattern unit used to create this tessellation?

b. What transformation(s) were used to create the pattern unit? Illustrate.

For **Extra Practice**, see page 619.

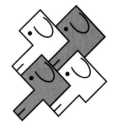

Mixed Review

18. *Geometry* Draw two similar right triangles. *(Lesson 5-6)*

19. **Standardized Test Practice** High temperatures for 12 cities on March 20 were: 40, 72, 74, 35, 58, 64, 40, 67, 40, 75, 68, 51. What is the range of this set of data? *(Lesson 4-5)*

A 75 **B** 51 **C** 40 **D** 11

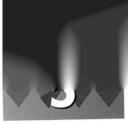

Vocabulary

After completing this chapter, you should be able to define each term, concept, or phrase and give an example or two of each.

Geometry
acute (p. 196)
alternate exterior angles (p. 189)
alternate interior angles (p. 189)
ASA congruence (p. 210)
complementary angles (p. 198)
congruent triangles (p. 210)
corresponding angles (p. 189)
corresponding parts (p. 210)
dilation (p. 214)
equilateral (p. 196)
isosceles (p. 196)
line symmetry (p. 206)
obtuse (p. 196)
parallel lines (p. 188)
parallelogram (p. 201)
perpendicular (p. 197)
polygon (p. 196)
quadrilateral (p. 201)
rectangle (p. 201)

reflection (p. 206)
rhombus (p. 201)
right (p. 196)
rotation (p. 207)
rotational symmetry (p. 207)
SAS congruence (p. 210)
scalene (p. 196)
similar triangles (p. 215)
square (p. 201)
SSS congruence (p. 210)
supplementary angles (p. 190)
tessellation (p. 220)
transformation (p. 220)
translation (p. 220)
transversal (p. 188)
trapezoid (p. 201)
triangle (p. 196)
vertical angles (p. 190)

Problem Solving
use a Venn diagram (p. 194)

Understanding and Using the Vocabulary

Choose the correct term or symbol to complete each sentence.

1. The symbol _____?_____ means *is similar to.*
2. _____?_____ are transformations that involve a turn about a given point.
3. _____?_____ are tilings formed by repeating a shape or shapes.
4. A figure that can be folded exactly in half is said to have _____?_____.
5. _____?_____ angles have measures greater than 90° and less than 180°.
6. The symbol _____?_____ means *is parallel to.*
7. A parallelogram that has four congruent sides is a(n) _____?_____.
8. A(n) _____?_____ is a line that intersects two other lines.

a. rhombus
b. line symmetry
c. transversal
d. acute
e. obtuse
f. rotations
g. tessellations
h. ≅
i. ~
j. ‖

In Your Own Words

9. *Tell* how to determine whether a figure has rotational symmetry.

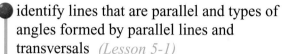

Objectives & Examples

Upon completing this chapter, you should be able to:

● identify lines that are parallel and types of angles formed by parallel lines and transversals *(Lesson 5-1)*

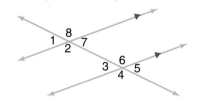

If $m\angle 6 = 135°$, then $m\angle 2 = 135°$.
If $m\angle 5 = 43°$, then $m\angle 6 = 137°$.
If $m\angle 8 = 118°$, then $m\angle 6 = 118°$.

● classify triangles by their angles and their sides and find the measures of missing angles in triangles *(Lesson 5-2)*

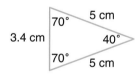

The triangle is acute and isosceles.

● classify quadrilaterals *(Lesson 5-3)*

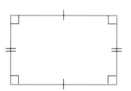

The names that describe this quadrilateral are parallelogram and rectangle.

Review Exercises

Use these exercises to review and prepare for the chapter test.

Refer to the figure below for Exercises 10–12.

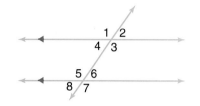

10. Find $m\angle 3$ if $m\angle 5 = 105°$.
11. Find $m\angle 1$ if $m\angle 2 = 53°$.
12. Find $m\angle 1$ if $m\angle 7 = 118°$.

Classify each triangle by its angles and by its sides.

13.   **14.**

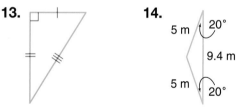

15. In $\triangle RST$, $m\angle R = 33°$ and $m\angle S = 72°$. What is $m\angle T$?

Sketch each figure. Let Q = quadrilateral, P = parallelogram, R = rectangle, S = square, RH = rhombus, and T = trapezoid. Write all of the letters that describe it inside the figure.

16.   **17.**

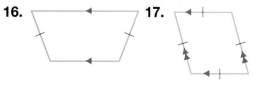

Objectives & Examples

Review Exercises

identify line symmetry and rotational symmetry *(Lesson 5-4)*

Draw all lines of symmetry for a square.

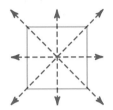

The figure also has rotational symmetry.

Trace each figure. Determine whether the figure has line symmetry. If so, draw the lines of reflection.

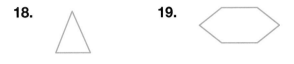

18. **19.**

20. Does the figure in Exercise 18 have rotational symmetry?

21. Does the figure in Exercise 19 have rotational symmetry?

verify congruent triangles by using SSS, ASA, and SAS *(Lesson 5-5)*

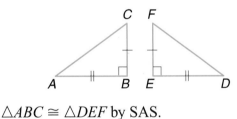

$\triangle ABC \cong \triangle DEF$ by SAS.

Determine whether each pair of triangles is congruent. If so, write a congruence statement and tell why the triangles are congruent.

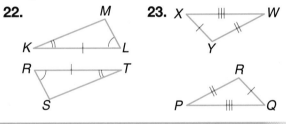

22. **23.**

identify similar triangles *(Lesson 5-6)*

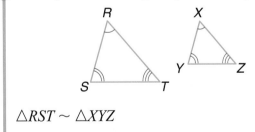

$\triangle RST \sim \triangle XYZ$

Tell whether each pair of triangles is *congruent, similar,* or *neither.* Justify your answer.

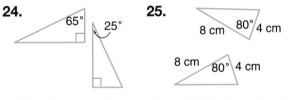

24. **25.**

create Escher-like drawings using translations and rotations *(Lesson 5-7)*

An example of a tessellation of squares is a floor covered by tiles. By modifying these squares, an Escher-like drawing can be made.

Make an Escher-like drawing for each pattern described.

26. **27.**

Applications & Problem Solving

28. *Traffic* The shape of a ROAD NARROWS sign is a parallelogram. How many pairs of sides of a ROAD NARROWS sign are parallel? *(Lesson 5-1)*

29. *Use a Venn Diagram* Oliver Middle School offers different activities before and after school. Twenty-one students have signed up for aerobics classes. Thirty students have signed up for basketball. Six students have signed up for both activities. What is the total number of students who have signed up for these two activities? *(Lesson 5-2A)*

30. *Construction* A 12-foot ladder is leaning against a house. The base of the ladder is 4 feet from the house. Classify the triangle formed by the house, the ground, and the ladder by its angles. *(Lesson 5-2)*

31. *Architecture* Frank Lloyd Wright was one of the most influential American architects. Identify the types of quadrilaterals Wright used in the roof of the home shown below. *(Lesson 5-3)*

Alternative Assessment

Open Ended

Suppose you are designing wallpaper for a friend. Your friend wants the design to contain a figure composed of any combination of two or more quadrilaterals or triangles. Your friend also wants the figure to have at least four lines of symmetry. Draw a figure to meet these requirements.

Your friend wants to know if your figure will tessellate. If it will, draw the tessellation. If it won't, explain why.

Completing the CHAPTER Project

Use the following checklist to make sure your school logo is complete.

☑ The original shape is shown with all of its lines of symmetry.

☑ Information about the rotational symmetry of the original shape is accurate.

☑ The logo is used to neatly fill the poster board.

PORTFOLIO   Select one of the words you learned in this chapter and place the word and its definition in your portfolio. Attach a note to it explaining why you selected it.

A practice test for Chapter 5 is provided on page 651.

Standardized Test Practice

Assessing Knowledge & Skills

Section One: Multiple Choice

There are nine multiple-choice questions in this section. Choose the best answer. If a correct answer is *not here,* choose the letter for **Not Here.**

1. The line plot shows the amount of allowance received by 30 students. How many students received $10 or less?

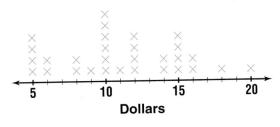

Dollars

A 15
B 10
C 5
D 3

2. How many lines of symmetry does an equilateral triangle have?

F 0
G 1
H 3
J 6

3. How many pairs of congruent triangles are formed by the diagonals of rectangle *ABCD*?

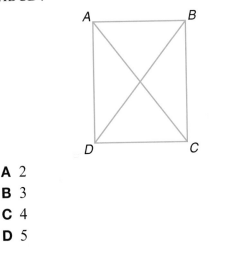

A 2
B 3
C 4
D 5

Please note that Questions 4–9 have five answer choices.

4. Mindy earns $7 each weekday for her paper route. She earns between $10 and $15 on Saturday and on Sunday. What is a reasonable amount for Mindy to earn during a seven-day week?

F less than $45
G between $45 and $55
H between $55 and $65
J between $65 and $75
K more than $75

5. The Venn diagram shows the relationship of five sets of eighth grade girls. Each set and the intersections shown have at least one member in it.

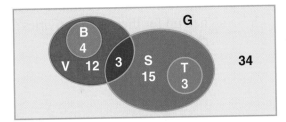

G = set of all eighth grade girls
B = set of all basketball players
V = set of all volleyball players
S = set of all softball players
T = set of all tennis players

Which is a valid conclusion concerning these sets?

A There is no volleyball player on the basketball team.
B There is no tennis player on the softball team.
C All softball players play volleyball.
D Some girls play both softball and volleyball.
E All tennis players play basketball.

6. The graph shows how the world's land is distributed. What percent of the world's land is *not* in the North America or South America?

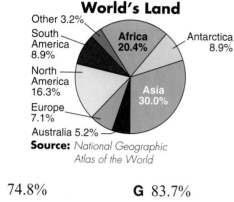

World's Land

Other 3.2%
South America 8.9%
North America 16.3%
Europe 7.1%
Australia 5.2%
Africa 20.4%
Antarctica 8.9%
Asia 30.0%

Source: *National Geographic Atlas of the World*

F 74.8%   **G** 83.7%
H 30.1%   **J** 25.2%
K 16.3%

7. Thomas bought a gym bag that was reduced 30%. If the bag originally cost $20, what was the final cost of the bag including 6% sales tax?
A $14.00   **B** $14.84
C $16.48   **D** $20.00
E $20.84

8. Moses wants to paint a wall that measures 12 feet by 15 feet. There are two windows on the wall that each measure 2 feet by 3 feet. How much surface area will be covered with paint?
F 180 ft²   **G** 178 ft²
H 168 ft²   **J** 152 ft²
K 150 ft²

9. During the day, the temperature rose 6°F every two hours. At 8 A.M., the temperature was −25°F. What was the temperature at 2 P.M.?
A 11°F   **B** −7°F
C −43°F   **D** −13°F
E Not Here

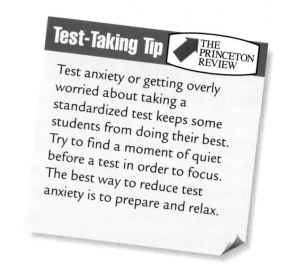

Test-Taking Tip | THE PRINCETON REVIEW

Test anxiety or getting overly worried about taking a standardized test keeps some students from doing their best. Try to find a moment of quiet before a test in order to focus. The best way to reduce test anxiety is to prepare and relax.

Section Two: Free Response

This section contains seven questions for which you will provide short answers. Write your answers on your paper.

10. Solve $5x + 32 < 42$. Show the solution on a number line.

11. Simplify $(3a + b) + (2a + 4b)$.

12. Solve $a = 20 \cdot (-2)$.

13. Find the absolute value of -28.

14. Find the value of x.

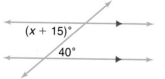

$(x + 15)°$
$40°$

15. Solve $b = -56 \div (-7)$.

16. Which point on the graph is in the third quadrant?

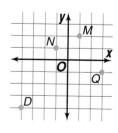

interNET CONNECTION Test Practice **For additional test practice questions, visit:**

www.glencoe.com/sec/math/mac/mathnet

Exploring Number Patterns

 you'll
learn in Chapter 6

- to solve problems by making a list,
- to find the greatest common factor and the least common multiple of two or more numbers,
- to identify, simplify, and compare rational numbers,
- to find the probability of a simple event, and
- to express numbers in scientific notation.

CHAPTER Project

TAKE ME OUT TO THE BALL GAME

In this project, you will write a news article on the record of a professional sports team for the past several years. You will include a comparison of the seasons in the article.

Getting Started

- Decide which professional league you would like to study. Two possibilities are a women's professional basketball association or a professional football league.

- Choose a sports team from your chosen league.

- Research your league and your team. For each of the past several years, find the total number of regular season games played, the number of games won, and team that won the league championship.

Technology Tips

- Use an **electronic almanac** to do your research.

- Use a **spreadsheet** to organize data and make calculations.

- Use a **word processor** to write your news article.

 **Data Update** For up-to-date information on professional sports, visit:

www.glencoe.com/sec/math/mac/mathnet

Working on the Project

You can use what you'll learn in Chapter 6 to help you write a news article about your chosen team.

Page	Exercise
248	33
252	41
264	32
271	Alternative Assessment

Divisibility Patterns

What you'll learn

You will learn to use divisibility rules for 2, 3, 4, 5, 6, 8, 9, and 10.

When am I ever going to use this?

Knowing how to use divisibility rules can help you determine which years are leap years.

Word Wise
divisible

LOOK BACK
You can refer to Lesson 1-2 to review factors.

Fina is the drum major for the Madison High School Band. The band has 144 marching members. How many ways can Fina arrange the band members so that they are marching in a rectangular formation? *This problem will be solved in Example 1.*

To answer this question, Fina must find all of the factor pairs for 144. If a number is a factor of a given number, we can also say the given number is **divisible** by the factor.

The following rules can help determine whether a number is divisible by 2, 3, 4, 5, 6, 8, 9, or 10.

A number is divisible by:

- 2 if the ones digit is divisible by 2.
- 3 if the sum of the digits is divisible by 3.
- 4 if the number formed by the last two digits is divisible by 4.
- 5 if the ones digit is 0 or 5.
- 6 if the number is divisible by 2 *and* 3.
- 8 if the number formed by the last three digits is divisible by 8.
- 9 if the sum of the digits is divisible by 9.
- 10 if the ones digit is 0.

Example
Real World APPLICATION

1 **School** Refer to beginning of the lesson. How many ways can Fina arrange the band members?

2: The ones digit, 4, is divisible by 2. So 144 is divisible by 2.

3: $1 + 4 + 4$ or 9 is divisible by 3. So 144 is divisible by 3.

4: 44 is divisible by 4. So 144 is divisible by 4.

5: The last digit is 4. So 144 is *not* divisible by 5.

6: 144 is divisible by 2 and 3. So 144 is divisible by 6.

8: 144 is divisible by 8.

9: $1 + 4 + 4$ or 9 is divisible by 9. So 144 is divisible by 9.

10: The last digit is 4. So 144 is *not* divisible by 10.

Fina also knows that $1 \times 144 = 144$ and $12 \times 12 = 144$.

The possible formations are 1 by 144, 2 by 72, 3 by 48, 4 by 36, 6 by 24, 8 by 18, 9 by 16, and 12 by 12.

How do you know these are all of the formations?

2 Determine whether 2,416 is divisible by 2, 3, 4, 5, 6, 8, 9, or 10.

2: Yes, 6 is divisible by 2.

3: No, 2 + 4 + 1 + 6 or 13 is *not* divisible by 3.

4: Yes, 16 is divisible by 4.

5: No, the ones digit is *not* a 0 or 5.

6: No, 2,416 is divisible by 2, but *not* by 3.

8: Yes, 416 is divisible by 8.

9: No, 2 + 4 + 1 + 6 or 13 is *not* divisible by 9.

10: No, the ones digit is *not* 0.

Therefore, 2,416 is divisible by 2, 4, and 8, but not by 3, 5, 6, 9 or 10.

INTEGRATION **3** **Geometry** The area of a rectangle is 4,329 square inches. The length and width of the rectangle are whole numbers. Find two possible dimensions of the rectangle.

The sum of the digits of 4,329 is 4 + 3 + 2 + 9, or 18, which is divisible by 3 and 9. So 4,329 is divisible by 3 and 9.

4329 ÷ 3 = *1443*   4329 ÷ 9 = *481*

Two possible dimensions of the rectangle are 3 inches by 1,443 inches and 9 inches by 481 inches.

CHECK FOR UNDERSTANDING

Communicating Mathematics

Read and study the lesson to answer each question.

1. *Explain* what is meant by the statement "*a* is divisible by *b*." Assume that *a* and *b* are whole numbers.

2. *Tell* why an odd number cannot be divisible by 6.

3. *You Decide* Alonzo says that any number that is divisible by 9 is also divisible by 3. Pamela disagrees. Who is correct? Explain.

Guided Practice

Determine whether each number is divisible by 2, 3, 4, 5, 6, 8, 9, or 10.

4. 48   5. 153   6. 2,470

7. Is 6 a factor of 198?   8. Is 795 divisible by 10?

9. Use mental math to find a four-digit number that is divisible by 3 and 10.

10. Find two ways to write 435 as a product of two whole number factors.

11. *School* In preparation for a pep assembly, the students are placing 40 chairs on the stage for the football team. If the chairs are to be placed in a rectangular array, what are the possible arrangements for the chairs?

Practice

Determine whether each number is divisible by 2, 3, 4, 5, 6, 8, 9, or 10.

12. 56   **13.** 165   **14.** 323   **15.** 918

16. 1,700   **17.** 2,865   **18.** 12,357   **19.** 16,084

20. Is 3 a factor of 777?   **21.** Is 5 a factor of 232?

22. Is 989 divisible by 9?   **23.** Is 2,348 divisible by 4?

24. Is 16,454 divisible by 8?   **25.** Is 2 a factor of 88,096?

Use mental math to find a number that satisfies the given conditions.

26. a three-digit number divisible by 3, 6, and 9

27. a four-digit number divisible by 2 but not by 6

28. a five-digit number *not* divisible by 2, 5 or 9

29. Find three ways to write 84 as a product of two whole number factors.

30. Find two different pairs of whole number factors of 14,121.

31. Find five different pairs of whole number factors of 17,184.

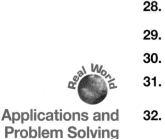

Applications and Problem Solving

32. *Calendar* Any year that is divisible by 4 and does *not* end in 00 is a leap year. Years ending in 00 are leap years only if they are divisible by 400.

     a. Was 1492 a leap year?

     b. Will 2015 be a leap year?

     c. Were you born in a leap year?

33. *Civics* Each star in the U.S. flag represents a state. If another state joins the Union, could the stars be arranged in a rectangular array? Explain.

34. *Critical Thinking* Write a rule to determine whether a number is divisible by 15.

Mixed Review

35. **Standardized Test Practice** Which of the following shapes will *not* tessellate? *(Lesson 5-7)*

     A pentagon

     B square

     C equilateral triangle

     D regular hexagon

36. *Statistics* Refer to the bar graph. How many times more students watch 0 to 5 hours than those who watch 15 to 17 hours? *(Lesson 4-1)*

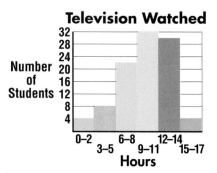

Television Watched

37. Express the ratio $\frac{\$2.88}{9 \text{ ounces}}$ as a unit rate. *(Lesson 3-1)*

For **Extra Practice,** see page 620.

38. Solve $p = 85 + (-47)$. *(Lesson 2-3)*

6-2 Prime Factorization

What you'll learn

You'll learn to find the prime factorization of a composite number.

When am I ever going to use this?

Knowing how to find the prime factorization of a number can help you make secret codes.

Word Wise

prime number
composite number
prime factorization
factor tree
Fundamental Theorem of Arithmetic

For over 2,000 years, mathematicians have been fascinated by numbers with exactly two factors. Using computers, today's mathematicians have discovered very large numbers with only two factors.

In the following Mini-Lab, you will investigate the number of factors for various numbers.

MINI-LAB

Work with a partner.

grid paper

Draw a rectangle with 2 square units on the grid paper. A 1-by-2 rectangle is the only way to draw such a rectangle using whole grid squares.

Try This

1. Draw as many rectangles as possible with an area of 3 square units.
2. Draw as many rectangles as possible with an area of 4 square units.
3. Continue to draw all possible rectangles for areas from 5 square units to 10 square units.

Talk About It

4. Which areas had only one rectangle?
5. Which areas had more than one rectangle?

Did you know In 1994, Paul Gage and David Slowinski used a supercomputer to find a prime number with 258,716 digits.

When a whole number greater than 1 has *exactly* two factors, 1 and itself, it is called a **prime number**. From the results of the Mini-Lab, you can see that the numbers 2, 3, 5, and 7 are prime numbers.

When a whole number greater than 1 has more than two factors, it is called a **composite number**. From the results of the Mini-Lab, you can see that the numbers 4, 6, 8, 9, and 10 are composite numbers.

The numbers 0 and 1 are *neither* prime *nor* composite. Notice that 0 has an endless number of factors and 1 has only one factor, itself.

A composite number can be expressed as a product of prime numbers. To find the prime factors of any composite number, begin by expressing the number as a product of two factors. Then continue to factor until all the factors are prime. When the number is expressed as a product of factors that are all prime, the expression is called the **prime factorization** of the number.

Each diagram shows a different way to find the prime factorization of 60. These diagrams are called **factor trees**. Use the divisibility rules and mental math to factor 60.

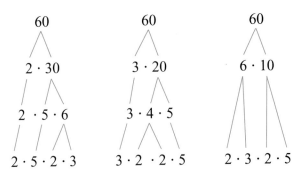

Notice that the bottom row of "branches" in each factor tree is the same except for the order in which the factors are written. Every number has a unique set of prime factors. This property of numbers is called the **Fundamental Theorem of Arithmetic**.

You can use a calculator and the divisibility rules to find the prime factorization.

Example ① Find the prime factorization of 1,530.

Divide by prime factors until the quotient is prime.

prime factors
$\downarrow$

$1{,}530 \div 2 = 765$ *0 is divisible by 2. So, 1,530 is divisible by 2.*

$765 \div 5 = 153$ *The last digit is 5. So, 765 is divisible by 5.*

$153 \div 3 = 51$ *1+5+3 or 9 is divisible by 3. 153 is divisible by 3.*

$51 \div 3 = 17$ *17 is a prime number.*

$1{,}530 = 2 \cdot 3^2 \cdot 5 \cdot 17$

Study Hint

Technology When you use a calculator to find prime factors, it is helpful to record them on paper as you find each one.

For centuries, mathematicians have tried to find an expression for calculating every prime number. While an expression may yield some prime numbers, none has been found to yield all prime numbers.

Example ② **INTEGRATION**

Algebra Show that the expression $n^2 - n + 41$ yields prime numbers for $n = 3$, 5, and 7.

$n = 3$: $n^2 - n + 41 = 3^2 - 3 + 41$ *Replace n with 3.*
$= 47$ *47 has only two factors, 1 and 47. So 47 is a prime number.*

$n = 5$: $n^2 - n + 41 = 5^2 - 5 + 41$ *Replace n with 5.*
$= 61$ *61 has only two factors, 1 and 61. So 61 is a prime number.*

$n = 7$: $n^2 - n + 41 = 7^2 - 7 + 41$ *Replace n with 7.*
$= 83$ *83 has only two factors, 1 and 83. So 83 is a prime number.*

Communicating Mathematics

Read and study the lesson to answer each question.

1. *Compare and contrast* prime and composite numbers.

2. *Draw* two different factor trees to find the prime factorization of 24.

HANDS-ON MATH

3. *Draw* as many rectangles as possible with areas of 11 square units, 12 square units, 13 square units, 14 square units, and 15 square units.
   a. Which areas have only one rectangle?
   b. Which areas have more than one rectangle?

Guided Practice

Determine whether each number is *prime, composite,* or *neither.*

4. 45   5. 23   6. 1

Find the prime factorization of each number.

7. 49   8. 32   9. 225

10. *Algebra* Is the value of $2a + 3b$ prime or composite if $a = 11$ and $b = 7$?

Practice

Determine whether each number is *prime, composite,* or *neither.*

11. 13	**12.** 27	**13.** 96	**14.** 37
15. 0	**16.** 177	**17.** 233	**18.** 507

Find the prime factorization of each number.

19. 25	**20.** 36	**21.** 75	**22.** 80
23. 117	**24.** 72	**25.** 4,900	**26.** 3,825

27. What is the least prime number that is a factor of 1,365?

28. What is the greatest prime number that is less than 60?

29. What is the least prime number that is greater than 60?

Applications and Problem Solving

30. *Number Theory* Primes that differ by two are called *twin primes.*
   a. Give three examples of twin primes.
   b. Are there any primes that differ by 1? Explain.

31. *Patterns* Consider the number pattern.
   a. What are the seventh, eighth, and ninth values in the pattern?

	1st	2nd	3rd	4th	5th	6th	7th	8th	9th
Pattern	2	4	8	16	32	64	?	?	?

   b. Write an algebraic expression you could use to find the *n*th value of the pattern. (*Hint:* Use the prime factorization.)

32. *Algebra* Find a whole number, *n*, for which the expression $n^2 - n + 41$ is not prime.

33. *Critical Thinking* Find the prime factorization of the perfect squares 9, 36, 100, and 144.

   a. What characteristics do the prime factorizations have?

   b. Is this true for all perfect squares? Explain.

34. **Standardized Test Practice** Ogima must arrange the chairs in the cafeteria in a rectangular pattern. If there are 486 chairs, in which of the following ways can they be arranged? *(Lesson 6-1)*

   A 27 rows of 19 chairs   **B** 18 rows of 26 chairs

   C 27 rows of 18 chairs   **D** 18 rows of 31 chairs

35. *Geometry* Tell whether the pair of figures are *congruent*, *similar*, or *neither*. Explain. *(Lesson 5-6)*

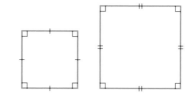

36. Express 38.5% as a decimal and as a fraction in simplest form. *(Lesson 3-4)*

37. *Algebra* Write an inequality for *Four times a number is less than 20.* Then solve the inequality. *(Lesson 1-9)*

For **Extra Practice,** see page 620.

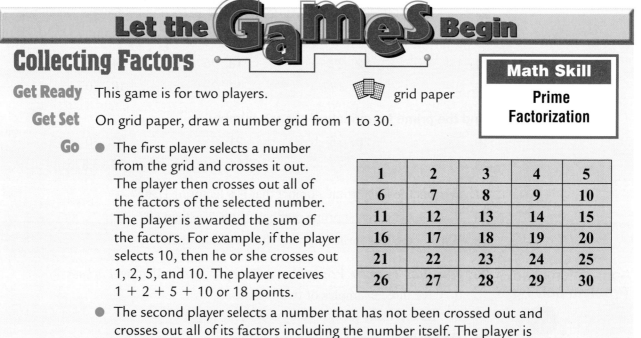

Let the Games Begin

Collecting Factors

Math Skill

Prime Factorization

Get Ready This game is for two players.   grid paper

Get Set On grid paper, draw a number grid from 1 to 30.

Go ● The first player selects a number from the grid and crosses it out. The player then crosses out all of the factors of the selected number. The player is awarded the sum of the factors. For example, if the player selects 10, then he or she crosses out 1, 2, 5, and 10. The player receives 1 + 2 + 5 + 10 or 18 points.

1	2	3	4	5
6	7	8	9	10
11	12	13	14	15
16	17	18	19	20
21	22	23	24	25
26	27	28	29	30

● The second player selects a number that has not been crossed out and crosses out all of its factors including the number itself. The player is awarded the sum of the factors that had not already been crossed out.

● If a player selects a number with no factors available other than itself, the player receives no points and loses his or her turn.

● The players continue to take turns choosing numbers until the numbers remaining have no other factors available.

● The player with the most points wins.

inter NET CONNECTION Visit www.glencoe.com/sec/math/mac/mathnet for more games.

HANDS-ON LAB

COOPERATIVE LEARNING

6-2B Basketball Factors

A Follow–Up of Lesson 6-2

colored pencils

tracing paper

Ms. Salgado coaches the eighth grade girls' basketball team. She is also a mathematics teacher. Sometimes she uses drills at basketball practice that reinforce concepts such as factors.

In one such drill, the ten players stand in a circle as indicated at the right. The player with the ball (B) calls out a number, such as 4, and passes the ball to the fourth player to her right. The person catching the ball continues the pattern by passing the ball to the fourth person to her right and so on.

Sometimes the player who starts the drill calls out a number that results in every player around the circle touching the ball once before it returns to her. When this happens, the team is given a water break.

TRY THIS

Work with a partner.

Step 1 Use tracing paper and colored pencils to record the different paths the ball takes when each of the numbers 1 through 9 is called.

Step 2 Copy and complete the table to determine when the team gets a water break.

Number of Players	Number Called	Water Break?
10	1	
10	2	
10	3	
10	4	
10	5	
10	6	
10	7	
10	8	
10	9	

ON YOUR OWN

1. What numbers can be called so the team can take a break?

2. What is the relationship between the factors of 10 and the factors of the numbers in the answer to Exercise 1?

3. What is the relationship between the factors of 10 and the factors of other numbers in the table?

4. *Reflect Back* Suppose there are 12 girls in the circle. Use factoring to determine which numbers will allow the girls to have a water break.

PROBLEM SOLVING

6-3A Make a List

A Preview of Lesson 6-3

Atepa and Curtis are helping to make and sell pizzas to raise money for their class trip. The students can add pepperoni, mushrooms, and/or green peppers to the basic cheese pizza. Atepa and Curtis are making posters to advertise the sale. Let's listen in!

Atepa

We can take orders faster if we list all of the different pizzas we can make.

Okay, let's start the list with a plain cheese pizza.

Then we can add the pizzas with just one topping to our list. Add pizza with pepperoni, pizza with mushrooms, and pizza with green peppers.

Curtis

We must also add pizzas with two or three toppings to our list.

Good! If that's all the different pizzas, we can tell how many choices there are on the posters.

THINK ABOUT IT

Work with a partner.

1. *Analyze* Atepa's and Curtis' thinking. Have they included all of the types of pizzas they can make?

2. *Make a list* of all the types of pizzas the students can make. How many different pizzas can Atepa and Curtis advertise on the posters?

3. Apply the **make a list** strategy to solve the following problem.

   Trish wants to buy a cookie from a vending machine. The cookie costs 45¢. If Trish uses exact change, how many different combinations of nickels, dimes, and quarters can she use?

For **Extra Practice,** see page 620.

ON YOUR OWN

4. The fourth step of the 4-step plan for problem solving tells you to *examine* the solution. **Tell** what things you should look for when examining a list used to solve a problem.

5. *Write a Problem* which you could solve using the make a list strategy.

6. *Look Ahead* Explain how the make a list strategy could be used to find what numbers are factors of both 12 and 18.

MIXED PROBLEM SOLVING

STRATEGIES

Look for a pattern.
Solve a simpler problem.
Act it out.
Guess and check.
Draw a diagram.
Make a chart.
Work backward.

Solve. Use any strategy.

7. *Entertainment* Paper cups come in packages of 40 or 75. Dimas needs 350 paper cups for the school party. How many packages of each size should he buy?

8. *Sports* In the World Series, two teams play each other until one team wins 4 games.

   a. What is the least number of games needed to determine a winner of the World Series?

   b. What is the greatest number of games needed to determine a winner?

   c. How many different ways can a team win the World Series in six games or less? (*Hint:* The team that wins the series must win the last game.)

9. *Camp* The camp counselor lists 21 chores on separate pieces of paper and places them in a basket. The counselor takes one piece of paper, and each camper takes one as it is passed around the circle. There is one piece of paper left when the basket returns to the counselor. How many people could be in the circle if the basket goes around the circle more than once?

10. *Critical Thinking* Copy and complete the factor tree.

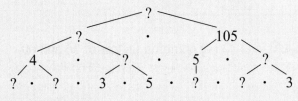

11. *Technology* Seven line segments are used to make the digits 0 to 9 on a digital clock. (The number 1 is made using the line segments on the right.)

   a. In forming these digits, which line segment is used most often?

   b. Which line segment is used the least?

12. *Hobbies* Gail says to Callie, "If you give me one of your football cards, I will have twice as many football cards as you have." Callie answers, "If you give me one of your cards, we will have the same numbers of cards." How many cards do each of the girls have?

13. *Standardized Test Practice* How many factors does 21 have?

   A exactly two factors

   B exactly three factors

   C exactly four factors

   D exactly five factors

Greatest Common Factor

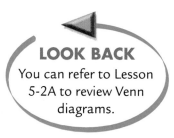

What you'll learn

You'll learn to find the greatest common factor of two or more numbers.

When am I ever going to use this?

Knowing how to find the greatest common factor can help you solve problems involving home improvements.

Word Wise

greatest common factor (GCF)

Different actors and actresses star in various movies. The Venn diagram shows the stars of Movie A and Movie B.

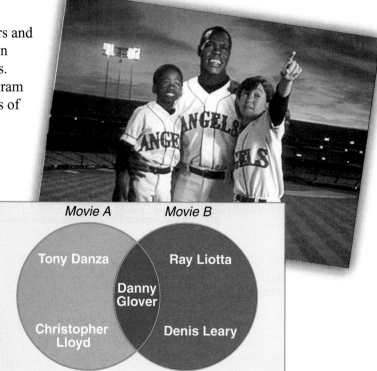

Danny Glover, Tony Danza, and Christopher Lloyd starred in Movie A. Danny Glover, Ray Liotta, and Denis Leary starred in Movie B. Notice that Danny Glover starred in both movies.

A Venn diagram is a visual way to examine common elements. Numbers may have common factors. The greatest of the factors common to two or more numbers is called the **greatest common factor (GCF)** of the numbers.

LOOK BACK

You can refer to Lesson 5-2A to review Venn diagrams.

Example

1 Find the greatest common factor of 36 and 60.

First, make a list of the factors of 36 and 60.

Factors of 36: 1, 2, 3, 4, 6, 9, 12, 18, 36
Factors of 60: 1, 2, 3, 4, 5, 6, 10, 12, 15, 20, 30, 60

Then, use a Venn diagram to summarize the information.

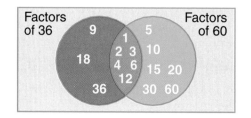

Notice that 1, 2, 3, 4, 6, and 12 are common factors of 36 and 60. The GCF is 12.

You can also use prime factorization to find the GCF.

② **Use prime factorization to find the GCF of 90, 54, and 81.**

Use a factor tree to find the prime factorization of each number.

$$90 \qquad\qquad 54 \qquad\qquad 81$$

$$9 \cdot 10 \qquad 6 \cdot 9 \qquad 9 \cdot 9$$

$$3 \cdot 3 \cdot 2 \cdot 5 \quad 2 \cdot 3 \cdot 3 \cdot 3 \quad 3 \cdot 3 \cdot 3 \cdot 3$$

90, 54, and 81 have 3 and 3 as common prime factors. The product of these prime factors, $3 \cdot 3$ or 9, is the GCF of 90, 54, and 81.

APPLICATION

③ **Interior Design** Toshiro wants to resurface the top of a table with ceramic tiles. The table is 30 inches long and 24 inches wide. What is the largest square tile that he can use without having to use any partial squares?

Explore You know the dimensions of the table. You want to know the largest square tile that he can use without having to use partial squares.

Plan The length of the side of the square must be a factor of both 30 and 24. You want the largest square, so you need to find the greatest common factor of these numbers.

Solve Write each number as a product of prime factors.

$30 = 2 \cdot 3 \cdot 5$
$24 = 2 \cdot 2 \cdot 2 \cdot 3$

Thus, the GCF is $2 \cdot 3$ or 6. The tile should be 6 inches on each side.

Examine Write all of the factors of 30 and 24 using a Venn diagram.

Factors of 30: 5, 10, 15, 30
1, 2, 3, 6
Factors of 24: 4, 8, 12, 24

The greatest common factor of 30 and 24 is 6. The answer is correct.

CHECK FOR UNDERSTANDING

Communicating Mathematics

Read and study the lesson to answer each question.

1. *Name* the common factors of 40 and 60. What is the greatest common factor of these numbers?

2. *Write* two numbers whose GCF is 16.

3. *Draw* a Venn diagram showing the factors of 30 and 42.
  a. Name the GCF of 30 and 42.
  b. Verify your answer by using factor trees to find the prime factorization of each number.

Guided Practice

Find the GCF of each set of numbers.

4. 45, 20   **5.** 27, 54   **6.** 26, 34, 64   **7.** 45, 105, 75

8. *Crafts* A quilt pattern calls for congruent square pieces. Some of the fabrics chosen for the quilt come in a width of 48 inches, and others come in a width of 60 inches. If no fabric is to be wasted, what is the greatest length that can be used for the sides of the squares?

EXERCISES

Practice

Find the GCF of each set of numbers.

9. 21, 28   **10.** 24, 48   **11.** 63, 84   **12.** 60, 80

13. 24, 36, 42   **14.** 36, 15, 45   **15.** 35, 49, 84   **16.** 36, 72, 90

17. 36, 60, 84   **18.** 210, 330, 150   **19.** 84, 126, 210   **20.** 510, 714, 306

21. What is the GCF of $2 \cdot 3^3 \cdot 5^2$ and $2^3 \cdot 3 \cdot 5^2$?

22. How do you know just by looking that two numbers will have 5 as a common factor?

23. Write three numbers that have a GCF of 7.

Applications and Problem Solving

24. *Interior Design* Refer to Example 3. How many 6-inch squares will be needed to cover the 30-inch by 24-inch table?

25. *Carpentry* Mali wants to cut two boards to make shelves. One board is 72 inches long, and the other is 54 inches long. Mali wants each shelf to be the same length and does not want to waste any of the wood.
  a. What is the longest possible length of the shelves?
  b. How many of these shelves will Mali have?

26. *Critical Thinking* Numbers that have a GCF of 1 are *relatively prime.* Use this definition to determine whether each statement is true or false and tell why.
  a. Any two prime numbers are relatively prime.
  b. If two numbers are relatively prime, one of them must be prime.

For **Extra Practice,** see page 621.

Mixed Review

27. Find the prime factorization of 56. *(Lesson 6-2)*

28. *Geometry* Determine whether the figure has rotational symmetry. *(Lesson 5-4)*

29. **Standardized Test Practice** Lucrecia is an experienced diver. During a dive, she descended from the surface at the rate of 6 meters per minute. What was her location at the end of 12 minutes? *(Lesson 2-7)*

  A -6 m   **B** -12 m   **C** -18 m   **D** -72 m   **E** Not Here

Rational Numbers

What you'll learn

You'll learn to identify and simplify rational numbers.

When am I ever going to use this?

Knowing how to simplify fractions can help you understand data such as 6 out of 30 games won.

Word Wise

rational number
simplest form

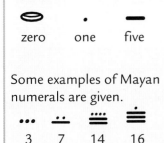

Cultural Kaleidoscope

The ancient Maya of Central America used place value and three symbols to represent numbers.

| zero | one | five |

Some examples of Mayan numerals are given.

| ••• | •• | •••• | ••••• |
| 3 | 7 | 14 | 16 |

About 5,000 years ago, Egyptians used hieroglyphics to represent numbers.

The Egyptian concept of fractions was mostly limited to fractions with numerators of 1. The hieroglyphics were placed under the symbol ⬯ to indicate the number was a denominator. Study the examples of Egyptian fractions.

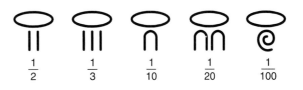

Fractions, their decimal equivalents, and integers are examples of **rational numbers**.

| **Rational Numbers** | Any number that can be expressed in the form $\frac{a}{b}$, where a and b are integers and $b \neq 0$, is called a rational number. |

You can use a Venn diagram to show how sets within the rational numbers are related.

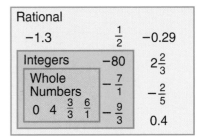

In the Venn diagram, notice that whole numbers and integers are included in the set of rational numbers. This is because integers, such as 4 and -7, can be written as $\frac{4}{1}$ and $\frac{-7}{1}$.

The set of rational numbers also includes decimals because they can be written as fractions with a power of ten for the denominator. For example, -0.29 can be written as $\frac{-29}{100}$.

Mixed numbers are also included in the set of rational numbers because any mixed number can be written as an improper fraction. For example, $5\frac{3}{4}$ can be written as $\frac{23}{4}$.

Name all the sets of numbers to which each number belongs.

1 25   25 is a whole number and an integer. Since it can be written as $\frac{25}{1}$, it is also a rational number.

2 -8 -8 is an integer. Since -8 can be written as $\frac{-8}{1}$ or $\frac{8}{-1}$, it is also a rational number.

3 $-7\frac{2}{3}$ Since $-7\frac{2}{3}$ can be written as $\frac{-23}{3}$ or $\frac{23}{-3}$, it is a rational number.

4 0.621   Since 0.621 can be written as $\frac{621}{1,000}$, it is a rational number.

When a rational number is represented as a fraction, it is often expressed in **simplest form**. A fraction is in simplest form when the GCF of the numerator and denominator is 1.

Examples

5 Write $\frac{30}{45}$ in simplest form.

Method 1 Divide by the GCF.

$30 = 2 \cdot 3 \cdot 5$ } *The GCF of 30*
$45 = 3 \cdot 3 \cdot 5$ } *and 45 is $3 \cdot 5$ or 15.*

$\frac{30}{45} \rightarrow \frac{30 \div 15}{45 \div 15} = \frac{2}{3}$

Method 2 Use prime factorization.

$\frac{30}{45} = \frac{2 \cdot \overset{1}{3} \cdot \overset{1}{5}}{\underset{1}{3} \cdot 3 \cdot \underset{1}{5}} = \frac{2}{3}$

The slashes indicate that the numerator and denominator are divided by $3 \cdot 5$, the GCF.

Since the GCF of 2 and 3 is 1, the fraction $\frac{2}{3}$ is in simplest form.

APPLICATION

Real World

6 **Space Exploration** On February 20, 1962, John Glenn became the first American to orbit Earth. He traveled around Earth 3 times. The trip took a total of 4 hours and 48 minutes. Express the length of time in terms of hours using a fraction in simplest form.

Write 48 minutes as hours and add to 4 hours.

48 minutes equals $\frac{48}{60}$ hour.

Write $\frac{48}{60}$ in simplest form.

$48 = 2 \cdot 2 \cdot 2 \cdot 2 \cdot 3$ } *The GCF of 48 and*
$60 = 2 \cdot 2 \cdot 3 \cdot 5$ } *60 is $2 \cdot 2 \cdot 3$ or 12.*

$\frac{48}{60} \rightarrow \frac{48 \div 12}{60 \div 12} = \frac{4}{5}$

John Glenn's trip took 4 plus $\frac{4}{5}$ or $4\frac{4}{5}$ hours.

Communicating Mathematics

Read and study the lesson to answer each question.

1. *Write* an example of a rational number that is not an integer.

2. *Explain* how to express $\frac{30}{50}$ in simplest form.

Guided Practice

Name all the sets of numbers to which each number belongs.

3. $1\frac{3}{4}$

4. -45

Write each fraction in simplest form.

5. $\frac{8}{72}$

6. $\frac{27}{45}$

7. $-\frac{60}{75}$

8. $\frac{36}{54}$

9. *Theater* The Antoinette Perry, or Tony, Award is given to exceptional plays and the people who make them happen. The award weighs 1 pound 10 ounces. Write the weight in pounds using a fraction in simplest form.

EXERCISES

Practice

Name all sets of numbers to which each number belongs.

10. 6.2

11. 0

12. $-5\frac{5}{7}$

13. 77

14. -9

15. -0.85

Write each fraction in simplest form.

16. $-\frac{3}{9}$

17. $\frac{15}{25}$

18. $\frac{36}{81}$

19. $-\frac{18}{54}$

20. $\frac{14}{66}$

21. $\frac{24}{54}$

22. $-\frac{15}{24}$

23. $\frac{48}{72}$

24. $\frac{24}{120}$

25. $-\frac{66}{88}$

26. $\frac{72}{98}$

27. $\frac{510}{2,159}$

28. Write two other names for –20.

29. Write 16 out of 40 as a fraction in simplest form.

30. Both numerator and denominator of a fraction are even. Can you tell whether the fraction is in simplest form? Explain.

For the latest weather information, visit: www/glencoe.com/sec/math/mac/mathnet

Applications and Problem Solving

31. *Weather* Yuma, Arizona, has an average of 175 days a year with temperatures of 90ºF or above. Using 365 days in a year, write a fraction representing these hot days as a part of a year. Express the fraction in simplest form.

32. *Geometry* Ordered pairs of rational numbers can be graphed on a coordinate plane in the same way as integers.

   a. Write the ordered pair for the coordinates of each point graphed on the coordinate plane at the right.

   b. Graph $R\left(\frac{1}{2}, -1\right)$, $S(0.75, 1.25)$, and $T\left(-\frac{1}{4}, 2\frac{1}{2}\right)$ on a coordinate plane.

33. *Working on the* **CHAPTER Project**

   Refer to the information you gathered about a sports team on page 231. For each year, write a fraction in simplest form representing the portion of regular season games won. Organize your results in a table.

34. *Critical Thinking* Does $\frac{4}{6.4}$ name a rational number? Explain.

Mixed Review

35. Find the GCF of 28, 126, and 56. *(Lesson 6-3)*

36. *Statistics* Would a pet store be a good location to find a representative sample for a survey of number of pets owned? Explain. *(Lesson 4-8)*

37. Solve $-42 \div (-14) = p$. *(Lesson 2-8)*

38. *Standardized Test Practice* If $3r + 8 = 35$, what is the value of r? *(Lesson 1-7)*

   A 52   **B** 43   **C** 14   **D** 9

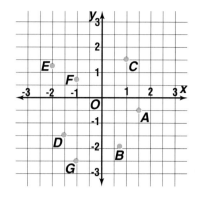

For **Extra Practice**, see page 621.

Mid-Chapter Self Test

Determine whether each number is divisible by 2, 3, 4, 5, 6, 8, 9, or 10. *(Lesson 6-1)*

1. 880

2. 12,321

Find the prime factorization of each number. *(Lesson 6-2)*

3. 68

4. 400

5. 63

Find the GCF for each set of numbers. *(Lesson 6-3)*

6. 28, 70

7. 12, 18, 30

Write each fraction in simplest form. *(Lesson 6-4)*

8. $\frac{12}{30}$

9. $-\frac{81}{99}$

10. *Weather* On the day of winter solstice, Hilo, Hawaii, has 10 hours and 55 minutes of sunshine. Express this length of time in terms of hours. *(Lesson 6-4)*

Rational Numbers and Decimals

The chart shows the recycling habits of Americans. How could you express the portion of iron and steel recycled as a decimal? How could you express the portion of paper recycled as a decimal?

Remember that a fraction is another way of writing a division problem. Any fraction can be expressed as a decimal by dividing the numerator by the denominator.

Portion of Trash Recycled

$\frac{1}{4}$ of all iron and steel

$\frac{1}{3}$ of all paper

$\frac{1}{10}$ of all wood

$\frac{1}{5}$ of all plastic

Source: *Vitality*, March, 1996

$\frac{1}{4}$ means $1 \div 4$.

$$\begin{array}{r} 0.25 \\ 4\overline{)1.00} \\ \underline{8} \\ 20 \\ \underline{20} \\ 0 \end{array}$$

Annex zeros to the numerator: 1 = 1.00.

$\frac{1}{4} = 0.25$

$\frac{1}{3}$ means $1 \div 3$.

$$\begin{array}{r} 0.333\ldots \\ 3\overline{)1.000} \\ \underline{9} \\ 10 \\ \underline{9} \\ 10 \\ \underline{9} \\ 1 \end{array}$$

The three dots means the three keeps repeating.

$\frac{1}{3} = 0.333\ldots$

The fraction $\frac{1}{4}$ can be expressed as the decimal 0.25. A decimal like 0.25 is called a **terminating decimal** because the division ends, or terminates, when the remainder is zero.

The fraction $\frac{1}{3}$ can be expressed as the decimal 0.333… . A decimal like 0.333… is called a **repeating decimal**. Since it is impossible to write all the digits, you can use **bar notation** to show that the 3 repeats.

$$0.33333\ldots = 0.\overline{3}$$

Examples

Express each decimal using bar notation.

① 0.454545...

The digits 45 repeat.

$0.454545\ldots = 0.\overline{45}$

② 0.1345345...

The digits 345 repeat.

$0.1345345\ldots = 0.1\overline{345}$

CONNECTION **③** **Geography** North America contains about $\frac{1}{6}$ of the dry land of Earth. Express this fraction as a decimal.

Use a calculator.

1 ÷ 6 = *0.166666667*

The calculator rounded the answer 0.166666666... to 0.166666667.

$\frac{1}{6} = 0.1\overline{6}$

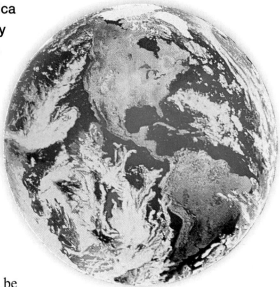

> **Study Hint**
>
> **Technology** Most calculators round answers, but some truncate answers. *Truncate* means to cut-off at a certain place-value position, ignoring the digits that follow.

Every terminating decimal can be expressed as a fraction with a denominator of 10, 100, 1,000, and so on. Thus, terminating decimals are rational numbers.

Examples

Express each decimal as a fraction or mixed number in simplest form.

④ 0.85

$0.85 = \frac{85}{100}$

$= \frac{17}{20}$ *Simplify. The GCF of 85 and 100 is 5.*

⑤ −4.125

$-4.125 = -4\frac{125}{1,000}$

$= -4\frac{1}{8}$ *Simplify.*

Repeating decimals can also be expressed as fractions. Thus, repeating decimals are rational numbers.

Example

INTEGRATION

⑥ **Algebra** Express $0.\overline{7}$ as a fraction.

Let $N = 0.\overline{7}$ or 0.777.... Then $10N = 7.777\ldots$ *Multiply N by 10, because 1 digit repeats.*

Subtract $N = 0.777\ldots$ to eliminate the repeating part, 0.777....

$\begin{aligned} 10N &= 7.777\ldots \\ -1N &= 0.777\ldots \qquad \textit{N = 1N} \\ \hline 9N &= 7 \\ N &= \frac{7}{9} \qquad \textit{Divide each side by 9.} \end{aligned}$

So, $0.\overline{7} = \frac{7}{9}$

Check: 7 ÷ 9 = *0.777777778* ✓

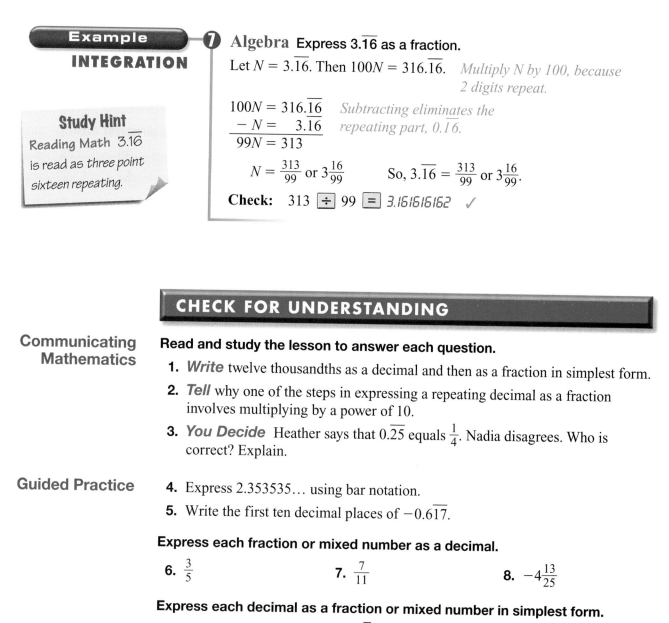

Example

INTEGRATION

7 Algebra Express $3.\overline{16}$ as a fraction.

Let $N = 3.\overline{16}$. Then $100N = 316.\overline{16}$. *Multiply N by 100, because 2 digits repeat.*

$$100N = 316.\overline{16} \quad \text{Subtracting eliminates the}$$
$$\underline{-N = \quad 3.\overline{16}} \quad \text{repeating part, } 0.\overline{16}.$$
$$99N = 313$$

$$N = \frac{313}{99} \text{ or } 3\frac{16}{99} \qquad \text{So, } 3.\overline{16} = \frac{313}{99} \text{ or } 3\frac{16}{99}.$$

Check: $313 \boxed{\div} 99 \boxed{=}$ *3.161616162* ✓

Study Hint

Reading Math $3.\overline{16}$ is read as *three point sixteen repeating.*

CHECK FOR UNDERSTANDING

Communicating Mathematics

Read and study the lesson to answer each question.

1. *Write* twelve thousandths as a decimal and then as a fraction in simplest form.

2. *Tell* why one of the steps in expressing a repeating decimal as a fraction involves multiplying by a power of 10.

3. *You Decide* Heather says that $0.\overline{25}$ equals $\frac{1}{4}$. Nadia disagrees. Who is correct? Explain.

Guided Practice

4. Express 2.353535… using bar notation.

5. Write the first ten decimal places of $-0.6\overline{17}$.

Express each fraction or mixed number as a decimal.

6. $\frac{3}{5}$ 7. $\frac{7}{11}$ 8. $-4\frac{13}{25}$

Express each decimal as a fraction or mixed number in simplest form.

9. 0.66 10. $0.\overline{2}$ 11. $-1.\overline{15}$

12. *Measurement* A micrometer measured the thickness of the wall of a plastic pipe as 0.084 inch. What fraction of an inch is this?

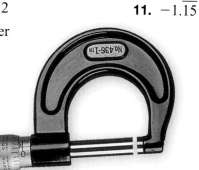

EXERCISES

Practice

Express each decimal using bar notation.

13. $0.255555\ldots$ 14. $-15.345345345\ldots$ 15. $7.0747474\ldots$

Write the first ten decimal places of each decimal.

16. $0.\overline{15}$ 17. $-0.\overline{305}$ 18. $0.3\overline{81}$

Express each fraction or mixed number as a decimal.

19. $\frac{3}{4}$ **20.** $\frac{4}{9}$ **21.** $-\frac{7}{25}$ **22.** $\frac{3}{22}$

23. $-5\frac{2}{3}$ **24.** $7\frac{5}{33}$ **25.** $12\frac{5}{8}$ **26.** $-8\frac{15}{44}$

Express each decimal as a fraction or mixed number in simplest form.

27. 0.88 **28.** $0.\overline{45}$ **29.** $0.8\overline{3}$ **30.** -0.225

31. $-1.\overline{5}$ **32.** 7.08 **33.** $-5.\overline{67}$ **34.** $2.1\overline{5}$

35. When $\frac{131}{200}$ is expressed as a decimal, is it a *terminating* or *repeating* decimal?

36. Suppose $N = 0.\overline{894}$. To change this number to a fraction, what power of ten should you multiply times N?

37. Express 0.38 and 0.383838 as fractions.

Applications and Problem Solving

38. *History* During the fourteenth and fifteenth centuries, printing presses used type cut from wood blocks. Each block was $\frac{7}{8}$ inch thick. Write this as a decimal.

39. *Life Science* An egg of a hummingbird weighs 0.013 ounce. Write this as a fraction of an ounce. Express the fraction in simplest form.

40. *Technology* A spreadsheet's display and two calculators' displays for $6 \div 9$ are listed below. How would each display the answer to $4 \div 11$?

     a. 0.7   **b.** 0.666666667   **c.** 0.666666666

41. *Working on the* **CHAPTER Project** Refer to the fractions that you found in Exercise 33 on page 248. Express each fraction as a decimal. Add these results to your table.

42. *Critical Thinking* A *unit fraction* is a fraction that has 1 as its numerator.

     a. Write the four greatest unit fractions that are terminating decimals. Express each fraction as a decimal.

     b. Write the four greatest unit fractions that are repeating decimals. Express each fraction as a decimal.

Mixed Review

43. Name all the sets of numbers to which -38 belongs. *(Lesson 6-4)*

44. *Geometry* Classify the triangle at the right according to its angles and its sides. *(Lesson 5-2)*

57 cm
29°   21°
27 cm 130° 36 cm

For **Extra Practice,** see page 621.

45. *Algebra* Solve $\frac{3}{17} = \frac{n}{68}$. *(Lesson 3-3)*

46. Without graphing, tell in which quadrant the point $M(6, -4)$ lies. *(Lesson 2-10)*

47. **Standardized Test Practice** Find the value of $2^3 \cdot 6^2$. *(Lesson 1-2)*

     A 48   **B** 144   **C** 288   **D** 864

Integration: Probability
Simple Events

What you'll learn

You'll learn to find the probability of a simple event.

When am I ever going to use this?

Knowing how to find the probability of a simple event can help you play some games wisely.

Word Wise

outcome
sample space
random
event
probability

An ancient Hebrew game is played with a four-sided top called a dreidel and tokens called gelt. Each side of the top has one of the following Hebrew letters.

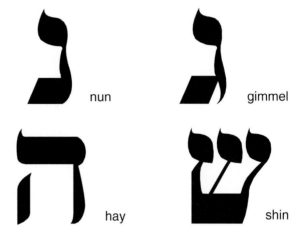

nun   gimmel

hay   shin

The players take turns spinning the dreidel. Depending on the spin, the player may add gelt to the center pot or take gelt from the center pot. For example, if a spin results in a hay, the player takes half of the gelt from the pot. What is the probability of spinning a hay?

When the dreidel is spun, there are four equally-likely results or **outcomes**. The list of all possible outcomes is called the **sample space**. The sample space of the dreidel is nun, gimmel, hay, and shin. Each outcome has the same chance of occurring. When all outcomes have an *equally-likely* chance of happening, we say that the outcomes happen at **random**.

An **event** is a specific outcome or type of outcome. In this case, the event is spinning a hay. **Probability** is the chance that an event will happen.

Probability	Probability $= \dfrac{\text{number of ways that an event can occur}}{\text{number of possible outcomes}}$

The probability of spinning a hay is $\frac{1}{4}$.

Use the scale below when you consider the probability of an event. When it is *impossible* for an event to happen, its probability is 0. An outcome that is *certain* to happen has a probability of 1.

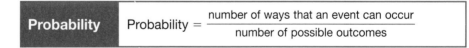

		50-50		
impossible		chance		certain
0		$\frac{1}{2}$		1

The probability of an outcome is written as a number from 0 to 1.

1 A bag contains 5 red marbles, 4 yellow marbles, and 3 blue marbles. Lashanda picks a marble without looking. What is the probability that she will pick a blue marble?

$$P(\text{blue}) = \frac{\text{blue marbles}}{\text{total number of marbles}}$$

$$= \frac{3}{12} \text{ or } \frac{1}{4} \text{ or } 0.25$$

INTEGRATION **2** **Geometry** A pin is dropped at random onto the square at the right. The point of the pin lands in one of the small regions.

a. What is the probability that the point lands inside a yellow region?

$$P(\text{yellow}) = \frac{\text{yellow squares}}{\text{total number of squares}}$$

$$= \frac{6}{25} \text{ or } 0.24$$

b. What is the probability that the point lands inside a blue or white region?

$$P(\text{blue or white}) = \frac{\text{blue squares} + \text{white squares}}{\text{total number of squares}}$$

$$= \frac{5 + 8}{25}$$

$$= \frac{13}{25} \text{ or } 0.52$$

c. What is the probability that the point lands inside a green region?

Since there are no green regions, the probability is 0.

CHECK FOR UNDERSTANDING

Communicating Mathematics

Read and study the lesson to answer each question.

1. *Write* an example of an outcome with a probability of 1.

2. *Draw* a spinner where the probability of an outcome of red is $\frac{1}{5}$.

HANDS-ON MATH

3. *Toss* a coin 50 times. Record the results as heads or tails. Are the results what you expected? Explain.

Guided Practice

State the probability of each outcome as a fraction and as a decimal.

4. A date picked at random is a Saturday.

5. This is a history book.

A number cube is rolled. Find each probability.

6. $P(3)$ 7. $P(\text{even})$ 8. $P(3 \text{ or } 5)$ 9. $P(\text{less than } 7)$

10. **Games** A board game has 100 tiles including 2 blank tiles, 9 tiles with the letter A, 12 tiles with E, 9 tiles with I, 8 tiles with O, and 4 tiles with U.

   a. What is the probability that the first tile picked is blank?

   b. What is the probability that the first tile picked is not blank?

   c. What is the probability that the first tile picked is A, E, I, O, or U?

Practice

State the probability of each outcome as a fraction and as a decimal.

11. A number cube is rolled and shows a number greater than 1.

12. A month picked at random starts with J.

13. A date picked at random is February 30.

14. A two-digit number picked at random is divisible by 10.

15. An integer picked at random is a rational number.

16. A positive one-digit number picked at random is prime.

The spinner at the right is used for a game. Find each probability.

17. $P(7)$

18. $P(2, 3, \text{or } 4)$

19. $P(\text{odd})$

20. $P(\text{less than } 6)$

21. $P(\text{not } 8)$

22. $P(\text{greater than } 2)$

The letters of the word "associative" are written one each on 11 identical slips of paper and shuffled in a hat. A blindfolded student draws one slip of paper. Find each probability.

23. $P(s)$

24. $P(\text{vowel})$

25. $P(\text{not } r)$

26. $P(s \text{ or } t)$

27. $P(\text{consonant})$

28. $P(\text{not } i)$

29. Suppose a person living in your home is picked at random. What is the probability that the person is a female?

30. Can a probability ever be greater than 1? Explain.

31. Can a probability ever be less than 0? Explain.

32. *Write a Problem* in which the answer will be a probability of $0.\overline{3}$.

Applications and Problem Solving

33. *Games* A card game has 25 red cards, 25 green cards, 25 yellow cards, 25 blue cards, and 8 wild cards.

  a. What is the probability that the first card dealt is a yellow card?

  b. Find the probability that the first card dealt is a wild card.

  c. There are two number 7 cards for each color. What is the probability that the first card dealt is a 7?

  d. There is only one number 0 card for each color. What is the probability that the first card dealt is a 0?

34. *History* The U.S. Bureau of the Census divides the United States into four regions. The table shows the population of these four regions in 1890 and 1990.

U.S. Population (thousands)		
Region	**1890**	**1990**
Northeast	17,407	50,809
Midwest	22,410	59,669
South	20,028	85,446
West	3,134	52,786

Source: U.S. Bureau of the Census

  a. Suppose a person living in the United States in 1890 was picked at random. What is the probability that the person lived in the West? Write your answer as a decimal to the nearest thousandth.

  b. Suppose a person living in the United States in 1990 was picked at random. What is the probability that the person lived in the West? Write your answer as a decimal to the nearest thousandth.

  c. How has the population of the West changed?

35. *Critical Thinking* If the probability that an event will occur is $\frac{3}{7}$, what is the probability that the event will not occur?

Mixed Review

36. *History* In 1864, Abraham Lincoln won the presidential election with about 0.55 of the popular vote. Write this as a fraction in simplest form. *(Lesson 6-5)*

37. *Geometry* In the figure, $m \parallel n$ and $m\angle 7 = 100°$. Find $m\angle 8$. *(Lesson 5-1)*

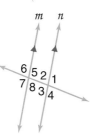

38. *Standardized Test Practice* To try to determine the most popular color for cars driven in the Minneapolis-St. Paul area, the colors of cars passing through an intersection in one hour were counted. The results are listed in the table.

Color	Frequency
black	ⵊⵊⵊ ⵊⵊⵊ ‖‖‖‖
white	ⵊⵊⵊ ⵊⵊⵊ ‖
red	ⵊⵊⵊ ⵊⵊⵊ ⵊⵊⵊ ⵊⵊⵊ ⵊⵊⵊ ‖‖
blue	ⵊⵊⵊ ⵊⵊⵊ ⵊⵊⵊ ⵊⵊⵊ ‖‖‖

Which color is the mode? *(Lesson 4-4)*

A black   **B** white   **C** red   **D** blue

For **Extra Practice,** see page 622.

39. *Algebra* Solve $-12 + t = -8$. *(Lesson 2-9)*

6-7 Least Common Multiple

What you'll learn

You'll learn to find the least common multiple of two or more numbers.

When am I ever going to use this?

Knowing how to find the LCM of integers can help you find when certain events such as the arrival of buses will coincide.

Word Wise

multiple
least common multiple (LCM)

Rebecca is buying hot dogs and hot dog buns for the photography club picnic. Hot dogs come in packages of 10. Hot dog buns came in packages of 8. Rebecca wants to have the same number of hot dogs and hot dog buns. What is the least number of hot dogs and hot dog buns she can buy?

You can use multiples and the problem-solving strategy *make a list* to answer this question. A **multiple** of a number is the product of that number and any whole number.

MINI-LAB

Work with a partner.   2 colored pencils

Try This

- List the numbers from 1 to 100 on a sheet of paper.
- Use divisibility rules to cross out all of the multiples of 10.
- Using a different color, cross out all of the multiples of 8.

Talk About It

1. Which numbers were crossed out by both colors?
2. How would you describe these numbers?
3. What is the least number crossed out by both colors?

In the Mini-Lab, you discovered some common multiples of 10 and 8. The least of the nonzero common multiples of two or more numbers is called the **least common multiple (LCM)** of the numbers. The LCM of 10 and 8 is 40. So, the least number of hot dogs and hot dog buns Rebecca can buy is 40.

Examples

1 List the first four multiples of 15.

multiples of 15

$0 \cdot 15 = 0$
$1 \cdot 15 = 15$
$2 \cdot 15 = 30$
$3 \cdot 15 = 45$

2 List the first four multiples of y.

multiples of y

$0 \cdot y = 0$
$1 \cdot y = y$
$2 \cdot y = 2y$
$3 \cdot y = 3y$

Example — **3** Find the LCM of 8, 9, and 12.

multiples of 8: 0, 8, 16, 24, 32, 40, 48, 56, 64, 72, 80, ...
multiples of 9: 0, 9, 18, 27, 36, 45, 54, 63, 72, 81, 90, ...
multiples of 12: 0, 12, 24, 36, 48, 60, 72, 84, 96, 108, 120, ...
The LCM of 8, 9, and 12 is 72. Remember that the LCM
is a nonzero number.

Prime factorization can be used to find the LCM of a set of numbers.
A common multiple contains *all* the prime factors of each number
in the set. The LCM contains *each* factor the greatest number of times
it appears in the set.

Examples — Use prime factorization to find the LCM of each set of numbers.

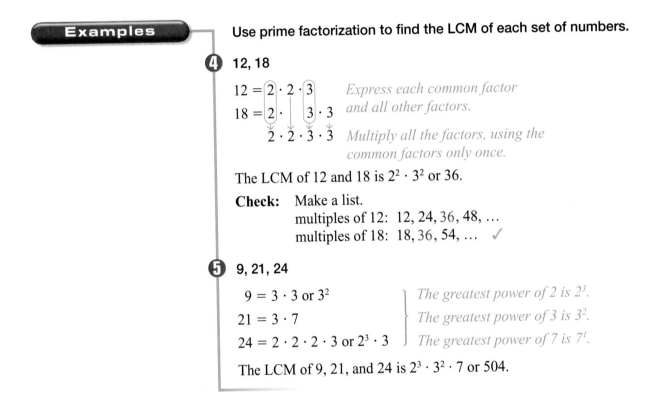

4 12, 18

$12 = 2 \cdot 2 \cdot 3$ *Express each common factor*
$18 = 2 \cdot \quad 3 \cdot 3$ *and all other factors.*
$\quad 2 \cdot 2 \cdot 3 \cdot 3$ *Multiply all the factors, using the common factors only once.*

The LCM of 12 and 18 is $2^2 \cdot 3^2$ or 36.

Check: Make a list.
multiples of 12: 12, 24, 36, 48, ...
multiples of 18: 18, 36, 54, ... ✓

5 9, 21, 24

$9 = 3 \cdot 3$ or 3^2 *The greatest power of 2 is 2^3.*
$21 = 3 \cdot 7$ *The greatest power of 3 is 3^2.*
$24 = 2 \cdot 2 \cdot 2 \cdot 3$ or $2^3 \cdot 3$ *The greatest power of 7 is 7^1.*

The LCM of 9, 21, and 24 is $2^3 \cdot 3^2 \cdot 7$ or 504.

CHECK FOR UNDERSTANDING

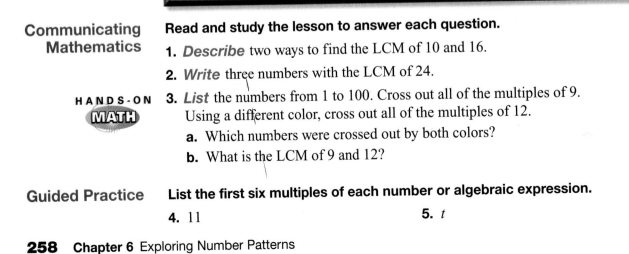

Communicating Mathematics

Read and study the lesson to answer each question.

1. *Describe* two ways to find the LCM of 10 and 16.

2. *Write* three numbers with the LCM of 24.

HANDS-ON MATH

3. *List* the numbers from 1 to 100. Cross out all of the multiples of 9.
Using a different color, cross out all of the multiples of 12.
   a. Which numbers were crossed out by both colors?
   b. What is the LCM of 9 and 12?

Guided Practice

List the first six multiples of each number or algebraic expression.

4. 11

5. *t*

Find the LCM of each set of numbers.

6. 8, 20   **7.** 15, 18   **8.** 2, 7, 8   **9.** 8, 28, 30

10. *Life Science* Cicadas emerge from the ground every 17 years. The number of one type of caterpillar peaks every 5 years. If the peak cycles of the caterpillar and the cicadas coincided in 1998, what will be the next year in which they coincide?

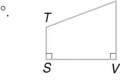

EXERCISES

Practice

List the first six multiples of each number or algebraic expression.

11. 7   **12.** 5   **13.** 14

14. 25   **15.** 150   **16.** $2x$

Find the LCM of each set of numbers.

17. 12, 16   **18.** 7, 12   **19.** 20, 50   **20.** 16, 24

21. 2, 3, 5   **22.** 4, 8, 12   **23.** 7, 21, 5   **24.** 10, 12, 14

25. 35, 25, 49   **26.** 24, 12, 6   **27.** 68, 170, 4   **28.** 45, 10, 6

29. Determine whether 285 is a multiple of 15.

30. When will the LCM of two numbers be one of the numbers?

31. When will the LCM of two numbers be the product of the numbers?

Applications and Problem Solving

32. *Civics* In the United States, a president is elected every four years. Members of the House of Representatives are elected every two years. Senators are elected every six years. If a voter had the opportunity to vote for a president, a representative, and a senator in 1996, what will be the next year the voter has a chance to make a choice for a president, a representative, and the same Senate seat?

33. *Bus Routes* Bus Route A reaches a particular bus stop every 12 minutes. Bus Route B reaches the same bus stop every 20 minutes. If a bus from Route A and a bus from Route B are at the bus stop at 7:00 A.M., what is the next time that a bus from each route will be at the bus stop?

34. *Critical Thinking* Find the LCM of $8xy$ and $6x^2yz$.

Mixed Review

35. *Standardized Test Practice* A box contains 7 orange marbles, 9 blue marbles, and 5 green marbles. If a marble is picked at random, what is the probability that the marble will be orange? *(Lesson 6-6)*

A 0   **B** $\frac{7}{20}$ **C** $\frac{1}{3}$ **D** 1

36. *Algebra* In trapezoid *STUV*, $m\angle S = 90°$, $m\angle T = 3x°$, $m\angle U = 60°$, and $m\angle V = 90°$. Find the value of x. *(Lesson 5-3)*

37. *Algebra* Solve $r + 125 = 483$. Check your solution. *(Lesson 1-4)*

For **Extra Practice,** see page 622.

38. *Manufacturing* A furniture company produces 15 rocking chairs in one hour. How long will it take to produce 45 chairs? *(Lesson 1-1)*

Lesson 6-7 Least Common Multiple   **259**

COOPERATIVE LEARNING

6-8A Density Property

A Preview of Lesson 6-8

 12-inch ruler

calculator

How many numbers are there between $-1\frac{3}{8}$ and $-1\frac{1}{4}$? Would you guess a few, many, or none? The *density property* states that between any two rational numbers, no matter how close they may seem, there is at least one other rational number.

TRY THIS

Work with a partner.

Step 1 Near the middle of a piece of paper, draw an eight-inch line segment parallel to the long side of the paper. Mark the ends and each one-inch interval with a vertical dash. Below the line, label the left endpoint 0, the right endpoint 2, and the middle point 1.

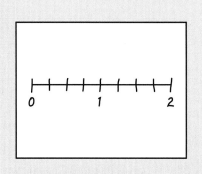

Step 2 Copy and complete the *Midpoint* column of the chart to find the midpoint between each pair of endpoints.

Step 3 Express the numbers in the *Endpoints* column as decimals. Find the mean of the two decimals. Record each result in the *Mean* column.

Endpoints	Midpoint of Endpoints	Mean of Endpoints
0 and 2		
1 and 2		
1 and $1\frac{1}{2}$		
$1\frac{1}{2}$ and $1\frac{1}{4}$		
$1\frac{3}{8}$ and $1\frac{1}{4}$		

ON YOUR OWN

1. What is the relationship between the numbers in the *Midpoint* column and the numbers in the *Endpoints* column?

2. What is the relationship between the numbers in the *Midpoint* column and the numbers in the *Mean* column?

3. Tell how you would find the midpoint between $\frac{3}{4}$ and $\frac{7}{8}$.

4. **Look Ahead** List two decimals with all but the last digits the same; for example, 4.26 and 4.27. Add a digit other than 0 to the lesser number. What is the relationship between the original numbers and the number formed by adding the digit?

Comparing and Ordering Rational Numbers

What you'll learn

You'll learn to compare and order rational numbers expressed as fractions and/or decimals.

When am I ever going to use this?

Knowing how to order rational numbers can help you determine which team has a better record.

Word Wise

least common denominator (LCD)

Lucille Treganowan is a grandmother and a car repair expert. She has her own weekly show on television.

Suppose a viewer incorrectly selects the $\frac{9}{16}$ socket while working on a car. The correct size is the next larger size. Which of the following sockets should be used?

$$\frac{11}{16}, \frac{1}{2}, \frac{5}{8}$$

One way to compare two rational numbers is to express them as fractions with like denominators. Any common denominator may be used, but the computation may be easier if the **least common denominator (LCD)** is used. The least common denominator is the LCM of the denominators.

Another way to compare these fractions is to express them as decimals and then compare the decimals.

Socket Size	Fraction Method	Decimal Method
$\frac{9}{16}$	$\frac{9}{16}$	0.5625
$\frac{11}{16}$	$\frac{11}{16}$	0.6875
$\frac{1}{2}$	$\frac{8}{16}$	0.5
$\frac{5}{8}$	$\frac{10}{16}$	0.625

Now it is easy to order the fractions. The socket order from least to greatest is $\frac{1}{2}, \frac{9}{16}, \frac{5}{8}$, and $\frac{11}{16}$. The viewer should use the $\frac{5}{8}$ socket since it is the next size larger than the $\frac{9}{16}$ socket.

Sports In the first round of the 1997 NCAA Baseball Tournament, the University of Florida played St. John's University. Before the tournament, Florida had won 38 of their 60 games. St. John's had won 35 of their 50 games. Which team had the better record?

Explore You are given the number of wins and the number of games for each team. You are asked to determine which team had a better record.

Plan Florida won $\frac{38}{60}$ of their games. St. John's won $\frac{35}{50}$ of their games. The LCD of these fractions is 300. Express each fraction as a fraction with a denominator of 300.

Solve Florida: $\frac{38}{60} = \frac{190}{300}$

St. John's: $\frac{35}{50} = \frac{210}{300}$

Since $\frac{190}{300} < \frac{210}{300}$, St. John's University had the better record.

Examine Express each record as a decimal.

Florida: 38 ÷ 60 = *0.633333333*

St. John's: 35 ÷ 50 = *0.7*

The decimals verify that St. John's had the better record.

When comparing or ordering rational numbers, it is usually easier and faster if all of the numbers are decimals.

Examples

② Replace ● with <, >, or = to make $-0.5 \bullet -\frac{3}{7}$ a true sentence.

First, express $-\frac{3}{7}$ as a decimal.

3 +○− ÷ 7 = -0.428571429

Now compare it to -0.5. In the tenths place, $-5 < -4$. Therefore $-0.5 < -\frac{3}{7}$.

Study Hint

Mental Math Since

$1.8 = 1.800$ and

$1.\overline{8} = 1.888\ldots,$

$1.8 < 1.\overline{8}.$

③ Order 1.8, 1.07, $\frac{17}{9}$, and $\frac{3}{2}$ from least to greatest.

17 ÷ 9 = *1.888888889*   3 ÷ 2 = *1.5*

$\frac{17}{9} = 1.\overline{8}$ $\frac{3}{2} = 1.5$

The order from least to greatest is 1.07, $\frac{3}{2}$, 1.8, and $\frac{17}{9}$.

INTEGRATION

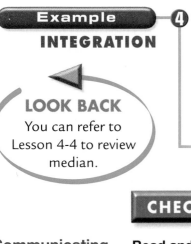

LOOK BACK
You can refer to Lesson 4-4 to review median.

④ Statistics Find the median of 17.4, 15, $17\frac{1}{2}$, $17\frac{3}{8}$, and 16.9.

To find the median, the numbers must be in order.

$$17\frac{1}{2} = 17.5 \qquad\qquad 17\frac{3}{8} = 17.375$$

The order from least to greatest is 15, 16.9, $17\frac{3}{8}$, 17.4, and $17\frac{1}{2}$. The median is $17\frac{3}{8}$ since it is the number in the middle.

CHECK FOR UNDERSTANDING

Communicating Mathematics

Read and study the lesson to answer each question.

1. *Draw* a 10-by-10 square on grid paper. Let the large square represent 1. Use the square to explain why $0.3 > 0.08$.

2. *Describe* two different ways to compare $\frac{5}{6}$ and $\frac{8}{9}$.

Math Journal

3. *Write* a fraction and a decimal that are between $\frac{2}{5}$ and $\frac{4}{7}$. Explain how you know these numbers are between the given numbers.

Guided Practice

4. Find the LCD for $\frac{7}{12}$ and $\frac{3}{8}$.

Replace each ● with <, >, or = to make a true sentence.

5. -6.2 ● -6.02 6. $\frac{6}{11}$ ● $0.5\overline{3}$ 7. 9.6 ● $9\frac{3}{5}$

8. Order $\frac{1}{2}$, $\frac{4}{5}$, $\frac{2}{5}$, $\frac{3}{4}$, and 0 from least to greatest.

9. Order $\frac{3}{8}$, 0.376, 0.367, and $\frac{2}{5}$ from greatest to least.

10. *Entertainment* Refer to the ride times for nine coasters at an amusement park in Canada. Find the median of the ride times.

Coaster	Ride Time
Dragon Fyre	$2\frac{1}{6}$ min
Mighty Canadian Minebuster	$2.\overline{6}$ min
Wilde Beast	2.5 min
Ghoster Coaster	$1\frac{5}{6}$ min
SkyRider	$2\frac{5}{12}$ min
Thunder Run	1.75 min
The Bat	$1\frac{5}{6}$ min
Vortex	1.75 min
TOP GUN	$2\frac{5}{12}$ min

EXERCISES

Practice

Find the LCD for each pair of fractions.

11. $\frac{5}{16}$, $\frac{3}{8}$ 12. $\frac{1}{6}$, $\frac{2}{9}$ 13. $\frac{5}{18}$, $\frac{4}{15}$

Replace each ● with <, >, or = to make a true sentence.

14. $3\frac{3}{7}$ ● $3\frac{4}{9}$

15. 1.4 ● 1.403

16. $\frac{1}{6}$ ● $0.1\overline{6}$

17. $-\frac{7}{2}$ ● $-\frac{9}{4}$

18. $\frac{5}{7}$ ● $\frac{9}{21}$

19. $11\frac{1}{8}$ ● 11.26

20. -5.2 ● $5\frac{1}{5}$

21. 0.77 ● $0.\overline{7}$

22. 5.92 ● $5\frac{23}{25}$

Order each set of rational numbers from least to greatest.

23. $\frac{1}{9}, \frac{1}{10}, -\frac{1}{3}, -\frac{1}{4}$

24. $0.6, \frac{6}{11}, \frac{1}{2}, 0.\overline{63}$

25. $-4.5, -4\frac{2}{3}, -4\frac{2}{5}, -4.1\overline{9}, -4.75$

26. $0.182, 0.182\overline{5}, 0.18\overline{2}, 0.\overline{18}$

27. Which is least, $\frac{7}{11}, 0.6, \frac{2}{3}, 0.6\overline{3}$, or $\frac{8}{13}$?

28. Using a number line, explain why $-6.7 < -6$.

29. Match each number with a point on the number line.

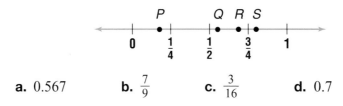

    a. 0.567
    b. $\frac{7}{9}$
    c. $\frac{3}{16}$
    d. 0.7

Applications and Problem Solving

30. *School* Marco scored 17 out of 20 on his history test and 21 out of 25 on his science test. Did Marco score better on his history test or his science test?

31. *Photography* The shutter speed on Natasha's camera is set at $\frac{1}{250}$ second. If Natasha wants to increase the shutter time, should she set the speed at $\frac{1}{500}$ second or $\frac{1}{125}$ second?

32. *Working on the* **CHAPTER Project** Refer to the table of fractions of games your chosen team won during the regular seasons. Order the fractions from least to greatest. Compare the seasons of the teams.

33. *Critical Thinking* Are there any rational numbers between $0.\overline{4}$ and $\frac{4}{9}$? Explain.

Mixed Review

34. **Standardized Test Practice** The #5 train runs every 8 minutes. The #3 train runs every 6 minutes. If they both leave the station at 8:12 A.M., when will be the next time the trains leave the station together? *(Lesson 6-7)*

    A 9:36 A.M.

    B 9:00 A.M.

    C 8:36 A.M.

    D 8:24 A.M.

35. Estimate 20.5% of 89. *(Lesson 3-6)*

36. Find the absolute value of -21. *(Lesson 2-1)*

37. Evaluate $4^2 + 6(12 - 8) - 15 \div 3$. *(Lesson 1-3)*

For **Extra Practice,** see page 622.

6-9 Scientific Notation

What you'll learn

You'll learn to express numbers in scientific notation.

When am I ever going to use this?

Knowing how to express numbers in scientific notation can help you in science classes.

Word Wise

scientific notation

In 1996, an amusement park in Anaheim, California, had about 15,000,000 visitors. The number 15,000,000 is in standard form. It can also be written as $1.5 \times 10,000,000$ or 1.5×10^7. The number 1.5×10^7 is in **scientific notation**. Large numbers like 15,000,000 are written in scientific notation to lessen the chance of omitting a zero or misplacing the decimal point.

When a number is expressed in scientific notation, it is written as a product of a factor and a power of 10. The factor must be greater than or equal to 1 and less than 10.

Multiplying by a positive power of 10 moves the decimal point to the right the same number of places as the exponent.

Example 1

Express 2.483×10^5 in standard form.

$$2.483 \times 10^5 = 2.483 \times 100,000$$
$$= 248,300$$

2.48300
5 places

Scientific notation is also used to express very small numbers. Study the pattern of products. Notice that multiplying by a negative power of 10 moves the decimal point to the left the same number of places as the absolute value of the exponent.

$4.16 \times 10^2 = 416$
$4.16 \times 10^1 = 41.6$
$4.16 \times 10^0 = 4.16$
$4.16 \times 10^{-1} = 0.416$
$4.16 \times 10^{-2} = 0.0416$
$4.16 \times 10^{-3} = 0.00416$
$4.16 \times 10^{-4} = 0.000416$

Example 2

CONNECTION

Life Science The diameter of a red blood cell is about 7.4×10^{-4} centimeter. Write this number in standard form.

$$7.4 \times 10^{-4} = 7.4 \times \frac{1}{10^4}$$
$$= 7.4 \times \frac{1}{10,000}$$
$$= 7.4 \times 0.0001$$
$$= 0.00074$$

0007.4
4 places

The diameter of a red blood cell is about 0.00074 centimeter.

Look at Examples 1 and 2. What is the relationship between the number of places the decimal point is moved and the absolute value of the exponent? This relationship allows you to use mental math to express any number in scientific notation.

Express each number in scientific notation.

③ 243,900,000

$243,900,000 = 2.439 \times 10^8$

The decimal point moves 8 places to the left.

④ 0.000000595

$0.000000595 = 5.95 \times 10^{-7}$

The decimal point moves 7 places to the right.

Example 5 shows how to enter a number in a calculator with too many digits to fit on the display screen in standard form.

⑤ Enter 0.00000000627 into a calculator.

First write the number in scientific notation.

$0.00000000627 = 6.27 \times 10^{-9}$

Then enter the number.

6.27 [EE] 9 [+◦−] *6.27 −09*

CHECK FOR UNDERSTANDING

Communicating Mathematics

Read and study the lesson to answer each question.

1. *Explain* why 36.2×10^3 and 0.362×10^{-4} are not written in scientific notation.

2. *Write* a number in scientific notation. Then write the same number in standard form.

Guided Practice

Express each number in standard form.

3. 4.882×10^5

4. 1.19×10^{-5}

5. 2.34×10^8

Express each number in scientific notation.

6. 2,085,000,000

7. 0.054

8. 0.000091

9. *Life Science* The Giganotosaurus weighed about 1.4×10^4 pounds. Write this number in standard notation.

EXERCISES

Practice

Express each number in standard form.

10. 3.1107×10^6

11. 8.08×10^9

12. 2.331×10^{-3}

13. 7.5×10^{-8}

14. 1.05×10^4

15. 2.52×10^{-5}

16. 6.81×10^{-2}

17. 2.021×10^2

18. 5.43×10^8

Express each number in scientific notation.

19. 0.00767

20. 5,750,000,000

21. 400,400

22. 0.0000051

23. 0.000033

24. 54,800,000

25. 7,600

26. 0.00083

27. 0.000000025

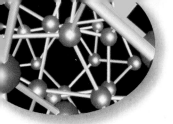

28. A googol is a number written with a 1 followed by 100 zeros. Write a googol in scientific notation.

29. Write the product of 0.00004 and 0.0008 in scientific notation.

Applications and Problem Solving

30. *Physical Science* An oxygen atom has a mass of 2.66×10^{-23} grams. Explain how to enter this number into a calculator.

31. *History* The United States purchased Alaska for $7,200,000. Write this dollar amount in scientific notation.

32. *Critical Thinking* Continue the following sequence.
$3^4 = 81$, $3^3 = 27$, $3^2 = 9$, $3^1 = 3$, $3^0 = 1$, $3^{-1} = ?$, $3^{-2} = ?$, $3^{-3} = ?$

Mixed Review

33. Order 7.35, $7\frac{2}{7}$, and $\frac{37}{5}$ from least to greatest. *(Lesson 6-8)*

34. *Statistics* Determine whether a scatter plot of the speed of a car and the stopping distance would show a positive, negative, or no relationship. *(Lesson 4-6)*

35. Solve $-319 - (-98) = w$. *(Lesson 2-5)*

36. **Standardized Test Practice** A board 42 inches long is cut in 2 pieces so that 1 piece is 7 inches longer than the other. Which equation could be used to find ℓ, the length of the shorter piece? *(Lesson 1-6)*

A $\ell + (\ell + 7) = 42$

B $2(\ell + 7) = 42$

C $\ell^2 + 7 = 42$

D $2\ell - 7 = 42$

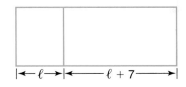

For **Extra Practice**, see page 623.

MATH IN THE MEDIA

Fox Trot

1. How did she determine 2,256 hours, 135,260 minutes, and 8,121,600 seconds?

2. Write 8,121,600 seconds in scientific notation.

3. There are 1,000,000,000 nanoseconds in a second. Use scientific notation to write the number of nanoseconds in 8,121,600 seconds.

4. Compute and express $\frac{(130,000)\,(0.0057)}{0.0004}$ and $\frac{(90,000)\,(0.0016)}{(200,000)\,(30,000)\,(0.00012)}$ in scientific notation.

inter NET
CONNECTION Chapter Review **For additonal lesson-by-lesson review, visit:**
www.glencoe.com/sec/math/mac/mathnet

Vocabulary

After completing this chapter, you should be able to define each term, concept, or phrase and give an example or two of each.

Number and Operations
bar notation (p. 249)
composite number (p. 235)
density property (p.260)
divisible (p. 232)
factor tree (p. 236)
Fundamental Theorem of Arithmetic (p. 236)
greatest common factor (GCF) (p. 242)
least common denominator (LCD) (p. 261)
least common multiple (LCM) (p. 257)
multiple (p. 257)
prime factorization (p. 235)
prime number (p. 235)
rational number (p. 245)

repeating decimal (p. 249)
scientific notation (p. 265)
simplest form (p. 246)
terminating decimal (p. 249)

Probability
event (p. 253)
outcome (p. 253)
probability (p. 253)
random (p. 253)
sample space (p. 253)

Problem Solving
make a list (p. 240)

Understanding and Using the Vocabulary

Choose the correct term or number to complete each sentence.

1. A (factor, multiple) of a whole number is the product of the number and any other whole number.

2. A whole number greater than 1 that has more than two factors is called a (prime, composite) number.

3. A fraction is in (bar notation, simplest form) when the GCF of the numerator and denominator is 1.

4. The least common denominator of two fractions is the (LCM, GCF) of the denominators.

5. An outcome that is certain to happen has a probability of (0, 1).

6. $0.4\overline{5}$ is an example of a (repeating, terminating) decimal.

7. The scientific notation for 32,000 is $(3.2 \times 10^4, 32 \times 10^3)$.

8. A number that can be expressed as a fraction where the numerator and denominator are both integers is called a (prime, rational) number.

In Your Own Words

9. *Explain* how to express a terminating decimal as a fraction.

Objectives & Examples

Upon completing this chapter, you should be able to:

● use divisibility rules for 2, 3, 4, 5, 6, 8, 9, and 10 *(Lesson 6-1)*

747 is divisible by 3 and 9 since $7 + 4 + 7 = 18$. Since 747 is not even and does not end in 5 or 0, it is *not* divisible by 2, 4, 5, 6, 8, or 10.

● find the prime factorization of a composite number *(Lesson 6-2)*

Write the prime factorization of 56.

$56 = 2 \cdot 28$

$\quad = 2 \cdot 2 \cdot 14$

$\quad = 2 \cdot 2 \cdot 2 \cdot 7$ or $2^3 \cdot 7$

● find the greatest common factor of two or more numbers *(Lesson 6-3)*

Find the GCF of 24 and 32.

factors of 24: 1, 2, 3, 4, 6, 8, 12, 24

factors of 32: 1, 2, 4, 8, 16, 32

The GCF of 24 and 32 is 8.

● identify and simplify rational numbers *(Lesson 6-4)*

Write $\frac{60}{140}$ in simplest form.

$\frac{60}{140} = \frac{\overset{1}{\cancel{2}} \cdot \overset{1}{\cancel{2}} \cdot 3 \cdot \overset{1}{\cancel{5}}}{\underset{1}{\cancel{2}} \cdot \underset{1}{\cancel{2}} \cdot \underset{1}{\cancel{5}} \cdot 7} = \frac{3}{7}$

● express rational numbers as decimals and decimals as fractions *(Lesson 6-5)*

Write $\frac{4}{11}$ as a decimal.

$\frac{4}{11} = 4 \div 11$

$\quad = 0.363636 \ldots$ or $0.\overline{36}$

Review Exercises

Use these exercises to review and prepare for the chapter test.

Determine whether each number is divisible by 2, 3, 4, 5, 6, 8, 9, or 10.

10. 533   11. 2,435

12. 332   13. 4,298

14. 16,548   15. 111

Find the prime factorization of each number.

16. 24   17. 63

18. 150   19. 202

20. 44   21. 81

Find the GCF for each set of numbers.

22. 16, 64   23. 14, 42

24. 14, 21, 42   25. 22, 33, 55

26. 160, 320, 480

Write each fraction in simplest form.

27. $\frac{15}{21}$ 28. $\frac{16}{20}$

29. $\frac{54}{63}$ 30. $-\frac{55}{66}$

Express each fraction or mixed number as a decimal.

31. $\frac{5}{8}$ 32. $\frac{16}{33}$

33. $2\frac{3}{5}$ 34. $5\frac{4}{9}$

Objectives & Examples

Review Exercises

Express $2.\overline{18}$ as a fraction.

Let $N = 2.\overline{18}$. Then $100N = 218.\overline{18}$.

$$100N = 218.\overline{18}$$
$$-N = 2.\overline{18}$$
$$99N = 216$$
$$N = \frac{216}{99} \text{ or } 2\frac{2}{11}. \text{ So, } 2.\overline{18} = 2\frac{2}{11}.$$

Express each decimal as a fraction or mixed number in simplest form.

35. 0.65 **36.** -0.075

37. $0.\overline{16}$ **38.** $6.\overline{3}$

39. 8.375 **40.** $-1.\overline{15}$

● find the probability of a simple event
(Lesson 6-6)

$$\text{Probability} = \frac{\text{number of ways an event can occur}}{\text{number of possible outcomes}}$$

A bag contains six red, three blue, and nine white marbles. A blindfolded student draws a marble. Find each probability.

41. $P(\text{white})$ **42.** $P(\text{red})$

43. $P(\text{not red})$ **44.** $P(\text{red or blue})$

● find the least common multiple of two or more numbers *(Lesson 6-7)*

Find the LCM of 6 and 9.

multiples of 6: 0, 6, 12, 18, 24, . . .
multiples of 9: 0, 9, 18, 27, 36, . . .

The LCM of 6 and 9 is 18.

Find the LCM for each set of numbers.

45. $15, 30$ **46.** $8, 10$

47. $21, 24$ **48.** $8, 12, 18$

49. $5, 10, 175$

● compare and order rational numbers expressed as fractions and/or decimals *(Lesson 6-8)*

Is $\frac{1}{3} > 0.33$?

$\frac{1}{3} = 0.\overline{3}$ and $0.\overline{3} > 0.33$. So, $\frac{1}{3} > 0.33$.

Replace each ● with $<$, $>$, or $=$ to make a true sentence.

50. -8.39 ● -8.4 **51.** $\frac{5}{12}$ ● $\frac{3}{4}$

52. $\frac{2}{3}$ ● 0.6 **53.** 0.5 ● $0.\overline{5}$

● express numbers in scientific notation
(Lesson 6-9)

Express 0.000165 in scientific notation.

$0.000165 \rightarrow 1.65 \times 10^{-4}$
4 places

Express each number in scientific notation.

54. $7,410,000$ **55.** 0.0000648

56. $17,500$ **57.** 0.00055

58. $87,500,000$ **59.** 0.00476

Applications & Problem Solving

60. *Make a List* Laura and Emilio are saving money to attend the state fair. Two tickets cost $30.00. Laura starts with $10.15 and saves $1.50 a week. Emilio starts with $8.50 and saves $1.75 a week. When will they save enough money? *(Lesson 6-3A)*

61. *Sports* A batting average is stated as a decimal rounded to the nearest thousandth. Ellis got two hits in eight times at bat. What is his batting average? *(Lesson 6-5)*

62. *Games* To finish the game shown at the right, the number cube must show the number equal to the number of spaces you have left to move. If you have one space left to move, what is the probability that the number cube will show a one? *(Lesson 6-6)*

63. *Geometry* A pin is dropped onto the rectangle below. If the point of the pin lands in one of the small square regions, what is the probability that the point lands inside a shaded region? *(Lesson 6-6)*

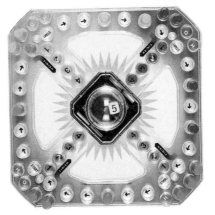

Alternative Assessment

Open Ended

Suppose your friend was absent when you learned how to express terminating and repeating decimals as fractions. You decide to use the decimals 0.83, 0.8$\overline{3}$, and 0.$\overline{83}$ to show your friend what she missed. How would you explain the differences among these decimals? How would you explain how to change these decimals to fractions?

Your friend was also absent when you learned about comparing and ordering rational numbers. How can you use decimals to show her how to order 0.83, 0.8$\overline{3}$, and 0.$\overline{83}$? How can you use the fractions to verify your answer?

A practice test for Chapter 6 is provided on page 652.

Completing the CHAPTER Project

Use the following checklist to make sure your news article is complete.

☑ The chart showing the fraction of games won is accurate.

☑ The seasons are compared in the article.

Add any finishing touches that you would like to make your article interesting.

PORTFOLIO Select one of the words you learned in this chapter. Place the word and its definition in your portfolio. Attach a note explaining why you selected the word.

Section One: Multiple Choice

There are thirteen multiple-choice questions in this section. Choose the best answer. If a correct answer is *not here*, choose the letter for Not Here.

1. What is the least prime factor of 42?
    A 4
    B 3
    C 2
    D 1

2. If $\triangle JKL$ is similar to $\triangle QRP$, what is the value of x?

    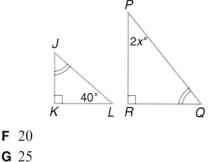

    F 20
    G 25
    H 40
    J 50

3. Express 0.0024 in scientific notation.
    A 2.4×10^3
    B 2.4×10^4
    C 2.4×10^{-3}
    D 2.4×10^{-4}

4. Write $0.\overline{24}$ as a fraction.
    F $\frac{8}{33}$
    G $\frac{1}{3}$
    H $\frac{24}{100}$
    J $\frac{6}{25}$

5. Which equation is equivalent to $x + 5 = 12$?
    A $x = 12$
    B $x + 5 - 5 = 12 - 5$
    C $x + 10 = 24$
    D $x + 5 - 5 = 12 + 5$

6. What is the probability of the spinner landing on a prime number?

    F 0.25
    G 0.5
    H 0.75
    J 1

7. If $\frac{y}{-2} = -4$, what is the value of y?
    A 2
    B -2
    C 8
    D -8

Please note that Questions 8–13 have five answer choices.

8. Carolyn makes $5.65 per hour for working her regularly scheduled hours. If she works any extra hours, she will earn $9.25 per hour. If r is the number of regularly scheduled hours and x is the number of extra hours, which number sentence can be used to find the total pay, p, that Carolyn makes in one week?
    F $p = 5.65 + 9.25x$
    G $p = 5.65r + 9.25x$
    H $p = 5.65r \times 9.25x$
    J $p = (5.25r + 9.25)x$
    K $p = 5.65r + 9.25$

9. Frank bought a suit for $149.99, a tie for $12.49, and a shirt for $14.99. What is the best estimate of the total amount he paid?
    A $160
    B $180
    C $200
    D $220
    E $240

10. Which integers are graphed on the number line?

$$-3\ -2\ -1\ 0\ 1\ 2\ 3$$

F $\{-2, 0, 2\}$

G $\{-3, 0, 3\}$

H $\{-2, 2\}$

J $\{-3, 3\}$

K $\{0\}$

11. Mirna paid $42.33 for a school jacket. If the price included $3.10 for sales tax, what was the price of the jacket before tax?

A $47.43

B $45.43

C $40.23

D $39.23

E $35.00

12. Rashid lost an average of two pounds a week on his diet. What was the change in his weight after 5 weeks?

F -5 lb

G -10 lb

H -15 lb

J -2 lb

K Not Here

13. Mr and Mrs. Dixon are having the ceiling of their rectangular living room painted. The living room is 14 feet wide and 20 feet long. A painter charges $0.24 for each square foot to be painted. How much will it cost the Dixons to have their ceiling painted?

A $33.60   **B** $53.40

C $67.20   **D** $81.80

E Not Here

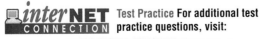

Test Practice For additional test practice questions, visit:

www.glencoe.com/sec/math/mac/mathnet

Test-Taking Tip   THE PRINCETON REVIEW

When taking a long test, work quickly through the problems. Skip those problems that take more than the average amount of time allotted for each problem. Mark these in the test booklet and come back to them after you have completed the problems that you could easily complete.

Section Two: Free Response

This section contains three questions for which you will provide short answers. Write your answers on your paper.

14. If the perimeter of the isosceles triangle is 45 units, what are the lengths of the sides?

a

$2a + 5$

a

15. Which two triangles are similar?

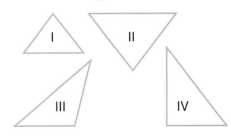

I   II

III   IV

16. What is the perimeter of the rectangle?

12.2 m

8.3 m

WHAT IN THE WORLD IS "WYSIWYG?"

Do you spend your spare time "on line"? Do you really get "bugged" when you "crash" and must "reboot"? Do you worry that some "hacker" will invade your "database"? If these terms are familiar to you, then you are probably one of the millions of people who use computers.

WYSIWYG (wizzywig) is an acronym for "What you see is what you get." It means that what you see on your computer screen is exactly what will be printed.

What You'll Do

In this investigation, you will research data about computer sales, display the results, and make predictions.

Materials almanacs or other factual books

 calculator

grid paper

Procedure

1. Work in pairs. Locate current information on yearly computer sales. Find the number of computers and the dollar sales of computers for the last ten years. Write the numbers in both standard and scientific notation.

2. Display the change in sales over the last ten years graphically.

3. Use the data to find the average cost per computer for each year.

4. Display the change in the cost of computers graphically.

5. Make some predictions about the sales of computers in the next ten years.

Technology Tips

• Surf the **Internet** to find information about computers.

• Use a **spreadsheet** to calculate the mean cost per computer.

• Use **graphing software** to display the data.

Making the Connection

Use the data collected about computers as needed to help in these investigations.

Language Arts

Write a newspaper article predicting the future of computers. Use the information that you gathered in the investigation. Be creative and describe the changes in computers that you think will occur. Explain how people will be affected.

Science

Research the impact of computers on the environment. Consider the amounts of paper and electricity used.

Music

Investigate how musicians can use computers in their work.

Go Further

- Survey 10 to 20 adults. Ask them how computers have changed their lives. Make a list of these changes.

- Investigate the binary number system. Find out how it works and why it is important to computers.

interNET **CONNECTION** Research **For current information on computers, visit:**

www.glencoe.com/sec/math/mac/mathnet

You may want to place your work on this investigation in your portfolio.

CHAPTER 7

Algebra: Using Rational Numbers

What you'll learn in Chapter 7

- to compute with fractions and mixed numbers,
- to solve problems by using patterns,
- to recognize and extend arithmetic and geometric sequences,
- to find the areas of triangles and trapezoids, and find the circumference of circles, and
- to solve equations and inequalities involving rational numbers.

CHAPTER Project

PATTERNS IN NATURE

In this project, you will explore one of the mathematical patterns found in nature. You will use what you find to create a poster or collage, or to present a slide show with narration.

Getting Started

- Collect a variety of pinecones. If these are not available, check with your science teacher. Identify the clockwise and counterclockwise spirals of scales from the bottoms of the cones. Count the number of individual spirals in each direction. Record the numbers of spirals you find.

- Study the arrangements of leaves on the stem of a variety of plants. If you look down on a plant, leaves are often arranged so that each one gets water and light. Make a drawing of each plant. Number the leaves from the bottom of the stem to the top. What is the difference between the numbers of the leaves that are aligned? How many times do you go around the stem to reach aligning leaves? Record these numbers for several different types of plants.

Technology Tips

- Use an **electronic encyclopedia** to do your research.
- Write portions of your poster or write your slide show script using **word processing** or **presentation** software.
- Record the information you gather about pinecone spirals or plant leaves in a **database**.

interNET CONNECTION Research For current information patterns and nature, visit:
www.glencoe.com/sec/math/mac/mathnet

Working on the Project

You can use what you'll learn in Chapter 7 to find patterns in nature.

Page	Exercise
298	29
311	26
325	Alternative Assessment

7-1 Adding and Subtracting Like Fractions

What you'll learn

You'll learn to add and subtract fractions with like denominators.

When am I ever going to use this?

Knowing how to add and subtract fractions can help you decorate your room.

Word Wise

mixed number

You can use rectangles to model adding fractions.

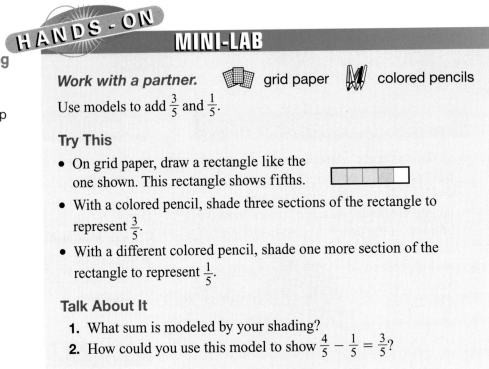

HANDS-ON MINI-LAB

Work with a partner. grid paper colored pencils

Use models to add $\frac{3}{5}$ and $\frac{1}{5}$.

Try This

- On grid paper, draw a rectangle like the one shown. This rectangle shows fifths.
- With a colored pencil, shade three sections of the rectangle to represent $\frac{3}{5}$.
- With a different colored pencil, shade one more section of the rectangle to represent $\frac{1}{5}$.

Talk About It

1. What sum is modeled by your shading?
2. How could you use this model to show $\frac{4}{5} - \frac{1}{5} = \frac{3}{5}$?

In the Mini-Lab, the rational numbers $\frac{3}{5}$ and $\frac{1}{5}$ have like denominators. Fractions with like denominators are also called *like fractions*. The results of the Mini-Lab suggest this rule for adding like fractions.

Adding Like Fractions	**Words:**	To add fractions with like denominators, add the numerators.
	Symbols: Arithmetic	$\frac{2}{5} + \frac{1}{5} = \frac{3}{5}$
	Algebra	$\frac{a}{c} + \frac{b}{c} = \frac{a+b}{c}, c \neq 0$

When the sum of two fractions is greater than 1, the sum is written as a **mixed number**, usually in simplest form. A mixed number is the sum of a whole number and a fraction.

Example ① Find $\frac{1}{2} + \frac{1}{2} + \frac{1}{2}$.

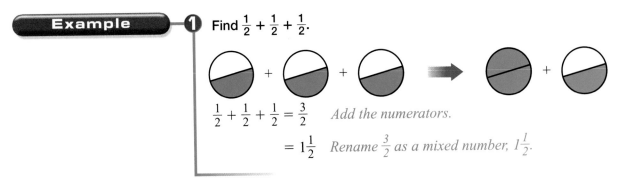

$\frac{1}{2} + \frac{1}{2} + \frac{1}{2} = \frac{3}{2}$ *Add the numerators.*

$\qquad\qquad = 1\frac{1}{2}$ *Rename $\frac{3}{2}$ as a mixed number, $1\frac{1}{2}$.*

Subtracting like fractions is similar to adding them.

Subtracting Like Fractions	**Words:**	To subtract fractions with like denominators, subtract the numerators.
	Symbols: **Arithmetic**	$\frac{2}{3} - \frac{1}{3} = \frac{1}{3}$
	Algebra	$\frac{a}{c} - \frac{b}{c} = \frac{a-b}{c}, c \neq 0$

Rational numbers include both positive and negative fractions. Use the rules for adding and subtracting integers to determine the sign of the sum or difference of two rational numbers.

Examples

2 Solve $x = -\frac{3}{8} - \left(-\frac{5}{8}\right)$.

$x = -\frac{3}{8} - \left(-\frac{5}{8}\right)$

$x = \frac{-3 - (-5)}{8}$ *Since the denominators are the same, subtract the numerators.*

$x = \frac{2}{8}$ or $\frac{1}{4}$ *Simplify.*

APPLICATION

Real World

3 **Home Decorating** Suppose you bought two storage blocks for your bedroom. One is $14\frac{3}{4}$ inches high. The other is $9\frac{3}{4}$ inches high. You want to stack them and fit them under the window, which is 24 inches from the floor. Will they fit?

Find the sum of the heights of the blocks.

$14\frac{3}{4} + 9\frac{3}{4} = 23\frac{6}{4}$ *Add the whole numbers and then the fractions.*

$= 24\frac{2}{4}$ or $24\frac{1}{2}$ *Rename the improper fraction.*

The sum of the blocks is higher than 24 inches. They will not fit under the window.

CHECK FOR UNDERSTANDING

Communicating Mathematics

Read and study the lesson to answer each question.

1. *Write* the subtraction sentence shown by the following model.

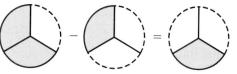

2. *Explain* how adding and subtracting rational numbers is similar to adding and subtracting integers.

HANDS-ON MATH

3. *Model* $\frac{3}{4} + \frac{3}{4}$ using a diagram and find the sum.

Guided Practice Solve each equation. Write the solution in simplest form.

4. $\frac{3}{8} + \frac{5}{8} = a$

5. $\frac{3}{4} + \left(-\frac{1}{4}\right) = d$

6. $k = -\frac{2}{3} - \frac{1}{3}$

7. $y = 2\frac{3}{5} - \left(-1\frac{4}{5}\right)$

8. *Algebra* Evaluate $a + b$ if $a = -\frac{3}{8}$ and $b = -2\frac{7}{8}$.

EXERCISES

Practice Solve each equation. Write the solution in simplest form.

9. $\frac{3}{8} - \frac{1}{8} = b$

10. $-\frac{5}{6} + \frac{1}{6} = f$

11. $m = -\frac{4}{5} - \frac{3}{5}$

12. $r = \frac{5}{6} - \left(-\frac{7}{6}\right)$

13. $-2\frac{1}{2} - \frac{1}{2} = d$

14. $\frac{1}{4} - \frac{3}{4} = z$

15. $c = -\frac{7}{12} - \frac{5}{12}$

16. $5\frac{2}{3} - \left(-2\frac{2}{3}\right) = x$

17. $-2\frac{5}{9} - \frac{5}{9} = y$

18. Subtract $-2\frac{1}{3}$ from $5\frac{1}{3}$.

19. What is the sum of $-\frac{5}{12}$ and $-\frac{1}{12}$?

Evaluate each expression if $a = -\frac{2}{5}$ and $b = \frac{3}{5}$.

20. $a + b$

21. $b - a$

22. $a - (a + b)$

Applications and Problem Solving

23. *Manufacturing* Plastic straps are often wound around large cardboard boxes to reinforce them during shipping. Suppose the ends of the strap must overlap $\frac{7}{16}$ inch to fasten. How long is the plastic strap around the box?

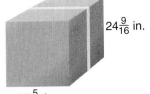

$24\frac{9}{16}$ in.

$28\frac{5}{16}$ in.

24. *Probability* A number cube is rolled.
  a. What is the probability of rolling a 2 or a 5?
  b. What is the probability of *not* rolling a 2 or a 5?
  c. Find the sum of the answers to parts a and b. Why is this true?

25. *Critical Thinking* A bookworm chewed its way from the first page of Volume I to the last page of Volume III. The books are on a bookshelf standing next to each other. If each volume is 2 inches thick and the covers are $\frac{3}{16}$-inch thick, how many inches did the bookworm chew?

Vol. I   Vol. II   Vol. III

Mixed Review

26. Express 9.2×10^{-4} in standard form. *(Lesson 6-9)*

27. **Standardized Test Practice** One centimeter is about 0.392 inch. What fraction of an inch is this? *(Lesson 6-5)*

A $\frac{49}{500}$ in.

B $\frac{98}{125}$ in.

C $\frac{49}{125}$ in.

D $\frac{392}{100}$ in.

For **Extra Practice**, see page 623.

28. *Geometry* Draw an isosceles right triangle. *(Lesson 5-2)*

Adding and Subtracting Unlike Fractions

What you'll learn

You'll learn to add and subtract fractions with unlike denominators.

When am I ever going to use this?

Knowing how to add and subtract fractions with unlike denominators is helpful when mixing paint.

Quoits (pronounced köits) was one of the five original games included in the ancient Greek pentathlon. It is similar to horseshoes, but is played with metal rings. The object of the game is to throw the quoit onto or as close as possible to a vertical metal pole.

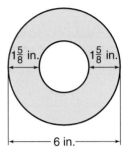

A quoit weighs about $3\frac{1}{2}$ pounds and is 6 inches across with a hole in the middle. If each section of metal is $1\frac{5}{8}$ inches wide, how large is the hole in the center of the quoit? *This problem will be solved in Example 2.*

In order to find the width of the hole in the center, you must subtract the total width of the metal, $1\frac{5}{8} + 1\frac{5}{8}$, from the width of the quoit, 6. The whole number 6 does not contain a fraction. To subtract, we must rename 6.

Adding and Subtracting Unlike Fractions	To find the sum or difference of two fractions with unlike denominators, rename the fractions with a common denominator. Then add or subtract and simplify, if necessary.

Example

INTEGRATION

The least common denominator (LCD) is the least common multiple (LCM) of the denominators. The LCD is helpful in renaming fractions for adding and subtracting.

1 Algebra Evaluate $m - n$ if $m = \frac{5}{6}$ and $n = -\frac{2}{3}$.

$$m - n = \frac{5}{6} - \left(-\frac{2}{3}\right) \quad \text{Replace } m \text{ with } \frac{5}{6} \text{ and } n \text{ with } -\frac{2}{3}.$$

$$= \frac{5}{6} - \left(-\frac{4}{6}\right) \quad \text{Use the LCM of 6 and 3 to rename } -\frac{2}{3} \text{ as } -\frac{4}{6}.$$

$$= \frac{5}{6} + \left(\frac{4}{6}\right) \quad \text{Subtract } -\frac{4}{6} \text{ by adding its inverse, } \frac{4}{6}.$$

$$= \frac{9}{6} \quad \text{Add the numerators.}$$

$$= 1\frac{3}{6} \text{ or } 1\frac{1}{2} \quad \text{Simplify.}$$

To subtract a fraction or mixed number from a whole number, you must rename the whole number with an improper fraction. To add or subtract mixed numbers with unlike denominators, first rename the fractions with a common denominator.

② Recreation Refer to the beginning of the lesson. Find the width of the hole in a quoit.

Explore   What do you know? The width of the quoit is 6 inches. Each section of metal is $1\frac{5}{8}$ inches wide.

Plan   Add the two metal widths and subtract from 6.

Solve   Find the total width of the metal.

$$1\frac{5}{8} + 1\frac{5}{8} = 2\frac{10}{8}$$
$$= 3\frac{2}{8} \text{ or } 3\frac{1}{4}$$

Now subtract $3\frac{1}{4}$ from 6.

$$6 - 3\frac{1}{4} = 5\frac{4}{4} - 3\frac{1}{4} \quad \textit{Rename 6 as } 5\frac{4}{4}.$$
$$= 2\frac{3}{4} \quad\quad\quad \textit{Simplify.}$$

The hole is $2\frac{3}{4}$ inches across.

Examine   See if the sum of the parts is equal to 6.

$$1\frac{5}{8} + 1\frac{5}{8} + 2\frac{3}{4} \stackrel{?}{=} 6$$
$$1\frac{5}{8} + 1\frac{5}{8} + 2\frac{6}{8} \stackrel{?}{=} 6 \quad \textit{Rename } \frac{3}{4} \textit{ as } \frac{6}{8}.$$
$$4\frac{16}{8} \stackrel{?}{=} 6$$
$$6 = 6 \quad \checkmark$$

③ Solve $6\frac{3}{8} - 2\frac{5}{6} = x$.

Estimate: 6 − 3 = 3

The LCM of 8 and 6 is 24.

$$
\begin{array}{lll}
6\frac{3}{8} & \rightarrow \quad 6\frac{9}{24} & \rightarrow \quad 5\frac{33}{24} \\
-2\frac{5}{6} & \quad -2\frac{20}{24} & \quad -2\frac{20}{24} \\
\hline
 & & \quad\quad 3\frac{13}{24}
\end{array}
$$

Rename $6\frac{9}{24}$ as $5\frac{33}{24}$. Why?

So, $3\frac{13}{24} = x$.   *Compare to the estimate.*

Communicating Mathematics

Read and study the lesson to answer each question.

1. *Write* the addition sentence shown by the model.

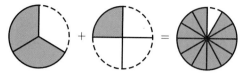

2. *Explain* why you might rename $5\frac{3}{8}$ as $4\frac{11}{8}$ in a subtraction problem.

HANDS-ON MATH

3. *Model* $\frac{7}{12} - \frac{1}{3}$ using the diagram. Then find the difference.

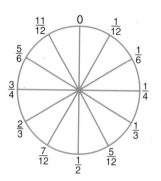

Guided Practice

Complete.

4. $5\frac{1}{3} = 4\frac{\blacksquare}{3}$

5. $3\frac{2}{5} = 2\frac{\blacksquare}{5}$

6. $4\frac{5}{8} = 3\frac{\blacksquare}{8}$

Solve each equation. Write the solution in simplest form.

7. $\frac{3}{4} + \frac{1}{2} = z$

8. $4\frac{3}{4} + \left(-\frac{1}{3}\right) = x$

9. $c = \frac{1}{4} - \left(-\frac{5}{8}\right)$

10. $h = -\frac{7}{10} + \frac{1}{5}$

11. $3\frac{7}{8} + 5\frac{5}{24} = d$

12. $5\frac{1}{3} - 3\frac{3}{4} = n$

13. *Algebra* Evaluate $c - d$ if $c = -\frac{3}{4}$ and $d = -12\frac{7}{8}$.

14. *Algebra* Simplify $1\frac{7}{8}a - \frac{1}{4}a$.

Practice

Solve each equation. Write the solution in simplest form.

15. $\frac{3}{8} + \frac{2}{3} = k$

16. $-\frac{5}{6} + \frac{1}{2} = g$

17. $m = -1\frac{4}{5} - \frac{3}{10}$

18. $y = -\frac{3}{4} + \left(-\frac{1}{3}\right)$

19. $9\frac{1}{3} - 2\frac{1}{2} = h$

20. $5 - 3\frac{1}{3} = f$

21. $s = 5 - 3\frac{3}{4}$

22. $j = -4\frac{1}{2} - \frac{3}{5}$

23. $b = -4\frac{1}{4} - 5\frac{5}{8}$

24. $d = 3\frac{2}{3} + \left(-5\frac{3}{4}\right)$

25. $7\frac{3}{4} - \left(-1\frac{1}{8}\right) = x$

26. $\frac{65}{187} - \frac{9}{136} = w$

27. Subtract $-6\frac{1}{4}$ from 9.

28. What is the sum of $-\frac{5}{8}$ and $-\frac{1}{2}$?

29. What is $2\frac{3}{8}$ less than $-8\frac{1}{5}$?

30. Find the sum of $\frac{1}{2}$, $\frac{1}{3}$, and $\frac{1}{5}$.

Evaluate each expression if $r = -\frac{5}{8}$, $s = 2\frac{5}{6}$, and $t = \frac{5}{9}$.

31. $r - s$ **32.** $r + s + t$ **33.** $s - (-r)$

Applications and Problem Solving

34. *Music* A waltz is written in $\frac{3}{4}$ (read as three-four) time. This means there are three beats in a measure and the quarter note gets one beat. The value of the notes can be expressed as fractions, and the total value of each measure is $\frac{3}{4}$. The first few measures of *The Laughing Song* are shown with the values of each note. What type of note must be used to finish the last measure?

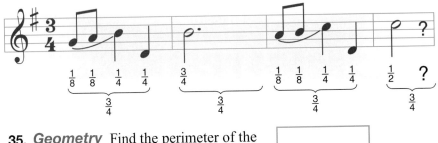

35. *Geometry* Find the perimeter of the rectangle if the width is $9\frac{3}{8}$ inches shorter than the length.

$21\frac{1}{4}$ in.

36. *Life Science* In frog jumping contests, the distance recorded is the combined lengths of three consecutive jumps. The longest jump in the United States was 21 feet $5\frac{3}{4}$ inches in 1986. However, the longest jump in the world was in South Africa in 1977 for a length of 33 feet $5\frac{1}{2}$ inches. How much farther is the world record than the U.S. record? (*Hint:* Change the lengths to inches.)

37. *Critical Thinking* The figure at the right is a magic square.
    a. Find the values of a and b so that each row, column, and diagonal has the same sum.
    b. Create your own magic square. Explain how you did it.

a	$\frac{7}{10}$	$1\frac{1}{5}$
$1\frac{3}{10}$	$\frac{9}{10}$	b
$\frac{3}{5}$	$1\frac{1}{10}$	1

38. *Critical Thinking* A bucket is put under two faucets. If one faucet is turned on alone, the bucket will be filled in 5 minutes; if the other is turned on, the bucket will be filled in 3 minutes. If both are turned on, how many seconds will it take to fill the bucket?

Mixed Review

39. Find the sum of $-5\frac{1}{6}$ and $2\frac{5}{6}$. *(Lesson 7-1)*

40. **Standardized Test Practice** José correctly answered 35 out of 50 questions on his math test. How would you describe his success as a fraction in simplest form? *(Lesson 6-4)*

  A $\frac{7}{10}$ **B** $\frac{18}{25}$ **C** $\frac{50}{35}$ **D** $\frac{35}{50}$

41. *Statistics* Determine whether a scatter plot of the number of base hits to the number of times at bat would show a *positive*, *negative*, or *no* relationship. *(Lesson 4-6)*

For **Extra Practice,** see page 623.

42. Solve $j = -7(15)(-10)$. *(Lesson 2-7)*

Fraction Track

Get Ready This game is for two to four players.

two number cubes   counters

Get Set
- Make the game board shown.
- Each person starts with a counter on each box at the left on the "fraction track."

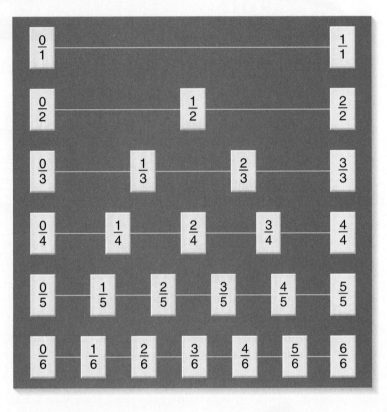

Go
- The first player rolls two number cubes and creates a fraction. The smaller number is always the numerator. For example, suppose a player rolls a 1 and a 6. The player's fraction is $\frac{1}{6}$.
- A player may move one counter the full value of the fraction or move any number of counters as long as they total exactly the value of the fraction rolled. For example, $\frac{1}{2}$ can be played as a $\frac{1}{3}$ and a $\frac{1}{6}$ since $\frac{1}{3} + \frac{1}{6} = \frac{1}{2}$.
- If the player rolls a fraction and cannot move, play moves to the next player.
- The winner is the first player to have all of the counters at the right.

inter NET CONNECTION Visit www.glencoe.com/sec/math/mac/mathnet for more games.

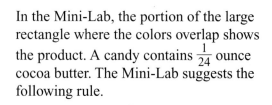

7-3 Multiplying Fractions

What you'll learn

You'll learn to multiply fractions.

When am I ever going to use this?

You will multiply fractions when you cut your brownie recipe in half.

Since the candy shown at the right was introduced on July 1, 1907, its size and shape has not changed. Each candy weighs $\frac{1}{6}$ ounce. If about $\frac{1}{4}$ of a candy is composed of cocoa butter, how much cocoa butter is in each candy?

You need to find $\frac{1}{4}$ of $\frac{1}{6}$. This means $\frac{1}{4} \times \frac{1}{6}$. You can use models to solve the problem.

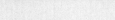

HANDS-ON MINI-LAB

Work with a partner.   ▦ lined paper   ◿ ruler

Try This   ✎ markers

- Draw a rectangle that is 6 rows long. Shade 1 row to represent $\frac{1}{6}$.
- Draw vertical lines that separate the rectangle into four equal columns. Using a different color, shade 1 column to represent $\frac{1}{4}$.

Talk About It

1. How many small rectangles are inside of the large rectangle?
2. How many of the small rectangles are shaded by two colors?
3. If the large rectangle has an area of 1, what fraction of the large rectangle is shaded by two colors?

In the Mini-Lab, the portion of the large rectangle where the colors overlap shows the product. A candy contains $\frac{1}{24}$ ounce cocoa butter. The Mini-Lab suggests the following rule.

Multiplying Fractions	**Words:**	To multiply fractions, multiply the numerators and multiply the denominators.
	Symbols: **Arithmetic**	$\frac{2}{5} \cdot \frac{2}{3} = \frac{4}{15}$
	Algebra	$\frac{a}{b} \cdot \frac{c}{d} = \frac{ac}{bd}, b \neq 0, d \neq 0$

Use the rules of signs for multiplying integers when you multiply rational numbers.

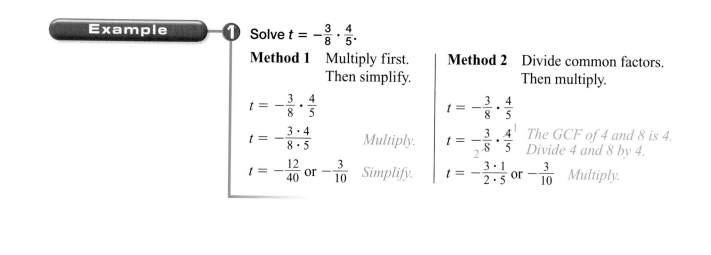

Example 1

Solve $t = -\frac{3}{8} \cdot \frac{4}{5}$.

Method 1 Multiply first. Then simplify.

$t = -\frac{3}{8} \cdot \frac{4}{5}$

$t = -\frac{3 \cdot 4}{8 \cdot 5}$ *Multiply.*

$t = -\frac{12}{40}$ or $-\frac{3}{10}$ *Simplify.*

Method 2 Divide common factors. Then multiply.

$t = -\frac{3}{8} \cdot \frac{4}{5}$

$t = -\frac{3}{\overset{}{\underset{2}{8}}} \cdot \frac{\overset{1}{4}}{5}$ *The GCF of 4 and 8 is 4.* *Divide 4 and 8 by 4.*

$t = -\frac{3 \cdot 1}{2 \cdot 5}$ or $-\frac{3}{10}$ *Multiply.*

To multiply mixed numbers, first rename them as improper fractions.

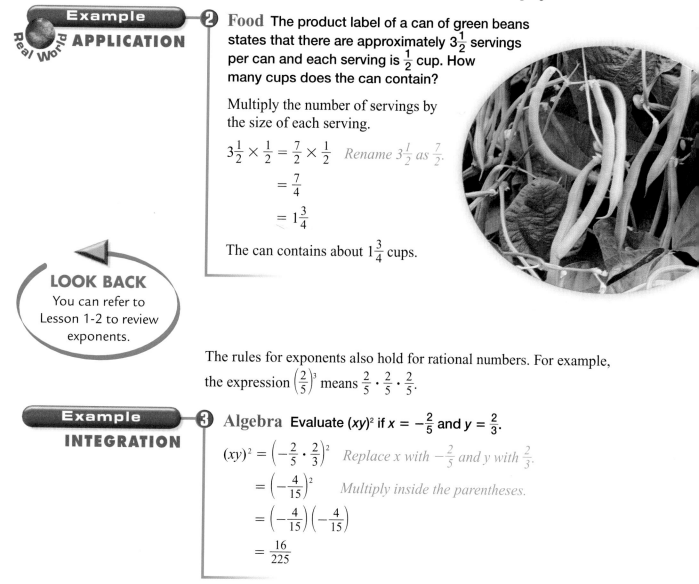

Example 2

Real World **APPLICATION**

Food The product label of a can of green beans states that there are approximately $3\frac{1}{2}$ servings per can and each serving is $\frac{1}{2}$ cup. How many cups does the can contain?

Multiply the number of servings by the size of each serving.

$3\frac{1}{2} \times \frac{1}{2} = \frac{7}{2} \times \frac{1}{2}$ *Rename $3\frac{1}{2}$ as $\frac{7}{2}$.*

$= \frac{7}{4}$

$= 1\frac{3}{4}$

The can contains about $1\frac{3}{4}$ cups.

LOOK BACK

You can refer to Lesson 1-2 to review exponents.

The rules for exponents also hold for rational numbers. For example, the expression $\left(\frac{2}{5}\right)^3$ means $\frac{2}{5} \cdot \frac{2}{5} \cdot \frac{2}{5}$.

Example 3

INTEGRATION

Algebra Evaluate $(xy)^2$ if $x = -\frac{2}{5}$ and $y = \frac{2}{3}$.

$(xy)^2 = \left(-\frac{2}{5} \cdot \frac{2}{3}\right)^2$ *Replace x with $-\frac{2}{5}$ and y with $\frac{2}{3}$.*

$= \left(-\frac{4}{15}\right)^2$ *Multiply inside the parentheses.*

$= \left(-\frac{4}{15}\right)\left(-\frac{4}{15}\right)$

$= \frac{16}{225}$

Communicating Mathematics

Read and study the lesson to answer each question.

1. **Tell** how the model shows the product of $\frac{3}{8}$ and $\frac{3}{4}$.

2. **Explain** how to evaluate ab^2 if $a = -\frac{1}{2}$ and $b = -\frac{2}{3}$.

HANDS-ON MATH

3. **Draw** a model that shows the product of $\frac{4}{5}$ and $\frac{3}{4}$. Then find the product.

Guided Practice

Solve each equation. Write the solution in simplest form.

4. $\frac{1}{3} \cdot \frac{3}{4} = z$

5. $\frac{4}{5} \cdot 3 = b$

6. $d = -2\left(-\frac{5}{8}\right)$

7. $x = \left(2\frac{2}{5}\right)^2$

8. **Algebra** Evaluate $c^2 d$ if $c = -\frac{2}{3}$ and $d = 2\frac{5}{8}$.

9. **Geometry** Find the perimeter and area of the parallelogram.

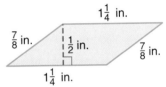

Practice

Solve each equation. Write the solution in simplest form.

10. $k = \frac{3}{8} \cdot \frac{4}{9}$

11. $f = -\frac{7}{8} \cdot \frac{5}{6}$

12. $\frac{3}{4} \cdot \left(-\frac{1}{3}\right) = y$

13. $-3\frac{3}{8} \cdot -\frac{2}{3} = j$

14. $g = -\frac{5}{6} \cdot 1\frac{4}{5}$

15. $n = 2\left(-\frac{3}{10}\right)$

16. $c = -4\frac{1}{4} \cdot \left(-3\frac{1}{3}\right)$

17. $p = 1\frac{1}{3} \cdot 5\frac{1}{2}$

18. $q = 9\left(-3\frac{2}{3}\right)$

19. $s = \left(-\frac{3}{4}\right)^2$

20. $t = \left(1\frac{1}{8}\right)^2$

21. $r = \left(-\frac{2}{3}\right)^3$

22. What is one-half of $-\frac{5}{8}$ times $-3\frac{1}{5}$?

23. Find the product of $\frac{1}{2}$, $\frac{2}{3}$, and $-\frac{3}{4}$.

Evaluate each expression if $w = 1\frac{1}{3}$, $x = -\frac{1}{2}$, $y = 3\frac{3}{4}$, and $z = -\frac{4}{5}$.

24. xz

25. $2y$

26. w^2

27. $x^3(-y)$

Applications and Problem Solving

28. **Food** Refer to Example 2 on page 287. Suppose another label on the can states that the can contained $9\frac{3}{4}$ ounces of beans before liquid was added for processing. If there are 8 ounces in a cup, does a serving of beans contain liquid, according to the information on the label? Explain.

29. *Bicycles* The saddle height is the distance between the height of the saddle (seat) and the lowest point of the pedal. To find your proper saddle height, multiply your inseam measure times $\frac{27}{25}$. The frame height is the perpendicular distance between the top of the frame and the line drawn between the centers of the wheels. To find your appropriate frame height, multiply your inseam by $\frac{2}{3}$ and add 2.

    a. Antonio's inseam is 25 inches. What saddle height should he use?

    b. What frame height should a person with a 30-inch inseam use?

30. *Critical Thinking* What fraction is halfway between $\frac{1}{4}$ and $\frac{1}{3}$?

Mixed Review

31. *Standardized Test Practice* Yana spent $\frac{2}{3}$ of an hour doing his homework on Monday. On Tuesday, he spent $1\frac{1}{2}$ hours doing homework. How much more time did he spend doing homework on Tuesday than on Monday? *(Lesson 7-2)*

    A $2\frac{1}{6}$ h   **B** $\frac{13}{6}$ h   **C** $\frac{5}{6}$ h   **D** $\frac{1}{4}$ h   **E** $\frac{1}{6}$ h

32. Find the LCM of 9 and 30. *(Lesson 6-7)*

33. *Algebra* Find the value of x in $\triangle DEF$ if $m\angle D = 4x°$, $m\angle E = 60°$, and $m\angle F = 72°$. *(Lesson 5-2)*

For Extra Practice, see page 624.

34. *Communication* The cost of a long distance phone call is $0.15 for the first 5 minutes and $0.10 for each additional minute. If Lenora makes a 30-minute long distance phone call, how much will she be charged? *(Lesson 1-1)*

MATH IN THE MEDIA

CLOSE TO HOME JOHN McPHERSON

© 1993 John McPherson/Dist. by Universal Press Syndicate

"I'll take a large pizza with half-onion, two-thirds olives, nine-fifteenths mushrooms, five-eighths pepperoni, one-eighth anchovies, and extra cheese on five-ninths of the onion half."

1. What fraction of the pizza should have extra cheese?

2. Is it possible to make the pizza ordered? Explain.

3. *Write a Problem* using the information in the comic.

Properties of Rational Numbers

Sojourner rover

What you'll learn

You'll learn to identify and use rational number properties.

When am I ever going to use this?

Understanding properties of rational numbers can help you determine weights on other planets.

Word Wise

inverse property of
    multiplication
multiplicative inverse
reciprocal

Did you know? The Sojourner rover was named after 19th-century African-American reformer Sojourner Truth.

On July 4, 1997, the Pathfinder landed on Mars. The next day, the Sojourner rover rolled out of Pathfinder.

Objects on Earth are about $2\frac{3}{5}$ times heavier than objects on Mars. If the Sojourner weighs about 9 pounds on Mars, how much did it weigh on Earth? *This problem will be solved in Example 2.*

All of the properties that were true for addition and multiplication of integers are also true for addition and multiplication of rational numbers.

Property	Arithmetic	Algebra
Commutative	$\frac{5}{8} + \frac{1}{3} = \frac{1}{3} + \frac{3}{8}$	$a + b = b + a$
	$\frac{1}{3} \cdot \frac{2}{5} = \frac{2}{5} \cdot \frac{1}{3}$	$a \cdot b = b \cdot a$
Associative	$\left(-\frac{1}{2} + \frac{2}{3}\right) + \frac{4}{5} = -\frac{1}{2} + \left(\frac{2}{3} + \frac{4}{5}\right)$	$(a + b) + c = a + (b + c)$
	$\left(-\frac{1}{2} \cdot \frac{2}{3}\right) \cdot \frac{4}{5} = -\frac{1}{2} \cdot \left(\frac{2}{3} \cdot \frac{4}{5}\right)$	$(a \cdot b) \cdot c = a \cdot (b \cdot c)$
Identity	$\frac{3}{4} + 0 = \frac{3}{4}$	$a + 0 = a$
	$-\frac{1}{2} \cdot 1 = -\frac{1}{2}$	$a \cdot 1 = a$

In Chapter 2, you learned about the inverse property of addition. A similar property that applies to the multiplication of rational numbers is called the **inverse property of multiplication**. Two numbers whose product is 1 are **multiplicative inverses**, or **reciprocals**, of each other. For example, $-\frac{5}{6}$ and $-\frac{6}{5}$ are multiplicative inverses because $-\frac{5}{6} \cdot \left(-\frac{6}{5}\right) = 1$.

Inverse Property of Multiplication	**Words:**	The product of a rational number and its multiplicative inverse is 1.
	Symbols: **Arithmetic**	$\frac{2}{5} \cdot \frac{5}{2} = 1$
	Algebra	$\frac{a}{b} \cdot \frac{b}{a} = 1$, where $a, b \neq 0$.

Example ❶ Name the multiplicative inverse of $2\frac{3}{4}$.

$2\frac{3}{4} = \frac{11}{4}$ *Rename the mixed number as an improper fraction.*

$\frac{11}{4} \times \square = 1$ *What number can you multiply by $\frac{11}{4}$ to get 1?*

$\frac{11}{4} \times \frac{4}{11} = 1$

The multiplicative inverse of $2\frac{3}{4}$ is $\frac{4}{11}$.

You can use the distributive property to find products involving mixed numbers.

Example ❷   **Physical Science** Refer to the beginning of the lesson. How much

CONNECTION   did the Sojourner rover weigh on Earth?

LOOK BACK
You can refer to Lesson 1-7 to review the distributive property.

Method 1 Use the distributive property.

$w = 2\frac{3}{5} \cdot 9$

$= 9 \cdot 2\frac{3}{5}$ *Commutative property*

$= 9\left(2 + \frac{3}{5}\right)$

$= (9 \cdot 2) + \left(9 \cdot \frac{3}{5}\right)$ *Distributive property*

$= 18 + \frac{27}{5}$

$= 18 + 5\frac{2}{5}$ or $23\frac{2}{5}$

Method 2 Multiply first. Then simplify.

$w = 2\frac{3}{5} \cdot 9$

$= \frac{13}{5} \cdot \frac{9}{1}$

$= \frac{117}{5}$

$= 23\frac{2}{5}$

The Sojourner rover weighed about 23 pounds on Earth.

CHECK FOR UNDERSTANDING

Communicating Mathematics

Read and study the lesson to answer each question.

1. **Write** the reciprocal of $\frac{5}{9}$.

2. **Identify** the property that allows you to compute $1\frac{1}{8} \cdot 8$ as $8 \cdot 1\frac{1}{8}$. Then state the product.

3. **Write** a sentence or two explaining how to find the multiplicative inverse of $-3\frac{7}{8}$.

Guided Practice

Name the multiplicative inverse of each of the following.

4. $-\frac{1}{5}$

5. $\frac{7}{8}$

6. $\frac{3}{4}$

Solve each equation using properties of rational numbers.

7. $\frac{3}{4} \cdot \left(-\frac{2}{3} \cdot -\frac{1}{2}\right) = x$

8. $y = -3\frac{3}{5} \cdot \frac{1}{3}$

9. **Algebra** Evaluate ab if $a = \frac{1}{2}$ and $b = -2\frac{4}{5}$.

Practice · **Name the multiplicative inverse of each rational number.**

10. 8   **11.** $-\frac{2}{3}$ **12.** $2\frac{3}{5}$ **13.** -0.2

14. -1 **15.** 2.25 **16.** $\frac{a}{b}$ **17.** $-x$

Solve each equation using properties of rational numbers.

18. $p = 6 \cdot 3\frac{5}{6}$ **19.** $\left(\frac{2}{3} \cdot -\frac{1}{2}\right) \cdot \frac{3}{8} = q$ **20.** $r = 16\frac{2}{9} \cdot \frac{3}{4}$

21. $x = -7 \cdot 8\frac{1}{2}$ **22.** $\frac{3}{5} \cdot 10\frac{1}{3} = t$ **23.** $w = -\frac{4}{5}\left(\frac{5}{8} \cdot \frac{10}{11}\right)$

24. Is $\frac{5}{6}$ the reciprocal of $1\frac{1}{5}$? Explain.

25. Are $-2\frac{1}{2}$ and $\frac{2}{5}$ reciprocals? Explain.

Evaluate each expression if $u = -1\frac{1}{6}$, $v = \frac{2}{3}$, and $w = -\frac{3}{4}$.

26. v^3 **27.** $4w + 3v$ **28.** $w^2(6 - v)$ **29.** uvw

Applications and Problem Solving

Real World

30. *Life Science* The average Alaskan brown bear is about $1\frac{1}{8}$ times as long as a grizzly bear. If the average grizzly bear is 8 feet long, how long is an Alaskan brown bear?

31. *Geometry* Find the area of the rectangle.

$14\frac{3}{4}$ in.

$2\frac{1}{2}$ in.

32. *Critical Thinking*
     a. Copy and complete the table.

n	1	2	3	4
n^2				
$\frac{1}{n^2}$				

For **Extra Practice**, see page 624.

     b. What pattern do you notice as you move across each row to the right?
     c. What relationship do you see between n^2 and $\frac{1}{n^2}$?

Mixed Review

33. Solve $-3\frac{5}{8}\left(-5\frac{1}{4}\right) = s$. *(Lesson 7-3)*

34. Find the GCF of 24 and 64. *(Lesson 6-3)*

35. Estimate 48% of 178. *(Lesson 3-6)*

36. **Standardized Test Practice** Brianna's Girl Scout troop sold 3,275 boxes of cookies last year. This year they sold 5,163 boxes. How many more boxes did they sell this year than last year? *(Lesson 2-5)*

     A 1,888   **B** 1,988   **C** 2,998   **D** 3,008   **E** Not Here

HEALTH

Dr. Vivian Pinn
PATHOLOGIST

Vivian Pinn is a pathologist and the director of the Office of Research on Women's Health at the National Institutes of Health. When Dr. Pinn entered the University of Virginia Medical School in 1963, she was the only African-American and the only woman.

Pathology is the study of the nature, cause, progression, and effects of diseases. A person who is interested in becoming a pathologist should consider an undergraduate degree in biology or chemistry with a strong emphasis in mathematics. Then the person must attend a medical school to complete a medical degree.

For more information:
American Medical Association
515 North State Street
Chicago, IL 60610

inter**NET**
CONNECTION
www.glencoe.com/sec/
math/mac/mathnet

I really enjoy my science class! I think that I would like to be a doctor someday!

Your Turn
The life of a doctor is very different from that shown on television dramas. Interview a doctor to find out about their experiences in completing medical training. Ask him or her about the importance of mathematics in their career. Write a paragraph about your findings.

7-5A Look for a Pattern

A Preview of Lesson 7-5

Santos and Becky are doing research on radiocarbon dating for a science report. Let's listen in!

Santos

This chart shows that half of the radiocarbon atoms decay every 5,700 years.

I see that each fraction is $\frac{1}{2}$ of the previous one.

So, if there is only $\frac{1}{64}$ of the original amount left, it is 34,200 years old.

I know how you came up with $\frac{1}{64}$. But how did you determine the number of years so fast?

I saw that the second number of years was double the first number. Then I noticed that the fourth number was double the second number. So, to get the sixth number, I doubled the third number.

Becky

Radiocarbon Atoms	
Original Amount Remaining	Age (yr)
$\frac{1}{2}$	5,700
$\frac{1}{4}$	11,400
$\frac{1}{8}$	17,100
$\frac{1}{16}$	22,800
$\frac{1}{32}$	28,500

THINK ABOUT IT

1. *Deductive reasoning* uses a rule to make a conclusion, and *inductive reasoning* makes a rule after seeing several examples. **Tell** whether Becky or Santos used inductive reasoning. Explain.

2. **Explain** how Becky and Santos knew the next fraction was $\frac{1}{64}$.

3. **Determine** the next two rows of the chart.

4. **Apply** the *look for a pattern* strategy to draw the next two figures in the pattern.

For **Extra Practice,** see page 624.

ON YOUR OWN

5. The first step of the 4-step plan for problem solving asks you to *explore*. **Describe** how looking for a pattern relates to this step.

6. *Write a Problem* that can be solved by finding a pattern. Then ask a classmate to describe the pattern and solve the problem.

7. *Look Ahead* Look for a pattern to determine the next two numbers in each list.

   a. 5, 10, 15, _?_ , _?_

   b. 39, 33, 27, _?_ , _?_

MIXED PROBLEM SOLVING

STRATEGIES

Look for a pattern.
Solve a simpler problem.
Act it out.
Guess and check.
Draw a diagram.
Make a chart.
Work backward.

Solve. Use any strategy.

8. *Pets* Mrs. Vallez loves cats and canaries. Altogether her pets have thirty heads and eighty legs. How many cats does she have?

9. *Physical Science* The Italian scientist Galileo discovered that there was a relationship between the time of the swing back and forth of a pendulum and its length.

Time of Swing	Length of Pendulum
1 second	1 unit
2 seconds	4 units
3 seconds	9 units
4 seconds	16 units

   a. How long is a pendulum with a swing of 5 seconds?

   b. How long is a pendulum with a swing of 1.5 seconds?

   c. Use the time as the *x*-coordinate and the length as the *y*-coordinate to graph the data in the table. Describe the graph.

10. *School* Judi was assigned some math exercises for homework. She answered half of them in a study period. After school, she answered 7 more exercises. If she still has 11 exercises to do, how many exercises were assigned?

11. *Geometry* What is the total number of rectangles, of any size, in the figure below?

12. *Money Matters* After a shopping trip, Abbi and Sophia counted their money to see how much they had left. Abbi said to Sophia, "If I had $4 more, I would have as much as you." Sophia replied, "If I had $4 more, I would have twice as much as you." How much does each girl have?

13. *Spreadsheets* The spreadsheet can be used to investigate bounce height when a ball is dropped from a height of 25 meters. The value in cell B2 tells you that each time the ball bounces it returns to a height that is $\frac{2}{5}$ or 0.4 times the previous height.

	A	B
1	Initial Height (m)	25
2	Bounce Factor	0.4
3	Height Number	Return Height
4	0	= B1
5	= A4 + 1	= B4*B2
6	= A5 + 1	= B5*B2

   a. What value will be in cell B5?

   b. A different ball will return to a height that is 0.5 times the previous height. How would you modify the spreadsheet to find heights for this ball?

14. *Standardized Test Practice* Which number is missing from this pattern?

   . . . , _?_ , 29, 22, 15, 8, . . .

   A 30   **B** 35   **C** 36   **D** 37

Integration: Patterns and Functions
Sequences

What you'll learn

You'll learn to recognize and extend arithmetic and geometric sequences.

When am I ever going to use this?

Sequences can be used to find the cost of a long-distance telephone call.

Word Wise

sequence
term
arithmetic sequence
common difference
geometric sequence
common ratio

Have you ever made yourself a hot bowl of instant oatmeal on a cold winter morning? If so, you know that it is made with a mix and boiling water. A box of oatmeal has the following table to determine the right amounts of water and oatmeal mix.

Servings	1	2	3
Water	1 c	$1\frac{3}{4}$ c	$2\frac{1}{2}$ c
Oats	$\frac{1}{2}$ c	1 c	$1\frac{1}{2}$ c

If you look at the amounts of water in order, the numbers form a **sequence**. A sequence is a list of numbers in a certain order, such as $1, 1\frac{3}{4}, 2\frac{1}{2}, ...$ or $\frac{1}{2}, 1, 1\frac{1}{2}, ...$. Each number is called a **term** of the sequence. When the difference between any two consecutive terms is the same, the sequence is called an **arithmetic sequence**. The difference is called the **common difference**.

Examples

APPLICATION

1 **Cooking** Refer to the beginning of the lesson. How many cups of water are needed for the next three numbers of servings of oatmeal?

Find the difference between consecutive terms.

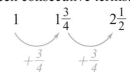

The difference between any two consecutive terms is $\frac{3}{4}$. So, this is an arithmetic sequence. Add $\frac{3}{4}$ to the last term of the sequence, and continue adding until the next three terms are found. The next three serving sizes will need $3\frac{1}{4}$ cups, 4 cups, and $4\frac{3}{4}$ cups of water.

2 State whether the sequence 15, 7, 0, −6, . . . is arithmetic. Then find the next three terms.

Since there is no common difference, the sequence is *not* arithmetic. But we can still identify a pattern to this sequence. Add −5 to the last term of the sequence, and continue adding the next greater integer until the next three terms are found. The next three terms are −11, −15, and −18.

Study Hint

Reading Math The three dots following a list of numbers are read as *and so on*.

The first term of a sequence can be represented by a_1, the second term a_2, the third term a_3, and so on up to the nth term, a_n. To find the nth term of an arithmetic sequence, you can use the formula $a_n = a_1 + (n - 1)d$, where n is the number of the term you want to find and d is the common difference.

Example

INTEGRATION

③ Algebra Find the fifteenth term, a_{15}, in the sequence 12.3, 13.8, 15.3, . . .

This sequence is arithmetic, and the common difference is 1.5.

$a_n = a_1 + (n - 1)d$

$a_{15} = 12.3 + (15 - 1)1.5$ *Replace n with 15, a_1 with 12.3, and d with 1.5.*

$a_{15} = 12.3 + (14)(1.5)$

$a_{15} = 12.3 + 21$

$a_{15} = 33.3$

So, the fifteenth term is 33.3.

Sometimes, the consecutive terms of a sequence are formed by multiplying by a constant factor. This type of sequence is called a **geometric sequence**. The factor is called the **common ratio**.

Examples

State whether each sequence is geometric. Then find the next three terms.

④ 3, −6, 12, −24, . . .

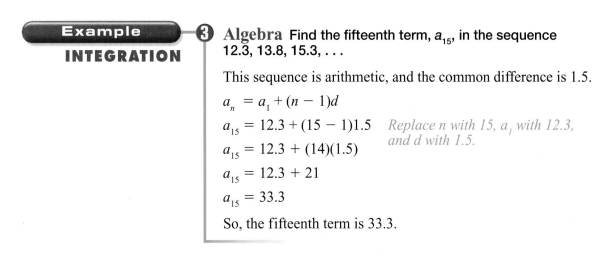

3   −6   12   −24

$\times(-2)$ $\times(-2)$ $\times(-2)$

Since there is a common ratio, −2, the sequence is a geometric sequence. Multiply the last term of the sequence by –2, and continue multiplying until the next three terms are found. The next three terms are 48, −96, and 192.

⑤ 96, 48, 12, 2, . . .

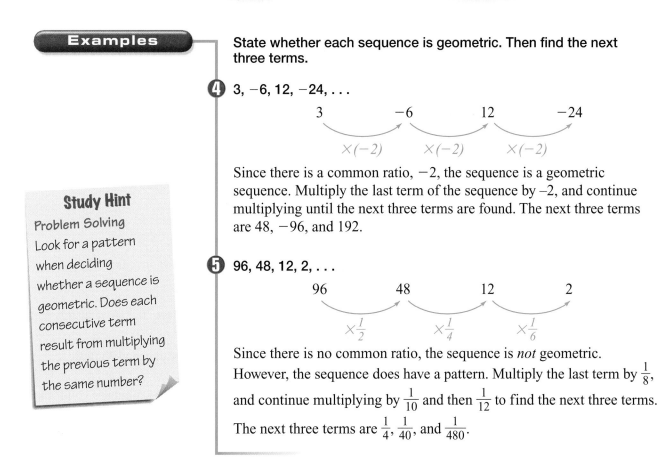

96   48   12   2

$\times\frac{1}{2}$ $\times\frac{1}{4}$ $\times\frac{1}{6}$

Since there is no common ratio, the sequence is *not* geometric. However, the sequence does have a pattern. Multiply the last term by $\frac{1}{8}$, and continue multiplying by $\frac{1}{10}$ and then $\frac{1}{12}$ to find the next three terms. The next three terms are $\frac{1}{4}$, $\frac{1}{40}$, and $\frac{1}{480}$.

Communicating Mathematics

Read and study the lesson to answer each question.

1. *Explain* how to determine whether a sequence is arithmetic.

2. *Write* a geometric sequence with a common ratio of 3.

3. *You Decide* Joanne says that the next term in the sequence 12, 8, 4, 0, ... is found by adding −4 to the last term. Marcie says that the next term is found by multiplying the last term by $\frac{1}{2}$. Who is correct? Explain.

Guided Practice

Identify each sequence as *arithmetic*, *geometric*, or *neither*. Then find the next three terms.

4. 2, 4, 6, 8, ...   5. 11, 4, −2, −7, ...   6. 10, 20, 40, 80, ...

7. 3, −6, 12, −24, ...   8. 1, 1, 2, 6, 24, ...   9. $4, 6\frac{1}{2}, 9, 11\frac{1}{2}, ...$

10. *Money Matters* Marianne has $5 in her bank. Suppose she puts $1.50 into her bank each week. If she does not take any money out, how much will be in the bank after 20 weeks?

Practice

Identify each sequence as *arithmetic*, *geometric*, or *neither*. Then find the next three terms.

11. 1, 3, 9, 27, ...

12. −6, −4, −2, 0, ...

13. $-4, -1, -\frac{1}{4}, -\frac{1}{16}, ...$

14. −5, 0, 4, 7, ...

15. 20, 24, 28, 32, ...

16. 20, 10, 5, 2.5, ...

17. 2, −6, 18, −54, ...

18. $1, \frac{1}{2}, \frac{1}{6}, \frac{1}{24}, ...$

19. $63, 21, 7, 2\frac{1}{3}, ...$

20. 1, 2, 5, 10, 17, ...

21. $4\frac{1}{2}, 4\frac{1}{6}, 3\frac{5}{6}, 3\frac{1}{2}, ...$

22. 156.55, 144.25, 131.95, 119.65,

23. 84, 72, 97, 85, 110, 98, 123, ...

24. −1, 1, −1, 1, ...

25. What are the first four terms in an arithmetic sequence with a common difference of $3\frac{1}{3}$ if the first term is 4?

26. Find the twentieth term in the sequence 18, 14, 10, 6, ...

27. What is the thirtieth term in the sequence 1.5, 4, 6.5, 9, ... ?

28. The sixth term of a geometric sequence is 81. The common ratio is $-\frac{1}{3}$. Find the first five terms.

Applications and Problem Solving

29. *Working on the* **CHAPTER Project** The sequence 1, 1, 2, 3, 5, 8, 13, 21, ... is called the *Fibonacci sequence.*

   a. Find the next three terms in the sequence. Is the sequence arithmetic, geometric, or neither? Explain.

   b. Compare the numbers in the Fibonacci sequence to the numbers you found in the pinecone spirals and plant leaves. Describe your findings.

   c. Find other plant materials and look for Fibonacci numbers. Add pictures and descriptions to your display or slide show.

30. Sports The graph shows how the area of the hitting surface of an average tennis racket has increased since 1976.

a. If this pattern continues, what will be the next area of the hitting surface of an average racket? Explain.

b. In what year would this increase be reported? Explain.

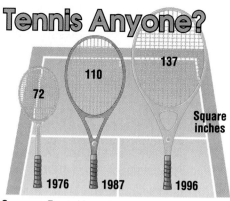

Tennis Anyone?

72 110 137

Square inches

1976 1987 1996

Source: *Tennis Magazine*

31. Money Matters Suppose a cellular phone service charges a base rate of $20 per month and $0.25 per minute for each call. If you talked for 55 minutes last month, how much would you be charged before taxes?

32. Critical Thinking The second term of an arithmetic sequence is 3, and the sixth term is 9.

a. What is the common difference? How did you find it?

b. Explain how to find the common difference of an arithmetic sequence if you know any two terms.

Mixed Review

33. Standardized Test Practice What number will make $\frac{4}{5} + \frac{2}{3} = \frac{2}{3} + g$ true? *(Lesson 7-4)*

    A $\frac{6}{8}$ **B** $\frac{4}{5}$ **C** $\frac{3}{5}$ **D** $\frac{2}{3}$

34. Find the LCD of $-\frac{5}{6}$ and $\frac{3}{8}$. *(Lesson 6-8)*

35. Probability A number cube is rolled. What is the probability it shows an odd number? *(Lesson 6-6)*

36. Statistics Find the interquartile range of the set {239, 226, 232, 212, 243, 216, 250}. *(Lesson 4-5)*

For **Extra Practice**, see page 625.

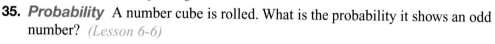

CHAPTER 7

Mid-Chapter Self Test

Solve each equation. Write the solution in simplest form.
(Lessons 7-1, 7-2, and 7-3)

1. $\frac{3}{8} + \left(-\frac{1}{8}\right) = d$ **2.** $z = \frac{1}{8} - \frac{7}{8}$ **3.** $12 - 5\frac{3}{5} = g$

4. $x = -7\frac{3}{4} + \left(-3\frac{4}{5}\right)$ **5.** $8\frac{3}{4}(-11) = b$ **6.** $k = \left(-\frac{5}{6}\right)^2$

7. Find the product of $\frac{1}{2}$ and $12\frac{4}{5}$ using the distributive property. *(Lesson 7-4)*

8. Construction Lavar is tiling the wall over a kitchen counter. The space requires 7 rows of tiles with $15\frac{1}{2}$ tiles in each row. How many tiles will be in the finished space? *(Lesson 7-4)*

Identify each sequence as *arithmetic*, *geometric*, or *neither*. Then find the next three terms. *(Lesson 7-5)*

9. 768, 192, 48, ...

10. 20, 23, 26, 29, ...

COOPERATIVE LEARNING

7-5B The Fibonacci Sequence

A Follow-Up of Lesson 7-5

grid paper

Leonardo was born about A.D. 1170 in the city of Pisa, Italy. This famous mathematician was also known by the name Fibonacci. Leonardo created story problems centered on a series of numbers that became known as the *Fibonacci sequence.* In this lab, you will discover this sequence.

TRY THIS

Work in groups of three.

Step 1 Using grid paper or dot paper, draw a "brick" that is 2 units long and 1 unit wide. If you build a "road" of grid paper bricks, there is only one way to build a road that is 1 unit long.

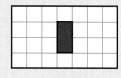

Step 2 Using two bricks from Step 1, draw all of the different roads possible that are 2 units long.

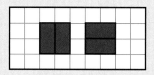

Step 3 Using three bricks, draw all of the different roads possible that are 3 units long.

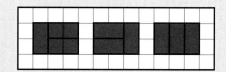

ON YOUR OWN

1. Draw all of the different roads possible that are 4 units long, 5 units long, and 6 units long using the indicated number of bricks.

2. Copy and complete the chart.

Length of Road	0	1	2	3	4	5	6
Number of Ways to Build the Road	1	1	2	3			

3. Look for a pattern in the numbers in the second row. How is each number related to the previous numbers?

4. Without making a drawing, how many ways are there to build a road that is 7 units long?

5. **Reflect Back** The numbers in the second row of the chart are the first numbers in the Fibonacci sequence. Write the ratios of consecutive terms of the Fibonacci sequence, $\frac{1}{1}, \frac{2}{1}, \frac{3}{2}, \frac{5}{3}, \frac{8}{5}, ..., \frac{89}{55}$, as decimals rounded to the nearest thousandth. Describe the pattern.

Integration: Geometry
Area of Triangles and Trapezoids

What you'll learn

You'll learn to find the areas of triangles and trapezoids.

When am I ever going to use this?

Many shapes can be subdivided into triangles or trapezoids to make it easier to determine their areas.

Word Wise

base
altitude
height
trapezoid

Sailboat races were conducted in 8 different classes of competition in the 1996 Olympics. In the Star Class, the mainsail area cannot exceed 20.6 square meters. The dimensions of a mainsail are shown in the diagram. Does it qualify for the Star Class? *This problem will be solved in Example 1.*

Any side of a triangle can be used as a **base**. A line segment perpendicular to the base from the opposite vertex is called the **altitude**. The length of the altitude is called the **height**.

In the triangle at the right, side $\overline{CD}$ is the base, and $\overline{EF}$ is the altitude.

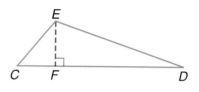

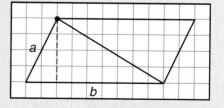
grid paper

HANDS-ON MINI-LAB

Work with a partner.

Try This

• Copy the triangles shown onto a piece of grid paper. Label as shown.

• Sketch the figure that is a result of moving each triangle as shown.

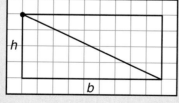
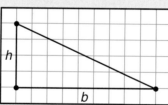

Talk About It

1. What is the length of the base and height of each parallelogram or rectangle?
2. Find the area of each parallelogram or rectangle.
3. Explain how the area of each parallelogram or rectangle is related to the product of the base and height of the corresponding triangle.

The Mini-Lab suggests the following rule.

| Area of a Triangle | **Words:** | The area of a triangle is equal to half the product of its base and height. |
| | **Symbols:** $A = \frac{1}{2}bh$ | **Model:** 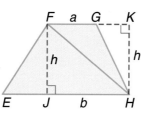 |

Example

Real World **APPLICATION**

1 **Sports** Refer to the beginning of the lesson. Does the mainsail qualify for the Star Class?

$A = \frac{1}{2}bh$

$A = 0.5 \boxed{\times} 3.5 \boxed{\times} 6.55$ *Replace b with 3.5 and h with 6.55.*

$A = 11.4625$

The area of the mainsail is 11.4625 square meters. Since 11.4625 is less than 20.6, the mainsail qualifies for the Star Class.

A **trapezoid** is a quadrilateral with exactly one pair of parallel sides. These parallel sides are its bases. When you draw a diagonal in a trapezoid, two triangles are formed. The area of the trapezoid is the sum of the areas of the triangles.

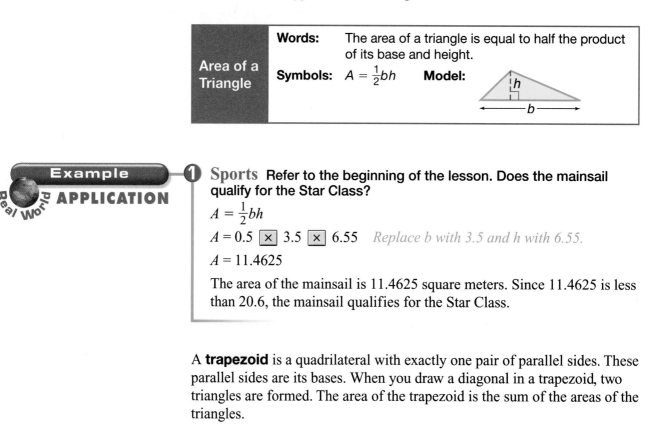

The triangles are $\triangle EFH$ and $\triangle FGH$. The altitudes of these triangles, $\overline{FJ}$ and $\overline{HK}$, are congruent. Both are h units long. The base of $\triangle FGH$ is a units long. The base of $\triangle EFH$ is b units long.

area of trapezoid $EFGH$ = area of $\triangle FGH$ + area of $\triangle EFH$

$\qquad\qquad = \qquad \frac{1}{2}ah \qquad + \qquad \frac{1}{2}bh$

$\qquad\qquad = \frac{1}{2}h(a + b)$ *Distributive property*

| Area of a Trapezoid | **Words:** | The area of a trapezoid is equal to the product of half the height and the sum of the bases. |
| | **Symbols:** $A = \frac{1}{2}h(a + b)$ | **Model:** 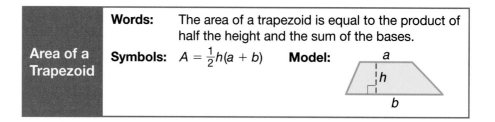 |

Example

2 **Find the area of the trapezoid.**

$A = \frac{1}{2}h(a + b)$

$A = \frac{1}{2} \cdot 3(12 + 6)$ *Replace h with 3, a with 12, and b with 6.*

$A = \frac{1}{2} \cdot 3 \cdot 18$ or 27   The area is 27 square feet.

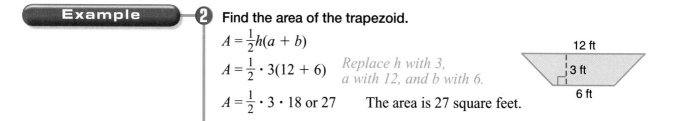

Communicating Mathematics

Read and study the lesson to answer each question.

1. *Draw* an isosceles triangle. Sketch and label the three possible altitudes.

2. *Explain* how to tell which two sides of a trapezoid are the bases.

HANDS-ON MATH

3. *Copy* the trapezoid on a piece of grid paper. Cut it out and then cut along the dashed line. Move the parts so they form a parallelogram. Explain how the height and length of the base of the parallelogram are related to the height and the sum of the lengths of the bases of the original trapezoid.

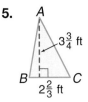

Guided Practice

State the measure of the base(s) and the height of each triangle or trapezoid. Then find the area of each figure.

4.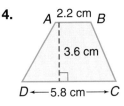

5.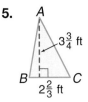

Find the area of each figure.

6. triangle: base, 7.5 cm; height, 3.75 cm

7. triangle: base, $3\frac{3}{4}$ in.; height, $2\frac{1}{8}$ in.

8. trapezoid: bases $7\frac{1}{6}$ yd and $5\frac{2}{3}$ yd; height, 8 yd

9. trapezoid: bases 30 ft and 22 ft; height, 9 ft

10. *Geography* The shape of the state of Virginia is close to a triangle.
    a. Estimate the area of Virginia.
    b. Research to find the actual area. Compare to your estimate.

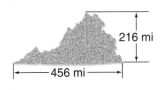

Practice

State the measure of the base(s) and the height of each triangle or trapezoid. Then find the area of each figure.

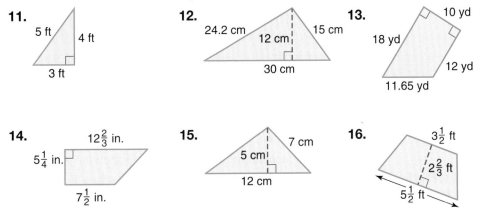

Find the area of each triangle.

	base	height
17.	5 in.	9 in.
19.	12 cm	5.5 cm
21.	$6\frac{1}{2}$ yd	2 yd

	base	height
18.	8 ft	5 ft
20.	3.75 m	2.2 m
22.	$9\frac{3}{4}$ in.	$12\frac{1}{8}$ in.

Find the area of each trapezoid.

	base (a)	base (b)	height
23.	6 yd	9 yd	5 yd
25.	4.5 m	5.33 m	6 m
27.	$3\frac{1}{2}$ ft	$5\frac{1}{2}$ ft	2 ft

	base (a)	base (b)	height
24.	4 km	9.5 km	3 km
26.	2.2 cm	5.8 cm	3.6 cm
28.	$4\frac{3}{4}$ in.	$5\frac{7}{8}$ in.	$2\frac{1}{4}$ in.

29. Draw and label a triangle that has an area of 9 square inches.

30. A trapezoid has an area of 81 square meters, and the lengths of its bases are 8 meters and 19 meters. Find the height of this trapezoid.

Applications and Problem Solving

31. *Algebra* The shorter base of a trapezoid is x units long. The longer base of the trapezoid is one and one-half times as long as the shorter base. Find the length of the bases if the area of the trapezoid is 150 square yards and the height is 10 yards.

32. *Landscaping* A diagram of Mr. Stone's yard is shown. The rectangles represent the house and driveway. He wants to fertilize his lawn. The label on a bag of fertilizer indicates that one bag will cover 2,000 square feet. How much fertilizer should he buy?

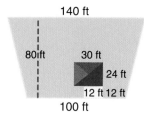

33. *Critical Thinking* What is the effect on the area of a trapezoid if the length of each base is doubled and the height is doubled?

Mixed Review

34. *Patterns* Write the next three terms in the sequence 80, 76, 72, 68, ... *(Lesson 7-5)*

35. Express 0.000084 in scientific notation. *(Lesson 6-9)*

36. **Standardized Test Practice** If △*EFG* is congruent to △*ABC*, then — *(Lesson 5-5)*

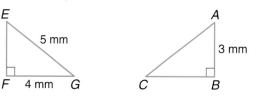

   A ∠*A* is a right angle.

   B the perimeter of △*EFG* is 11 millimeters.

   C $\overline{FG} \cong \overline{AB}$.

   D the measure of side $\overline{BC}$ is 4 millimeters.

   E $m\angle E = m\angle C$.

37. *Algebra* Evaluate $5a^2 - (b + c)$ if $a = 3$, $b = 8$, and $c = 10$. *(Lesson 1-3)*

For **Extra Practice,** see page 625.

7-6B Area and Pick's Theorem

A Follow-Up of Lesson 7-6

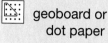 geoboard or dot paper

💿 spreadsheet software

Pick's Theorem is a formula that you can use to find the area of a shape formed on a geoboard. You can investigate this formula by using a spreadsheet.

Pick's Theorem is: $\dfrac{Boundary\ Points}{2} + Interior\ Points - 1$.

TRY THIS

Work in groups of four.

1 Make these shapes on your geoboard. *If you don't have a geoboard, draw them on dot paper.*

- a triangle with an area of 1.5 square units
- a rectangle with an area of 2 square units
- a triangle with an area of 3 square units
- a rectangle with an area of 4 square units
- a triangle with an area of 4.5 square units
- a rectangle with an area of 6 square units

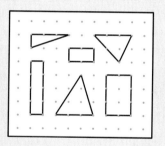

2 Copy the spreadsheet.

	A	B	C	D	
1	Area	Boundary Points	Interior Points	Pick's Theorem	⬆
2	1.5			= B2/2 + C2 – 1	
3	2			= B3/2 + C3 – 1	
4	3			= B4/2 + C4 – 1	
5	4			= B5/2 + C5 – 1	
6	4.5			= B6/2 + C6 – 1	
7	6			= B7/2 + C7 – 1	

- Count the number of dots on the boundary of each shape. Then count the number of dots in the interior of each shape. Record these numbers in columns B and C of the spreadsheet.

- Use the spreadsheet formula in column D to find the results of using Pick's Theorem.

1. Compare your results in column D to the area of each shape. What do you notice?
2. If *b* represents the number of dots on the boundary of a figure, *i* represents the number of dots in its interior, and *A* represents its area, write a formula for Pick's Theorem.
3. Do you think Pick's Theorem would work for a square? Show several squares on your geoboard to support your answer.
4. Do you think Pick's Theorem would work for trapezoids? Show examples to support your answer.

5. Make each of the figures below on your geoboard or dot paper.

I.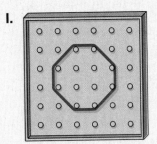

II.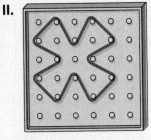

III.

IV.

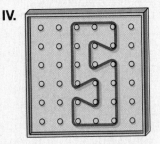

a. How could you find the area of each figure without using Pick's Theorem? (*Hint:* Look for ways to divide each figure into rectangles, triangles, or trapezoids.)
b. Investigate Pick's Theorem with each figure. With which types of figures can you use Pick's Theorem?

7-7A Graphing Pi

A Preview of Lesson 7-7

grid paper

circular objects

tape

Press $\boxed{\pi}$ on your calculator. Have you ever wondered why π is such a special number? Before you investigate π, let's study some properties related to circles.

A *circle* is a set of points in a plane that are the same distance from a given point in the plane, called the *center*. The distance across the circle through its center is its *diameter*. The *circumference* of a circle is the distance around the circle.

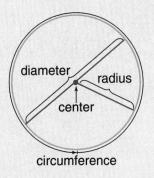

TRY THIS

Work in groups of three or four.

Step 1 Tape several pieces of grid paper together to form a large coordinate grid. Draw the horizontal and vertical axes so that the origin is 1 or 2 units from the lower left-hand corner.

Step 2 Place the bottom of a can or other circular object so that it touches the origin and the *x*-axis cuts the circle in half. Make a mark on the can where it crosses the *x*-axis. Then trace the circumference.

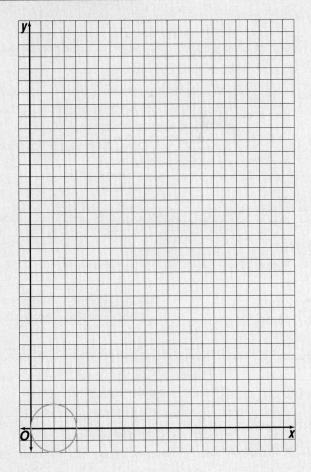

Step 3 Roll the can up one rotation, keeping the path of its edge on the *y*-axis. Draw a point on the graph where the can completes one rotation.

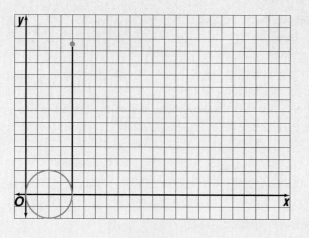

Step 4 Repeat Steps 2 and 3 for three to five more circular objects.

Step 5 From the origin, draw a curve that connects the points. Describe the curve.

The steepness of the blue line is called the *slope* of the line. Slope is the ratio of the *rise,* or vertical change, to the *run,* or horizontal change, as you move from one point on the line to another.

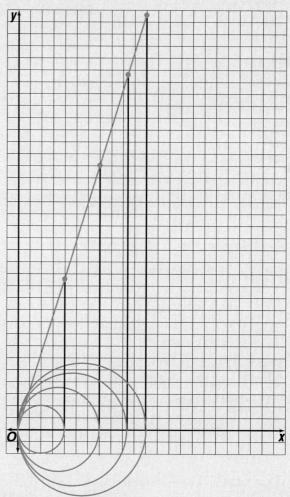

ON YOUR OWN

1. Find the *x*-coordinate and *y*-coordinate of each of the points you graphed.
   a. What does the *x*-coordinate represent?
   b. What does the *y*-coordinate represent?
   c. Determine the ratio of the *y*-coordinate to the *x*-coordinate for each point.
   d. What is the mean of the ratios?

2. *Look Ahead* The Greek letter π (pi) represents the ratio of the circumference of a circle to its diameter. What do you think would be a good approximation for π? Explain how π is related to the slope of the blue line.

Integration: Geometry
Circles and Circumference

Is the Ferris wheel one of your favorite amusement park rides? The largest Ferris wheel was built in Yokohama City, Japan in 1985. The wheel is 328 feet in diameter. What distance would you travel in one revolution if you rode on the largest Ferris wheel? *This problem will be solved in Example 2.*

A **circle** is a set of points in a plane that are the same distance from a given point in the plane. The given point is called the **center**. The distance from the center to any point on the the circle is called the **radius (r)**. The distance across the circle through the center is its **diameter (d)**. The **circumference (C)** of a circle is the distance around the circle. The diameter of a circle is twice its radius, or $d = 2r$.

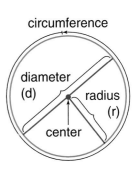

The relationship you discovered in Lesson 7-7A is true for all circles. The circumference of a circle divided by its diameter is always 3.1415926…. The Greek letter π(pi) represents this number. The numbers 3.14 and $\frac{22}{7}$ are often used as approximations for π.

Circumference	**Words:**	The circumference of a circle is equal to its diameter times π, or 2 times its radius times π.	
	Symbols:	$C = \pi d$ or $C = 2\pi r$	**Model:**

Example ① Find the circumference of a circle with a radius of 14 meters.

Method 1 Use $\frac{22}{7}$ for π.

$C = 2\pi r$

$C = 2 \cdot \frac{22}{7} \cdot 14$

$C = 2 \cdot \frac{22}{\underset{1}{7}} \cdot \frac{\overset{2}{14}}{1}$

$C \approx 88$ m

The circumference is about 88 meters.

Method 2 Use 3.14 for π.

$C = 2\pi r$

$C = 2 \cdot 3.14 \cdot 14$

$C \approx 87.92$ m

Most calculators have a π key that gives a more precise approximation of pi.

Example 2
Real World APPLICATION

Entertainment Refer to the beginning of the lesson. What distance would you travel in one revolution if you rode on the Ferris wheel?

$C = \pi d$

$C = \pi \cdot 328$ *Replace d with 328.*

[π] [×] 328 [=] *1030.44239*

You would travel about 1,030 feet in one revolution.

CHECK FOR UNDERSTANDING

Communicating Mathematics

Read and study the lesson to answer each question.

1. *Identify* two numbers that are approximations for π.

2. *Tell* how to estimate the circumference of a circle if you know the radius.

3. *Write* a few sentences explaining how you can find the diameter of a circle that has a circumference of 22 inches.

Guided Practice

Find the circumference of each circle to the nearest tenth. Use $\frac{22}{7}$ or 3.14 for π.

4.

14 in.

5.

9 in.

6.

7.5 cm

7. The radius is 17 centimeters.

8. The diameter is $\frac{1}{2}$ foot.

9. *Archaeology* Stonehenge is a circular array of giant stones in England. The diameter of Stonehenge is 30.5 meters. Find the approximate distance it is around Stonehenge.

EXERCISES

Practice

Find the circumference of each circle to the nearest tenth. Use $\frac{22}{7}$ or 3.14 for π.

10.
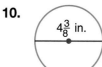
$4\frac{3}{8}$ in.

11.
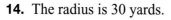
10.5 mm

12.
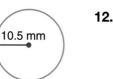
12.56 m

13.

29.5 cm

14. The radius is 30 yards.

15. The diameter is 18 feet.

16. The diameter is 13.3 meters.

17. The radius is $5\frac{1}{4}$ inches.

18. The radius is 4.5 centimeters.

19. The diameter is 19.33 meters.

20. The diameter is $6\frac{1}{2}$ feet.

21. The radius is $6\frac{3}{4}$ inches.

22. The circumference of a circle is π units. What is the diameter of the circle?

23. The circumference of a circle is 2.041 meters. Find its radius.

24. *Sports* Three tennis balls are packaged one on top of the other in a can. Which measure is greater, the height of the can or the circumference of the can? Explain.

25. *Manufacturing* The label that goes around a jar of peanut butter overlaps itself by $\frac{3}{8}$ inch to allow for sealing the label on the jar. If the diameter of the jar is 3 inches, what is the length of the label?

26. *Working on the* **CHAPTER Project**

You can model a spiral using the Fibonacci numbers and circles.

    a. Use a ruler to draw two squares with sides 1 centimeter long as shown. Then add a square with sides of 2 centimeters, 3 centimeters, 5 centimeters, and so on. Use a compass to draw quarter circles in each square so that a continuous curve results. The result is a *Fibonacci spiral.*

        b. Add descriptions and pictures of spirals in nature to your project.

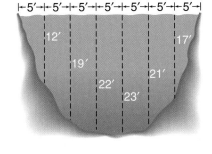

27. *Critical Thinking* A *central angle* has the center of a circle as its vertex and separates a circle into a *major arc* and a *minor arc.* (A 90° central angle intersects a minor arc of 90° and a major arc of 270°.) A *chord* is a line segment with its endpoints on a circle. If a diameter is perpendicular to a chord, then it bisects the chord. An *inscribed angle* has its vertex on the circle and sides that contains chords. (The measure of an inscribed angle equals one-half measure of its intercepted arc.) Find the measures of the following angles and arcs.

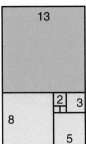

    a. minor arc *LN*   **b.** angle *x*

    c. major arc *NL*   **d.** angle *y*

    e. arc *KJ*   **f.** arc *KL*

    g. arc *LM*   **h.** arc *MN*

28. *Earth Science* The sketch of the cross section of a stream was made by taking a depth reading every 5 feet. Estimate the area of the cross section. *(Lesson 7-6)*

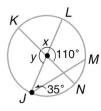

29. *Standardized Test Practice* What number will make $\frac{3}{4} \cdot \frac{7}{8} = n \cdot \frac{7}{8}$ true? *(Lesson 7-4)*

  A $\frac{4}{8}$ **B** $\frac{3}{4}$ **C** $\frac{10}{12}$ **D** $\frac{7}{8}$

For **Extra Practice,** see page 625.

30. *Number Sense* Use divisibility rules to determine whether 284 is divisible by 6. Explain why or why not. *(Lesson 6-1)*

Dividing Fractions

What you'll learn

You'll learn to divide fractions.

When am I ever going to use this?

Knowing how to divide fractions can help you find the diameter of a circle when the circumference is known.

Have you thought of a career as a teacher? If so, you will need to earn a bachelor's degree in education. In 1993, about 108,000 people earned bachelor's degrees in education in the United States. Men earned a little more than $\frac{1}{5}$ of those degrees. Approximately how many men earned a bachelor's degree in education in 1993?

You could find this number in two ways.

Method 1 Divide by 5.

$$108000 \boxed{\div} \; 5 \; \boxed{=} \; \textit{21600}$$

Method 2 Multiply by $\frac{1}{5}$.

$$108000 \boxed{\times} \; 0.2 \; \boxed{=} \; \textit{21600} \qquad \tfrac{1}{5} = 0.2$$

About 21,600 men earned bachelor's degrees in education in 1993.

In the example above, notice that dividing by 5 and multiplying by $\frac{1}{5}$ gave you the same result. Also note that 5 and $\frac{1}{5}$ are multiplicative inverses.

Dividing Fractions	**Words:**	To divide by a fraction, multiply by its multiplicative inverse.
	Symbols: Arithmetic	$\frac{5}{6} \div \frac{2}{3} = \frac{5}{6} \cdot \frac{3}{2}$
	Algebra	$\frac{a}{b} \div \frac{c}{d} = \frac{a}{b} \cdot \frac{d}{c}$, where $b, c, d \neq 0$.

Example ❶ Solve $a = \dfrac{-6}{-1\frac{1}{2}}$.

$a = \dfrac{-6}{-1\frac{1}{2}}$ *Estimate:* $-6 \div (-2) = 3$

$a = -6 \div \left(-1\frac{1}{2}\right)$ *Remember that the fraction bar means division.*

$a = -\dfrac{6}{1} \cdot \left(-\dfrac{2}{3}\right)$ *Dividing by $-1\frac{1}{2}$ or $-\frac{3}{2}$ is the same as multiplying by $-\frac{2}{3}$.*

$a = -\dfrac{\overset{2}{\cancel{6}}}{1} \cdot \left(-\dfrac{2}{\underset{1}{\cancel{3}}}\right)$

$a = 4$

Solve each equation.

② $\frac{3}{4} \div \frac{5}{6} = y$

$\frac{3}{4} \div \frac{5}{6} = y$

$\frac{3}{4} \cdot \frac{6}{5} = y$

$\frac{3}{\underset{2}{4}} \cdot \frac{\overset{3}{6}}{5} = y$

$\frac{9}{10} = y$

③ $f = -3\frac{1}{4} \div 2\frac{1}{2}$

$f = -3\frac{1}{4} \div 2\frac{1}{2}$

$f = -\frac{13}{4} \div \frac{5}{2}$

$f = -\frac{13}{\underset{2}{4}} \cdot \frac{\overset{1}{2}}{5}$

$f = -\frac{13}{10} \text{ or } -1\frac{3}{10}$

④ **Food** A major hamburger chain in the United States is famous for its square burgers. The burgers are $2\frac{1}{2}$ inches on a side. How many hamburgers can fit across a grill that is 30 inches wide?

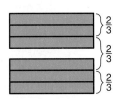

Did you know? The ground beef hamburger was first made in Hamburg, Germany. Hamburgers were introduced in the United States at the 1904 World's Fair in St. Louis, Missouri.

$30 \div 2\frac{1}{2} = \frac{30}{1} \div \frac{5}{2}$ *Rename 30 as $\frac{30}{1}$ and $2\frac{1}{2}$ as $\frac{5}{2}$.*

$= \frac{\overset{6}{30}}{1} \cdot \frac{2}{\underset{1}{5}}$

$= \frac{12}{1} \text{ or } 12$

You can fit 12 hamburgers across a grill that is 30 inches wide.

CHECK FOR UNDERSTANDING

Communicating Mathematics

Read and study the lesson to answer each question.

1. *Explain* how the model shows the quotient of 2 and $\frac{2}{3}$.

2. *Tell* how to estimate the quotient of $10\frac{7}{8} \div \left(-\frac{1}{4}\right)$.

$\left. \right\} \frac{2}{3}$

$\left. \right\} \frac{2}{3}$

$\left. \right\} \frac{2}{3}$

Guided Practice

Solve each equation. Write the solution in simplest form.

3. $\frac{2}{3} \div \frac{3}{4} = a$

4. $\frac{1}{2} \div 6 = d$

5. $r = -\frac{3}{8} \div \frac{9}{10}$

6. $q = 2\frac{1}{2} \div 1\frac{1}{4}$

7. What is the value of $\frac{2\frac{2}{3}}{4}$?

8. *Cooking* How many slices of pepperoni, each $\frac{1}{16}$ inch thick, can be cut from a stick 8 inches long?

Practice   **Solve each equation. Write the solution in simplest form.**

9. $12 \div -3\frac{1}{3} = j$ **10.** $\frac{2}{3} \div \frac{5}{6} = x$ **11.** $r = \frac{9}{10} \div (-6)$

12. $y = 1\frac{1}{2} \div \frac{1}{2}$ **13.** $-2\frac{1}{4} \div \left(-\frac{3}{8}\right) = t$ **14.** $\frac{7}{8} \div 2\frac{4}{5} = b$

15. $4\frac{1}{2} \div \frac{3}{4} = h$ **16.** $f = -3\frac{3}{4} \div \left(-2\frac{1}{2}\right)$ **17.** $c = -12\frac{1}{4} \div 4\frac{2}{3}$

18. *Algebra* Evaluate $\frac{d}{q}$ if $d = -3\frac{3}{4}$ and $q = 6\frac{2}{3}$.

19. What is the quotient of $-3\frac{3}{5}$ and -10?

20. *Algebra* Evaluate $\frac{a^2}{b^2}$ if $a = -\frac{4}{5}$ and $b = \frac{2}{3}$.

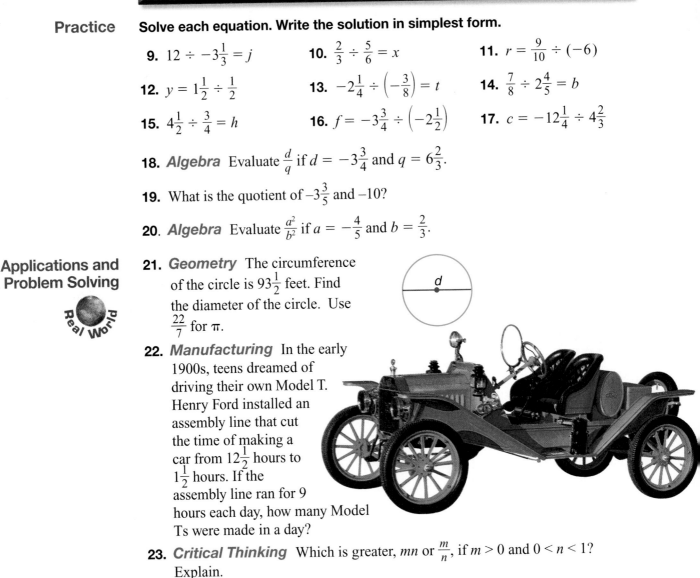

Applications and Problem Solving

21. *Geometry* The circumference of the circle is $93\frac{1}{2}$ feet. Find the diameter of the circle. Use $\frac{22}{7}$ for π.

22. *Manufacturing* In the early 1900s, teens dreamed of driving their own Model T. Henry Ford installed an assembly line that cut the time of making a car from $12\frac{1}{2}$ hours to $1\frac{1}{2}$ hours. If the assembly line ran for 9 hours each day, how many Model Ts were made in a day?

23. *Critical Thinking* Which is greater, mn or $\frac{m}{n}$, if $m > 0$ and $0 < n < 1$? Explain.

Mixed Review

24. *Geometry* Find the circumference of a circle with a radius of 2.7 millimeters. *(Lesson 7-7)*

25. **Standardized Test Practice** Paige has been jogging every day for the past four weeks. At the beginning of each week, she has been increasing the length of her daily jog. If she continues jogging according to the pattern shown, how many minutes will she spend jogging each day during her fifth week of jogging? *(Lesson 7-5)*

Week	Time Jogging (min)
1	8
2	16
3	24
4	32
5	?

A 32 min   **B** 40 min   **C** 48 min   **D** 56 min

26. *Geometry* Make an Escher-like drawing for the pattern shown. Use a tessellation of two rows of five equilateral triangles as your base. *(Lesson 5-7)*

For **Extra Practice,** see page 626.

Solving Equations

What you'll learn

You'll learn to solve equations involving rational numbers.

When am I ever going to use this?

Knowing how to solve equations can help you find the length of rectangle when you know the area and width.

Would you pay $121,000 for an ink pen? The Meisterstück Solitaire Royal fountain pen is 24K or 24-karat gold and is encased with 4,810 diamonds.

The amount of gold in jewelry is usually expressed using the karat system. One karat means that the metal is $\frac{1}{24}$ pure gold. Thus, pure gold is 24-karat. If there are $2\frac{1}{2}$ ounces of gold in a gold bracelet that weighs 6 ounces, how should the bracelet be advertised using karats?

You can apply the skills you have learned for rational numbers to solve equations containing rational numbers. You can find the amount of gold in jewelry, G, using the formula $G = \frac{k}{24} \cdot w$, where k is the number of karats and w is the weight.

$$G = \frac{k}{24} \cdot w$$

$$2\frac{1}{2} = \frac{k}{24} \cdot 6 \qquad \textit{Replace G with } 2\frac{1}{2} \textit{ and w with 6.}$$

$$\frac{5}{2} \cdot \frac{1}{6} = \frac{k}{24} \cdot 6 \cdot \frac{1}{6} \qquad \textit{Multiply each side by } \frac{1}{6}, \textit{ the multiplicative inverse of 6}$$

$$\frac{5}{12} = \frac{k}{24}$$

$$24\left(\frac{5}{12}\right) = 24 \cdot \frac{k}{24} \qquad \textit{Multiply each side by 24.}$$

$$10 = k$$

The bracelet should be advertised as 10-karat gold.

Example 1

Solve $-\frac{5}{8}x = \frac{3}{4}$. Check your solution.

$$-\frac{5}{8}x = \frac{3}{4}$$

$$-\frac{8}{5} \cdot -\frac{5}{8}x = -\frac{\overset{2}{8}}{5} \cdot \frac{3}{\underset{1}{4}} \qquad \textit{Multiply each side by } -\frac{8}{5}. \textit{ Why?}$$

$$x = -\frac{6}{5} \text{ or } -1\frac{1}{5}$$

Check: $\qquad -\frac{5}{8}x = \frac{3}{4}$

$$-\frac{\overset{1}{5}}{\underset{4}{8}}\left(-\frac{\overset{3}{6}}{\underset{1}{5}}\right) \overset{?}{=} \frac{3}{4} \qquad \textit{Replace x with } -\frac{6}{5}.$$

$$\frac{3}{4} = \frac{3}{4} \checkmark$$

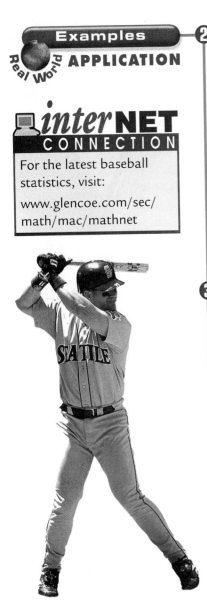

2 Sports In baseball, a player's batting average is determined by dividing the number of hits by the number of at-bats. On August 22, 1997, Edgar Martinez' batting average was 0.327, and he had 447 at-bats. How many hits did he have?

$$0.327 = \frac{h}{447}$$

$$447 \cdot 0.327 = \frac{447}{1} \cdot \frac{h}{447} \quad \textit{Multiply each side by 447.}$$

$$146.169 = h$$

Mr. Martinez had 146 hits.

Check: 146 $\boxed{\div}$ 447 $\boxed{=}$ *0.326621924*

Since batting averages are rounded to the thousandths place, the answer is correct.

3 Solve $\frac{b}{1.5} - 13 = -2.2$. Check your solution.

$$\frac{b}{1.5} - 13 = -2.2$$

$$\frac{b}{1.5} - 13 + 13 = -2.2 + 13 \quad \textit{Add 13 to each side.}$$

$$\frac{b}{1.5} = 10.8$$

$$1.5 \cdot \frac{b}{1.5} = 1.5 \cdot 10.8 \quad \textit{Multiply each side by 1.5.}$$

$$b = 16.2$$

Check: $\frac{b}{1.5} - 13 = -2.2 \quad \textit{Replace b with 16.2.}$

16.2 $\boxed{\div}$ 1.5 $\boxed{-}$ 13 $\boxed{=}$ *2.2* ✓

CHECK FOR UNDERSTANDING

Communicating Mathematics

Read and study the lesson to answer each question.

1. **Write** an equation that is solved by using the multiplicative inverse of $1\frac{1}{4}$.

2. **Explain** how to solve $3n + \frac{1}{4} = -\frac{7}{8}$.

3. **You Decide** Esohe says that to solve the equation $7m - \frac{2}{3} = 1\frac{2}{3}$, the first step is to add $\frac{2}{3}$ to each side. Priya says the first step is to divide each side by 7. Whose method would you use and why?

Guided Practice

Solve each equation. Check your solution.

4. $\frac{3}{4}v = -\frac{7}{8}$

5. $2.3y = 6.9$

6. $-4.2 = \frac{c}{7}$

7. $t + 0.25 = -4.125$

8. $\frac{1}{3}x - \frac{3}{8} = -\frac{1}{2}$

9. $-1.6w + 3.5 = 0.48$

10. **Physical Science** Use the formula $F = \frac{9}{5}C + 32$ to find the Celsius temperature C for the temperature of 68°F.

Practice

Solve each equation. Check your solution.

11. $-12 = -\dfrac{z}{5}$

12. $1.1 + d = -4.4$

13. $2\dfrac{1}{2}x = 5\dfrac{3}{4}$

14. $u - (-0.03) = 3.2$

15. $\dfrac{n}{2} = -1.6$

16. $2a = -12$

17. $\dfrac{t}{3.2} = -4.5$

18. $1.8 = 0.6 - g$

19. $-c + \dfrac{1}{6} = \dfrac{3}{5}$

20. $\dfrac{2}{3h} - (-3) = 6$

21. $\dfrac{x - 10}{5} = 2.5$

22. $-12 + \dfrac{w}{4} = 9$

23. $4.7 = 1.2 - 7m$

24. $\dfrac{c}{2} + (-3) = 5$

25. $1.6 = \dfrac{-8 + k}{-7}$

26. Find the solution of $-2 = j - \dfrac{2}{5}$.

27. What is the solution of $7p - \left(-\dfrac{2}{9}\right) = \dfrac{5}{9}$?

Applications and Problem Solving

28. *Geometry* The volume of a cone is given by the formula $V = \dfrac{1}{3}\pi r^2 h$ where r is the radius and h is the height of the cone. The volume of a cone is 346.5 cubic inches. If the cone has a radius of 3.5 inches, what is its height?

29. *Air-Conditioning* An estimate of the size of a room air conditioner needed is given by the formula, $B = 27\ell w$, where B is the number of British Thermal Units per hour, ℓ is the length of the room, and w is the width of the room. A room air conditioner is rated at 4,000 BTUs. What is the maximum length room it can cool if the room is $11\dfrac{1}{2}$ feet wide?

30. *Safety* The time that a traffic light remains yellow is given by the formula $t = \dfrac{1}{2}s + 1$, where t is the time in seconds and s is the speed limit. What is the yellow time for a traffic light on a street with a speed limit of 25 miles per hour?

31. *Critical Thinking* Using the formula in Exercise 10, find the temperature that is the same on both the Fahrenheit and Celsius scales.

Mixed Review

32. Solve $a = \dfrac{7}{12} \div \dfrac{3}{4}$. *(Lesson 7-8)*

33. *Number Sense* Find the prime factorization of 48. *(Lesson 6-2)*

34. **Standardized Test Practice** The graph shows how Melissa spends her day. What percent of Melissa's day is not spent in school or studying? *(Lesson 3-4)*

   A 70%

   B 62%

   C 38%

   D 30%

   E Not Here

For **Extra Practice,** see page 626.

35. Find the value of $4^2 - 2^3$. *(Lesson 1-2)*

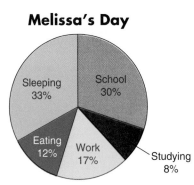

Melissa's Day

Sleeping 33%

School 30%

Eating 12%

Work 17%

Studying 8%

7-10 Solving Inequalities

What you'll learn

You'll learn to solve inequalities involving rational numbers and graph their solutions.

When am I ever going to use this?

You can use inequalities to find the minimum number of signatures needed for a petition.

Have you ever ordered something from a catalog? Then you know the price listed in the catalog is not the total cost of your order! Usually, taxes and shipping and handling fees are added.

Shipping and handling fees are often determined by the total price of the items you order. Consider the table at the right. Regardless of how much you order, the shipping and handling charges, s, can be expressed as the inequality $s \le \$11.98$.

AVOID DELAY—PLEASE INCLUDE SHIPPING & HANDLING

Orders up to $10.00	add $3.98
From $10.01 to $20.00	add $4.98
From $20.01 to $30.00	add $5.98
From $30.01 to $40.00	add $6.98
From $40.01 to $50.00	add $7.98
From $50.01 to $60.00	add $8.98
From $60.01 to $70.00	add $9.98
From $70.01 to $80.00	add $10.98
Orders over $80.00	add $11.98

In Lesson 1-9, you learned to solve inequalities by solving the related equations. You can solve inequalities involving rational numbers in the same way.

Example

1 Solve $\frac{1}{2}x - 6 > 5\frac{1}{4}$. **Graph the solution on a number line.**

Solve the related equation $\frac{1}{2}x - 6 = 5\frac{1}{4}$.

$$\frac{1}{2}x - 6 = 5\frac{1}{4}$$

$$\frac{1}{2}x - 6 + 6 = 5\frac{1}{4} + 6 \quad \textit{Add 6 to each side.}$$

$$\frac{1}{2}x = 11\frac{1}{4} \quad \textit{Simplify.}$$

$$2\left(\frac{1}{2}x\right) = 2\left(11\frac{1}{4}\right) \quad \textit{Multiply each side by 2.}$$

$$x = 22\frac{2}{4} \text{ or } 22\frac{1}{2}$$

The solution must be either numbers greater than $22\frac{1}{2}$ or numbers less than $22\frac{1}{2}$. Let's test two numbers, one greater than $22\frac{1}{2}$ and one less than $22\frac{1}{2}$, to see which is correct.

Try 24.

$$\frac{1}{2}(24) - 6 \overset{?}{>} 5\frac{1}{4} \quad \textit{Replace x with 24.}$$

$$12 - 6 \overset{?}{>} 5\frac{1}{4}$$

$$6 > 5\frac{1}{4} \quad \textit{true}$$

Try 10.

$$\frac{1}{2}(10) - 6 \overset{?}{>} 5\frac{1}{4} \quad \textit{Replace x with 10.}$$

$$5 - 6 \overset{?}{>} 5\frac{1}{4}$$

$$-1 > 5\frac{1}{4} \quad \textit{false}$$

Numbers greater than $22\frac{1}{2}$ make up the solution set. So, $x > 22\frac{1}{2}$. Graph the solution on a number line.

20 21 22 23 24 25

Since $22\frac{1}{2}$ is not included, an open circle is used.

---

2 Solve $\frac{d - 2.5}{1.25} \le -3.8$. Graph the solution on a number line.

Solve the related equation $\frac{d - 2.5}{1.25} = -3.8$.

$$\frac{d - 2.5}{1.25} = -3.8$$

$$(1.25)\left(\frac{d - 2.5}{1.25}\right) = (1.25)(-3.8) \quad \textit{Multiply each side by 1.25.}$$

$$d - 2.5 = -4.75 \quad \textit{Simplify.}$$

$$d - 2.5 + 2.5 = -4.75 + 2.5 \quad \textit{Add 2.5 to each side.}$$

$$d = -2.25$$

The solution includes -2.25. *Why?* Now test two numbers to determine whether the solution also includes numbers greater than -2.25 or numbers less than -2.25.

Try 0.

$$\frac{0 - 2.5}{1.25} \stackrel{?}{<} -3.8$$

$$\frac{-2.5}{1.25} \stackrel{?}{<} -3.8$$

$$-2 < -3.8 \quad \textit{false}$$

Try -2.5.

$$\frac{-2.5 - 2.5}{1.25} \stackrel{?}{<} -3.8$$

$$\frac{-5}{1.25} \stackrel{?}{<} -3.8$$

$$-4 < -3.8 \quad \textit{true}$$

Numbers less than 2.25 are included in the solution. So, $d \le -2.25$. The number line shows $d \le -2.25$.

$$\text{Since } -2.25 \text{ is included,}$$
$$\text{a solid dot is used.}$$

CONNECTION **3** **Civics** Mrs. Ramirez needs 340 valid signatures on a petition before she can enter the race for a city council position. The elections board told her that about $\frac{3}{20}$ of the signatures usually prove to be invalid. How many signatures should she get to ensure that she will qualify for the ballot?

If $\frac{3}{20}$ of the signatures are invalid, then $\frac{17}{20}$ will be valid. Let n = the number of signatures Mrs. Ramirez should get. So $\frac{17}{20}$ of n must be greater than or equal to 340. This can be written as $\frac{17}{20}n \ge 340$.

Solve the related equation $\frac{17}{20}n = 340$.

$$\frac{17}{20}n = 340$$

$$\frac{20}{17} \cdot \frac{17}{20}n = \frac{20}{17} \cdot 340 \quad \textit{Multiply each side by } \frac{20}{17}.$$

$$n = 400 \quad \textit{Simplify.}$$

If you test numbers greater than and less than 400, you find the solution set includes numbers greater than 400. So, $n \ge 400$.

Mrs. Ramirez needs at least 400 signatures to ensure her petition.

In Examples 1, 2, and 3, notice that the inequality symbol in the solution is the same as the one in the original inequality. Is this always true?

4 Solve $-4x + 5 < 11$. Graph the solution on a number line.

Solve the related equation $-4x + 5 = 11$.

$$-4x + 5 = 11$$
$$-4x + 5 - 5 = 11 - 5 \quad \textit{Subtract 5 from each side.}$$
$$-4x = 6$$
$$\left(-\frac{1}{4}\right)(-4x) = \left(-\frac{1}{4}\right)6 \quad \textit{Multiply each side by the reciprocal of } -4, -\frac{1}{4}.$$
$$x = -\frac{3}{2} \text{ or } -1\frac{1}{2}$$

The solution includes numbers greater than $-1\frac{1}{2}$ or less than $-1\frac{1}{2}$. Test two numbers to see which is correct.

Try 2.

$$-4(2) + 5 \overset{?}{<} 11$$
$$-8 + 5 \overset{?}{<} 11$$
$$-3 < 11 \quad \textit{true}$$

Try -2.

$$-4(-2) + 5 \overset{?}{<} 11$$
$$8 + 5 \overset{?}{<} 11$$
$$13 < 11 \quad \textit{false}$$

Numbers greater than $-1\frac{1}{2}$ are included in the solution. So, $x > -1\frac{1}{2}$. The number line shows the solution set.

Notice that the inequality sign of the solution in Example 4 is the opposite of that in the original inequality. Also notice that in solving the related equation, you multiplied by a negative number. Whenever you multiply or divide by a negative number in solving the related equation, the solution will have the opposite inequality sign from that given in the original inequality.

CHECK FOR UNDERSTANDING

Communicating Mathematics

Read and study the lesson to answer each question.

1. *Compare* solving inequalities involving rational numbers with solving inequalities involving whole numbers.

2. *Tell* whether each statement is *true* or *false*. If false, explain how to change the inequality to make it true.
   a. If $\frac{1}{2}x > -7$, then $x < -14$.
   b. If $3x > -15$, then $x > -5$.

Guided Practice

Solve each inequality. Graph the solution on a number line.

3. $-6x > 18$

4. $\frac{3}{8}a < -\frac{5}{8}$

5. $3.3 + y \geq -13.2$

6. $n + 3\frac{7}{8} \leq 6\frac{1}{4}$

7. $\frac{2}{3}h - (-3) > 6$

8. $-2x - 1\frac{1}{2} < 3$

9. *Money Matters* Theo wants to earn at least $20 this week to go to the fair. His father said he would pay him $9 for mowing the lawn and $2.75 an hour to weed the flower bed. Theo mows the lawn. Use the inequality $\$9 + \$2.75h \geq \$20$ to find the number of hours he needs to weed to earn at least $20.

Practice

Solve each inequality. Graph the solution on a number line.

10. $2w > -12$

11. $p + 0.03 < 3.2$

12. $\frac{8}{5}r \geq 4$

13. $-12 \leq -\frac{1}{7}y$

14. $3.6 > \frac{f}{0.9}$

15. $-\frac{3}{5}k > \frac{2}{3}$

16. $z - \frac{2}{5} \geq -2$

17. $t + \frac{1}{4} < -4\frac{1}{8}$

18. $-6 > \frac{1}{3}m$

19. $2\frac{1}{2}d > 5\frac{3}{4}$

20. $5 + 8x > 59$

21. $-11h + 15 \geq 12\frac{1}{2}$

22. $-\frac{5}{8}c + \frac{1}{6} \leq \frac{3}{5}$

23. $\frac{2}{1.5}b - 13 \geq 2.2$

24. $\frac{2x - 15}{3} > 3$

25. $\frac{1}{2}g + 3 > -1\frac{3}{5}$

26. $-3.2q + 7 < 0.96$

27. $\frac{n}{2.3} - 1.3 > -5.2$

28. Find the solution of $4p - \frac{1}{5} > -7\frac{2}{5}$.

29. Choose the inequality for which $h \leq -5\frac{1}{2}$ is a solution.

   a. $-4h + 7 > 15$ **b.** $-\frac{4}{5}h - \frac{7}{5} \geq 3$ **c.** $-\frac{4}{5}h - \frac{7}{5} \leq 3$

Applications and Problem Solving

Real World

30. *Food* Do you ever grab some cookies for an afternoon snack? Nutritionists recommend that you select cookies that contain 3 grams or less of fat per serving and no more than 8 to 12 grams of sugar. Which cookies meet these standards?

Cookie (cookies per serving)	Grams per Serving	
	Fat	Sugar
Iced gingersnaps (6)	1	8
Reduced fat creme sandwich cookies (2)	2.5	10
Reduced fat chocolate sandwich cookies (3)	3.5	13
Cinnamon baby grahams (25)	4	8

31. *Transportation* Car Safe offers a special rate for long term parking. It charges $5.50 for the first day and $4 for each additional day not to exceed $50 for a 15-day period. Use the inequality $5.50 + $4d \leq $50 to find out when you will owe $50 for your stay.

32. *Critical Thinking* The expression *the numbers between* $-2\frac{1}{2}$ *and* $3\frac{1}{3}$ can be written as $-2\frac{1}{2} < n < 3\frac{1}{3}$. What does the algebraic sentence $-1\frac{1}{3} < x + \frac{1}{2} < 2\frac{5}{6}$ represent? Give examples of these numbers.

Mixed Review

33. *Algebra* Solve $\frac{2}{3}x + 4 = -6$. *(Lesson 7-9)*

34. *Geometry* Find the area of a triangle whose base is 2.5 inches and whose height is 3.5 inches. *(Lesson 7-6)*

35. *Probability* Two number cubes are rolled. Find the probability that the sum of the numbers showing is 7. *(Lesson 6-6)*

36. *Standardized Test Practice* Rick took his father out to dinner for his birthday. When the bill came, Rick's father reminded him that it is customary to tip the server 15% of the bill. If the bill was for $19.60, a good estimate for the tip is — *(Lesson 3-6)*

  A $6.   **B** $5.   **C** $4.   **D** $3.   **E** $2.

For **Extra Practice**, see page 626.

CHAPTER 7

Study Guide and Assessment

interNET
CONNECTION Chapter Review For additonal lesson-by-lesson review, visit:
www.glencoe.com/sec/math/mac/mathnet

Vocabulary

After completing this chapter, you should be able to define each term, concept, or phrase and give an example or two of each.

Algebra
arithmetic sequence (p. 296)
common difference (p. 296)
common ratio (p. 297)
Fibonacci sequence (p. 300)
geometric sequence (p. 297)
sequence (p. 296)
term (p. 296)

Numbers and Operations
inverse property of multiplication (p. 290)
mixed number (p. 278)
multiplicative inverse (p. 290)
reciprocal (p. 290)

Geometry
altitude (p. 301)
base (pp. 301, 302)
center (pp. 307, 309)
circle (pp. 307, 309)
circumference (pp. 307, 309)
diameter (pp. 307, 309)
height (p. 301)
Pick's Theorem (p. 305)
radius (p. 309)
trapezoid (p. 302)

Problem Solving
deductive reasoning (p. 294)
inductive reasoning (p. 294)
look for a pattern (pp. 294–295)

Understanding and Using the Vocabulary

Choose the letter of the term or number that best matches each phrase.

1. the sum of a whole number and a fraction

2. the LCM of 3 and 9

3. a sequence whose terms increase or decrease by a constant factor

4. a sequence having the same difference between any two consecutive terms

5. a quadrilateral with exactly one pair of parallel sides

6. the distance across a circle through the center

7. the distance around a circle

8. the number you would multiply each side of the equation $\frac{2}{3}t = -\frac{4}{9}$ by to solve it

a. $\frac{2}{3}$

b. $\frac{3}{2}$

c. 9

d. 18

e. diameter

f. circumference

g. mixed number

h. geometric sequence

i. arithmetic sequence

j. trapezoid

In Your Own Words

9. *Describe* how to multiply two mixed numbers.

Upon completing this chapter, you should be able to:

● add and subtract fractions with like denominators *(Lesson 7-1)*

Solve $b = \frac{1}{5} - \frac{3}{5}$.

$b = \frac{1-3}{5}$ *Subtract the numerators.*

$b = \frac{-2}{5}$ or $-\frac{2}{5}$

● add and subtract fractions with unlike denominators *(Lesson 7-2)*

Solve $h = \frac{2}{3} + \frac{3}{4}$.

$h = \frac{8}{12} + \frac{9}{12}$ *Rename each fraction.*

$h = \frac{17}{12}$ or $1\frac{5}{12}$ *Simplify.*

● multiply fractions *(Lesson 7-3)*

Solve $f = -\frac{5}{8} \cdot \frac{1}{3}$.

$f = \frac{-5 \cdot 1}{8 \cdot 3}$ *Multiply the numerators and the denominators.*

$f = -\frac{5}{24}$

● identify and use rational number properties *(Lesson 7-4)*

Commutative:	$a + b = b + a$
	$a \cdot b = b \cdot a$
Associative:	$(a + b) + c = a + (b + c)$
	$(a \cdot b) \cdot c = a \cdot (b \cdot c)$
Identity:	$a + 0 = a$
	$a \cdot 1 = a$
Multiplicative Inverse:	$\frac{a}{b} \cdot \frac{b}{a} = 1$, where $a, b \neq 0$.

Use these exercises to review and prepare for the chapter test.

Solve each equation. Write the solution in simplest form.

10. $\frac{2}{9} - \frac{5}{9} = n$

11. $w = -\frac{1}{8} - \frac{3}{8}$

12. $x = \frac{5}{11} + \frac{6}{11}$

Solve each equation. Write the solution in simplest form.

13. $m = -\frac{3}{5} + \frac{2}{3}$

14. $z = 5 - 1\frac{2}{5}$

15. $-4\frac{1}{2} + \left(-6\frac{2}{3}\right) = t$

Solve each equation. Write the solution in simplest form.

16. $p = \left(-\frac{1}{6}\right)\left(-\frac{4}{5}\right)$ 17. $2\frac{2}{3}\left(-3\frac{1}{8}\right) = s$

18. $-\frac{3}{7} \cdot \frac{5}{7} = k$ 19. $g = \frac{8}{9} \cdot 3\frac{1}{4}$

Name the multiplicative inverse of each rational number.

20. $\frac{3}{4}$ 21. $-5\frac{1}{3}$

Solve each equation using properties of rational numbers.

22. $p = 4 \cdot 6\frac{1}{4}$ 23. $\left(\frac{3}{5}\right)\left(-2\frac{1}{6}\right) = b$

● recognize and extend arithmetic and geometric sequences *(Lesson 7-5)*

State whether the sequence 2, 4, 8, 16, . . . is geometric. Then write the next three terms.

Since there is a common ratio, 2, the sequence is geometric. The next three terms are 32, 64, and 128.

Identify each sequence as *arithmetic, geometric,* or *neither.* Then find the next three terms.

24. $-7, -4, -1, 2, \ldots$

25. $32, 16, 8, 4, \ldots$

26. $6, 14, 21, 27, \ldots$

27. $80, 68, 56, 44, \ldots$

● find the areas of triangles and trapezoids *(Lesson 7-6)*

Find the area of the triangle.

$A = \frac{1}{2}bh$

$A = \frac{1}{2}(3.4)(2.6)$

$A = 4.42 \text{ cm}^2$

2.6 cm

3.4 cm

Find the area of each figure described below.

28. triangle: base, $2\frac{1}{4}$ in.; height, $3\frac{1}{2}$ in.

29. trapezoid: bases, 10.5 m and 7 m; height, 6.4 m

● find the circumference of circles *(Lesson 7-7)*

6 in.

$C = \pi d$

$C \approx 3.14(6)$

$C \approx 18.84 \text{ in}^2$

Find the circumference of each circle described to the nearest tenth.

30. The diameter is $4\frac{1}{3}$ feet.

31. The radius is 2.6 meters.

32. The radius is 18 yards.

33. The diameter is 6.5 inches.

● divide fractions *(Lesson 7-8)*

Solve $r = \frac{7}{12} \div \left(-\frac{2}{3}\right)$.

$r = \frac{7}{\overset{4}{\cancel{12}}} \cdot -\frac{\overset{1}{\cancel{3}}}{2}$ *Use the multiplicative inverse.*

$r = -\frac{7}{8}$ *Simplify.*

Solve each equation. Write the solution in simplest form.

34. $j = -\frac{7}{9} \div \left(-\frac{1}{3}\right)$

35. $4\frac{2}{5} \div 2 = f$

36. $2\frac{1}{7} \div \left(-3\frac{3}{5}\right) = d$

37. $q = \frac{3}{10} \div 2\frac{2}{5}$

● solve equations involving rational numbers *(Lesson 7-9)*

Solve $\frac{3}{4}n = \frac{7}{8}$.

$\frac{4}{3} \cdot \frac{3}{4}n = \frac{4}{3} \cdot \frac{7}{8}$ *Multiply each side by $\frac{4}{3}$.*

$n = \frac{7}{6} \text{ or } 1\frac{1}{6}$

Solve each equation. Check your solution.

38. $\frac{x}{4} = -2.2$ **39.** $-7.2 = \frac{r}{1.6}$

40. $2b - \frac{6}{7} = 5\frac{2}{7}$ **41.** $t - (-0.8) = 4$

● solve inequalities involving rational numbers and graph their solutions. *(Lesson 7-10)*

Solve $-3.7x < 11.4$.

$-3.8x = 11.4$ *Solve the related equation.*

$\dfrac{-3.8x}{-3.8} = \dfrac{11.4}{-3.8}$ *Divide each side by -3.8.*

$x = -3$

Test numbers greater than -3 and less than -3 to find that the solution is $x > -3$.

Solve each inequality. Graph the solution on a number line.

42. $-5 < \frac{x}{4}$

43. $-\frac{d}{7} \le 3$

44. $36 + \frac{k}{5} \ge 51$

45. $2.6v + 9 < 28.76$

Applications & Problem Solving

46. ***Farming*** A farmer needs to fence a rectangular pasture. The length of the pasture is $20\frac{3}{4}$ yards, and the width is $20\frac{1}{3}$ yards. How many feet of fence does he need to buy? *(Lesson 7-2)*

47. ***Look for a Pattern*** Bo rides in the May bike marathon. To train for the event, he rides 4 miles the first day, 8 miles the second day, and 12 miles the third day. If he continues this pattern, how many miles will Bo have ridden in all by the end of the fourth day? *(Lesson 7-5A)*

48. ***Flooring*** Ms. Herr owns a carpet and tile store. She needs to special order some tile for the floor area shown below. What is the area of the tile needed? *(Lesson 7-6)*

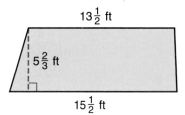

$13\frac{1}{2}$ ft

$5\frac{2}{3}$ ft

$15\frac{1}{2}$ ft

Alternative Assessment

● ***Open Ended***

You are designing a patio that is shaped like a trapezoid. You want the area of the patio to be between 130 and 150 square feet. One of the bases of the trapezoid must be 8 feet long. Make a sketch of a possible trapezoid. Label your sketch with the dimensions. What is the area of your patio?

You want to place a round table with a circumference of about 18.84 feet in the middle of your patio. Will the table fit on your patio? Explain.

A practice test for Chapter 7 is provided on page 653.

● *Completing the* **CHAPTER Project**

Use the following checklist to make sure your display or slide show is complete.

☑ You have included a clear, well-organized description of the Fibonacci sequence and the Fibonacci spiral.

☑ Include pictures of examples of Fibonacci numbers in nature.

● **PORTFOLIO**   Select one of the equations you solved in this chapter and place it in your portfolio. Attach a note to it explaining why you selected it.

Section One: Multiple Choice

There are eleven multiple-choice questions in this section. Choose the best answer. If a correct answer is *not here*, mark the letter for Not Here.

1. A four-sided figure with exactly one pair of parallel sides is a —

A rectangle.

B rhombus.

C parallelogram.

D trapezoid.

2. Express 42.8% as a decimal.

F 42.8

G 4.28

H 0.428

J 0.0428

3. If $\frac{x}{5} + 4 = 32$, what is the value of x?

A 5.6

B 7.2

C 28

D 140

4. What is the greatest common factor of 35 and 42?

F 6

G 7

H 14

J 1,470

5. What is the probability that Tashima will draw a blue jelly bean from a jar containing 3 black jelly beans, 2 red jelly beans, and 5 yellow jelly beans?

A 0

B $\frac{1}{10}$

C $\frac{1}{3}$

D 1

6. The histogram shows the quiz scores of the students in Mr. Guerrero's class.

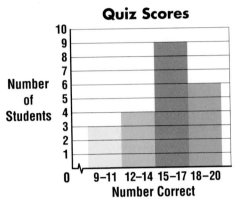

Quiz Scores

How many students had scores less than 15?

F 4

G 7

H 16

J 22

Please note that Questions 7–11 have five answer choices.

7. Carla averages 3 miles per hour walking. Jorge averages 7 miles per hour on his bicycle. They start at the same point and go in opposite directions on a straight road. Which equation could be used to find the number of hours h, when they would be 12 miles apart?

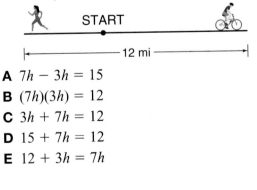

A $7h - 3h = 15$

B $(7h)(3h) = 12$

C $3h + 7h = 12$

D $15 + 7h = 12$

E $12 + 3h = 7h$

8. Luther works 20 hours each week. So far this week, he has worked 6.5 hours on Tuesday and 4.75 hours on Thursday. Which sentence could be used to find h, the number of hours remaining for Luther to work this week?

   F $6.5 + 4.75 = h + 20$

   G $6.5 + 4.75 = h \times 20$

   H $h = 20 - (6.5 + 4.75)$

   J $h - 6.5 + 4.75 = 20$

   K $h = 20 - 6.5 + 4.75$

9. Jaya works 21 hours a week and earns $122.85. How much money does Jaya earn per hour?

   A $6.50

   B $5.85

   C $5.05

   D $4.11

   E Not Here

10. Nora's classmates collected 4,656 aluminum cans last year. This year they collected 6,024 cans. How many more cans did they collect this year than last year?

    F 1,368

    G 1,468

    H 2,478

    J 2,632

    K Not Here

11. If an airplane travels 365 miles per hour, how many miles will it travel in 4 hours?

    A 1,240 mi

    B 1,260 mi

    C 1,440 mi

    D 1,460 mi

    E Not Here

 **inter**NET **CONNECTION** Test Practice For additonal test practice questions, visit:

www.glencoe.com/sec/math/mac/mathnet

Test-Taking Tip THE PRINCETON REVIEW

Most standardized tests have a time limit, so you must use your time wisely. Some questions will be easier than others. If you cannot answer a question quickly, go on to another. If there is time remaining when you are finished, go back to the questions you skipped.

Section Two: Free Response

This section contains eight questions for which you will provide short answers. Write your answers on your paper.

12. Which point is in the first quadrant?

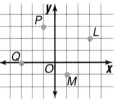

13. State the measure of the bases and the height of the trapezoid. Then find the area.

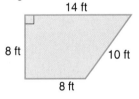

14. What is the value of $35 \cdot (-2)$?

15. Evaluate the expression ab if $a = -44$ and $b = 11$.

16. What is the multiplicative inverse of $-8\frac{1}{4}$?

17. Express 3.49×10^{-3} in standard form.

18. State the greater of the numbers, 7.089 or 7.09.

19. Does an isosceles triangle have line symmetry? If so, draw the lines of reflection.

What you'll learn in Chapter 8

- to solve problems by using proportions,
- to solve problems involving percents,
- to solve problems by first solving a simpler problem,
- to find missing measures of similar polygons, and
- to graph dilations on a coordinate plane.

TV WATCH

INSIDE:
- ORIGINAL "EMMIE" COMIC
- REVIEW OF "SPACE PARK"
- CLASSIC COMEDY VIDEOS

TV's Most Popular Young Stars!

THE VIEWERS CHOOSE THE BEST AND BRIGHTEST OF THE NEW SEASON

THIS MONTH'S NETWORK PREMIERE MOVIE

$1.20

CHAPTER Project

STAY TUNED!

In this project, you will write a newspaper article discussing the popularity of television programs. You will poll your classmates to determine how your peers rate several programs. You will also analyze the media ratings for one of these programs.

Getting Started

- Each person in the class should pick a television program to follow in the Nielsen ratings.
- As a class, make a list of all of the shows picked.
- Poll the class members to determine how many watch each program. Record the results of the poll in the form of a fraction. For example, if there are 25 students in your class and 6 of them watch a certain program, record $\frac{6}{25}$ for that program.
- Media ratings are expressed in percents. For example, a rating of 10 indicates that 10% of the 97 million households with a television watched the program. Record the media ratings for your program for three consecutive weeks.

Technology Tips

- Use the **Internet** to search for the media ratings for your television program.
- Use a **spreadsheet** to organize the class ratings and the media ratings.
- Use a **word processor** to write your article.

 **interNET** **CONNECTION** Data Update **For up-to-date information on the Nielsen ratings, visit:**

www.glencoe.com/sec/math/mac/mathnet

Working on the Project

You can use what you'll learn in Chapter 8 to change fractions to percents and to compare the ratings.

Page	Exercise
338	45
351	37
377	Alternative Assessment

Using Proportions

What you'll learn

You'll learn to solve problems by using proportions.

When am I ever going to use this?

Knowing how to use proportions can help you find an actual length represented on a scale drawing.

LOOK BACK

You can refer to Lesson 3-3 for information on proportions and cross products.

Are you a "channel-surfer" during commercials? A typical 30-minute TV program in the United States has about 8 minutes of commercials. At that rate, how many commercial minutes are shown during a 3-hour TV movie?

This problem can be solved using a proportion. A 3-hour TV movie is 3×60 or 180 minutes long. In the proportion below, c represents the number of commercial minutes during a 3-hour TV movie.

commercial minutes $\rightarrow$ $\dfrac{8}{30} = \dfrac{c}{180}$ $\leftarrow$ *commercial minutes*
program length $\rightarrow$ $\leftarrow$ *program length*

Cross products can be used to solve this proportion. *You will solve this problem in Exercise 5.*

Example ① **Health** To burn off the Calories from a 12-ounce can of regular cola, you would have to cycle 24 minutes at 13 miles per hour. How long would you have to cycle at the same speed to burn off the Calories from a 16-ounce bottle of cola?

CONNECTION

Explore   You must cycle for 24 minutes to burn off the Calories from a 12-ounce can of cola. You want to know how long you would need to cycle to burn off the Calories from a 16-ounce bottle of cola.

Plan   Let m represent the number of minutes needed to burn the Calories from a 16-ounce bottle of cola. Write a proportion.

minutes of cycling $\rightarrow$ $\dfrac{24}{12} = \dfrac{m}{16}$ $\leftarrow$ *minutes of cycling*
ounces of cola $\rightarrow$ $\leftarrow$ *ounces of cola*

Solve   Solve the proportion.

$$\frac{24}{12} = \frac{m}{16}$$

$24 \cdot 16 = 12 \cdot m$ *Find the cross products.*

$$384 = 12m$$

$$\frac{384}{12} = \frac{12m}{12} \quad \textit{Divide each side by 12.}$$

$$32 = m$$

You will need to cycle 32 minutes to work off the Calories from a 16-ounce bottle of regular cola.

Study Hint

Mental Math
Sometimes you can solve a proportion mentally by using equivalent fractions.
$$\frac{3}{11} = \frac{12}{x}$$
THINK: $3 \times 4 = 12$
$11 \times 4 = 44$
So, $x = 44$.

Examine Check your solution by substituting 32 for *m* in your original ratios. Simplify each ratio.

$$\frac{24}{12} = \frac{2}{1} \qquad \frac{32}{16} = \frac{2}{1}$$

(÷12 and ÷16 arrows shown)

Since both ratios are equivalent to the same fraction, $\frac{2}{1}$, they form a proportion. So, 32 is correct.

Sometimes you need to round your answer so it makes sense.

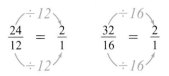

❷ **Life Science** About 1 out of every 5 people is left handed. If there are 28 students in a class, how many would you expect to be left handed?

Let *s* represent the number of left-handed students.

$$\begin{array}{ll} \textit{left handed} \rightarrow & \frac{1}{5} = \frac{s}{28} \leftarrow \textit{left handed} \\ \textit{size of group} \rightarrow & \qquad\quad \leftarrow \textit{size of group} \end{array}$$

$$1 \cdot 28 = 5 \cdot s \qquad \textit{Find the cross products.}$$

$$28 = 5s$$

$$\frac{28}{5} = \frac{5s}{5} \qquad \textit{Divide each side by 5.}$$

$$5.6 = s$$

You would expect to find 5 or 6 left-handed students out of 28 students.

CHECK FOR UNDERSTANDING

Communicating Mathematics

Read and study the lesson to answer each question.

1. **Tell** which proportions from the list below could be used to solve the problem. Be prepared to defend your answer.

   If a 15-minute long-distance call costs $2.10, how much does a similar 12-minute call cost?

   a. $\frac{15}{2.10} = \frac{12}{c}$ **b.** $\frac{15}{2.10} = \frac{c}{12}$ **c.** $\frac{c}{12} = \frac{2.10}{15}$

2. **Write a Problem** that could be solved by using a proportion. Then write a proportion that could be used to solve the problem.

Guided Practice

Write a proportion to solve each problem. Then solve.

3. If 16 ounces equals a pound, then 56 ounces equals *p* pounds.

4. If $\frac{1}{4}$ inch on a scale drawing represents 1 foot, then 3 inches on the drawing represent *f* feet.

Solve each problem.

5. *Advertising* Refer to the beginning of the lesson. How many commercial minutes are shown during a 3-hour TV movie?

6. *Photography* A photograph is 3 inches wide and 5 inches long. If the width of an enlargement of this photograph is 9 inches, find the length of the enlargement.

7. *Life Science* A microscope slide shows 35 red blood cells out of 60 blood cells. How many red blood cells would be expected in a blood sample of 75,000 blood cells?

EXERCISES

Practice

Write a proportion to solve each problem. Then solve.

8. If 1 foot equals 12 inches, 8 feet equals x inches.

9. If 100 centimeters equals a meter, then 170 centimeters equals m meters.

10. If a roll of 12-exposure film costs $4.20 to develop, then each picture costs x cents to develop.

11. If 12 eggs cost 96¢, then 4 eggs cost x cents.

12. If Trevor can type 2 pages in 15 minutes, then he can type 35 pages in m minutes.

13. If 35 copies are made in 2.5 minutes, then 420 copies will take m minutes.

Solve each problem.

14. If Maka can bicycle 25 miles in 2 hours, how far can she bicycle in 3 hours?

15. How many feet are in $2\frac{3}{4}$ miles? (*Hint:* 1 mile = 5,280 feet)

16. *Money Matters* If it costs $90 to feed a family of 3 for one week, how much will it cost to feed a family of 5 for a week?

17. If a car can travel 536 miles on 16 gallons of gasoline, how far can it travel on 10 gallons of gasoline?

18. *Fund-raising* The student council at Maple Springs Middle School decided to sell school sweatshirts as a fund-raiser. They surveyed 75 students. Of those students, 53 said they would buy a sweatshirt. If there are 1,023 students in the school, about how many students can you expect to buy a sweatshirt?

19. *Earth Science* The ratio of a person's weight on the moon to his or her weight on Earth is 1 to 6.
    a. If a person weighs 126 pounds on Earth, how much would he or she weigh on the moon?
    b. Find your weight on the moon.

Applications and Problem Solving

20. *Health* The average human heart beats 72 times in 60 seconds.
    a. How many times does it beat in 15 seconds?
    b. Take your pulse. Record the number of beats in 30 seconds.
    c. Use a proportion to determine how many times your heart beats in 45 seconds.

21. *Lawn Care* The amount of grass seed needed to plant a new lawn is directly proportional to the area of the lawn. If 3 pounds of seed are needed to plant a 2,000 square foot lawn, how many pounds are needed to plant a lawn that is 3,500 square feet?

22. *Life Science* An area of 2,500 square feet of grass produces enough oxygen for a family of 4.

   a. Find the area of grass needed to supply a family of 5 with oxygen.

   b. Find the number of people that could be supplied with oxygen by a lawn that covers 3,800 square feet.

23. *Critical Thinking* Write four different proportions that could be used to solve this problem.

   If 15 gallons of punch are needed for 120 people, how many people will 25 gallons serve?

Mixed Review

24. **Standardized Test Practice** What is the solution of $2r + 3 > 11$? *(Lesson 7-10)*

   A $r > 4$ **B** $r < 3$ **C** $r > 7$ **D** $r < 7$

25. Express $2.\overline{44}$ as a mixed number in simplest form. *(Lesson 6-5)*

For **Extra Practice**, see page 627.

26. *Inventory* A restaurant serves 300 cups of coffee each day. If one pound of coffee is used to make 100 cups, how many pounds will the restaurant need for a week? *(Lesson 1-1)*

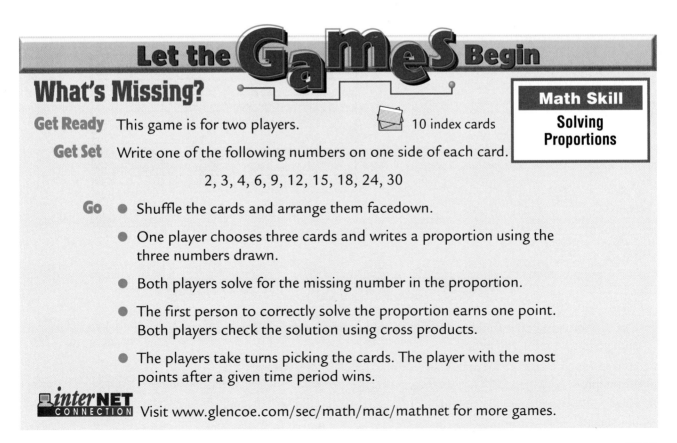

Let the Games Begin

What's Missing?

Math Skill	Solving Proportions

Get Ready   This game is for two players.   10 index cards

Get Set   Write one of the following numbers on one side of each card.

2, 3, 4, 6, 9, 12, 15, 18, 24, 30

Go
- Shuffle the cards and arrange them facedown.
- One player chooses three cards and writes a proportion using the three numbers drawn.
- Both players solve for the missing number in the proportion.
- The first person to correctly solve the proportion earns one point. Both players check the solution using cross products.
- The players take turns picking the cards. The player with the most points after a given time period wins.

interNET CONNECTION Visit www.glencoe.com/sec/math/mac/mathnet for more games.

8-1B Proportions

A Follow-Up of Lesson 8-1

computer

spreadsheet software

The Elegant Eatery caters food for various functions. Their chili dip recipe is shown at the right. A spreadsheet can be used to determine how much of each ingredient to use for various sized groups. The proportion below will give the number of batches needed.

> **Chili Dip**
>
> 1 c. cottage cheese $\frac{1}{4}$ c. chili sauce
>
> $\frac{1}{4}$ t. onion powder $\frac{1}{4}$ c. skim milk
>
> 3 T. grated Parmesan cheese
>
> Mix all ingredients in blender until smooth. Chill. Serves 20.

$$\frac{number\ of\ people\ expected}{number\ of\ servings\ per\ batch} = \frac{number\ of\ batches\ needed}{1\ batch}$$

Since B1 is the number of people and the number of servings per batch is 20, we can rewrite the proportion $\frac{B1}{20} = \frac{x}{1}$. The formula in cell B2 uses B1 ÷ 20 to find the number of batches that need to be made.

TRY THIS

Work with a partner.

Use the spreadsheet to determine the recipe for a function with 120 people by making the following substitution.

$$B1 = 120$$

CHILI DIP RECIPE

	A	B	C
1	People To Serve	B1	
2	Batches Needed	= B1/20	
3	INGREDIENT	NUMBER	
4	cottage cheese	= B2	cups
5	chili sauce	= B2/4	cups
6	onion powder	= B2/4	t.
7	skim milk	= B2/4	cups
8	parmesan cheese	= B2 * 3	T.

ON YOUR OWN

1. Use the spreadsheet to find the amount of each ingredient needed to make enough chili dip for 180 people.

2. How could you change the spreadsheet if one batch of chili dip served 12 people?

3. How could you change the spreadsheet if the recipe included $\frac{1}{8}$ teaspoon of hot sauce?

The Percent Proportion

What you'll learn

You'll learn to solve problems using the percent proportion.

When am I ever going to use this?

Knowing how to use the percent proportion can help you determine the amount of tax you must pay.

Word Wise

percentage
base
rate
percent proportion

About 29% of the United States is forest or woodland. The symbol % is a common symbol for the word percent. A *percent* is a ratio that compares a number to 100. Percent also means *hundredths,* or *per hundred.* So 29% of the United States means that 29 square miles out of every 100 square miles or $\frac{29}{100}$ of the land is forest.

The models below represent 50%, 75%, and 10%.

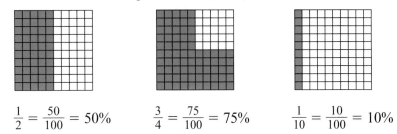

$$\frac{1}{2} = \frac{50}{100} = 50\% \qquad \frac{3}{4} = \frac{75}{100} = 75\% \qquad \frac{1}{10} = \frac{10}{100} = 10\%$$

You can use a proportion to find a percent.

Example 1

What percent of the circle is shaded?

Let *x* represent the percent of the circle that is shaded.

$$\begin{array}{ll} \text{parts shaded} & \rightarrow \\ \text{parts in whole} & \rightarrow \end{array} \quad \frac{5}{8} = \frac{x}{100} \quad \begin{array}{l} \leftarrow \text{ percent shaded} \\ \leftarrow \text{ percent in whole} \end{array}$$

$$5 \cdot 100 = 8 \cdot x \quad \textit{Find the cross products.}$$

$$500 = 8x$$

$$\frac{500}{8} = \frac{8x}{8} \quad \textit{Divide each side by 8.}$$

$$62.5 = x$$

So, 62.5% of the circle is shaded.

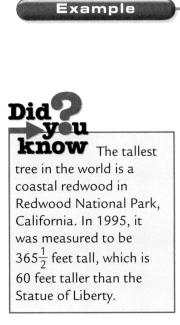

Did you know The tallest tree in the world is a coastal redwood in Redwood National Park, California. In 1995, it was measured to be $365\frac{1}{2}$ feet tall, which is 60 feet taller than the Statue of Liberty.

In Example 1, the number of parts shaded, 5, is called the **percentage (P)**. The total number of parts, 8, is called the **base (B)**. The ratio $\frac{62.5}{100}$ is called the **rate (r)**.

$$\frac{5}{8} = \frac{62.5}{100} \quad \rightarrow \quad \frac{\text{Percentage}}{\text{Base}} = \text{Rate}$$

If *r* represents the percent, the proportion can be written as $\frac{P}{B} = \frac{r}{100}$. This proportion is called the **percent proportion**.

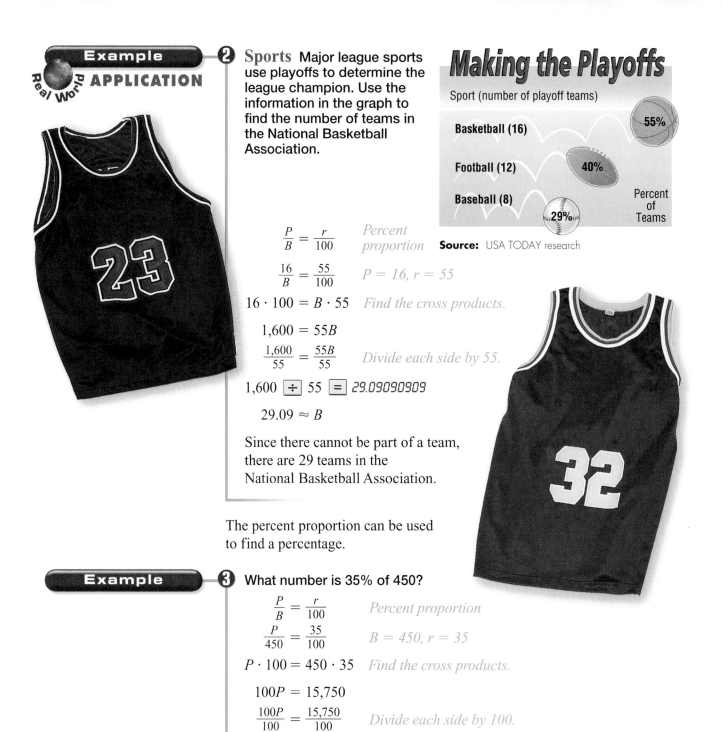

Sports Major league sports use playoffs to determine the league champion. Use the information in the graph to find the number of teams in the National Basketball Association.

Making the Playoffs

Sport (number of playoff teams)

Basketball (16)   55%

Football (12)   40%

Baseball (8)   29%

Percent of Teams

Source: USA TODAY research

$\frac{P}{B} = \frac{r}{100}$ *Percent proportion*

$\frac{16}{B} = \frac{55}{100}$ *P = 16, r = 55*

$16 \cdot 100 = B \cdot 55$ *Find the cross products.*

$1,600 = 55B$

$\frac{1,600}{55} = \frac{55B}{55}$ *Divide each side by 55.*

$1,600 \; \boxed{\div} \; 55 \; \boxed{=} \; 29.09090909$

$29.09 \approx B$

Since there cannot be part of a team, there are 29 teams in the National Basketball Association.

The percent proportion can be used to find a percentage.

Example   **What number is 35% of 450?**

$\frac{P}{B} = \frac{r}{100}$ *Percent proportion*

$\frac{P}{450} = \frac{35}{100}$ *B = 450, r = 35*

$P \cdot 100 = 450 \cdot 35$ *Find the cross products.*

$100P = 15,750$

$\frac{100P}{100} = \frac{15,750}{100}$ *Divide each side by 100.*

$P = 157.5$

157.5 is 35% of 450.

CHECK FOR UNDERSTANDING

Communicating Mathematics

Read and study the lesson to answer each question.

1. **Explain** why the value of r in $\frac{P}{B} = \frac{r}{100}$ represents a percent.

2. **Tell** why the following statement is true.

   40% of the figure is shaded.

3. The percent proportion can be used to find the percentage, the base, or the percent. *Write* an example of each type of problem. Include the percent proportion needed to solve each problem.

Guided Practice

4. Write a percent to represent the shaded portion of the figure.

Express each fraction as a percent.

5. $\frac{1}{20}$ **6.** $\frac{3}{10}$ **7.** $\frac{19}{50}$ **8.** $\frac{7}{40}$

Write a percent proportion to solve each problem. Then solve. Round to the nearest tenth.

9. Find 15% of 60.

10. 70 is what percent of 280?

11. 15 is what percent of 45?

12. 2 is 16% of what number?

13. *Taxes* Mrs. Reeder stayed at a hotel during a mathematics teachers' conference. Her bill listed the room tax as $19.46 and the room rate as $139.00. Write the tax as a percent of the room rate.

EXERCISES

Practice

Write a percent to represent the shaded portion of each figure.

14.   **15.**   **16.**

Express each fraction as a percent.

17. $\frac{1}{5}$ **18.** $\frac{7}{10}$ **19.** $\frac{13}{25}$ **20.** $\frac{9}{20}$ **21.** $\frac{41}{50}$ **22.** $\frac{7}{8}$

23. $\frac{1}{2}$ **24.** $\frac{33}{40}$ **25.** $\frac{3}{16}$ **26.** $\frac{24}{25}$ **27.** $\frac{7}{28}$ **28.** $\frac{11}{40}$

Write a percent proportion to solve each problem. Then solve. Round to the nearest tenth.

29. Find 25% of 66.

30. Find 13% of 80.

31. 16 is what percent of 48?

32. 22 is what percent of 110?

33. 5 is 26% of what number?

34. 8 is 13% of what number?

35. 7 is what percent of 20?

36. What is 27% of 70?

37. 60 is 15% of what number?

38. Find 47% of 65.

39. 45 is 35% of what number?

40. 9 is what percent of 15?

41. *Geometry* What percent of the area of the square is shaded?

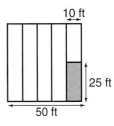

Applications and Problem Solving

42. *Jewelry* Jewelers measure gold in karats. Pure gold is 24 karat gold.

a. In the United States, most jewelry is 18 karat gold. What percent of the 18-karat jewelry is gold?

b. In the Netherlands, most jewelry is 22 karat gold. What percent of the 22-karat jewelry is gold?

43. *Royalties* Consuela Reyes writes rap songs and receives a royalty of 8% of the sales of her songs. Last month she received a royalty check for $3,896.00. How much money was spent on her songs?

44. *Education* Jamal scored a 92% on his science test. If there were 25 questions on the test, how many questions did Jamal answer correctly?

45. *Working on the* **CHAPTER Project** Refer to the class poll that you took on television programs on page 329.

a. Change each fraction to a percent, and identify which television program is the most popular in your class.

b. Make a graph to display your data.

46. *Critical Thinking* Hao made 56% of his free throws in the first half of the basketball season. If he makes 7 shots out of the next 13 attempts, will it help or hurt his average? Explain.

Mixed Review

47. *Baking* A recipe calls for 4 cups of flour for 64 cookies. How much flour is needed for 96 cookies? *(Lesson 8-1)*

48. Solve $g = \frac{5}{6} \div \frac{4}{3}$. *(Lesson 7-8)*

49. Write $\frac{54}{81}$ in simplest form. *(Lesson 6-4)*

50. **Standardized Test Practice** What is the value of x in the figure? *(Lesson 5-1)*

A $14°$

B $42°$

C $98°$

D $140°$

$140°$

$(5x - 70)°$

For **Extra Practice,** see page 627.

51. Express *240 shrimp in 6 pounds* as a unit rate. *(Lesson 3-1)*

52. Find $|-20|$. *(Lesson 2-1)*

Integration: Algebra
The Percent Equation

What you'll learn

You'll learn to solve problems using the percent equation.

When am I ever going to use this?

Knowing how to use the percent equation can help you understand grades given in percents.

A candy company gave consumers the opportunity to vote for a new color of candy to replace the tan-colored ones. Blue won with over 54% of the vote. The table shows the new mix of colors. In a 16-ounce bag of the candy, there are about 525 pieces of candy. How many would you expect to be red?

You have learned to find 20% of 525 by using the percent proportion. Another way to solve this problem is to express 20% as a decimal and then multiply.

Plain Chocolate Candies	
Color	**Percent**
brown	30%
yellow	20%
red	20%
orange	10%
green	10%
blue	10%

$0.20 \times 525 = 105$ *Estimate: $\frac{1}{5}$ of 500 is 100.*

In a 16-ounce bag of the candy, there should be about 105 red ones. In this example, 20% is the rate, 525 is the base, and 105 is the percentage.

You can use an equation to describe how the quantities are related.

Percent Equation	**Words:**	The rate (R) times the base (B) is equal to the percentage (P).
	Symbols:	$RB = P$

In the percent equation, the rate is usually expressed as a decimal.

Examples

1 **Find 4% of $625.** *Estimate: 1% of 600 is 6; $4 \times 6 = 24$*

$$RB = P \quad \text{\textit{Percent equation}}$$
$$0.04 \times 625 = P \quad \text{\textit{Replace R with 4\% or 0.04 and B with 625.}}$$
$$25 = P$$

So, 4% of $625 is $25.   *Compare to the estimate.*

Study Hint

Technology To find 4% of $625 with your calculator, press 4 [2nd] [%] [×] 625 [=] 25.

2 **12 is 60% of what number?** *Estimate: $60\% \approx \frac{1}{2}$; 12 is $\frac{1}{2}$ of 24.*

$$RB = P \quad \text{\textit{Percent equation}}$$
$$0.6B = 12 \quad \text{\textit{Replace R with 60\% or 0.6 and P with 12.}}$$
$$\frac{0.6B}{0.6} = \frac{12}{0.6} \quad \text{\textit{Divide each side by 0.6.}}$$
$$12 \div 0.6 = 20$$

So, 12 is 60% of 20.   *Check your answer by finding 60% of 20.*

Real World APPLICATION

③ Money Matters Betty bought supplies at the hardware store that cost $28.40. If she paid sales tax of $1.42, what was the sales tax rate?

Explore   You know the amount of the purchases and the sales tax. You need to find the rate.

Plan   To find the rate, use the percent equation. Let B represent the base, $28.40, and let P represent the percentage, $1.42.

Solve

$$RB = P \qquad \text{\textit{Percent equation}}$$

$$R \cdot 28.40 = 1.42 \qquad \text{\textit{Replace B with 28.40 and P with 1.42.}}$$

$$\frac{R \cdot 28.40}{28.40} = \frac{1.42}{28.40} \qquad \text{\textit{Divide each side by 28.40.}}$$

$$1.42 \;\boxed{\div}\; 28.40 \;\boxed{=}\; 0.05$$

So, the sales tax rate is 0.05 or 5%.

Examine   5% of $28.40 is 1.42. The answer checks.

CHECK FOR UNDERSTANDING

Communicating Mathematics

Read and study the lesson to answer each question.

1. *Write* an equation to find the percent of questions answered correctly if you correctly answered 28 out of 34 questions on a history test.

2. *Tell* what number you would use for R in the equation $RB = P$ if the rate is 35%.

Guided Practice

Write an equation in the form $RB = P$ for each problem. Then solve.

3. What number is 47% of 52?

4. What number is 56% of 80?

5. What percent of 90 is 36?

6. $17 is what percent of $51?

7. $48 is 30% of what amount?

8. Fifty is 10% of what number?

9. *Money Matters* The price of a home video game system is $198. If the sales tax on the system is $8.91, what is the sales tax rate?

EXERCISES

Practice

Write an equation in the form $RB = P$ for each problem. Then solve.

10. Find 16.5% of 60.

11. Fifty-five is what percent of 66?

12. 75 is 50% of what number?

13. 15% of what number is 30?

14. 18 is what percent of 60?

15. Find 24% of 72.

16. Find $33\frac{1}{3}$% of 420.

17. 16 is $66\frac{2}{3}$% of what number?

18. 45 is what percent of 150?

19. 6 is what percent of 300?

20. Find 5% of 3,200.

21. What percent of 80 is 25?

22. Find 28% of $231.

23. $6 is what percent of $50?

24. 15 is 75% of what number?

25. What percent of 80 is 70?

26. $1.47 is 7% of what amount?

27. $54 is 8% of what amount?

28. There are 35 students. Eight of them are wearing green. What percent are wearing green?

29. Twenty percent of what amount is $5,000?

Applications and Problem Solving

30. *Life Science* Earth's age is estimated at 4.6 billion years. If life began 3.5 billion years ago, for what percent of Earth's existence has life been present?

31. *Music* In a six-month period, Americans ages 12 to 54 buy an average of 11 CDs, tapes, or records. The graph below shows the reasons they give for making a purchase. Suppose you survey 25 people. Predict how many of those surveyed bought music because they saw the video.

With a family member, find the nutrition label on a box of cereal. What information does it provide? How are percents used on the label?

Why We Buy Music

Heard on radio	80%
Saw video	43%
Word-of-mouth, store display (tie)	36%
Fan loyalty	28%
Was on sale	24%
Saw TV performance	23%
Read review, saw ad (tie)	15%
Attended concert	13%

Note: May name more than one reason

Source: Strategic Record Research

32. *Critical Thinking* Todd adds 10% of a number to the number. Then he subtracts 10% of the total. Is the result equal to his original number? Explain.

Mixed Review

33. *Recycling* Americans recycled 35.4% of the aluminum products discarded in a recent year. If Americans discarded 3,000,000 tons of aluminum products, how many tons were recycled? *(Lesson 8-2)*

34. Express $\frac{7}{8}$ as a percent. *(Lesson 3-4)*

35. **Standardized Test Practice** Order the integers 7, 12, -61, 23, -22, -6, and 0 from greatest to least. *(Lesson 2-2)*

A -61, 23, -22, 12, 7, -6, 0

B 0, -6, 7, 12, -22, 23, -61

C -61, -22, -6, 0, 7, 12, 23

D 23, 12, 7, 0, -6, -22, -61

For **Extra Practice,** see page 627.

36. *Algebra* Solve $q + 6.8 = 15.2$. Check your solution. *(Lesson 1-4)*

PROBLEM SOLVING

8-3B Solve a Simpler Problem

A Follow-Up of Lesson 8-3

Shaniqua and Michael are on the holiday social committee. They need to order cups of ice cream for the 300 students who will be attending the holiday dance. Let's listen in!

It would take too much time to ask everyone what flavor of ice cream he or she wants. How will we know how many of each flavor to order?

I took this survey of 20 of our classmates. We could use the results of the survey to determine how many of each flavor to order.

Michael

Since 45% prefer chocolate chip ice cream, let's write the percent equation $RB = P$ and solve for P.

In this case, R is 45% or 0.45, and B is 300. Since 0.45 × 300 = 135, we should order about 135 cups of chocolate chip ice cream.

Shaniqua

Flavor	Vanilla	Chocolate	Strawberry	Chocolate Chip	Peanut Butter
Number	1	5	3	9	2
Percent	5%	25%	15%	45%	10%

THINK ABOUT IT

Work with a partner.

1. *Explain* why Michael chose to survey 20 students instead of 18 or 22 students. What other number might have been a good number to survey?

2. *Describe* how Shaniqua and Michael could use proportions to decide how many cups of chocolate chip ice cream to order.

3. *Apply* the **solve a simpler problem** strategy to determine how many cups of each flavor of ice cream to order.

For **Extra Practice,** see page 628.

ON YOUR OWN

4. The second step of the 4-step plan for problem solving asks you to *plan.* **Explain** why this step is important when using the strategy of solving a simpler problem.

5. *Write a Problem* that can be solved by solving a simpler problem.

6. *Refer* to the beginning of Lesson 8-3. How many of each color of candy would you expect in a 16-ounce bag of the candy?

MIXED PROBLEM SOLVING

STRATEGIES

Look for a pattern.
Solve a simpler problem.
Act it out.
Guess and check.
Draw a diagram.
Make a chart.
Work backward.

Solve. Use any strategy.

7. *Life Science* A paramecium, which is 0.23 millimeter long, is pictured in a science book. If the paramecium in the picture is 6 millimeters long, about how many times has the paramecium been magnified for the book?

8. *Geography* Minnesota is known for its lakes. In fact, 5.75% of its area is covered with water. If Minnesota has 4,854 square miles of inland water, find the total area of Minnesota.

9. *Laundry* Two clothespins are needed to hang one towel on a clothesline. One clothespin can be used on a corner of one towel and a corner of the towel next to it. What is the least number of clothespins you need to hang 8 towels?

10. *Geometry* The perimeter of a rectangle is 22 meters. Its area is 24 square meters. What are the dimensions of the rectangle?

11. *Health* A frozen lean dinner entree contains 200 Calories. About 25% of the Calories are from fat. How many Calories are from fat?

12. *Statistics* The graph shows three of the major concerns of adults in the United States today.

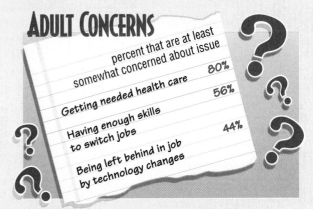

ADULT CONCERNS

percent that are at least somewhat concerned about issue

Getting needed health care — 80%

Having enough skills to switch jobs — 56%

Being left behind in job by technology changes — 44%

Source: KRC Research & Consulting for *U.S News*

a. If there are about 142,146,000 people between 20 and 59 years of age in the United States, how many people in this age group have each concern?

b. What do you think are the major concerns of teenagers today? How do these concerns compare with those of adults?

13. **Standardized Test Practice** If the probability of a certain event is 100%, which statement is true?

A The probability is more than 1.

B The probability is equal to 1.

C The probability is less than 1.

D The probability is equal to 0.

Large and Small Percents

What you'll learn

You'll learn to express percents greater than 100 or less than 1 as fractions and decimals.

When am I ever going to use this?

Knowing how to express percents greater than 100 can help you understand the meaning of a sales increase of 110%.

Farmers from the United States produce 50% of the world's soybeans, 40% of the world's corn, 25% of the world's beef, and 15% of the world's cotton. However, farmers who live in the United States represent only 0.3% of the total number of farmers in the world. What does 0.3% mean? *You will answer this question in Exercise 1.*

You can use grid paper to investigate percents less than 1 such as 0.3% and percents greater than 100 such as 125%.

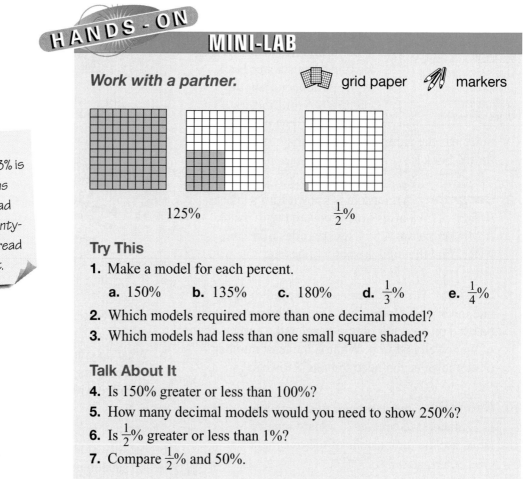

Study Hint

Reading Math 0.3% is read as *three tenths percent*. 125% is read as *one hundred twenty-five percent*. $\frac{1}{2}$% is read as *one-half percent*.

Work with a partner. ⬚⬚ grid paper ✏️ markers

125% $\frac{1}{2}$%

Try This

1. Make a model for each percent.
   a. 150%   **b.** 135%   **c.** 180%   **d.** $\frac{1}{3}$%   **e.** $\frac{1}{4}$%
2. Which models required more than one decimal model?
3. Which models had less than one small square shaded?

Talk About It

4. Is 150% greater or less than 100%?
5. How many decimal models would you need to show 250%?
6. Is $\frac{1}{2}$% greater or less than 1%?
7. Compare $\frac{1}{2}$% and 50%.

To express a percent greater than 100 or less than 1 as a fraction or a decimal, you can use the same procedure you used for percents from 1 to 100.

Express each percent as a fraction or mixed number in simplest form.

1 **0.8%**

$0.8\% = \dfrac{0.8}{100}$ *Multiply the numerator and denominator by 10.*

$= \dfrac{8}{1,000}$ or $\dfrac{1}{125}$

LOOK BACK

You can refer to Lesson 7-8 for information on dividing fractions.

2 **145%**

$145\% = \dfrac{145}{100}$ *Change the improper fraction to a mixed number.*

$= 1\dfrac{45}{100}$ or $1\dfrac{9}{20}$

3 $\dfrac{1}{5}\%$

$\dfrac{1}{5}\% = \dfrac{\frac{1}{5}}{100}$

$= \dfrac{1}{5} \div 100$ *The fraction bar indicates division.*

$= \dfrac{1}{5} \times \dfrac{1}{100}$ *To divide by 100, multiply by its multiplicative inverse, $\dfrac{1}{100}$.*

$= \dfrac{1}{500}$

Express each percent as a decimal.

4 **137%**

$137\% = \dfrac{137}{100}$

$= 1.37$

5 **0.7%**

$0.7\% = \dfrac{0.7}{100}$

$= \dfrac{7}{1,000}$ or 0.007

APPLICATION

Real World

6 **Business** The five states with the greatest increases in female-owned companies from 1987 to 1996 are shown in the graph. Order these states according to their percent increase starting with the state with the greatest increase.

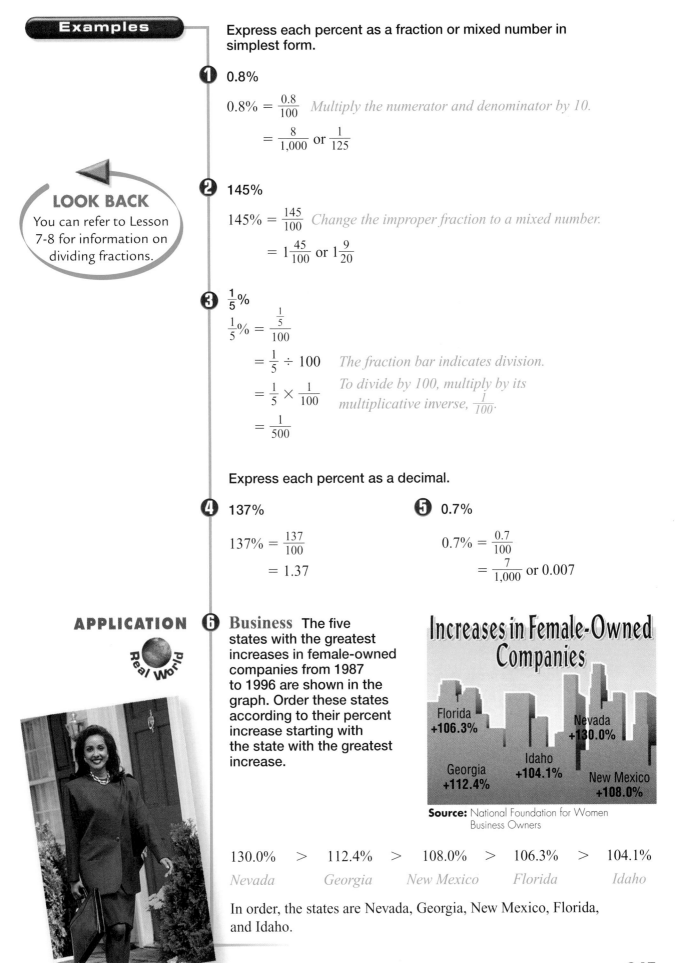

Increases in Female-Owned Companies

Florida +106.3%

Nevada +130.0%

Idaho +104.1%

Georgia +112.4%

New Mexico +108.0%

Source: National Foundation for Women Business Owners

$130.0\% > 112.4\% > 108.0\% > 106.3\% > 104.1\%$
Nevada   Georgia   New Mexico   Florida   Idaho

In order, the states are Nevada, Georgia, New Mexico, Florida, and Idaho.

Communicating Mathematics

Read and study the lesson to answer each question.

1. *Tell* what is meant by the following statement.

   Farmers who live in the United States represent only 0.3% of the total number of farmers in the world.

2. *Compare* 125% and 1 using $>$, $<$, or $=$.

HANDS-ON MATH

3. *Make a model* for each percent.

   a. 110%   **b.** 0.25%

Guided Practice

Express each percent as a fraction or mixed number in simplest form.

4. 0.04%   **5.** 175%   **6.** $\frac{1}{4}$%

Express each percent as a decimal.

7. 155%   **8.** 0.25%   **9.** 123.5%

10. Order 25%, $\frac{1}{4}$%, 125%, and 1 from least to greatest.

11. *Taxes* The percent of tax returns audited by the Internal Revenue Service varies from year to year. The percent of returns audited for five years are given in the table.

Year	1990	1991	1992	1993	1994
Percent of Audits	1.17%	1.06%	0.92%	1.08%	2.21%

   a. Write each percent as a decimal.

   b. Which year had the greatest percent of audits?

   c. Which year had the least percent of audits?

Practice

Express each percent as a fraction or mixed number in simplest form.

12. 0.5%   **13.** 115%   **14.** $\frac{1}{8}$%   **15.** 0.06%   **16.** 136.5%

17. 245%   **18.** $\frac{3}{4}$%   **19.** 13,447%   **20.** $\frac{4}{25}$%   **21.** $33\frac{1}{3}$%

Express each percent as a decimal.

22. 118%   **23.** 0.03%   **24.** 225%   **25.** 10.25%   **26.** 0.009%

27. 0.079%   **28.** $\frac{3}{10}$%   **29.** $\frac{2}{5}$%   **30.** 2,355.2%   **31.** $110\frac{2}{5}$%

32. Order $\frac{1}{5}$%, 20%, 200%, and 1 from least to greatest.

33. Write 1.04%, 0.4%, 104%, 0.04%, and 1 in order from greatest to least.

34. Which of the following numbers is the greatest? the least?

   $0.8, 67\%, 7, \frac{3}{8}\%$

35. ***Movies*** The average cost of a ticket at a movie theater in the United States increased about 181% from 1970 to 1995. Write this percent as a decimal.

36. ***Education*** The graph shows the percent of high school students studying various languages.

   a. Write each percent as a fraction in simplest form.

   b. Which language do the greatest percent of students study?

   c. Which language do the least percent of students study?

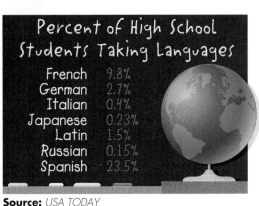

Percent of High School Students Taking Languages

French	9.8%
German	2.7%
Italian	0.4%
Japanese	0.23%
Latin	1.5%
Russian	0.15%
Spanish	23.5%

Source: *USA TODAY*

37. ***Critical Thinking*** The average cost of four years of college is expected to double in about 20 years. Explain whether this means that the cost is expected to be 200% of the current cost.

Mixed Review

38. ***Sports*** Eleven of the 48 members of the football team are on the field. What percent of the team members are playing? *(Lesson 8-3)*

39. ***Geometry*** Find the circumference of the circle. *(Lesson 7-7)*

8.2 cm

40. Find the GCF of 36, 108, and 180. *(Lesson 6-3)*

41. **Standardized Test Practice** Two angles of a triangle have the same measure, and the third angle measures 68°. Let *m* represent the measure in degrees of each of the congruent angles. Which equation could be used to find *m*? *(Lesson 5-2)*

   A $m + 68 = 180$

   B $m + 68 = 90$

   C $2m + 68 = 180$

   D $2m - 68 = 180$

   E $2m + 68 = 360$

42. Solve $u = -8(10)(-12)$. *(Lesson 2-7)*

For **Extra Practice,** see page 628.

Percent of Change

You can express an increase or a decrease in terms of a percent.

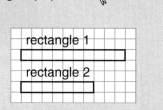

HANDS-ON MINI-LAB

Work with a partner.   grid paper   scissors

Try This

1. Draw a rectangle 10 units long and 1 unit wide. Then draw another rectangle 7 units long and 1 unit wide.

    rectangle 1
    rectangle 2

2. Cut out the second rectangle and place it over the first rectangle. Align one end of the rectangles. Shade the squares that are not covered.

3. Write a percent that represents the shaded squares in the first rectangle using the first rectangle as the base for the percent.

Talk About It

4. In Step 3, you found the percent of change in the area from the first rectangle to the second rectangle. Does this percent of change represent an increase or a decrease?

5. Cut out a third rectangle that is 12 units long and 1 unit wide. Find the percent of change in the area from the first to the third rectangle. Does this change represent an increase or a decrease?

When an increase or decrease is expressed as a percent, the percent is called the **percent of change**.

Examples

Find each percent of change. Round to the nearest percent.

1 original: 25
new: 28

First, find the amount of change.

$28 - 25 = 3$

Then, find the percent using the original number, 25, as the base.

$$\frac{3}{25} = \frac{r}{100}$$

$3 \cdot 100 = 25 \cdot r$

$300 = 25r$

$$\frac{300}{25} = \frac{25r}{25}$$

$12 = r$

The percent of change is 12%.

2 original: 45
new: 30

First, find the amount of change.

$45 - 30 = 15$

Then, find the percent using the original number, 45, as the base.

$$\frac{15}{45} = \frac{r}{100}$$

$15 \cdot 100 = 45 \cdot r$

$1,500 = 45r$

$$\frac{1,500}{45} = \frac{45r}{45} \quad \textit{Use a calculator.}$$

$33.3 \approx r$

The percent of change is 33.3%.

In Example 1, the new number is greater than the original number, so the percent of change is a **percent of increase**. In Example 2, the new number is less than the original number, so the percent of change is a **percent of decrease**.

Do you shop at sales? The amount by which the regular price is reduced is called the **discount**. The percent of the discount is a percent of decrease. You can find the sale price by subtracting the discount.

 **Example 3**

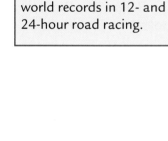

Real World APPLICATION

Shopping The Just for Fun Sporting Goods store advertises that all in-line skates are on sale for 20% off the regular price. Find the sale price of a pair of in-line skates that cost $135.

Method 1	Method 2
First, use the percent equation to find the amount of the discount. $RB = P$ $0.20 \cdot 135 = P$ $27 = P$ Then, subtract the $27 discount from the regular price, $135. $135 - 27 = 108$	First, find the percent paid. $100\% - 20\% = 80\%$ Then, use the percent equation to find the sale price of the skates. $RB = P$ $0.80 \cdot 135 = P$ $108 = P$

The sale price of the in-line skates is $108.

Did you know The first world championships of in-line skating were held in 1992. Derek Parra holds two overall world championship titles. Kimberly Ames holds the world records in 12- and 24-hour road racing.

A store sells an item for more than it paid for that item. The extra money is used to cover the expenses of the store and to make a profit. The increase in the price is called the **markup**. The percent of a markup is a percent of increase. The amount the customer pays for an item is called the **selling price**.

Example 4

Real World APPLICATION

Business The Just for Fun Sporting Goods store prices items 30% over the price paid by the store. If the store purchases a tennis racket for $65, find the selling price of the racket.

Method 1	Method 2
Use the percent proportion to find the markup. $$\frac{m}{65} = \frac{30}{100}$$ $m \cdot 100 = 65 \cdot 30$ $100m = 1{,}950$ $$\frac{100m}{100} = \frac{1{,}950}{100}$$ $m = 19.50$ Add the markup to the price paid by the store. $65.00 + 19.50 = 84.50$	The customer will pay 100% of the price plus an extra 30% of the price. Find $100\% + 30\%$ or 130% of the price paid by the store. $$\frac{s}{65} = \frac{130}{100}$$ $s \cdot 100 = 65 \cdot 130$ $100s = 8{,}450$ $$\frac{100s}{100} = \frac{8{,}450}{100}$$ $s = 84.50$

The selling price of the racket for the customer is $84.50.

Communicating Mathematics

Read and study the lesson to answer each question.

1. *State* the first step in finding the percent of change.

2. *Explain* how you know whether a percent of change is a percent of increase or a percent of decrease.

HANDS-ON MATH

3. *Make a model* to show a decrease of 20%.

Guided Practice

Find each percent of change. Round to the nearest percent.

4. original: 5
   new: 4

5. original: 325
   new: 400

Find the sale price of each item to the nearest cent.

6. $29.95 jeans, 25% off

7. $300.00 stereo, $33\frac{1}{3}$% off

Find the selling price for each item given the amount paid by the store and the markup. Round to the nearest cent.

8. $150.00 skis, 35% markup

9. $12.00 shirt, 40% markup

10. *Advertising* In a television commercial, Sally's Burgers claimed that their Big Beefy was 75% larger than Mickey's Big Burger. The Big Beefy contains $\frac{1}{3}$ pound of ground beef, while the Big Burger contains $\frac{1}{4}$ pound of ground beef. Does the Big Beefy contain 75% more beef by weight than the Big Burger? Explain.

EXERCISES

Find each percent of change. Round to the nearest percent.

Practice

11. original: 10
    new: 6

12. original: 50
    new: 67

13. original: 12
    new: 20

14. original: 80
    new: 55

15. original: 775
    new: 825

16. original: 835
    new: 900

Find the sale price of each item to the nearest cent.

17. $14.50 CD, 10% off

18. $39.95 sweater, 25% off

19. $3.59 tennis balls, $\frac{1}{4}$ off

20. $119.50 lamp, $\frac{1}{3}$ off

21. $13.00 book, 20% off

22. $18.00 shirt, 30% off

Find the selling price for each item given the amount paid by the store and the markup. Round to the nearest cent.

23. $1,200.00 computer, 20% markup

24. $15.00 belt, 35% markup

25. $14.00 video tapes, 40% markup

26. $87.00 coat, $33\frac{1}{3}$% markup

27. $59.00 camera, 25% markup

28. $45.00 dress, 45% markup

29. A math class had 25 students. If 2 more students enrolled in the class, what is the percent of change?

30. Find the discount rate on a $25 watch that regularly sells for $30.

31. What is the sale price of a $250 bicycle on sale at 10% off?

32. Find the markup rate on a $60 jacket that sells for $75.

33. A store has a markup of 30%. If a store buys a sweatshirt for $15, what is the selling price?

Applications and Problem Solving

34. *Shopping* The retail price of a software program was $50.00. At one store, the program was on sale for 10% off the retail price. Jason had a coupon for an additional 10% off the sale price. At another store, Gabrielle bought the same program for 20% off the retail price. Did the two people pay the same price?

35. *Publishing* In 1998, the circulation of a boy's magazine was 1,242,722. In 1999, its circulation was 1,229,052.
   a. Find the percent of change in the circulation of the magazine.
   b. Is this a percent of *increase* or a percent of *decrease*?

36. *Population* Between the 1980 census and the 1990 census, the United States population increased by approximately 22,164,873 or 9.8%.
   a. What was the census population in 1980?
   b. What was the census population in 1990?
   c. Assume the percent of change remains the same. Predict the population for the 2000 census.

37. *Working on the* **CHAPTER Project** Refer to the media ratings you collected on page 329.
   a. Compute the percent of change from the first week to the second week.
   b. Compute the percent of change from the second week to the third week.
   c. How do the media ratings compare with your class ratings? Draw a double-line graph to show these results.

38. *Critical Thinking* Ron bought a computer, listed at $x,$ at a 15% discount. He also paid a 5% sales tax. After 6 months, he decided to sell the computer secondhand for $y,$ which was 55% of what he paid originally. Express y as a function of x.

Mixed Review

39. Express $\frac{2}{5}\%$ as a fraction in simplest form. *(Lesson 8-4)*

40. **Standardized Test Practice** A deck is constructed in the shape of a trapezoid. The parallel sides of the deck are 24 feet and 18 feet, and the height of the trapezoid is 47 feet. What is the area of the deck? *(Lesson 7-6)*
   A 987 sq ft   **B** 432 sq ft   **C** 164 sq ft   **D** 89 sq ft

41. Use divisibility rules to determine whether 672 is divisible by 3. *(Lesson 6-1)*

42. *Statistics* Refer to the line plot. Find the median and mode of the data. *(Lesson 4-4)*

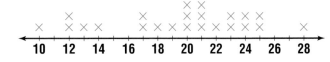

For **Extra Practice,** see page 628.

43. Find the value of $9^2 - 2^3$. *(Lesson 1-2)*

SPREADSHEETS

8-5B Discounts

A Follow-Up of Lesson 8-5

- computer
- spreadsheet software

Suppose you are the manager of the casual clothes department of a department store. The store has frequent sales when many of the items are discounted by the same percent. You use the spreadsheet below to generate sale signs for each item. The discount rate is entered into cell B1. Then the formulas in the cells in column C determine the sale prices.

TRY THIS

Work with a partner.

Use the spreadsheet to determine the price of each item during the Midnight Madness Sale when the prices are discounted 25%. To find these prices, substitute 25 for B1.

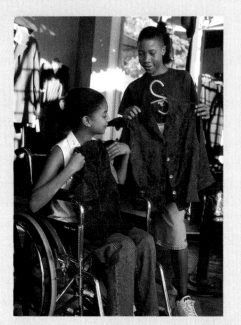

	A	B	C
1	Discount Rate	B1	
2	Item	Original Price	Sale Price
3	Cotton Sweaters	29.99	= (100-B1)/100 * B3
4	Denim Jackets	36.29	= (100-B1)/100 * B4
5	Team Sweatshirts	24.89	= (100-B1)/100 * B5
6	Sport Socks 3-Pack	6.59	= (100-B1)/100 * B6
7	T-Shirts	7.99	= (100-B1)/100 * B7

ON YOUR OWN

1. **List** the prices of the items during the Midnight Madness Sale.
2. **Explain** why the formula in cell C3 correctly finds the sale price of cotton sweaters.
3. Use the spreadsheet to find the sale price of a denim jacket if the discount rate is 35%.
4. What is the discount on a T-shirt if the discount rate is 40%?
5. Suppose you wanted to add a row 8 to the spreadsheet for a $99.59 suede jacket. List each of the cell entries (A8, B8, and C8) that you would enter.

8-6 Simple Interest

What you'll learn

You'll learn to solve problems involving simple interest.

When am I ever going to use this?

Knowing how to solve problems involving simple interest can help you determine the interest on your savings account.

Word Wise

interest
principal

Study Hint

Reading Math *I = prt* means interest equals principal times rate times time.

Darius wants to buy a car. He now has $750. He would like to deposit his money in a bank savings account. The newspaper listed several interest rates for certificates of deposit (CDs) as shown at the right. If he puts his money in a five-year CD, how much will he have at the end of the 5 years?

Interest is the amount paid or earned for the use of money. The formula $I = prt$ is used to solve problems involving simple interest. In the formula, I represents the interest, p represents the amount of money invested, called the **principal**, r represents the annual interest *rate*, and t represents *time* in years.

CERTIFICATE OF DEPOSIT INTEREST RATES

6-month	1-year	$2\frac{1}{2}$-year	5-year
4.80%	5.12%	5.37%	5.69%

$$I = prt$$
$$I = 750 \cdot 0.0569 \cdot 5 \quad p = 750, r = 5.69\% \text{ or } 0.0569, t = 5$$

750 ⊗ .0569 ⊗ 5 ⊜ *213.375*

$$I = 213.375$$

To the nearest cent, Darius will earn $213.38 in interest. He will have $750.00 + $213.38 or $963.38 in five years.

Example REAL WORLD APPLICATION

1 **Savings** Lindsey sold stationery to her family and her mother's friends. She deposited the $125 she earned in a savings account. The account earns 5.18% interest annually. If she does not deposit or withdraw any money for 18 months, how much will she have in her account?

Since there are 12 months in a year, 18 months equals $\frac{18}{12}$ or $\frac{3}{2}$ years. Use the simple interest formula to find the amount of interest earned during this time.

$$I = prt$$
$$I = 125 \cdot 0.0518 \cdot \frac{3}{2} \quad p = 125, r = 5.18\% \text{ or } 0.0518, t = \frac{3}{2}$$

125 ⊗ .0518 ⊗ 3 ⊘ 2 ⊜ *9.7125*

$$I = 9.7125$$

To the nearest cent, the interest will be $9.71. Lindsey will have $125.00 + $9.71 or $134.71 in her account in 18 months.

Interest is charged to you when you borrow money. When this happens, the principal is the amount borrowed.

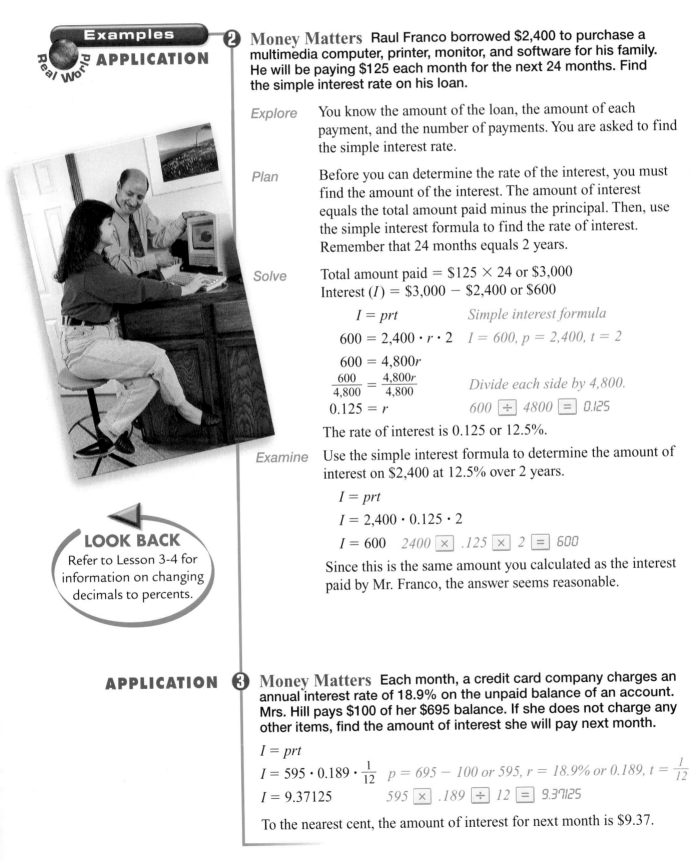

Examples

Real World APPLICATION

② **Money Matters** Raul Franco borrowed $2,400 to purchase a multimedia computer, printer, monitor, and software for his family. He will be paying $125 each month for the next 24 months. Find the simple interest rate on his loan.

Explore You know the amount of the loan, the amount of each payment, and the number of payments. You are asked to find the simple interest rate.

Plan Before you can determine the rate of the interest, you must find the amount of the interest. The amount of interest equals the total amount paid minus the principal. Then, use the simple interest formula to find the rate of interest. Remember that 24 months equals 2 years.

Solve Total amount paid = $125 × 24 or $3,000
Interest (I) = $3,000 − $2,400 or $600

$I = prt$ *Simple interest formula*
$600 = 2,400 \cdot r \cdot 2$ *$I = 600, p = 2,400, t = 2$*
$600 = 4,800r$
$\dfrac{600}{4,800} = \dfrac{4,800r}{4,800}$ *Divide each side by 4,800.*
$0.125 = r$ *600 ÷ 4800 = 0.125*

The rate of interest is 0.125 or 12.5%.

Examine Use the simple interest formula to determine the amount of interest on $2,400 at 12.5% over 2 years.

$I = prt$
$I = 2,400 \cdot 0.125 \cdot 2$
$I = 600$ *2400 × .125 × 2 = 600*

Since this is the same amount you calculated as the interest paid by Mr. Franco, the answer seems reasonable.

LOOK BACK
Refer to Lesson 3-4 for information on changing decimals to percents.

APPLICATION **③** **Money Matters** Each month, a credit card company charges an annual interest rate of 18.9% on the unpaid balance of an account. Mrs. Hill pays $100 of her $695 balance. If she does not charge any other items, find the amount of interest she will pay next month.

$I = prt$
$I = 595 \cdot 0.189 \cdot \dfrac{1}{12}$ *$p = 695 − 100$ or 595, $r = 18.9\%$ or 0.189, $t = \dfrac{1}{12}$*
$I = 9.37125$ *595 × .189 ÷ 12 = 9.37125*

To the nearest cent, the amount of interest for next month is $9.37.

Communicating Mathematics

Read and study the lesson to answer each question.

1. *Write* the simple interest formula. Explain in your own words the meaning of each value.

2. *Explain* the advantages and disadvantages of using credit to buy the computer items in Example 2.

3. *You Decide* Janay and Sonia are trying to find the interest on a bank account of $360 for 8 months if the rate of interest is 5.5%. Each student's work is shown below. Who is correct? Explain your reasoning.

Janay	**Sonia**
$I = 360 \cdot 0.055 \cdot \frac{2}{3}$	$I = 360 \cdot 0.055 \cdot 8$
$I = 13.20$	$I = 158.40$

Guided Practice

Find the simple interest to the nearest cent.

4. $500 at 6.5% for 3 years

5. $230 at 12% for 8 months

Find the total amount in each account to the nearest cent.

6. $660 at 5.25% for 2 years

7. $685 at 6.3% for 21 months

8. *History* In 1626, Peter Minuit paid the natives of Manhattan Island $24 in trade goods for their island. If they had taken the money and invested it at a simple interest rate of 9.8%, what would the balance be today?

Practice

Find the simple interest to the nearest cent.

9. $300 at 7.5% for 5 years

10. $770 at 16% for 6 months

11. $668 at 9.25% for 15 months

12. $285 at 8.5% for $2\frac{1}{2}$ years

13. $360 at 18.5% for 2 years

14. $175 at 5.45% for 30 months

Find the total amount in each account to the nearest cent.

15. $800 at 7.5% for 8 months

16. $460 at 4.5% for 2 years

17. $235 at 8.5% for 3 years

18. $240 at 6.35% for 28 months

19. $385 at 12.6% for 9 months

20. $190 at 5.45% for $1\frac{1}{2}$ years

21. Suppose $1,250 is placed in a savings account for 2 years. Find the simple interest if the interest rate is 4.5%.

22. A savings account starts with $980. If the simple interest rate is 5%, find the total amount in the account after 9 months.

23. Estrella's bank statement listed a balance of $328.80. She originally opened the account with a $200 deposit and a simple interest rate of 4.6%. If there were no deposits or withdrawals, how long ago was the account opened?

Applications and Problem Solving

24. *Money Matters* Antonio Lopez can get a small loan from his credit union at 12% simple interest. He can also use his credit card to get a cash advance at a 1.25% per month interest rate. If he intends to pay back the loan within a month, which is the better rate? Explain.

25. *Credit Cards* Recently, the worldwide charges on a certain credit card totaled $390,000,000,000. Assume that these charges remained unpaid for one year. Using 18% as the interest rate, find the amount of interest due on these charges.

26. *Critical Thinking* Refer to the beginning of the lesson. Suppose Darius had put his money in a 30-month CD. After 30 months or $2\frac{1}{2}$ years, he redeposited the original $750 plus interest into another 30-month CD with the same interest rate.

For **Extra Practice**, see page 629.

    a. Would he have more or less money than with the 5-year CD?

    b. How would future changes in interest rates affect which investment would be better?

    c. *Compound interest* is paid on the initial deposit and on interest earned in the past. Suppose Darius deposited $750 in a bank CD that pays 5.69% interest compounded semiannually. How much interest will there be in 5 years?

Mixed Review

27. *Fund-raising* The Roosevelt Middle School Pep Club is selling sweatshirts. They purchase the sweatshirts for $19.00 and sell them for $25.00. What is the percent of markup? *(Lesson 8-5)*

28. *Standardized Test Practice* What is the prime factorization of 756? *(Lesson 6-2)*

    A $2 \cdot 3 \cdot 7$ **B** $2^2 \cdot 3^3 \cdot 7$ **C** $2 \cdot 3^5 \cdot 7$ **D** $2 \cdot 3 \cdot 7^5$

29. Solve $n = -282 + 41$. *(Lesson 2-3)*

CHAPTER 8 — Mid-Chapter Self Test

1. *Agriculture* A 4-acre field has a yield of 112 bushels of wheat. What yield can be expected from a 42-acre field? *(Lesson 8-1)*

Write a percent proportion to solve each problem. Then solve. *(Lesson 8-2)*

2. Find 35% of 700.

3. 63 is what percent of 84?

Write an equation in the form of *RB = P* for each problem. Then solve. *(Lesson 8-3)*

4. 9 is 45% of what number?

5. Find 55% of 86.

Express each percent as a fraction or mixed number in simplest form. *(Lesson 8-4)*

6. 0.6%

7. 285%

8. *Shopping* What is the sale price of a $79 pair of boots on sale at 25% off? *(Lesson 8-5)*

Find the simple interest to the nearest cent. *(Lesson 8-6)*

9. $750 at 6.25% for 3 years

10. $430 at 13.5% for 6 months

Integration: Geometry
Similar Polygons

What you'll learn

You'll learn to identify corresponding parts of similar polygons and to find missing measures of similar polygons.

When am I ever going to use this?

Knowing how to find missing measures of similar polygons can help you find actual sizes of figures shown in a scale drawing.

Word Wise

similar polygons
pentagon

Artists often use geometric shapes in ways that illustrate their mathematical properties. Josef Albers' *Homage to the Square: Terra Caliente* shows squares of different sizes. These are examples of similar figures.

In mathematics, a polygon is a simple, closed figure in a plane formed by three or more sides. A quadrilateral is a polygon with four sides. The quadrilaterals shown below have the same shape, but differ in size. These quadrilaterals are **similar polygons**.

LOOK BACK
You can refer to Lesson 5-6 for information on similar triangles.

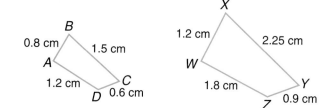

quadrilateral $ABCD$ ~ quadrilateral $WXYZ$

Recall that the symbol ~ means "is similar to."

Corresponding angles of similar polygons are congruent. *You can use a protractor to make sure the angles have the same measure.*

Using corresponding angles as a guide, you can easily identify the corresponding sides.

$$\angle A \cong \angle W \qquad \angle B \cong \angle X$$

$$\angle C \cong \angle Y \qquad \angle D \cong \angle Z$$

$$\overline{AB} \leftrightarrow \overline{WX} \qquad \overline{BC} \leftrightarrow \overline{XY}$$

$$\overline{CD} \leftrightarrow \overline{YZ} \qquad \overline{DA} \leftrightarrow \overline{ZW}$$

A special relationship also exists among the corresponding sides of the polygons. Compare the ratios of the lengths of the corresponding sides.

$$\frac{AB}{WX} = \frac{0.8}{1.2} \text{ or } \frac{2}{3} \qquad \frac{BC}{XY} = \frac{1.5}{2.25} \text{ or } \frac{2}{3} \qquad \frac{CD}{YZ} = \frac{0.6}{0.9} \text{ or } \frac{2}{3} \qquad \frac{DA}{ZW} = \frac{1.2}{1.8} \text{ or } \frac{2}{3}$$

As you can see, the ratios of the lengths of the corresponding sides all equal $\frac{2}{3}$. Since the ratios are all equivalent, you can form proportions using corresponding sides.

Similar Polygons	Two polygons are similar if their corresponding angles are congruent and their corresponding sides are in proportion.

HANDS-ON MINI-LAB

Work in groups of 3 or 4.   ☐ transparency   🔦 flashlight

📏 ruler   📐 protractor

Try This

- Draw a scalene triangle on a transparency.
- Hold a flashlight 6 inches directly behind the transparency and project the triangle's image onto a paper held against a wall 12 inches in front of the transparency. Trace the projected image of the triangle onto the paper.
- Measure the angles of each triangle. Record your results.
- Measure the lengths of the sides of the new triangle. Find the ratio of the length of each side of the new triangle to the corresponding side of the original triangle.

Talk About It

1. Are the two triangles similar? Explain your answer.
2. If the flashlight were held at an angle below the triangle on the overhead transparency, would the projected triangle be similar to the original triangle? Test your answer.

Proportions are useful in finding the missing length of a side in any pair of similar polygons.

Example ①

A **pentagon** is a polygon with five sides. If pentagons *ABCDE* and *HIJKL* are similar, find the length of side $\overline{IJ}$.

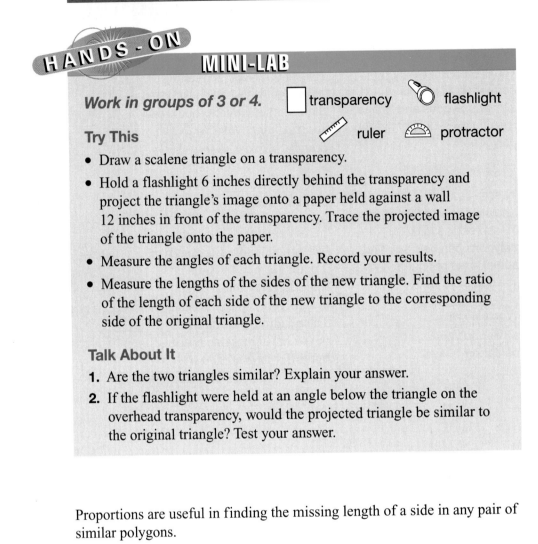

$\overline{AB}$ corresponds to $\overline{HI}$, and $\overline{BC}$ corresponds to $\overline{IJ}$. So you can write a proportion.

$$\frac{AB}{HI} = \frac{BC}{IJ}$$

$$\frac{5}{4} = \frac{7}{x} \qquad AB = 5,\ HI = 4,\ BC = 7,\ IJ = x$$

$$5 \cdot x = 4 \cdot 7 \qquad \textit{Find the cross products.}$$

$$5x = 28$$

$$\frac{5x}{5} = \frac{28}{5} \qquad \textit{Divide each side by 5.}$$

$$x = 5.6$$

The length of side $\overline{IJ}$ is 5.6 meters.

Study Hint

Estimation You can often tell which angles of similar polygons are corresponding by estimating their angle measure instead of measuring each angle.

Example **2**
APPLICATION

Design Ashley designed a logo for her school. The logo, which is 5 inches wide and 8 inches long, will be enlarged and used on a school sweatshirt. If the logo will be $12\frac{1}{2}$ inches wide on the sweatshirt, find its length.

Let x represent the length. Use a proportion to solve for x.

$$\begin{array}{ll} width \rightarrow \\ length \rightarrow \end{array} \quad \frac{5}{8} = \frac{12\frac{1}{2}}{x} \quad \begin{array}{l} \leftarrow \ width \\ \leftarrow \ length \end{array}$$

$$5 \cdot x = 8 \cdot 12\frac{1}{2} \quad \textit{Find the cross products.}$$

$$5x = 100$$

$$\frac{5x}{5} = \frac{100}{5} \quad \textit{Divide each side by 5.}$$

$$x = 20$$

The logo will be 20 inches long.

CHECK FOR UNDERSTANDING

Communicating Mathematics

Read and study the lesson to answer each question.

1. **Tell** how you can determine whether two polygons are similar.

2. **Compare and contrast** similar polygons and congruent polygons.

HANDS-ON MATH

3. **Draw** a parallelogram on an overhead transparency. With the help of some of your classmates, use a flashlight to draw a similar parallelogram.

Guided Practice

4. Tell whether the rectangles are similar. Explain your reasoning.

5. In the figure at the right, $\triangle ABC \sim \triangle EBD$. Write a proportion to find the missing measure x. Then find the value of x.

6. Refer to Example 1 on page 358. Find the length of $\overline{HL}$.

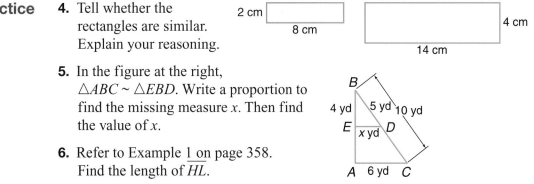

7. **Publishing** Portrait proofs for the yearbook are 4 inches by 5 inches. The yearbook staff must reduce the proofs so that they can fit 3 photographs across the page. To do this, the ratio of the original to the reduced print is 8:5. Find the dimensions of the pictures as they will appear in the yearbook.

Practice

Tell whether each pair of polygons is similar. Explain your reasoning.

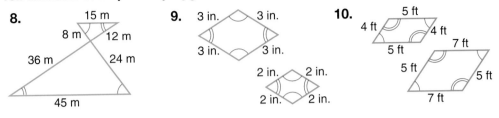

8.

15 m
8 m   12 m
36 m   24 m
45 m

9.

3 in.   3 in.
3 in.   3 in.
2 in.   2 in.
2 in.   2 in.

10.

5 ft
4 ft   4 ft
5 ft   7 ft
5 ft
7 ft   5 ft

Each pair of polygons is similar. Write a proportion to find the missing measure x. Then find the value of x.

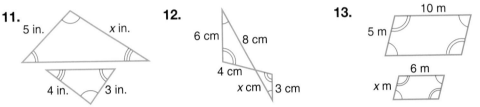

11.

5 in.   x in.
4 in.   3 in.

12.

6 cm   8 cm
4 cm
x cm   3 cm

13.

10 m
5 m
6 m
x m

Refer to Example 1 on page 358.

Applications and Problem Solving

14. Find the length of $\overline{CD}$.

15. Find the length of $\overline{DE}$.

16. *Hobbies* The world's smallest miniature model railroad with a scale of 1:1,400 was made by Bob Henderson of Gravenhurst, Ontario. If the engine measures $\frac{5}{16}$ inch, how long is the real engine?

17. *Art* Refer to the picture at the beginning of the lesson. *Regular polygons* have all sides equal and all angles equal.

    a. Are the squares in the picture regular polygons?

    b. Draw two squares of different sizes than the ones in the picture. Are they similar to the ones in the picture?

    c. Are all squares similar polygons? Explain.

18. *Critical Thinking* A rectangle is 5 centimeters by 7 centimeters. The sides of a similar rectangle are twice as long.

    a. What is the ratio of the perimeter of the first rectangle to the perimeter of the second rectangle?

    b. What is the ratio of the area of the first rectangle to the area of the second rectangle?

Mixed Review

19. *Money Matters* Shala's savings account earned $4.56 in 6 months at a simple interest rate of 4.75%. How much was in her account? *(Lesson 8-6)*

20. **Standardized Test Practice** What is the multiplicative inverse of $-\frac{1}{a}$? *(Lesson 7-4)*

    A $\frac{1}{a}$ **B** $-a$ **C** a **D** -1

21. Express $9\frac{3}{5}$ as a decimal. *(Lesson 6-5)*

22. Solve $\frac{-108}{12} = a$. *(Lesson 2-8)*

For **Extra Practice,**
see page 629.

8-8 Indirect Measurement

What you'll learn

You'll learn to solve problems involving similar triangles.

When am I ever going to use this?

Knowing how to solve problems involving similar triangles can help you find the height of trees or other tall structures.

Word Wise

indirect measurement

The rope-stretchers, or surveyors, of ancient Egypt used a technique called shadow reckoning to determine the heights of tall objects. The height of a staff and the length of its shadow are proportional to the height of an object and its shadow. The objects and their shadows form two sides of similar triangles from which a proportion can be written.

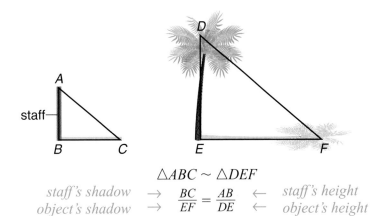

$$\triangle ABC \sim \triangle DEF$$

$$\text{staff's shadow} \rightarrow \frac{BC}{EF} = \frac{AB}{DE} \leftarrow \text{staff's height}$$
$$\text{object's shadow} \rightarrow \qquad\qquad \leftarrow \text{object's height}$$

Using proportions to find a measurement is called **indirect measurement**.

Example
APPLICATION

1

Tourism While touring Egypt, Isabel visited the Great Pyramid of Cheops. At the same time of day that Isabel's shadow was 0.6 meter, the pyramid's shadow was 56 meters. If Isabel is 1.5 meters tall, how high is the Great Pyramid of Cheops?

Study Hint

Estimation Isabel's actual height is about 3 times her shadow, so the pyramid's actual height is about 3 times its shadow.

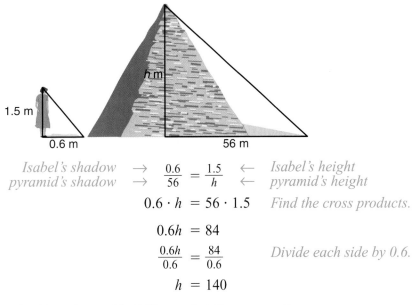

$$\text{Isabel's shadow} \rightarrow \frac{0.6}{56} = \frac{1.5}{h} \leftarrow \text{Isabel's height}$$
$$\text{pyramid's shadow} \rightarrow \qquad\qquad \leftarrow \text{pyramid's height}$$

$$0.6 \cdot h = 56 \cdot 1.5 \qquad \text{Find the cross products.}$$

$$0.6h = 84$$

$$\frac{0.6h}{0.6} = \frac{84}{0.6} \qquad \text{Divide each side by 0.6.}$$

$$h = 140$$

The Great Pyramid of Cheops is 140 meters tall.

You can also use similar triangles that do not involve shadows to find missing measurements.

2 Engineering Delaware and Elmwood Avenues are parallel. Elmwood now ends at Forest Avenue, but the city is planning to extend Elmwood to Military Avenue. The triangles are similar triangles. Find the length of the Elmwood extension.

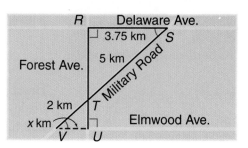

$$\triangle RST \sim \triangle UVT$$

$$\frac{RS}{UV} = \frac{ST}{VT}$$

$$\frac{3.75}{x} = \frac{5}{2} \qquad RS = 3.75,\ UV = x,\ ST = 5,\ VT = 2$$

$$3.75 \cdot 2 = x \cdot 5 \qquad \textit{Find the cross products.}$$

$$7.5 = 5x$$

$$\frac{7.5}{5} = \frac{5x}{5} \qquad \textit{Divide each side by 5.}$$

$$1.5 = x$$

The part of Elmwood Avenue that is under construction is 1.5 kilometers long.

CHECK FOR UNDERSTANDING

Communicating Mathematics

Read and study the lesson to answer each question.

1. ***Tell*** what is meant by indirect measurement.

2. ***Draw and label*** a diagram for the following problem. Then write an appropriate proportion.

   A staff's shadow is 9 feet and a tree's shadow is 15 feet. If the staff is 6 feet tall, how tall is the tree?

   Math Journal

3. ***Write a Problem*** that requires indirect measurement. Explain how to solve the problem.

Guided Practice

Write a proportion for each problem and then solve it. Assume the triangles are similar.

4. Find the distance between Lewiston and Buffalo. Estimate to check your answer.

5. A guy wire is attached to the top of a telephone pole and goes to the ground 9 feet from its base. When Honovi stands under the guy wire so that his head touches it, he is 2 feet 3 inches from where the wire goes into the ground. If Honovi is 5 feet tall, how tall is the telephone pole? (*Hint:* Make a drawing and identify two similar triangles.)

6. *Entertainment* The *Texas Star* at Fair Park in Dallas, Texas, is the tallest Ferris wheel in the United States. A $2\frac{1}{2}$-foot tall child standing next to the Ferris wheel casts a 2-foot shadow at the same time the Ferris wheel's shadow is 170 feet. How tall is the *Texas Star?*

EXERCISES

Practice

Write a proportion for each problem and then solve it. Assume the triangles are similar.

7. Find the distance across Otter Lake.

8. A stop sign casts a shadow 7.5 meters long, while a bush planted nearby casts a shadow 5.4 meters long. If the stop sign is 3 meters high, how tall is the bush?

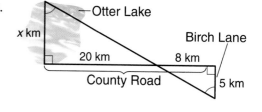

9. The state highway department is investigating the possibility of building a tunnel through the mountain from Millersburg to Pleasant Valley. How long would the tunnel be?

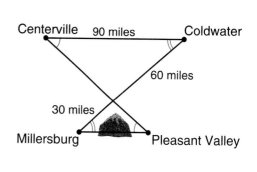

10. An office building in New York City casts a shadow $10\frac{1}{2}$ feet long. At the same time, the Empire State Building casts a $156\frac{1}{4}$-foot shadow. If the office building is 84 feet high, how tall is the Empire State Building? Estimate to check your answer.

11. A lakefront building that is 26 feet high casts a shadow on the water. How long is that shadow if a 10-foot high truck parked nearby casts a 7-foot shadow?

12. From the shoreline, the ground slopes down under the water at a constant incline. If the water is 2 feet deep at a distance of 5 feet from the shore, how deep will it be 100 feet from the shore?

Applications and Problem Solving

13. *Architecture* The Gateway to the West Arch in St. Louis, Missouri, casts a 210-foot shadow. At the same time, a nearby tourist who is 6 feet tall casts a 2-foot shadow. How tall is the arch?

14. *Forestry* The base of a ranger lookout station must be at least 10 feet above the tallest tree in order for the rangers to be able to oversee the surrounding area. Ranger Hwan is $5\frac{1}{4}$ feet tall. At the same time that her shadow is 3 feet long, Ranger Hwan measured the tree's shadow to be 28 feet. How high should the base of the station be built?

15. *Astronomy* An astronomer cuts a $\frac{1}{4}$ inch square hole in a piece of cardboard. Placing it 30 inches from his eye, the circular shape of the moon fits exactly into the square hole. Using 240,000 miles as the distance to the moon, find the diameter of the moon.

16. *Critical Thinking* What factors might cause you to get an incorrect measurement when using the indirect measurement method?

Mixed Review

17. *Architecture* The Pentagon, located outside of Washington, D.C., consists of five similar pentagons one inside of the other. A courtyard in the center is also in the shape of a pentagon. Each side of the outside pentagon is 921.6 feet, and each side of the inside courtyard is 360.8 feet. Write a ratio relating a side of the courtyard to a side of the exterior in simplest form. *(Lesson 8-7)*

18. Standardized Test Practice In the following list of data, which number represents the median? *(Lesson 4-4)*

27, 13, 26, 26, 17, 14, 15, 26, 16

A 16   **B** 17   **C** 20   **D** 26

For **Extra Practice**, see page 629.

MATH ⟩ IN THE MEDIA

B.C.

1. How is the caveman measuring the distance to the Sun?

2. Is he using appropriate proportional reasoning?

3. What must be true to measure a distance using indirect measurement?

COOPERATIVE LEARNING

8-8B Trigonometry

A Follow-Up of Lesson 8-8

protractor

metric ruler

calculator

Mathematicians have been studying the relationships among the sides of right triangles since before 2000 B.C. The study of these relationships is called *trigonometry*.

TRY THIS

Work in groups of three.

Step 1 Each person should draw a right triangle *ABC* in which $m\angle A = 30°$, $m\angle B = 60°$, and $m\angle C = 90°$.

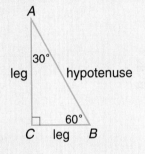

Step 2 Use your ruler to measure the leg opposite the 30° angle. Record the measurement to the nearest millimeter.

Step 3 The leg adjacent to an angle is the side of the angle that is not the hypotenuse. For example, in this triangle, the leg adjacent to the 30° angle is $\overline{AC}$. Measure the leg adjacent to the 30° angle and record the measurement. Measure the hypotenuse and record the measurement.

Step 4 Use your measurements and your calculator to find each ratio for the 30° angle.

ratio 1: $\dfrac{\text{opposite leg}}{\text{adjacent leg}}$ ratio 2: $\dfrac{\text{opposite leg}}{\text{hypotenuse}}$ ratio 3: $\dfrac{\text{adjacent leg}}{\text{hypotenuse}}$

Compare your ratios with the others in your group.

Step 5 Repeat Steps 2–4 for the 60° angle.

ON YOUR OWN

1. ***Make a conjecture*** about the ratios of the sides of any 30°–60° right triangle.

2. Are all 30°–60° right triangles similar to each other? Explain.

3. ***Draw*** a 45°–45° right triangle. Find the value of each of the ratios for a 45° angle.

4. ***Reflect Back*** Suppose you know the length of one side of a 30°–60° right triangle. How could you apply what you have learned to find the length of the other sides without measuring it directly?

What you'll learn

You'll learn to solve problems involving scale drawings.

When am I ever going to use this?

Knowing how to solve problems involving scale drawings can help you read maps.

Word Wise

scale drawing
scale model
scale

The famous composer Wolfgang Amadeus Mozart had a model of each of his opera stage sets built before approving its design. These models can be seen at the Mozart Museum in Salzburg, Austria, Mozart's birthplace.

A **scale drawing** or **scale model** is used to represent an object that is too large or too small to be drawn or built at actual size. The **scale** is determined by the ratio of a given length on the drawing or model to its corresponding length in reality.

One of the most common types of scale drawings is a map. Maps are very useful when planning a trip, whether it is across town or across the country.

Example 1 — Real World APPLICATION

Travel The Quitana family is planning to drive from Dallas, Texas, to San Antonio, Texas, to visit some friends. On a map with a scale of 1 inch = 105 miles, the distance between Dallas and San Antonio is 3 inches. If they travel 60 miles per hour, how long will it take the family to make the trip?

Use the scale and the distance given to write a proportion.

$$\frac{1 \text{ inch}}{105 \text{ miles}} = \frac{3 \text{ inches}}{d \text{ miles}}$$

$$1 \cdot d = 105 \cdot 3$$

$$d = 315$$

The distance between the cities is 315 miles.

$$315 \text{ miles} \div 60 \text{ miles per hour} = 5.25 \text{ hours}$$

The trip will take 5.25 hours or 5 hours and 15 minutes.

Some people like to collect scale models of cars or trains. Scale models are frequently used in the movie industry.

Example

Real World APPLICATION

2 Movies In a recent movie, a child who was 28 inches tall was "blown up" to be 112 feet tall. Find the scale needed for the props in the movie.

Write the ratio of the height of the child to the height he appeared to be in the movie. Then simplify the ratio.

$$\frac{28 \text{ inches}}{112 \text{ feet}} = \frac{1 \text{ inch}}{4 \text{ feet}}$$

The scale is 1 inch = 4 feet.

Some designers prefer that both values in the scale be in the same units.

$$\frac{1 \text{ inch}}{4 \text{ feet}} = \frac{1 \text{ inch}}{(4 \times 12) \text{ inches}} = \frac{1 \text{ inch}}{48 \text{ inches}} = \frac{1}{48}$$

The scale is 1:48. *Note that if the scale uses the same units, it is not necessary to include them.*

CHECK FOR UNDERSTANDING

Communicating Mathematics

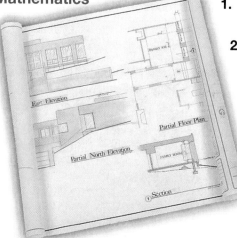

Read and study the lesson to answer each question.

1. *Describe* why it is sometimes necessary to use scale drawings and models.

2. *Name* three occupations or careers that use scale drawings or models. Tell how they use scale drawings or models.

3. *You Decide* On a blueprint, 1 inch represents 3 feet. Renee says the scale is 1:3, but Sierra disagrees. Sierra says the scale is 1:36. Who is correct? Explain.

Guided Practice

4. The figure at the right is a scale drawing of the west side of a cabin. In the drawing, the side of each square represents 2 feet. Find the actual size of each of the following.

   a. the width of the cabin
   b. the height of the cabin
   c. the width of the door
   d. the height of the door

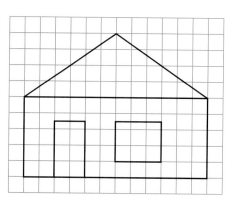

5. If the actual distance between two airports is 500 kilometers and the distance on a map is 4 centimeters, what is the scale for the map?

6. *Movies* One of the models of the gorilla used in the filming of a 1933 movie was only 18 inches tall. In the movie, the gorilla was seen as 24 feet high. What was the scale used?

EXERCISES

Practice

7. The figure at the right is a scale drawing of a house plan. In the drawing, the side of each square represents 4 feet. Find the length and width of each room.
   a. living room
   b. bedroom 1
   c. kitchen
   d. bedroom 3

8. On a map, the scale is 1 inch = 90 miles. Find the actual distance for each map distance.

	From	To	Map Distance
a.	Tampa Bay, Florida	Buffalo, New York	$14\frac{1}{2}$ inches
b.	Columbus, Ohio	Chicago, Illinois	4 inches
c.	Montgomery, Alabama	Savannah, Georgia	$3\frac{1}{2}$ inches

9. A scale model of the Statue of Liberty is 10 inches high. If the Statue of Liberty is 305 feet tall, find the scale of the model.

10. A model of a sports car is 10 inches long. If the actual car is 14 feet, find the scale of the model.

11. The distance between Little Rock, Arkansas, and Memphis, Tennessee, is 140 miles. If the scale on a map is 1 inch = 50 miles, find the distance between the two cities on the map.

12. *Life Science* An amoeba is pictured in a science book. If it is 2.4 centimeters long in the picture and the scale is 200:1, find the actual size of the amoeba.

Applications and Problem Solving

13. *Geography* Madurodam is a tourist attraction in the Netherlands. It is a miniature town representing the country. It includes typical houses, historic buildings, windmills, oil rigs, and tulips. The scale of Madurodam is 1:25. If a typical tulip is 65 centimeters tall, how tall are the tulips in Madurodam?

Madurodam

14. *Earth Science* On a weather map, a hurricane is shown $2\frac{1}{4}$ inches southeast of Puerto Rico. The hurricane is traveling northwest at approximately 20 miles per hour. If the scale of the map is $\frac{3}{5}$ inch = 300 miles, how long before the hurricane should reach Puerto Rico?

15. *Hobbies* A doll house is being designed as a replica of a townhouse. The scale is 1 inch = 12 inches. If the actual townhouse has the outside dimensions of 25 feet by 35 feet, find the outside dimensions of the doll house.

16. *Critical Thinking* Kansas is approximately the shape of a rectangle measuring 480 miles by 267 miles. As a social studies project, you must make a scale drawing of the state on a piece of $8\frac{1}{2}$-inch by 11-inch paper. What should 1 inch on the drawing represent so that the drawing is as large as possible?

Mixed Review

For **Extra Practice,** see page 630.

17. *Geography* The Pyramid of the Sun at Teotihuacán near Mexico City casts a shadow 13.3 meters long. At the same time, a 1.83-meter tall tourist casts a shadow 0.4 meters long. How tall is the Pyramid of the Sun? *(Lesson 8-8)*

18. **Standardized Test Practice** Solve $t = 5\frac{1}{6} + 4\frac{2}{9}$. Write the solution in simplest form. *(Lesson 7-2)*

A $9\frac{1}{3}$　　　**B** $9\frac{7}{18}$　　　**C** $10\frac{1}{3}$　　　**D** $10\frac{7}{18}$

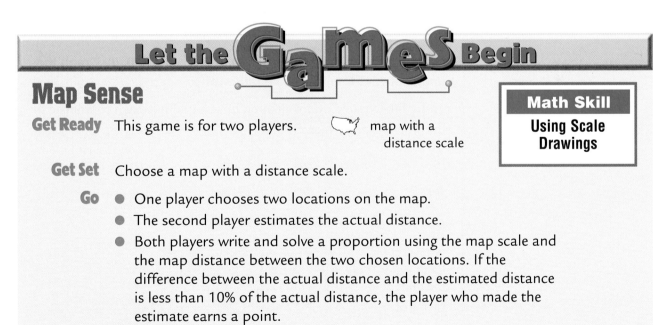

Let the Games Begin

Map Sense

Math Skill
Using Scale Drawings

Get Ready This game is for two players. 　map with a distance scale

Get Set Choose a map with a distance scale.

Go ● One player chooses two locations on the map.
● The second player estimates the actual distance.
● Both players write and solve a proportion using the map scale and the map distance between the two chosen locations. If the difference between the actual distance and the estimated distance is less than 10% of the actual distance, the player who made the estimate earns a point.
● Players exchange roles.
● The player with the most points wins the game.

*inter*NET CONNECTION Visit www.glencoe.com/sec/math/mac/mathnet for more games.

Integration: Geometry
Dilations

The Renaissance architect Sebastian Serlio designed a stage set in 1545 that created the illusion of distance and depth using **dilations**. The arched doorways on the right are the same shape, but vary in size. Can you find other parts of the set design that show similar shapes?

In mathematics, the process of enlarging or reducing an image is called a dilation. Since the dilated image has the same shape as the original, the two images are similar. The ratio of the dilated image to the original is called the **scale factor**.

Example ❶ Graph △**ABC** with vertices **A(2, 4)**, **B(6, 2)**, and **C(8, 6)**. Graph the image of △**ABC** with a scale factor of $\frac{3}{2}$.

To find the vertices of the dilation image, multiply each coordinate in the ordered pairs by $\frac{3}{2}$.

$$A(2, 4) \rightarrow \left(2 \cdot \tfrac{3}{2}, 4 \cdot \tfrac{3}{2} \right) \rightarrow A'(3, 6)$$

$$B(6, 2) \rightarrow \left(6 \cdot \tfrac{3}{2}, 2 \cdot \tfrac{3}{2} \right) \rightarrow B'(9, 3)$$

$$C(8, 6) \rightarrow \left(8 \cdot \tfrac{3}{2}, 6 \cdot \tfrac{3}{2} \right) \rightarrow C'(12, 9)$$

A dilated image is usually named using the same letters as the original figure, but with primes, as in △*ABC* ~ △*A'B'C'.*

Now graph A', B', and C'. Connect them to form △$A'B'C'$. △$A'B'C'$ is the dilation of △ABC with a scale factor of $\frac{3}{2}$.

To check your graph, draw lines through the origin (0, 0) and each of the vertices of the original figure. If the vertices of the dilated figure do not lie on those same lines, you have made a mistake.

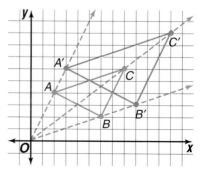

In art, the point where the lines connecting corresponding points intersect is called the *vanishing point*. In Example 1, the origin (0, 0) is the vanishing point. Can you find the vanishing point in Sebastian Serlio's drawing?

Sometimes you need to find the scale factor of a dilation.

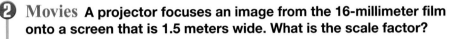

Example

Real World APPLICATION

2 **Movies** **A projector focuses an image from the 16-millimeter film onto a screen that is 1.5 meters wide. What is the scale factor?**

Write a ratio of the width of the image to the width of the film. Change meters to millimeters. Then write the scale factor in simplest form.

$$\frac{1.5 \text{ meters}}{16 \text{ millimeters}} = \frac{(1.5 \times 1,000) \text{ millimeters}}{16 \text{ millimeters}}$$

$$= \frac{1,500 \text{ millimeters}}{16 \text{ millimeters}}$$

$$= \frac{375}{4}$$

The scale factor is $\frac{375}{4}$ or 93.75.

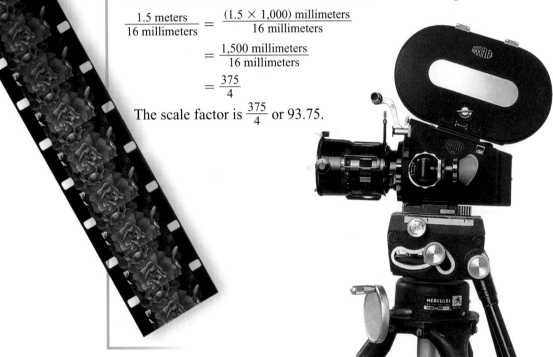

What happens if the scale factor of the dilation is less than 1, or perhaps equal to 1?

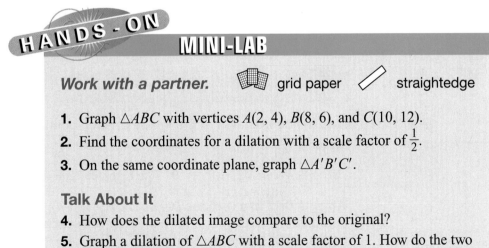

HANDS-ON

MINI-LAB

Work with a partner. grid paper straightedge

1. Graph △*ABC* with vertices *A*(2, 4), *B*(8, 6), and *C*(10, 12).
2. Find the coordinates for a dilation with a scale factor of $\frac{1}{2}$.
3. On the same coordinate plane, graph △*A'B'C'*.

Talk About It

4. How does the dilated image compare to the original?
5. Graph a dilation of △*ABC* with a scale factor of 1. How do the two triangles compare?

Communicating Mathematics

Read and study the lesson to answer each question.

1. *Describe* in your own words the meaning of dilation.

2. *Write* a general rule for finding the new coordinates of any ordered pair (x, y) for a dilation with a scale factor of k.

HANDS-ON MATH

3a. *Graph* $\triangle RST$ with vertices $R(3, 6)$, $S(6, 3)$, and $T(9, 12)$.

   b. *Write* the coordinates of a dilation with scale factor of $\frac{2}{3}$.

   c. *Graph* $\triangle R'S'T'$ on the same coordinate plane.

   d. *Compare* the two triangles.

Guided Practice

4. Find the coordinates of the image of $A(4, 12)$ for a dilation with a scale factor of 2.5.

5. In the figure at the right, the green triangle is a dilation of the blue triangle. Find the scale factor.

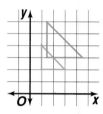

6. Triangle XYZ has vertices $X(3, 6)$, $Y(6, 9)$ and $Z(12, 6)$. Find the coordinates of its image for a dilation with a scale factor of $\frac{4}{3}$. Graph $\triangle XYZ$ and its dilation.

7. *Photocopying* Some photocopiers have various settings to produce copies that are larger or smaller than the original.

   a. Suppose an enlargement is required. Should the machine be set on 50% or 200%? Explain.

   b. If the machine is set on 75%, what is the scale factor of the dilation?

Practice

Find the coordinates of the image of each point for a dilation with a scale factor of $\frac{3}{4}$.

8. $A(8, 4)$ 9. $B(10, 16)$ 10. $C(6, 4)$

Triangle DRT has vertices $D(-4, 12)$, $R(-2, -4)$, and $T(8, 6)$. Find the coordinates of its image for a dilation with each given scale factor. Graph $\triangle DRT$ and each dilation.

11. 2 12. $\frac{1}{4}$ 13. $\frac{5}{4}$

In each figure, the green figure is a dilation of the blue figure. Find each scale factor.

14.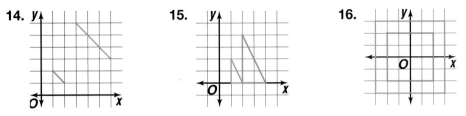

15.

16.

17. Draw a rectangle and its dilated image if the scale factor is 1.5.

18. *Art* Examine Van Gogh's painting of Montmartre, Paris. How did he create the illusion of depth in the two-dimensional painting?

19. *Geometry* Graph polygon *ABCD* with vertices *A*(2, 3), *B*(8, 3), *C*(8, 7), and *D*(2, 7).

   a. Find the perimeter and area of polygon *ABCD*.

   b. Graph the image of polygon *ABCD* for a dilation with a scale of 3.

   c. Find the perimeter and area of polygon $A'B'C'D'$.

   d. Write a ratio in simplest form comparing the perimeter of $A'B'C'D'$ to the perimeter of *ABCD*.

   e. Write a ratio in simplest form comparing the area of $A'B'C'D'$ to the area of *ABCD*.

   f. Make a conjecture about your findings.

Montmartre, Paris

20. *Critical Thinking* Examine the triangle.

   a. How many different sizes of triangles do you see?

   b. What is the scale factor of each size as compared to the large triangle?

   c. How many triangles are there altogether?

For **Extra Practice,** see page 630.

Mixed Review

21. *Scale Drawings* On a scale drawing of a house with a scale of $\frac{1}{4}$ inch = 1 foot, the length of the house is 14 inches. Find the length of the actual house. *(Lesson 8-9)*

22. **Standardized Test Practice** Express 52,380,000 in scientific notation. *(Lesson 6-9)*

   A 52.38×10^6 **B** 5.238×10^8 **C** 5.238×10^6 **D** 5.238×10^7

23. *Geometry* True or *false*? Every square is a rhombus. *(Lesson 5-3)*

CHAPTER 8

Study Guide and Assessment

*inter*NET
CONNECTION Chapter Review **For additonal lesson-by-lesson review, visit:**
www.glencoe.com/sec/math/mac/mathnet

Vocabulary

After completing this chapter, you should be able to define each term, concept, or phrase and give an example or two of each.

Number and Operations
base (p. 335)
discount (p. 349)
interest (p. 353)
markup (p. 349)
percent of change (p. 348)
percent of decrease (p. 349)
percent of increase (p. 349)
percent proportion (p. 335)
percentage (p. 335)

principal (p. 353)
rate (p. 335)
selling price (p. 349)

Measurement
indirect measurement (p. 361)

Problem Solving
solve a simpler problem
    (p. 342)

Geometry
dilation (p. 370)
pentagon (p. 358)
scale (p. 366)
scale drawing (p. 366)
scale factor (p. 370)
scale model (p. 366)
similar polygons (p. 357)
trigonometry (p. 365)

Understanding and Using the Vocabulary

Choose the letter of the term that best matches each phrase.

1. value represented by P in the equation $RB = P$
2. polygons that have the same shape but may differ in size
3. process of enlarging or reducing an image in mathematics
4. percent of increase or decrease
5. finding a measurement using proportions
6. drawing that represents a larger or smaller object
7. ratio of a dilated image to the original

a. indirect measurement

b. similar polygons

c. scale drawing

d. percentage

e. percent of change

f. dilation

g. scale factor

In Your Own Words

8. *Explain* how you would find the simple interest on $500 invested at 4.5% for 1 year.

Objectives & Examples

Upon completing this chapter, you should be able to:

● solve problems by using proportions *(Lesson 8-1)*

3 tapes cost $29.97. How much do 5 tapes cost?

$\frac{3}{29.97} = \frac{5}{x}$ → $3x = 29.97(5)$ → $x = 49.95$

5 tapes cost $49.95.

Review Exercises

Use these exercises to review and prepare for the chapter test.

Write a proportion to solve each problem. Then solve.

9. Luis earns $80 in 10 hours. How much will he earn in 15 hours?

10. A turtle can move 5 inches in 4 minutes. How far can the turtle move in 10 minutes?

solve problems using the percent proportion *(Lesson 8-2)*

75 is what percent of 250?

$\frac{75}{250} = \frac{r}{100}$ *Percent proportion*

$7,500 = 250r$ *Find the cross products.*

$30 = r$ *Divide each side by 250.*

So, 75 is 30% of 250.

solve problems using the percent equation *(Lesson 8-3)*

Find 15% of 82.

$RB = P$ *Percent equation*

$0.15 \cdot 82 = P$ *R = 0.15 and B = 82.*

$12.3 = P$

So, 15% of 82 is 12.3.

express percents greater than 100 or less than 1 as fractions and decimals *(Lesson 8-4)*

Express 130% as a mixed number.

$130\% = \frac{130}{100} = 1\frac{30}{100} = 1\frac{3}{10}$

find and use the percent of increase or decrease *(Lesson 8-5)*

Find the percent of change of the price of a sweater that was $37.50 and is now on sale for $30.00.

$\$37.50 - \$30 = \$7.50$

$\frac{7.50}{37.50} = \frac{r}{100}$ *Percent proportion*

$750 = 37.50r$ *Find the cross products.*

$20 = r$ *Divide each side by 37.50.*

The percent of change is 20%.

Write a percent proportion to solve each problem. Then solve. Round to the nearest tenth.

11. Find 45% of 18.

12. 18 is what percent of 27?

13. What is 86% of 80?

14. 21 is 14% of what number?

Write an equation in the form $RB = P$ for each problem. Then solve.

15. What is 66% of 7,000?

16. 7 is what percent of 56?

17. 15 is 30% of what number?

18. 60 is what percent of 500?

Express each percent as a fraction or mixed number in simplest form.

19. 215%   **20.** $\frac{1}{7}\%$

Express each percent as a decimal.

21. 0.6%   **22.** 126.5%

Find each percent of change. Round to the nearest percent.

23. original: 8   **24.** original: 10
    new: 10   new: 15

25. original: 18   **26.** original: 900
    new: 12   new: 725

27. Find the percent of change if the original number of members in the club was 35 and now there are 49 members.

solve problems involving simple interest
(Lesson 8-6)

Find the simple interest for $350 at 5% for 3 years.

$$I = 350 \cdot 0.05 \cdot 3$$

$$= 52.50 \quad \text{The interest is \$52.50.}$$

Find the simple interest to the nearest cent.

28. $100 at 8.5% for 2 years

29. $780 at 6% for 8 months

30. $125 at 5.55% for 2.5 years

31. $260 at 17.5% for 18 months

identify corresponding parts of similar polygons and find missing measures of similar polygons
(Lesson 8-7)

Two polygons are similar if their corresponding angles are congruent and their corresponding sides are in proportion.

In the figure, $\triangle ABC \sim \triangle RST$. Find the length of each side.

32. $\overline{BC}$

33. $\overline{RS}$

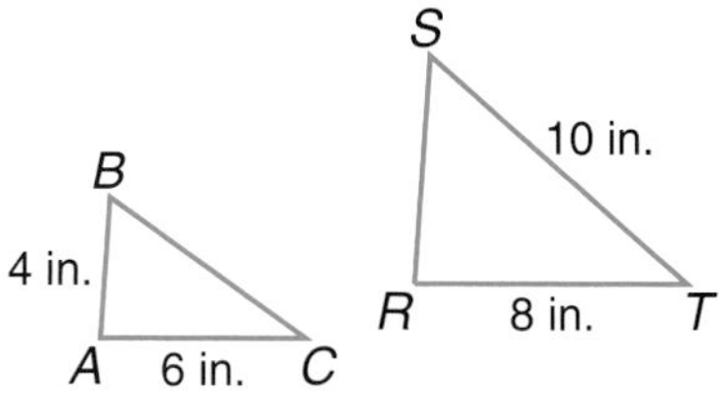

solve problems involving similar triangles
(Lesson 8-8)

A mailbox casts a shadow 6.5 meters long, while a nearby tree casts a shadow 52 meters long. If the mailbox is 2 meters high, how tall is the tree?

mailbox shadow → $\dfrac{6.5}{2} = \dfrac{52}{x}$ ← *tree shadow*
mailbox height →   ← *tree height*

$$6.5x = 104$$

$$x = 16$$

The tree is 16 meters tall.

Solve.

34. A house casts a shadow $5\frac{1}{2}$ feet long. At the same time, a light post casts a $2\frac{1}{4}$-foot shadow. If the light post is 6 feet high, how tall is the house?

35. A flagpole casts a shadow 9 feet long at the same time that an office building casts a 15-foot shadow. If the office building is 60 feet tall, how tall is the flagpole?

solve problems involving scale drawings
(Lesson 8-9)

In a drawing, 3 cm = 45 m. Find the length of a line representing 75 meters.

$$\frac{3 \text{ centimeters}}{45 \text{ meters}} = \frac{x \text{ centimeters}}{75 \text{ meters}}$$

$$225 = 45x$$

$$5 = x$$

The line is 5 centimeters long.

On a map, 2 inches = 5 miles. Find the actual distance for each map distance.

36. 4 inches

37. 5.5 inches

38. 12 inches

39. 9 inches

graph dilations on a coordinate plane
(Lesson 8-10)

In a dilation with a scale factor of k, the ordered pair (x, y) becomes the ordered pair (kx, ky).

Find the coordinates of the image of each point for a dilation with a scale factor of 4.

40. $A(6, 9)$

41. $B(15, 3)$

42. $C(2, 5)$

43. $D(1, 4)$

Applications & Problem Solving

44. *Entertainment* Domingo hired a rock band to play at the school dance. The band charges $3,000 per performance. The manager required a 20% deposit. How much did Domingo pay for the deposit? *(Lesson 8-2)*

45. *Solve a Simpler Problem* How many cuts are needed to separate a long board into 19 shorter pieces with the same width as the original board? *(Lesson 8-3B)*

46. *Money Matters* Emily wants to buy a CD player that costs $189.95. If she waits to buy the CD player until it is on sale for $132.97, what is the percent of savings? *(Lesson 8-5)*

47. *Travel* A map of the roads between four cities is shown below. How far is it from Main City to Carlton to the nearest mile? *(Lesson 8-8)*

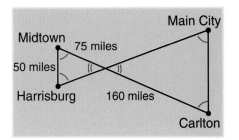

Alternative Assessment

Open Ended

Suppose you earn $42 for working 8 hours. A friend says that you can use the following proportion to find out how much you would earn for 24 hours.

$$\frac{42}{8} = \frac{24}{h}$$

Is your friend correct? Why or why not? If not, write the correct proportion. Use the correct proportion to solve the problem.

Determine how to find out how much you would earn for 24 hours of work without using proportions.

A practice test for Chapter 8 is provided on page 654.

Completing the CHAPTER Project

Use the following checklist to make sure your newspaper article is complete.

☑ The class rating for each program is given both as a fraction and as a percent.

☑ The Nielsen ratings for your program are recorded for three consecutive weeks and the percent of change is computed.

☑ The class ratings and the Nielsen ratings are compared.

PORTFOLIO Select one of the problems you solved in this chapter and place it in your portfolio. Attach a note explaining how your problem illustrates one of the important concepts covered in this chapter.

Standardized Test Practice

Assessing Knowledge & Skills

Section One: Multiple Choice

There are eleven multiple-choice questions in this section. Choose the best answer. If a correct answer is *not here,* choose the letter for Not Here.

1. What is the probability that Kinsey will draw a green marble from a jar containing 3 white marbles, 2 blue marbles, 5 orange marbles, and no other marbles?

A 0

B $\frac{1}{10}$

C $\frac{1}{3}$

D 1

2. If $2x - 5 = -3$, what is the value of x?

F -4

G -1

H 1

J 4

3. $6a + 5a - 2a =$

A $9a$

B $13a$

C $11a - 2$

D $9a^3$

4. $0.3\% =$

F 3

G 0.3

H 0.03

J 0.003

5. How many lines of symmetry does a square have?

A none

B 1

C 2

D 4

Please note that Questions 6–11 have five answer choices.

6. The Venn diagram shows the relationship of four sets of students in the music program at Jones Middle School.

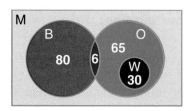

M = the set of all students in the school music program.

B = the set of all students in the band.

O = the set of all students in the orchestra.

W = the set of all students in the woodwinds.

Which is a valid conclusion concerning these sets?

F There is no orchestra student in the band.

G All woodwind students are in the band.

H All woodwind students are in the orchestra.

J There are students in the band that are not in the music program.

K All orchestra students play woodwinds.

7. The student senate surveyed the 8th grade class to determine what kind of decorations to have for the winter dance. Of the 320 students surveyed, 100 preferred streamers. If the same survey were conducted among the entire middle school population of 1,280 students, what is a reasonable number of students in the entire middle school that would prefer streamers?

A 300

B 400

C 500

D 600

E 700

8. If it costs approximately $150 per week to feed a family of 5, how much should it cost to feed a family of 7?

    F $30   **G** $150

    H $180   **J** $210

    K $250

9. Bananas are on sale for $0.32 per pound. If Iggy's Ice Cream Parlor pays $30.08 for a shipment of bananas, how many pounds were purchased?

    A 94 lb   **B** 99 lb

    C 104 lb   **D** 127 lb

    E Not Here

10. Suppose $840 is deposited in a savings account that earned 4.5% interest. If there are no deposits or withdrawals, how much was in the account at the end of 6 months?

    F $1,066.80   **G** $858.90

    H $226.80   **J** $18.90

    K Not Here

11. Including tax, the cost of renting a movie on videotape for 2 days is $2.50 and $2.00 for each day after that. What is the cost for renting a movie for 7 days?

    A $10.00   **B** $12.50

    C $14.00   **D** $16.50

    E Not Here

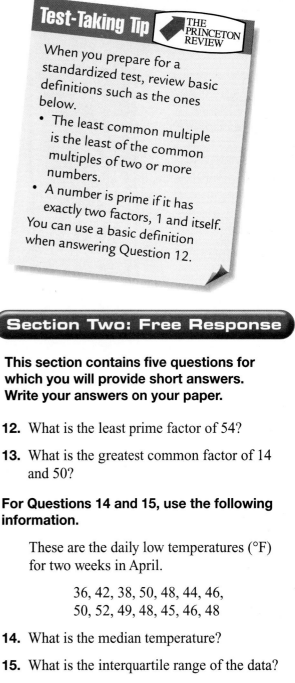

Test-Taking Tip THE PRINCETON REVIEW

When you prepare for a standardized test, review basic definitions such as the ones below.

- The least common multiple is the least of the common multiples of two or more numbers.

- A number is prime if it has exactly two factors, 1 and itself. You can use a basic definition when answering Question 12.

Section Two: Free Response

This section contains five questions for which you will provide short answers. Write your answers on your paper.

12. What is the least prime factor of 54?

13. What is the greatest common factor of 14 and 50?

For Questions 14 and 15, use the following information.

These are the daily low temperatures (°F) for two weeks in April.

36, 42, 38, 50, 48, 44, 46,
50, 52, 49, 48, 45, 46, 48

14. What is the median temperature?

15. What is the interquartile range of the data?

16. An architectural drawing measuring 9 inches wide by 15 inches long is enlarged. The new enlargement is 36 inches wide. How long is the enlargement?

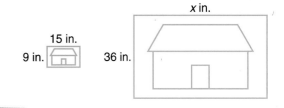

What you'll
learn in Chapter 9

- to estimate and find square roots,
- to identify and classify numbers,
- to use the Pythagorean Theorem,
- to solve problems by drawing a diagram, and
- to find the distance between points in the coordinate plane.

PROCEED WITH CAUTION

Traffic accidents can happen at any time and in any weather condition. Police officers often reconstruct the details of the accident. If an officer doesn't actually see the accident, how can he or she know what happened? Officers look for certain clues at the accident scene. In this project, you will use square roots, real numbers, and distance on a coordinate plane to write a traffic accident report for the accident scene shown in the diagram.

Getting Started

- Research how to investigate an accident scene and how to write an accident report. What types of information should go into an accident report?

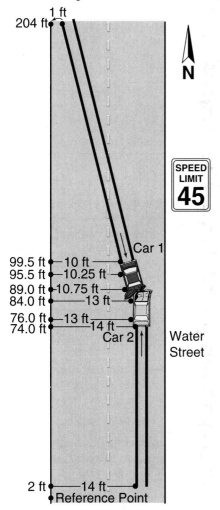

Technology Tips

- Use a **word processor** to write your accident report.
- Surf the **Internet** for more information.

inter NET CONNECTION Data Update **For up-to-date information on traffic, visit:**

www.glencoe.com/sec/math/mac/mathnet

Working on the Project

You can use what you'll learn in Chapter 9 to help you write a traffic report.

Page	Exercise
388	25
393	32
413	21
421	Alternative Assessment

Square Roots

The Pythagorean school of ancient Greece focused on the study of philosophy, mathematics, and natural science. The students, called Pythagoreans, made many advances in these fields. One of their studies was *figurate numbers*. By drawing pictures of various numbers, patterns can be discovered. For example, some whole numbers can be represented by drawing dots arranged in squares.

1	4	9	16	64
1 by 1	2 by 2	3 by 3	4 by 4	8 by 8

Numbers that can be pictured in squares of dots are called **perfect squares**. The number of dots in each row or column in the square is called a **square root** of the perfect square. The perfect square 9 has a square root of 3 because there are 3 rows and 3 columns. We say that 8 is a square root of 64 because $64 = 8 \times 8$ or 8^2.

Square Root	If $x^2 = y$, then x is a square root of y.

Perfect squares also include decimals and fractions like 0.09 and $\frac{4}{9}$ since $(0.3)^2 = 0.09$ and $\left(\frac{2}{3}\right)^2 = \frac{4}{9}$.

It is also true that $(-8)^2 = 64$ and $(-12)^2 = 144$. So we can say that -8 is also a square root of 64 and -12 is a square root of 144.

The positive square root of a number is called the **principal square root**. The symbol $\sqrt{}$, called a **radical sign**, is used to indicate the principal square root.

$$\sqrt{25} = 5 \quad \textit{principal square root}$$

$$-\sqrt{25} = -5 \quad \textit{negative square root}$$

Examples

1 Find $\sqrt{225}$.

The symbol $\sqrt{}$ indicates the principal square root.

Since $15^2 = 225$, $\sqrt{225} = 15$.

2 Find $-\sqrt{49}$.

The $-\sqrt{}$ symbol indicates the negative square root.

Since $(-7)^2 = 49$, $-\sqrt{49} = -7$.

Study Hint

Reading Math $\sqrt{225}$ is read as *the square root of 225.*

Some equations that involve squares can be solved by taking the square root of each side of the equation.

3 **Algebra** Find a number that when squared equals 361.

Let n be the number.

Then $n^2 = 361$.

$$\sqrt{n^2} = \sqrt{361} \qquad \textit{Take the square root of each side.}$$

361 [2nd] [$\sqrt{\ }$] *19*   *Use a calculator.*

$n = 19$ or $n = -19$

The number could be either 19 or -19.

4 **Design** Ms. Lewis is creating several cubicle work spaces for her consulting business. One square workspace needs to have a floor area of 576 square feet. Draw a diagram and describe how she could find the length of each side for the cubicle.

Use the formula for the area of a square.

$$A = s^2$$

$576 = s^2$ *Replace A with 576.*

$\sqrt{576} = s$ *Take the square root of each side.*

576 [2nd] [$\sqrt{\ }$] *24*   *Use a calculator.*

$s = 24$ or $s = -24$

Therefore, each side of the workspace must be 24 feet.

−24 does not make sense because length is never negative.

s ft	
576 square feet	s ft

CHECK FOR UNDERSTANDING

Communicating Mathematics

Read and study the lesson to answer each question.

1. *Write* the symbol for the negative square root of 100.

2. *Tell* why $\frac{36}{81}$ is a perfect square.

3. *Write* a description of the relationship between a perfect square and its two square roots. Give a specific example and include a drawing of dots arranged in a square.

Guided Practice

Find each square root.

4. $\sqrt{100}$ 5. $\sqrt{81}$ 6. $\sqrt{121}$ 7. $-\sqrt{64}$

8. **Algebra** Solve $x^2 = 144$.

9. **History** The Great Pyramid at Giza, built around 2600 B.C., is one of the "Seven Wonders of the Ancient World." It has a square base with an area of 602,176 square feet. How long is each side of the base?

Practice

Find each square root.

10. $\sqrt{25}$ 11. $-\sqrt{9}$ 12. $\sqrt{256}$ 13. $-\sqrt{36}$

14. $\sqrt{400}$ 15. $\sqrt{\frac{4}{9}}$ 16. $\sqrt{0.16}$ 17. $-\sqrt{\frac{64}{100}}$

18. $\sqrt{441}$ 19. $-\sqrt{196}$ 20. $\sqrt{\frac{121}{625}}$ 21. $-\sqrt{2.89}$

Solve each equation.

22. $x^2 = 81$ 23. $0.25 = x^2$ 24. $x^2 = \frac{16}{25}$

25. Find two square roots of 169. Identify the principal square root.

26. *Geometry* If the area of a square is 1.69 square meters, what is the length of each of its sides?

Applications and Problem Solving

27. *Agriculture* Pecan trees are planted in square patterns to take advantage of land space and for ease of harvesting. If you wanted to plant 289 trees, how many rows and how many trees in each row would you plant?

28. *Design* The designs of tile floors are often planned on a piece of grid paper. Suppose you have 100 1-inch square blue tiles.

   a. What is the largest square you could make with these tiles?

   b. If you wanted to make two square designs that were the same size, what is the largest that each square could be? How many tiles would be left over?

   c. Describe three different-sized squares that you could make at the same time out of your 100 tiles. How many tiles are left over?

29. *Critical Thinking* Is the product of two perfect squares always a perfect square? Explain why or why not.

Mixed Review

30. *Geometry* Graph $\overline{EF}$ with endpoints $E(2, 6)$ and $F(4, -4)$. Then graph its image for a dilation with a scale factor of 2. *(Lesson 8-10)*

31. *Geometry* Find the area of a trapezoid with bases 9 and 12 centimeters long and a height of 11 centimeters. *(Lesson 7-6)*

32. **Standardized Test Practice** A ladder leaning against the side of a building forms a 75° angle with the ground. Approximately what is the measure of the angle between the ground and the other side of the ladder? *(Lesson 5-1)*

   A 60°   **B** 75°   **C** 105°   **D** 255°

For **Extra Practice,** see page 630.

33. *Statistics* Find the mean, median, and mode of the data 6, 4, 6, 12, 10, 8, 7, 12, 11, 9. *(Lesson 4-4)*

COOPERATIVE LEARNING

9-2A Estimating Square Roots

A Preview of Lesson 9-2

algebra tiles or
base-ten blocks

calculator

Suppose you try to arrange 50 tiles into a square. You discover that it's impossible. This suggests that 50 is not a perfect square. In this lab, you will estimate the square roots of numbers that are not perfect squares.

TRY THIS

Work with a partner.

Use tiles to estimate the square root of 50.

- Arrange 50 tiles into the largest square possible.

  The largest possible square has 49 tiles, with one left over.

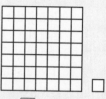

$\sqrt{49} = 7$

- Add tiles until you have the next larger square.

  You need to add 7 tiles on top and 7 tiles on the side, and then the leftover tile from above can be placed in the upper right corner. Therefore, you added 14 new tiles to make a square that has 64 tiles.

- The square root of 49 is 7, and the square root of 64 is 8. Therefore, the square root of 50 is between the whole numbers 7 and 8. $\sqrt{50}$ is closer to 7 because 50 is closer to 49 than 64.

  $\sqrt{64} = 8$

- Verify the estimate with a calculator. 50 [2nd] [√] 7.071067812

ON YOUR OWN

For each number, arrange tiles or base-ten blocks into the largest square possible. Then add tiles until you have the next larger square. To the nearest whole number, estimate the square root of each number.

1. 20   **2.** 76   **3.** 133   **4.** 150

5. *Look Ahead* Describe another method that could be used to find the square root of a number by squaring numbers to estimate, rather than by taking square roots to estimate.

Estimating Square Roots

CLOSE TO HOME JOHN McPHERSON

What you'll learn

You'll learn to estimate square roots.

When am I ever going to use this?

Knowing how to estimate square roots can help you approximate how far you can see to the horizon.

Could you count off by $\sqrt{7}$? The class is stumped since 7 is not a perfect square.

We can estimate the square roots of numbers that are not perfect squares. For example, the number 130 is not a perfect square. However, we know that 130 is between two perfect squares, 121 and 144. So the square root of 130 is between 11 and 12.

Deep down inside, Coach Knott had always wanted to be a math teacher.

Study the tiled squares at the right. Notice that the 11×11 square contains 121 squares, and the 12×12 square contains 144 squares. If you shade 130 squares, you can see that you are closer to 121 squares than you are to 144 squares.

Since 130 is closer to 121 than 144, the best whole number estimate for the square root of 130 is 11.

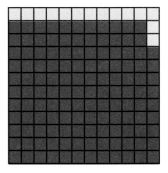

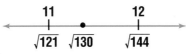

Examples

1 Estimate $\sqrt{78}$.

$64 < 78 < 81$ *64 and 81 are perfect squares.*

$8^2 < 78 < 9^2$

$8 < \sqrt{78} < 9$

Since 78 is closer to 81 than 64, the best whole number estimate for $\sqrt{78}$ is 9.

INTEGRATION **2** **Geometry** Heron, an Egyptian mathematician, created a formula that finds the area of a triangle if the length of each side is known.

If the measures of the sides of a triangle are a, b, and c, then the area of the triangle equals $\sqrt{s(s-a)(s-b)(s-c)}$ where s equals $\frac{1}{2}$ of the perimeter of the triangle.

Find the area of the triangular piece of land shown at the right.

6 miles

5 miles

3 miles

$a = 3, b = 5, c = 6,$ and $s = \frac{1}{2}(3 + 5 + 6)$ or 7.

$A = \sqrt{s(s - a)(s - b)(s - c)}$ *Substitute the values for a, b, c,*

$\quad = \sqrt{7(7 - 3)(7 - 5)(7 - 6)}$ *and s into the formula.*

$\quad = \sqrt{7(4)(2)(1)}$

$\quad = \sqrt{56}$

The area is $\sqrt{56}$. Now, estimate $\sqrt{56}$.

$$49 < 56 < 64$$
$$7^2 < 56 < 8^2$$
$$7 < \sqrt{56} < 8$$

Since $49 < 56 < 64$, we know that the best whole number estimate for $\sqrt{56}$ is between 7 and 8. Because 56 is closer to 49 than to 64, $\sqrt{56}$ is closer to 7 than 8. Therefore, the area of the triangular piece of land is approximately 7 square miles.

CHECK FOR UNDERSTANDING

Communicating Mathematics

Read and study the lesson to answer each question.

1. *Tell* how the drawing at the right can be used to estimate $\sqrt{12}$.

2. *Draw* a figure that can be used to explain why the square root of 75 is between 8 and 9.

Guided Practice

Estimate to the nearest whole number.

3. $\sqrt{68}$ 4. $\sqrt{29}$ 5. $\sqrt{135}$ 6. $\sqrt{11}$

7. *Physical Science* The distance you can see to the horizon can be estimated by using the formula $d = 1.22 \times \sqrt{h}$. In the formula, d represents the distance you can see, in miles, and h represents the height your eyes are from the ground, in feet. If you are standing on the ground, suppose your eyes are about 5 feet above ground level. About how far can you see to the horizon?

EXERCISES

Practice

Estimate to the nearest whole number.

8. $\sqrt{23}$ 9. $\sqrt{15}$ 10. $\sqrt{44}$ 11. $\sqrt{13.5}$

12. $\sqrt{113}$ 13. $\sqrt{200}$ 14. $\sqrt{408}$ 15. $\sqrt{31}$

16. $\sqrt{23.5}$ 17. $\sqrt{5}$ 18. $\sqrt{250}$ 19. $\sqrt{136}$

20. Find the nearest whole number estimate for $\sqrt{65.25}$.

21. *Algebra* Estimate the solution of $y = \sqrt{78}$ to the nearest whole number.

22. *Algebra* If $t = \sqrt{221}$, what is an estimate for the value of t?

23. *Landscaping* A landscape architect is preparing costs for a rectangular side yard that contains a triangular-shaped pond. The planner needs to know how many square feet will be planted in grass. Using the diagram at the right and Heron's formula in Example 2, what would be a good estimate for the grassy area?

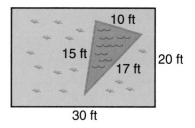

24. *Physical Science* Suppose you are visiting Washington, D.C., and you climb the 898 steps to the top of the Washington Monument. Your eyes are approximately 520 feet above the ground. Using the formula in Exercise 7, about how far can you see to the horizon?

25. *Working on the* CHAPTER Project

When investigating a car accident, police officers can estimate how fast a car was going by measuring the length of the skid marks. The formula $s = \sqrt{30df}$ gives the minimum speed s of the car in miles per hour, where d is the distance in feet the car skidded after its brakes were applied and f is a drag factor that depends on the road surface. Some typical drag factors are listed in the table.

Road Surface	Drag Factor (*f*)
dry concrete	0.82
packed gravel	0.66
wet concrete	0.57

a. A car left skid marks for 60 feet on a dry, concrete road surface. If the speed limit on the road were 35 mph, was the car speeding when the brakes were applied? If so, by how much?

b. Choose ten different values for d and evaluate the formula for a wet, concrete road surface. Graph the resulting ordered pairs on a coordinate grid and describe the graph.

26. *Critical Thinking* Without a calculator, determine which is greater, $\sqrt{358}$ or 19. Explain your reasoning.

Mixed Review

27. Find the principal square root of 900. *(Lesson 9-1)*

28. *Money Matters* Eleven-year old Currito regularly performs some of the jobs listed in the chart below. *(Lesson 8-5)*

Pay Rates			
Job	Unit of Pay	Ages 9-11	Ages 12-14
Baby-sitting	per hour	$3	$3.50
Mowing lawns	per lawn	$5	$6
Other yard work	per job	$2.88	$5
Pet care	per hour	$1	$3
Washing cars	per car	$4.50	$5

a. If in the course of one week, Currito mows 3 lawns, takes care of the neighbor's dog each day for 1 hour, and washes 2 cars, how much money will he earn for the week?

b. What is the percent of change in earnings if he were 12 years old?

29. Standardized Test Practice The Venn diagram shows the relationship of four sets of students at Natalie's school. Which is a valid conclusion? *(Lesson 5-2)*

U = set of all students at Natalie's middle school

M = set of students in Natalie's music class

G = set of all eighth-grade girls

B = set of all eighth-grade basketball players

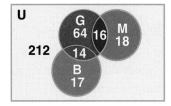

A No eighth-grade girl is in Natalie's music class.

B All eighth-grade girls are in Natalie's music class.

C All students in Natalie's music class are eighth-grade basketball players.

D All students in Natalie's music class are eighth-grade girls.

E No eighth-grade basketball players are in Natalie's music class.

For **Extra Practice,**
see page 631.

Let the Games Begin

Estimate and Eliminate

📇 40 index cards   ✏️ 4 markers

Math Skill
Estimating Square Roots

Get Ready   This game is for four players.

Get Set   Each player is given 10 index cards. Player 1 writes one of each of the whole numbers 1 – 10 on each card. Player 2 writes the square of one of each of the whole numbers 1 – 10 on each card. Player 3 writes a different whole number between 11-50, that is not a perfect square, on each card. Player 4 writes a different whole number between 51-99, that is not a perfect square, on each card.

Go   ● Mix all 40 cards together. The dealer deals all 40 cards to the four players, one at a time.

● In turn, moving clockwise, each player lays down any pair(s) of a perfect square and its square root that they have in their hand. The two cards should be laid perpendicular to each other.

● After the first round, any player, during their turn: (1) lays down a square and a square root pair, or (2) lays down a card on the card that is perpendicular to the best approximate square root if it is already on the table in front of a player. For example, a 21 should be laid on a 25 if the 5 and 25 pair is already on the table. A player makes as many plays as possible during their turn.

● After each complete round, each player passes one card to their left, and the play resumes.

● The winner is the first person who places all of their 10 cards on the table.

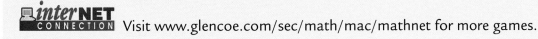

interNET
CONNECTION Visit www.glencoe.com/sec/math/mac/mathnet for more games.

The Real Number System

What you'll learn

You'll learn to identify and classify numbers in the real number system.

When am I ever going to use this?

Knowing how to use real numbers can help you estimate distance and time.

Word Wise

irrational number
real number

Have you ever gotten caught out in a thunderstorm and wished you knew how long it would be before it's over? Meteorologists use the formula $t^2 = \frac{d^3}{216}$ to estimate the amount of time that a thunderstorm will last. In this formula, t is the time in hours, and d is the diameter of the storm in miles.

Numbers such as 6, 8.4, and $10\frac{1}{2}$ are common replacements for d in $t^2 = \frac{d^3}{216}$.

These numbers are from sets of numbers you've studied.

Whole Numbers	$\{0, 1, 2, 3, 4, \ldots\}$
Integers	$\{\ldots, -2, -1, 0, 1, 2, \ldots\}$
Rational Numbers	$\{$all numbers that can be expressed in the form $\frac{a}{b}$, where a and b are integers and $b \neq 0\}$

LOOK BACK
You can refer to Lesson 6-4 to review rational numbers.

The Venn diagram represents a summary of the classification of rational numbers.

Remember, terminating or repeating decimals are rational numbers since they can be expressed as fractions. The square roots of perfect squares are also rational numbers. For example, $\sqrt{0.09}$ is a rational number because $\sqrt{0.09} = 0.3$, a rational number.

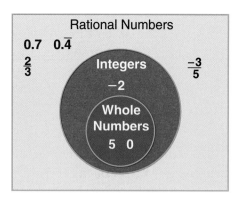

Where do numbers like $\sqrt{2}$ and $\sqrt{5}$ fit into the Venn diagram? Notice what happens when you find $\sqrt{2}$ and $\sqrt{5}$ with your calculator.

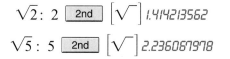

$\sqrt{2}: \ 2$ [2nd] $[\sqrt{\ }]$ 1.414213562

$\sqrt{5}: \ 5$ [2nd] $[\sqrt{\ }]$ 2.236067978

The numbers appear to not terminate nor is there any obvious pattern of repeating digits. These numbers are not rational numbers. Numbers like $\sqrt{2}$ and $\sqrt{5}$ are called **irrational numbers**.

Irrational Number	An irrational number is a number that cannot be expressed as $\frac{a}{b}$, where a and b are integers and b does not equal 0.

Examples

Determine whether each number is rational or irrational.

① **0.16666666 . . .**

This decimal ends in a repeating pattern. It is a rational number and can be expressed as $\frac{1}{6}$.

② **0.16116111611116111116 . . .**

This decimal ends in a pattern that will never repeat or terminate. It is an irrational number.

③ **π**

$\pi = 3.1415926 \ldots$
This decimal does not repeat or terminate. It is an irrational number.

You have graphed rational numbers on a number line. If you graph all of the rational numbers, you would still have some "holes" in the number line. The irrational numbers "fill in" the number line. The set of rational and irrational numbers

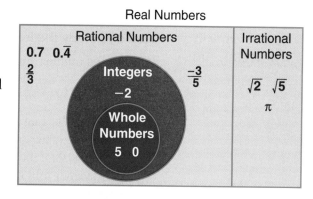

combine to form the set of **real numbers**. The graph of all real numbers is the entire number line without any "holes."

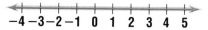

To graph irrational numbers, you can use a calculator or a table of squares and square roots to find approximate square roots in decimal form.

Example

④ Graph $\sqrt{5}$, π, and $-\sqrt{2}$ on a number line.

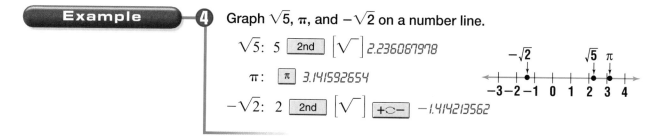

So far in this textbook, you have solved equations that have rational number solutions. However, some equations have solutions that are irrational numbers.

Example **APPLICATION**

⑤ Meteorology Refer to the beginning of the lesson. Use the formula to determine how long a thunderstorm will last if it is 7.8 miles wide.

$$t^2 = \frac{d^3}{216}$$

$$t^2 = \frac{7.8^3}{216} \quad \textit{Replace d with 7.8.}$$

$$\sqrt{t^2} = \sqrt{2.197} \quad \textit{Take the square root of each side.}$$

2.197 [2nd] $\left[\sqrt{\ } \right]$ *1.482228053* *Use a calculator.*

$$t \approx 1.5$$

The thunderstorm will last about 1.5 hours.

CHECK FOR UNDERSTANDING

Communicating Mathematics

Read and study the lesson to answer each question.

1. *Draw* a number line and graph $\sqrt{15}$.

2. *Explain* why $\sqrt{\frac{25}{36}}$ is a rational number.

3. *You Decide* Lan and Sharon disagree about whether a square root of a number is greater or less than the original number. Lan claims that the square root is less than the number, and Sharon says that the square root is greater. Their teacher told them that they were both correct. Who is really correct? Explain and give some examples.

Guided Practice

Let R = real numbers, Q = rational numbers, Z = integers, W = whole numbers, and I = irrational numbers. Name the set or sets of numbers to which each real number belongs.

4. $-\sqrt{9}$

5. $0.\overline{27}$

Find an estimate for each square root. Then graph the square root on a number line.

6. $\sqrt{7}$

7. $\sqrt{20}$

Solve each equation. Round solutions to the nearest tenth.

8. $x^2 = 148$

9. $y^2 = 59$

10. *Carpentry* Hotah and his aunt are building a deck. In one space, they need a board that measures $\sqrt{110}$ feet. About how long is the board?

EXERCISES

Practice

Let R = real numbers, Q = rational numbers, Z = integers, W = whole numbers, and I = irrational numbers. Name the set or sets of numbers to which each real number belongs.

11. 36

12. $\sqrt{31}$

13. $0.121121112\ldots$

14. $\dfrac{5}{8}$

15. $-\sqrt{121}$

16. $\dfrac{\sqrt{10}}{\sqrt{49}}$

Find an estimate for each square root. Then graph the square root on a number line.

17. $\sqrt{6}$

18. $-\sqrt{27}$

19. $\sqrt{50}$

20. $-\sqrt{99}$

21. $\sqrt{108}$

22. $\sqrt{300}$

Solve each equation. Round solutions to the nearest tenth.

23. $x^2 = 69$

24. $y^2 = 12$

25. $x^2 = 75$

26. $n^2 = 41$

27. $x^2 = 925$

28. $t^2 = 0.28$

29. Which is greater, $\sqrt{\dfrac{25}{4}}$ or $\dfrac{8}{3}$?

Applications and Problem Solving

Real World

30. *Algebra* In the geometric sequence 4, 12, _?_, 108, 324, the missing number is called the *geometric mean* of 12 and 108. It can be found by simplifying $\sqrt{ab}$ where a and b are the numbers on either side of the geometric mean. Find the missing number.

31. *Physical Science* The formula $d = 16t^2$ represents the distance, d, in feet, that an object falls in t seconds. Copy and complete the table of a ball falling from different distances. Do you need the negative square root? What pattern can you find?

Distance (ft)	Time (s)
128	
64	
32	
16	
8	

32. *Working on the* CHAPTER Project Police officers often use a rolling device known as a trundle wheel to measure distances at an accident scene.

    a. If a trundle wheel has a radius of 8 inches, what distance on the ground has been measured when the wheel has made one complete revolution? Name the set of numbers to which this distance belongs.

    b. Make a table of the distances measured in 1, 2, 3, …, 10 revolutions. Then graph the ordered pairs and describe the graph.

33. *Critical Thinking* The area of each of the smallest squares at the right is 10 square units.

    a. What is the length of each side of the smallest squares to the nearest tenth?

    b. Find the length of a diagonal of the largest square to the nearest tenth.

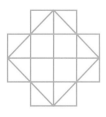

34. Find the nearest whole number estimate for $\sqrt{60}$. Then check your estimate by using a calculator. *(Lesson 9-2)*

For the latest color survey, visit:

www.glencoe.com/ sec/math/mac/ mathnet

35. *Statistics* What is your favorite color? Pantone Color Institute surveyed people about color preferences. Of the 1,000 women surveyed, 280 said blue was their favorite. If next year's survey was increased to include 2,500 women, how many would you predict say blue is their favorite color? *(Lesson 8-1)*

36. *Forestry* The ranger at Crosswoods Nature Preserve recorded the heights of several trees in an area of the preserve. Make a histogram of this data. *(Lesson 4-1)*

Height (ft)	Number of Trees
11-20	4
21-30	10
31-40	14
41-50	22
51-60	28
61-70	18

37. *Standardized Test Practice* Eighth-grade students who were randomly sampled gave the following responses when asked in how many extracurricular activities they were involved: 1, 3, 2, 0, 4, 3, 1, 5, 2, 3, 4, 2, 0, 1, 2. Which expression shows how to determine what percent of the sampled students are involved in 3 or more extracurricular activities? *(Lesson 3-4)*

A $\frac{3}{6} \times 100$

B $\frac{3}{12} \times 100$

C $\frac{3}{15} \times 100$

D $\frac{6}{9} \times 100$

E $\frac{6}{15} \times 100$

For **Extra Practice,** see page 631.

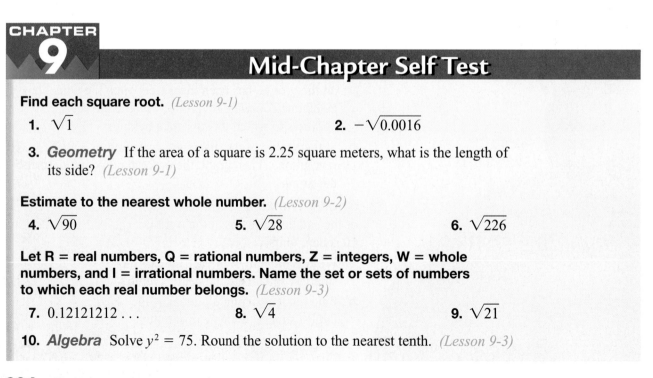

CHAPTER 9

Mid-Chapter Self Test

Find each square root. *(Lesson 9-1)*

1. $\sqrt{1}$

2. $-\sqrt{0.0016}$

3. *Geometry* If the area of a square is 2.25 square meters, what is the length of its side? *(Lesson 9-1)*

Estimate to the nearest whole number. *(Lesson 9-2)*

4. $\sqrt{90}$

5. $\sqrt{28}$

6. $\sqrt{226}$

Let R = real numbers, Q = rational numbers, Z = integers, W = whole numbers, and I = irrational numbers. Name the set or sets of numbers to which each real number belongs. *(Lesson 9-3)*

7. 0.12121212 . . .

8. $\sqrt{4}$

9. $\sqrt{21}$

10. *Algebra* Solve $y^2 = 75$. Round the solution to the nearest tenth. *(Lesson 9-3)*

LAW ENFORCEMENT

> Someday, I'd like to be a highway patrol trooper like Trooper Quinonez.

Evelio H. Quinonez
POLICE OFFICER

Across the nation, highway patrol troopers like Evelio Quinonez enforce traffic laws, respond to traffic accidents, and assist motorists with car trouble. Troopers also enforce criminal laws and assist local police and sheriffs in crime investigation. Trooper Quinonez was a co-recipient of the Florida Highway Patrol Trooper of the Month award in January 1997.

To be a highway patrol trooper, you should be a U.S. citizen, have a high school education, and be in good physical condition. You should also enjoy working with people. For highway patrol work, you should have a solid background in mathematics and have good communication skills. Some departments prefer candidates with some college. Many colleges offer degrees in law enforcement and criminal justice.

For more information:
Qualifications for highway patrols vary from state to state. To find out how to become a trooper in your state, contact your state patrol agency. For example,

Florida Highway Patrol
1011 NW 111th Ave.
Miami, FL 33172

inter NET CONNECTION
www.glencoe.com/sec/
math/mac/mathnet

Your Turn
Traffic safety is a primary concern for state highway patrol troopers. Investigate a safety issue such as the number of accidents that occur at different speeds. Use a statistical graph to design a poster about the need for safe driving speeds that could be used in a traffic safety campaign.

HANDS-ON LAB

geoboard

dot paper

geobands

COOPERATIVE LEARNING

9-4A The Pythagorean Theorem

A Preview of Lesson 9-4

In the previous lessons, you learned about the relationship between the area of a square and the length of its side. You will use this relationship when you investigate a famous rule about right triangles. This rule is called the Pythagorean Theorem.

TRY THIS

Work with a partner.

Step 1 Build a triangle like the one shown on the geoboard. Remember, because it has one right angle, it is called a *right triangle*.

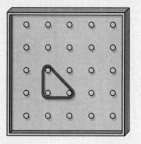

Step 2 Using the longest side of the triangle, build a square. Each side of the square has the same length as the longest side of the triangle. As you can see below, the area of this square is 2 square units.

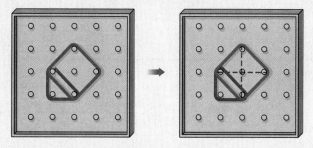

Step 3 Now build squares on the two shorter sides. Each square has an area of 1 square unit.

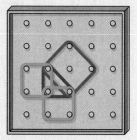

Step 4 Copy the table below and enter the results from this lab in the first row.

Area of Square on Short Side 1	Length of Short Side 1	Area of Square on Short Side 2	Length of Short Side 2	Area of Square on Longest Side	Length of Longest Side

ON YOUR OWN

On dot paper, draw each triangle below and then build squares on each side. Record your results in your table.

1.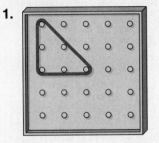

2.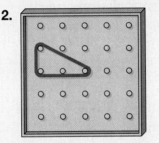

3.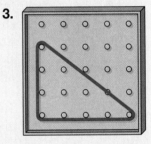

4. In a certain right triangle, the two shorter sides are 2 units and 3 units long. Without making a drawing, what is the area of the square built on the longest side?

5. In another right triangle, the longest side is 6 units long, and one of the other sides is 2 units long. What is the area of the square built on the remaining side?

6. Can you build a square on the longest side of a right triangle with an area of 10 square meters? If you think of 10 as the sum of 1 and 9, you can build squares like the ones shown. The square that is built on the longest side has an area of 10 square units. On dot paper, build squares on the longest side of a right triangle with areas of 13, 17, 40, and 50 square units

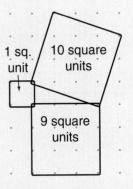

7. *Look Ahead* Use the results in your table to look for a relationship between the area of the two smaller squares and the area of the larger square. Write a statement that describes this relationship.

The Pythagorean Theorem

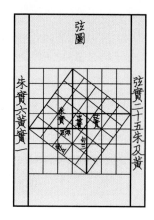

What you'll learn

You'll learn to use the Pythagorean Theorem.

When am I ever going to use this?

You can use the Pythagorean Theorem to find how far baseball players throw a ball to reach certain positions.

Word Wise

hypotenuse
legs
Pythagorean Theorem
converse

Chou-pei Suan-king is an ancient Chinese book that was written sometime during 1200–600 B.C. It contains a diagram named *hsuan-thu,* which offers an interesting puzzle showing the relationship among the sides of a right triangle.

The sides of each right triangle have lengths of 3, 4, and 5 units. The longest side of each triangle, called the **hypotenuse**, is opposite the right angle. The sides that form the right angles are called the **legs**.

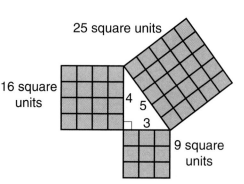

Many years after the Chinese, in the fifth century B.C., a Greek mathematician, Pythagoras, and the students in his school studied the 3-4-5 triangle. They noticed that if they built a square on each of the three sides, the total area of the two smaller squares was equal to the area of the large square.

Today, we call this relationship the **Pythagorean Theorem**. It is true for *any* right triangle.

Pythagorean Theorem	**Words:** In a right triangle, the square of the length of the hypotenuse is equal to the sum of the squares of the lengths of the legs.
	Symbols: $c^2 = a^2 + b^2$ **Model:** 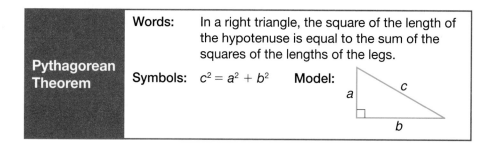

Example ① **Find the length of the hypotenuse in the triangle.**

$c^2 = a^2 + b^2$ *Pythagorean Theorem*

$c^2 = 9^2 + 12^2$ *Replace a with 9 and b with 12.*

$c^2 = 81 + 144$

$c^2 = 225$

$c = \sqrt{225}$ *You can ignore $-\sqrt{225}$ because it is*

$c = 15$ *not reasonable to have a negative length.*

The length of the hypotenuse is 15 feet.

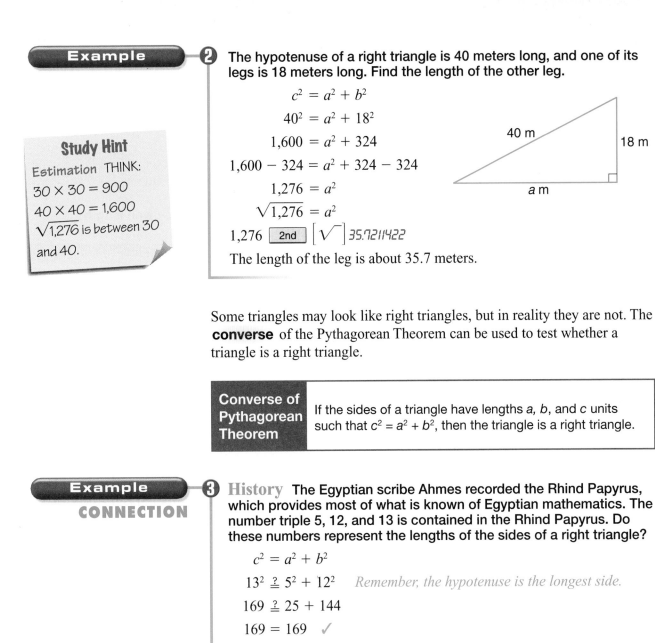

Example **2**

The hypotenuse of a right triangle is 40 meters long, and one of its legs is 18 meters long. Find the length of the other leg.

$$c^2 = a^2 + b^2$$
$$40^2 = a^2 + 18^2$$
$$1,600 = a^2 + 324$$
$$1,600 - 324 = a^2 + 324 - 324$$
$$1,276 = a^2$$
$$\sqrt{1,276} = a^2$$

1,276 [2nd] [$\sqrt{\ }$] *35.7211422*

The length of the leg is about 35.7 meters.

Study Hint

Estimation THINK:
$30 \times 30 = 900$
$40 \times 40 = 1,600$
$\sqrt{1,276}$ is between 30 and 40.

Some triangles may look like right triangles, but in reality they are not. The **converse** of the Pythagorean Theorem can be used to test whether a triangle is a right triangle.

Converse of Pythagorean Theorem	If the sides of a triangle have lengths *a*, *b*, and *c* units such that $c^2 = a^2 + b^2$, then the triangle is a right triangle.

Example **3**

CONNECTION

History The Egyptian scribe Ahmes recorded the Rhind Papyrus, which provides most of what is known of Egyptian mathematics. The number triple 5, 12, and 13 is contained in the Rhind Papyrus. Do these numbers represent the lengths of the sides of a right triangle?

$$c^2 = a^2 + b^2$$
$$13^2 \stackrel{?}{=} 5^2 + 12^2 \quad \textit{Remember, the hypotenuse is the longest side.}$$
$$169 \stackrel{?}{=} 25 + 144$$
$$169 = 169 \quad \checkmark$$

The numbers represent the sides of a right triangle because the Pythagorean Theorem holds true for the lengths of the sides.

CHECK FOR UNDERSTANDING

Communicating Mathematics

Read and study the lesson to answer each question.

1. **Tell** the area of the shaded square shown at the right.

2. **Draw** a right triangle and label the right angle, the hypotenuse, and the legs.

Math Journal

3. Use centimeter grid paper to determine, and then **write** why a triangle with sides 6, 12, and 14 cannot be a right triangle.

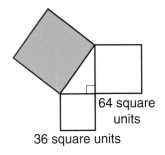

64 square units

36 square units

Write an equation you could use to find the length of the missing side of each right triangle. Then find the missing length. Round to the nearest tenth.

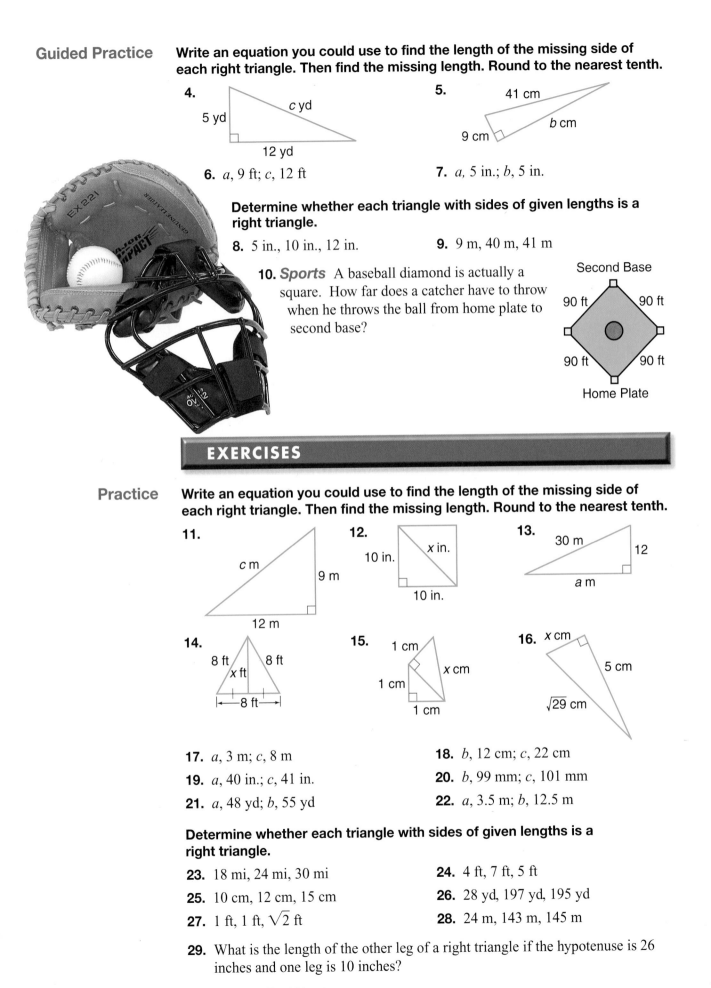

4.

5 yd
c yd
12 yd

5.

41 cm
b cm
9 cm

6. *a*, 9 ft; *c*, 12 ft

7. *a*, 5 in.; *b*, 5 in.

Determine whether each triangle with sides of given lengths is a right triangle.

8. 5 in., 10 in., 12 in.

9. 9 m, 40 m, 41 m

10. *Sports* A baseball diamond is actually a square. How far does a catcher have to throw when he throws the ball from home plate to second base?

Second Base
90 ft   90 ft
90 ft   90 ft
Home Plate

EXERCISES

Practice Write an equation you could use to find the length of the missing side of each right triangle. Then find the missing length. Round to the nearest tenth.

11.

c m
9 m
12 m

12.

10 in.
x in.
10 in.

13.

30 m
12
a m

14.

8 ft   8 ft
x ft
8 ft

15.

1 cm
x cm
1 cm
1 cm

16.

x cm
5 cm
√29 cm

17. *a*, 3 m; *c*, 8 m

18. *b*, 12 cm; *c*, 22 cm

19. *a*, 40 in.; *c*, 41 in.

20. *b*, 99 mm; *c*, 101 mm

21. *a*, 48 yd; *b*, 55 yd

22. *a*, 3.5 m; *b*, 12.5 m

Determine whether each triangle with sides of given lengths is a right triangle.

23. 18 mi, 24 mi, 30 mi

24. 4 ft, 7 ft, 5 ft

25. 10 cm, 12 cm, 15 cm

26. 28 yd, 197 yd, 195 yd

27. 1 ft, 1 ft, $\sqrt{2}$ ft

28. 24 m, 143 m, 145 m

29. What is the length of the other leg of a right triangle if the hypotenuse is 26 inches and one leg is 10 inches?

30. *Geometry* A circle with a diameter of 10 centimeters is inscribed in a square. What is the length of the diagonal of the square?

10 cm

31. *Design* A woodworking company makes entertainment centers that include a space for a television. Televisions are sized by the length of the diagonal of the rectangular screen. The designer wants to accommodate a television with a 27-inch screen, but TVs vary in width and length. The screen widths for a 27-inch TV range from 18 to 20 inches. What is the range of the heights?

32. *Write a Problem* about the photograph at the right that requires using the Pythagorean Theorem in the solution.

33. *Critical Thinking* About 2000 B.C., Egyptian engineers discovered a way to make a right triangle using a rope with 12 evenly spaced knots tied in it. They attached one end of the rope to a stake in the ground. At what knot locations should the other two stakes be placed in order to form a right triangle? Draw a diagram.

Mixed Review

34. *Algebra* Solve $x^2 = 1.70$. Round the solution to the nearest tenth. *(Lesson 9-3)*

35. At 3:00 P.M., Vinita's shadow was 2.5 feet long. The shadow of a tree was 21 feet long. If Vinita is 5.25 feet tall, estimate the height of the tree. *(Lesson 8-7)*

36. *Geometry* Roberto became lost when driving from his house to a restaurant. From his house, he drove west 3 blocks, north 2 blocks, east 5 blocks, south 4 blocks, east 3 blocks, and north 6 blocks before arriving at the restaurant. If Roberto's house has coordinates $(0, 0)$, what are the coordinates of the restaurant? *(Lesson 2-10)*

37. **Standardized Test Practice** Alisa works at the Dairy Dream 20 hours each week. So far this week, she has worked $4\frac{1}{2}$ hours on Monday and $7\frac{1}{4}$ hours on Wednesday. Which sentence could be used to find h, the number of hours remaining for Alisa to work this week? *(Lesson 1-6)*

A $4\frac{1}{2} + 7\frac{1}{4} = h + 20$

B $4\frac{1}{2} + 7\frac{1}{4} = h \times 20$

C $h = 20 - \left(4\frac{1}{2} + 7\frac{1}{4}\right)$

D $h - 4\frac{1}{2} - 7\frac{1}{4} = 20$

E $h = 20 - 4\frac{1}{2} + 7\frac{1}{4}$

For **Extra Practice,**
see page 631.

9-5A Draw a Diagram

A Preview of Lesson 9-5

Lonzo zoomed up on his in-line skates yelling, "I've got a great puzzle for you. You'll never figure it out!" Paul rolled his eyes, but he was ready to take Lonzo's challenge. Let's listen in!

My invisible pet ant must travel from one corner to the opposite corner of a 2-inch cube. My ant knows the shortest way, do you?

That's pretty easy. It would need to go 2 inches up, 2 inches across, and 2 inches back. That's 6 inches.

Good try! But there's a shorter way.

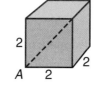

OK, use the diagonal across the face of the cube, which is the hypotenuse of a right triangle. This length can be found by solving $h^2 = 2^2 + 2^2$ or $h^2 = 8$. That means that $h = \sqrt{8}$, which is a little less than 3. So, now you have a little less than 3 inches plus the length of the edge to the back that is 2 inches. This makes the path length slightly less than 5 inches.

Lonzo

That's better, but it's still not the shortest way. Remember, the shortest distance between two points is a straight line.

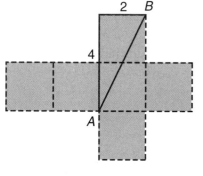

Maybe a picture would help. A cube has 6 faces and they all fit together like this. You need to use a hypotenuse, but the right triangle you use has a 2 inch leg and a 4 inch leg. Therefore, $h^2 = 2^2 + 4^2$ or $h^2 = 20$. That means that $h = \sqrt{20}$, which is about 4.5 inches. This must be the shortest path.

Paul

For **Extra Practice,** see page 632.

THINK ABOUT IT

Work with a partner.

1. *Compare and contrast* the three solutions. Are all three paths feasible for an ant to travel?

2. *Draw* a different diagram that could be used to solve this problem. How is it different? What path would the ant travel and how long is that path?

3. *Draw a diagram* to solve the following problem.

   A section of a theater is arranged so that each row has the same number of seats. Annie is seated in the fifth row from the front and the third row from the back. Her seat is sixth from the left and second from the right. How many seats are in this section?

ON YOUR OWN

4. The first step of the 4-step plan for problem solving asks you to *explore* the problem. *Explain* how you determined what information was given, what was needed, and if there was too much information given for Exercise 3.

5. *Write a Problem* that can be solved by drawing a diagram. Exchange problems with another student and solve each others' problem.

6. *Look Ahead* Draw a diagram that could be used for solving Exercise 10 on page 407.

MIXED PROBLEM SOLVING

STRATEGIES

Look for a pattern.
Solve a simpler problem.
Act it out.
Guess and check.
Draw a diagram.
Make a chart.
Work backward.

Solve. Use any strategy.

7. *Recreation* Ms. Llarena's dance class is evenly spaced in a circle. If the sixth person is directly opposite the sixteenth person, how many people are in the circle?

8. *Number Theory* When a number is decreased by 4, the result is –23. What is the number?

9. *Geometry* Two sides of a triangle have the same length. The third side is 2 meters long. If the perimeter of the triangle is 20 meters, find the lengths of the sides.

10. *Food* Of the 30 members in a cooking club, 20 like to mix salads, 17 prefer baking desserts, and 8 like to do both.
    a. How many like to mix salads, but not bake desserts?
    b. How many do not like either baking desserts or mixing salads?

11. *Money* Matsuko has some quarters and nickels in her pocket. She has 8 more nickels than quarters. The total value is $3.70. How many of each coin does Matsuko have?

12. *Entertainment* The graph shows the numbers of types of outdoor grills sold.
    a. How does the number of charcoal grills compare to the number of gas grills?
    b. Make a circle graph of the information.

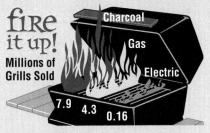

Source: Barbecue Industry Association

13. *Standardized Test Practice* There are eight people at a business meeting. Each person shakes hands with everyone else exactly once. How many handshakes occur?

    A 8   **B** 16   **C** 28   **D** 64

Lesson 9-5A **THINKING LAB**   **403**

Using the Pythagorean Theorem

The Anasazi Indians built pueblos in the southwest United States around A.D. 900, long before the European Americans reached that area. Bandelier National Monument in New Mexico has preserved some of their ancient ruins including the amazing pueblos and cave dwellings.

The stone walls are rugged, making it difficult to measure the height of the caves from the ground. However, using the Pythagorean Theorem, archeologists can easily calculate the heights.

Example

Real World APPLICATION

① **Archeology** A 10-foot ladder that is placed 4 feet from the base of the vertical rock wall just reaches the entrance of the cave. How high is the cave entrance from the ground?

Explore Draw a diagram. The wall, the ground, and the ladder form a right triangle and the ladder is the hypotenuse. You need to know the height of the cave.

Plan Let c = the height of the cave, g = the length on the ground from the wall to the bottom of the ladder, and ℓ = the length of the ladder. Find the value of c by using the Pythagorean Theorem.

Solve
$$c^2 + g^2 = \ell^2$$
$$c^2 + 4^2 = 10^2 \quad \text{Substitute the values for } g \text{ and } \ell$$
$$c^2 + 16 = 100 \quad \text{into the Pythagorean Theorem.}$$
$$c^2 + 16 - 16 = 100 - 16 \quad \text{Subtract 16 from each side.}$$
$$c^2 = 84$$
$$c = \sqrt{84}$$

84 [2nd] [$\sqrt{\ }$] $9.1651513\vphantom{9}$ *Use a calculator.*

The cave is about 9.2 feet above the ground.

Examine Check to see if the measures satisfy the Pythagorean Theorem.

$$4^2 + 9.2^2 \stackrel{?}{=} 10^2$$
$$16 + 84.64 \stackrel{?}{=} 100$$
$$100.64 \approx 100 \quad \checkmark \quad \text{\textit{The numbers aren't exactly the}}$$
$$\text{\textit{same because 9.2 was rounded.}}$$

APPLICATION

Real World

② **Sports** Drew Bledsoe and Terry Glenn play for the Patriots. On a play, the announcer said, "Bledsoe drops back to the 10-yard line and fires the ball to Glenn. Glenn catches it at the 50-yard sideline. What a catch!" If Bledsoe was 25 yards from the sideline when he dropped back to the 10-yard line, how far did he throw the ball?

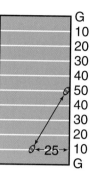

$a = 25$ yards and
$b = 50 - 10$ or
    40 yards

$a^2 + b^2 = c^2$ *Pythagorean Theorem*

$25^2 + 40^2 = c^2$

$625 + 1{,}600 = c^2$

$2{,}225 = c^2$

$\sqrt{2{,}225} = c$

2225 [2nd] [√] 47.16990566

Drew Bledsoe's pass was just over 47 yards.

You know that a triangle with sides 3, 4, and 5 units is a right triangle because these numbers satisfy the Pythagorean Theorem. Such numbers are called **Pythagorean triples**. By using multiples of a primitive set of Pythagorean triples, you can create additional triples. *The triple 3-4-5 is called a primitive Pythagorean triple because the numbers are relatively prime.*

Example

INTEGRATION

③ **Number Patterns**

a. Multiply the triple 3-4-5 by the numbers 2, 3, 4, and 10 to find more Pythagorean triples.

b. Compare the ratio of the areas of a 3-4-5 triangle and a 30-40-50 triangle to the ratio of the sides of the two triangles.

a. You can organize your answers in a table. Multiply each Pythagorean triple entry by the same number and then check the Pythagorean relationship.

Study Hint

Reading Math ×2 means multiply by 2, ×3 means multiply by 3, and so on.

	a	b	c	Check: $a^2 + b^2 = c^2$
original	3	4	5	$9 + 16 = 25$ ✓
× 2	6	8	10	$36 + 64 = 100$ ✓
× 3	9	12	15	$81 + 144 = 225$ ✓
× 4	12	16	20	$144 + 256 = 400$ ✓
× 10	30	40	50	$900 + 1{,}600 = 2{,}500$ ✓

b. Find the area of each triangle.

$A = \frac{1}{2}bh$ *Formula for area of a triangle*

3-4-5 triangle: 30-40-50 triangle:

$A = \frac{1}{2}(3)(4)$ $A = \frac{1}{2}(30)(40)$

$A = 6$ $A = 600$

(continued on the next page)

$$\text{ratio of areas} = \frac{6}{600} \text{ or } \frac{1}{100}$$

$$\text{ratio of sides} = \frac{3}{30} \text{ or } \frac{1}{10}, \frac{4}{40} \text{ or } \frac{1}{10}, \text{ and } \frac{5}{50} \text{ or } \frac{1}{10}$$

Each pair of sides has a ratio of $\frac{1}{10}$, and the ratio of the areas is $\frac{1}{100}$.

CHECK FOR UNDERSTANDING

Communicating Mathematics

Read and study the lesson to answer each question.

1. *Write* two Pythagorean triples other than 5-12-13 that are in the 5-12-13 family.

2. *Explain* why you can use any two sides of a right triangle to find the third side.

Guided Practice

Write an equation that can be used to answer the question. Then solve. Round to the nearest tenth.

3. How long is the lake?

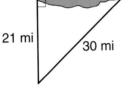

21 mi   30 mi

4. *Geography* The state of Wyoming is shaped like a rectangle that is about 360 miles by 280 miles.

Wyoming

   a. What is the longest distance you could go in a straight line and still remain in Wyoming?

   b. Estimate the longest distance in your state.

EXERCISES

Practice

Write an equation that can be used to answer each question. Then solve. Round to the nearest tenth.

5. How far apart are the planes?

6. How high does the ladder reach?

7. How long is each rafter?

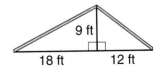

5 mi   8 mi

15 ft   3 ft

9 ft   18 ft   12 ft

8. Name the primitive Pythagorean triple to which 16-30-34 belongs.

9. **Safety** For safety reasons, the base of a 24-foot ladder should be at least 8 feet from the wall. How high can a 24-foot ladder safely reach?

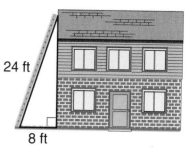

24 ft

8 ft

10. **Sports** At 13 years old, Dominique Moceanu was the youngest gymnast ever to win the U.S. Senior National Championship. Using the diagonal of the floor mat for her run, flip, and finish, gives her more distance. The routine is performed on a 40-foot by 40-foot mat. How much more room does she get by using the diagonal rather than by using a side?

11. **Geography** The flags of Papua New Guinea and the Solomon Islands are shown below. How much longer is the diagonal of the Solomon Islands flag than the diagonal of the Papua New Guinea flag?

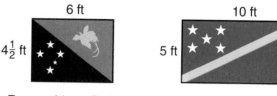

6 ft

$4\frac{1}{2}$ ft

Papua New Guinea

10 ft

5 ft

Solomon Islands

12. **Recreation** The acceleration ramp for a skateboard competition is 20 meters long, and it extends 15 meters from the base of the starting point. How high is the ramp?

13. **Geography** Jewel likes to hike in the San Miguel Mountains at Bandelier National Monument. The park boundaries are two legs of a right triangle. One leg is 6 kilometers long, and the other leg is 11.8 kilometers long. The trail from the Rio Grande River to Jewel's campsite is slightly longer than the hypotenuse of the triangle. Approximate the length of her hike.

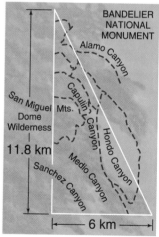

BANDELIER NATIONAL MONUMENT

Alamo Canyon

San Miguel Mts. Dome Wilderness

Capulin Canyon

Hondo Canyon

11.8 km

Sanchez Canyon

Medio Canyon

6 km

14. **Critical Thinking** What is the distance from the center of a circle of radius 4 to a chord of length 6 centimeters?

For **Extra Practice,** see page 632.

Mixed Review

15. **Standardized Test Practice** Triangle *DEF* is a right triangle. Its hypotenuse is 15 inches long, and one leg is 12 inches long. How long is the other leg? *(Lesson 9-4)*

    A 11 in.   **B** 9 in.   **C** 4 in.   **D** 3 in.

16. Factor 162 completely. *(Lesson 6-2)*

9-5B Graphing Irrational Numbers

A Follow-Up of Lesson 9-5

compass

straightedge

Lewis Howard Latimer (1848-1928), an African-American engineer and drafter, made the drawings needed to get a patent for the telephone, which was invented by Alexander Graham Bell.

Drafters, inventors, and engineers used to use a compass and straightedge to copy measurements. Today, they use computers to generate duplicate measurements.

You already know how to graph integers and rational numbers on a number line. How would you graph an irrational number like $\sqrt{5}$?

TRY THIS

Work with a partner.

1 Graph $\sqrt{5}$.

- Find two numbers whose squares have a sum or difference of 5. One pair that works is 1 and 2, since $1^2 + 2^2 = 5$.

- Draw a number line. At 2 on the number line, construct a perpendicular line segment 1 unit long.
  You can construct a perpendicular segment by folding the number line at 2, making sure the two parts of the number line align. The fold is a segment perpendicular to the number line.

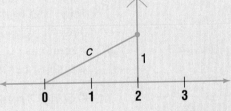

- Draw the line segment shown from 0 to the top of the 1 unit segment. Label it c.

- The Pythagorean Theorem can be used to show that c is $\sqrt{5}$ units long.

  $c^2 = a^2 + b^2$

  $c^2 = 1^2 + 2^2$ *Replace a with 1 and b with 2.*

  $c^2 = 5$

  $c = \sqrt{5}$

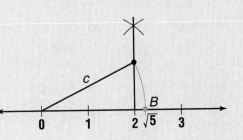

- Open the compass to the length of c. With the tip of the compass at 0, draw an arc that intersects the number line at B. The distance from 0 to B is $\sqrt{5}$ units.

❷ Graph $\sqrt{7}$.

- Find two numbers whose squares have a sum or difference of 7. One pair that works is 4 and 3, since $4^2 - 3^2 = 7$.

- Draw a number line. At 3 on the number line, construct a perpendicular line segment. Put the tip of the compass at 0. With the compass at 4 units, construct an arc that intersects the perpendicular line segment. Label the perpendicular leg a.

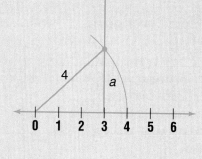

- The Pythagorean Theorem can be used to show that a is $\sqrt{7}$ units long.

  $c^2 = a^2 + b^2$

  $4^2 = a^2 + 3^2$ *Replace c with 4 and b with 3.*

  $16 = a^2 + 9$

  $7 = a^2$

  $\sqrt{7} = a$

- Open the compass to the length of a. With the tip of the compass at 0, draw an arc that intersects the number line at D. The distance from 0 to D is $\sqrt{7}$ units.

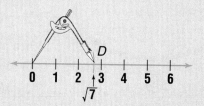

ON YOUR OWN

Graph each number on a number line.

1. $\sqrt{2}$ 2. $\sqrt{3}$ 3. $\sqrt{8}$ 4. $\sqrt{10}$

5. Explain how you graphed $\sqrt{2}$.

6. Describe two different ways to graph $\sqrt{8}$.

7. Explain how to graph $-\sqrt{5}$.

8. Explain how the graph of $\sqrt{2}$ can be used to locate the point that represents $\sqrt{3}$.

9. *Reflect Back* For each number from 11-20 that is not a perfect square, write the equation that is the sum or difference of two squares that could be used in order to graph the square root of that number.

Integration: Geometry
Distance on the Coordinate Plane

What you'll learn

You'll learn to find the distance between points in the coordinate plane.

When am I ever going to use this?

Knowing how to find the distance between points in the coordinate plane can help you estimate the distance between places on a map.

In 1791, Pierre L'Enfant was hired to create a plan for the nation's capital city, Washington, D.C. He placed the Capitol at the center of a grid-style layout for the city. Below is a map of Capitol Hill and its surrounding area with a centimeter grid laid over top.

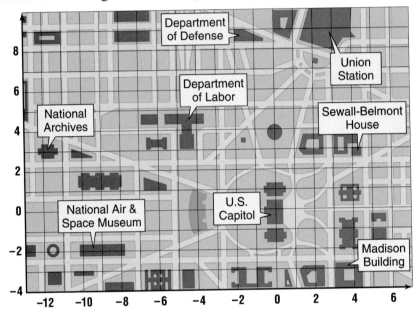

When points are graphed on a coordinate grid, you can use ordered pairs to determine the distance between two points.

LOOK BACK
You can refer to Lesson 2-10 to review the coordinate plane.

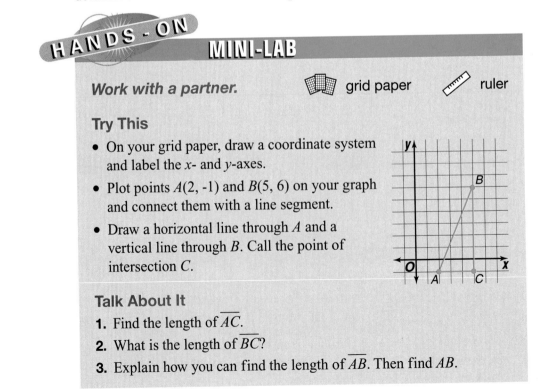

HANDS-ON MINI-LAB

Work with a partner. grid paper   ruler

Try This

- On your grid paper, draw a coordinate system and label the *x*- and *y*-axes.

- Plot points $A(2, -1)$ and $B(5, 6)$ on your graph and connect them with a line segment.

- Draw a horizontal line through A and a vertical line through B. Call the point of intersection C.

Talk About It

1. Find the length of $\overline{AC}$.

2. What is the length of $\overline{BC}$?

3. Explain how you can find the length of $\overline{AB}$. Then find AB.

1 Graph the ordered pairs (2, 0) and (5, −4). Then find the distance between the points.

Let c = the distance between the two points, $a = 3$, and $b = 4$.

Use the Pythagorean Theorem.

$$c^2 = a^2 + b^2$$
$$c^2 = 3^2 + 4^2$$
$$c^2 = 9 + 16$$
$$c^2 = 25$$
$$c = \sqrt{25}$$
$$c = 5$$

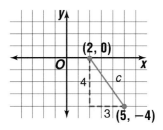

The points are 5 units apart.

APPLICATION

2 **Travel** Refer to the map at the beginning of the lesson. Suppose that the Capitol is located at the origin. How many miles is it from the National Air and Space Museum to the Department of Defense if a unit on the grid is about 0.05 mile?

The coordinates of the National Air and Space Museum are (−8, −2) and the coordinates of the Department of Defense are (−2, 9).

Let c = the distance between the National Air and Space Museum and the Department of Defense. Then $a = 6$ and $b = 11$.

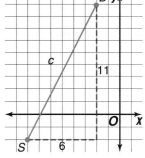

$$c^2 = a^2 + b^2 \quad \textit{Pythagorean Theorem}$$
$$c^2 = 6^2 + 11^2$$
$$c^2 = 36 + 121$$
$$c^2 = 157$$
$$c = \sqrt{157}$$

157 [2nd] $\left[\sqrt{} \right]$ *12.52996409*   *Use a calculator.*

They are about 13 units apart.

Multiply 13 units by 0.05 mile per unit to find the number of miles.

$$13 \times 0.05 = 0.65$$

The National Air and Space Museum is about 0.65 mile from the Department of Defense.

 Did you know Suppose all of the Smithsonian treasures were lined up in one long exhibit. If a person spent 1 second looking at each item nonstop for 24 hours a day, it would take that person more than 2.5 years to see all of them.

CHECK FOR UNDERSTANDING

Communicating Mathematics

Read and study the lesson to answer each question.

1. *Draw* the triangle that you can use to find the distance between points at (−2, 1) and (−5, −3) on the coordinate plane.

2. *Write* the steps you could use to find the distance between points at (5, 5) and (2, 2).

Lesson 9-6 Integration: Geometry   Distance on the Coordinate Plane   **411**

3. Find the distance from the Capitol to the Sewall-Belmont House and from the Capitol to the Madison Building. Are the distances the same or different? Explain.

Guided Practice

4. Find the distance between the pair of points whose coordinates are given. Round to the nearest tenth.

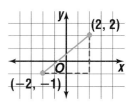

Graph each pair of ordered pairs. Then find the distance between the points. Round to the nearest tenth.

5. $(-2, 4), (3, -1)$ **6.** $(1, 4), (6, 2)$

7. *Maps* On a street map, the side of each square represents 1 mile. The Wright Ball Park is located at (1, 2), and the Weaver Rollerdome is located at (6, 10). A diagonal street runs directly between the two locations. Approximately how far is it from Wright Ball Park to Weaver Rollerdome?

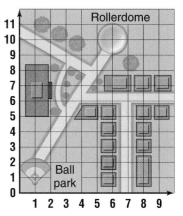

EXERCISES

Practice

Find the distance between each pair of points whose coordinates are given. Round to the nearest tenth.

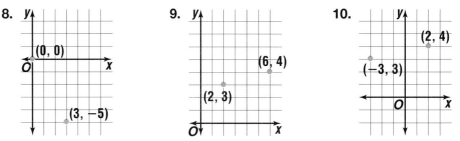

8. (0, 0), (3, −5) **9.** (2, 3), (6, 4) **10.** (−3, 3), (2, 4)

Graph each pair of ordered pairs. Then find the distance between the points. Round to the nearest tenth.

11. $(3, -2), (2, 5)$ **12.** $(-1, 0), (2, 7)$ **13.** $(1, 5), (3, 1)$

14. $(-6, 3), (2, -5)$ **15.** $(-5, 0), (2, 0)$ **16.** $(7, -4), (-3, 5)$

17. The coordinates of points *R* and *S* are (4, 3) and (1, 6). What is the distance between the points?

Applications and Problem Solving

18. *Geometry* A triangle on the coordinate plane has vertices $A(2, -1)$, $B(-2, 2)$, and $C(-6, 14)$.

  a. Draw the triangle.

  b. Find the perimeter of the triangle.

19. *Geometry* A circle goes through the points A(3, 0), B(6, 3), C(9, 0), and D(6, −3). Desiree connected those points and determined that they formed a square inside the circle. What is the length of a side of the square?

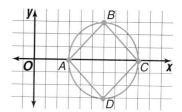

20. *Graphing Calculator* The graphing calculator program will find the distance between two points in the coordinate plane. To use the program, first enter it into the calculator's memory. To access the program memory, press [PRGM] [▶] [▶] [ENTER]. Enter all the program instructions at the right. Then press [2nd] [QUIT] to return to the Home screen. To run the program, press [PRGM] and choose the program from the list by pressing the number next to its name then [ENTER]. Enter the coordinates of the points as the program asks for them. *X1 refers to the x-coordinate of the first point.* The calculator automatically rounds the distance to the nearest tenth of a unit.

```
Prgm 1: DISTANCE
:  Fix 1
:  Disp "ENTER X1"
:  Input A
:  Disp "ENTER Y1"
:  Input B
:  Disp "ENTER X2"
:  Input C
:  Disp "ENTER Y2"
:  Input D
:  √((A-C)^2+(B-D)^2)→E
:  Disp "THE DISTANCE IS"
:  Disp E
```

Use the program to find the distance between each pair of points.

a. (3, 3), (−7, −1)   **b.** (−12, 1), (15, −5)   **c.** (9, 1), (−2, 1)

d. (−8, 3), (−2, −4)   **e.** (−47, 21), (125, 72)   **f.** (2.4, 6.1), (0.2, 0.3)

21. *Working on the* CHAPTER Project
Police officers may use a technique called the *coordinate method* to diagram an accident scene. An officer chooses a reference point such as a mile post or a telephone pole. For each item to be diagramed, the officer places a marker along the side of the road at a point that is directly across from the item.

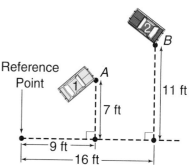

a. Place the accident scene diagram above on a coordinate plane and determine how far point A is from point B.

b. Simulate an accident scene in the school parking lot. Make a diagram on a coordinate grid and write an accident report.

22. *Critical Thinking* The distance between points A and B is 17 units. Find the value of x if the coordinates of A and B are (−3, x) and (5, 2).

Mixed Review

23. Standardized Test Practice What is the height of the tower? *(Lesson 9-5)*

A 8 feet   **B** 31.5 feet

C 992 feet   **D** 49.9 feet

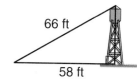

24. Express $8.\overline{72}$ as a mixed number. *(Lesson 6-5)*

25. Solve $f = 8(-3)(-4)^2$. *(Lesson 2-7)*

For **Extra Practice**, see page 632.

Integration: Geometry
Special Right Triangles

What you'll learn

You'll learn to find missing measures in 30°-60° right triangles and 45°-45° right triangles.

When am I ever going to use this?

Knowing how to find missing measures in special right triangles can help you determine roof span length.

The sides of right triangles with 30° and 60° angles have a special relationship.

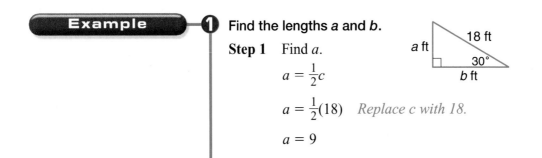

HANDS-ON MINI-LAB

Work with a partner.

✎ compass   ⌓ protractor

📏 ruler   ✂ scissors

Try This

- Construct and cut out an equilateral triangle.
- Fold the triangle in half and cut along the fold line.
- Measure each side and each angle. Measure the length to the nearest tenth of a centimeter.
- Repeat the above steps for at least two other equilateral triangles.

Talk About It

What is the relationship between the length of the hypotenuse and the length of the side opposite the 30° angle?

◄ **LOOK BACK**

You can refer to Lesson 5-1A to review how to measure and construct line segments and angles.

The relationship you discovered in the Mini-Lab is always true in a 30°-60° right triangle. The length of the hypotenuse is twice the length of the side opposite the 30° angle.

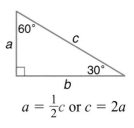

$$a = \frac{1}{2}c \text{ or } c = 2a$$

Therefore, the ratio $\dfrac{\text{length of hypotenuse}}{\text{length of side opposite the 30° angle}}$ is $\dfrac{2}{1}$.

Example ——❶ **Find the lengths a and b.**

Step 1   Find a.

$$a = \frac{1}{2}c$$

$$a = \frac{1}{2}(18) \quad \textit{Replace c with 18.}$$

$$a = 9$$

18 ft a ft   30° b ft

Step 2 Find b.

$$a^2 + b^2 = c^2 \quad \textit{Pythagorean Theorem}$$
$$9^2 + b^2 = 18^2 \quad \textit{Replace a with 9 and c with 18.}$$
$$81 + b^2 = 324$$
$$b^2 = 243$$
$$b = \sqrt{243}$$

243 [2nd] [$\sqrt{}$] 15.58845727

$$b \approx 15.6$$

The length a is 9 feet, and the length b is about 15.6 feet.

There is also a special relationship between the sides of a triangle with two 45° angles.

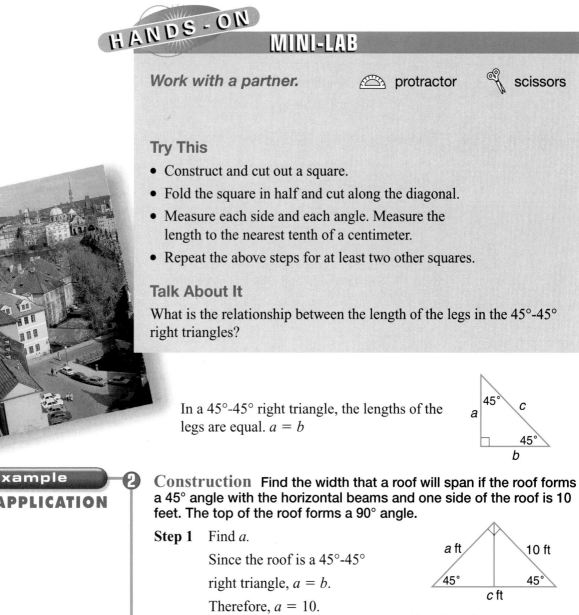

HANDS-ON MINI-LAB

Work with a partner. 📐 protractor ✂ scissors

Try This

- Construct and cut out a square.
- Fold the square in half and cut along the diagonal.
- Measure each side and each angle. Measure the length to the nearest tenth of a centimeter.
- Repeat the above steps for at least two other squares.

Talk About It

What is the relationship between the length of the legs in the 45°-45° right triangles?

In a 45°-45° right triangle, the lengths of the legs are equal. $a = b$

Example 2

🌎 **Real World APPLICATION**

Construction Find the width that a roof will span if the roof forms a 45° angle with the horizontal beams and one side of the roof is 10 feet. The top of the roof forms a 90° angle.

Step 1 Find a.

Since the roof is a 45°-45° right triangle, $a = b$.

Therefore, $a = 10$.

(continued on the next page)

Step 2 Find c.

$c^2 = a^2 + b^2$ *Pythagorean Theorem*

$c^2 = 10^2 + 10^2$ *Replace a and b with 10.*

$c^2 = 100 + 100$

$c^2 = 200$

200 [2nd] [$\sqrt{}$] *14.14213562*

$c \approx 14.1$

The roof spans about 14.1 feet.

CHECK FOR UNDERSTANDING

Communicating Mathematics

Read and study the lesson to answer each question.

1. **Write** a sentence describing the relationship between the hypotenuse of a 30°-60° right triangle and the leg opposite the 30° angle.

2. **Tell** why a triangle with two 45° angles is an isosceles right triangle.

HANDS-ON MATH

3. **Draw** an acute triangle and an altitude. Describe the triangles formed by the sides and the altitude of the original triangle. Describe what you know about the two new triangles.

Guided Practice

Find the missing lengths. Round decimal answers to the nearest tenth.

4.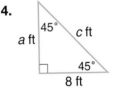

5.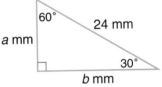

6. **Signs** A stop sign is cut from a square piece of sheet metal. If each side of the sign is 1 foot long and all of the angles are equal, how long were the sides of the original square piece of metal?

EXERCISES

Practice

Find the missing lengths. Round decimal answers to the nearest tenth.

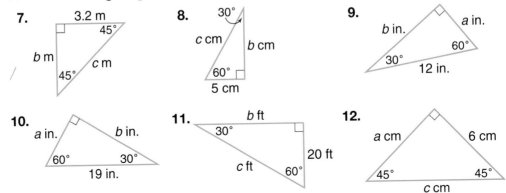

7. 3.2 m, 45°, b m, 45°, c m

8. 30°, c cm, b cm, 60°, 5 cm

9. a in., b in., 30°, 60°, 12 in.

10. a in., b in., 60°, 30°, 19 in.

11. b ft, 30°, 20 ft, c ft, 60°

12. a cm, 6 cm, 45°, 45°, c cm

13. The length of the hypotenuse of a 30°-60° right triangle is 7.5 inches. Find the length of the side opposite the 30° angle.

Applications and Problem Solving

14. *Geometry* Draw a 30°-60° right triangle whose hypotenuse is 16 units. Draw the height to the hypotenuse. In each of the two new triangles, draw the height to the hypotenuse. In each of the four new triangles, draw the height to the hypotenuse. Find and label the measure of each angle and find the length of as many sides as you can without using the Pythagorean Theorem.

15. *Geometry* Use the diagram at the right.
   a. Find the length of the hypotenuse in each of the four smallest triangles.
   b. Are there any 45°-45° right triangles? Are there any 30°-60° right triangles?
   c. Look at the pattern. What do you think is the length of the hypotenuse of the next triangle? Verify your answer by calculating the length.

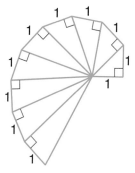

16. *Sports* Did you have your first taste of team sports playing Tee-ball? Tee-ball is similar to softball, but is played without a pitcher and on a smaller field. The baselines on a Tee-ball diamond are 40 feet long. How far is it from home plate to second base on a Tee-ball diamond?

17. *Geometry* Solve the following problems.
   a. Draw a diagram to help you find the area and perimeter of a regular hexagon inscribed in a circle of radius 4.
   b. What is the side length of an isosceles right triangle with hypotenuse $\sqrt{162}$?

18. *Critical Thinking* Determine how much material is needed to make the sail shown at the right.

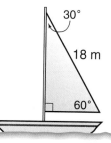

Mixed Review

19. **Standardized Test Practice** What is the distance between points $Q(1, -3)$ and $R(6, 9)$? *(Lesson 9-6)*

   A 9 units   **B** 5 units   **C** 12 units   **D** 13 units

20. *Money* The cost of making a $1 bill and each coin is listed in the chart. What percent of the $1 bill's value and each coin's value is used to make it? *(Lesson 8-3)*

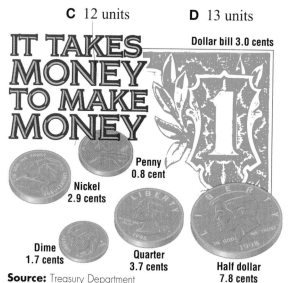

IT TAKES MONEY TO MAKE MONEY

Dollar bill 3.0 cents
Penny 0.8 cent
Nickel 2.9 cents
Dime 1.7 cents
Quarter 3.7 cents
Half dollar 7.8 cents

Source: Treasury Department

For **Extra Practice**, see page 633.

Vocabulary

After completing this chapter, you should be able to define each term,
concept, or phrase and give an example or two of each.

Number and Operations
irrational number (p. 391)
perfect square (p. 382)
principal square root (p. 382)
radical sign (p. 382)
real numbers (p. 391)
square root (p. 382)

Geometry
converse (p. 399)
hypotenuse (p. 398)
legs (p. 398)
Pythagorean Theorem (pp. 396-398)
Pythagorean triple (p. 405)

Problem Solving
draw a diagram (p. 402)

Understanding and Using the Vocabulary

Choose the letter of the term that best matches each statement or phrase.

1. the symbol $\sqrt{}$
2. the square of a rational number
3. the combined set of rational numbers and irrational numbers
4. can always be expressed as a terminating or repeating decimal
5. one of the sides that forms the right angle in a right triangle
6. The Pythagorean Theorem is true for this type of triangle.
7. In a 30°-60° right triangle, the length of the side opposite the 30° angle is one-half the length of this side.
8. the name of x in relation to y if $x^2 = y$

a. triangle
b. radical sign
c. real numbers
d. rational number
e. irrational number
f. square root
g. perfect square
h. right triangle
i. hypotenuse
j. leg

In Your Own Words

9. *Explain* the difference between a rational number and an irrational number. Give an example of each.

Objectives & Examples

Upon completing this chapter, you should be able to:

● find square roots of perfect squares
(Lesson 9-1)

Find $\sqrt{64}$.

Since $8^2 = 64$, $\sqrt{64} = 8$.

● estimate square roots *(Lesson 9-2)*

Estimate $\sqrt{42}$.

$36 < 42 < 49$

$6^2 < 42 < 7^2$

$6 < \sqrt{42} < 7$

Since 42 is closer to 36 than to 49, the best whole number estimate is 6.

● identify and classify numbers in the real number system *(Lesson 9-3)*

Determine whether 7.43 is a rational or irrational number.

7.43 is a terminating decimal.
Therefore, 7.43 is a rational number.

● use the Pythagorean Theorem *(Lesson 9-4)*

Find the missing length of the right triangle.

$c^2 = a^2 + b^2$

$c^2 = 5^2 + 8^2$

$c^2 = 25 + 64$

$c^2 = 89$

$c = \sqrt{89}$

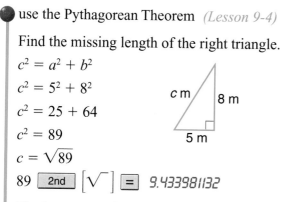

$89 \boxed{\text{2nd}} \left[\sqrt{\ }\right] \boxed{=} \;\; 9.433981132$

The hypotenuse is about 9.4 meters long.

Review Exercises

Use these exercises to review and prepare for the chapter test.

Find each square root.

10. $\sqrt{225}$ **11.** $\sqrt{6.25}$

12. $-\sqrt{\dfrac{4}{9}}$ **13.** $\sqrt{\dfrac{36}{81}}$

14. $-\sqrt{100}$ **15.** $\sqrt{3.24}$

Estimate to the nearest whole number.

16. $\sqrt{135}$ **17.** $\sqrt{50.1}$

18. $\sqrt{696}$ **19.** $\sqrt{320}$

20. $\sqrt{19.25}$ **21.** $\sqrt{230}$

Let R = real numbers, Q = rational numbers, Z = integers, W = whole numbers, and I = irrational numbers. Name the set or sets of numbers to which each real number belongs.

22. $\sqrt{32}$ **23.** -12

24. $-0.171171117\ldots$ **25.** $0.333333\ldots$

Write an equation you could use to find the length of the missing side of each right triangle. Then find the missing length. Round to the nearest tenth.

26.   **27.**

Objectives & Examples

solve problems using the Pythagorean Theorem *(Lesson 9-5)*

How tall is the tree? Let h represent the height of the tree.

$$h = a^2 + b^2$$
$$62^2 = 40^2 + h^2$$
$$3{,}844 = 1{,}600 + h^2$$
$$2{,}244 = h^2$$
$$\sqrt{2{,}244} = h$$
$$47.4 \approx h$$

The tree is about 47.4 feet tall.

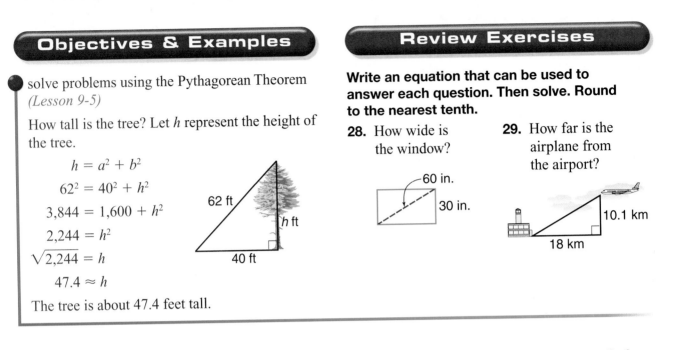

find the distance between points in the coordinate plane *(Lesson 9-6)*

Find c.

$$c^2 = a^2 + b^2$$
$$c^2 = 4^2 + 2^2$$
$$c^2 = 16 + 4$$
$$c^2 = 20$$
$$c = \sqrt{20}$$
$$c \approx 4.5$$

The distance is about 4.5 units.

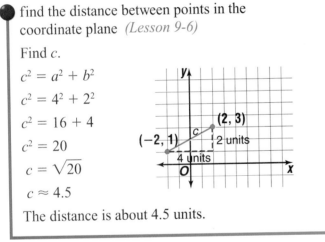

find the missing measures in 30°-60° right triangles and 45°-45° right triangles *(Lesson 9-7)*

In a 30°-60° right triangle, $a = \frac{1}{2}c$ or $c = 2a$.

In a 45°-45° right triangle, $a = b$.

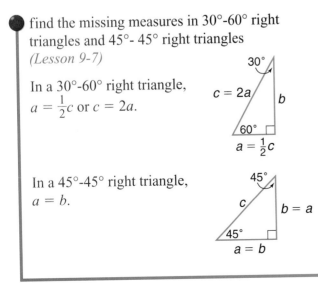

Review Exercises

Write an equation that can be used to answer each question. Then solve. Round to the nearest tenth.

28. How wide is the window?

29. How far is the airplane from the airport?

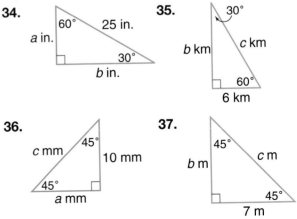

Graph each pair of ordered pairs. Then find the distance between each pair of points. Round to the nearest tenth.

30. $(3, 4), (2, 7)$

31. $(-1, 2), (4, 8)$

32. $(0, -3), (5, 5)$

33. $(-6, 2), (-4, 5)$

Find the missing lengths. Round decimal answers to the nearest tenth.

34. **35.** **36.** **37.**

Page 420, Chapter 9 Algebra: Exploring Real Numbers

Applications & Problem Solving

38. *Design* Albert is an interior designer. His client wants him to decorate a square room that has an area of 484 square feet. What is the length of each side of the room? *(Lesson 9-1)*

39. *Gardening* A diagram of Nina's garden is shown below. How long is the walkway to the nearest tenth of a foot? *(Lesson 9-4)*

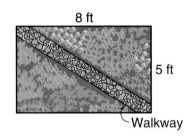

8 ft

5 ft

Walkway

40. *Draw a Diagram* Ashland Middle School's band director arranges the band so that the same number of students are in each row. Charmaine is in the third row from the front and the fourth row from the back. Her place in the row is fifth from the left and third from the right. How many students are in the band? *(Lesson 9-5A)*

41. *Construction* Meda needs to reach a window that is 8 feet above ground level. When she placed a ladder so that it reached the bottom of the window, the base of the ladder was 6 feet from the house. How long is the ladder? *(Lesson 9-5)*

Alternative Assessment

Open Ended

You have four landscape timbers. They are 13 feet long, 10 feet long, 9 feet long, and 12 feet long. You want to use three of the timbers to make a flower garden shaped like a right triangle. You must use all of the timbers, but you can cut one of the timbers in half and use one or both halves. Is it possible to build the garden? Explain why or why not.

Suppose you decide that one of the legs of the right triangle must be 9 feet long. Can you create your garden using the landscape timbers you have? Explain why or why not.

Completing the CHAPTER Project

Now that you have studied the concepts in this chapter, you should have the skills to complete your chapter project.

☑ Summarize your research of traffic accident investigations and reports. Explain why it is important to carefully measure and record the location of each item at the accident scene and pertinent landmarks.

☑ Refer to the traffic accident diagram on page 381. Write an accident report for the accident scene. Be sure to include all appropriate information, including the estimated speed of each car.

PORTFOLIO   Select one of the assignments you completed in this chapter and place it in your portfolio. Attach a note explaining why you selected it.

A practice test for Chapter 9 is provided on page 655.

Section One: Multiple Choice

There are eleven multiple-choice questions in this section. Choose the best answer. If a correct answer is *not here*, choose the letter for Not Here.

1. A red blood cell is about 0.00075 centimeter long. How is this measure expressed in scientific notation?

   A 0.75×10^{-3}

   B 7.5×10^{-4}

   C 7.5×10^{-5}

   D 75×10^{-5}

2. $\triangle RST$ is a right triangle. What is the length of $\overline{RT}$?

   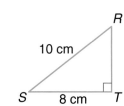

   F 4 cm

   G 6 cm

   H 8 cm

   J 10 cm

3. $-\sqrt{0.81} =$

   A -0.09

   B -0.9

   C 0.09

   D 0.9

4. What ordered pair represents the intersection of line t and line m?

   F $(2, 3)$

   G $(-2, -3)$

   H $(2, -3)$

   J $(-2, 3)$

   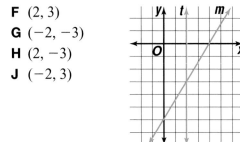

Please note that Questions 5-11 have five answer choices.

5. To the nearest inch, how long is the diagonal of a 20-inch by 16-inch window?

   A 12 in.

   B 24 in.

   C 25 in.

   D 26 in.

   E 27 in.

6. Arturo wants to wallpaper a wall that measures 10 feet by 15 feet. A window that is 3 feet by 4 feet is in the center of the wall. What is the wallpapered area?

   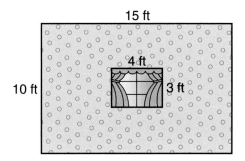

   F 110 square feet

   G 120 square feet

   H 138 square feet

   J 144 square feet

   K 150 square feet

7. A drawing measures 4.5 centimeters long and 5.5 centimeters wide. An enlargement is similar to the original and its length is 36 centimeters. What is the width of the enlargement?

   A 10 cm

   B 24 cm

   C 44 cm

   D 64 cm

   E 124 cm

8. Margi worked for a total of 33 hours in a 4-day period. How much did she earn before deductions, if she was paid $5.25 per hour?

   F $132.00

   G $135.00

   H $171.60

   J $173.25

   K Not Here

9. Felica's bedroom is 15 feet by 15 feet. If the ceiling is 8 feet high and a can of paint will cover 120 square feet, how many cans will be needed to cover only the walls?

   A 4

   B 5

   C 6

   D 7

   E 8

10. Jill runs on a 400-meter track. Last week she ran 4.5 kilometers on Monday, 3.5 kilometers on Wednesday, and 4.75 kilometers on Friday. What was the average length of Jill's runs on those 3 days?

    F 3.25 km

    G 4 km

    H 4.75 km

    J 12.75 km

    K Not Here

11. Foothill Middle School sold 2,053 rolls of wrapping paper last year. This year the school sold 1,837 rolls. The profit that the school makes on each roll of paper sold is $2.25. How much more profit did they have last year than this year?

    A $216

    B $405

    C $486

    D $4133.25

    E Not Here

Test-Taking Tip THE PRINCETON REVIEW

Most of the basic formulas that you need in order to answer the questions on a standardized test are usually given to you in the test booklet. However, it is a good idea to review the formulas before the test. For example, if your test includes the Pythagorean Theorem, familiarize yourself with the formula and its uses.

Section Two: Free Response

This section contains six questions for which you will provide short answers. Write your answers on your paper.

12. To use a calculator to divide 8 by $3\frac{2}{5}$, what decimal number should you enter for $3\frac{2}{5}$?

13. If $x + 2\frac{2}{3} = -4\frac{1}{3}$, what is the value of x?

14. What is $0.\overline{39}$ written as a fraction?

15. Which figure could contain exactly one right angle: a rectangle, an isosceles triangle, an acute triangle, or an obtuse triangle?

16. $-(8)^3 =$

17. To the nearest tenth, what is the distance between points A and B?

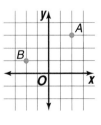

 **Test Practice** For additional test practice questions, visit:

www.glencoe.com/sec/math/mac/mathnet

MATH AT THE MALL

What is your favorite store at the mall? Maybe it's the bakery, the music store, or the photo shop. Did you know that they use math at all of these stores?

When you go shopping, you have dozens of choices. How do you decide which is the best use of your hard-earned money?

What You'll Do

In this investigation, you will calculate and compare the ratio of areas of similar polygons to help you choose which picture frame or cake to buy.

Materials   grid paper   tape measure

            unlined paper   calculator

Procedure

1. Work in groups of four. Your first stop at the mall is the photo shop. Suppose you have a photograph you took on your family vacation that you would like to enlarge and frame. The original photograph is 6 inches by 8 inches. There are frames available in the following standard sizes.

8" by 10"	9" by 12"	10" by 12"
11" by 14"	12" by 16"	14" by 18"
16" by 20"	18" by 24"	20" by 24"
22" by 28"	24" by 36"	30" by 40"

Use proportions and similar figures to determine which standard size frames are suitable for an enlargement of your photograph if you do not wish to crop any of the image.

The photo shop also offers custom-made frames in any size you specify. List three possible custom frame sizes you could use.

2. Your next stop is at the bakery to check out the prices for different sizes of cakes. Draw a diagram of each cake. How do the sizes and the areas of the cakes compare? Write the ratio of each cake's area to its price. Is one of the cakes less expensive per square inch? What other things are there to consider when choosing which cake to order?

Schmidt's Bakery

Custom cakes for every occasion. Choose chocolate, white, or marble.

	Size	Price
Small	9" x 13"	$7.99
Medium	13" x 18"	$15.99
Large	18" x 26"	$29.99

Making the Connection

Use the information about the frame and cake sizes as needed to help in these investigations.

Language Arts

Use what you learned about the area of similar polygons to decide whether different-sized enlargements at a local photo shop are priced fairly. Write a paragraph supporting your opinion.

Art

Find a small drawing or cartoon. Make an enlargement that has 9 times the area. Write a paragraph explaining how you know the ratio of the areas is 1 to 9.

Social Studies

Choose a lake on a map. Approximate the actual area of the lake by using the scale provided on the map.

Go Further

- Draw a polygon such as a rectangle. Multiply the measures of the sides of the polygon by 2, 3, 4, and 5, and draw the enlargements. Investigate how the area of the enlarged polygons compare to the area of the original polygon.

- Most copiers do not enlarge by 300%. How could you enlarge a photograph by 300%?

Technology Tips

- Use **geometry software** to draw your figures and calculate the areas.

- Use a **spreadsheet** to make a table and calculate the ratios and areas.

inter NET CONNECTION Research For current information on math at the mall, visit:
www.glencoe.com/sec/math/mac/mathnet

PORTFOLIO

You may want to place your work on this investigation in your portfolio.

Algebra: Graphing Functions

What you'll learn in Chapter 10

- to complete function tables and graph linear and quadratic functions,

- to find solutions of equations with two variables,

- to solve systems of linear equations by graphing,

- to solve problems by using a graph, and

- to graph translations, reflections, and rotations on a coordinate plane.

CHAPTER Project

GAMES PEOPLE PLAY

In this project, you will use functions to design a computer game. After word processing, playing games is the most popular use for computers. A good computer game has a storyline and characters, a goal that the player is trying to achieve, obstacles that the player must face, and rewards for overcoming the obstacles. Many computer games use sophisticated graphics and animation. Movement of everything on the screen is controlled by mathematical formulas.

Getting Started

- Research computer graphics to find out how a grid like the coordinate plane is used to position pictures on a computer screen.
- Research computer animation to find out how objects appear to move on a computer screen.
- Brainstorm a theme or story for your game.

Technology Tips

- Use an **electronic encyclopedia** to do your research.
- Use a **word processor** to write out your game plan.
- Use a **software drawing program** to sketch a screen.

inter NET CONNECTION Data Update **For up-to-date information on designing computer games, visit:**

www.glencoe.com/sec/math/mac/mathnet

Working on the Project

You can use what you'll learn in Chapter 10 to help you design your computer game.

Page	Exercise
444	21
449	22
467	18
471	Alternative Assessment

10-1 Functions

What you'll learn

You'll learn to complete function tables.

When am I ever going to use this?

You will evaluate functions when you study motion in physical science.

Word Wise

function
function table
domain
range

LOOK BACK

Refer to Lesson 1-7 for information on solving two-step equations.

In a recent movie, Wayne Szalinski's invention shrinks four adults to a height of just $\frac{3}{4}$ inch! Of course, the movie is make believe. But did you know that after a person turns 30, he or she shrinks about 0.06 centimeter per year? At that rate, how many centimeters shorter would a person be when he or she is 35, 42, 57, 61, and 83 years old? *This problem will be solved in Example 3.*

The amount of height loss depends upon a person's age. A relationship like this where one thing depends upon another is called a **function**. In a function, one or more operations are performed on one number to get another. So, the second number depends on, or is a function of, the first.

Functions are often written as equations. Suppose the variable n is used for the first number, or the *input*. The notation for the second number, or the *output*, is $f(n)$. This is read *the function of n,* or more simply *f of n.* The operations performed in the function are sometimes called the *rule.* You can organize the input, rule, and output of a function into a **function table**.

Example

1 Copy and complete the function table to find the function values of $\{-1, 0, 1, 2, 3\}$ for $f(n) = n - 3$.

Substitute each value of n, or input, into the function rule. Then simplify to find the output.

$$f(n) = n - 3$$
$$f(-1) = (-1) - 3 \text{ or } -4 \quad \textit{Replace n with } -1.$$
$$f(0) = (0) - 3 \text{ or } -3$$
$$f(1) = (1) - 3 \text{ or } -2$$
$$f(2) = (2) - 3 \text{ or } -1$$
$$f(3) = (3) - 3 \text{ or } 0$$

Input n	Rule $n - 3$	Output $f(n)$
-1		
0		
1		
2		
3		

Input n	Rule $n - 3$	Output $f(n)$
-1	-1 - 3	-4
0	0 - 3	-3
1	1 - 3	-2
2	2 - 3	-1
3	3 - 3	0

The set of input values in a function is called the **domain**. The set of output values is called the **range**. If the domain contains all values of n, then the range contains all values of $f(n)$.

2 Make a function table to find the range of $f(n) = 2n - 6$ if the domain is $\{-2, -1, 0, \frac{1}{2}, 14\}$.

First make a function table. List the domain, or input values.

Next, substitute each member of the domain into the function to find the range, or output values, of $f(n)$.

Domain n	$2n - 6$	Range $f(n)$
-2	2(-2)-6	-10
-1	2(-1)-6	-8
0	2(0)-6	-6
$\frac{1}{2}$	$2\left(\frac{1}{2}\right)-6$	-5
14	2(14)-6	22

The range is $\{-10, -8, -6, -5, 22\}$.

CONNECTION

3 **Life Science** Refer to the beginning of the lesson. If a person shrinks 0.06 centimeter per year, how many centimeters shorter would the person be when he or she is 35, 42, 57, 61, and 83 years old?

Each year, the person shrinks 0.06 cm. To find the total height loss, multiply the number of years after age 30 by 0.06.

$$f(n) = 0.06(n - 30)$$

The person would be 0.3 cm shorter at 35, 0.72 cm shorter at 42, 1.62 cm shorter at 57, 1.86 cm shorter at 61, and 3.18 cm shorter at 83 years old.

n	$0.06(n - 30)$	$f(n)$
35	0.06(35 − 30)	0.3
42	0.06(42 − 30)	0.72
57	0.06(57 − 30)	1.62
61	0.06(61 − 30)	1.86
83	0.06(83 − 30)	3.18

CHECK FOR UNDERSTANDING

Communicating Mathematics

Read and study the lesson to answer each question.

1. *State* the mathematical names for the input values and the output values.

2. *Define* function in your own words.

3. *You Decide* Raul says that $f(6)$ for $f(n) = 2n^2 - 18$ is 54. Candace says that $f(6) = 126$. Who is correct and why?

Guided Practice

4. Copy and complete the function table.

Find each function value.

5. $f(6)$ if $f(n) = 2n - 8$

6. $f(-3)$ if $f(n) = \frac{1}{2}n + 4$

n	$3n$	$f(n)$
-2		
0		
2		
$3\frac{1}{2}$		
4.3		

7. **Life Science** Veterinarians have used the rule that one year of a dog's life is equivalent to seven years of human life. This can be written as $f(n) = 7n$, where n represents the dog's age and $f(n)$ represents the equivalent human age. If Wendy's dog Goldie is 6 years old, what is Goldie's human equivalent age?

EXERCISES

Practice

Copy and complete each function table.

8.

n	$2n - 3$	$f(n)$
-4		
-1		
0		
1.5		
$4\frac{1}{2}$		

9.

n	$-5n$	$f(n)$
-4		
-2		
0		
$2\frac{1}{4}$		
5.7		

10.

n	$-0.5n + 1$	$f(n)$
-2		
0		
2.5		
8		
14.9		

Find each function value.

11. $f(4)$ if $f(n) = n - 6$

12. $f(-2)$ if $f(n) = 4n + 1$

13. $f(-8)$ if $f(n) = 3n + 24$

14. $f\left(\frac{5}{8}\right)$ if $f(n) = n + \frac{1}{6}$

15. $f(2.2)$ if $f(n) = n^2 + 1$

16. $f(-3.4)$ if $f(n) = 2n^2 - 5$

17. If $f(n) = -5n - 4$, find $f\left(\frac{4}{5}\right)$.

18. Find $f(-3.3)$ if $f(n) = 8.5 - 2n$.

Applications and Problem Solving

19. **Earth Science** Have you ever noticed that you see the lightning before you hear the thunder? You can use that time difference to estimate the distance in feet between you and a lightning strike. If n is the number of seconds between the lightning and the thunder and $f(n)$ is the number of feet between you and the lightning, then $f(n) = 1,100n$.

a. Copy and complete the function table.

Time (seconds) n	Distance (feet) $f(n)$
2	
5	
11	
18	

b. If you are two miles from a lightning strike, in how many seconds would you hear thunder? Round to the nearest tenth of a second.

20. **Sports** Each team in a bowling league has an equal chance of winning because team members are given handicaps based on their bowling averages. Enrico's handicap is 30, so the function used to determine his handicapped score is $f(n) = n + 30$, where n is the game score. Make a function table to show Enrico's handicapped scores if he bowled 153, 144, 161, 163, and 166 in the first five matches of the season.

21. **Write a Problem** that can be solved using a function.

22. **Critical Thinking** A diagonal of a polygon connects vertices that are not adjacent. A triangle has no diagonals and a rectangle has 2.

   a. Sketch polygons with 3, 4, 5, 6, 7, and 8 sides. Include the diagonals.

   b. If there are n sides in a polygon, write an expression for the number of diagonals that can be drawn from one vertex.

   c. Use $f(n)$ to write an equation for the relationship between n, the number of sides in a polygon, and $f(n)$, the number of diagonals.

For **Extra Practice,** see page 633.

Mixed Review

23. **Geometry** The length of a leg of a 45°-45° right triangle is 6.4 feet. Find the length of the hypotenuse to the nearest tenth. *(Lesson 9-7)*

24. Find $\sqrt{\frac{9}{16}}$. *(Lesson 9-1)*

25. **Standardized Test Practice** Fifteen out of the 60 eighth-graders at Seabring Junior High are on the track team. What percent of the eighth-graders are on the track team? *(Lesson 8-2)*

   A 15%   **B** 25%   **C** 45%   **D** 60%   **E** Not Here

---

Let the Games Begin

Guess My Rule

Math Skill
Functions

Get Ready This game is for two players.

Get Set Make a function table like the one at the right. Player A creates a function and secretly writes its equation on a piece of scrap paper.

n	f(n)

Go
- Player B writes a value for n in the function table. Then Player A writes the corresponding value for $f(n)$ in the table.
- Player B may guess the function rule at any time. If the guess is correct, the round ends and Player B receives 10 points. If the guess is incorrect, Player A receives 5 points and Player B may guess again or continue writing values of n.
- After Player B guesses the function, Player A receives one point for each value of n Player B wrote.
- Exchange positions for the next round.
- The winner is the player with the most points after four rounds.

 Visit www.glencoe.com/sec/math/mac/mathnet for more games.

GRAPHING CALCULATORS

10-1B Function Tables

A Follow-Up of Lesson 10-1

graphing calculator

You can use a graphing calculator to create function tables. When you enter the function and the domain values, the calculator will find the corresponding range values.

TRY THIS

Work with a partner.

Make a function table to find the range of $f(n) = 4n + 2$ if the domain is $\{-5, -3, 0, 0.5, 12\}$.

The graphing calculator uses X for domain values and Y for range values. So the function can be represented by $Y = 4X + 2$.

Step 1 Enter the function into the Y=list.
Press [Y=] to access the function list. Then enter the function by pressing 4 [X,T,θ,n] [+] 2.

Step 2 Next, set up your table. Press [2nd] [TBLSET] to display the table setup screen. Use the arrow and [ENTER] keys to highlight Indpnt: ASK and Depend: Auto. *These selections allow you to enter values of X and have the calculator find the values for Y.*

Step 3 Access the table by pressing [2nd] [TABLE]. The calculator will display an empty function table.

Step 4 Now input the domain values. Enter each value and press [ENTER]. The calculator will display each range value.

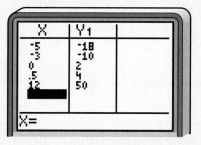

ON YOUR OWN

Use a graphing calculator to complete each function table.

1.
X	Y = 15X
−3	
−2	
0	
1.5	
4.8	

2.
X	Y = 5X − 4
−8	
−1.7	
0.3	
3	
5.4	

3.
X	Y = 3 − 2X
−14	
−3.3	
1.3	
4.5	
9.4	

4. Study the table you created for Exercise 3. If you entered a value of X that is greater than 10, would the value of Y be greater or less than −16? Explain.

10-2 Using Tables to Graph Functions

What you'll learn

You'll learn to graph functions by using function tables.

When am I ever going to use this?

You can compare prices by graphing functions.

LOOK BACK

You can refer to Lesson 2-10 for information on graphing points.

You may have heard of the *Titanic*, the "unsinkable" luxury ship that sank April 15, 1912. Robert Ballad visited the wreck of the *Titanic*. He located the shipwreck using sonar. Sonar units find distances to objects using the time it takes to reflect sound. The function used is $f(n) = 727n$, where $f(n)$ is the distance to the object in meters and n is the time in seconds.

You have made function tables to represent functions. Another way to represent a function is by graphing. When a function is graphed, the domain and range values are written as ordered pairs.

To form an ordered pair, let the *x*-coordinate be the value of *n*. The *y*-coordinate is the value of $f(n)$. The table shows ordered pairs for the function $f(n) = 2n + 1$. Each ordered pair is graphed on the coordinate grid. Notice that the *x*-axis is labeled *n* and the *y*-axis is labeled $f(n)$.

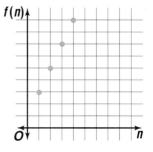

What pattern do the points suggest?

n	2n + 1	f(n)	(n, f(n))
1	2(1) + 1	3	(1, 3)
2	2(2) + 1	5	(2, 5)
3	2(3) + 1	7	(3, 7)
4	2(4) + 1	9	(4, 9)

Example ①

CONNECTION

History Refer to the beginning of the lesson. Draw a graph of the sonar function $f(n) = 727n$.

Step 1 Make a function table.

Step 2 Choose several values of *n* and find the value of $f(n)$ for each.

n	727n	f(n)	(n, f(n))
0	727(0)	0	(0, 0)
5	727(5)	3,635	(5, 3,635)
10	727(10)	7,270	(10, 7,270)
15	727(15)	10,905	(15, 10,905)

Step 3 Write the ordered pairs.

Step 4 Graph the ordered pairs. Draw the line that the points suggest.

Lesson 10-2 Using Tables to Graph Functions **433**

Not all functions have graphs that are straight lines. The graphs of many functions also lie in more than the first quadrant.

Example ❷ Graph the function $f(n) = 2n^2 + 1$.

Make a function table for different values of n.

Then graph the ordered pairs and connect the points to complete the graph. There are infinitely many values possible for n. If you graphed all of them, the points would form a smooth curve. Draw the curve suggested by these points.

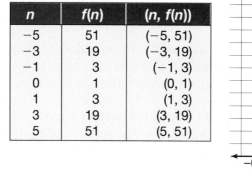

n	$f(n)$	$(n, f(n))$
-5	51	$(-5, 51)$
-3	19	$(-3, 19)$
-1	3	$(-1, 3)$
0	1	$(0, 1)$
1	3	$(1, 3)$
3	19	$(3, 19)$
5	51	$(5, 51)$

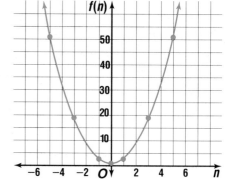

CHECK FOR UNDERSTANDING

Communicating Mathematics

Read and study the lesson to answer each question.

1. Write in your own words how you can use a function table to graph a function.

2. Tell what $f(0)$ means in terms of the sonar described in Example 1.

Guided Practice

Copy and complete each function table. Then graph the function.

3. $f(n) = n + 5$

n	$f(n)$	$(n, f(n))$
-2		
-1		
0		
1		
2		

4. $f(n) = \dfrac{4}{n}$

n	$f(n)$	$(n, f(n))$
$\frac{1}{2}$		
1		
2		
4		
16		

5. *Driver Safety* A "tailgater" risks an accident by following cars too closely. The function $f(n) = 3.48n - 20$ estimates the stopping distance $f(n)$ in feet for a car traveling n miles per hour.

a. Copy and complete the function table.

b. Graph the function.

n	$f(n)$	$(n, f(n))$
10		
20		
30		
40		
50		

Practice

Copy and complete each function table. Then graph the function.

6. $f(n) = n + 4$

n	f(n)	(n, f(n))
−3		
−1		
1		
3		
5		

7. $f(n) = -8n$

n	f(n)	(n, f(n))
−2		
0		
0.5		
1		
2		

8. $f(n) = 2n + 3$

n	f(n)	(n, f(n))
−4		
−1		
0		
2		
5		

9. $f(n) = 8 - n$

n	f(n)	(n, f(n))
−4		
−1		
0		
6		
10		

10. $f(n) = \dfrac{8}{n}$

n	f(n)	(n, f(n))
$\frac{1}{2}$		
2		
4		
8		
16		

11. $f(n) = n^2 + 3$

n	f(n)	(n, f(n))
−3		
−2		
−1		
0		
1		
2		
3		

12. Choose values for n and graph $f(n) = 0.4n + 5$.

13. Choose values for n and graph $f(n) = -(n^2)$.

Applications and Problem Solving

14. *Sports* In October 1995, Joe Dengler and his team attempted the first rafting expedition on China's Shuiluo river. The 150-mile river cuts through the mountains descending 3,000 feet. At the start of the trip, Joe estimated that the river was flowing at 1,200 cubic feet per second.

   a. Let n represent the number of seconds and $f(n)$ represent the number of cubic feet of water that flows past a point in n seconds. Choose values for n and graph $f(n) = 1,200n$.

   b. Team member Beth Rypins hiked ahead to check the rapids. How much water passed by in the 8 minutes Beth was gone?

15. *Life Science* A person's recommended weight is a function of his or her height. The function is $w = \dfrac{11(h - 40)}{2}$ for a weight of w pounds and a height of h inches.

   a. Choose values of h and graph the function using the h values as the x-coordinates and the values of w as the y-coordinates.

   b. Monifa is 5 feet 4 inches tall. According to this formula, what is Monifa's recommended weight?

16. *Critical Thinking* Water is poured at a constant rate into a 2-liter soda bottle. Draw a graph of the water level as a function of time.

For **Extra Practice**, see page 633.

Mixed Review

17. *History* Refer to the beginning of the lesson. How far beneath the surface is the *Titanic* if it takes 5.24 seconds for sound to reflect? *(Lesson 10-1)*

18. **Standardized Test Practice** What is the GCF of 72, 132, and 24? *(Lesson 6-3)*

   A 2   **B** 3   **C** 6   **D** 12

COMPUTER PROGRAMMING

Roberta Williams
COMPUTER GAME DESIGNER

Roberta Williams co-founded Sierra On-Line with her husband in their kitchen in 1979. Today, Sierra is one of the world's top computer game sellers. Ms. Williams' popular *King's Quest* adventure game series has sold over 7 million copies! As a game designer, she invents a plot and characters for a game and then works closely with a team of computer programmers and artists to create the game.

To be a computer programmer, you should have a solid background in mathematics and science and take courses in keyboarding, computer skills, and business. Although some computer programmers can get jobs with a degree from a junior or technical college, most programmers have a bachelor's degree.

For more information:
The Association for
   Computing Machinery
1515 Broadway
New York, NY 10036

inter NET CONNECTION
www.glencoe.com/sec/math/mac/mathnet

How can I get to be a successful computer game designer like Roberta Williams?

Your Turn
Interview a computer programmer about the wide variety of work programmers do. Be sure to ask them how they use mathematics in their job.

Equations with Two Variables

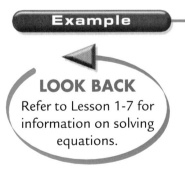

What you'll learn

You'll learn to find solutions of equations with two variables.

When am I ever going to use this?

Knowing how to solve equations with two variables can help you to find how long a car trip will take.

What are your career plans? For many people, a college degree is a big part of their future plans. One way to estimate future college costs is to multiply the current cost of the college of your choice by the inflation factor. An estimated inflation factor for 4 years is 1.34. So if the cost of a college is n dollars, the cost in four years will be $f(n) = 1.34n$.

Harvard University, Cambridge, MA

Sometimes functions do not use the $f(n)$ notation. Instead, they use two variables — one to represent the input and one to represent the output. For example, you could write the college cost function as $y = 1.34c$ if y is your cost in four years and c is the current cost. Other functions you know are written in this way.

Finding the perimeter of a square	$P = 4s$
The distance you travel at 65 miles per hour	$d = 65t$
Changing Celsius to Fahrenheit	$F = \frac{9}{5}C + 32$

Values for the two variables that make the equation true form an ordered pair that is a solution of the equation. Most equations with two variables have an infinite number of solutions.

One variable in an equation with two variables represents the input. The other variable represents the output of the function. When x and y are used, x usually represents the input.

Example

LOOK BACK

Refer to Lesson 1-7 for information on solving equations.

1 Find four solutions of $2x + 3 = y$. Write the solutions as a set of ordered pairs.

First, make a function table to find the ordered pairs. Use x for the input and y for the output.

Choose any four values for x. Then complete the table.

x	$2x + 3$	y	(x, y)
-3	$2(-3) + 3$	-3	$(-3, -3)$
0	$2(0) + 3$	3	$(0, 3)$
1	$2(1) + 3$	5	$(1, 5)$
3	$2(3) + 3$	9	$(3, 9)$

Four solutions of the equation $2x + 3 = y$ are $\{(-3, -3), (0, 3), (1, 5), (3, 9)\}$.

In Example 1, we found four ordered pairs that are solutions of $2x + 3 = y$.
There are many more solutions. *Can you name another?*

Example

CONNECTION

2

Earth Science Crickets
and katydids can both tell
you the temperature in degrees
Fahrenheit. If you are listening
to a katydid, subtract 19 from
the number of chirps in a
minute. Then divide the result
by 3 and add 60. This is a good
estimate of the temperature.
Determine which ordered pairs in the
set {(28, 63), (37, 69), (40, 69), (52, 71)}
represent solutions of this function.

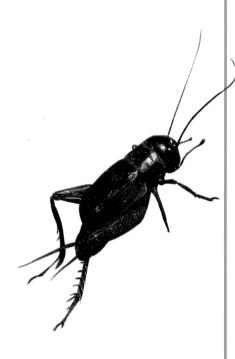

Did?you know The hot
humid days of summer,
when the crickets and
katydids are chirping, are
called the "dog days"
because Sirius, the dog
star, is visible then.

First, translate the verbal sentence into an equation.
Let x = the number of chirps and y = the temperature.

temperature	equals	19 less than the number of chirps divided by 3	plus	60
y	$=$	$\dfrac{x - 19}{3}$	$+$	60

Try each ordered pair in the equation.

Try (28, 63).

$y = \dfrac{x - 19}{3} + 60$

$63 \stackrel{?}{=} \dfrac{28 - 19}{3} + 60$

$63 \stackrel{?}{=} \dfrac{9}{3} + 60$

$63 \stackrel{?}{=} 3 + 60$

$63 = 63$ ✓

(28, 63) is a solution.

Try (37, 69).

$69 = \dfrac{x - 19}{3} + 60$

$69 \stackrel{?}{=} \dfrac{37 - 19}{3} + 60$

$69 \stackrel{?}{=} \dfrac{18}{3} + 60$

$69 \stackrel{?}{=} 6 + 60$

$69 \neq 66$

(37, 69) is not a solution.

Try (40, 69).

$y = \dfrac{x - 19}{3} + 60$

$69 \stackrel{?}{=} \dfrac{40 - 19}{3} + 60$

$69 \stackrel{?}{=} \dfrac{21}{3} + 60$

$69 \stackrel{?}{=} 7 + 60$

$69 \neq 67$

(40, 69) is not a solution.

Try (52, 71).

$y = \dfrac{x - 19}{3} + 60$

$71 \stackrel{?}{=} \dfrac{52 - 19}{3} + 60$

$71 \stackrel{?}{=} \dfrac{33}{3} + 60$

$71 \stackrel{?}{=} 11 + 60$

$71 = 71$ ✓

(52, 71) is a solution.

The ordered pairs (28, 63) and
(52, 71) are solutions. The
ordered pair (28, 63)
represents 28 chirps and a
temperature of 63° F.
*Check by graphing the function
and the points.*

Communicating Mathematics

Read and study the lesson to answer each question.

1. *Find* four ordered pairs that are solutions of $2x + 3 = y$ other than those found in Example 1.

2. *Show* how you could use the method in Example 2 to check the solutions in Example 1.

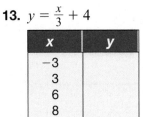

3. *Write* two examples of equations with two variables that you have used in your math or science classes.

Guided Practice

Copy and complete the table for each equation.

4. $y = 4x - 1$

x	y
−1	
0	
2	
4	

5. $0.5x = y$

x	y
−4	
0	
5	
16	

6. Find four solutions of $y = \frac{1}{4}x + 5$.

7. *Geometry* The equation $s = 180(n - 2)$ relates the sum of the measures of the angles s formed by the sides of a polygon to the number of sides n. Find four ordered pairs *(n, s)* that are solutions of the equation.

EXERCISES

Practice

Copy and complete the table for each equation.

8. $y = x + 1$

x	y
−3	
−1	
1	
2	

9. $y = -1.5x$

x	y
−6	
−1	
3	
16	

10. $y = 5x - 2$

x	y
$-\frac{4}{5}$	
2	
$\frac{7}{15}$	

11. $y = 7 + 0.2x$

x	y
−5	
0	
5	
10	

12. $1 - 5x = y$

x	y
0	
−2	
4	
10	

13. $y = \frac{x}{3} + 4$

x	y
−3	
3	
6	
8	

Find four solutions of each equation.

14. $y = 4x + 3$

15. $y = -x + 1$

16. $y = -2x + 3$

17. List four solutions of $y = 15x - 57$.

18. *Finances* Refer to the beginning of the lesson. In the regional science fair, Latanya won a scholarship that will pay $1,000 per year for college.

  a. Write an equation with two variables for Latanya's college costs in four years.

  b. Use the current costs listed in the table to find Latanya's annual cost for each of the schools she is considering.

College	Current Annual Cost
Texas A & M	$11,192
Purdue University	$15,156
The University of Virginia	$18,441

19. *Physical Science* Chemists often use the Kelvin scale to measure temperature. The equation $1.8K = F + 459.67$ relates degrees Fahrenheit F to kelvins K. Write three ordered pairs (F, K) that represent equivalent Fahrenheit and Kelvin temperatures.

20. *Critical Thinking* The equations $y = 4x - 2$ and $8x - 2y = 4$ have the same solution set.

  a. Find four solutions of the first equation. Then verify that the ordered pairs are solutions of the second equation.

  b. For which equation do you think it is easier to find solutions? Explain.

21. *Algebra* Choose values for n and graph $f(n) = \frac{n}{2} - 1$. *(Lesson 10-2)*

22. Standardized Test Practice Which of the following does not describe the figure? *(Lesson 5-3)*

  A parallelogram   **B** square

  C trapezoid   **D** rhombus

For **Extra Practice,**
see page 634.

23. *Music* The number of pages in a magazine in the last nine issues is 196, 188, 184, 200, 168, 176, 192, 160, and 180. Find the median, upper quartile, and lower quartile of this data. *(Lesson 4-5)*

MATH IN THE MEDIA

1. If the small candles are 1 year, how old is Herman?

2. Let x represent the number of large candles and y represent the number of small candles. Write an equation for the numbers of candles that can be used to show an age of 60 years.

3. Use your equation to find all of the different ways to show an age of 60 years.

HERMAN®

2-5   © 1991 Jim Unger

"The big ones are 15 years."

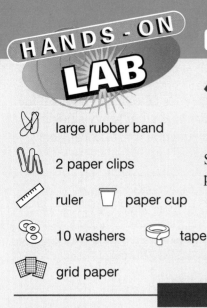
10-4A Graphing Relationships

A Preview of Lesson 10-4

large rubber band

2 paper clips

ruler   paper cup

10 washers   tape

grid paper

Scientists and engineers often graph relationships to look for patterns in the points they graph.

TRY THIS

Work in groups of four.

Step 1   Use a pencil to punch a small hole in the bottom of the cup. Place one paper clip onto the rubber band. Push the other end of the rubber band through the hole in the cup. Attach the second paper clip to the other end of the rubber band to keep it from coming back out of the hole.

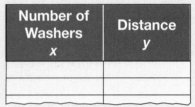

Step 2   Copy the table at the right.

Step 3   Tape the top paper clip to the edge of a desk. Measure and record the distance from the bottom of the desk to the bottom of the cup. Drop one washer into the cup. Measure and record the new distance.

Number of Washers x	Distance y

Step 4   Keep dropping washers in the cup and recording distances until you have ten sets of distances.

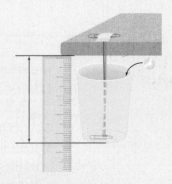

ON YOUR OWN

1. Refer to your table of recorded data. Let *x* represent the number of washers and *y* the distance from the paper clip to the bottom of the cup. Do you think that the distance is a function of the number of washers? Explain.

2. On a coordinate grid, graph the ordered pairs formed by your data. What pattern do the points suggest?

3. ***Look Ahead***   Combine your data with the rest of the class. Graph the ordered pairs. A line that is drawn to represent the data closely is called a *line of best fit*. Draw a line of best fit for the data.

Graphing Linear Functions

What you'll learn

You'll learn to graph linear functions by plotting points.

When am I ever going to use this?

Accountants graph linear functions to show costs of manufactured items.

Word Wise

linear function

Experts estimate that smoking a pack of cigarettes takes a half hour off the average person's life. That may not sound like much, but heavy smokers can take years off their lives. A graph of the equation $y = 0.5x$ can show how much time is lost to smoking.

HANDS-ON MINI-LAB

Work with a partner.   grid paper   ruler

Try This

• Make a function table to find ten solutions of the equation $y = 0.5x$.

• The solutions include the ordered pairs (0, 0), (1, 0.5), (2, 1), and (4, 2). *(4, 2) means 4 packs of cigarettes take 2 hours off a person's life.*

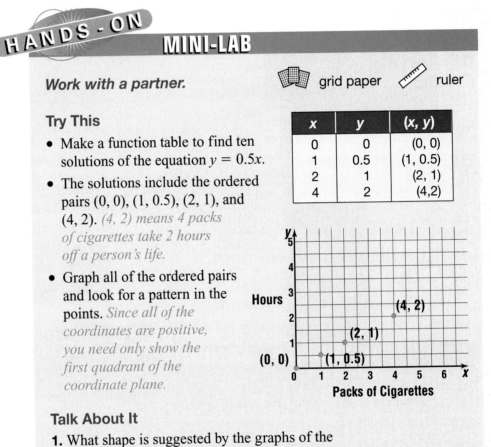

x	y	(x, y)
0	0	(0, 0)
1	0.5	(1, 0.5)
2	1	(2, 1)
4	2	(4,2)

• Graph all of the ordered pairs and look for a pattern in the points. *Since all of the coordinates are positive, you need only show the first quadrant of the coordinate plane.*

Talk About It

1. What shape is suggested by the graphs of the ordered pairs?

2. Sketch the figure.

3. Find four more solutions of the equation. Do the graphs of these ordered pairs lie on the figure you sketched?

4. What conclusion can you make about other solutions of $y = 0.5x$?

In the Mini-Lab, you saw that all solutions of the equation suggested a straight line. A function in which the graphs of the solutions form a line is called a **linear function**.

The graphs of linear functions that interpret real-life activities often lie in the first quadrant. However, most linear functions lie in at least two quadrants of the coordinate plane.

Examples

1 Graph $y = \frac{x}{2} + 1$.

Begin by making a function table. List at least three values for x.

Graph each ordered pair. Connect the points with a line. Add arrows to the ends of the line to show that the line continues indefinitely.

x	y	(x, y)
−4	−1	(−4, −1)
0	1	(0, 1)
4	3	(4, 3)

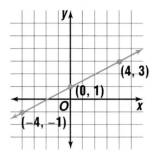

Study Hint

Estimation Estimate how long the axes should be before you graph a function. The graph may be more manageable if each unit on the axis represents 2, 3, 5, or some other number greater or less than 1.

 **APPLICATION**

2 **Sports** In 1989, Mark Wellman became the first paraplegic to climb Yosemite's 3,595-foot *El Capitan.* As he and his partner climbed, they had to prepare for lower temperatures at the top. If the temperature at sea level is 78° F, the function $t = 78 - 5.4h$ describes the temperature t at a height of h thousand feet above sea level. Graph the temperature function.

Make a function table. Then graph the points and complete the line.

h	t	(h, t)
0	78.0	(0, 78.0)
2	67.2	(2, 67.2)
5	51.0	(5, 51.0)

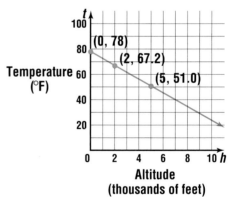

Mark Wellman

CHECK FOR UNDERSTANDING

Communicating Mathematics

Read and study the lesson to answer each question.

1. *Explain* why you should use at least three ordered pairs to graph a linear function.

2. *Tell* why only points in the first quadrant make sense for the graph in the Mini-Lab.

HANDS-ON MATH

3. *Extend* the graph in the Mini-Lab to find the amount of time taken off the life of a person who smokes ten packs of cigarettes.

Graph each function.

    4. $y = 3x$ **5.** $y = 10 - x$ **6.** $y = 6x + 4$

    7. *History* American Indians used wampum beads for jewelry and money when English settlers came to New England. Wampum beads were made of polished clam shell. Three black beads could be traded for one English penny. Graph the function $p = 3w$.

EXERCISES

Practice **Graph each function.**

    8. $y = 5x$ **9.** $y = 25 - x$ **10.** $y = 3x + 1$

    11. $y = 15 - 3x$ **12.** $y = -4x$ **13.** $y = 1.5x + 2.5$

    14. $y = \frac{1}{3}x + 5$ **15.** $y = 10 - \frac{x}{4}$ **16.** $y = 60 - 5x$

    17. Draw a graph of $y = 75 - 3x$.

    18. Graph the linear function $y = -\frac{1}{2}x + 6$.

Applications and Problem Solving

    19. *Food* Maple syrup is made by boiling the sap of maple trees. The function $y = 40x$ describes the relationship between the number of gallons of sap y, used to make x gallons of maple syrup. Graph the function.

    20. *Transportation* Because of their tolerance of hot weather, camels have been used for transportation in Africa and in the Middle East for centuries. A thirsty camel can drink up to 25 gallons of water in 10 minutes.

        a. Suppose a camel is drinking from a 25-gallon tank. Let x be the number of minutes he has been drinking and y be the number of gallons of water remaining in the tank. Write two ordered pairs that relate the gallons of water in the tank y to the number of minutes x.

        b. Graph the linear function that contains these two points.

        c. How could you use the graph to determine the number of gallons in the tank at any time?

    21. *Working on the* **CHAPTER Project** A computer game includes an animated insect that moves across the screen. The insect is at (x, y). Graph the insect's path as it moves along $y = 2x + 5$.

    22. *Critical Thinking*

        a. Graph $y = 2x$ and $y = -\frac{1}{2}x$ on the same coordinate plane.

        b. At what point do the graphs intersect?

        c. Describe how these lines are related.

For **Extra Practice**, see page 634.

Mixed Review

    23. *Algebra* Find four solutions of $y = 3x + 10$. *(Lesson 10-3)*

    24. *Patterns* Name the eighth term in the sequence 81, 27, 9, ... *(Lesson 7-5)*

    25. **Standardized Test Practice** There are 5,280 feet in one mile. Jun-Ko and Lorie rode their bicycles 3 miles. How many feet is this? *(Lesson 2-7)*

        A 1,760 ft   **B** 5,280 ft   **C** 10,560 ft   **D** 15,840 ft

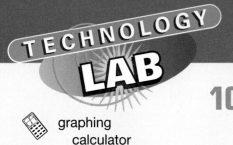
GRAPHING CALCULATORS

10-4B Linear Functions

A Follow-Up of Lesson 10-4

graphing
calculator

A graphing calculator can be a valuable tool to study functions. You need to set an appropriate range before you can create a graph. For example, a viewing window of $[-10, 10]$ by $[-10, 10]$ with a scale factor of 1 on both axes means that the graph will show x-coordinates from -10 to 10 and y-coordinates from -10 to 10. The scale factor of 1 indicates that the tick marks on both axes are one unit apart. This is called the *standard viewing window.* *You can choose a viewing window other than the standard by pressing* WINDOW *and entering your values.*

TRY THIS

Work with a partner.

Graph $y = 3x + 1$ in the standard viewing window.

Step 1 Choose the standard viewing window by pressing ZOOM 6.

Step 2 Enter and graph the equation. Press Y= to access the function list. Then enter the equation by pressing 3 X,T,θ,n + 1. Complete the graph by pressing GRAPH.

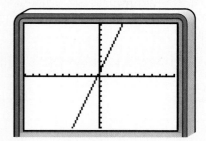

You will need to clear the graphics screen before you can graph another function. To do this, press Y= and clear any equations in the list.

ON YOUR OWN

Graph each linear function on a graphing calculator. Then sketch the graph on a piece of paper.

1. $y = x - 4$

2. $y = 4 - x$

3. $y = 5$

4. $y = 3x$

5. $y = 3x + 5$

6. $y = -3x$

7. Describe the graphs of $y = x - 4$ and $y = 4 - x$.

8. What will the graph of $y = c$ look like if c is any number?

Working students may find their mailboxes stuffed with offers for credit cards. But how do you know which one is the best choice? One way is by comparing your spending habits with each offer.

Ana Torres is comparing two credit card offers. The Card A offer has no annual fee and an annual interest rate of 18%. Card B has a $25 annual fee and an annual interest rate of 16%. Which offer is better for Ana? *This problem will be solved in Example 2.*

A set of two or more equations is called a **system of equations**. When you find an ordered pair that is a solution of all of the equations in the system, you have solved the system. The ordered pair for the point where the graphs of the equations intersect is the solution of the system of equations.

Example

① Graph the system of equations $y = 2x - 5$ and $y = 1 - x$. Then find the solution of the system.

Step 1 Make a function table for each equation.

$y = 2x - 5$	
x	y
0	−5
1	−3
3	1

$y = 1 - x$	
x	y
−1	2
1	0
4	−3

Step 2 Graph each equation on one coordinate grid. Then locate the point where the graphs intersect.

Step 3 The lines intersect at the point whose coordinates are $(2, -1)$. So, the solution of the system of equations is $(2, -1)$.

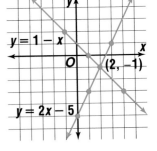

Step 4 Check the solution in *both* equations.

$y = 2x - 5$
$-1 \stackrel{?}{=} 2(2) - 5$
$-1 \stackrel{?}{=} 4 - 5$
$-1 = -1$ ✓

Substitute 2 for x and −1 for y.

$y = 1 - x$
$-1 \stackrel{?}{=} 1 - 2$
$-1 = -1$ ✓

The solution checks in both equations.

2 **Money Matters** Refer to the beginning of the lesson. Which card is the better choice for Ana?

Let x represent the balance and y represent the annual cost. The total paid for Card A can be represented by $y = 0.18x$. For Card B, the cost is represented by $y = 25 + 0.16x$.

Graph both equations on the same coordinate plane. You can use a calculator to find the y-values for each value of x quickly.

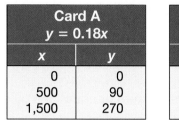

Card A $y = 0.18x$		Card B $y = 25 + 0.16x$	
x	y	x	y
0	0	0	25
500	90	500	105
1,500	270	1,500	265

Graph each function. The graphs intersect at (1,250, 225).

If Ana carries a balance of less than $1,250, Card A is the better choice. If her balance is more than $1,250, Ana should choose Card B.

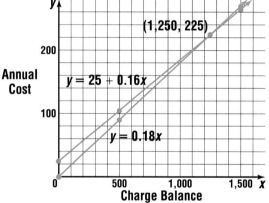

Not all systems of equations can be solved by graphing.

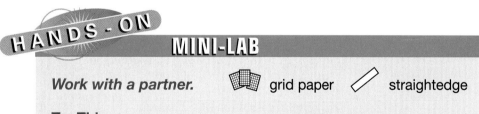

HANDS-ON   MINI-LAB

Work with a partner.   grid paper   straightedge

Try This

Graph the system of equations $y = x + 2$ and $y = x - 5$.

- Make a function table for each equation.
- Graph the ordered pairs and draw the lines.

Talk About It

1. What type of lines are these?
2. Where do the lines meet?
3. What conclusion can you make about the solution of this system of equations?

Communicating Mathematics

Read and study the lesson to answer each question.

1. *Write* in your own words what is meant by a system of equations.

2. *Identify* the solution of the system of equations graphed at the right.

HANDS-ON MATH

3. *Sketch* a graph for a system of equations that has no solution.

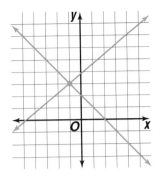

Guided Practice

Solve each system of equations by graphing.

4. $y = 3 + x$
   $y = 2x + 5$

5. $y = x$
   $y = 4 - x$

6. $y = -2x - 6$
   $y = -2x - 3$

7. *Economics* In March 1997, American exports were $71.4 billion, down $1.1 billion from February. Imports were $80.5 billion, up $1.2 billion from February. If the trend continues, exports for a month could be represented by $y = 71.4 - 1.1x$ and imports by $y = 80.5 + 1.2x$ where x is the number of months since March. Would there be a time after March when exports and imports are equal? If so, when?

EXERCISES

Practice

Solve each system of equations by graphing.

8. $y = 2x + 1$
   $y = 7 - x$

9. $y = 4 - 2x$
   $y = 2x$

10. $y = x - 4$
    $y = -4x + 16$

11. $y = 3x + 1$
    $y = 3x - 1$

12. $x + y = 7$
    $3x - y = 5$

13. $y = 4x - 15$
    $y = x + 3$

14. $y = -x - 1$
    $y = -2x + 4$

15. $y = 4x - 13$
    $y = -\frac{1}{2}x + 5$

16. $2x + y = 5$
    $x - y = 1$

17. Graph the system $y = 4x - 1$ and $x + 2y = 16$. Find the solution of the system.

18. Find the solution of the system $y = 3.4x - 5$ and $11x - 5y = -5$.

19. *Write a Problem* involving a system of equations whose solution is $(2, 8)$.

Applications and Problem Solving

20. *Business* Part of making a business plan is finding the *break-even point*. The break-even point is the point where the cost of doing business and the revenue is the same. The break-even point is the intersection of the functions that describe the cost and the revenue. Suppose the cost function for a company is $y = 1,365 + 2.5x$ and the revenue function is $y = 6x$.

    a. What is the break-even point?

    b. If profit is the revenue minus the cost, what is the profit at the break-even point?

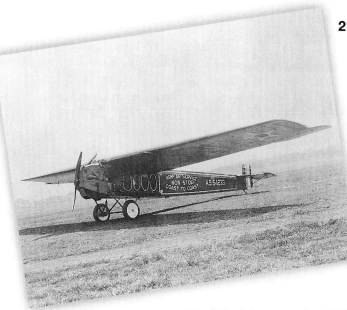

21. **Aviation** The first nonstop coast to coast flight across the U.S. was made by a Fokker T-2 in 1923. It took about 27 hours. In 1987, the same flight took just over 4 hours in a Boeing 747. The Fokker averaged 93 miles per hour, and the Boeing plane averaged about 600 miles per hour. Suppose the flights had been made on the same day with a 5-hour lead for the Fokker.

    a. Let y represent miles and x represent hours. The Fokker could be represented by $y = 93x$. The Boeing could be represented by $y = 600(x - 5)$. Graph the system of equations.

    b. When would the Boeing pass the Fokker? Use the graph to justify your answer.

22. **Working on the CHAPTER Project** In a computer game, if you cross the path of the monkey, it will steal the bananas you have collected. The monkey is moving along $y = -3x + 23$. If the player is traveling along $y = 2x + 3$, will the monkey steal the bananas? If so, where?

23. **Critical Thinking** Graph $5x + y = 4$ and $10x + 2y = 8$ on the same grid.

    a. Describe the graphs.

    b. Does this system of equations have a solution? If so, what is the solution?

Mixed Review

24. **Standardized Test Practice** Which is the graph of $y = -3x$? *(Lesson 10-4)*

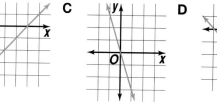

A   B   C   D

25. Compute $\frac{5}{6} \cdot 3\frac{3}{7}$ mentally. *(Lesson 7-3)*

For **Extra Practice**, see page 634.

26. **Geography** There are 5,828,987 people who live in El Salvador. The area of the country is 8,124 square miles. Find the average number of people per square mile in El Salvador. *(Lesson 3-1)*

CHAPTER 10

Mid-Chapter Self Test

1. Copy and complete the function table for $f(n) = 3n + 5$. *(Lesson 10-1)*

n	3n + 5	f(n)
−2		
−1		
0		
3		
5		

2. Use the table you made in Exercise 1 to graph $f(n) = 3n + 5$. *(Lesson 10-2)*

3. Find four solutions of $y = 2.5x + 1$. *(Lesson 10-3)*

4. Graph $y = 6x + 5$. *(Lesson 10-4)*

5. Find the solution of the system $y = 14x - 23$ and $4x + 2y = 18$. *(Lesson 10-5)*

10-6A Use a Graph

A Preview of Lesson 10-6

Wesley

Wesley and Kate are lab partners in their earth science class. Their assignment is to write a paper about the effects of cutting down the rainforests. Let's listen in!

This article I found says that the rain forests are being cleared for lumber production, cattle ranches, farms, dams, and highways.

Kate

We should have something about each of those topics in our paper. Mr. Hatano likes it when we back up what we say with numbers or graphs. Let's start with the lumber.

Okay. There's a graph in here that shows the lumber production for different parts of the world.

Let's put the graph in the paper and point out how much the annual lumber production has grown.

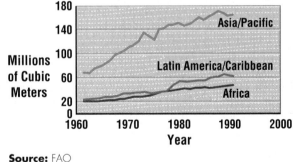

Lumber Production in the Tropics

Millions of Cubic Meters

Asia/Pacific
Latin America/Caribbean
Africa

Source: FAO

The graph shows that in 1962 it was about 70 million cubic meters. In 1991, it was up to about 160 million cubic meters.

So in 29 years, the production increased 90 million cubic meters. If we write that as a percent, it's 90 divided by 70 or a 129% increase!

THINK ABOUT IT

Work with a partner.

1. *Tell* why the graphs of lumber production are not straight lines.

2. *Write* a sentence about the change of lumber production in Africa from 1962 to 1991.

3. *Find* a graph in a newspaper, a magazine, or on the Internet. Write a sentence explaining the information contained in the graph.

For **Extra Practice,** see page 635.

ON YOUR OWN

4. The first step of the 4-step plan for problem solving is to *explore* the problem. **Describe** two things you could observe about lumber production by examining the graph.

5. Look Ahead Refer to the graph in Example 1 on page 452. Which grows faster as you move along the graph, the *x* values or the *y* values?

6. *Write a Problem* that can be answered using the graph on teen credit.

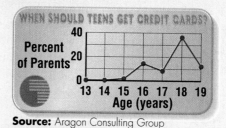

Source: Aragon Consulting Group

MIXED PROBLEM SOLVING

STRATEGIES

Look for a pattern.
Solve a simpler problem.
Act it out.
Guess and check.
Draw a diagram.
Make a chart.
Work backward.

Solve. Use any strategy.

7. *Money Matters* A 1997 auction of items owned by members of the *Beatles* was held in Tokyo, Japan. Paul McCartney's birth certificate was sold for $14,613 and sold again for $73,064. How much more did the second purchaser pay than the first?

8. *Earth Science* Graciela and Lisa's science paper is on chlorofluorocarbon (CFC) production. They found the graph below.

Chlorofluorocarbon Production

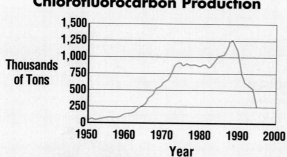

Source: Du Pont, Worldwatch estimates

a. About how much greater was the CFC production in 1995 than in 1950?

b. How much was the CFC production reduced from 1988 to 1995?

9. *Patterns* A number is doubled and then −19 is added to it. If the result is 11, what was the original number?

10. *Health* The rate of death among liver cancer patients has dropped gradually. Choose the graph that shows a gradual decrease over time.

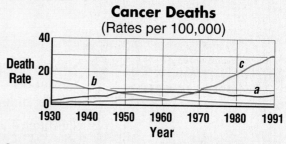

Source: National Center for Health Statistics

11. *Fashion* Eric, Julian, and Louisa each chose different stones for their class rings. Their stones are black onyx, blue sapphire, and purple amethyst. Julian's stone is not amethyst. Eric wished he had chosen onyx. Louisa's stone is blue. What stone does each student have?

12. *History* The Lighthouse of Pharos was one of the Seven Wonders of the World. It was 500 feet tall. The tallest structure in the United States is a television tower in Blanchard, North Dakota, which is 2,063 feet tall. About how many times taller is the television tower than the lighthouse?

13. Standardized Test Practice Approximately 36,460,000 people attended college football games recently. What is this number rounded to the nearest million?

A 36,000,000   **B** 1,000,000

C 40,000,000   **D** 36,500,000

What you'll learn

You'll learn to graph quadratic functions.

When am I ever going to use this?

You will use quadratic functions when you study gravity in physical science.

Word Wise

quadratic function

Do you dream of hitting the track in a Hot Rod? You may not have to wait. The National Hot Rod Association sponsors Junior Drag Racing Championships for up and coming drivers. Cars half the size of the pros' compete on a one-eighth mile track.

In 1997, the *Texas Raceway* team took the championship. The distance their car traveled can be described by the function $d(t) = \frac{1}{2}at^2$, where a is the rate of acceleration and t is time. This is a **quadratic function**. In a quadratic function, the greatest power is 2.

If you knew the acceleration and wanted to know how far the team's car had traveled at each second, you could evaluate $d(t)$ for several values of t. A better way is to graph the function.

Example APPLICATION

1 **Sports** Suppose the *Texas Raceway* car accelerated at a rate of 4.5 feet per second per second. Graph $d(t) = \frac{1}{2}at^2$ to find the number of feet that the car traveled after 6 seconds.

Substitute 4.5 for a in the function.

$$d(t) = \frac{1}{2}at^2$$

$$d(t) = \frac{1}{2}(4.5)t^2$$

$$d(t) = 2.25t^2$$

Now make a function table.

t	$d(t)$	$(t, d(t))$
0	$2.25(0)^2 = 0$	(0, 0)
2	$2.25(2)^2 = 9$	(2, 9)
4	$2.25(4)^2 = 36$	(4, 36)
6	$2.25(6)^2 = 81$	(6, 81)

Now graph the ordered pairs. Notice that the points suggest a curve. Sketch a smooth curve to connect the points. At that rate of acceleration, the car will have traveled 81 feet after 6 seconds.

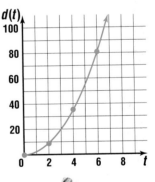

Often the graphs of quadratic functions that represent real-life situations are only in the first quadrant because negative numbers are unreasonable. But the graph of a quadratic function or equation can be in any quadrant.

Example

2 **Graph $y = -3x^2 + 6$.**

Make a table and graph the ordered pairs.

x	y	(x, y)
-2	$-3(-2)^2 + 6 = -6$	$(-2, -6)$
-1.5	$-3(-1.5)^2 + 6 = -0.75$	$(-1.5, -0.75)$
-1	$-3(-1)^2 + 6 = 3$	$(-1, 3)$
0	$-3(0)^2 + 6 = 6$	$(0, 6)$
1	$-3(1)^2 + 6 = 3$	$(1, 3)$
1.5	$-3(1.5)^2 + 6 = -0.75$	$(1.5, -0.75)$
2	$-3(2)^2 + 6 = -6$	$(2, -6)$

The graphs of the ordered pairs suggest a downward curve. Connect the points with a smooth curve.

A graph with this shape is called a parabola. A parabola can also curve up, to the left, or to the right.

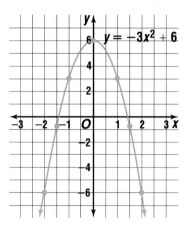

CHECK FOR UNDERSTANDING

Communicating Mathematics

Read and study the lesson to answer each question.

1. *State* the highest power in a quadratic function.

2. *Explain* why the graphs of some real-life functions such as in Example 1 use only the first quadrant of the graph.

3. *You Decide* Nicole says that $y = \dfrac{x^2}{3}$ is a quadratic function. Eduardo disagrees. Who is correct and why?

Guided Practice

4. Copy and complete the function table for $y = x^2 - 1$. Then graph the function.

x	y	(x, y)
-2		
-1		
0		
1		
2		

Graph each quadratic function.

5. $y = x^2$ 6. $f(x) = -2x^2$ 7. $y = 2x^2 + 1$

8. **Physical Science** If you have stood on a beach and watched a boat become tiny as it approaches the horizon, you may have wondered how far you could see. Standing on the shore, you can see d miles if your eyes are $f(d)$ feet above the ground. This is described by the function $f(d) = \frac{2}{3}d^2$.

    a. Graph the function $f(d) = \frac{2}{3}d^2$.

    b. Find how many feet above the ground your eyes would need to be in order to see a friend waving from a small boat 4 miles from shore.

EXERCISES

Practice

Copy and complete each function table. Then graph the function.

9. $y = -x^2$

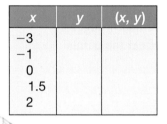

x	y	(x, y)
−3		
−1		
0		
1.5		
2		

10. $f(x) = -1.5x^2$

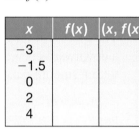

x	f(x)	(x, f(x))
−3		
−1.5		
0		
2		
4		

11. $f(n) = 2n^2 - 4$

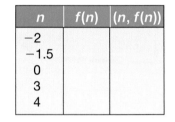

n	f(n)	(n, f(n))
−2		
−1.5		
0		
3		
4		

Graph each quadratic function.

12. $y = 3x^2$

13. $f(x) = -2x^2 + 10$

14. $y = 5x^2 + 1$

15. $f(n) = 1.5n^2 - 1$

16. $f(x) = \frac{1}{2}x^2 + 2$

17. $y = -1.5x^2 - 1$

18. $y = -2x^2 + 6$

19. $y = 10 - x^2$

20. $f(n) = n^2 + n$

21. Consider the function $y = 3x^2 - 4$. Which ordered pairs from the set $\{(-2, 8), (-1, -7), (0, -4), (1, -1), (2, -8)\}$ are solutions?

Applications and Problem Solving

22. **Fireworks** Is going to see fireworks one of your family's favorite Independence Day activities? You can estimate the height of a fireworks rocket using a quadratic function. If the rocket is fired from the ground with an initial velocity of 40 meters per second, then the height of the rocket after t seconds is $h = 40t - 4.9t^2$.

    a. Graph the height function.

    b. Estimate the height of the rocket after 3.2 seconds using the graph and the equation. Compare your graph estimate to your equation estimate.

    c. What is the maximum height the rocket reaches?

23. **Aviation** The formula $a = 0.66d^2$ represents the number of miles d that can be seen when flying at a height of a feet.

    a. Graph the function.

    b. In 1988, the *Nashua Number One* set a record for the most people on an untethered balloon flight. If the balloon was 328 feet above the ground, use your graph to estimate how far the passengers and crew could see.

24. **Critical Thinking** The length of a rectangle is twice its width. If the width is x, write the equation of the area of the rectangle. Then graph the area as a function of x.

25. Critical Thinking Equations like $y = x^3$ and $y = 2x^3$ are examples of *cubic equations.* Since the graphs of quadratic and cubic equations are not straight lines, they are called *nonlinear equations.* Graph each set of equations on the same coordinate plane.

  a. Quadratic Equations: How do the graphs of $y = x^2 + 2$ and $y = x^2 + (-2)$ differ from the graph of $y = x^2$? Write a statement comparing the graph of $y = x^2 + c$ to the graph of $y = x^2$ depending on the value of c.

  b. Cubic Equations: Repeat part a for $y = x^3$, $y = x^3 + 2$, and $y = x^3 + (-2)$.

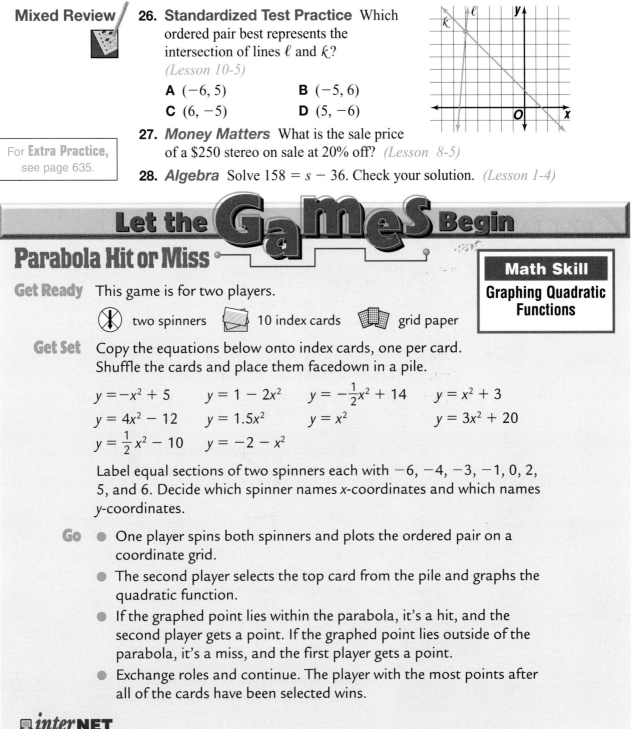

Mixed Review

26. Standardized Test Practice Which ordered pair best represents the intersection of lines ℓ and k? *(Lesson 10-5)*

  A $(-6, 5)$ **B** $(-5, 6)$

  C $(6, -5)$ **D** $(5, -6)$

27. Money Matters What is the sale price of a $250 stereo on sale at 20% off? *(Lesson 8-5)*

For **Extra Practice,** see page 635.

28. Algebra Solve $158 = s - 36$. Check your solution. *(Lesson 1-4)*

Let the GameS Begin

Parabola Hit or Miss

Math Skill

Graphing Quadratic Functions

Get Ready This game is for two players.

⊗ two spinners   📇 10 index cards   🗂 grid paper

Get Set Copy the equations below onto index cards, one per card. Shuffle the cards and place them facedown in a pile.

$y = -x^2 + 5$ $y = 1 - 2x^2$ $y = -\frac{1}{2}x^2 + 14$ $y = x^2 + 3$

$y = 4x^2 - 12$ $y = 1.5x^2$ $y = x^2$ $y = 3x^2 + 20$

$y = \frac{1}{2}x^2 - 10$ $y = -2 - x^2$

Label equal sections of two spinners each with -6, -4, -3, -1, 0, 2, 5, and 6. Decide which spinner names x-coordinates and which names y-coordinates.

Go ● One player spins both spinners and plots the ordered pair on a coordinate grid.

  ● The second player selects the top card from the pile and graphs the quadratic function.

  ● If the graphed point lies within the parabola, it's a hit, and the second player gets a point. If the graphed point lies outside of the parabola, it's a miss, and the first player gets a point.

  ● Exchange roles and continue. The player with the most points after all of the cards have been selected wins.

inter NET CONNECTION Visit www.glencoe.com/sec/math/mac/mathnet for more games.

Integration: Geometry
Translations

What you'll learn

You'll learn to graph translations on a coordinate plane.

When am I ever going to use this?

Animators use translations to create moving images for movies.

Word Wise

translation

LOOK BACK

Refer to Lesson 5-7 for more information on translations.

If you play video games, you have probably seen an application of *motion geometry*. The game designers use computers to move points on the screen, or pixels, to create the illusion of movement. In Chapter 5, you learned that a sliding motion is called a **translation**.

HANDS-ON MINI-LAB

Work with a partner.   grid paper   straightedge

 colored pencils   unlined paper   scissors

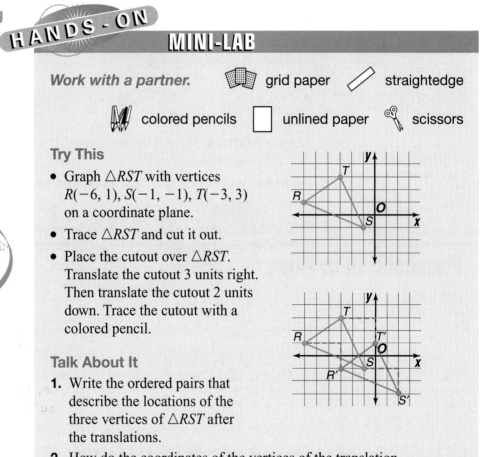

Try This

• Graph △RST with vertices R(−6, 1), S(−1, −1), T(−3, 3) on a coordinate plane.

• Trace △RST and cut it out.

• Place the cutout over △RST. Translate the cutout 3 units right. Then translate the cutout 2 units down. Trace the cutout with a colored pencil.

Talk About It

1. Write the ordered pairs that describe the locations of the three vertices of △RST after the translations.

2. How do the coordinates of the vertices of the translation compare to the coordinates of the original vertices?

In a coordinate plane, a translation down or to the left is negative. A translation up or to the right is positive. The translation in the Mini-Lab could be written as (3, −2). If we add 3 to the *x*-coordinate and −2 to the *y*-coordinate of each vertex of the original triangle, the results are the coordinates of the vertices of the translated triangle.

Translation	**Words:**	To translate a point as described by an ordered pair, add the coordinates of the ordered pair to the coordinates of the point.
	Symbols:	**Arithmetic:** (4, −3) translated by (−1, 2) becomes (3, −1). **Algebra:** (*x*, *y*) translated by (*a*, *b*) becomes (*x* + *a*, *y* + *b*).

1 The vertices of △ABC are A(4, −3), B(0, 2), and C(5, 1). Graph △ABC. Then graph the triangle after a translation 4 units left and 3 units up.

The location of a point after being moved is often written using a prime. So the new coordinates of A are written as A', B as B', and C as C'.

Study Hint

Reading Math A' is read as A prime. If point A has been moved twice, the new point is labeled A'', which is read as A double prime.

vertex		4 left, 3 up	translation	
$A(4, -3)$	$+$	$(-4, 3)$	$\rightarrow$	$A'(0, 0)$
$B(0, 2)$	$+$	$(-4, 3)$	$\rightarrow$	$B'(-4, 5)$
$C(5, 1)$	$+$	$(-4, 3)$	$\rightarrow$	$C'(1, 4)$

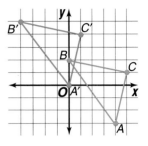

The coordinates of the vertices of the translated triangle are $A'(0, 0)$, $B'(-4, 5)$, and $C'(1, 4)$.

Graph A', B', and C' and draw the translated triangle.

Sometimes you may need to know how a figure was moved. You can use your knowledge of translations and algebra to find the ordered pair that describes the translation.

2 **Algebra** Parallelogram MNPQ has vertices M(−3, 0), N(−1, −1), P(0, 2), and Q(−2, 3). Use an equation to find the ordered pair that describes the translation if M' has coordinates (2, −1). Then graph parallelogram M'N'P'Q'.

Explore   You know the coordinates of the vertices of MNPQ and the coordinates of M'. You need to find the ordered pair for the translation and graph M'N'P'Q'.

Plan   Use an equation to compare each coordinate of M and M'. Find the ordered pair for the translation. Then use the ordered pair to find the coordinates of N', P', and Q'.

Solve   Let (a, b) represent the ordered pair for the translation.

M has coordinates (−3, 0) and M' has coordinates (2, −1), so we can say M(−3, 0) + (a, b) → M'(2, −1). Write an equation for each coordinate.

x-coordinate	*y-coordinate*
$-3 + a = 2$	$0 + b = -1$
$a = 5$	$b = -1$

The ordered pair for the translation is $(5, -1)$. a is positive, so the first move is 5 units right. b is negative, so the next move is 1 unit down.

(continued on the next page)

Find the coordinates of N', P', and Q'.

$N(-1, -1) + (5, -1) \rightarrow N'(4, -2)$

$P(0, 2) + (5, -1) \qquad \rightarrow P'(5, 1)$

$Q(-2, 3) + (5, -1) \quad \rightarrow Q'(3, 2)$

Now graph $M'N'P'Q'$.

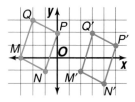

Examine Check your work by graphing $MNPQ$ and translating each vertex by $(5, -1)$. Do the points correspond to $M'N'P'Q'$?

CHECK FOR UNDERSTANDING

Communicating Mathematics

Read and study the lesson to answer each question.

1. *Describe* the type of movement that results from a translation.

2. *Write* a sentence to describe the translation named by $(-3, 4)$.

HANDS-ON LAB

3. *Graph* triangle ABC with vertices $A(1, 2)$, $B(3, 1)$, and $C(3, 4)$. Show the position of the triangle after a translation of $(-2, 1)$.

Guided Practice

Find the coordinates of the vertices of each figure after the translation described. Then graph the figure and its translation.

4. $\triangle BCD$ with vertices $B(5, 2)$, $C(-2, 4)$, and $D(-1, 1)$, translated by $(1, -3)$

5. rectangle $HJKL$ with vertices $H(1, 3)$, $J(4, 0)$, $K(3, -1)$, and $L(0, 2)$, translated by $(3, -4)$

6. Triangle RST has vertices $R(-2, 3)$, $S(-3, 0)$, and $T(1, 1)$. When translated, R' has coordinates $(3, 5)$.
   a. Describe the translation using an ordered pair.
   b. Find the coordinates of S' and T'.

7. *Geometry* If a figure can be translated onto another so that all of the vertices correspond, the figures are congruent. To determine whether two figures are congruent, find the ordered pair that translates each vertex of one figure to the corresponding vertex of the other. If all of these ordered pairs are the same, the figures are congruent. Is $\triangle ABC$ with vertices $A(-2, 5)$, $B(9, 0)$, and $C(-2, -2)$ congruent to $\triangle XYZ$ with vertices $X(4, 4)$, $Y(15, -1)$, and $Z(4, -3)$?

EXERCISES

Practice

Find the coordinates of the vertices of each figure after the translation described. Then graph the figure and its translation.

8. $\triangle EFG$ with vertices $E(-5, -2)$, $F(-2, 3)$, and $G(2, -3)$, translated by $(6, 3)$

9. $\triangle PQR$ with vertices $P(0, 0)$, $Q(-3, -4)$, and $R(1, 3)$, translated by $(-6, 3)$

10. square $SQAR$ with vertices $S(2, 1)$, $Q(4, 3)$, $A(2, 5)$, and $R(0, 3)$, translated by $(-1, 3)$

11. rectangle $WXYZ$ with vertices $W(-4, 1)$, $X(2, 4)$, $Y(3, 2)$, and $Z(-3, -1)$, translated by $(-1, 4)$

12. parallelogram *FGHJ* with vertices *F*(7, 5), *G*(5, 2), *H*(7, 0), and *J*(9, 3), translated by (−4, −2)

13. pentagon *ABCDE* with vertices *A*(−2, −1), *B*(0, −1), *C*(1, 1), *D*(−1, 3), and *E*(−3, 1), translated by (−2, 1)

14. Triangle *NMP* has vertices *M*(4, 2), *N*(−8, 0), and *P*(6, 7). When translated, *M′* has coordinates (−2, 4).

  a. Describe the translation using an ordered pair.

  b. Find the coordinates of *N′* and *P′*.

15. The vertices of parallelogram *QRST* are *Q*(−10, 2), *R*(−4, 0), *S*(6, 2), and *T*(0, 4). *S′* has coordinates (8, −3).

  a. Describe the translation using an ordered pair.

  b. Find the coordinates of *Q′*, *R′*, and *T′*.

16. Hexagon *ABCDEF* has vertices *A*(0, 2), *B*(−5, 0), *C*(−4, −4), *D*(0, −4), *E*(6, −2), and *F*(3, 1). When translated, *E′* has coordinates (8, −5).

  a. Describe the translation using an ordered pair.

  b. Find the coordinates of *A′*, *B′*, *C′*, *D′* and *F′*.

Applications and Problem Solving

17. *Demographics* The center of population is the point around which a nation's population is equally distributed. That is, half of the population lives north of this point and half lives south; half lives east of it and half lives west. Describe the translations of the center of the U.S. population.

Source: *Statistical Abstract of the United States*

18. *Music* Composers refer to *transposing* music from one key to another. When music is transposed, each note is translated the same distance up or down the musical scale.

  a. The music at the top is from Mozart's *Clarinet Quintet in A Major* in the key of C major. Copy the music at the bottom and translate the music to the key of A major. The first note is done for you.

  b. How would you describe the transformation in part a mathematically?

For **Extra Practice,** see page 635.

19. *Critical Thinking* A figure is translated by (4, −1). Then the result is translated by (−4, 1). Without graphing, what is the final position of the figure?

Mixed Review

20. *Algebra* Graph $y = -3x^2 + 1$. *(Lesson 10-6)*

21. **Standardized Test Practice** Which is the best estimate for $\sqrt{317}$? *(Lesson 9-2)*

  A 15   **B** 16   **C** 17   **D** 18   **E** 19

22. Express $35.70 for 6 pounds as a unit rate. *(Lesson 3-1)*

Lesson 10-7 Integration: Geometry Translations   **459**

10-8

Integration: Geometry
Reflections

What you'll learn

You'll learn to graph reflections on a coordinate plane.

When am I ever going to use this?

You will study reflections when you study light and sound in physical science.

LOOK BACK

Refer to Lesson 5-4 for more information on reflections.

The techniques for making Hopi pottery have been passed down from mother to daughter for centuries. Many of the patterns are made by repeated transformations. Dextra Quotskuyva designed the seed jar at the right to be *symmetric*.

Something is symmetric if you can fold the design so that the halves coincide. In Chapter 5, you learned that this fold line is called the *line of symmetry*. One half of the design is the *reflection* of the other.

Every point of the original figure has a corresponding point on the other side of the line of symmetry.

HANDS-ON MINI-LAB

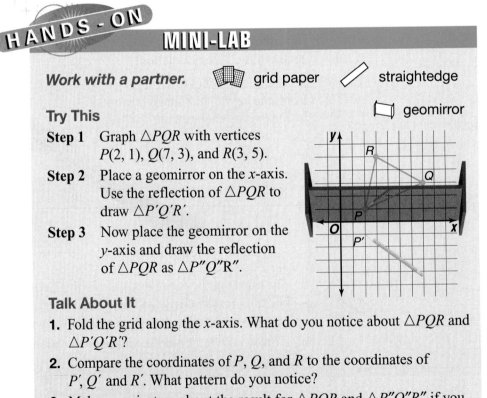

Work with a partner. 🗐 grid paper ╱ straightedge

▭ geomirror

Try This

Step 1 Graph △*PQR* with vertices *P*(2, 1), *Q*(7, 3), and *R*(3, 5).

Step 2 Place a geomirror on the *x*-axis. Use the reflection of △*PQR* to draw △*P'Q'R'*.

Step 3 Now place the geomirror on the *y*-axis and draw the reflection of △*PQR* as △*P"Q"R"*.

Talk About It

1. Fold the grid along the *x*-axis. What do you notice about △*PQR* and △*P'Q'R'*?

2. Compare the coordinates of *P*, *Q*, and *R* to the coordinates of *P'*, *Q'* and *R'*. What pattern do you notice?

3. Make a conjecture about the result for △*PQR* and △*P"Q"R"* if you fold the grid again, this time along the *y*-axis. Fold to verify your conjecture.

4. Write a sentence comparing the coordinates of *P*, *Q*, and *R* to the coordinates of *P"Q"* and *R"*.

In the Mini-Lab, you reflected $\triangle PQR$ over the x-axis to draw $\triangle P'Q'R'$. For these figures, the x-axis is the line of symmetry. The coordinates of the vertices have a special relationship.

Reflection over the x-axis	**Words:**	To reflect a point over the x-axis, use the same x-coordinate and multiply the y-coordinate by -1.
	Symbols: **Arithmetic:**	$(4, -3)$ becomes $(4, 3)$.
	Algebra:	(x, y) becomes $(x, -y)$.

Example 1 Reflect parallelogram $ABCD$ with vertices $A(1, 2)$, $B(0, 0)$, $C(-5, 0)$, and $D(-4, 2)$ over the x-axis.

Find the coordinates of each vertex after the reflection.

$$A(1, 2) \quad \rightarrow \quad (1, 2 \cdot -1) \quad \rightarrow \quad A'(1, -2)$$
$$B(0, 0) \quad \rightarrow \quad (0, 0 \cdot -1) \quad \rightarrow \quad B'(0, 0)$$
$$C(-5, 0) \quad \rightarrow \quad (-5, 0 \cdot -1) \quad \rightarrow \quad C'(-5, 0)$$
$$D(-4, 2) \quad \rightarrow \quad (-4, 2 \cdot -1) \quad \rightarrow \quad D'(-4, -2)$$

Graph $ABCD$ and $A'B'C'D'$.

In the Mini-Lab, you also investigated the relationship for the coordinates of points reflected over the y-axis.

Reflection over the y-axis	**Words:**	To reflect a point over the y-axis, multiply the x-coordinate by -1 and use the same y-coordinate.
	Symbols: **Arithmetic:**	$(4, -3)$ becomes $(-4, -3)$.
	Algebra:	(x, y) becomes $(-x, y)$.

You can use a reflection over the y-axis to help graph some quadratic functions.

Example 2

INTEGRATION

Algebra The graph of $y = x^2 + 4$ is symmetric about the y-axis. Complete a function table and reflect the points over the y-axis to complete the graph.

Find each function value.

x	y	(x, y)
0	$(0)^2 + 4$	$(0, 4)$
1	$(1)^2 + 4$	$(1, 5)$
2	$(2)^2 + 4$	$(2, 8)$
3	$(3)^2 + 4$	$(3, 13)$
4	$(4)^2 + 4$	$(4, 20)$

(continued on the next page)

Now reflect each point over the *y*-axis.

$(0, 4)$ → $(0 \cdot -1, 4)$ → $(0, 4)$

$(1, 5)$ → $(1 \cdot -1, 5)$ → $(-1, 5)$

$(2, 8)$ → $(2 \cdot -1, 8)$ → $(-2, 8)$

$(3, 13)$ → $(3 \cdot -1, 13)$ → $(-3, 13)$

$(4, 20)$ → $(4 \cdot -1, 20)$ → $(-4, 20)$

Finally graph each point and its reflection. Complete by drawing a smooth curve.

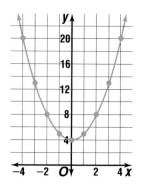

CHECK FOR UNDERSTANDING

Communicating Mathematics

Read and study the lesson to answer each question.

1. ***Write*** a sentence explaining how a mirror reflection is related to a geometric reflection.

2. ***Develop*** a way to remember how to reflect a point over one of the axes.

HANDS-ON LAB

3. ***Sketch*** the figure at the right. Then use a geomirror to draw all of the lines of symmetry.

Guided Practice

4. Name the line of symmetry for the pair of figures.

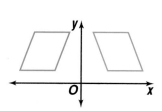

5. Graph $\triangle ABC$ with vertices $A(-2, 3)$, $B(5, 1)$, and $C(4, 5)$.
   a. Reflect $\triangle ABC$ over the *x*-axis
   b. Reflect $\triangle ABC$ over the *y* axis.
   c. Is $\triangle ABC$ congruent to its reflections? That is, are the lengths of corresponding sides in the ratio 1 to 1? Justify your answer.

6. ***Architecture*** Georgian architecture was the major architectural style in England and the American colonies during the 1700s. Study the photo of the Georgian style Blenheim Palace at the right. Does it exhibit symmetry? If so, sketch the building and draw the line of symmetry.

Practice **Name the line of symmetry for each pair of figures.**

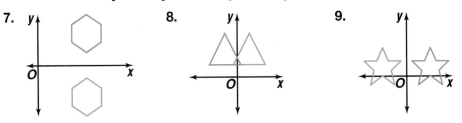

7. **8.** **9.**

Graph each figure. Then draw its reflections over the *x*-axis and over the *y*-axis.

10. △*MNP* with vertices *M*(−8, 4), *N*(−3, 8), and *P*(−2, 2)

11. square *WXYZ* with vertices *W*(−3, 2), *X*(1, −2), *Y*(−3, −6), and *Z*(−7, −2)

12. trapezoid *PQRS* with vertices *P*(4, −2), *Q*(8, −2), *R*(8, 5), and *S*(2, 5)

13. Graph rectangle *RECT* with vertices *R*(−3, 3), *E*(3, 3), *C*(3, −3), and *T*(−3, −3).
  a. Reflect *RECT* over the *x*-axis.
  b. Reflect *RECT* over the *y*-axis on the same coordinate plane.
  c. What do you observe about the three graphs?

14. Graph *A*(2, 6) and *B*(5, 5).
  a. Reflect *A* and *B* over the *y*-axis and graph *A*′ and *B*′.
  b. Complete *A*′*ABB*′. What type of figure is formed?

Applications and Problem Solving

15. *Art* Refer to the beginning of the lesson. Draw your own pottery design using one or more reflections.

16. *Life Science* If you look at your face in the mirror and imagine a vertical line drawn between your eyes, you will find that most of your features are symmetric. Studies show that people of all different cultures find faces that are symmetric to be attractive. Find a photograph of a friend or a famous person. What features of his or her face are symmetric? Are there features that are not symmetric?

17. *Critical Thinking* Determine the vertices of a triangle, whose vertices were originally at (−1, −1), (1, −2), and (5, 1) after it is translated 1 unit left and 2 units right and then reflected across the graph of *y* = 2*x* − 1. (*Hint:* Use a geomirror.)

Mixed Review

18. **Standardized Test Practice** What are the coordinates of *W*(−6, 3) if it is translated by (2, −1)? *(Lesson 10-7)*
  A (−4, 2)   **B** (−8, 4)   **C** (−4, 4)   **D** (−7, 5)

19. *Advertising* How many times do you see commercials before you buy the product? The results of a survey on how many times people watch infomercials before they buy are shown. Make a circle graph of this data. *(Lesson 4-2)*

Times Seen Before We Buy	Percent
Once	27%
Twice	31%
Three times	18%
Four times	9%
Five or more times	15%

Source: National Infomercial Marketing Assn.

For **Extra Practice**, see page 636.

Integration: Geometry
Rotations

Rotations

What you'll learn

You'll learn to graph rotations on a coordinate plane.

When am I ever going to use this?

You can use rotations to create quilt patterns.

Word Wise

rotation

Have you ever studied vexillology *(vek suh LAHL uh jee)*? Vexillology is the study of the history and symbolism of flags. Egyptian art shows us that the first flags were symbols attached to the tops of poles and carried into battle. Cloth flags were first made in China about 3000 B.C. The colors and symbols of national flags are often chosen to represent heritage or an event in history.

The symbols on the flags of Anguilla and the Isle of Man can be made using a **rotation**. A rotation moves a figure about a central point.

Anguilla

Isle of Man

Cultural Kaleidoscope

Nations that have a common history often share the same colors in their flag. For example, the flags of the Arab nations contain black, green, red, and white.

MINI-LAB

Work in groups of two or three.

protractor

Try This

- The graph models the blades of a fan. Record the coordinates of each lettered point.
- Measure ∠MON, ∠NOP, ∠POQ, and ∠QOM. Record your measurements.
- As the fan blades turn, each point of one blade will occupy the previous location of the corresponding point on another blade. Make a list of the corresponding points as the blade rotates.

Talk About It

1. Which direction is the fan turning?
2. How many degrees did the blade rotate to move from point M to point N?
3. How do the coordinates of M compare to the coordinates of N?
4. How many degrees did the blade rotate to move from point M to point P?
5. Compare the coordinates of M and P.

As you discovered in the Mini-Lab, the coordinates of points rotated 90° and 180° are related.

Rotation of 90° counterclockwise	**Words:**	To rotate a figure 90° counterclockwise about the origin, switch the coordinates of each point and then multiply the new first coordinate by −1.
	Symbols:	**Arithmetic:** $P(6, 2) \rightarrow P'(-2, 6)$ **Algebra:** $P(x, y) \rightarrow P'(-y, x)$
Rotation of 180°	**Words:**	To rotate a figure 180° about the origin, multiply both coordinates of each point by −1.
	Symbols:	**Arithmetic:** $P(6, 2) \rightarrow P'(-6, -2)$ **Algebra:** $P(x, y) \rightarrow P'(-x, -y)$

Examples

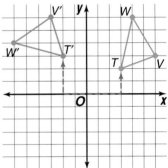

1 Triangle *TVW* has vertices *T*(3, 2), *V*(6, 3), and *W*(4, 6). Graph △*TVW* and rotate it 90° counterclockwise about the origin. Then graph △*T′V′W′*.

To rotate △*TVW* 90°, switch the coordinates of each vertex and multiply the first by −1.

$T(3, 2) \quad \rightarrow \quad T'(-2, 3)$

$V(6, 3) \quad \rightarrow \quad V'(-3, 6)$

$W(4, 6) \quad \rightarrow \quad W'(-6, 4)$

Graph △*T′V′W′*.

Study Hint

Estimation You can verify a 90° rotation using a corner of a piece of paper. If you can place the paper so that the corner is on the origin and *T* and *T′* are on the edges, you have probably rotated correctly.

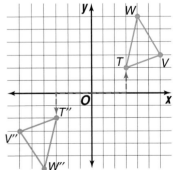

2 Use △*TVW* from Example 1. Rotate it 180° about the origin. Then graph △*T″V″W″*.

To rotate △*TVW* 180°, multiply the coordinates of each vertex by −1.

$T(3, 2) \quad \rightarrow \quad T''(-3, -2)$

$V(6, 3) \quad \rightarrow \quad V''(-6, -3)$

$W(4, 6) \quad \rightarrow \quad W''(-4, -6)$

Graph △*T″V″W″*.

You know that some figures have line symmetry. Other figures have *rotational* or *point symmetry*. If a figure has rotational symmetry, when you turn it around its center point, there is at least one other position in which the figure looks the same as it did originally. *Any figure will look the same after a 360° turn, so a figure must look the same after a rotation of less than 360° to have rotational symmetry.*

❸ Flags Refer to the beginning of the lesson. Find the measures of the counterclockwise turns that make the symbol on the flag of the Isle of Man look the same.

Label the symbol to keep track of the rotations.

 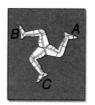 

| Original | 120° turn | 240° turn | 360° turn |

The parts of the symbol are in different positions, but it looks the same after each turn. A 360° turn returns the figure to its original position. So divide 360° into thirds to find that the counterclockwise turns are 120° and 240°.

CHECK FOR UNDERSTANDING

Communicating Mathematics

Read and study the lesson to answer each question.

1. **State** the quadrant in which a triangle will lie if it is in the first quadrant and is rotated 90° counterclockwise.

2. **Explain** why a figure and its rotation image are congruent figures.

3. **Write** three examples of rotating objects you see every day.

Guided Practice

4. Determine whether the figures represent a rotation.

5. Graph $\triangle ABC$ with vertices $A(1, 3)$, $B(6, 7)$, and $C(9, 1)$.
   a. Rotate the triangle 90° counterclockwise and graph $\triangle A'B'C'$.
   b. Rotate the triangle 180° and graph $\triangle A''B''C''$.

6. **Geometry** An equilateral triangle has rotational symmetry. Draw three other figures that have rotational symmetry.

EXERCISES

Practice

Determine whether each pair of figures represents a rotation. Write *yes* or *no*.

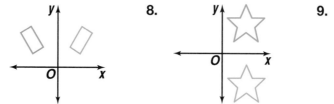

7.

8.

9.

10. Graph $\triangle XYZ$ with vertices $X(-6, -9)$, $Y(-1, -3)$, and $Z(-8, -5)$.
    a. Rotate the triangle 90° counterclockwise and graph $\triangle X'Y'Z'$.
    b. Rotate the triangle 180° and graph $\triangle X''Y''Z''$.

11. Rectangle *RECT* has vertices $R(-5, 4)$, $E(-5, -2)$, $C(-3, -2)$, and $T(-3, 4)$. Graph *RECT*.

    a. Graph $R'E'C'T'$, the rotation of *RECT* 90° counterclockwise .

    b. Rotate the rectangle 180° and graph $R''E''C''T''$.

12. Graph trapezoid *ABCD* with vertices $A(3, -2)$, $B(7, -2)$, $C(9, -7)$, and $D(-1, -7)$.

    a. Rotate the trapezoid 90° counterclockwise and graph $A'B'C'D'$.

    b. Rotate the trapezoid 180° and graph $A''B''C''D''$.

13. After a triangle is rotated 180°, its vertices are at $(-4, 1)$, $(-1, 4)$, and $(-5, -8)$. What were the coordinates of the vertices before the rotation?

14. A figure is rotated 180° and then its image is rotated 180°, what is the result?

Applications and Problem Solving

Family Activity

Look at items in your home that have corporate logos. Do any of them have rotational symmetry?

15. *Marketing* Marketers create corporate logos to identify products from their company. Some corporate logos have rotational symmetry.

    a. Which of these logos exhibit rotational symmetry?

    b. For each logo that has rotational symmetry, find the degree turns that show the symmetry.

16. *Art* How many degrees is hedgehog A rotated to create hedgehog B in this piece of African art?

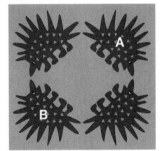

17. *Entertainment* In a deck of cards, the eight of diamonds has rotational symmetry. What other cards have rotational symmetry?

18. *Working on the* **CHAPTER Project**
A computer game has a spinning windmill. One blade has vertices $A(-6, 4)$, $B(-11, 2)$, $C(-10, -2)$, $D(-6, -2)$, $E(0, 0)$, and $F(-3, 3)$. After 6 equal rotations, A' has coordinates $(-4, -6)$.

    a. How many degrees does the blade move on each rotation?

    b. Find the coordinates of the other vertices of the blade after 6 rotations.

19. *Critical Thinking* If the point at (x, y) is rotated 90° *clockwise,* what are the new coordinates?

Mixed Review

20. *Geometry* Graph $\triangle RST$ with vertices $R(3, 3)$, $S(0, 0)$, and $T(6, -1)$. *(Lesson 10-8)*

    a. Reflect $\triangle RST$ over the *x*-axis.

    b. Reflect $\triangle RST$ over the *y*-axis.

21. **Standardized Test Practice** It is 326 kilometers from Milford to Loveland. If there are 1,000 meters in a kilometer, use scientific notation to write the distance from Milford to Loveland in meters. *(Lesson 6-9)*

    A 3.26×10^6 **B** 32.6×10^5 **C** 326×10^5 **D** 3.26×10^5

For **Extra Practice,** see page 636.

22. Solve $q = 360 \div (-40)$. *(Lesson 2-8)*

*inter*NET
CONNECTION Chapter Review **For additional lesson-by-lesson review, visit:**
www.glencoe.com/sec/math/mac/mathnet

Vocabulary

After completing this chapter, you should be able to define each term, concept, or phrase and give an example or two of each.

Patterns and Functions
domain (p. 428)
function (p. 428)
function table (p. 428)
linear function (p. 442)
quadratic function (p. 452)
range (p. 428)
standard viewing window
(p. 445)
system of equations
(p. 446)

Geometry
line of symmetry (p. 460)
reflection (p. 460)
rotation (p. 464)
symmetric (p. 460)
translation (p. 456)

Problem Solving
use a graph (p. 450)

Understanding and Using the Vocabulary

Choose the correct term or number to complete each sentence.

1. The (domain, range) is the set of input values of a function.

2. The range is the set of (input, output) values of a function.

3. When you find a common solution of two or more equations, you are solving a (function, system of equations).

4. A function in which the graphs of the solutions form a line is called a (linear, quadratic) function.

5. A function in which the greatest power is (two, three) is called a quadratic function.

6. The movement of a figure 2 units right and 4 units down is a (rotation, translation).

7. $A(2, 1) \rightarrow A'(-2, 1)$ describes a (reflection, translation) over the y-axis.

8. In a (function, rotation), one or more operations are performed on one number to get another.

In Your Own Words

9. **Tell** how to determine whether an ordered pair is a solution of a function.

Upon completing this chapter, you should be able to:

● complete function tables *(Lesson 10-1)*

Find $f(5)$ if $f(n) = 2n + 6$.

$f(5) = 2(5) + 6$ *Replace n with 5.*

$f(5) = 10 + 6$

$f(5) = 16$

Use these exercises to review and prepare for the chapter test.

10. Copy and complete the function table for $f(n) = 1 - 3n$.

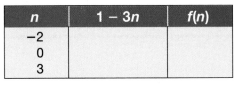

n	1 − 3n	f(n)
−2		
0		
3		

● graph functions by using function tables *(Lesson 10-2)*

Graph $f(n) = n - 1$.

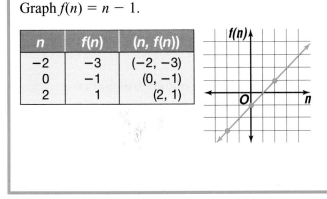

n	f(n)	(n, f(n))
−2	−3	(−2, −3)
0	−1	(0, −1)
2	1	(2, 1)

11. Copy and complete the function table for $f(n) = 3n + 1$. Then graph the function.

n	f(n)	(n, f(n))
−3		
0		
1		

Choose values for n and graph each function.

12. $f(n) = 0.5n - 4$

13. $f(n) = -2n$

● find solutions of equations with two variables *(Lesson 10-3)*

Find a solution of $y = 5x + 2$.

Let $x = 3$.

$y = 5(3) + 2$

$y = 15 + 2$

$y = 17$

A solution of $y = 5x + 2$ is (3, 17).

Copy and complete the table for each equation.

14. $y = 1 - 1.5x$

x	y
−4	
0	
2	

15. $y = x - 4$

x	y
−2	
4	
6	

16. Find four solutions of $y = 5x - 10$.

● graph linear functions by plotting points *(Lesson 10-4)*

Make a function table and choose at least three values for *x*. Then graph the ordered pairs and connect the points with a line.

Graph each function.

17. $y = -5x$

18. $y = 4 + 2x$

19. $y = 1\frac{1}{2} - 2\frac{1}{2}x$

20. $y = \frac{x}{2} - 2$

● solve systems of linear equations by graphing *(Lesson 10-5)*

The coordinates of the point where the graphs of two linear equations intersect is the solution of the system of equations.

Solve each system of equations by graphing.

21. $y = 5x$
$y = x + 4$

22. $y = 3x - 6$
$y = x - 4$

23. Graph the system $y = 6x$ and $y = x + 5$. Find the solution of the system.

● graph quadratic functions *(Lesson 10-6)*

Make a function table. Then graph the ordered pairs and draw a smooth curve connecting the points.

Graph each quadratic function.

24. $y = 0.5x^2 + 4$

25. $y = x^2 - 2$

26. $f(n) = 5 - n^2$

27. $y = -\frac{1}{2}x^2 - 3$

● graph translations on a coordinate plane *(Lesson 10-7)*

To translate a point by *(a, b)*, add *a* to the *x*-coordinate and add *b* to the *y*-coordinate.

Graph each figure and its translation.

28. $\triangle XYZ$ with vertices $X(2, 2)$, $Y(3, 5)$, and $Z(5, 3)$, translated by $(-2, -4)$

29. rectangle $ABCD$ with vertices $A(-3, -1)$, $B(0, -1)$, $C(0, 1)$, and $D(-3, 1)$, translated by $(3, 2)$

● graph reflections on a coordinate plane *(Lesson 10-8)*

reflection over the *x*-axis
(x, y) becomes $(x, -y)$.

reflection over the *y*-axis
(x, y) becomes $(-x, y)$.

30. Graph square $EFGH$ with vertices $E(2, 5)$, $F(4, 5)$, $G(4, 3)$, and $H(2, 3)$ and its reflection over the *x*-axis.

31. Graph $\triangle TUV$ with vertices $T(-4, -5)$, $U(-3, -3)$, and $V(-5, -2)$ and its reflection over the *y*-axis.

● graph rotations on a coordinate plane *(Lesson 10-9)*

rotation 90° counterclockwise
(x, y) becomes $(-y, x)$.

rotation 180°
(x, y) becomes $(-x, -y)$.

32. Graph rectangle $IJKL$ with vertices $I(1, 4)$, $J(3, 6)$, $K(6, 3)$, and $L(4, 1)$ and its rotation of 90° counterclockwise.

33. Graph $\triangle QRS$ with vertices $Q(2, 2)$, $R(4, -1)$, and $S(2, -3)$ and its 180° rotation.

Applications & Problem Solving

34. Sports Golfers are given a handicap based on their average. Tatanka's handicap is 15, so $f(n) = n - 15$, where n is his actual score, is used to determine his handicapped score. Make a function table of his handicapped scores if Tatanka had actual scores of 85, 87, and 89. *(Lesson 10-1)*

35. Communication Code flags can send messages between ships at sea. Tell whether each code flag displays symmetry. If a flag has symmetry, describe it as line or rotational symmetry. *(Lessons 10-8 and 10-9)*

36. Business One company's profit is described by the equation $p = 100x - 200$. Another company's profit is described by $p = 100x + 200$. They plan to merge when their profit is the same. At what point will that occur? Explain your answer. *(Lesson 10-5)*

37. Use a Graph The graph below shows the number of farms in the United States from 1940 to 1995. *(Lesson 10-6A)*

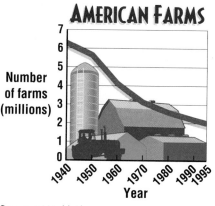

Source: *World Almanac*

a. About how many more farms were there in 1950 than in 1980?

b. During which ten-year period was the decrease in the number of farms the greatest?

Alternative Assessment

● **Open Ended**

Ofelia drew a graph of the cost of renting videos at two stores near her home using x as the number of videos and y as the cost. One line passes through $(0, 5)$ and $(5, 15)$. The other line passes through the origin and $(5, 15)$. Graph the lines. Describe the situations shown by the lines.

One of the stores decided to change its prices. Ofelia drew a new graph and found that the two lines she drew did not intersect. Describe how the prices could have changed.

● **Completing the** **CHAPTER Project**

Use the following checklist to make sure that your project is complete.

☑ Outline your computer game plan.

☑ Use a grid to design the background for a scene from your game.

☑ Design the characters in your game.

☑ Use functions to describe how the characters move on the screen.

PORTFOLIO Select one of the assignments from this chapter and place it in your portfolio. Attach a note to it explaining why you selected it.

A practice test for Chapter 10 is provided on page 656.

Section One: Multiple Choice

There are eleven multiple-choice questions in this section. Choose the best answer. If a correct answer is *not here,* choose the letter for Not Here.

1. What number will make the number sentence true?

$$\frac{2}{3} + \frac{5}{9} = \frac{5}{9} + h$$

A $\frac{7}{12}$

B $\frac{2}{3}$

C $\frac{3}{9}$

D $\frac{4}{3}$

2. $0.9 =$

F 9%

G 0.9%

H 0.09%

J 90%

3. An electronics store is having a going out of business sale. The store advertises 35% off the price of a $210 VCR. To find the amount saved on the cost of the VCR, multiply $210 by —

A 0.035.

B 0.35.

C 0.65.

D 0.065.

4. If $\frac{3}{8} = \frac{c}{12}$, which statement is true?

F $8c > 3(12)$

G $8c < 3(12)$

H $8c = 3(12)$

J Not enough information

5. Hakeem has been increasing the length of his walks at the beginning of each week for three weeks.

Week	Number of Miles
1	1.4 miles
2	2.8 miles
3	4.2 miles
4	

If Hakeem continues at the same rate, how many miles will he walk during the fourth week of training?

A 5.0 miles

B 5.2 miles

C 5.4 miles

D 5.6 miles

Please note that Questions 6–11 have five answer choices.

6. A store sold $22,057 worth of merchandise in the first three weeks of the month. In the fourth week, they sold $6,900 worth of merchandise. A reasonable conclusion would be that the store sold —

F less than $2,500 per week.

G between $2,500 and $5,000 per week.

H between $5,000 and $7,500 per week.

J between $7,500 and $10,000 per week.

K more than $10,000 per week.

7. Which set are solutions of the equation $y = 2x - 4$?

A $\{(-1, -6), (1, -2)\}$

B $\{(-6, -1), (-2, 1)\}$

C $\{(1, -2), (3, -2)\}$

D $\{(-2, 1), (2, 3)\}$

E $\{(9, 14), (-2, 0)\}$

8. In a scale drawing of a room, 1 unit = 6 inches. What are the scale dimensions of a 40 in. by 60 in. table?

  F 4 units by 6 units

  G $6\frac{2}{3}$ units by 10 units

  H 8 units by 10 units

  J 24 units by 36 units

  K 240 units by 360 units

9. Tasha bought three bags of apples for apple bobbing. The bags weighed 5.5 pounds, 8.4 pounds, and 7.35 pounds. What was the total weight of the apples?

  A 20.45 lb   **B** 21.25 lb

  C 23.60 lb   **D** 25.55 lb

  E Not Here

10. If Mike worked a total of 31 hours over a four-day period and is paid $7.25 per hour, how much did he earn before deductions?

  F $29.00   **G** $124.00

  H $224.75   **J** $899.00

  K Not Here

11. Four friends are saving money to buy a computer game that costs $64.95. Lydia has saved $17.94, Owen $12.55, Tomás $16.85, and Alma $15.90. What is the total that the four friends have saved?

  A $61.24   **B** $62.84

  C $64.86   **D** $64.94

  E Not Here

 **Test Practice For additional test practice questions, visit:**

www.glencoe.com/sec/math/mac/mathnet

Section Two: Free Response

This section contains six questions for which you will provide short answers. Write your answers on your paper.

12. Which figure has rotational symmetry?

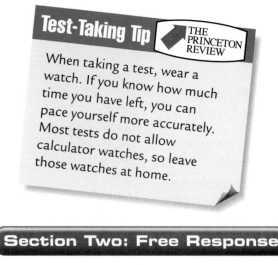

13. What is the least prime factor of 24?

14. What is the solution of the system of equations?

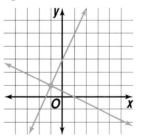

15. What are the coordinates of the rectangle *MATH* translated by $(2, -1)$ if $M(-3, 1)$, $A(2, 6)$, $T(6,2)$, and $H(1, 3)$?

16. To the nearest inch, how long is the diagonal of a 20 in. by 50 in. window?

17. Find the value of x in the square.

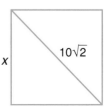

$10\sqrt{2}$

x

CHAPTER 11

Geometry: Using Area and Volume

What you'll learn in Chapter 11

- to find the areas of circles,
- to solve problems by making a model,
- to find the volumes of prisms, cylinders, pyramids, and cones,
- to find the surface areas of prisms and cylinders, and
- to analyze measurements.

CHAPTER Project

UP, UP, AND AWAY

In this project, you will design a three-dimensional kite. You will make a three-dimensional scale drawing of the kite, compute the amount and cost of materials you will need to make the kite, and determine the instrument(s) you will need to make the appropriate measurements to build the kite. You will place your plans and information on your kite in a folder.

Getting Started

- Research different styles and shapes of three-dimensional kites. Choose a type to design.
- Research the cost of the materials you would need to make the kite.

Technology Tips

- Use an **electronic encyclopedia** to do your research.
- Use a **drawing program** to sketch your design.
- Use a **spreadsheet** to record the cost of materials.
- Use a **word processor** to write the information about your kite.

 interNET **CONNECTION** Research **For up-to-date information on kites, visit:**

www.glencoe.com/sec/math/mac/mathnet

Working on the Project

You can use what you learn in Chapter 11 to draw your kite and determine what you will need to make the kite.

Page	Exercise
485	16
502	16
507	18
511	Alternative Assessment

Area of Circles

What you'll learn

You'll learn to find the areas of circles.

When am I ever going to use this?

Knowing how to find the areas of circles can help you compare the size of pizzas and choose the better buy.

The Floral Clock located in Niagara Falls, Canada, is a favorite tourist attraction. Each year over 19,000 plants are arranged to form the circular face of the working clock which has a diameter of 40 feet. If $1\frac{1}{2}$ ounces of fertilizer are needed for each square foot of planting area, how many pounds of fertilizer should be used for the clock? *You will solve this problem in Example 2.*

In order to solve this problem, you need to find the area of the clock. Finding the area of a circle can be related to finding the area of a parallelogram. Draw several equally-spaced radii of a circle. Cut the circle along the radii to form wedge-like pieces. Rearrange the pieces to form a parallelogram-shaped figure.

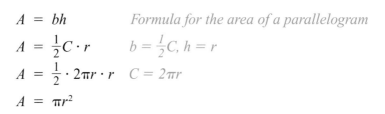

LOOK BACK

You can refer to Lesson 7-7 to review circles.

The length of each wedge from the point to its curved side is the same as the radius (r) of the circle. So the height of the parallelogram is r. The curved sides of the wedges form the circumference (C) of the circle. The base of the parallelogram is made of half of these curves, or half of the circumference of the circle.

$$A = bh \qquad \textit{Formula for the area of a parallelogram}$$
$$A = \frac{1}{2}C \cdot r \qquad b = \frac{1}{2}C,\ h = r$$
$$A = \frac{1}{2} \cdot 2\pi r \cdot r \qquad C = 2\pi r$$
$$A = \pi r^2$$

So, the formula for the area of a circle is $A = \pi r^2$.

Area of a Circle	**Words:**	The area (A) of a circle equals π times the radius (r) squared.
	Symbols: $A = \pi r^2$	**Model:** 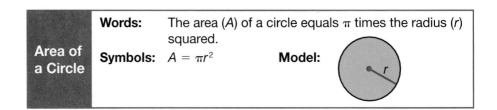

1 Find the area of a circle with a radius of 6 inches to the nearest square inch.

Method 1 Use paper and pencil.

$A = \pi r^2$

$A \approx 3.14 \cdot 6^2$ *Use 3.14 for π.*

$A \approx 3.14 \cdot 36$

$A \approx 113.04$

Method 2 Use a calculator.

$A = \pi r^2$

$A = \pi \cdot 6^2$

$\boxed{\pi} \;\; \boxed{\times} \; 6 \; \boxed{x^2} \;\; \boxed{=} \; \mathit{113.0973355}$

$A \approx 113$

The area of the circle is about 113 square inches.

APPLICATION

2 **Horticulture** Refer to the beginning of the lesson. How many pounds of fertilizer should be used for the clock?

Explore   You know the diameter of the circular clock and the amount of fertilizer needed for each square foot. You must find how many pounds of fertilizer are needed for the clock.

Plan   Since the diameter of the clock is 40 feet, the radius is half of 40 or 20 feet. Use the radius to find the area of the clock. Then calculate the number of ounces of fertilizer needed and change this answer to pounds.

Solve $A = \pi r^2$ *Formula for the area of a circle*

       $A = \pi \cdot 20^2$ *r = 20*

       $A \approx 1{,}257$ $\boxed{\pi} \;\; \boxed{\times} \; 20 \; \boxed{x^2} \;\; \boxed{=} \; \mathit{1256.637061}$

Multiply 1,257 by $1\frac{1}{2}$ to find how many ounces of fertilizer are needed.

$$1{,}257 \times 1\tfrac{1}{2} = \tfrac{3{,}771}{2} \text{ or } 1{,}885\tfrac{1}{2}$$

Divide by 16 to find the number of pounds of fertilizer needed.

$$1{,}885\tfrac{1}{2} \div 16 = \tfrac{3{,}771}{32} \text{ or about } 118 \quad \textit{1 lb = 16 oz}$$

About 118 pounds of fertilizer are needed.

Examine   Estimate the area of the clock.

$$3 \times 20^2 = 3 \times 400 \text{ or } 1{,}200 \text{ square feet}$$

Since $1{,}200 \times 1\frac{1}{2} = 1{,}800$, the number of ounces needed seems reasonable.

Remember that the probability of an event is defined as the ratio of the number of ways something can happen to the total possible outcomes. Probability can also be related to the area of a figure.

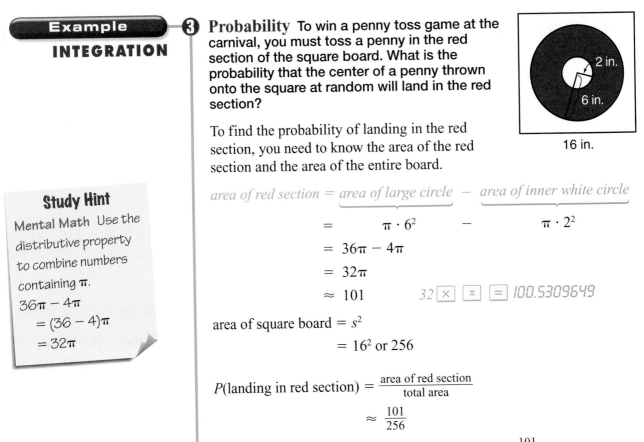

3 **Probability** To win a penny toss game at the carnival, you must toss a penny in the red section of the square board. What is the probability that the center of a penny thrown onto the square at random will land in the red section?

To find the probability of landing in the red section, you need to know the area of the red section and the area of the entire board.

area of red section = area of large circle − area of inner white circle

$$= \qquad \pi \cdot 6^2 \qquad - \qquad \pi \cdot 2^2$$
$$= 36\pi - 4\pi$$
$$= 32\pi$$
$$\approx 101 \qquad 32 \times \pi = 100.5309649$$

area of square board = s^2
$$= 16^2 \text{ or } 256$$

$P(\text{landing in red section}) = \dfrac{\text{area of red section}}{\text{total area}}$

$$\approx \dfrac{101}{256}$$

The probability of landing in the red section is about $\dfrac{101}{256}$ or about 39.5%.

Study Hint

Mental Math Use the distributive property to combine numbers containing π.
$36\pi - 4\pi$
$= (36 - 4)\pi$
$= 32\pi$

CHECK FOR UNDERSTANDING

Communicating Mathematics

Read and study the lesson to answer each question.

1. **Tell** how you can estimate the area of a circle whose radius is 10 meters.

2. **Explain** whether a person is likely to win the penny toss game in Example 3.

3. **You Decide** Alisa says that the area of the semicircle equals $\frac{1}{2}\pi \cdot 5^2$. Megan says that it equals $\frac{1}{2}\pi \cdot 10^2$. Cheryl says that it equals $\pi \cdot 5^2$. Who is correct? Explain.   ← 10 m →

Guided Practice

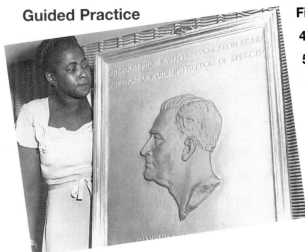

Find the area of each circle to the nearest tenth.

4. diameter, 24 millimeters

5. radius, 13 feet

6. Find the area of the shaded region to the nearest tenth.   8 ft

7. **Art** Sculptor Selma Burke sculpted the profile of former president Franklin D. Roosevelt used on the dime first minted in 1946. The diameter of a dime is 17.91 millimeters. Find the area of one side of a dime.

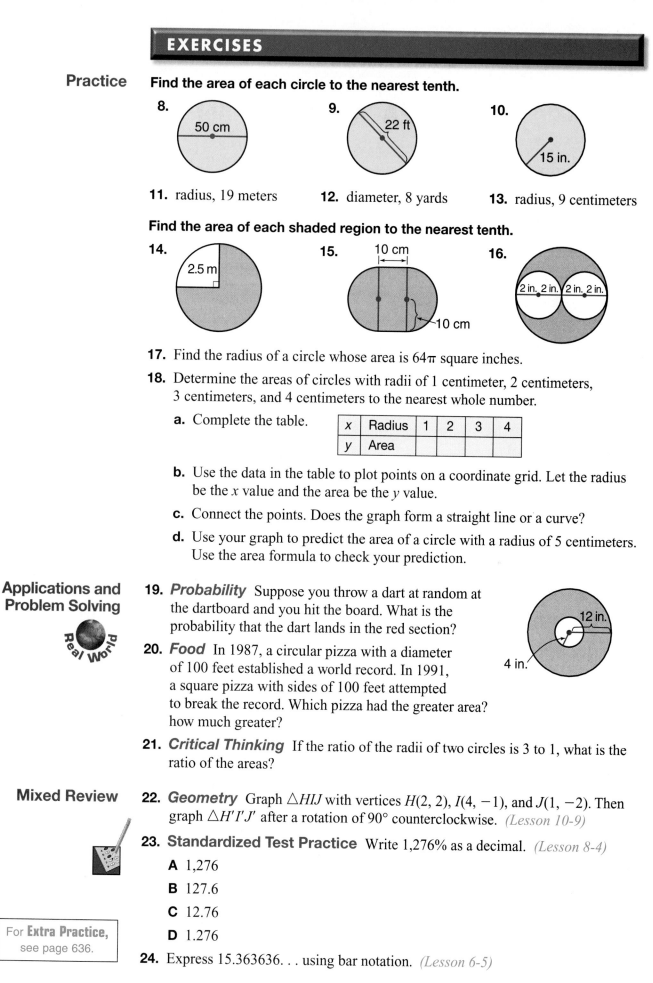

Practice

Find the area of each circle to the nearest tenth.

8. 50 cm

9. 22 ft

10. 15 in.

11. radius, 19 meters

12. diameter, 8 yards

13. radius, 9 centimeters

Find the area of each shaded region to the nearest tenth.

14. 2.5 m

15. 10 cm / 10 cm

16. 2 in. 2 in. 2 in. 2 in.

17. Find the radius of a circle whose area is 64π square inches.

18. Determine the areas of circles with radii of 1 centimeter, 2 centimeters, 3 centimeters, and 4 centimeters to the nearest whole number.

   a. Complete the table.

x	Radius	1	2	3	4
y	Area				

   b. Use the data in the table to plot points on a coordinate grid. Let the radius be the x value and the area be the y value.

   c. Connect the points. Does the graph form a straight line or a curve?

   d. Use your graph to predict the area of a circle with a radius of 5 centimeters. Use the area formula to check your prediction.

Applications and Problem Solving

Real World

19. *Probability* Suppose you throw a dart at random at the dartboard and you hit the board. What is the probability that the dart lands in the red section?

12 in. / 4 in.

20. *Food* In 1987, a circular pizza with a diameter of 100 feet established a world record. In 1991, a square pizza with sides of 100 feet attempted to break the record. Which pizza had the greater area? how much greater?

21. *Critical Thinking* If the ratio of the radii of two circles is 3 to 1, what is the ratio of the areas?

Mixed Review

22. *Geometry* Graph $\triangle HIJ$ with vertices $H(2, 2)$, $I(4, -1)$, and $J(1, -2)$. Then graph $\triangle H'I'J'$ after a rotation of 90° counterclockwise. *(Lesson 10-9)*

23. **Standardized Test Practice** Write 1,276% as a decimal. *(Lesson 8-4)*

   A 1,276

   B 127.6

   C 12.76

   D 1.276

24. Express 15.363636. . . using bar notation. *(Lesson 6-5)*

For **Extra Practice,** see page 636.

11-2A Make a Model

A Preview of Lesson 11-2

For their community living project, Lorena and Jeffrey suggest building a children's play area using pre-fabricated cubical units. Lorena and Jeffrey study the plans showing each side of the building. Let's listen in!

Lorena

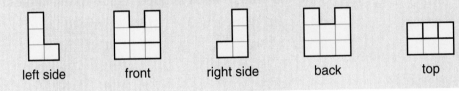

left side   front   right side   back   top

If we use blocks and our plans to build a model, we will know how many pre-fabricated units we need. Remember, the dark lines indicate a break in the surface.

From the top view, I can see we need 6 blocks to form the base of our building.

From the left and right views, I can see we need to build a three-story tower in each back corner.

Jeffrey

There is just one story along the front, but there are two stories in the middle of the back of the building.

THINK ABOUT IT

Work with a partner.

1. **Tell** what each block in the model represents.

2. **Make a model** of the building. How many pre-fabricated cubical units are needed?

3. **Explain** how to build the model by starting with one of the side views.

4. **Apply** the **make a model** strategy to determine how many pre-fabricated cubical units are needed for the following plans.

left side   front   right side

back   top

For **Extra Practice,** see page 637.

ON YOUR OWN

5. The last step of the 4-step plan for problem solving asks you to *examine* your solution. *Explain* what you need to check if you are using a model to solve a problem.

6. *Write a Problem* that can be solved by making a model.

7. *Look Ahead* The figure is called a rectangular prism. Use cubes to make a model of a rectangular prism that is 3 units long, 2 units wide, and 4 units high.

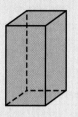

MIXED PROBLEM SOLVING

STRATEGIES

Look for a pattern.
Solve a simpler problem.
Act it out.
Guess and check.
Draw a diagram.
Make a chart.
Work backward.

Solve. Use any strategy.

8. *Education* Ricardo is reading a 216-page book for his book report. He needs to read twice as many pages as he has already read to finish the book. How many pages has he read so far?

9. **Standardized Test Practice** Four cubes are glued together. If you could pick them up and look at them from all sides, which of these arrangements shows the fewest squares?

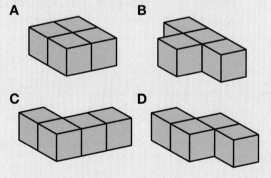

A

B

C

D

10. *Design* Edu-Toys is designing a new package to hold a set of 30 alphabet blocks. Each block is a cube with each edge of the cube being 2 inches long. Give two possible dimensions for the box.

11. *Architecture* The Sears Tower in Chicago is one of the tallest structures in the world. It is actually a building of varying heights. The diagram shows a top view of the building with the number of stories for each section indicated. Use grid paper to draw the view of the Sears Tower from each side.

50 stories	89 stories	66 stories
110 stories	110 stories	89 stories
66 stories	89 stories	50 stories

12. *Money Matters* Dion paid $45 for a jacket that was on sale at 40% off. What was the original price of the jacket?

13. **Standardized Test Practice** Which of the following is *not* a top or side view of the figure?

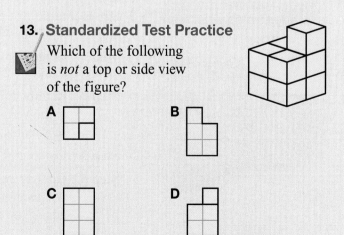

A

B

C

D

11-2 Three-Dimensional Figures

What you'll learn

You'll learn to identify and sketch three-dimensional figures.

When am I ever going to use this?

Knowing how to sketch three-dimensional figures can help you make drawings in art class.

Word Wise

solid
prism
face
edge
vertex
base
pyramid

Geologists classify crystals by the number of flat surfaces and the angles at which the surfaces meet.

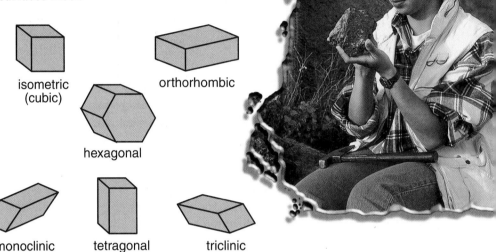

isometric (cubic)

orthorhombic

hexagonal

monoclinic

tetragonal

triclinic

In geometry, these three-dimensional figures are called **solids**. Some common geometric solids are shown below

rectangular prism   triangular prism   pyramid   cone   cylinder

You can use isometric dot paper to draw geometric solids.

Example 1

Use isometric dot paper to sketch a rectangular prism that is 3 units high, 1 unit long, and 2 units wide.

Step 1 Lightly draw the bottom of the prism 1 unit by 2 units.

Step 2 Lightly draw the vertical segments at the vertices of the base. Each segment is 3 units high.

Step 3 Complete the top of the prism.

Step 4 Go over your lines. Use dashed lines for the edges of the prism you cannot see from your perspective and solid lines for the edges you can see.

You could draw several other perspectives of the same prism.

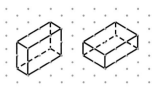

Prisms, like the ones above, have flat surfaces. The surfaces of a prism are called **faces**. The faces meet to form the **edges** of the prism. The edges meet at corners called **vertices**.

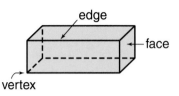

All prisms have at least one pair of faces that are parallel and congruent. These are called **bases** and are used to name the prism. The prisms shown above are *rectangular prisms*.

A **pyramid** is a solid that has a polygon for a base and triangles for sides. It is named according to its base.

 **Examples**

② Describe the solid in the drawing. Include its name and the number of faces, edges, and vertices.

The base of the figure is a hexagon. The other faces are triangles. The figure is a hexagonal pyramid. It has a total of 7 faces. It has 12 edges and 7 vertices.

APPLICATION

③ **Architecture** An architect's sketch shows the plans for a pedestal for a sculpture in front of a library.

a. What are the dimensions of the bottom of the base?

The base is 8 units by 3 units.

b. What are the dimensions of the top of the center section?

The top is 2 units by 3 units.

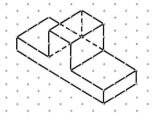

c. What is the area of the top of the center section?

The area is 2 × 3 or 6 square units.

Communicating Mathematics

Read and study the lesson to answer each question.

1. *Explain* which lines should be solid and which lines should be dashed when drawing a solid figure.

2. *Name* each figure.

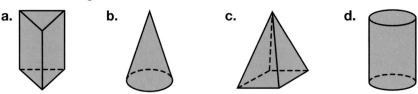

a.   b.   c.   d.

3. *Write* a paragraph comparing a pentagonal prism and a pentagonal pyramid. Include a drawing of each solid.

Guided Practice

4. Use isometric dot paper to draw a triangular pyramid with an isosceles triangle as its base.

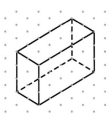

5a. Name the solid at the right.
  b. Give the dimensions of the solid.
  c. How many faces does the solid have?
  d. How many edges does the solid have?
  e. How many vertices does the solid have?
  f. Draw the solid from a different perspective.

6. *Pets* The Roosevelt Middle School science classes have a pet lizard that lives in an aquarium with a hexagonal base and height of 4 units. Sketch the aquarium.

Practice   **Use isometric dot paper to draw each solid.**

7. a rectangular prism that is 3 units high, 4 units long, and 5 units deep

8. a triangular prism that is 5 units high

9. a pyramid with a square base

10a. Name the solid at the right.
   b. What is the height of the solid?
   c. How many faces does the solid have?
   d. How many edges does the solid have?
   e. How many vertices does the solid have?
   f. Draw the solid from a different perspective.

11. Write a few sentences to tell how you would draw a cylinder. Then draw one.

12. Write a few sentences to tell how you would draw a cone. Then draw one.

13. On grid paper, draw the front view, back view, two side views, and top view of the figure.

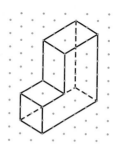

14. *History* Solids with faces that are polygons are called *polyhedrons.* During the 1700s, Swiss mathematician Leonard Euler discovered a relationship among the vertices (*V*), faces (*F*), and edges (*E*), of polyhedrons.

   a. Copy and complete the table.

Solid	Vertices (V)	Faces (F)	Edges (E)	V + F − E
triangular prism	6	5	9	6 + 5 − 9 = 2
triangular pyramid				
rectangular prism				
rectangular pyramid				
pentagonal prism				

Leonard Euler

   b. What is the relationship among the vertices, faces, and edges of polyhedrons?

   c. If a prism has 10 faces and 24 edges, how many vertices does it have? Name the prism.

15. *Interior Design* Toshiko is arranging storage cubes to form a wall unit in her bedroom. Sketch two designs that can be made using 6 cubes.

16. *Working on the* **CHAPTER Project** Refer to the kite you chose on page 475. Make a three-dimensional scale drawing of your kite. Label the measurements.

17. *Critical Thinking* *True* or *False*? If true, give an example. If false, explain why. Three planes in three-dimensional space can:

   a. intersect in a line.

   b. intersect in a point.

   c. have no intersection at all.

Mixed Review

18. *Sports* During an attempted foul shot, the shooter stands behind the foul line. All other players must stay out of the shaded region. What is the area of this region? *(Lesson 11-1)*

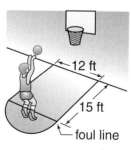
12 ft
15 ft
foul line

For **Extra Practice,** see page 637.

19. **Standardized Test Practice** The formula $A = s^2$ is used to find the area of a square. If the area of a square is 40 mm², what is the approximate length of one side of the square? *(Lesson 9-3)*

   A 6.3 mm   **B** 7.5 mm   **C** 18.6 mm   **D** 25.2 mm

Volume of Prisms and Cylinders

What you'll learn

You'll learn to find the volumes of prisms and cylinders.

When am I ever going to use this?

Knowing how to find volume can help you determine how much ice a cooler can hold.

Word Wise

volume
circular cylinder

SHOE

In the cartoon, the word **volume** has two meanings. Volume can mean one book in a set. However in geometry, volume is the measure of the space occupied by a solid. It is measured in cubic units. Two common units of measure for volume are cubic centimeter (cm^3) and cubic inch (in^3).

HANDS-ON MINI-LAB

Work with a partner.   12 cubes

Try This

- Model the prism at the right.
- Model at least three other rectangular prisms.
- Assume each edge of the cubes represents one unit and each cube represents one cubic unit. Copy and complete the following table.

Prism	Area of Base	Height	Volume
1	4 square units	2 units	8 cubic units
2			
3			
4			

Talk About It

1. Describe how the area of the base and the height of a prism are related to its volume.
2. Write a formula for computing the volume of a prism.

Study Hint

Reading Math
Remember that the *B* in the formula $V = Bh$ represents the area of the base.

Volume of a Prism	**Words:** The volume (*V*) of a prism is the area of the base (*B*) times the height (*h*).	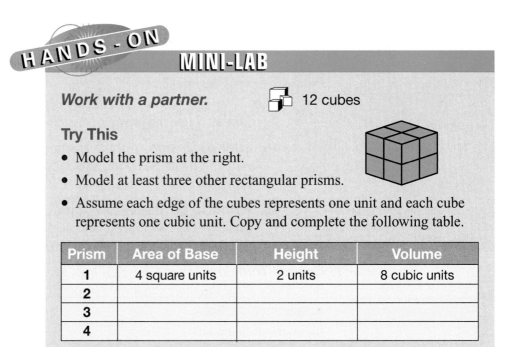
	Symbols: $V = Bh$ **Model:**	

In the case of a rectangular prism, the area of the base (B) equals the length (ℓ) times the width (w). For a rectangular prism, the formula $V = Bh$ becomes $V = \ell wh$.

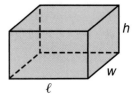

Examples

Find the volume of each prism.

1

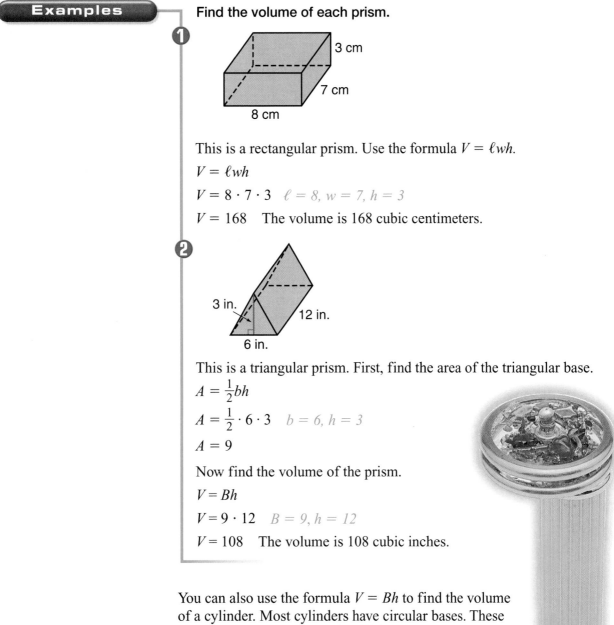

3 cm

7 cm

8 cm

This is a rectangular prism. Use the formula $V = \ell wh$.

$V = \ell wh$

$V = 8 \cdot 7 \cdot 3$ $\ell = 8, w = 7, h = 3$

$V = 168$ The volume is 168 cubic centimeters.

2

3 in.

12 in.

6 in.

This is a triangular prism. First, find the area of the triangular base.

$A = \frac{1}{2}bh$

$A = \frac{1}{2} \cdot 6 \cdot 3$ $b = 6, h = 3$

$A = 9$

Now find the volume of the prism.

$V = Bh$

$V = 9 \cdot 12$ $B = 9, h = 12$

$V = 108$ The volume is 108 cubic inches.

You can also use the formula $V = Bh$ to find the volume of a cylinder. Most cylinders have circular bases. These cylinders are called **circular cylinders**. For a circular cylinder, the area of the base is the area of a circle (πr^2). So, the formula for finding the volume of a cylinder is $V = \pi r^2 h$.

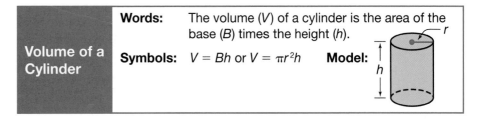

| **Volume of a Cylinder** | **Words:** | The volume (V) of a cylinder is the area of the base (B) times the height (h). |
| | **Symbols:** | $V = Bh$ or $V = \pi r^2 h$ **Model:** |

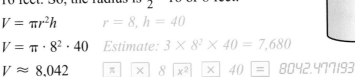

Example

APPLICATION

Real World

3 **Agriculture** The diagram shows the dimensions of a silo. What is the volume of the grain that can be stored in the silo?

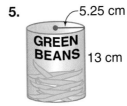

─16 ft─→

40 ft

The diameter of the base of the cylinder is 16 feet. So, the radius is $\frac{1}{2} \cdot 16$ or 8 feet.

$V = \pi r^2 h$ *r = 8, h = 40*

$V = \pi \cdot 8^2 \cdot 40$ *Estimate: $3 \times 8^2 \times 40 = 7,680$*

$V \approx 8,042$ [π] [×] 8 [x²] [×] 40 [=] *8042.477193*

The volume of the grain is about 8,042 cubic feet.

CHECK FOR UNDERSTANDING

Communicating Mathematics

Read and study the lesson to answer each question.

1. *Compare and contrast* the formula used for finding the volume of a rectangular prism, a triangular prism, and a cylinder.

2. *Draw* a cube. Let *s* represent the length of each side of the cube. Write a formula for the volume of a cube.

HANDS-ON MATH

3. *Model* a rectangular prism using 18 cubes.
   a. Find the area of the base of the prism.
   b. Find the height of the prism.
   c. Find the product of the area of the base times the height of the prism. How does this product relate to the volume of the prism?

Guided Practice

Find the volume of each solid to the nearest tenth.

4.

JUICE   4 in.

3 in. $1\frac{1}{2}$ in.

5.   ─5.25 cm
GREEN BEANS   13 cm

6. *Energy* Strategic petroleum reserves at Bryan Mound, Texas, are connected to port and terminal facilities at Freeport, Texas, by two pipelines. Each pipeline is $2\frac{1}{2}$ feet in diameter and 4 miles long. Find the maximum volume of the petroleum that can be in two pipelines.

EXERCISES

Practice

Find the volume of each solid to the nearest tenth.

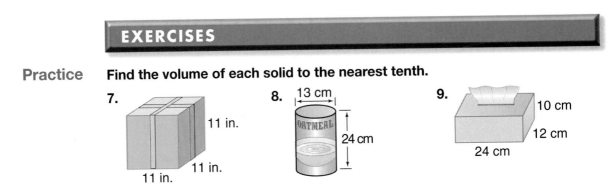

7.
11 in.

11 in.

11 in.

8.   13 cm
OATMEAL   24 cm

9.
10 cm

12 cm

24 cm

10.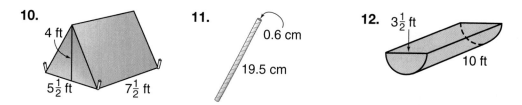

4 ft

$5\frac{1}{2}$ ft $7\frac{1}{2}$ ft

11. 0.6 cm

19.5 cm

12. $3\frac{1}{2}$ ft

10 ft

13. Estimate the volume of a cylinder that is 10 feet tall and whose diameter is 14 feet.

14. For a rectangular prism, $\ell = 3$, $w = 2$, and $h = 5$.
   a. Find the volume of the prism.
   b. Suppose the height of the prism is doubled. What is the volume of the new prism?
   c. What is the ratio of the volume of the new prism to the volume of the original prism?
   d. What happens to the volume of a prism if the height is doubled?

15. For a rectangular prism, $\ell = 4$, $w = 1$, and $h = 6$.
   a. Find the volume of the prism.
   b. Suppose all three of the dimensions of the prism are doubled. What is the volume of the new prism?
   c. What is the ratio of the volume of the new prism to the volume of the original prism?
   d. What happens to the volume of a prism if all three dimensions are doubled?

16. What happens to the volume of a cylinder if the radius is doubled?

Applications and Problem Solving

Real World

17. *Health* The inside of a refrigerator in a medical laboratory measures 17 inches by 18 inches by 42 inches.
   a. Tanisha Adams estimates she needs at least 8 cubic feet to refrigerate the samples from the lab. Is the refrigerator large enough for the samples? Explain.
   b. Tell how you would change the dimensions of the refrigerator to double its volume.

18. *Environment* The cylindrical flue of an incinerator has a diameter of 30 inches and a height of 120 feet. If it takes 15 minutes for the contents of the flue to be expelled into the air, what is the volume of the substances being expelled each hour?

19. *Critical Thinking* Write the nonlinear function for each prism. Then graph your function.
   a. The volume, V, of a cube as a function of the edge length a.
   b. The volume, V, of a rectangular prism as a function of a fixed height of 5 and a square base of length s.
   c. The volume, V, of a triangular prism as a function of a fixed height of 4 and an equilateral triangular base of length b.

Mixed Review

20. Draw a hexagonal prism that is 5 units tall. *(Lesson 11-2)*

21. **Standardized Test Practice** Doralina is enclosing a circular area of her yard that measures 31 feet in diameter. How much fencing will she need to buy? *(Lesson 7-7)*

   A 48.69 ft   **B** 97.39 ft   **C** 754.77 ft   **D** 3,019.07 ft

22. *Algebra* Solve $2x - 8 = 12$. *(Lesson 1-7)*

For **Extra Practice,** see page 637.

Volume of Pyramids and Cones

What you'll learn

You'll learn to find the volumes of pyramids and cones.

When am I ever going to use this?

Knowing how to find the volume of cones can help you determine the volume of a mound of sand.

Word Wise

circular cone
altitude

Do you like to buy ice cream in a cone? A sugar cone used for ice cream is an example of the geometric solid called a **circular cone**.

A segment that goes from the vertex of the cone to its base and is perpendicular to the base is called the **altitude**. The height of a cone is measured along the altitude.

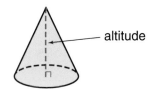

altitude

HANDS-ON MINI-LAB

Work with a partner. 📁 construction paper   📖 rice

⊚ compass   📏 ruler   ✂ scissors   🎞 tape

Try This

- Draw and cut out a circle with a radius of $1\frac{1}{2}$ inches. Draw and cut out a rectangle that is $9\frac{3}{4}$ inches by $2\frac{5}{8}$ inches. Wrap and tape the rectangle around the circle to form a cylinder with an open top.

- Draw and cut out a circle with a radius of 3 inches. Fold the circle into quarters and open it to form a cone.

- Set the cylinder and cone on your desk with the base of each figure down. Are the solids about the same height?

- Place the base of the cone on top of the cylinder. Are the circles that form the bases of the cylinder and cone the same size?

- Fill the cone with rice. Slide the ruler across the top of the rice to make sure the cone is full. Pour the rice into the cylinder. Repeat until the cylinder is filled.

Talk About It

1. How many times did you fill the cone in order to fill the cylinder?

2. How is the area of the base of the cone related to the area of the base of the cylinder?

3. A cone and a cylinder have the same base and height. What is the ratio of the volume of the cone to the volume of the cylinder?

4. Write a formula for finding the volume of a cone.

The results of the Mini-Lab suggest the formula for finding the volume of a cone.

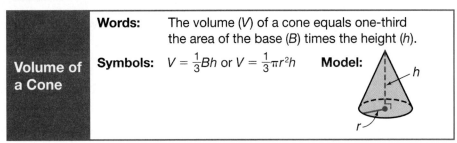

Volume of a Cone	**Words:** The volume (V) of a cone equals one-third the area of the base (B) times the height (h).
	Symbols: $V = \frac{1}{3}Bh$ or $V = \frac{1}{3}\pi r^2 h$ **Model:**

Example

1 **Business** Manuel fills parfait glasses with vanilla custard and strawberry whip at the Harvest Restaurant. The glasses are in the shape of a cone. If each glass is 7 centimeters across and 15 centimeters tall, what is the volume of each dessert?

If the glass is 7 centimeters across, the radius of the base of the cone is $\frac{1}{2} \times 7$ or 3.5 centimeters.

$V = \frac{1}{3}(\pi r^2)h$ *Replace r with 3.5 and h with 15.*

$V = \frac{1}{3}(\pi \cdot 3.5^2) \cdot 15$ *Estimate: $\frac{1}{3} \cdot 3 \cdot 4^2 \cdot 15 = 240$*

$V \approx 192.4$ 1 $\div$ 3 $\times$ π $\times$ 3.5 x^2 $\times$ 15 $=$ *192.42255*

The volume of each dessert is about 192.4 cubic centimeters.

The relationship between a pyramid and a prism is similar to the relationship between a cylinder and a cone. If you have a pyramid and a prism with the same base and height, the ratio of the volume of the pyramid to the volume of the prism is 1 to 3.

Volume of a Pyramid	**Words:** The volume (V) of a pyramid equals one-third the area of the base (B) times the height (h).
	Symbols: $V = \frac{1}{3}Bh$ **Model:**

Example

2 A tetrahedron is a pyramid with a triangular base. Find the volume of the tetrahedron.

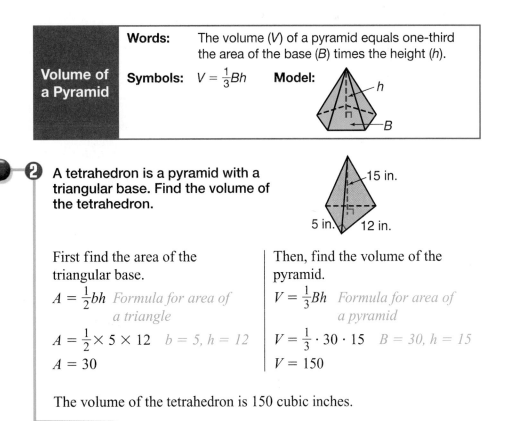

First find the area of the triangular base.

$A = \frac{1}{2}bh$ *Formula for area of a triangle*

$A = \frac{1}{2} \times 5 \times 12$ *b = 5, h = 12*

$A = 30$

Then, find the volume of the pyramid.

$V = \frac{1}{3}Bh$ *Formula for area of a pyramid*

$V = \frac{1}{3} \cdot 30 \cdot 15$ *B = 30, h = 15*

$V = 150$

The volume of the tetrahedron is 150 cubic inches.

Communicating Mathematics

Read and study the lesson to answer each question.

1. ***Explain*** why $V = \frac{1}{3}Bh$ and $V = \frac{1}{3}\pi r^2 h$ can both be used to find the volume of a cone.

2. ***Compare and contrast*** the formulas for the volume of a prism and the volume of a pyramid.

HANDS-ON MATH

3a. Draw and cut out five squares that are 2 inches on each side. Tape the squares together to form a cube with an open top.

   b. Draw and cut out four isosceles triangles with a base of 2 inches and a height of $2\frac{1}{4}$ inches. Tape the triangles together to form a square pyramid without the base.

   c. Compare the bases and heights of the prism and the pyramid.

   d. Fill the pyramid with rice. Slide a ruler across the top to make sure the pyramid is full. Pour the rice into the prism. Repeat until the prism is filled. How many times did you need to fill the pyramid to fill the prism?

   e. Suppose you have a pyramid and a prism with the same base and height. What do you think would be the ratio of the volume of the pyramid to the volume of the prism?

Guided Practice

Find the volume of each solid to the nearest tenth.

4.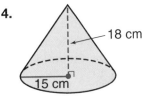
18 cm
15 cm

5.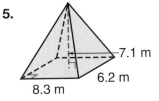
7.1 m
6.2 m
8.3 m

6. ***Physical Science*** A model of a volcano constructed for science class has a height of 10 inches and a diameter of 9 inches. What is the volume of the volcano?

Practice

Find the volume of each solid to the nearest tenth.

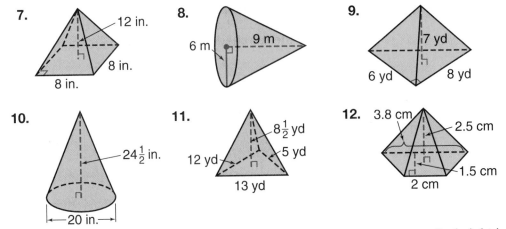

7. 12 in.
8 in.
8 in.

8. 6 m
9 m

9. 7 yd
6 yd
8 yd

10. $24\frac{1}{2}$ in.
20 in.

11. $8\frac{1}{2}$ yd
5 yd
12 yd
13 yd

12. 3.8 cm
2.5 cm
1.5 cm
2 cm

13. The area of the base of a hexagonal pyramid is 45.3 square meters. Its height is 19.7 meters. Estimate the volume of the pyramid.

14. The radius of a cone is 5 inches. The height is 8 inches.
   a. Find the volume of the cone.
   b. Suppose the radius of the cone is doubled and the height remains the same. What is the volume of the new cone?
   c. What is the ratio of the volume of the new cone to the volume of the original cone?
   d. What happens to the volume of a cone if the radius is doubled?

15. What happens to the volume of a pyramid if the height is doubled?

16. A cone and a cylinder each have a diameter of 10 centimeters. A prism and a pyramid each have square bases that are 10 centimeters on a side. All of the solids have a height of 8 centimeters. Without calculating the volume of the solids, order the solids according to their volume starting with the solid with the least volume.

Applications and Problem Solving

17. *Architecture* The Great American Pyramid in Memphis, Tennessee, was built as a memorial to American music. It is 321 feet high and has a base that is about 296,000 square feet. Find the approximate volume of this pyramid.

18. *Highway Maintenance* A mixture of salt and sand is used to melt snow and ice on highways. A mound of this mixture is in a conical shape. It has a diameter of 18 feet and a height of 12 feet.
   a. Find the volume of the salt and sand.
   b. If 1 cubic foot of this mixture is needed for each 500 square feet of highway, how many square feet of highway can be treated with the salt and sand in this mound?

Great American Pyramid

19. *Critical Thinking* Suppose the radius of a cone is doubled. How could you change the height so that the volume will remain the same?

Mixed Review

20. **Standardized Test Practice** What is the volume of a rectangular solid that measures 4 feet by 7 feet by 3 feet? *(Lesson 11-3)*
   A 14 ft³   **B** 28 ft³   **C** 84 ft³   **D** 128 ft³

21. *Money Matters* Find the percent of change if last month's electric bill was $52.50 and this month's bill is $57.20. *(Lesson 8-5)*

22. Order 23, −8, 0, −16, 51, −51, and −30 from greatest to least. *(Lesson 2-2)*

For **Extra Practice,** see page 638.

CHAPTER 11   Mid-Chapter Self Test

1. *Sports* In track, a shot-putter must stay inside a circle with a diameter of 7 feet. What is the area in which the athlete has to move in this competition? *(Lesson 11-1)*

2. Use isometric dot paper to draw a rectangular prism that is 2 units high, 5 units long, and 3 units deep. *(Lesson 11-2)*

Find the volume of each solid to the nearest tenth. *(Lessons 11-3 and 11-4)*

3.   4 cm   8 cm   16 cm

4. 14 in.   6 in.

5. 5 m   2 m

COOPERATIVE LEARNING

11-5A Nets

A Preview of Lesson 11-5

Every solid with at least one flat surface can be formed from a two-dimensional pattern called a *net*.

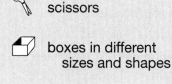 scissors

boxes in different sizes and shapes

ruler

TRY THIS

Work with a partner.

Step 1 Each person should pick at least one empty box. You can choose a cereal box, a pasta box, or any other type of box. Tape the box shut.

Step 2 Flatten each box by cutting along some of the edges. You should have only one piece for each box when it is flattened. Try to cut each box in a different way.

Step 3 Trace or sketch each flattened box showing each of the shapes used for the sides of the box. Your drawing is called a net.

Step 4 Cut out each net. Fold the net along the interior lines to form a box.

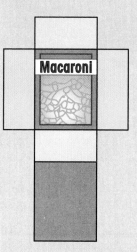

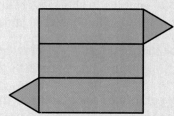

ON YOUR OWN

1. **Describe** each shape that makes up each net.
2. **Tell** how many shapes are needed for each net.
3. **Identify** any congruent shapes in each net.
4. **Describe** the solid formed by the net shown at the right. Sketch the solid.

5. **Look Ahead** Use one of your boxes.
   a. Measure the length, width, and height of the box.
   b. Estimate the total area of the sides of the box.
   c. Compute the area of each side of the box. Record the area of each surface on your net.
   d. Find the sum of the areas of the sides of the box.

Surface Area of Prisms

What you'll learn

You'll learn to find the surface areas of rectangular and triangular prisms.

When am I ever going to use this?

Knowing how to find surface area can help you determine the amount of canvas needed to construct a tent.

Word Wise

surface area

The faces of the three-dimensional figures that form Tony Smith's modern sculpture *Wandering Rocks* are constructed of stainless steel. Mr. Smith had to know the **surface area**

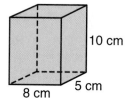

Wandering Rocks

of the sculpture to determine the amount of stainless steel to order. Surface area is the sum of the areas of all faces or surfaces of a solid.

Study Hint

Reading Math
Remember that volume is given in cubic units and surface area is given in square units.

HANDS-ON MINI-LAB

Work with a partner.

grid paper   scissors

Try This

tape

- Draw the pattern on grid paper and cut it out. *This pattern is an example of a net.*

- Fold the pattern along the red lines and tape the edges to form a rectangular prism.

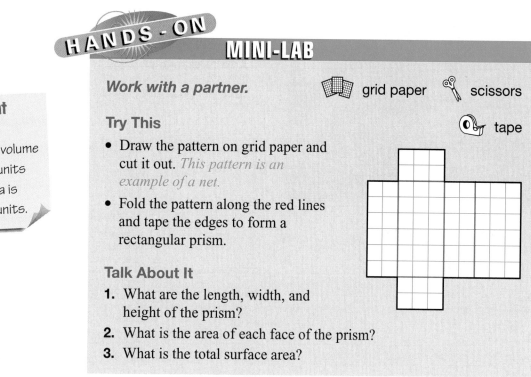

Talk About It

1. What are the length, width, and height of the prism?
2. What is the area of each face of the prism?
3. What is the total surface area?

To find the surface area of a rectangular prism, you must first determine the area of each of its six faces.

Example ①

Find the surface area of the rectangular prism.

Faces	Area
top and bottom	8×5 or 40
front and back	8×10 or 80
right and left sides	5×10 or 50

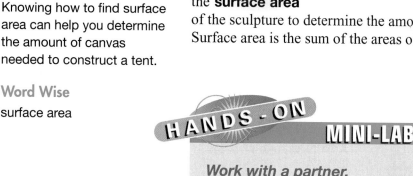

10 cm

8 cm   5 cm

Add the areas.
$2(40) + 2(80) + 2(50) = 340$
The surface area of the rectangular prism is 340 square centimeters.

Camping The Kennedy Middle School Camping Club has designed a pup tent. The tent has a canvas floor as well as canvas sides. How much canvas will the students need to construct the tent?

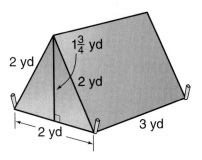

Explore   You know the dimensions of a tent that is in the shape of a triangular prism. You need to find the amount of canvas needed to construct the tent.

Plan   To find the amount of canvas needed, you must find the surface area of a triangular prism. To find the surface area, you must find the area of each face of the prism.

Solve   Two faces of the tent are triangles, and the area of each triangle is $\frac{1}{2} \times 2 \times 1\frac{3}{4}$ or $1\frac{3}{4}$ square yards. Three faces of the tent are rectangles, and the area of each rectangle is 3×2 or 6 square yards.

    area of triangular faces $2 \times 1\frac{3}{4}$ or $3\frac{1}{2}$

    area of rectangular faces 3×6 or <u>18 </u>

                                             Total $21\frac{1}{2}$

The students will need $21\frac{1}{2}$ square yards of canvas.

Examine

Faces	Area		
front	$\frac{1}{2} \times 2 \times 1\frac{3}{4}$	or	$1\frac{3}{4}$
back	$\frac{1}{2} \times 2 \times 1\frac{3}{4}$	or	$1\frac{3}{4}$
floor	3×2	or	6
right side	3×2	or	6
left side	3×2	or	<u>6 </u>
		Total	$21\frac{1}{2}$ ✓

CHECK FOR UNDERSTANDING

Communicating Mathematics

Read and study the lesson to answer each question.

1. *Explain* why the expression $2\ell h + 2\ell w + 2wh$ could be used to find the surface area of the rectangular prism.

2. *Write* an expression for the surface area of a cube if each edge is s units long.

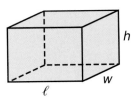

3. *Draw* a net for a rectangular prism that is 2 units by 3 units by 4 units on grid paper. Draw a net for a rectangular prism that is 4 units by 6 units by 8 units on grid paper.

   a. Find the surface area of each prism.

   b. Find the ratio of the length of each side of the first prism to the length of the corresponding side of the second prism.

   c. Write a ratio comparing the surface area of the first prism to the surface area of the second prism.

   d. Do the ratios from part b and c form a proportion? Explain.

Guided Practice

Family
Activity

Look for rectangular containers in your home. Choose one and work with a family member to design and cut out all the possible nets that will completely cover the solid without overlapping.

Find the surface area of each prism to the nearest tenth.

4. 8 yd   10 yd   6 yd   20 yd

5. 15 in.   12 in.   5 in.

6. Estimate the surface area of a rectangular prism with a length of 2.06 cm, a width of 1.89 cm, and a height of 0.91 cm.

7. *Sports* A water skiing ramp is built in the shape of a triangular prism. Salali is planning to paint it with water repellent paint. Find the surface area to be painted.

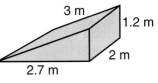
3 m   1.2 m   2 m   2.7 m

EXERCISES

Practice   **Find the surface area of each prism to the nearest tenth.**

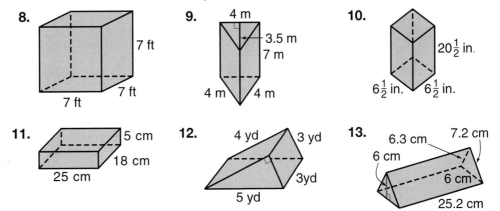

8. 7 ft   7 ft   7 ft

9. 4 m   3.5 m   7 m   4 m   4 m

10. $20\frac{1}{2}$ in. $6\frac{1}{2}$ in. $6\frac{1}{2}$ in.

11. 5 cm   18 cm   25 cm

12. 4 yd   3 yd   3yd   5 yd

13. 6.3 cm   7.2 cm   6 cm   6 cm   25.2 cm

14. Find the surface area of a triangular prism whose bases are right triangles with legs 6 feet and 8 feet and whose height is 10 feet.

15. Estimate the surface area of a cube with each edge 9.7 meters long.

16. Suppose the length of each edge of a cube is doubled. Find the ratio of the surface area of the original cube to the surface area of the new cube.

17. *Write a Problem* that requires computing the surface area of a prism.

Real World

Applications and Problem Solving

18. *Horticulture* Find the area of the glass needed to cover the roof and sides of the greenhouse.

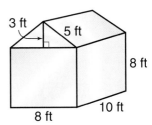
3 ft   5 ft   8 ft   8 ft   10 ft

19. *Physical Science* The rate that ice melts depends on the amount of surface area. The greater the surface area, the faster it will melt. Which of the following blocks of ice will be the last to melt? Explain.

    block 1: 1 inch by 2 inches by 32 inches

    block 2: 4 inches by 8 inches by 2 inches

    block 3: 16 inches by 4 inches by 1 inch

    block 4: 4 inches by 4 inches by 4 inches

20. *Critical Thinking* The length of each side of a cube is 3 inches. Suppose the cube is painted and then it is cut into 27 smaller congruent cubes.

    a. How many of the smaller cubes will be painted on three sides?

    b. How many of the smaller cubes will be painted on two sides?

    c. How many of the smaller cubes will be painted on one side?

    d. How many of the smaller cubes will not be painted at all?

For **Extra Practice**, see page 638.

21. *Critical Thinking* Draw a diagram to find the volume and surface area of a square-based pyramid whose lateral faces are equilateral triangles with each side equal to 6.

Mixed Review

22. *Algebra* Solve the system $y = -x + 5$ and $y = 2x - 4$ by graphing. *(Lesson 10-5)*

23. **Standardized Test Practice** What is the GCF of 84 and 36? *(Lesson 6-3)*

    A 2   **B** 3   **C** 6   **D** 12

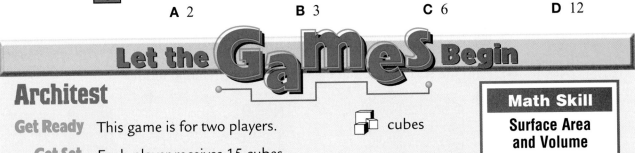

Let the Games Begin

Architest

Math Skill

Surface Area and Volume

Get Ready This game is for two players. [cubes]

Get Set Each player receives 15 cubes.

Go
- Each player designs a structure with some of his or her cubes. The player then draws the top, front, back, and side views of the structure and computes its volume and surface area. Assume each cube represents one cubic unit.

- Player A tries to guess Player B's structure. Player A does this by asking Player B for information about the structure. Pieces of information include the volume, surface area, or drawing of one view of the structure. Player A receives 7 points for correctly guessing Players B's structure after receiving one piece of information, 6 points for correctly guessing after two pieces of information, and so on. If Player A cannot guess Player B's structure after receiving all 7 pieces of information, then Player B receives 3 points.

- Player B now tries to guess Player A's structure, according to the same rules.

- Play continues with different structures. The player with the most points at the end of the game wins.

*inter*NET CONNECTION Visit www.glencoe.com/sec/math/mac/mathnet for more games.

Surface Area of Cylinders

What you'll learn

You'll learn to find the surface areas of cylinders.

When am I ever going to use this?

Knowing how to find the surface area of cylinders can help you determine the amount of sheet metal needed to make a can.

People in the United States consume more soft drinks than any other type of beverage. Soft drinks are usually sold in containers in the shape of cylinders because they are easy to hold, relatively easy to manufacture, and can be easily sealed. The amount of material needed to manufacture a beverage container depends on the surface area of the cylinder.

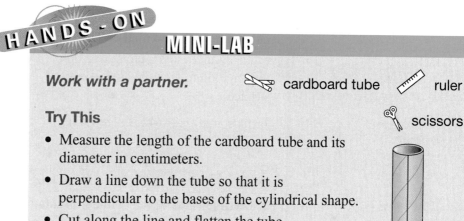

As with prisms, you find the surface area of a cylinder by finding the area of the two bases and adding the area of the side. However, the "side" of a cylinder is one curved surface.

HANDS-ON MINI-LAB

Work with a partner.   cardboard tube   ruler   scissors

Try This

- Measure the length of the cardboard tube and its diameter in centimeters.
- Draw a line down the tube so that it is perpendicular to the bases of the cylindrical shape.
- Cut along the line and flatten the tube.

Talk About It

1. Describe the shape of the flattened tube. What part of the cylinder does the flattened tube represent?
2. Measure the dimensions of the flattened tube in centimeters.
3. Find the circumference of either base of the original tube. How is this measure related to the dimensions of the flattened tube?
4. How is the height of the tube related to the dimensions of the flattened tube?
5. What is the area of the flattened tube?
6. Write an expression for the area of the curved surface of a cylinder using d for the diameter of the cylinder and h for the height of the cylinder.

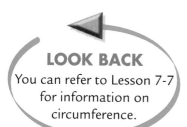

LOOK BACK
You can refer to Lesson 7-7 for information on circumference.

In the Mini-Lab, you see that the curved side of the cylinder can be flattened into a rectangle. The rectangle has the same height (h) as the cylinder. Its length is the same measure as the circumference (C) of the base. So, the area of the curved surface is Ch or πdh or $2\pi rh$.

The area of each base of a cylinder is πr^2. So the sum of the areas of the bases is $\pi r^2 + \pi r^2$, or $2\pi r^2$.

The surface area of a cylinder is the sum of the areas of the two circular bases and the area of the curved surface.

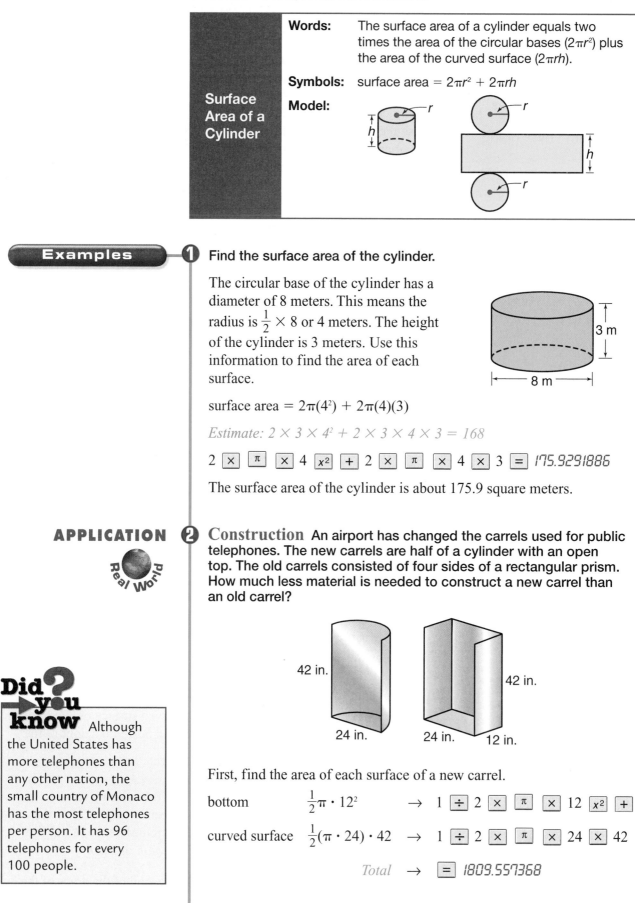

	Words:	The surface area of a cylinder equals two times the area of the circular bases ($2\pi r^2$) plus the area of the curved surface ($2\pi rh$).
Surface Area of a Cylinder	**Symbols:**	surface area $= 2\pi r^2 + 2\pi rh$
	Model:	

Examples

1 Find the surface area of the cylinder.

The circular base of the cylinder has a diameter of 8 meters. This means the radius is $\frac{1}{2} \times 8$ or 4 meters. The height of the cylinder is 3 meters. Use this information to find the area of each surface.

3 m

8 m

surface area $= 2\pi(4^2) + 2\pi(4)(3)$

Estimate: $2 \times 3 \times 4^2 + 2 \times 3 \times 4 \times 3 = 168$

2 $\boxed{\times}$ $\boxed{\pi}$ $\boxed{\times}$ 4 $\boxed{x^2}$ $\boxed{+}$ 2 $\boxed{\times}$ $\boxed{\pi}$ $\boxed{\times}$ 4 $\boxed{\times}$ 3 $\boxed{=}$ *175.9291886*

The surface area of the cylinder is about 175.9 square meters.

APPLICATION

Real World

2 **Construction** An airport has changed the carrels used for public telephones. The new carrels are half of a cylinder with an open top. The old carrels consisted of four sides of a rectangular prism. How much less material is needed to construct a new carrel than an old carrel?

42 in.

42 in.

24 in.

24 in.

12 in.

First, find the area of each surface of a new carrel.

bottom $\frac{1}{2}\pi \cdot 12^2$ →   1 $\boxed{\div}$ 2 $\boxed{\times}$ $\boxed{\pi}$ $\boxed{\times}$ 12 $\boxed{x^2}$ $\boxed{+}$

curved surface $\frac{1}{2}(\pi \cdot 24) \cdot 42$ →   1 $\boxed{\div}$ 2 $\boxed{\times}$ $\boxed{\pi}$ $\boxed{\times}$ 24 $\boxed{\times}$ 42

Total → $\boxed{=}$ *1809.557368*

Then, find the area of each surface of an old carrel.

bottom	24×12	$\rightarrow$	24 $\boxed{\times}$ 12 $\boxed{+}$
back	24×42	$\rightarrow$	24 $\boxed{\times}$ 42 $\boxed{+}$
right side	12×42	$\rightarrow$	12 $\boxed{\times}$ 42 $\boxed{+}$
left side	12×42	$\rightarrow$	12 $\boxed{\times}$ 42
	Total	$\rightarrow$	$\boxed{=}$ *2304*

About $2{,}304 - 1{,}810$ or 494 square inches of material are saved.

CHECK FOR UNDERSTANDING

Communicating Mathematics

Read and study the lesson to answer each question.

1. ***Explain*** why the expression $2\pi r^2 + 2\pi rh$ could be used to find the surface area of the cylinder.

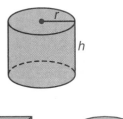

2. ***Write a Problem*** where you need to find the surface area of a cylinder.

3. ***You Decide*** The two solids have about the same volume. Geraldo says that the surface area of the cylinder is less than the surface area of the rectangular solid. Katie disagrees. Who is correct? Explain.

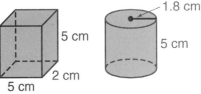

HANDS-ON MATH

4. ***Make*** a net of a cylinder. Trace the bases of a can on grid paper. Cut out the circles. Measure the height of the can. Cut a long strip of grid paper so that its width is the height of the can. Wrap the strip around the can. Cut the excess paper off so that the strip just fits around the can.

   a. Describe the shape of each of the three pieces you have created

   b. Assume each square on the grid represents one square unit. Use the three pieces to estimate the surface area of the can.

Guided Practice

5. Find the surface area of the cylinder to the nearest tenth.

6. A cylinder has a diameter of 7.88 centimeters and a height of 6.04 centimeters. Estimate the area of the curved surface of the cylinder.

7. ***Community Services*** A community is building a cylindrical tank to store water. The diameter of the tank will be 20 feet, and its height will be 24 feet. The tank will be lined with a material to prevent corrosion and contamination. How much of this material will be needed to line the tank?

Practice **Find the surface area of each cylinder to the nearest tenth.**

8.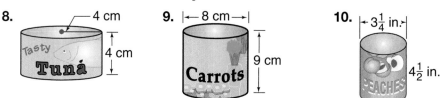
4 cm

4 cm

Tasty
Tuna

9. |←— 8 cm —→|

9 cm

Carrots

10. ←$3\frac{1}{4}$ in.→

$4\frac{1}{2}$ in.

PEACHES

11. Find the area of the curved surface of a cylinder with a radius of 3 feet and a height of 9 feet.

12. Find the surface area of a cylinder with a diameter of 4.6 meters and a height of 4.6 meters.

13. Estimate the surface area of a cylinder with a diameter of 10 inches and a height of 10 inches.

Applications and Problem Solving

14. *Maintenance* The toddler pool at the community center is cylindrical. It has a radius of 12 feet and is 2 feet deep. If one gallon of paint covers 24 square feet, how many gallons of paint need to be purchased to paint the toddler pool?

15. *Mail* Find the amount of metal needed to construct the mailbox.

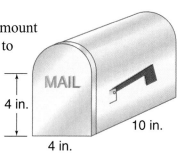

MAIL

4 in.

10 in.

4 in.

16. *Working on the* **CHAPTER Project** Refer to the kite you designed on page 485.

   a. Determine the amount of materials you will need to make your kite. Some of your materials are sold by the length, and some of the materials are sold by the square unit.

   b. Determine the cost of these materials.

17. *Critical Thinking* Will the surface area of a cylinder increase more if you double the height or double the radius of the base? Explain.

Mixed Review

18. *Geometry* Find the surface area of a rectangular prism that is 5 inches long, 3 inches wide, and 2 inches tall. *(Lesson 11-5)*

19. **Standardized Test Practice** Haloke invited 20 people to a skating party. The cost of food and skating is $5.17 per person. Which is the best estimate of the cost of the party? *(Lesson 3-6)*

   A less than $40   **B** between $50 and $60

   C between $70 and $80   **D** between $100 and $110

   E more than $130

For **Extra Practice,** see page 638.

COOPERATIVE LEARNING

11-6B Surface Area and Volume

A Follow-Up of Lesson 11-6

calculator

The average cylindrical soft drink can holds 355 milliliters of beverage. In the metric system, 1 milliliter = 1 cubic centimeter. Therefore, the volume of one of these cans is about 355 cubic centimeters. A manufacturer uses a can that has a radius of 3.3 centimeters and a height of 10.4 centimeters, but is considering changing the shape of the can. The manufacturer would like the cans to have the least surface area possible, unless it would hurt sales.

TRY THIS

Work with a partner.

Step 1 Design three more cylinders that have a volume between 354 and 356 cubic centimeters. Use a calculator to help determine appropriate measurements. Record the dimensions and volume of each cylinder in a chart. Include at least one cylinder whose radius is less than 3.3 centimeters and at least one whose radius is more than 3.3 centimeters.

Cylinder	Height	Radius	Volume	Surface Area
A	10.4 cm	3.3 cm	355.8 cm³	
B				
C				
D				

Step 2 Compute the surface area of each cylinder.

Step 3 Design three rectangular prisms that have a volume between 354 and 356 cubic centimeters. Record the dimensions and volume of each prism in a chart.

Prism	Length	Width	Height	Volume	Surface Area
E					
F					
G					

Step 4 Compute the surface area of each prism.

ON YOUR OWN

1. Which container resulted in the greatest surface area?

2. Which container resulted in the least surface area?

3. *Write* a paragraph recommending a container for the manufacturer to use. Give your reasons. Consider the appearance as well as the surface area.

4. *Reflect Back* In general, what type of container would hold the most liquid for the least amount of surface area?

Integration: Measurement
Precision and Significant Digits

What you'll learn

You'll learn to analyze measurements.

When am I ever going to use this?

Knowing how to analyze measurements can help you know how precise of an instrument you need for a particular measurement.

Word Wise

precision
significant digits
greatest possible error
relative error

In the 1996 Olympics, Beth Botsford won the gold medal in the 100-meter backstroke. Her time was recorded as 61.19 seconds.

No measurement can be any more exact than the scale being used to make the measurement. The **precision** of a measurement depends on the unit of measure being used. The timing device used in the Olympics had a precision of 0.01 second. *How does the precision of this instrument compare with the precision of your watch?*

The digits you record when you measure are **significant**. These digits indicate the precision of the measurement. In Ms. Botsford's Olympic time, there are four significant digits.

Ms. Botsford may have taken 0.005 second more or less than the 61.19 seconds. The instrument used to record her time could not distinguish a more precise time. The **greatest possible error** is half the smallest unit used to make the measurement. For her time, the greatest possible error is 0.005 second.

The **relative error** of measurement is found by comparing the greatest possible error with the measurement itself.

$$\text{relative error} = \frac{\text{greatest possible error}}{\text{measurement}}$$

The relative error of Ms. Botsford's time can be determined by dividing 0.005 by 61.19.

$$\text{relative error} = \frac{0.005}{61.19} \text{ or about } 0.00008$$

Compare the two measurements of a paper clip.

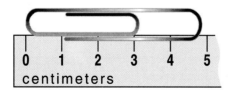
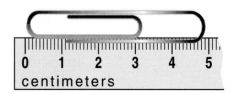

measurement: 5 cm
precision: 1 cm
significant digits: 1
greatest possible error: 0.5 cm
relative error: $\frac{0.5}{5}$ or 0.1

measurement: 4.9 cm
precision: 0.1 cm
significant digits: 2
greatest possible error: 0.05 cm
relative error: $\frac{0.05}{4.9}$ or about 0.01

Physical Science The mass of a substance is determined to be 0.0045 kilogram. Analyze this measurement.

- The mass of the substance has been measured to the nearest 0.0001 kilogram.

- There are two significant digits. The zeros in 0.0045 are used to show only the place value of the decimal and are not counted as significant digits.

- The greatest possible error is 0.00005 kilogram.

- The relative error is $\frac{0.00005}{0.0045}$ or about 0.011.

Significant digits and greatest possible error also apply to rounded data.

Example
APPLICATION

Entertainment The graph shows the number of people who attended the largest state fairs.

a. **Are the numbers exact?**

No, the measurements are to the nearest 0.1 million or 100,000 people.

b. **How many significant digits are used in these numbers?**

The numbers have two significant digits.

FAIR ATTENDANCE
(millions)

State Fair of Texas	3.4
Ohio State Fair	1.8
State Fair of Oklahoma	1.7
Minnesota State Fair	1.6
New Mexico State Fair	1.5

Source: International Association of Fairs and Expositions

c. **What is the greatest possible error?**

The greatest possible error is 0.05 million or 50,000 people.

d. **Can you be sure that 100,000 more people attended the Ohio State Fair than the State Fair of Oklahoma?**

No, 1,760,000 could have attended the Ohio State Fair and 1,740,000 could have attended the State Fair of Oklahoma. In this case, the difference would only be 20,000 people.

Cultural Kaleidoscope

A world's fair is an international exposition. The first world's fair was held in London in 1851. The Eiffel Tower was built for the world's fair in Paris in 1889.

CHECK FOR UNDERSTANDING

Communicating Mathematics

Read and study the lesson to answer each question.

1. *Tell* whether each statement is true or false.

   a. One yard is about 1 meter.

   b. One centimeter is about 3 inches long.

   c. One kilometer is about 6 miles.

   d. One kilogram is about 2.2 pounds.

   e. One liter is about 1 quart.

   f. One millimeter is about the height of a person.

   g. One gram is about the weight of a paper clip.

2. *Describe* a situation where two measurements have the same greatest possible error, but different relative errors.

3. *Write* a paragraph describing an instrument used for measuring length. How precise is the instrument? Give two examples of measurements for which the instrument would be used.

Guided Practice **Analyze each measurement. Give the precision, significant digits if appropriate, greatest possible error, and relative error to two significant digits.**

4. $6\frac{1}{4}$ in.

5. 76.4 km

6. *Clothing* The sizes of men's hats are given as the diameter of a circle before the hat is made more oval. Hat sizes start at $6\frac{1}{4}$ and go up by $\frac{1}{8}$ inch increments. How precise are the hat sizes?

EXERCISES

Practice **Analyze each measurement. Give the precision, significant digits if appropriate, greatest possible error, and relative error to two significant digits.**

7. 7 mm

8. 0.052 kg

9. $9\frac{1}{2}$ lb

10. 87.23 m

11. 47 min

12. 42.8 cm

13. Which would be the most precise measurement for a can of tomatoes: 2 pounds, 34 ounces, or 34.3 ounces? Explain.

14. Which is the more precise measurement for the height of the Statue of Liberty: 100 yards or 305 feet? Explain.

15. The lengths of two poles are 17 inches and 78 inches. Which measurement has the least relative error? Explain.

Applications and Problem Solving

16. *Entertainment* The graph shows the amount of money spent on professional team logo merchandise.

    a. Are the numbers exact? Explain.

    b. What is the greatest possible error?

    c. What is the total amount of money spent on the logos for these four professional sports?

Sale of Logo Merchandise (billions)

Hockey	$1.00
Baseball	$1.90
Basketball	$2.65
Football	$3.15

Source: Sporting Goods Manufacturers Association

17. **Technology** A calculator gives the value of π as 3.141592654. How many significant digits does this value for π have?

18. **Working on the** **CHAPTER Project** Refer to the kite you designed on page 485. Determine the appropriate instruments you will need to make the appropriate measurements to build the kite.

19. **Critical Thinking** To be precise when multiplying measurements, your rounded answer should have the same number of significant digits as the least precise measurement. With that in mind, find the area of a circle whose radius is 4.2 centimeters.

Mixed Review

20. **Geometry** Find the surface area of a cylinder 9 centimeters tall with a radius of 4 centimeters. *(Lesson 11-6)*

21. **Standardized Test Practice** Which equation describes the function represented by the table? *(Lesson 10-1)*

n	f(n)
−2	−7
0	−3
2	1
4	5

A $f(n) = 2n - 3$ **B** $f(n) = n + 4$ **C** $f(n) = n - 3$
D $f(n) = 2n + 3$ **E** $f(n) = n + 3$

22. **Probability** A cereal box stated that it may contain a prize. A consumer protection agency bought 50 boxes of the cereal. Five of them contained a prize. Estimate the probability of winning a prize when you buy a box of that brand of cereal. *(Lesson 6-6)*

For **Extra Practice**, see page 639.

MATH IN THE MEDIA

Hi and Lois

1. What precision is the daughter, Dot, using?

2. What precision might Dot's mother use to show the pieces are the same size?

3. Can you cut two pieces exactly the same size? Explain.

Lesson 11-7 Integration: Measurement   Precision and Significant Digits   **507**

CHAPTER 11

Study Guide and Assessment

inter NET
CONNECTION Chapter Review **For additional lesson-by-lesson review, visit:**
www.glencoe.com/sec/math/mac/mathnet

Vocabulary

After completing this chapter, you should be able to define each
term, concept, or phrase and give an example or two of each.

Geometry
altitude (p. 490)
base (p. 483)
circular cone (p. 490)
circular cylinder (p. 487)
edge (p. 483)
face (p. 483)
net (p. 494)
prism (p. 483)
pyramid (p. 483)
solid (p. 482)
surface area (p. 495)
vertex (p. 483)
volume (p. 486)

Measurement
greatest possible error (p. 504)
precision (p. 504)
relative error (p. 504)
significant digits (p. 504)

Problem Solving
make a model (p. 480)

Understanding and Using the Vocabulary

**Choose the letter of the term that best matches each
statement or phrase.**

1. any three-dimensional figure
2. the measure of the space occupied by a solid
3. a flat surface of a prism
4. a figure that has an area of π times the radius squared
5. the greatest possible error divided by the measurement
6. a figure that has two parallel, congruent circular bases
7. a figure that has a volume that is one-third the area of the
   base times the height
8. the segment used to measure the height of a pyramid and
   a cone
9. half the smallest unit used to make a measurement

a. pyramid
b. cylinder
c. rectangular prism
d. circle
e. precision
f. solid
g. face
h. surface area
i. volume
j. greatest possible error
k. altitude
l. relative error

In Your Own Words

10. *Explain* how to find the surface area of a rectangular prism.

Objectives & Examples

Upon completing this chapter, you should be able to:

● find the areas of circles *(Lesson 11-1)*

Find the area of a circle with a radius of 5 feet.

$A = \pi r^2$

$A = \pi \cdot 5^2$ ⬚π ⬚× 5 ⬚x² ⬚= *78.53981634*

$A \approx 78.5$

The area of the circle is about 78.5 square feet.

● identify and sketch three-dimensional figures *(Lesson 11-2)*

Make the edges you can see solid lines and the edges you can't see dashed lines.

● find the volumes of prisms and cylinders *(Lesson 11-3)*

Find the volume of the rectangular prism.

$V = \ell wh$

$V = 3 \cdot 4 \cdot 6$ or 72

The prism has a volume of 72 cubic meters.

6 m
4 m
3 m

Review Exercises

Use these exercises to review and prepare for the chapter test.

Find the area of each circle to the nearest tenth.

11.

7 m

12.

12 in.

Use isometric dot paper to draw each solid.

13. a rectangular prism that is 4 units long, 5 units wide, and 2 units high

14. a hexagonal pyramid

15. a rectangular prism with all edges 3 units long

Find the volume of each solid to the nearest tenth.

16.

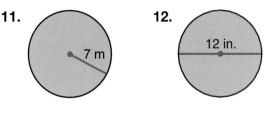

9 in.

12 in.

17.

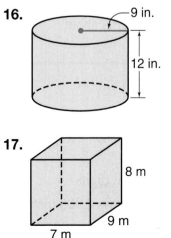

8 m

9 m

7 m

Objectives & Examples

Review Exercises

● find the volumes of pyramids and cones
(Lesson 11-4)

Find the volume of each solid to the nearest tenth.

Find the volume of the square pyramid.

$$V = \frac{1}{3}Bh$$

$$V = \frac{1}{3} \cdot 4^2 \cdot 9$$

$$V = 48$$

The volume of the square pyramid is 48 cubic feet.

18. **19.**

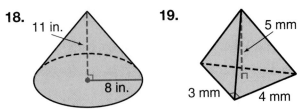

● find the surface areas of rectangular and triangular prisms *(Lesson 11-5)*

Find the surface area of each prism to the nearest tenth.

To find the surface area of a prism, find the area of each face. Then add all the areas.

20. **21.**

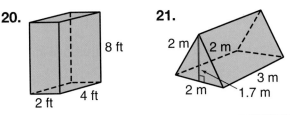

● find the surface areas of cylinders
(Lesson 11-6)

Find the surface area of each cylinder to the nearest tenth.

Find the surface area of a cylinder with a radius of 2 feet and a height of 5 feet.

surface area = $2\pi r^2 + 2\pi r h$

surface area = $2\pi(2^2) + 2\pi(2)(5)$

surface area ≈ 88.0

The surface area of cylinder is about 88.0 square feet.

22 **23.**

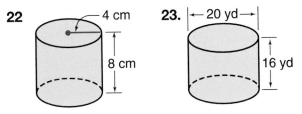

● analyze measurements *(Lesson 11-7)*

Analyze 3.08 meters.

• The measurement is to the nearest 0.01 meter.

• There are three significant digits.

• The greatest possible error is 0.005 meter.

• The relative error is $\frac{0.005}{3.08}$ or about 0.00162.

Analyze each measurement. Give the precision, significant digits if appropriate, greatest possible error, and relative error to two significant digits.

24. 0.06 millimeter

25. 12 feet

26. $2\frac{1}{3}$ yards

27. 24.2 meters

Applications & Problem Solving

28. *Music* Find the area of the top of a compact disc if its diameter is 12 centimeters and the diameter of the hole is 1.5 centimeters. *(Lesson 11-1)*

29. *Make a Model* In the game of pool, a rack is used to organize the balls. There is one ball in the first row, two balls in the next row, three balls in the next row, and so on, until 5 rows are completed. How many balls can be placed in a pool rack? *(Lesson 11-2A)*

30. *Food* A soup can is in the shape of a circular cylinder. The top has a diameter of 6.5 centimeters, and the can is 9.5 centimeters tall. Find the volume of the soup can. *(Lesson 11-3)*

31. *Maintenance* The diagram shows the design of the trash cans in the school cafeteria. The cans need to be painted.

  a. Find the surface area of each trash can.

  b. The paint covers 200 square feet per gallon. Approximately how many trash cans can be covered with 1 gallon of paint? *(Lesson 11-6)*

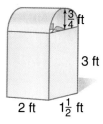

$\frac{3}{4}$ ft

3 ft

2 ft $1\frac{1}{2}$ ft

Alternative Assessment

Open Ended

You have a circular swimming pool with a radius of 9 meters and a volume of about 1,017.36 cubic meters. Your friend has a circular swimming pool with a diameter that is the same as the radius of your pool. The depth of your friend's pool is the same as the depth of your pool. Explain how to find the volume of your friend's pool. Then find the volume. Compare the volume of the two pools.

You and your friend decide to drain the water from your pools and paint the inside surface. Your friend says that he needs 3 cans of paint for his pool. How can you use this information to determine the number of cans you will need to paint your pool?

Completing the CHAPTER Project

Use the following checklist to make sure your kite design is complete.

☑ The three-dimensional drawing of the kite is accurate.

☑ The computations involving the amount and cost of materials needed for the kite are accurate.

☑ A statement about the appropriate measuring instruments needed to construct the kite is included.

PORTFOLIO   Select one of the assignments from this chapter and place it in your portfolio. Attach a note to it explaining why you selected it.

A practice test for Chapter 11 is provided on page 657.

Section One: Multiple Choice

There are ten multiple-choice questions in this section. Choose the best answer. If a correct answer is *not here,* choose the letter for Not Here.

1. A ladder was leaning up against a building. The ladder formed a 60° angle with the ground.

What is the measure of the angle between the ground and the other side of the ladder?

  A 30°

  B 60°

  C 100°

  D 120°

2. If an architectural drawing measures 6 inches wide by 8 inches long, how wide is an enlargement that is 32 inches long?

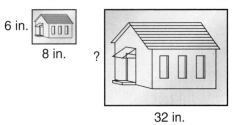

  F 14 inches

  G 48 inches

  H 24 inches

  J 18 inches

3. Find the volume of a 5-meter by 6-meter by 2-meter rectangular prism.

  A 10 cubic meters

  B 12 cubic meters

  C 30 cubic meters

  D 60 cubic meters

4. △LMN is a right triangle. What is the length of $\overline{LM}$?

  F 2 inches

  G 3 inches

  H 4 inches

  J 5 inches

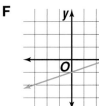

Please note that Questions 5–10 have five answer choices.

5. A circular clock has a diameter of 14 inches. Which expression represents the area of the face of the clock?

  A $\pi \cdot 7 \cdot 7$

  B $\pi \cdot 28$

  C $\pi \cdot 14$

  D $\pi \cdot 14 \cdot 14$

  E $2 \cdot \pi \cdot 14$

6. Which is the graph of $y = 3x - 3$?

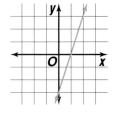

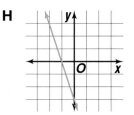

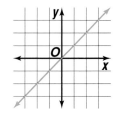

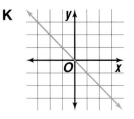

7. A flower bed is 10 feet by 3 feet. To fill it with soil 6 inches deep, how much topsoil is needed?

A 30 cubic feet

B 15 cubic feet

C 120 cubic feet

D 150 cubic feet

E 180 cubic feet

8. On the first play of the Cowboys' game, the quarterback was sacked and lost 18 yards. On the second and third plays, the offense advanced the football a total of 12 yards. What is the total number of yards gained or lost?

F 6 yards lost

G 6 yards gained

H 30 yards lost

J 30 yards gained

K Not Here

9. Kwan-Yong bought $6\frac{1}{2}$ pounds of ground beef for $10.27 and 2 dozen cookies for $3.78. What was the cost per pound of the ground beef?

A $1.89

B $1.58

C $1.08

D $2.16

E Not Here

10. Consuela is participating in a walk-a-thon. Khalid is sponsoring her at $0.10 per mile, and Debbie is sponsoring her at $0.11 per mile. If Consuela walks 9 miles, what is the total amount she will collect from Khalid and Debbie?

F $0.90

G $0.99

H $1.89

J $1.90

K $1.99

Test-Taking Tip
THE PRINCETON REVIEW

Remember, on most tests, you get as much credit for correctly answering the easy questions as you do for correctly answering the difficult ones. Answer the easy questions first and then spend time on the more challenging questions.

Section Two: Free Response

This section contains three questions for which you will provide short answers. Write your answers on your paper.

11. The bases of a triangular prism are right triangles with sides of 3 feet, 4 feet, and 5 feet. The height of the prism is 10 feet. What is the surface area?

12. The curved part of a can will be covered by a label. What is this area to the nearest whole number?

13. Rectangle *ABCD* has vertices *A*(1, 2), *B*(1, −3), *C*(−5, −3), and *D*(−5, 2). What are the coordinates of the rectangle after a translation 1 unit right and 2 units down?

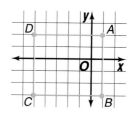

interNET **CONNECTION**   Test Practice **For additional test practice questions, visit:**

www.glencoe.com/sec/math/mac/mathnet

Investigating Discrete Math and Probability

What you'll
learn in Chapter 12

- to count outcomes by using a tree diagram or the Counting Principle,
- to find the number of permutations and combinations of objects,
- to find theoretical and experimental probability,
- to solve problems by acting them out, and
- to predict the actions of a large group by using a sample.

CHAPTER Project

CONSIDER THE PROBABILITIES

In this project, you will write a news article about what it takes to compete at a high level in a sport. You will compute the probability that an athlete will reach the higher levels of competition in a sport of your choice. You will make a graph showing how many players out of 100,000 would be expected to play at the higher levels.

Getting Started

- Choose a sport that interests you.
- Find out how many players in your community are involved in the sport at the little league level or some other level.
- Interview a middle school or high school coach in the sport. Find out how many students play the sport for the school. Find out how many of his or her players have gone on to play in college. Ask if any of his or her players have competed in the Olympics or have played on professional teams.

Technology Tips

- Use a **spreadsheet** to organize data and make calculations.
- Use **computer software** to make the graph.
- Use a **word processor** to write your article.

interNET **CONNECTION** Data Update **For up-to-date** information on sports participation, visit:

www.glencoe.com/sec/math/mac/mathnet

Working on the Project

You can use what you'll learn in Chapter 12 to help you calculate

Page	Exercise
543	11
548	11
553	Alternative Assessment

probabilities and make predictions.

12-1A Fair and Unfair Games

A Preview of Lesson 12-1

2 number cubes

Many people enjoy playing games they believe are fair. A *fair game* is defined as one in which each player has an equal chance of winning. In an *unfair game,* players do *not* have an equal chance of winning.

Have you ever played Scissors, Paper, Stone? This ancient game, also known as Hic, Hacec, Hoc, is played all over the world. On the count of three, two players simultaneously display one hand with either two fingers forming a V (scissors), an open hand (paper), or a fist (stone).

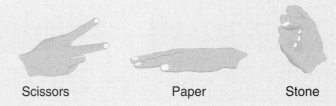

Scissors   Paper   Stone

The winner of the game is decided by the following rules.

- Scissors cut paper.
- Paper wraps stone.
- Stone breaks scissors.

If both players pick the same object, the round is a draw.

TRY THIS

Work with a partner.

1 Play 20 rounds of Scissors, Paper, Stone. Tally the number of times each player wins in a table like the one below.

Winner		
Player A	**Player B**	**Draw**

ON YOUR OWN

1. To analyze the game, make a list of all of the possible outcomes.
2. How many different outcomes are possible?
3. How many ways can Player A win?
4. How many ways can Player B win?
5. How many outcomes are a draw?
6. Is each outcome equally likely?
7. Is Scissors, Paper, Stone a fair game? Explain.

Work with a partner.

❷ Roll two number cubes. If the product of the numbers is even, Player 1 wins. If the product of the numbers is odd, Player 2 wins. Play 20 rounds of this game. Tally the number of times each player wins in a table like the one below.

Winner	
Player 1	Player 2

ON YOUR OWN

8. To analyze the game, make a list of all of the possible outcomes.

9. How many different outcomes are possible?

10. How many ways can Player 1 win?

11. How many ways can Player 2 win?

12. Is it possible to have a draw in this game?

13. Is each outcome equally likely?

14. Is this game *fair* or *unfair?* Explain.

15. **Look Ahead** Two players each spin one of the spinners. Player X wins if the spinners show the same letter. Player Y wins if the spinners show different letters. Is this game fair or unfair? Be prepared to defend your answer.

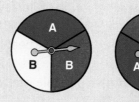

12-1 Counting Outcomes

What you'll learn

You'll learn to count outcomes by using a tree diagram or the Counting Principle.

When am I ever going to use this?

Knowing how to count outcomes can help you determine the number of sandwich choices.

Word Wise

tree diagram
outcome
Counting Principle

Cultural Kaleidoscope

The Chinese wrote about the mathematics of combinations 4,000 years ago.

The games you played in Lesson 12-1A involved counting outcomes by listing them. In this lesson, you will learn two other ways to count outcomes.

Ernesto is making his lunch for school. He has a choice of wheat or rye bread. He can use ham, turkey, or beef. He can use Swiss cheese or American cheese. How many different sandwiches can he make using one bread, one meat, and one cheese?

You can draw a **tree diagram** to find the number of possible combinations or **outcomes**.

Bread	Meat	Cheese	Outcome
wheat (W)	ham (H)	Swiss (S)	WHS
		American (A)	WHA
	turkey (T)	Swiss (S)	WTS
		American (A)	WTA
	beef (B)	Swiss (S)	WBS
		American (A)	WBA
rye (R)	ham (H)	Swiss (S)	RHS
		American (A)	RHA
	turkey (T)	Swiss (S)	RTS
		American (A)	RTA
	beef (B)	Swiss (S)	RBS
		American (A)	RBA

There are twelve possible sandwiches, or outcomes.

Example 1 — Real World APPLICATION

Sports Members of a soccer team have red shorts and black shorts. They also have white jerseys, red jerseys, and black jerseys. How many different uniforms can they wear?

Use a tree diagram to find all of the possible uniforms.

Shorts	Jerseys	Outcome
red	white	red and white
	red	red and red
	black	red and black
black	white	black and white
	red	black and red
	black	black and black

There are six different uniforms.

You can also find the total number of outcomes by multiplying. This principle is known as the **Counting Principle**.

Counting Principle	If event *M* can occur in *m* ways and is followed by event *N* that can occur in *n* ways, then the event *M* followed by the event *N* can occur in $m \cdot n$ ways.

The number of possible sandwiches that Ernesto can make can be determined using this principle.

number of choices for bread	×	*number of choices for meat*	×	*number of choices for cheese*	=	*number of possible sandwiches*
2	×	3	×	2	=	12

Example 2

There are six different numbers on a number cube. Suppose you roll a red number cube, a green number cube, and a black number cube. What is the probability of rolling a 1 on the red number cube, a 2 on the green number cube, and a 3 on the black number cube?

Use the Counting Principle to determine the number of possible outcomes.

number of outcomes on red number cube	×	*number of outcomes on green number cube*	×	*number of outcomes on black number cube*	=	*total number of possible outcomes*
6	×	6	×	6	=	216

LOOK BACK
You can refer to Lesson 6-6 to review probability of simple events.

There are 216 possible outcomes. Each outcome is equally likely. Only one of these outcomes shows a 1 on the red number cube, a 2 on the green number cube, and a 3 on the black number cube. The probability of this roll is $\frac{1}{216}$.

CHECK FOR UNDERSTANDING

Communicating Mathematics

Read and study the lesson to answer each question.

1. *Write a Problem* that corresponds to the tree diagram.

2. *Use* the Counting Principle to write a mathematical equation that corresponds to the tree diagram.

Math Journal

3. *Write* a paragraph comparing the use of a tree diagram and the use of the Counting Principle to determine the number of outcomes.

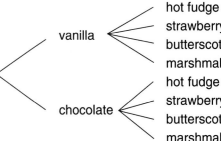

4. Suppose a number cube is rolled and a coin is tossed. Draw a tree diagram to find the number of possible outcomes.

5. Suppose you can buy a souvenir T-shirt in long or short sleeve. Each type of T-shirt comes in 5 colors and 4 sizes. How many different T-shirts are available?

6. *Games* In a board game, there are 6 suspects, 6 weapons, and 9 rooms. Find the number of possible solutions to the game's question of who, how, and where.

EXERCISES

Practice

Draw a tree diagram to find the number of possible outcomes for each situation.

7. The spinner is spun three times.

8. Four coins are tossed.

9. Kana has a choice of a floral, plaid, or striped blouse with a choice of tan, black, navy, or white skirt.

State the number of possible outcomes for each event.

10. A restaurant offers a choice of three types of pasta with five types of sauce. Each pasta entrée comes with or without a meatball.

11. There are four choices for each of five multiple-choice questions on a science quiz.

12. A car comes with two or four doors, a four or six-cylinder engine, and eight exterior colors.

Applications and Problem Solving

13. *Games* The Hopi Indians invented a game of chance called Totolospi. Players use three cane dice and a counting board. Each die can land either round side up (R) or flat side up (F).
   a. Draw a tree diagram to show the possible outcomes for the three cane dice.
   b. How many outcomes are possible?

14. *Sports* The Silvercreek Ski Resort has four ski lifts up the mountain and eleven trails down the mountain. How many different ways can a skier go up and down the mountain?

15. *Communication* In the United States, radio and television stations use call letters that start with K or W. How many different call letters with four letters are possible?

For **Extra Practice**, see page 639.

16. *Critical Thinking* If *x* coins are tossed, write an algebraic expression for the number of possible outcomes.

Mixed Review

17. *Measurement* How many significant digits are in the measurement 14.4 centimeters? *(Lesson 11-7)*

18. **Standardized Test Practice** The area of a square playground is 361 square feet. What is the perimeter of the playground? *(Lesson 9-1)*

   A 19 ft   **B** 76 ft   **C** 127 ft   **D** 361 ft

What you'll learn

You'll learn to find the number of permutations of objects.

When am I ever going to use this?

Knowing how to find the number of permutations can help you find the number of ways you can order floats in a parade.

Word Wise

permutation
factorial

The Environmental Club at Greenway Junior High School has decided to complete the following projects during the school year.

- Clean up the creek that runs behind the school.
- Collect and recycle old phone books.
- Plant trees around the school on Arbor Day.

The club president plans to appoint a different chairperson for each of the projects. Seven students have volunteered to chair the projects. How many possible ways can the president appoint the chairpersons?

1st Project Any of the 7 students can chair the first project.
2nd Project One of the 6 remaining students can chair the second project.
3rd Project One of the 5 remaining students can chair the third project.

Using the Counting Principle, there will be $7 \times 6 \times 5$ or 210 possible ways to pick the chairpersons from just 7 volunteers.

An arrangement or listing in which order is important is called a **permutation**. In the above example, the symbol $P(7, 3)$ represents the number of permutations of 7 volunteers taken 3 at a time.

$P(n, r)$	**Words:**	$P(n, r)$ means the number of permutations of n things taken r at a time.
	Symbols:	
	Arithmetic	$P(7, 3) = 7 \cdot 6 \cdot 5$
	Algebra	$P(n, r) = n \cdot (n - 1) \cdot (n - 2) \cdot \ldots \cdot (n - r + 1)$

MINI-LAB

Work in groups of three or four.

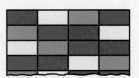 4 different colored pencils

Try This

- Draw four columns on notebook paper. Use the lines of the paper to complete a grid so that each row has four units.
- On the first row, color each unit a different color.
- On the next row, use the same four colors to color the units, but do not repeat the pattern you used in the first row.
- Continue coloring the units in each row until you have created all possible arrangements of the four colors.

The number of permutations in the Mini-Lab can be expressed as $P(4, 4)$. Notice that $P(4, 4)$ means the number of permutations of 4 things taken 4 at a time.

$$P(4, 4) = 4 \cdot 3 \cdot 2 \cdot 1$$

The mathematical notation 4! also means $4 \cdot 3 \cdot 2 \cdot 1$. The symbol 4! is read *four **factorial***. *n*! means the product of all counting numbers beginning with *n* and counting backward to 1. We define 0! as 1.

Examples

CONNECTION

1 Geography Each summer, Ms. McGatha likes to visit a different state in the United States. Over the next five years, she wants to visit the five largest states; Alaska, Texas, California, Montana, and New Mexico. How many different ways can she arrange the trips?

You must find the number of permutations of 5 states taken 5 at a time.

$$P(5, 5) = 5!$$
$$= 5 \cdot 4 \cdot 3 \cdot 2 \cdot 1$$
$$= 120$$

She can arrange the trips in 120 ways.

APPLICATION

Real World

2 Pets The local Humane Society needs to select 2 out of 6 dogs to feature in a newspaper advertisement. One will appear in the Saturday paper, and the other will appear in the Sunday paper. How many ways can the dogs be selected?

You must find the number of permutations of 6 dogs taken 2 at a time.

$$P(6, 2) = 6 \cdot 5$$
$$= 30$$

There are 30 different ways for the dogs to be selected. *You can draw a tree diagram to check the answer.*

CHECK FOR UNDERSTANDING

Communicating Mathematics

Read and study the lesson to answer each question.

1. *Write a Problem* that could be solved by finding the value of $P(5, 2)$.

2. *Compare* 8! and $P(8, 3)$.

3. *Draw* three columns on notebook paper. Use the lines of the paper to complete a grid so that each row has three units. On the first row, color each unit a different color. On the next row, use the same three colors to color the units, but do not repeat the pattern you used in the first row. Continue coloring the units in each row until you have created all possible arrangements of the three colors.

   a. How many different arrangements are possible?

   b. How does your number of arrangements compare with 3!?

Guided Practice

Find each value.

4. $P(5, 3)$ 5. $P(9, 2)$ 6. 6!   7. 2!

8. How many different ways can you arrange the letters in the word *math*?

9. *Games* In the game Tic Tac Toe, players place an X or an O in any of the 9 locations that is empty. How many different ways can the first 3 moves of the game occur?

EXERCISES

Practice

Find each value.

10. $P(7, 4)$ 11. $P(8, 3)$ 12. $P(10, 5)$ 13. $P(6, 6)$

14. 3!   15. 7!   16. 0!   17. 10!

18. $P(25, 3)$ 19. $P(14, 4)$ 20. $P(7, 5)$ 21. 8!

22. How many 3-digit whole numbers can you write using the digits 1, 3, 5, 7, and 9 if no digit can be used twice?

23. How many different ways can you arrange the letters in the word *equals*?

24. A family with 5 members is going to the mall in a car. Two people can sit in the front seat and 3 can sit in the back. If the mother is driving, how many ways can the members of the family be seated in the car?

Applications and Problem Solving

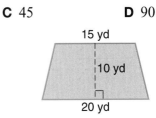

25. *Recreation* There were 17 floats in a parade in New York City. If the Santa float must be last, how many ways could the parade organizer pick the first two floats in the parade?

26. *Sports* There are 9 players on a baseball team. How many ways can the coach pick the first 4 batters?

27. *Critical Thinking* Compare $P(n, n)$ and $P(n, n - 1)$, where n is any whole number greater than one. Explain.

Mixed Review

28. **Standardized Test Practice** A pizza shop has 3 different crusts, 3 different meat toppings, and 5 different vegetables. If Shalema wants a pizza with one meat and one vegetable, how many different pizzas can she order? *(Lesson 12-1)*

    A 11   **B** 15   **C** 45   **D** 90

29. **Geometry** State the measure of the bases and the height of the trapezoid. Then find the area of the trapezoid. *(Lesson 7-6)*

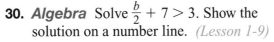

15 yd

10 yd

20 yd

For **Extra Practice**, see page 639.

30. *Algebra* Solve $\frac{b}{2} + 7 > 3$. Show the solution on a number line. *(Lesson 1-9)*

Combinations

What you'll learn

You'll learn to find the number of combinations of objects.

When am I ever going to use this?

Knowing how to find the number of combinations can help you determine the number of 4-topping pizzas you could order.

Word Wise

combination

Ms. Manzueta has been collecting books and information on butterflies for her science students to use to complete their projects. She currently has information on the Tiger Swallowtail, Giant Swallowtail, Clouded Sulfur, Viceroy, Monarch, Meadow Fritillary, and Dog Face. Each student must pick three butterflies to study. How many different groups of butterflies can the students study if they use Ms. Manzueta's resources? *This problem will be solved in Example 1.*

An arrangement or listing, like the one above, where order is not important is called a **combination**.

MINI-LAB

Work with a partner.   6 index cards

Try This

- Mark each of six index cards with one of the letters A, B, C, D, E, and F.
- Select any three cards. Record your selection.
- Select another three cards, but not the same group you chose before. Record your selection.
- Continue selecting groups of three cards until all possible groups are recorded.

Talk About It

1. How many groups did you record?
2. In how many different orders can three letters be arranged?
3. Find $P(6, 3)$.
4. What is the relationship between the number of groups, the number of arrangements of three cards, and $P(6, 3)$?

Study Hint

Reading Math $C(6, 3)$ is read as *the number of combinations of 6 things taken 3 at a time.*

A quick way to find the number of groupings, or combinations, of 6 cards taken 3 at a time, $C(6, 3)$, is to divide the number of permutations, $P(6, 3)$, by the number of orders 3 cards can be arranged, which is 3!.

$$C(6, 3) = \frac{P(6, 3)}{3!} \quad \textit{Dividing by 3! eliminates the combinations that are the same.}$$

$$= \frac{6 \cdot 5 \cdot 4}{3 \cdot 2 \cdot 1}$$

$$= \frac{120}{6} \text{ or } 20$$

There are 20 groups.

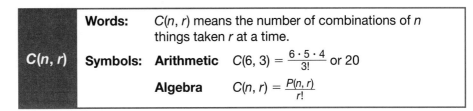

	Words:	$C(n, r)$ means the number of combinations of n things taken r at a time.
$C(n, r)$	Symbols: **Arithmetic**	$C(6, 3) = \frac{6 \cdot 5 \cdot 4}{3!}$ or 20
	Algebra	$C(n, r) = \frac{P(n, r)}{r!}$

Examples

CONNECTION

1 **Life Science** Refer to the beginning of the lesson. How many different groups of butterflies can the students study if they use Ms. Manzueta's resources?

Ms. Manzueta has information about 7 different butterflies. You need to find the number of combinations of 7 butterflies taken 3 at a time.

Study Hint

Mental Math You can simplify $\frac{7 \cdot 6 \cdot 5}{3 \cdot 2 \cdot 1}$ before multiplying.

$$\frac{7 \cdot \overset{2}{6} \cdot 5}{\underset{1}{3} \cdot 2 \cdot \underset{1}{1}} = 35$$

$$C(7, 3) = \frac{P(7, 3)}{3!}$$
$$= \frac{7 \cdot 6 \cdot 5}{3 \cdot 2 \cdot 1}$$
$$= \frac{210}{6} \text{ or } 35$$

There are 35 different groups of three butterflies each, that the students can study.

INTEGRATION **2** **Geometry** Six points are located on a circle. How many line segments can be drawn with these points as endpoints?

Explore There are 6 points on the circle. You want to know how many line segments can be drawn using these points as endpoints.

Plan You need to find the number of combinations of 6 things taken 2 at a time.

Solve $C(6, 2) = \frac{P(6, 2)}{2!}$
$$= \frac{6 \cdot 5}{2 \cdot 1}$$
$$= \frac{30}{2} \text{ or } 15$$

There are 15 possible line segments that can be drawn.

Examine Draw all of the possible line segments. There are 15 possible line segments. The answer checks.

Communicating Mathematics

Read and study the lesson to answer each question.

1. *Write* an expression to represent the number of six-person committees that could be formed from a group of 20 people.

2. *Tell*, without calculating, which is greater, $P(8, 3)$ or $C(8, 3)$. Explain.

3. *You Decide* Bernice says that picking a class president, a class vice president, and a secretary from all of the members of the class is an example of a combination. Montega says it is an example of a permutation. Who is correct? Explain.

HANDS-ON MATH

4. *Mark* each of five index cards with one of the letters A, B, C, D, and E. Select any two cards. Record your selection. Continue selecting groups of two cards until all possible groups are recorded.
   a. How many groups did you record?
   b. How does your number of groups compare with the value of $\frac{P(5, 2)}{2!}$?

Guided Practice

Find each value.

5. $C(8, 3)$ 6. $C(9, 2)$ 7. $C(5, 2)$

Determine whether each situation is a *permutation* or a *combination*.

8. placing 9 model cars in a line

9. choosing a team of 5 players from 11 students

10. How many different combinations of 3 flowers can you choose from 12 different flowers?

11. *Business* A pizza shop has 12 different toppings from which to choose. This week, if you buy a 2-topping pizza, you get 2 more toppings free. How many different ways can the special pizza be ordered?

Practice

Find each value.

12. $C(5, 4)$ 13. $C(7, 5)$ 14. $C(14, 2)$ 15. $C(10, 4)$
16. $C(8, 4)$ 17. $C(9, 5)$ 18. $C(20, 3)$ 19. $C(13, 3)$

Determine whether each situation is a *permutation* or a *combination*.

20. choosing a committee of 7 from 25 members

21. five students forming a line

22. writing a 3-digit number using no digit more than once

23. packing 5 out of 15 outfits for a trip

24. buying 3 CDs from a selection of 18 CDs

25. placing 3 out of 18 CDs in the CD player

26. How many different starting squads of 6 players can be picked from 10 volleyball players?

27. How many different 5-card hands can be dealt from a standard deck of 52 cards?

28. There are 20 runners in a race. In how many ways can the runners take first, second, and third place?

29. In a group of 23 people, there is about a 50% chance that two people have the same birthday. How many different combinations of 2 people could have the same birthday in the group?

Applications and Problem Solving

30. *Business* An ice cream store has 31 flavors of ice cream.

　　a. Lamel wants to buy three pints of ice cream. If each pint of ice cream is a different flavor, how many different purchases can he make?

　　b. Maria wants to buy a three-scoop ice cream cone. If each scoop is a different flavor and the order of scoops in the cone is important to her, how many different cones can she order?

31. *Games* In the game of cribbage, a player gets two points for each combination of cards that totals 15. How many points could a player get with the hand at the right?

32. *Critical Thinking* How many ways can a committee of 2 boys and 3 girls be formed from a class of 18 boys and 12 girls?

Mixed Review

33. How many 4-digit whole numbers can you write using the digits 1, 2, 3, 4, 5, 6, and 7 if no digit can be used twice? *(Lesson 12-2)*

34. **Standardized Test Practice** What is the area of the shaded region? *(Lesson 11-1)*

　　A 107.5 cm²

　　B 77.7 cm²

　　C 59.7 cm²

　　D 27.5 cm²

35. *Geography* The distance between two cities on a map is $2\frac{3}{8}$ inches. Find the actual distance between the cities if the scale on the map is 1 inch:300 miles. *(Lesson 8-9)*

For **Extra Practice**, see page 640.

36. *Statistics* Find the mean, median, and mode for the data set. *(Lesson 4-4)*

　　14, 3, 6, 8, 11, 9, 3, 2, 7

Pascal's Triangle

What you'll learn

You'll learn to identify patterns in Pascal's Triangle.

When am I ever going to use this?

Knowing how to identify patterns in Pascal's Triangle can help you determine the number of combinations.

Word Wise

Pascal's Triangle

For many years, mathematicians have been interested in a special pattern called **Pascal's Triangle**. This triangle is named for a French mathematician Blaise Pascal (1623–1662). However, knowledge of the triangle is believed to have existed in China and Persia as early as the twelfth century. Published references to the triangle were made by Chinese mathematician Chu Shih-Chieh in 1303. Study the first five rows of Pascal's Triangle.

```
            1
          1   1
        1   2   1
      1   3   3   1
    1   4   6   4   1
```

Example 1 Find the pattern in Pascal's Triangle and use the pattern to complete the next two rows.

The right and left sides of the triangle are formed by ones. Notice the relationship of each inside number and the two numbers above it.

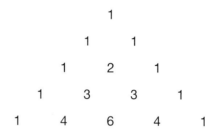

To find each inside number, add the two numbers above it.
The rows in Pascal's Triangle are numbered beginning with the top row, which is row 0.

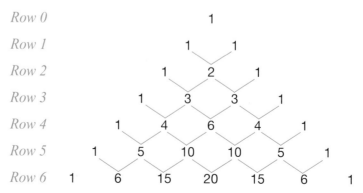

Row 0   1
Row 1   1   1
Row 2   1   2   1
Row 3   1   3   3   1
Row 4   1   4   6   4   1
Row 5   1   5   10   10   5   1
Row 6   1   6   15   20   15   6   1

Work with a partner.

Try This

- Copy Pascal's Triangle formed in Example 1 and add two rows on your copy.
- Now, draw four tree diagrams to represent the possible outcomes for tossing 2, 3, 4, and 5 coins.

Talk About It

1. Which row of Pascal's Triangle has a sum equal to the number of outcomes for tossing 2 coins? 3 coins? 4 coins? 5 coins?
2. When tossing 4 coins, how many ways can the outcome of 2 heads and 2 tails result? Where is the answer located in Pascal's Triangle?
3. How many possible outcomes are there when you toss 8 coins? How many ways can you get 4 heads and 4 tails?

Pascal's Triangle can be used to answer questions involving combinations. To find the number of combinations of 6 things taken 4 at a time, refer to row 6 of Pascal's Triangle.

Number Taken at a Time	0	1	2	3	4	5	6
Row 6	1	6	15	20	15	6	1

There are 15 combinations of 6 things taken 4 at a time.

Example

CONNECTION

2 **Life Science** Mr. Santiago is cross breeding 8 different types of tomato plants in order to find a plant that produces larger tomatoes. If he wants to cross every possible pair, how many pairs will he have?

Explore   Mr. Santiago has 8 different plants. He wants to know how many pairs he can form from these plants.

Plan   Use Pascal's Triangle to find the number of combinations of 8 things taken 2 at a time.

Solve

Number Taken at a Time	0	1	2	3	4	5	6	7	8
Row 8	1	8	28	56	70	56	28	8	1

There are 28 pairs.

Examine   Find the number of combinations of 8 things taken 2 at a time.

$$C(8, 2) = \frac{P(8, 2)}{2!}$$

$$= \frac{8 \cdot 7}{2 \cdot 1} \text{ or } 28 \quad \checkmark$$

Lesson 12-4 Pascal's Triangle   **529**

Communicating Mathematics

Read and study the lesson to answer each question.

1. *Show* how to find row 9 of Pascal's Triangle.

2. *Tell* how the sum of the numbers in row 5 compares to the total number of possible ways to answer a true-or-false quiz with 5 questions.

HANDS-ON MATH

3. *Draw* a tree diagram to represent the possible outcomes of tossing 6 coins.

   a. How many ways can you toss 6 heads? 5 heads? 4 heads? 3 heads? 2 heads? 1 head? 0 heads?

   b. What row of Pascal's Triangle corresponds with the results of tossing 6 coins?

Guided Practice

Use Pascal's Triangle to find each value.

4. $C(6, 2)$ 5. $C(4, 3)$

Use Pascal's Triangle to answer each question.

6. How many different committees of 3 members can be chosen from 8 people?

7. Five coins are tossed. What is the probability that all 5 coins will show heads?

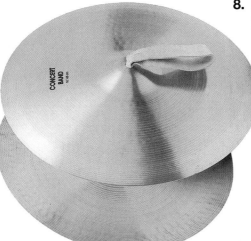

8. *Fund-raising* The Howard Middle School Band is making pizzas to raise money for new percussion instruments. The members can add pepperoni, mushrooms, onions, and/or green peppers to the basic cheese pizzas.

   a. How many 2-item pizzas can the band prepare?

   b. How many 3-item pizzas can the band prepare?

   c. How many different pizzas can the band prepare?

Practice

Use Pascal's Triangle to find each value.

9. $C(8, 3)$ 10. $C(3, 2)$ 11. $C(5, 2)$
12. $C(6, 5)$ 13. $C(9, 5)$ 14. $C(7, 5)$

Use Pascal's Triangle to answer each question.

15. How many combinations of 6 things taken 3 at a time are possible?

16. How many different 4-question quizzes can be formed from 7 possible questions?

17. How many branches will there be in the last column of a tree diagram showing the outcomes of tossing 6 coins?

18. Six coins are tossed. What is the probability that 3 will show heads and 3 will show tails?

19. What does the 6 in row 4 of Pascal's Triangle mean?

20. In row 6 of Pascal's Triangle, which combinations have the same number of possibilities?

Applications and Problem Solving

21. *Life Science* A bee is in cell A and wants to go to cell B. The bee must travel downward from one cell directly into another.

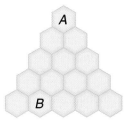

    a. How many ways can a bee get from cell A to cell B?

    b. Explain how this problem relates to Pascal's Triangle.

22. *Probability* Five coins are tossed. Miriam wins if there are more heads than tails. Cleveland wins if there are more tails than heads. Who is more likely to win? Explain.

23. *Critical Thinking* Compare $C(9, n)$ and $C(9, 9 - n)$, where n is a whole number and $0 \leq n \leq 9$. Explain.

Mixed Review

24. Find $C(9, 3)$. *(Lesson 12-3)*

25. *Algebra* Graph the quadratic function $y = 2x^2 - 8$. *(Lesson 10-6)*

26. **Standardized Test Practice** Joe Marsh and three of his co-workers ate lunch at Old Town Café. They plan to leave a 20% tip for the waiter. Two of his co-workers had turkey sandwiches, one had soup and salad, and Mr. Marsh had pasta. What other information is necessary to determine how much to leave for a tip? *(Lesson 3-5)*

    A the cost of the pasta

    B the cost of the 4 meals

    C what day they had lunch

    D the soup of the day

    E the name of the waiter

For **Extra Practice,** see page 640.

CHAPTER 12

Mid-Chapter Self Test

1. Sweatshirts are available in 3 colors (black, white, and gray) and in 4 sizes (small, medium, large, and extra large). Draw a tree diagram that illustrates the outcomes. *(Lesson 12-1)*

2. The school cafeteria offers a lunch special. A person can pick one of six different sandwiches, one of three beverages, and one of three types of fruit. How many different lunch specials are possible? *(Lesson 12-1)*

3. *Music* Five band members play the trumpet. How many ways can these members be chosen for the first, second, and third chairs of the trumpet section? *(Lesson 12-2)*

4. How many ways can 2 student council members be elected from 7 candidates? *(Lesson 12-3)*

5. Use Pascal's Triangle to find the number of combinations that is possible for 7 things taken 3 at a time. *(Lesson 12-4)*

COOPERATIVE LEARNING

12-4B Patterns in Pascal's Triangle

A Follow-Up of Lesson 12-4

hexagonal grid

highlighter

calculator

A famous sequence of numbers is the Fibonacci sequence, 1, 1, 2, 3, 5, 8, ... The sequence begins with 1, 1, and each number that follows is the sum of the previous two numbers. *Refer to Lesson 7-5B to review the Fibonacci sequence.*

The Fibonacci numbers and numerous number patterns can be found in Pascal's Triangle.

TRY THIS

Work with a partner.

1 • Copy Pascal's Triangle at the right and add another two rows to your copy.

• Draw diagonals beginning at each 1 along the left-hand side (passing under the 1 just above it) and extend it to the far side of the triangle as shown.

• Find the sum of the numbers through which each diagonal passes.

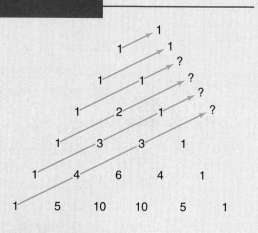

ON YOUR OWN

1. What sequence do the sums form?
2. What should the sum along the next diagonal be?

TRY THIS

Work with a partner.

2 • Find the product of the shaded ring of numbers in the Pascal's Triangle at the right.

• Find the product of any two other rings of the same shape and size in Pascal's Triangle.

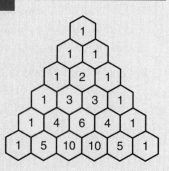

ON YOUR OWN

3. Determine the square root of the product for each ring.
4. What conclusion can you make about the product of rings in Pascal's Triangle?

Work with a partner.

3 • Using a hexagonal grid triangle like the one shown below, highlight all of the numbers that are multiples of 2. Highlight as many rows as necessary to determine a geometric pattern.

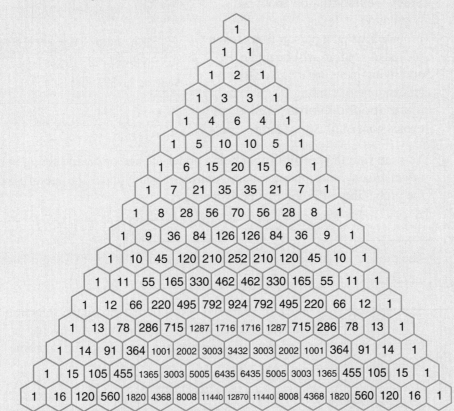

• On another triangle, highlight all of the numbers that are multiples of 3. Determine a geometric pattern.

• On a third triangle, highlight all of the numbers that are multiples of 4. Determine a geometric pattern.

• On a fourth triangle, highlight all of the numbers that are multiples of 7. Determine a geometric pattern.

ON YOUR OWN

5. What general visual pattern is common in all four triangles? Is this true for multiples of all numbers? Explain.

6. Which multiples of the numbers 2, 3, 4, and 7 have patterns that are symmetrical at each of the three vertices of the triangle?

7. *Reflect Back* Write a paragraph about some of the patterns found in Pascal's Triangle.

Probability of Compound Events

What you'll learn

You'll learn to find the probability of independent and dependent events.

When am I ever going to use this?

Knowing how to find probability can help you determine whether something is likely to happen in a game.

Word Wise

compound events
independent events
dependent events

LOOK BACK

You can refer to Lesson 3-4 to review writing fractions as percents.

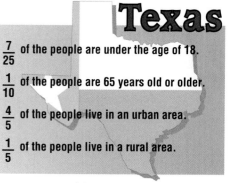

Texas

$\frac{7}{25}$ of the people are under the age of 18.

$\frac{1}{10}$ of the people are 65 years old or older.

$\frac{4}{5}$ of the people live in an urban area.

$\frac{1}{5}$ of the people live in a rural area.

Source: Bureau of the Census

The United States Bureau of the Census keeps statistical information about each of the states.

Study the information about the population of Texas. What is the probability that a person from Texas picked at random will be under 18 and living in an urban area? This is an example of finding the probability of **compound events**. Compound events consist of two or more events.

You can find the probability of being under the age of 18 and living in an urban area by multiplying the probability of being under the age of 18 by the probability of living in an urban area.

$$\begin{array}{r}\textit{number of ways it can occur} \rightarrow \\ \textit{total numbers of possible outcomes} \rightarrow\end{array} \frac{7}{25} \cdot \frac{4}{5} = \frac{28}{125}$$

The probability of a person in Texas being under the age of 18 and living in an urban area is $\frac{28}{125}$ or 22.4%.

Selecting a person under 18 years old does not depend on the selection of a person living in an urban area. We call these compound events **independent**. The outcome of one event does not affect the outcome of the other event.

Probability of Two Independent Events	**Words:**	The probability of two independent events can be found by multiplying the probability of the first event by the probability of the second event.
	Symbols:	$P(A \text{ and } B) = P(A) \cdot P(B)$

Example **APPLICATION**

① **Census** Refer to the beginning of the lesson. Find the probability that a person from Texas picked at random is 65 years old or older and lives in an urban area.

$$P(65 \text{ or older}) = \frac{1}{10} \qquad P(\text{lives in urban area}) = \frac{4}{5}$$

$$P(65 \text{ or older and lives in urban area}) = \frac{1}{10} \cdot \frac{4}{5} = \frac{4}{50} \text{ or } \frac{2}{25}$$

The probability that the two events will occur is $\frac{2}{25}$ or 8%.

Example 2

Suppose a number cube is rolled twice. What is the probability that an even number will show both times?

$$P(\text{even}) = \frac{1}{2}$$

$$P(\text{even and even}) = \frac{1}{2} \cdot \frac{1}{2} \text{ or } \frac{1}{4}$$

The probability of rolling two even numbers is $\frac{1}{4}$ or 25%.

Sometimes the outcome of one event affects the outcome of another.

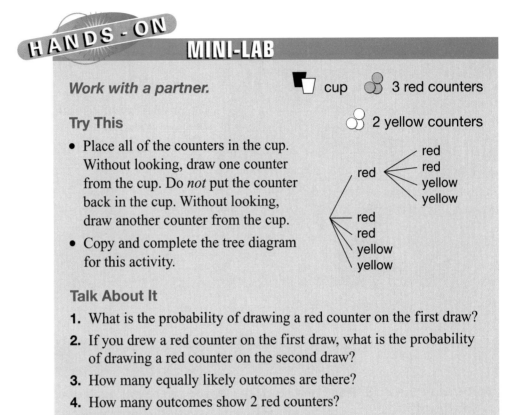

HANDS-ON MINI-LAB

Work with a partner.

🔲 cup ⚫ 3 red counters

🟡 2 yellow counters

Try This

- Place all of the counters in the cup. Without looking, draw one counter from the cup. Do *not* put the counter back in the cup. Without looking, draw another counter from the cup.
- Copy and complete the tree diagram for this activity.

red < red, red, yellow, yellow

red, red, yellow, yellow

Talk About It

1. What is the probability of drawing a red counter on the first draw?
2. If you drew a red counter on the first draw, what is the probability of drawing a red counter on the second draw?
3. How many equally likely outcomes are there?
4. How many outcomes show 2 red counters?
5. If the first counter is replaced, is the number of outcomes affected? If so, how?

If the outcome of one event affects the outcome of another event, these compound events are called **dependent**. In the Mini-Lab, a counter was removed from the cup on the first draw. For the second draw, there were fewer counters from which to choose. These events are dependent.

Like independent events, the probability of two dependent events is found by multiplying probabilities. In this case, you multiply the probability that the first event occurs times the probability that the second event follows the first one.

Probability of Two Dependent Events	**Words:**	If two events, *A* and *B*, are dependent, then the probability of both events occurring is the product of the probability of *A* and the probability of *B* after *A* occurs.
	Symbols:	$P(A \text{ and } B) = P(A) \cdot P(B \text{ following } A)$

Example 3

Study Hint

Reading Math In dependent events, you always assume you were successful on the first event.

Games In a board game, there are 44 cards. Each set of cards consists of four of each denomination (1, 2, 3, 4, 5, 7, 8, 10, 11, and 12) and four other cards. Cards are placed faceup after being drawn. A card with a 2 allows a player to draw again. What is the probability that the first player will draw a 2 followed by another 2?

This is an example of dependent events since the first draw affects the second draw.

First draw: $P(2) = \frac{4}{44}$ or $\frac{1}{11}$

Second draw: $P(\text{second } 2) = \frac{\text{number of 2s left}}{\text{number of cards left}}$

$= \frac{4-1}{44-1}$ or $\frac{3}{43}$

$P(\text{two 2s}) = \frac{1}{11} \cdot \frac{3}{43}$

$= \frac{3}{473}$

The probability of a 2 followed by another 2 is $\frac{3}{473}$ or about 0.6%.

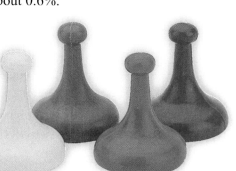

CHECK FOR UNDERSTANDING

Communicating Mathematics

Read and study the lesson to answer each question.

1. **Explain** whether the probability of $\frac{2}{25}$ or 8% found in Example 1 indicates the person picked at random is likely to be 65 years old or older and living in an urban area.

2. **Describe** how to find the probability of the second of two dependent events.

3. **You Decide** Nancy says that the probability of spinning a vowel on each of two consecutive spins is $\frac{2}{5} \cdot \frac{1}{4}$ or $\frac{1}{10}$. Tonya says the probability is $\frac{2}{5} \cdot \frac{2}{5}$ or $\frac{4}{25}$. Who is correct? Explain.

HANDS-ON MATH

4. Suppose there are 4 red and 2 yellow counters in a cup. One counter is drawn at random and is *not* returned to the cup. A second counter is drawn at random from the cup.
   a. Draw a tree diagram showing all of the possible outcomes.
   b. What is the probability of drawing 2 red counters?
   c. What is the probability of drawing 2 yellow counters?

Guided Practice

A coin is tossed and a number cube is rolled. Find each probability.

5. $P(\text{heads and } 5)$

6. $P(\text{tails and even number})$

There are 3 white marbles, 7 red marbles, and 5 blue marbles in a bag. Once a marble is selected, it is not replaced. Find the probability of each outcome.

7. two red marbles

8. a white marble and then a blue marble

9. *Games* A standard Western domino set has 28 tiles. Seven of these tiles have the same number of dots on each side and are called doubles. To start the game, each player draws a tile. What is the probability that the first and second players each draw a double if the tiles are not replaced?

EXERCISES

Practice

Each spinner is spun once. Find each probability.

10. $P(1 \text{ and } A)$

11. $P(2 \text{ and } B)$

12. $P(\text{even and } C)$

13. $P(\text{odd and } B)$

14. $P(\text{even and vowel})$

15. $P(\text{odd and consonant})$

Two cards are drawn from a deck of ten cards numbered 1 though 10. Once a card is selected, it is not replaced. Find the probability of each outcome.

16. a 5 and then a 3

17. two even numbers

18. two numbers greater than 4

19. a 6 and then an odd number

20. an odd number and then an even number

21. a number greater than 7 and then a number less than 6

22. What is the probability of tossing a coin three times and getting heads each time?

23. What is the probability of rolling a number cube three times and getting numbers greater than 4 each time?

Applications and Problem Solving

24. *Statistics* In the United States, $\frac{3}{5}$ of all households have some kind of pet, and $\frac{1}{3}$ of all households have at least one child. What is the probability that a household picked at random will have a pet and one or more children?

25. *Games* In a board game, there are 16 Community Box cards. One is a You're Grounded card, and one is a You're Grounding Is Over card. A player keeps the You're Grounding Is Over card. What is the probability that the first card picked will say You're Grounding Is Over and the next card drawn will say You're Grounded?

For **Extra Practice,** see page 640.

26. *Critical Thinking* During his career, the probability that Babe Ruth would get a hit was 0.342. What is the probability that Babe Ruth would *not* get a hit in two consecutive times at bat?

27. Use Pascal's Triangle to find $C(6, 2)$. *(Lesson 12-4)*

Mixed Review

28. **Standardized Test Practice** Find the amount of discount for a pair of $89 shoes that are on sale at 30% off. *(Lesson 8-5)*

A $17.80 **B** $26.70 **C** $35.60 **D** $62.30 **E** Not Here

29. Find the LCM of 8, 15, and 12. *(Lesson 6-7)*

PROBLEM SOLVING

12-6A Act It Out

A Preview of Lesson 12-6

Miko and Kelsey are going on a hot air balloon ride for Miko's birthday. The girls have checked with Rise Above It All, the local air balloon company, and found that there are many considerations involved in taking the ride. Let's listen in!

I really want to ride in the one that is painted to look like the planet Earth. I think it is so cool! But the company says they have 6 balloons, and I can't pick the one I want.

Kelsey

I want to fly over our school, but that depends on how the wind is blowing the day of the ride. The manager at Rise Above It All says that we have a one in four chance of having the right wind to travel over the school.

Miko

What do you think about the chances of both of us getting our wishes?

I don't know. Let's conduct an experiment to answer that question.

That's a good idea. We can use a number cube to simulate which balloon we will get. The number one will represent the one I want. We can use a spinner with four equal sections to simulate the direction of the wind. One section will represent the wind we need to go over the school.

THINK ABOUT IT

Work with a partner.

1. *Explain* why you could use Miko's experiment, or simulation, to estimate the probability that both events will occur. Try the simulation.

2. *Compute* the probability that the girls will get the balloon that Miko wants and will travel over the school.

3. *Tell* whether the results of the simulation tried 100 times will be exactly the same as computing the results. Explain.

4. *Select* another model to simulate the possible outcomes of the balloon ride.

For **Extra Practice,** see page 641.

ON YOUR OWN

5. The second step of the 4-step plan for problem solving asks you to *plan*. **Explain** why this step is important when using the **act it out** strategy.

6. *Write a Problem* that can be solved by acting it out.

7. *Look Ahead* Explain why the act it out strategy might sometimes be called experimental probability.

MIXED PROBLEM SOLVING

STRATEGIES

Look for a pattern.
Solve a simpler problem.
Act it out.
Guess and check.
Draw a diagram.
Make a chart.
Work backward.

Solve. Use any strategy.

8. *Sports* There are 16 tennis players in a tournament. If each losing player is eliminated from the tournament, how many tennis matches will be played during the tournament?

9. *Money Matters* Devin bought a jacket and a shirt. The total cost, not including tax, was $62.50. The jacket cost four times as much as the shirt. How much did Devin spend on each item?

10. *Fashion* Namel has 6 ties. On Thursday through Sunday, Namel works at the mall. Each work day, he chooses a tie at random to wear for his job. Find the probability that Namel wears the same tie more than once in his four-day week. Select an appropriate model and use the act it out strategy.

11. *Statistics* In a recent survey of 120 students, 50 students said they play baseball, and 60 said they play soccer. If 20 students play both sports, how many students do not play either baseball or soccer?

12. *Physical Science* The light in the circuit will turn on if one or more switches are closed. How many combinations of open and closed switches will result in the light being on?

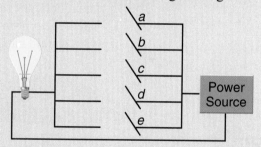

13. *Business* At a certain restaurant, toys are given with each child's meal. During the spring promotional, three different toys are given at random.
  a. Act it out to determine how many meals must be purchased in order to get all three toys.
  b. What number would guarantee that you would receive all three toys? Explain.

14. *Standardized Test Practice* There are 5 red marbles, 2 yellow marbles, and 1 blue marble in a bag. One marble is selected at random, and it is not replaced. Then, a second marble is selected. Find the probability that both marbles are red.

  A $\frac{5}{14}$ **B** $\frac{25}{64}$ **C** $\frac{5}{16}$ **D** $\frac{25}{56}$

Experimental Probability

What you'll learn

You'll learn to find experimental probability.

When am I ever going to use this?

Knowing how to find experimental probability can help you find the probability that you will make a free throw.

Word Wise

simulation
experimental probability
theoretical probability

Do you like to play basketball? What is the probability that you will make a basket from the free-throw line?

Sometimes you can't tell what the probability of an event is until you conduct an experiment. For example, suppose you try 50 free throws and count the number of times you make a basket. The results of this experiment allow you to estimate the probability that you will make a basket from the free-throw line.

Probabilities that are based on frequencies obtained by conducting an experiment, or doing a **simulation** as you did in Lesson 12-6A, are called **experimental probabilities**. Experimental probabilities may vary when an experiment is repeated.

Probabilities based on known characteristics or facts are called **theoretical probabilities**. For example, you can compute the theoretical probability of rolling a 6 on a number cube or spinning green on a spinner. Theoretical probability tells you *approximately* what should happen in an experiment.

Did you know

Around 1900, the English statistician Karl Pearson tossed a coin 24,000 times and got 12,012 heads.

HANDS-ON MINI-LAB

Work with a partner.

📄 paper bag containing 10 colored marbles

Try This

- Draw one marble from the bag, record its color, and replace it in the bag. Repeat this 10 times.

- Find the experimental probability for each color of marble.

  experimental probability = $\dfrac{\text{number of times color was drawn}}{\text{total number of draws}}$

- Repeat both steps described above for 20, 30, 40, and 50 draws.

Talk About It

1. Is it possible to have a certain color marble in the bag and never draw that color?

2. Open the bag and compute the theoretical probability of drawing each color of marble.

3. Compare the experimental probability and theoretical probability.

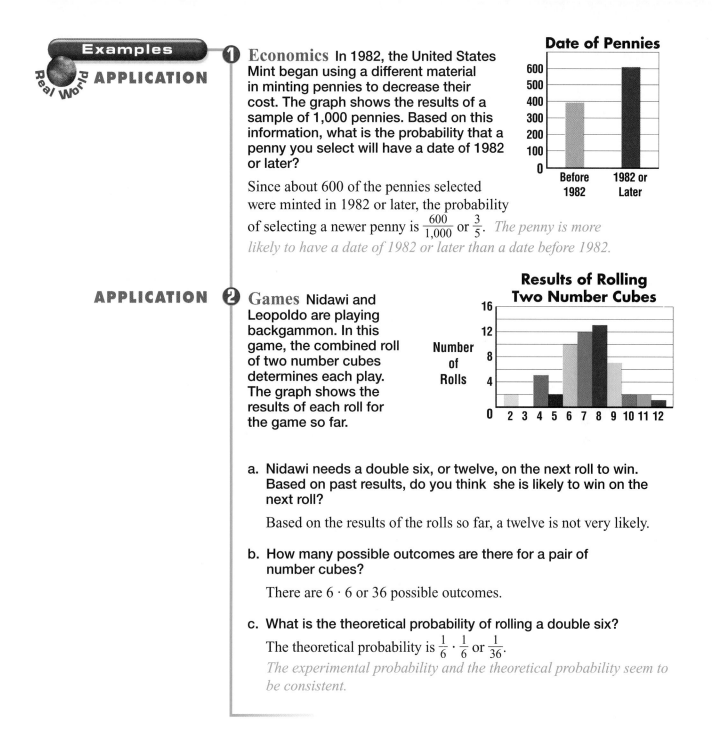

APPLICATION

① Economics In 1982, the United States Mint began using a different material in minting pennies to decrease their cost. The graph shows the results of a sample of 1,000 pennies. Based on this information, what is the probability that a penny you select will have a date of 1982 or later?

Date of Pennies

Since about 600 of the pennies selected were minted in 1982 or later, the probability of selecting a newer penny is $\frac{600}{1,000}$ or $\frac{3}{5}$. *The penny is more likely to have a date of 1982 or later than a date before 1982.*

APPLICATION ② **Games** Nidawi and Leopoldo are playing backgammon. In this game, the combined roll of two number cubes determines each play. The graph shows the results of each roll for the game so far.

Results of Rolling Two Number Cubes

a. Nidawi needs a double six, or twelve, on the next roll to win. Based on past results, do you think she is likely to win on the next roll?

   Based on the results of the rolls so far, a twelve is not very likely.

b. How many possible outcomes are there for a pair of number cubes?

   There are $6 \cdot 6$ or 36 possible outcomes.

c. What is the theoretical probability of rolling a double six?

   The theoretical probability is $\frac{1}{6} \cdot \frac{1}{6}$ or $\frac{1}{36}$.
   The experimental probability and the theoretical probability seem to be consistent.

CHECK FOR UNDERSTANDING

Communicating Mathematics

Read and study the lesson to answer each question.

1. *Tell* why you might not be able to determine the theoretical probability of an event.

2. *Explain* why you would not expect the theoretical probability and the experimental probability to always be the same.

HANDS-ON MATH

3. In conducting your experiment for the Mini-Lab, could you accurately predict the probability of each color after 10 draws? What suggestion might you make to others who would conduct a similar experiment?

4. Roll a number cube 50 times.
   a. Record the results.
   b. Based on your results, what is the probability of rolling a 2?
   c. Based on your results, what is the probability of rolling an even number?
   d. Are the probabilities in parts b and c examples of *experimental* or *theoretical* probabilities? Explain.
   e. Suppose you rolled a number cube 10,000 times. Predict the number of times you would roll a 2 based on your results in part a.

5. *Entertainment* A local video store has advertised that one out of every four customers will receive a free box of popcorn with their video rental. So far, 15 of the first 75 customers have won the popcorn.
   a. Based on the results so far, what is the experimental probability that a customer will win?
   b. What is the theoretical probability that a customer will win?

EXERCISES

Practice

Find a game in your house that uses a spinner, number cubes, or cards. Then design and carry out an experiment to find the experimental probability for one particular outcome. Compare the experimental probability to the theoretical probability.

6. Toss a paper cup 50 times and count the number of times it lands up.
   a. What is the experimental probability that the paper cup will land up?
   b. Can you determine the theoretical probability that the paper cup will land up? Explain.

7. Toss two coins 50 times.
   a. Record the results.
   b. Based on your results, what is the probability that both coins will show heads?
   c. Based on your results, what is the probability that one coin will show heads and the other will show tails?
   d. Draw a tree diagram to show all of the possible outcomes of tossing two coins.
   e. What is the theoretical probability that both coins will show heads?
   f. What is the theoretical probability that one coin will show heads and the other will show tails?

8. *Graphing Calculator* You can use the random number generator function on a graphing calculator to simulate a probability experiment. The graphing calculator program will generate 30 random numbers between 1 and the number you enter for T. In order to use the program, you must first enter the program into the calculator's memory. To access the program memory, use the following keystrokes.

Enter: [PRGM] [▶] [▶] [ENTER]

Run the program. Press PRGM and choose the program from the list by pressing the number next to its name. Enter a number for T. Press enter thirty times. The calculator will display a number each time you press enter.

PROGRAM:RANDOM
: Input T
: For (N, 1, 30)
: int((T*rand) + 1) → D
: Disp D
: Pause
: End

   a. Run the program to simulate 30 rolls of a number cube.
   b. Run the program to simulate 30 spins of a spinner that has 7 equal sections.
   c. How do the results compare to the theoretical probabilities of these events?

Applications and Problem Solving

9. *Agriculture* Over the last eight years, the probability that corn seeds planted by Mr. Simon produce corn is $\frac{5}{6}$.

   a. Is this probability an experimental or a theoretical probability? Explain.

   b. If Mr. Simon wants to have 10,000 corn bearing plants, how many seeds should he plant?

10. *Contests* The PTO at Grassland Middle School announced that "One out of every 25 students will win a movie pass." They will draw a name from each homeroom. Ambrosia complained that she does not have a $\frac{1}{25}$ chance of winning since there are 32 students in her homeroom. Is she correct? Explain.

11. *Working on the* **CHAPTER Project** Refer to the information about sports you gathered on page 515. Find the theoretical probabilities that a little league player picked at random will go on to play in higher level sports.

12. *Critical Thinking* Alfonso believes a coin is weighted to give an advantage to one side. He asked ten of his friends to toss the coin 40 times and record their results in the table. Do you think the coin is unfair? Explain.

For **Extra Practice**, see page 641.

Student	1	2	3	4	5	6	7	8	9	10
Number of Heads	21	22	18	26	21	21	19	22	18	29
Number of Tails	19	18	22	14	19	19	21	18	22	11

Mixed Review

13. **Standardized Test Practice** Marcus tossed a coin and rolled a number cube. What is the probability that he will toss tails and roll a multiple of 3? *(Lesson 12-5)*

   A $\frac{1}{2}$ B $\frac{1}{3}$ C $\frac{1}{4}$ D $\frac{1}{6}$

14. *Algebra* Solve the system $y = 2x - 7$ and $y = -2x + 9$. *(Lesson 10-5)*

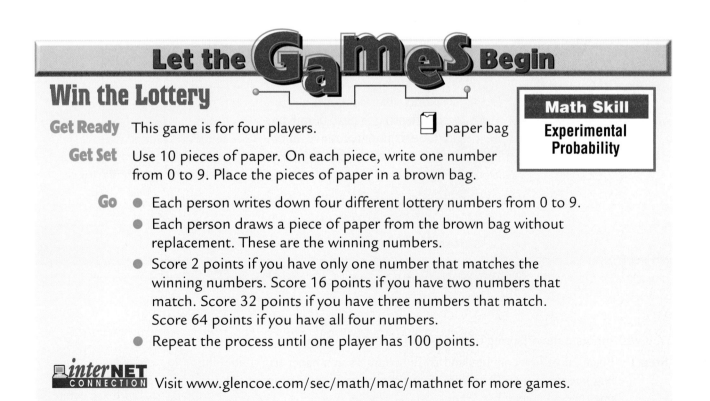

Let the Games Begin

Win the Lottery

Math Skill
Experimental Probability

Get Ready This game is for four players.   📄 paper bag

Get Set Use 10 pieces of paper. On each piece, write one number from 0 to 9. Place the pieces of paper in a brown bag.

Go
- Each person writes down four different lottery numbers from 0 to 9.
- Each person draws a piece of paper from the brown bag without replacement. These are the winning numbers.
- Score 2 points if you have only one number that matches the winning numbers. Score 16 points if you have two numbers that match. Score 32 points if you have three numbers that match. Score 64 points if you have all four numbers.
- Repeat the process until one player has 100 points.

interNET CONNECTION Visit www.glencoe.com/sec/math/mac/mathnet for more games.

COOPERATIVE LEARNING

12-7A Punnett Squares

A Preview of Lesson 12-7

40 yellow counters

40 red counters

2 paper bags

In the early 1900s, Reginald Punnett developed a model to show the possible ways genes can combine at fertilization. You can use a *Punnett square* to simulate how random fertilization works.

In a Punnett square, dominant genes are shown with capital letters. Recessive genes are shown with lowercase letters. Letters representing the parent's genes are placed on the left and upper sides of the Punnett square. Letters inside the boxes of the square show the possible gene combinations for their offspring.

Suppose pea plants are being crossed. Let **T** represent the dominant gene for tallness. Let **t** represent the recessive gene for shortness. Offspring receive one gene from each parent. A pea plant with **TT** genes is pure dominant and is tall. A pea plant with **tt** genes is pure recessive and is short. A pea plant with **Tt** genes is hybrid and is tall because it has a dominant gene. Notice that the capital letter goes first in hybrid genes.

The Punnett square at the right represents a cross between two hybrid tall pea plants.

Notice that 1 out of 4 offspring is pure dominant, 1 is pure recessive, and 2 are hybrid.

	T	t
T	TT	Tt
t	Tt	tt

TRY THIS

Work with a partner.

You will simulate the offspring of two hybrid parents.

Step 1 Place 20 yellow counters and 20 red counters in a paper bag. Label the bag *female parent*. This represents a parent with one dominant gene and one recessive gene.

Step 2 Place 20 yellow counters and 20 red counters in a second paper bag. Label this bag *male parent*. This represents a parent with one dominant gene and one recessive gene.

Step 3 Make a Punnett square to show the expected offspring of these parents. Use **R** for the dominant gene, red. Use **r** for the recessive gene, yellow.

Step 4 Copy the table.

Trial	Color of Female Parent Counter	Color of Male Parent Counter	Gene Pairs
1			
2			
3			
⋮			
40			

Step 5 Shake the bags. Reach into each bag and, without looking, remove one counter. Record the colors of the counters in your table. Put each counter back into its original bag.

Step 6 Repeat Step 5 thirty-nine more times.

ON YOUR OWN

1. Refer to Step 3. Out of 40 offspring, how many would you expect to be pure dominant (**RR**)? hybrid (**Rr**)? pure recessive (**rr**)?

2. Which of the three gene combinations did you expect to occur most often?

3. What is the theoretical probability of an offspring having pure dominant genes? hybrid genes? pure recessive genes?

4. Find each experimental probability. Describe how it compares to the corresponding theoretical probability.

5. In humans, free earlobes is a dominant trait over attached earlobes. Let **F** represent free earlobes and **f** represent attached earlobes. Draw a Punnett square for each parent combination below. Find the theoretical probability for each possible offspring.

   a. FF, Ff   **b. FF, FF**   **c. ff, ff**   **d. Ff, ff**

6. *Look Ahead* Suppose 100 babies have two parents who both have the hybrid earlobe trait. How many babies would you expect to have free earlobes? attached earlobes?

Integration: Statistics
Using Sampling to Predict

What you'll learn

You'll learn to predict the actions of a larger group by using a sample.

When am I ever going to use this?

Knowing how to predict using a sample can help you plan the type of refreshments needed for a large party.

Word Wise

sample
population

In the spring of 1997, water in the Red River in Minnesota overflowed its banks by as much as 14 miles. More than 60,000 people had to be evacuated from their homes. There was over 1 billion dollars in property damage from the flooding. Unfortunately, only about 1 out of every 15 homes had flood insurance.

As insurance companies began to determine how much money they might eventually pay in claims, they took a survey of a small group of residents in the area. This group is called a **sample**. A sample is representative of a larger group called the **population**. A sample is selected at random.

Example
Real World APPLICATION

① **Marketing** A video store is thinking about opening a new section in the store that offers only animated movies. They decide to take a survey to determine whether there is enough interest in this type of movie. For two Saturdays, they ask people at the mall if they would rent an animated movie at least every four months. The table gives the results.

a. How many people were in the survey?

$98 + 75 + 187 + 116 = 476$

476 people were surveyed.

	Yes	No
First Saturday	98	75
Second Saturday	187	116

b. What percent said they would rent an animated movie?

$98 + 187 = 285$ *285 people would rent an animated movie.*

285 ÷ 476 = 0.598739496

About 60% of the people said they would rent an animated movie at least every four months.

LOOK BACK

Refer to Lessons 3-2 and 8-2 to review percents.

c. Out of 2,000 customers, how many should the video store expect to rent animated movies?

$0.60 \times 2,000 = 1,200$

1,200 customers would be expected to rent an animated movie at least every four months.

d. Would the survey have a representative sample if they asked people coming out of a movie theater that is showing an animated movie?

No, because they would tend to like animated movies.

Example 2

Weather The experimental probability that various cities will have snow on December 25 is given in the chart. Suppose Cecilia lives to be 80 years old and spends every December in Cheyenne, Wyoming. How many times should she enjoy snow on December 25?

Find 35% of 80.

$0.35 \times 80 = 28$

Cecilia should enjoy snow 28 times.

Snow on December 25?

Boston, MA	23%
Cheyenne, WY	35%
Denver, CO	42%
Detroit, MI	50%
Minneapolis, MN	73%
Seattle, WA	5%

Source: Northeast Regional Climate Center at Cornell University

---

CHECK FOR UNDERSTANDING

Communicating Mathematics

Read and study the lesson to answer each question.

1. *Tell* how a sample group survey can be used to predict the actions of a whole population.

2. *Compare* taking a survey and finding an experimental probability.

Math Journal

3. *Write* a paragraph about the importance of using a random sample to make predictions. Tell how you could find an appropriate sample to determine the favorite sport of the students at your school. Give an example of an inappropriate sample for the same question.

Guided Practice

4. The school bookstore sells 3-ring binders. All incoming sixth-grade students will need a binder. The binders come in 4 colors; red, green, blue, or yellow. The students who run the store decide to survey 50 sixth graders to find out their favorite color. They will use this information to order 450 binders to sell in the fall.

   a. From an alphabetical list, the students survey every ninth student. Is this a good sample? Explain.

   b. Of the students surveyed, 25 chose red, 10 chose green, and 2 chose yellow. How many chose blue?

   c. For the 450-binder order, how many of each color should be ordered?

5. *Entertainment* As people leave a concert, 50 people are surveyed at random. Six people say they bought a concert T-shirt.

   a. If 6,330 people attended the concert, how many people would you predict bought T-shirts at the concert?

   b. What could T-shirt sellers learn from this survey?

EXERCISES

Practice

6. A survey is being taken to determine how much money the average family in the United States spends to heat their home. A sample of people from Arizona is used for the survey. Is this a good sample? Explain.

7. Three students are running for class president. Marcy surveyed some of her classmates during lunch. Seven students said they were voting for Pedro, twelve said they were voting for Jasmine, and six said they were voting for Tim.

   a. What is the size of the sample?

   b. What percent said they were voting for Jasmine?

   c. What fraction of the students said they were voting for Tim?

   d. There are 180 students in the class. Based on this survey, how many votes do you think Pedro will receive?

8. A music store surveyed some of their customers. Use the results to answer each question.

   a. What is the size of the sample?

   b. What fraction chose country?

   c. What percent chose oldies?

   d. If the management is going to order 1,000 new CDs, how many of each type would you recommend be purchased?

Favorite Type of Music	
rock	168
jazz	42
rhythm and blues	78
country	152
alternative	23
oldies	44
other	33

Applications and Problem Solving

9. *Advertising* The Sunshine Orange Juice Company conducted a taste test. Out of 600 people who compared Sunshine Orange Juice to the competitor's brand, 327 said they preferred Sunshine Orange Juice. Which of the following statements can the company correctly make in their advertising? Explain.

   a. "Consumers prefer Sunshine Orange Juice 2 to 1."

   b. "Over 50% of the people surveyed prefer Sunshine Orange Juice."

   c. "More people always choose Sunshine Orange Juice over our competitor's brand."

10. *Life Science* Miss Thompson is a park ranger. She wants to estimate the number of fish in a small lake. She captures and marks 50 fish. She returns them to the lake. Several weeks later, she captures another 50 fish. She notices that 15 of the fish are some of the ones she marked.

   a. Write a proportion that Miss Thompson could use to predict the number of fish in the lake.

   b. Estimate the number of fish in the lake.

11. *Working on the* **CHAPTER Project** Refer to the probabilities of playing higher level sports that you calculated on page 543. Use these probabilities to predict how many out of 100,000 little league players will play at these levels. Make a graph showing your predictions.

For **Extra Practice**, see page 641.

12. *Critical Thinking* How could things such as the wording of a question or the tone of voice of the interviewer affect a survey? Explain.

Mixed Review

13. *Probability* Define experimental probability. *(Lesson 12-6)*

14. **Standardized Test Practice** A hiker walked 22 miles north and then walked 17 miles west. How far is the hiker from the starting point? *(Lesson 9-5)*

   A 374 mi   **B** 112.6 mi   **C** 39 mi   **D** 27.8 mi

Debra Wein
NUTRITIONIST

Debra Wein is the founder of a nutrition consulting firm in Boston called The Sensible Nutrition Connection (SnaC). She has made presentations for several national sports and fitness organizations including the USA Track & Field Association, the American College of Sports Medicine (ACSM), and the International Dance Exercise Association (IDEA). She is a member of the Massachusetts Governor's Council on Physical Fitness & Sports. Ms. Wein has degrees in nutrition and applied physiology.

To be a nutritionist, you'll need a bachelor's degree in dietetics, food and nutrition, or food service systems management. Courses in biology, mathematics, statistics, psychology, and sociology are required for these degrees.

For more information:
American Dietetic Association
216 West Jackson Boulevard
Chicago, IL 60606

inter NET CONNECTION
www.glencoe.com/sec/
math/mac/mathnet

Someday, I'd like to work with athletes to help them reach their full potential.

Your Turn
Find out what types of foods an athlete should eat in the 24 hours before competition. List some examples of each of these types of foods. How many combinations of the foods listed could an athlete eat before competition? List these combinations.

Study Guide and Assessment

inter NET
CONNECTION Chapter Review **For additional lesson-by-lesson review, visit:**
www.glencoe.com/sec/math/mac/mathnet

Vocabulary

After completing this chapter, you should be able to define each term, concept, or phrase and give an example or two of each.

Statistics and Probability

combination (p. 524)
compound events (p. 534)
Counting Principle (p. 519)
dependent events (p. 535)
experimental probability (p. 540)
factorial (p. 522)
independent events (p. 534)
outcome (p. 518)
Pascal's Triangle (p. 528)

permutation (p.521)
population (p. 546)
sample (p. 546)
simulation (p. 540)
theoretical probability (p. 540)
tree diagram (p. 518)

Problem Solving

act it out (p. 538)

Understanding and Using the Vocabulary

Choose the letter of the term that best matches each phrase.

1. an arrangement or listing in which order is important
2. an arrangement or listing in which order is not important
3. the product of all counting numbers beginning with n and counting back to 1
4. a diagram used to find the total number of outcomes of an event
5. a smaller group representative of a larger group
6. probabilities based on frequencies obtained in an experiment
7. probabilities based on known facts
8. two or more events for which the outcome of one event affects the outcome of another event
9. two or more events for which the outcome of one event does not affect the outcome of the other event

a. experimental probabilities
b. theoretical probabilities
c. population
d. sample
e. combination
f. permutation
g. tree diagram
h. $n!$
i. dependent events
j. independent events
k. simulation

In Your Own Words

10. *Write* a definition of the Counting Principle.

Objectives & Examples

Upon completing this chapter, you should be able to:

● count outcomes by using a tree diagram or the Counting Principle *(Lesson 12-1)*

Two coins are tossed. How many outcomes are possible?

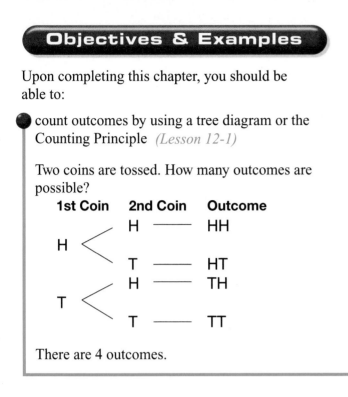

There are 4 outcomes.

● find the number of permutations of objects *(Lesson 12-2)*

The symbol $P(n, r)$ represents the number of permutations of n things taken r at a time.

$P(n, r) =$

$n \cdot (n - 1) \cdot (n - 2) \cdot ... \cdot (n - r + 1)$

● find the number of combinations of objects *(Lesson 12-3)*

The symbol $C(n, r)$ represents the number of combinations of n things taken r at a time.

$$C(n, r) = \frac{P(n, r)}{r!}$$

Review Exercises

Use these exercises to review and prepare for the chapter test.

State the number of possible outcomes for each event.

11. A restaurant offers 4 types of soft drinks in two different sizes.

12. A car comes in 3 models and 4 colors. There is a choice of a standard or automatic transmission.

13. Phones come in wall or desk models with straight or coiled cords. They come in three colors—black, neutral, and white.

Find each value.

14. $P(7, 2)$

15. $P(5, 5)$

16. How many different ways can 4 different books be arranged on a shelf?

17. How many 3-digit whole numbers can you write using the digits 1, 2, 3, 4, 5, and 6 if no digit can be used twice?

Find each value.

18. $C(9, 5)$

19. $C(12, 2)$

20. How many different pairs of puppies can be selected from a litter of 8?

21. How many different groups of 3 marbles can be chosen from a box containing 20 marbles?

Chapter 12   Study Guide and Assessment

Objectives & Examples

identify patterns in Pascal's Triangle
(Lesson 12-4)

How many different groups of 2 people can be taken from a group of 5 people?

Number Taken at a Time	0	1	2	3	4	5
Row 5	1	5	10	10	5	1

There are 10 different groups.

find the probability of independent and dependent events *(Lesson 12-5)*

The probability of two events can be found by multiplying the probability of the first event by the probability of the second event.

find experimental probability *(Lesson 12-6)*

Experimental probabilities are probabilities based on frequencies obtained by conducting an experiment.

predict the actions of a larger group by using a sample *(Lesson 12-7)*

If 30 out of a sample of 90 students prefer grape juice over orange juice, how many cartons of grape juice does the cafeteria need for 750 students?

$$\text{grape sample} \rightarrow \frac{30}{90} = \frac{g}{750} \leftarrow \text{grape order}$$
$$\text{total sample} \rightarrow \quad\quad\quad\quad \leftarrow \text{total order}$$

30 $\boxed{\times}$ 750 $\boxed{\div}$ 90 $\boxed{=}$ *250*

The cafeteria needs to order 250 cartons of grape juice.

Review Exercises

Use Pascal's Triangle to answer each question.

22. Find the value of $C(9, 3)$.

23. How many combinations are possible when 8 objects are chosen 4 at a time?

24. How many different kinds of 3-topping pizzas can be made when 7 toppings are offered?

25. A number cube is rolled twice. What is the probability that an odd number will show both times?

26. A bag contains 7 white marbles and 3 blue marbles. Once a marble is selected, it is not replaced. What is the probability of picking 2 blue marbles?

27. Toss two coins 20 times.
    a. Record the results.
    b. What is the experimental probability of no tails? of one head and one tail?
    c. What are the theoretical probabilities for part b?

The Superfast Computer Company conducted a survey. Participants in the survey were asked to choose which computer they would buy. Of those surveyed, 48 said they would buy a Superfast computer, 34 said they would buy computer X, and 18 said they would buy computer Y.

28. What is the size of the sample?

29. What fraction chose computer Y?

30. For 8,000 people, how many would buy a Superfast computer?

Applications & Problem Solving

31. *Food* County Line Restaurant has a Saturday breakfast special that offers the choices shown in the menu below. Draw a tree diagram to find the number of possible outcomes for a meal consisting of one choice from each column. *(Lesson 12-1)*

32. *Sports* There are 10 batters on a softball team. How many ways can the coach pick the first 4 batters? *(Lesson 12-2)*

33. *Business* Midtown Music Store has 20 different CDs on a sale table. These CDs are priced 3 for $30. How many combinations of 3 CDs can be purchased? *(Lesson 12-3)*

34. *Act It Out* Each morning, Allison walks past 4 park benches. Based on her observation, she estimates that any one bench is occupied 0.75 of the time. Find the probability that all 4 benches will be occupied as Allison walks past this morning. *(Lesson 12-6A)*

Alternative Assessment

● Open Ended

Your friend says that he is playing a game in which the probability of event A occurring is 30% and the probability of event B occurring is 40%. Your friend says that events A and B are independent. Give an example of a game in which these conditions are met. What is P(A and then B)?

Suppose events A and B are dependent. Explain how to change your game so this is true.

● Completing the CHAPTER Project

Use the following checklist to make sure your news article is complete.

☑ The probabilities that a player will play at higher levels are based on information you have collected.

☑ The predictions about how many out of 100,000 little league players will play sports at a higher level are accurate.

☑ The graph showing how many players you expect to play at higher levels is clear and concise.

● PORTFOLIO   Select one of the assignments you completed in this chapter and place it in your portfolio. Attach a note explaining why you selected it.

A practice test for Chapter 12 is provided on page 658.

Section One: Multiple Choice

There are eleven multiple choice questions in this section. Choose the best answer. If a correct answer is *not here,* choose the letter for Not Here.

1. What is the probability of choosing a yellow marble from a bag containing 4 blue marbles, 3 red marbles, and 5 black marbles?

   A 0

   B $\frac{1}{7}$

   C $\frac{4}{15}$

   D 1

2. Samantha is redecorating her room. She has a choice of 4 colors of paint, 3 colors of carpet, and 2 colors of curtains. How many combinations of paint, carpet, and curtain colors can she use?

   F 9

   G 12

   H 24

   J 48

3. The back of the couch forms a 115° angle with its seat. Which type of angle is this?

   A right

   B acute

   C obtuse

   D isosceles

4. Write $-2.\overline{8}$ as a fraction or mixed number.

   F $-2\frac{4}{5}$

   G $-2\frac{8}{9}$

   H $-\frac{28}{10}$

   J $-\frac{28}{9}$

Please note that Questions 5–11 have five answer choices.

5. Which statement is correct concerning the probabilities of reaching into the jars without looking and pulling out a blue marble?

   A greater for Jar 1 than Jar 2

   B greater for Jar 2 than Jar 1

   C equal for both jars

   D both are greater than 1

   E cannot be determined

6. Which mathematical expression represents the number of ways the four shapes can be arranged in a row?

   F $P(4, 1)$

   G $P(4, 4)$

   H $C(4, 1)$

   J $C(4, 4)$

   K $\frac{4!}{C(4, 4)}$

7. One centimeter is about 0.3937 inch. If Henry is 68 inches tall, estimate his height in centimeters.

   A less than 140 centimeters

   B between 140 and 150 centimeters

   C between 150 and 160 centimeters

   D between 160 and 170 centimeters

   E more than 170 centimeters

8. A swimming pool filled with water is losing water through a hole at a rate of 3 gallons per minute. At this rate, in how many minutes will the pool lose 78 gallons?

F 18 minutes

G 24 minutes

H 26 minutes

J 30 minutes

K Not Here

9. One drinking cup holds 8.8 ounces. What is the largest number of drinking cups that can be filled with a 80 ounce bottle?

A 7 cups

B 8 cups

C 9 cups

D 10 cups

E 11 cups

10. Jackson Auto Rental charges $20 per day and $0.15 per mile to rent a car. Find the cost of renting a car for two days and driving 100 miles.

F $35

G $45

H $55

J $65

K Not Here

11. A paycheck has 23% deducted for federal income tax and 4.5% deducted for state income tax. What is the total percent deducted from the paycheck for these taxes?

A 18.5%

B 22.5%

C 27.5%

D 100%

E Not Here

Test-Taking Tip THE PRINCETON REVIEW

When taking a standardized test, you may be required to fill in an answer sheet. Your answers may be machine scored. It is important to make your marks completely and fill in circles darkly. Answer each question only once, and make sure you check frequently that you are matching the correct problem number with the correct answer blank.

Section Two: Free Response

This section contains five questions for which you will provide short answers. Write your answers on your paper.

12. What is the least common multiple of 33 and 36?

13. What is the greatest common factor of 24 and 36?

14. What are the prime factors of 65?

15. List the factors of 72.

16. If $\triangle ABC \sim \triangle DEF$, write three ratios relating the corresponding sides.

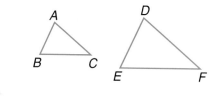

www.glencoe.com/sec/math/mac/mathnet

Interdisciplinary Investigation

EXTRA! EXTRA! NEWSPAPERS MAY TAKE OVER THE PLANET!

Does your community have a recycling center? Do you recycle unwanted items? Did you know that over 40% of what we throw away in the United States is paper and paperboard? Many trees are wasted each week as a result of not recycling newspapers.

What You'll Do

In this investigation, you will collect data about a daily newspaper. You will investigate the volume of newspapers discarded by just one person and discover how that volume can accumulate over many years.

Materials   a week's supply of newspapers   ruler

 calculator   grid paper

Procedure

1. Work in a group. Collect daily editions of a newspaper for a week. Make a stack containing the newspapers. Find the volume of the newspapers.

2. Suppose you bought and saved this newspaper every day for a year. What would be the total volume of the newspapers? Mark off a space in your classroom that has approximately the same volume.

3. Under certain conditions, newspapers in a landfill can be intact after 20 or 30 years. Make a function table showing the accumulation of one person's newspapers from 1 to 20 years. Write an equation in two variables describing the function. Graph the function.

Technology Tips

- Use a **spreadsheet** to calculate the accumulation of newspapers.

- Use **graphing software** to graph your function.

- Surf the **Internet** to find information on recycling.

Making the Connection

Use the data collected about newspapers as needed to help in these investigations.

Language Arts

Write an article to encourage the recycling of newspapers. Give specific information and data about recycling newspapers.

Science

Research the conditions needed for newspapers to decompose in a landfill. Investigate the advantages and disadvantages of burning paper instead of dumping it in landfills.

Art

Find a recipe for making recycled paper. Make some recycled paper and use it for an art project.

Go Further

- Estimate the volume of another type of garbage that you discard. What would be the accumulated volume of this garbage in one year?

- Collect facts about the types and amounts of discarded items in the United States. Prepare a graph, table, or informational pamphlet displaying your findings.

interNET **CONNECTION** Research For current information on recycling, visit the following website.

Data Collection and Comparison To share and compare your data with other students in the U.S., visit:

www.glencoe.com/sec/math/mac/mathnet

You may want to place your work on this investigation in your portfolio.

Algebra: Exploring Polynomials

What you'll
learn in Chapter 13

- to represent and simplify polynomials using algebra tiles,

- to add, subtract, and factor polynomials using algebra tiles,

- to multiply monomials and polynomials using algebra tiles, and

- to solve problems by guess and check.

CHAPTER Project

HIT OR MISS

In this project, you will create geometric and algebraic models that describe the number of baskets a player can expect to make when he or she shoots free throws. You will write a report explaining your models and how they can be used.

Getting Started

- Find the free throw percent of at least two of your favorite college or professional basketball players.

- Convert these percents to probabilities. For example, if the percent is 90%, then the probability is 0.9.

- What is the probability of each of these players hitting a free throw? What is the probability of their missing a free throw?

- Suppose the players are to shoot two free throws in a row. What is the probability of their hitting both free throws? missing both free throws? hitting the first and missing the second? missing the first and hitting the second?

- As you work through the lessons in the chapter, you will make area models to describe the probabilities you have just found. You will use these models to write a polynomial that you can use to find the number of points each player can expect to score if he or she attempts two free throws.

Technology Tips

- Use a **spreadsheet** to keep track of the probabilities.

- Use **computer software** to draw your geometric models.

- Use a **word processor** to write your report.

 **Data Update For up-to-date information on sports statistics, visit:**

www.glencoe.com/sec/math/mac/mathnet

Working on the Project

You can use what you'll learn in Chapter 13 to help you create your models.

Page	Exercise
564	34
585	21
591	25
595	Alternative Assessment

COOPERATIVE LEARNING

13-1A Algebra Tiles

A Preview of Lesson 13-1

yellow, red, blue, and green construction paper

scissors

Throughout this text, you have used rectangles to show multiplication problems. For example, the figure at the right shows 4×5 as a rectangle that is 4 units wide and 5 units long. Its area is 20 square units.

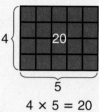

$4 \times 5 = 20$

In this lab, you will make models called *algebra tiles*.

TRY THIS

Work with a partner.

Step 1 From yellow construction paper, cut out a square that is 1 unit long and 1 unit wide.

Step 2 Draw a line segment that is longer than 1 unit. Since the line segment can be any length, we will say that it is x units long. From green construction paper, cut out a rectangle that is 1 unit wide and x units long.

Step 3 Using the same measure for x, cut out a square that is x units long and x units wide from blue construction paper.

ON YOUR OWN

1. Find the area of each shape and write the area on each tile.
2. For each tile, write a sentence that tells the relationship among the length, width, and area.
3. From construction paper, cut out four more sets of tiles of the same sizes and colors as the first set. Label each with its area.
4. From red construction paper, cut out five sets of tiles with the same dimensions as the yellow, green, and blue tiles. Label the areas -1, $-x$, and $-x^2$.
5. *Look Ahead* Use your tiles to model $4x$.

560 **Chapter 13** Algebra: Exploring Polynomials

Modeling Polynomials

What you'll learn

You'll learn to represent polynomials with algebra tiles.

When am I ever going to use this?

Knowing how to represent polynomials with algebra tiles will help you recognize like terms.

Word Wise

monomial
polynomial

interNET CONNECTION

For the latest statistics on allowances, visit: www.glencoe.com/sec/math/mac/mathnet

Do you get an allowance? A survey of 784 kids ages 9 through 14 was taken. Of those surveyed, 361 receive an allowance. Kids 13 and older were less likely to get an allowance than those younger.

Models are often used to help us visualize numbers. The graph models the amount of kids' allowances for ages 13 and 14.

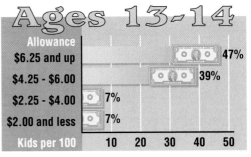

Source: *Zillions,* Jan./Feb., 1997

The number 361 can be modeled as follows.

$$361 = (3 \times 10^2) + (6 \times 10) + 1$$

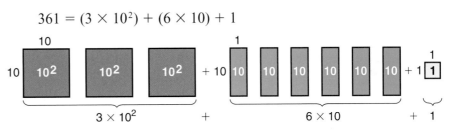

In the previous Hands-On Lab, you made algebra tiles. Algebra tiles can be used to model **monomials**.

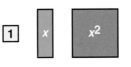

The expressions 1, x, and x^2 are examples of monomials. A monomial is a number, a variable, or a product of a number and one or more variables.

Examples

Model each monomial using drawings or algebra tiles.

1 5

To model this expression, you need 5 yellow 1-tiles.

2 $-2x$

Use red tiles to model negative values. To model this expression, you need 2 red x-tiles.

3 $3x^2$

To model this expression, you need 3 blue x^2-tiles.

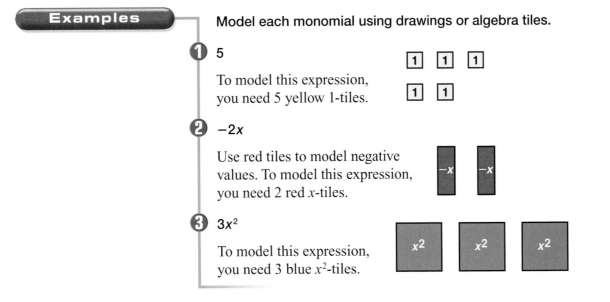

Algebraic expressions that contain more than one monomial are called **polynomials**. A polynomial is the sum or difference of two or more monomials. You can also model polynomials with algebra tiles.

Examples

Model each polynomial using drawings or algebra tiles.

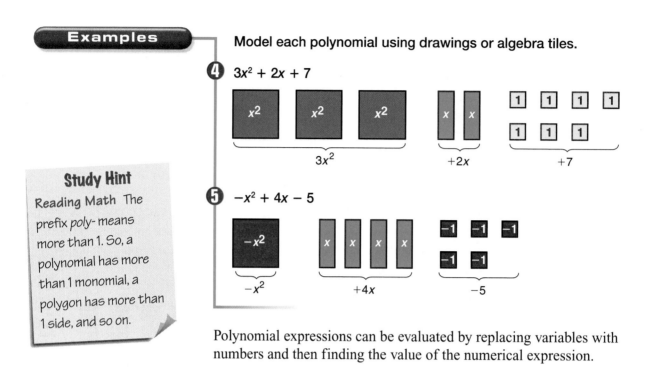

4 $3x^2 + 2x + 7$

5 $-x^2 + 4x - 5$

Study Hint

Reading Math The prefix *poly-* means more than 1. So, a polynomial has more than 1 monomial, a polygon has more than 1 side, and so on.

Polynomial expressions can be evaluated by replacing variables with numbers and then finding the value of the numerical expression.

Example

CONNECTION

6 **Physical Science** The expression $-16x^2 + 56x$ describes the height of an arrow after x seconds if it is shot upward at 56 feet per second. Find the height of the arrow after 2 seconds by evaluating the expression for $x = 2$.

$$-16x^2 + 56x = -16(2)^2 + 56(2) \quad \textit{Replace x with 2.}$$
$$= -16(4) + 56(2)$$
$$= -64 + 112$$
$$= 48$$

After 2 seconds, the arrow is 48 feet high.

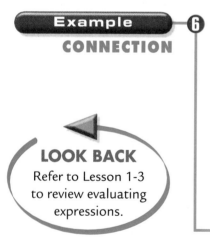

LOOK BACK

Refer to Lesson 1-3 to review evaluating expressions.

CHECK FOR UNDERSTANDING

Communicating Mathematics

Read and study the lesson to answer each question.

1. *Tell* whether each expression is a monomial or a polynomial.

   a. $x + 2$ **b.** $3x$ **c.** $7xy$ **d.** $2x^2 + x - 5$

Math
Journal

2. *Tell* how to model a monomial like $-3x$.

3. *Write* a polynomial and use a drawing or algebra tiles to represent it.

Guided Practice

Write a monomial or polynomial for each model.

4.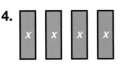

5.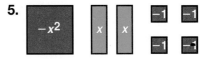

Model each monomial or polynomial using drawings or algebra tiles.

6. $2x^2$

7. $3x - 1$

Evaluate each expression.

8. $5x - 2$, if $x = 6$

9. $x^2 + 4x$, if $x = -1$

10. $x^2 - 9x + 14$, if $x = 7$

11. *Life Science* Stories about sea monsters may have come from sightings of giant squid. The expression $x - 50$ represents the length of a giant squid in feet. Evaluate the expression for $x = 120$ to find the length of a giant squid.

EXERCISES

Practice

Write a monomial or polynomial for each model.

12.

13.

14.

15.

16.

17.

Model each monomial or polynomial using drawings or algebra tiles.

18. $-3x^2$

19. $5x$

20. $-4x + 2$

21. $2x^2 - 3x$

22. $x^2 - 2x + 1$

23. $-2x^2 + x - 4$

Evaluate each expression.

24. $12x + 4$, if $x = 3$

25. $6x^2 + 3x$, if $x = 2$

26. $x^2 - 7x - 8$, if $x = 1$

27. $x^2 - 5x$, if $x = -2$

28. $-x^2 + 4x - 5$, if $x = 4$

29. $2x^2 - x - 7$, if $x = -1$

30. Find the value of $9x^2 + x$ if $x = 5$.

31. What is the sum of $3x^2$, $4x$, and -10 if $x = -3$?

Real World

32. *Sports* If Juan Gonzalez of the Detroit Tigers hits a baseball straight up with a speed of 150 feet per second, the height of the ball after x seconds is given by $-16x^2 + 150x$.

Height of Ball	
Time (s)	**Height (ft)**
0	
2	
4	
6	
8	
9	

  a. Tell how you would use algebra tiles to model the expression $-16x^2 + 150x$.

  b. Copy and complete the table.

  c. Graph the ordered pairs and connect the points with a smooth curve.

  d. Use the graph to estimate the maximum height of the ball.

33. *Geometry* Refer to the figure at the right.

  a. Write an expression for the surface area of the box.

  b. Find the surface area when $x = 2$.

  c. Is the box a cube or a rectangular prism? Explain.

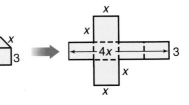

34. *Working on the* CHAPTER *Project*

  Refer to the probabilities you found on page 559.

  a. Use a product mat to represent the probabilities of hitting (h) or missing (m) a first free throw on the horizontal axis. Then model the probabilities of hitting or missing a second free throw on the vertical axis.

  b. Use the percents of each of your players to complete the area model for the product of the probabilities of making both, missing the first then hitting the second, hitting the first then missing the second, or missing both free throws.

35. *Critical Thinking* Suppose 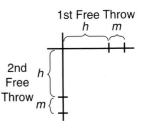 represents an area of a and b represents an area of b. Draw a model that represents an area of ab.

Mixed Review

36. *Statistics* Refer to the favorite soft drink survey. For 6,300 people, how much ginger ale should the Band Boosters order? *(Lesson 12-7)*

Favorite Soft Drink	Number of Responses
Lemon Lime	17
Cola	25
Root Beer	10
Fruit	12
Ginger Ale	8

For **Extra Practice**, see page 642.

37. *Geometry* Find the distance between $P(-3, 4)$ and $Q(5, -2)$. *(Lesson 9-6)*

38. **Standardized Test Practice** The back of a chair forms an angle of $112°$ with its seat. What type of angle is this? *(Lesson 5-2)*

  A obtuse   **B** right   **C** straight   **D** acute

39. What percent of 70 is 42? *(Lesson 3-6)*

Simplifying Polynomials

From 1987 to 1996, the number of businesses owned by minority women increased 153%. How many businesses did minority women own in 1996? This number can be found by adding the numbers in the minority groups. The total is 382,400 + 305,700 + 405,200 or 1,093,300.

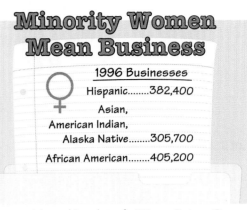

Minority Women Mean Business

1996 Businesses
Hispanic........382,400
Asian, American Indian, Alaska Native........305,700
African American........405,200

Source: National Foundation for Women Business Owners

In the same way, the polynomial $7x + 5x$ can be simplified by adding the number of x's. We can model $7x + 5x$ with algebra tiles.

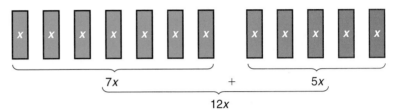

$7x$ + $5x$

$12x$

Pearline Motley

The polynomial $7x + 5x$ contains two monomials. Each monomial in a polynomial is called a **term**. The monomials $7x$ and $5x$ are called **like terms** because they have the same variable to the same power. When you use algebra tiles, you can recognize like terms because they have the same size and shape.

The model shown above suggests that you can simplify polynomials that have like terms. An expression that has no like terms is in **simplest form**. In simplest form, $7x + 5x$ is $12x$.

Example

1 Simplify $x^2 + 4x^2 + 3x$.

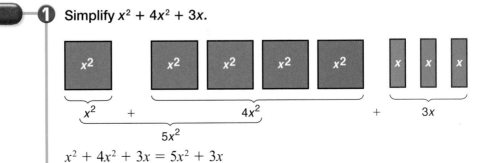

x^2 + $4x^2$

$5x^2$

+ $3x$

$x^2 + 4x^2 + 3x = 5x^2 + 3x$

What happens when you have both positive and negative terms in a polynomial? Let's use algebra tiles to find out.

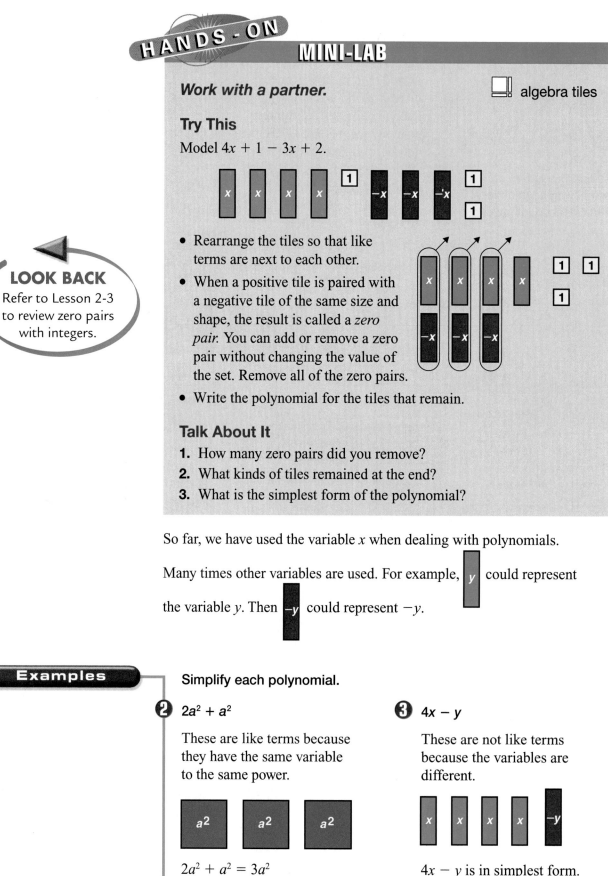

HANDS-ON

MINI-LAB

Work with a partner.

algebra tiles

Try This
Model $4x + 1 - 3x + 2$.

LOOK BACK
Refer to Lesson 2-3 to review zero pairs with integers.

- Rearrange the tiles so that like terms are next to each other.
- When a positive tile is paired with a negative tile of the same size and shape, the result is called a *zero pair*. You can add or remove a zero pair without changing the value of the set. Remove all of the zero pairs.
- Write the polynomial for the tiles that remain.

Talk About It
1. How many zero pairs did you remove?
2. What kinds of tiles remained at the end?
3. What is the simplest form of the polynomial?

So far, we have used the variable x when dealing with polynomials.

Many times other variables are used. For example, y could represent the variable y. Then $-y$ could represent $-y$.

Examples

Simplify each polynomial.

❷ $2a^2 + a^2$

These are like terms because they have the same variable to the same power.

$2a^2 + a^2 = 3a^2$

❸ $4x - y$

These are not like terms because the variables are different.

$4x - y$ is in simplest form.

Example

4 **Food** Matt brought 2 boxes of chocolate cookies and 1 box of peanut butter cookies to the class picnic. Tara brought 2 boxes of chocolate cookies and 3 boxes of peanut butter cookies. If c represents the number of chocolate cookies in a box, and p represents the number of peanut butter cookies in a box, then the total number of cookies is $2c + p + 2c + 3p$. Simplify the polynomial to find the total number of cookies.

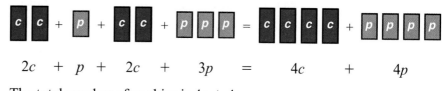

$$2c \ + \ p \ + \ 2c \ + \ 3p \ = \ 4c \ + \ 4p$$

The total number of cookies is $4c + 4p$.

CHECK FOR UNDERSTANDING

Communicating Mathematics

Read and study the lesson to answer each question.

1. **Tell** the like terms in $2x^2 - y - 3x^2 + 2y^2 + 5y$.

2. **Write** a polynomial in simplest form to represent the model at the right.

HANDS-ON MATH

3. **Draw** a model of $-2x^2 + x^2 - 2x + x - 3$. Then write a polynomial in simplest form to represent the model.

Guided Practice

Name the like terms in each list of terms.

4. $6x, -4x^2, -11x$

5. $7x, 8y, 9z$

6. Simplify $-x^2 + 4x + 2x + x^2$ using the model.

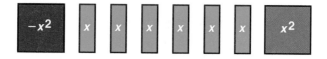

Simplify each polynomial. Use drawings or algebra tiles if necessary.

7. $7x + 1 - 3x$

8. $-3y^2 - 2y^2 - 4y + 3$

Simplify each expression. Then evaluate if $x = -1$ and $y = 6$.

9. $5x - 2y + 3x + y$

10. $x + 2x - 7y + 4x$

11. **Money** Miyoko found three quarters, five dimes, and two nickels in her backpack. In her pocket, she had one quarter, three dimes, and three nickels.

   a. Let $q, d,$ and n represent the value of a quarter, a dime, and a nickel, respectively. Represent all of the coins Miyoko has as a polynomial expression in simplest form.

   b. Evaluate the simplified expression to determine how much money Miyoko has.

Practice

Name the like terms in each list of terms.

12. $4, 3m^2, -5, 2m$

13. $3x^2, 4x, 10, -2x^2$

14. $15y^2, 2y, 8$

15. $7a, 6b, 10a, 14b$

16. $-a^2, 3a^2, -4x^2$

17. $4y, 8, 9, -2y, -3y$

Simplify each polynomial using the model.

18. $3a^2 - 2a^2 + 3a$

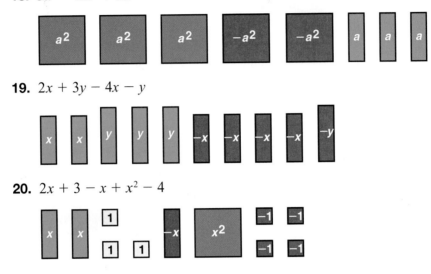

19. $2x + 3y - 4x - y$

20. $2x + 3 - x + x^2 - 4$

Simplify each polynomial. Use drawings or algebra tiles if necessary.

21. $2a + 1 + 3a + 4$

22. $2x^2 + 3 + 4x - 7$

23. $y^2 - 5y - y^2 - 2y$

24. $10x + 3y - 8x + 5y$

25. $4x^2 + 6 - x^2$

26. $-a^2 + a + 10a + 9a^2$

Simplify each expression. Then evaluate if $a = 4$ and $b = -2$.

27. $3a + 9b + 14a + 2b$

28. $7a + 9b + 1a + 5b$

29. $a + 5b - 2a + 4b$

30. $-4a + 7a + 10b - 7b$

31. Find the value of $4x^2 + x - x^2$ if $x = 4$.

32. Find the sum of $8a, 7b, -2a,$ and b if $a = 3$ and $b = 2$.

Applications and Problem Solving

33. *Money Matters* Shana receives $50 each birthday from her uncle. Her parents put this money in a savings account with an interest rate of r. The table gives the account balance after each birthday.

Birthday	Balance ($)
1	50
2	$50 + 50r + 50$
3	$50 + 100r + 50r^2$ $+ 50 + 50r + 50$

a. Write the amount Shana has in her account on her second birthday in simplest form. If r is 6% or 0.06, how much is in her account?

b. Write the balance of Shana's account after her third birthday in simplest form.

34. Geometry A new playground is to be constructed as shown at the right.

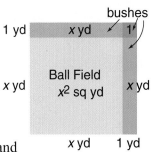

a. Write the area of the playground in simplest form.

b. Find the area if x is 60.

35. School As a fund-raiser, the school band is selling fruit. The sales for the first two weeks are shown in the table.

a. Using O and G to represent the selling prices of a basket of oranges and a basket of grapefruit respectively, write a polynomial expression for the total sales.

Baskets of Fruit Sold		
Week	Oranges	Grapefruit
1	52	39
2	62	57

b. If oranges cost $12 a basket and grapefruit cost $9 a basket, what was the total amount of sales?

36. Critical Thinking Write the polynomial modeled at the right in simplest form.

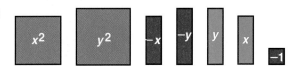

Mixed Review

37. Algebra Evaluate $5x^2 - 3x + 1$ if $x = 2$. *(Lesson 13-1)*

38. Functions Find the value of $f(3)$ if $f(n) = 2n^3 - 6$. *(Lesson 10-1)*

39. Standardized Test Practice A photograph measures 4 inches wide by 6 inches long. An enlargement is made that is 18 inches long. How wide is the enlargement? *(Lesson 8-9)*

4 in.

6 in.

? in.

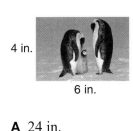

18 in.

A 24 in.

B 12 in.

C 28 in.

D 36 in.

For **Extra Practice**, see page 642.

40. Express 8.99×10^4 in standard form. *(Lesson 6-9)*

Adding Polynomials

Where is the best place for your savings? A bank Certificate of Deposit (CD) is very safe, but the earnings may be low. Investing money in a mutual fund is more risky, but can produce better profits. The graph shows the earnings of a $500 investment in a CD and of a mutual fund for a recent year.

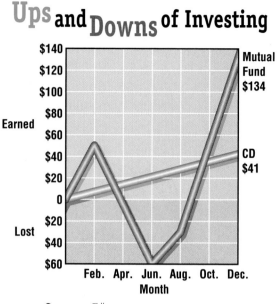

Ups and Downs of Investing

Source: *Zillions*

The mutual fund had lost $60 in June but finished the year earning a total of $134. It was worth $500 plus $134 growth.

$$
\begin{array}{rcl}
\$500 & \rightarrow & (5 \times 10^2) + (0 \times 10) + 0 \\
+134 & \rightarrow & (1 \times 10^2) + (3 \times 10) + 4 \\
\hline
\$634 & \leftarrow & (6 \times 10^2) + (3 \times 10) + 4
\end{array}
$$

The total value of the mutual fund was $634.

To find the total value, "like terms"—1s, 10s, and 100s—were added. Two or more polynomials can be added in a similar way.

HANDS-ON MINI-LAB

Work with a partner. algebra tiles

Try This

Add $2x^2 + 3x - 3$ and $x^2 - 4x + 2$.

• Model each polynomial.

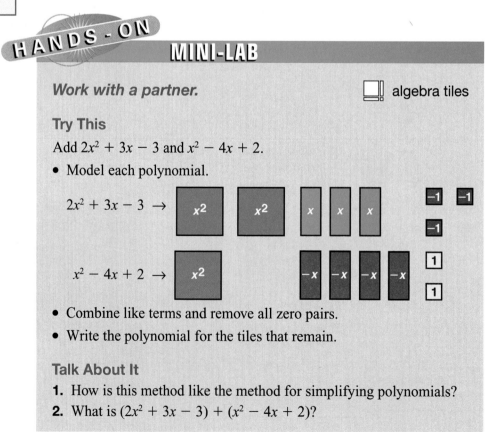

$2x^2 + 3x - 3 \rightarrow$

$x^2 - 4x + 2 \rightarrow$

• Combine like terms and remove all zero pairs.
• Write the polynomial for the tiles that remain.

Talk About It

1. How is this method like the method for simplifying polynomials?
2. What is $(2x^2 + 3x - 3) + (x^2 - 4x + 2)$?

1 Write the two polynomials represented below. Then find their sum.

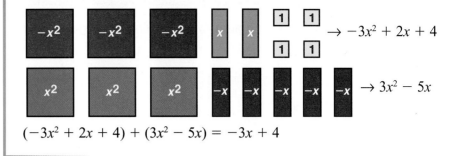

$$\rightarrow -3x^2 + 2x + 4$$

$$\rightarrow 3x^2 - 5x$$

$$(-3x^2 + 2x + 4) + (3x^2 - 5x) = -3x + 4$$

One way to add polynomials is to write them vertically and line up like terms.

2 Find $(x^2 - 4x + 3) + (2x^2 - 2x - 4)$. Then evaluate the sum if $x = -2$.

$$\begin{array}{r} x^2 - 4x + 3 \\ + 2x^2 - 2x - 4 \\ \hline 3x^2 - 6x - 1 \end{array}$$
Arrange like terms in columns. Then add.

Now evaluate for $x = -2$.

$$\begin{aligned} 3x^2 - 6x - 1 &= 3(-2)^2 - 6(-2) - 1 \quad \textit{Replace x with } -2. \\ &= 3(4) + 12 - 1 \\ &= 12 + 12 - 1 \text{ or } 23 \end{aligned}$$

CHECK FOR UNDERSTANDING

Communicating Mathematics

Read and study the lesson to answer each question.

1. *Tell* how to add two polynomials.

2. *Write* the two polynomials represented at the right. Then find their sum.

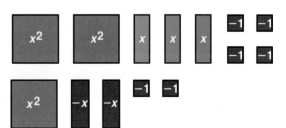

HANDS-ON MATH

3. *Show* how to find the sum of $x^2 - 2x - 3$ and $-2x^2 + 2x + 4$ using drawings or algebra tiles.

Guided Practice

Find each sum. Use drawings or algebra tiles if necessary.

4. $\begin{array}{r} 2x^2 + x + 4 \\ + 3x^2 + 3x + 1 \\ \hline \end{array}$

5. $\begin{array}{r} 5a + 3b \\ + 4a + 2b \\ \hline \end{array}$

6. $(-2y^2 + y + 8) + (6y^2 + 3y + 2)$

7. $(-3x + 1) + (2x + 5)$

8. Find the sum $(8r - 3s) + (r + 5s)$. Then evaluate the sum if $r = 2$ and $s = 1$.

9. *Geometry* Refer to the figure.
   a. Write and simplify an expression for its perimeter.
   b. Find the perimeter if $x = 4$.

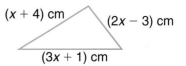

$(x + 4)$ cm $(2x - 3)$ cm $(3x + 1)$ cm

Practice

Find each sum. Use drawings or algebra tiles if necessary.

10. $\quad x^2 + 2x + 1$
$\quad + x^2 - 3x - 2$

11. $\quad 5y^2 - 2y + 5$
$\quad + y^2 + 4y - 3$

12. $\quad -4a^2 + 3a - 2$
$\quad + 6a^2 - 5a - 7$

13. $\quad 3x^2 - 3x + 5$
$\quad + -2x^2 + 3x$

14. $\quad 4a + 6b + c$
$\quad + 5a - 3b - 2c$

15. $15q + 2r - 1$
$\quad + q - 3r + 2$

16. $(2x^2 - 6x) + (x^2 - 4x)$

17. $(9a + 5b) + (3a + 7b)$

18. $(5r - 7s) + (3r + 8s)$

19. $(3a^2 + 2a) + (7a^2 - 3)$

20. $(-2x^2 + 4x - 6) + (-5x^2 - 3x + 7)$ **21.** $(3x^2 - 5x - 7) + (-x^2 + 2x - 3)$

Find each sum. Then evaluate if $g = 5$ and $h = 4$.

22. $(6g - 3h) + (2g + 5h)$

23. $(3g + 2h - 1) + (4g + h + 5)$

24. $(-3g + 5h) + (6g - 6h)$

25. $(14g + 3h + 2) + (g - 2h + 9)$

26. Find the sum of $-4x + 2$ and $6x^2 - 3$.

27. What is the value of $(p^2 - 4p) + (2p^2 + 3p)$ if $p = 3$?

Applications and Problem Solving

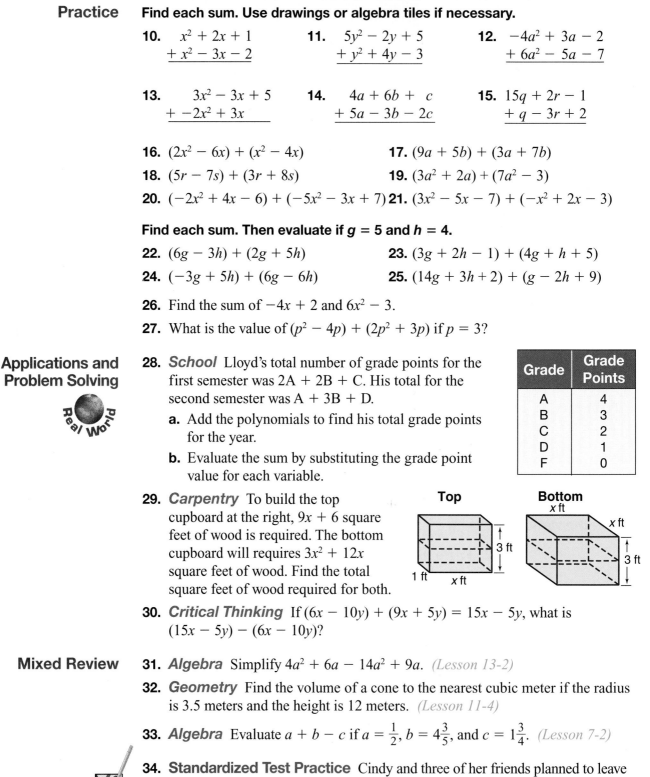

28. *School* Lloyd's total number of grade points for the first semester was $2A + 2B + C$. His total for the second semester was $A + 3B + D$.

  a. Add the polynomials to find his total grade points for the year.

  b. Evaluate the sum by substituting the grade point value for each variable.

Grade	Grade Points
A	4
B	3
C	2
D	1
F	0

29. *Carpentry* To build the top cupboard at the right, $9x + 6$ square feet of wood is required. The bottom cupboard will requires $3x^2 + 12x$ square feet of wood. Find the total square feet of wood required for both.

Top

Bottom

30. *Critical Thinking* If $(6x - 10y) + (9x + 5y) = 15x - 5y$, what is $(15x - 5y) - (6x - 10y)$?

Mixed Review

31. *Algebra* Simplify $4a^2 + 6a - 14a^2 + 9a$. *(Lesson 13-2)*

32. *Geometry* Find the volume of a cone to the nearest cubic meter if the radius is 3.5 meters and the height is 12 meters. *(Lesson 11-4)*

33. *Algebra* Evaluate $a + b - c$ if $a = \frac{1}{2}$, $b = 4\frac{3}{5}$, and $c = 1\frac{3}{4}$. *(Lesson 7-2)*

34. **Standardized Test Practice** Cindy and three of her friends planned to leave a 20% tip for the waiter when they had lunch. Two of her friends had hamburgers, one had soup and salad, and Cindy had pasta. What other information is necessary to determine how much to leave for the tip? *(Lesson 3-6)*

  A the cost of the hamburger
  B where they had lunch
  C the cost of the four meals
  D what day they had lunch
  E the salary of the waiter

For **Extra Practice,**
see page 642.

13-4

Subtracting Polynomials

What you'll learn

What you'll learn

You'll learn to subtract polynomials by using algebra tiles.

When am I ever going to use this?

Knowing how to subtract polynomials can help you solve problems involving interest rates and finance.

You can use algebra tiles to subtract polynomials.

HANDS-ON MINI-LAB

Work with a partner.

algebra tiles

Try This

Find $(3x + 6) - (-1x + 3)$.

- Model the polynomial $3x + 6$.
- To subtract $(-1x + 3)$, remove 1 negative x-tile and 3 1-tiles. You can remove the 1-tiles, but there are no negative x-tiles, so you can't remove $-1x$.
- Add a zero pair, and then remove the negative x-tile.

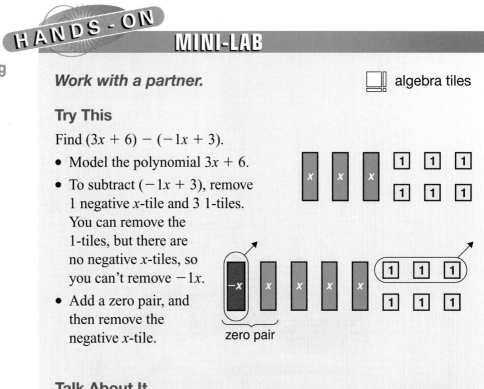

zero pair

Talk About It

1. What kinds of tiles remain?
2. What is $(3x + 6) - (-1x + 3)$ in simplest form?
3. Why can you add or remove a zero pair?

To subtract an integer, it is usually easier to add its opposite. For example, $3 - 8 = 3 + (-8) = -5$. In a similar manner, to subtract a polynomial, it may be easier to add the opposite of each term of the polynomial.

Examples

Find each difference.

1

$$\begin{array}{r} 6x - 4 \\ -(2x - 1) \\ \hline \end{array}$$
The opposite of $2x$ is $-2x$.
The opposite of -1 is 1.

$$\begin{array}{r} 6x - 4 \\ + (-2x + 1) \\ \hline 4x - 3 \end{array}$$

2 $(3x^2 - 2x - 1) - (5x^2 - 3x + 4)$

$$\begin{array}{r} 3x^2 - 2x - 1 \\ - (5x^2 - 3x + 4) \\ \hline \end{array} \rightarrow \begin{array}{r} 3x^2 - 2x - 1 \\ + (-5x^2 + 3x - 4) \\ \hline -2x^2 + x - 5 \end{array}$$

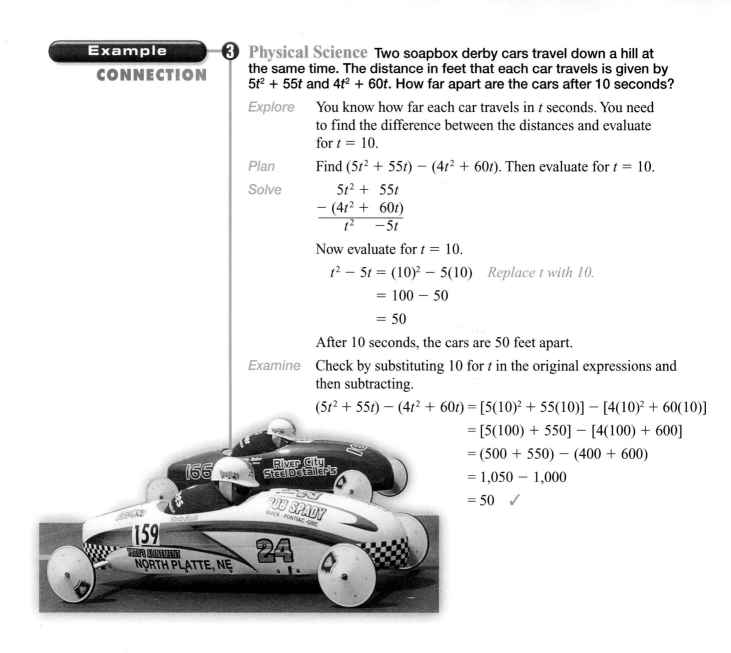

Example ▶ ❸ **Physical Science** Two soapbox derby cars travel down a hill at the same time. The distance in feet that each car travels is given by $5t^2 + 55t$ and $4t^2 + 60t$. How far apart are the cars after 10 seconds?

CONNECTION

Explore   You know how far each car travels in t seconds. You need to find the difference between the distances and evaluate for $t = 10$.

Plan   Find $(5t^2 + 55t) - (4t^2 + 60t)$. Then evaluate for $t = 10$.

Solve
$$\begin{array}{r} 5t^2 + 55t \\ - (4t^2 + 60t) \\ \hline t^2 - 5t \end{array}$$

Now evaluate for $t = 10$.

$$t^2 - 5t = (10)^2 - 5(10) \quad \textit{Replace t with 10.}$$
$$= 100 - 50$$
$$= 50$$

After 10 seconds, the cars are 50 feet apart.

Examine   Check by substituting 10 for t in the original expressions and then subtracting.

$$(5t^2 + 55t) - (4t^2 + 60t) = [5(10)^2 + 55(10)] - [4(10)^2 + 60(10)]$$
$$= [5(100) + 550] - [4(100) + 600]$$
$$= (500 + 550) - (400 + 600)$$
$$= 1{,}050 - 1{,}000$$
$$= 50 \quad ✓$$

CHECK FOR UNDERSTANDING

Communicating Mathematics

Read and study the lesson to answer each question.

1. *Tell* the opposite of $2x$.

2. *Write* the subtraction problem modeled at the right. What is the difference?

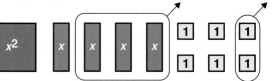

HANDS-ON MATH

3. *Show* how to find $(2x^2 - 3x + 3) - (x^2 + 1)$ using algebra tiles or drawings.

Guided Practice   **Find each difference. Use drawings or algebra tiles if necessary.**

4.
$$\begin{array}{r} 9s - 1 \\ - (7s + 2) \\ \hline \end{array}$$

5.
$$\begin{array}{r} 4p^2 - 3p + 1 \\ - (2p^2 - 2p) \\ \hline \end{array}$$

6. $(a^2 + 2a) - (2a^2 + 2a)$

7. $(5x^2 - 3x + 2) - (3x^2 - 3)$

8. Find $(7r + 5s) - (r + 4s)$. Then evaluate the difference if $r = 1$ and $s = -2$.

9. **Food** Valerie ordered 6 tacos and 3 drinks from a fast-food drive-thru. When she got home, she discovered that the bag of food contained 4 tacos and 4 drinks. Find $(6t + 3d) - (4t + 4d)$ and evaluate for the values shown in the table to find how much she was overcharged.

Item	Cost ($)
taco (t)	0.89
drink (d)	0.75

EXERCISES

Practice

Find each difference. Use drawings or algebra tiles if necessary.

10. $7x + 5$
    $- (3x + 4)$

11. $5y^2 + 9$
    $- (4y^2 + 9)$

12. $-4a^2 - 3a - 2$
    $-(2a^2 + 2a + 7)$

13. $3r^2 - 3rt + t^2$
    $- (2r^2 + 5rt - t^2)$

14. $(5x + 3) - (2x + 1)$

15. $(-4a + 5) - (a - 1)$

16. $(10c - 2d) - (6c + 3d)$

17. $(4a^2 - 3a - 2) - (2a^2 + 2a + 7)$

18. $(-3x^2 + 2x + 1) - (x^2 + 3x - 1)$

19. $(6x^2 + 2x + 9) - (3x^2 + 5x + 9)$

Find each difference. Then evaluate if $c = -2$ and $d = 5$.

20. $(3c + 4d) - (2c + 5d)$

21. $(4c - 6d - 1) - (2c + 5d + 3)$

22. $(-2c + 7d) - (15c + 9d)$

23. $(10c + 4d - 2) - (c - 2d + 8)$

24. Find $(7a + 3b - c)$ minus $(4a - 2b + 2c)$.

25. What is $(3p^2 + 5pq - q^2)$ decreased by $(p^2 + 3pq - 2q^2)$?

Applications and Problem Solving

26. **Money Matters** Alan borrowed $200 from his father each year for college expenses. The amount he owes his father at the beginning of his second and third years is $(400 + 200r)$ and $(600 + 600r + 200r^2)$ respectively, where r is the interest rate.

    a. Find how much his debt increased between his second and third years.

    b. Evaluate the increase for $r = 8\%$.

27. **Sports** Katrina's scoring total for Game 1 was $5A + 8B + 7C$. Her scoring total for Game 2 was $3A + 12B + 10C$.

Game	A 3 points	B 2 points	C 1 point
1	5	8	7
2	3	12	10

    a. Subtract to find how many more points she scored in Game 2 than in Game 1.

    b. Evaluate the difference for A = 3, B = 2, and C = 1.

28. **Critical Thinking** Mentally find $(3x^2 + 7x + 9) - (5x^2 - 3x + 2)$ if $x = 0$. Explain.

Mixed Review

29. Algebra Find the sum $(6y^2 + 7y - 5) + (8y + 8)$. *(Lesson 13-3)*

30. Probability Two number cubes are rolled. Find the probability that a prime number is rolled on one number cube and an odd number is rolled on the other number cube. *(Lesson 12-5)*

31. Algebra Solve $3x - 5 = -6$. *(Lesson 7-9)*

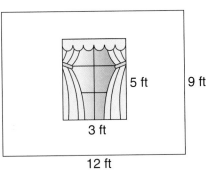

32. Standardized Test Practice Miguel wants to wallpaper a wall that measures 9 feet by 12 feet. A window that is 5 feet by 3 feet is in the center of the wall. How much surface area will be covered with wallpaper? *(Lesson 1-8)*

    A 36 sq ft

    B 93 sq ft

    C 108 sq ft

For **Extra Practice,** see page 643.

    D 15 sq ft

    E Not Here

CHAPTER 13

Mid-Chapter Self Test

Model each monomial or polynomial using drawings or algebra tiles.
(Lesson 13-1)

1. $-5x$

2. $x^2 - 2x + 4$

3. Simplify $x^2 + x - 2x + 4 - 1$ using the model.
(Lesson 13-2)

Simplify each polynomial. Use drawings or algebra tiles if necessary. *(Lesson 13-2)*

4. $6x - 4 + x$

5. $y^2 - 5y + 3 + 2y^2$

Find each sum or difference. Use drawings or algebra tiles if necessary.
(Lessons 13-3 and 13-4)

6. $(-3y^2 + 4y) + (-1y^2 - 3y)$

7. $(5x^2 + 6x + 7) - (x^2 + 3x - 4)$

Find each sum or difference. Then evaluate if $a = 1$ and $b = 3$.
(Lessons 13-3 and 13-4)

8. $(-2a + 7b) + (8a - 8b)$

9. $(3a - 5b + 6) - (a + 2b - 1)$

10. Geometry Write the simplified expression for the perimeter of the figure. *(Lesson 13-3)*

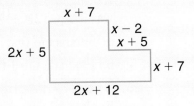

COOPERATIVE LEARNING

13-5A Modeling Products

A Preview of Lesson 13-5

algebra tiles

product mat

The algebra tiles that you have been using are based on the fact that the area of a rectangle is the product of the width and length.

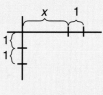

In this lab, you will use these algebra tiles to build more complex rectangles. These rectangles will help you understand how to find the product of simple polynomials. The width and length each represent a polynomial being multiplied; the area of the rectangle represents their product.

TRY THIS

Work with a partner.

Step 1 Make a rectangle with a width of 2 units and a length of $x + 1$ units. This rectangle has an area of $2(x + 1)$. Use your algebra tiles to mark off the dimensions on a product mat.

Step 2 Using the marks as a guide, fill in the rectangle with algebra tiles.

Step 3 The area of the rectangle is $x + x + 1 + 1$. In simplest form, the area is $2x + 2$. Therefore, $2(x + 1) = 2x + 2$.

ON YOUR OWN

Find each product using algebra tiles.

1. $2(x + 2)$

2. $x(x + 3)$

3. $3(2x + 1)$

4. *Look Ahead* Find $3(n + 4)$.

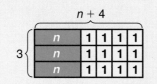

13-5 Multiplying Monomials and Polynomials

What you'll learn

You'll learn to multiply monomials and polynomials.

When am I ever going to use this?

Knowing how to multiply polynomials can help you solve problems involving surface area.

You can use a calculator to find patterns and see how rules are developed for exponents.

TECHNOLOGY

MINI-LAB

Work with a partner.

scientific calculator

Try This

Copy the table at the right. Then use a calculator to find each product and complete the table.

Factors	Product	Product Written as a Power
$10^1 \cdot 10^1$		
$10^1 \cdot 10^2$		
$10^1 \cdot 10^3$		
$10^1 \cdot 10^4$		
$10^1 \cdot 10^5$		

Talk About It

1. Compare the exponents of the factors to the exponent in the products. What do you observe?
2. Write a rule for determining the exponent of the product when you multiply powers. Test your rule by multiplying $2^2 \cdot 2^4$ using a calculator.

The pattern you observed in the Mini-Lab leads to the following rule for multiplying powers that have the same base.

Product of Powers	**Words:**	You can multiply powers that have the same base by adding their exponents.
	Symbols:	For any number a and integers m and n, $a^m \cdot a^n = a^{m+n}$.

Monomials that are powers with the same base can be multiplied using the rule for the product of powers.

Examples

1 Find $6^3 \cdot 6^4$.

$6^3 \cdot 6^4 = 6^{3+4}$ or 6^7

Check: $6^3 \cdot 6^4 = (6 \cdot 6 \cdot 6)(6 \cdot 6 \cdot 6 \cdot 6)$
$= 6 \cdot 6 \cdot 6 \cdot 6 \cdot 6 \cdot 6 \cdot 6$ or 6^7 ✓

2 Find $a^2 \cdot a^3$.

$a^2 \cdot a^3 = a^{2+3}$ or a^5

You can use models or the distributive property to multiply a polynomial by a monomial. Recall that the distributive property allows you to multiply the factor *outside* the parentheses by each term *inside* the parentheses.

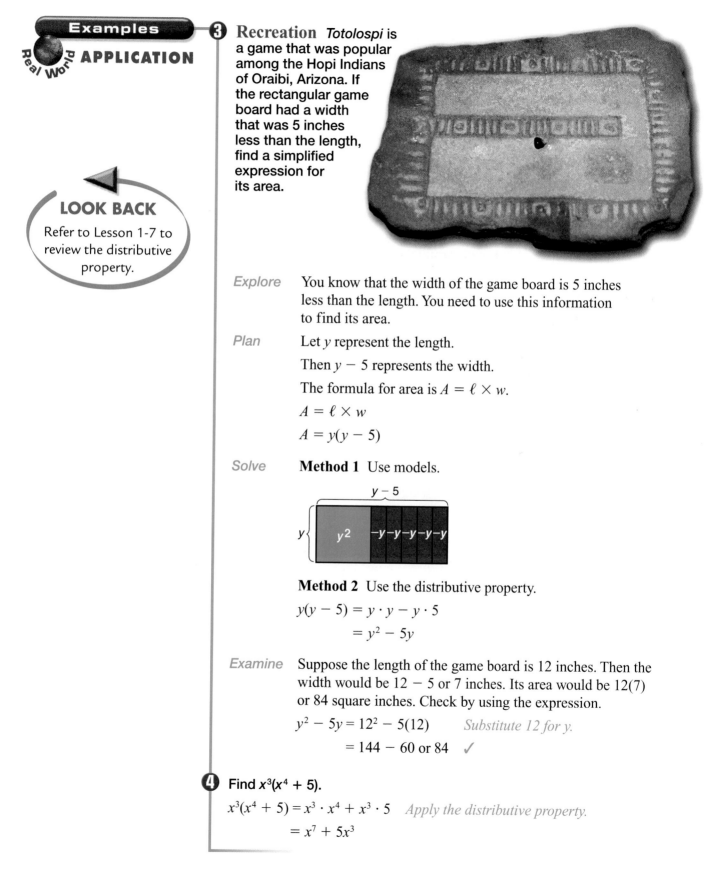

Examples

Real World APPLICATION

LOOK BACK

Refer to Lesson 1-7 to review the distributive property.

③ Recreation *Totolospi* is a game that was popular among the Hopi Indians of Oraibi, Arizona. If the rectangular game board had a width that was 5 inches less than the length, find a simplified expression for its area.

Explore You know that the width of the game board is 5 inches less than the length. You need to use this information to find its area.

Plan Let y represent the length.

Then $y - 5$ represents the width.

The formula for area is $A = \ell \times w$.

$A = \ell \times w$

$A = y(y - 5)$

Solve **Method 1** Use models.

$y - 5$

$y \{ \quad y^2 \quad -y -y -y -y -y$

Method 2 Use the distributive property.

$y(y - 5) = y \cdot y - y \cdot 5$

$= y^2 - 5y$

Examine Suppose the length of the game board is 12 inches. Then the width would be $12 - 5$ or 7 inches. Its area would be $12(7)$ or 84 square inches. Check by using the expression.

$y^2 - 5y = 12^2 - 5(12)$ *Substitute 12 for y.*

$= 144 - 60$ or 84 ✓

④ Find $x^3(x^4 + 5)$.

$x^3(x^4 + 5) = x^3 \cdot x^4 + x^3 \cdot 5$ *Apply the distributive property.*

$= x^7 + 5x^3$

Communicating Mathematics

Read and study the lesson to answer each question.

1. *Tell* how to use the distributive property to find $3x(5x + 2)$.

2. *Use* a calculator to multiply $5^4 \cdot 5^6$.

HANDS-ON

3. *Draw* a rectangle to model $2x(x + 4)$. Then write the product.

Guided Practice

Find each product. Express the answer in exponential form.

4. $3^4 \cdot 3^2$

5. $n^3 \cdot n^5$

6. Find $x(x + 6)$.

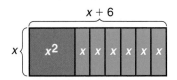

Find each product.

7. $b(b - 2)$

8. $j^4(j^2 + 7)$

9. **Gardening** A square garden plot measures x feet on each side. Suppose you double the length of the plot and increase the width by 3 feet.

   a. Draw the new garden.

   b. Write two expressions for the area of the new plot.

   c. If the original plot was 10 feet on a side, what is the area of the new plot?

Practice

Find each product. Express the answer in exponential form.

10. $4 \cdot 4^3$

11. $2^6 \cdot 2^4$

12. $x^2 \cdot x^7$

13. $p^5 \cdot p^9$

Find each product.

14. $x(x + 3)$

15. $z(z - 1)$

16. $2x(x + 2)$

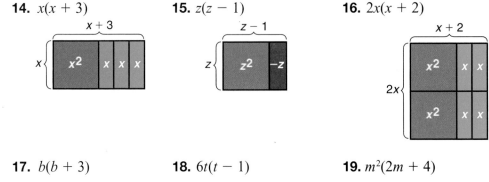

17. $b(b + 3)$

18. $6t(t - 1)$

19. $m^2(2m + 4)$

20. $3y(2 + y)$

21. $d^3(d^5 + 15)$

22. $2x^4(2x - 1)$

23. Find the product of a^8 and a^3.

24. Multiply $(z - 4)$ by $3z^2$.

25. *Geometry* Find the measure of the area of the shaded region in simplest form.

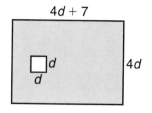

26. *Geometry* A rectangle is 5 feet longer than it is wide. Find the area of the rectangle.

27. *Money Matters* Tanisha's neighbors pay her $2 an hour to baby-sit their baby, but have promised her a raise. If her raise is x dollars, how much will she earn for 3 hours of baby-sitting?

28. *Critical Thinking* Copy the table. Use a calculator to find each quotient and complete the table.

a. Compare the exponents of the division expressions to the exponents in the quotients. What pattern do you observe?

Division	Quotient	Quotient Written as a Power
$10^5 \div 10^1$		
$10^4 \div 10^1$		
$10^3 \div 10^1$		
$10^2 \div 10^1$		
$10^1 \div 10^1$		

b. Write a rule for determining the exponent in the quotient when you divide powers. Test your rule by dividing 7^5 by 7^3 on a calculator.

c. Find each quotient.

i. $\dfrac{x^7}{x^7}$ **ii.** $\dfrac{x^5}{x^7}$ **iii.** $\dfrac{x^7}{x^5}$

iv. $\dfrac{(60a^6b^3)}{(15a^2b^7)}$ **v.** $\dfrac{x^{-9}}{x^{-4}}$ **vi.** $\dfrac{(a^5b^{-3}c^{-1})}{(a^3b^8c^{-3})}$

Mixed Review

29. *Algebra* Find $(8c + d) - (3c - 2d)$. Then evaluate if $c = 3$ and $d = 2$. *(Lesson 13-4)*

30. *Probability* A test has five multiple-choice questions. Each question has three choices. How many outcomes for giving answers to the five questions are possible? *(Lesson 12-1)*

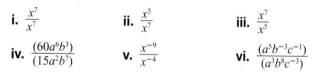

31. *Standardized Test Practice* A can of green beans is 11 centimeters tall and has a diameter of 7.5 centimeters. How many square centimeters of paper will it take to make a label for the can? (Remember: The label does not cover the top or bottom of the can.) *(Lesson 11-6)*

A 347.6 cm²   **B** 82.5 cm²

C 164.3 cm²   **D** 259.1 cm²

For **Extra Practice,** see page 643.

32. *Architecture* In the afternoon, a building casts a shadow 137 feet long. At the same time, a nearby street sign that is 10.4 feet high casts a shadow 1.3 feet long. How tall is the building? *(Lesson 8-8)*

SPORTS MANAGEMENT

Jerry Colangelo
FORMER NBA SCOUT

Jerry Colangelo began his career in sports management as the head scout and director of merchandising for the Chicago Bulls in 1966. Then, in 1987 he bought the Phoenix Suns and has since served as president and CEO of the franchise.

A professional sports scout evaluates the athletic skills of athletes to determine their potential. A person who is interested in becoming a professional sports scout should consider obtaining a college degree or attending a special school for the chosen sport. A good understanding of mathematics, especially statistics, is necessary in analyzing data on specific players.

For more information:
Athletic Institute
200 Castlewood Drive
North Palm Beach, FL 33408

interNET
CONNECTION
www.glencoe.com/sec/
math/mac/mathnet

I love basketball! I think it would be great to be a scout for a professional team!

Your Turn
• Choose three college basketball players who play the same position. Find and list their current statistics. If you were an NBA scout, which of these players would you encourage your team to draft? Write a paragraph describing why the player you have chosen would be the best for the team. Be sure to discuss the comparison of the statistics that you found.

Multiplying Binomials

What you'll learn

You'll learn to multiply binomials by using algebra tiles.

When am I ever going to use this?

Knowing how to multiply binomials is useful in landscaping and building design.

Word Wise

binomial

What do the words *bicycle* and *binomial* have in common? They both begin with the prefix *bi-*, meaning two. A bicycle is a cycle with two wheels. A **binomial** is a polynomial with two terms. Some examples of binomials are $x + 2$, $3y - 4$, and $r + s$. You can find the product of simple binomials by using algebra tiles.

HANDS-ON MINI-LAB

Work with a partner. ▭ algebra tiles ▯ product mat

Try This

Find $(x + 1)(x + 3)$.

- Make a rectangle with a width of $x + 1$ and a length of $x + 3$. Use algebra tiles to mark off the dimensions on a product mat.
- Using the marks as a guide, fill in the rectangle with algebra tiles.

Talk About It

1. How is this method like the method you used when multiplying a monomial by a polynomial?
2. What is $(x + 1)(x + 3)$?

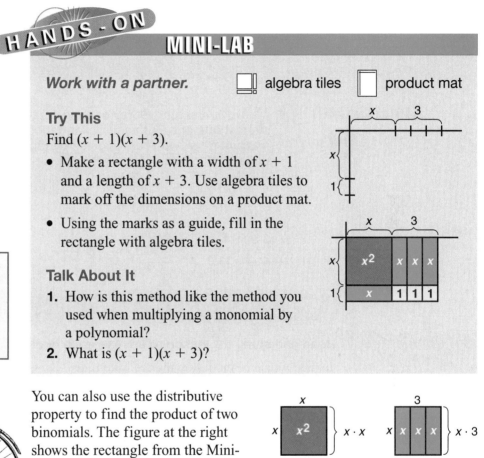

Did you know The first modern bicycle appeared in England in 1885. It was called the Rover safety bicycle.

You can also use the distributive property to find the product of two binomials. The figure at the right shows the rectangle from the Mini-Lab, separated into four parts. Notice that each term from the first parentheses $(x + 1)$ is multiplied by each term from the second parentheses $(x + 3)$.

Example

1 Find $(y + 2)(2y + 1)$.

Method 1 Use models.

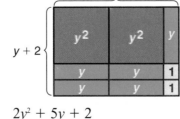

$2y^2 + 5y + 2$

Method 2 Use the distributive property.

$$(y + 2)(2y + 1) = y(2y + 1) + 2(2y + 1)$$
$$= 2y^2 + 1y + 4y + 2$$
$$= 2y^2 + 5y + 2 \ \textit{Simplify.}$$

Example ➋ Find $(x + 3)(x + 2)$.

$$(x + 3)(x + 2) = x(x + 2) + 3(x + 2)$$
$$= x^2 + 2x + 3x + 6$$
$$= x^2 + 5x + 6 \quad \textit{Simplify.}$$

You can also multiply binomials vertically.

Examples ➌ Find $(3t + 1)(2t + 3)$.

$(3t + 1)(2t + 3) \rightarrow$

$$\begin{array}{r} 2t + 3 \quad \textit{Multiply as with whole numbers.} \\ \times\ 3t + 1 \\ \hline 2t + 3 \quad \textit{Multiply by 1.} \\ 6t^2 + \ 9t \quad \textit{Multiply by 3t.} \\ \hline 6t^2 + 11t + 3 \quad \textit{Add like terms.} \end{array}$$

APPLICATION

Real World

➍ **Money Matters** Corey opened a savings account paying an interest rate of r. For each dollar deposited, the amount in the account after two years is given by the formula $A = (1 + r)(1 + r)$. Write the formula in simplest form.

$A = (1 + r)(1 + r) \quad A = (1 + r)^2$
$A = 1 \cdot (1 + r) + r \cdot (1 + r)$
$A = 1 + r + r + r^2$
$A = 1 + 2r + r^2 \quad \textit{Simplify.}$

CHECK FOR UNDERSTANDING

Communicating Mathematics

Read and study the lesson to answer each question.

1. **Tell** which of the following are binomials.

   a. $2x$ **b.** $x + 5$ **c.** $3y^2 + 4x$ **d.** $x^2 + 2x + 8$

2. **Write** the product shown at the right.

HANDS-ON MATH

3. **Draw** a rectangle to model $(2x + 2)(2x + 3)$. Then write the product.

Guided Practice

4. Find $(x + 1)(x + 2)$.

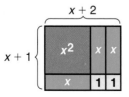

Find each product. Use drawings or algebra tiles if necessary.

5. $(x + 3)(x - 1)$ 6. $(3a + 1)(2a + 3)$

7. *Geometry* A square has dimensions of x feet by x feet. A rectangle is 4 feet longer and 3 feet wider than the square. Find the area of the rectangle.

EXERCISES

Practice

Find each product.

8. $(x + 3)(x + 3)$

9. $(2x + 1)(x + 3)$

10. $(x + 4)(2x + 1)$

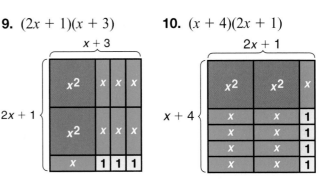

Find a rectangular area in your home or a friend's home that has square tiles. Sketch the area. Then record the dimensions of the area using x to represent the length of a side of one tile.

Find each product. Use drawings or algebra tiles if necessary.

11. $(b + 6)(b + 4)$

12. $(x + 1)(2x - 3)$

13. $(2m + 2)(2m + 1)$

14. $(2x + 1)(x + 5)$

15. $(d + 3)(d + 3)$

16. $(3z - 1)(z + 2)$

17. Find the product of $(3a + 1)$ and $(a - 3)$.

18. Multiply $(z - 5)$ by $(2z + 4)$.

Applications and Problem Solving

19. *Recreation* Refer to the diagram of the pool area.
  a. Find the area of the pool.
  b. Find the area of the cement path around the pool.

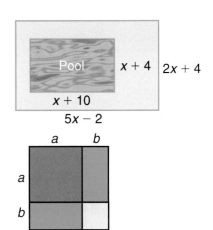

20. *Geometry* The model represents the square of a binomial.
  a. What product does this model represent?
  b. What is the area of each small square and rectangle?
  c. Write the area of the square as a polynomial.

21. *Working on the CHAPTER Project* Refer to the model you drew in Exercise 34 on page 564. Write a polynomial represented by the model. (*Hint*: hm and mh are like terms.) Does this model work for all players? Explain.

22. *Critical Thinking* The length and width of a rectangle are $3x + 1$ and $x + 4$. Give the dimensions of a rectangle that has twice the area. Check your answer by finding each area.

For **Extra Practice,** see page 643.

Mixed Review

23. *Algebra* Find $t(t - 4)$. *(Lesson 13-5)*

24. Standardized Test Practice Which of the following letters does not have rotational symmetry? *(Lesson 10-9)*

   A H   **B** I   **C** M   **D** O

25. Find the LCM of 18, 34, and 6. *(Lesson 6-7)*

PROBLEM SOLVING

13-7A Guess and Check

A Preview of Lesson 13-7

A soft drink company is offering free bikes to people who collect enough points by buying bottles or cans of soda. Ramón and Sheila are figuring out how they can get a free bike. Let's listen in!

To get a free bike, all you have to do is save up 915 points.

Ramón

A two-liter bottle is worth 5 points and a six-pack of soda is worth 10 points. So how much soda do we need to buy?

It will be a lot — I would guess around 40 bottles and 40 six-packs.

I think it would be more like 30 bottles and 70 six-packs.

Let's see who's guess is closer. Forty times 5 is 200 and 40 times 10 is 400. That's only 600 points total!

Sheila

Your guess would give us 30 times 5, which is 150, plus 70 times 10, which is 700. That's still only 850 points. You're closer, but I don't know if we'll ever be able to drink all that soda.

We better check to see how long this promotion is supposed to last!

THINK ABOUT IT

Work with a partner.

1. **Compare and contrast** Sheila's and Ramón's guesses. How would you have guessed differently?

2. **Use** the **guess–and–check** strategy to find a better estimate of how much soda you would need to drink to win a bike.

3. **Apply** what you have learned to solve the following problem.

   The product of two consecutive odd integers is 899. What are the numbers?

For **Extra Practice,** see page 644.

ON YOUR OWN

4. The third step of the 4-step plan for problem solving asks you to *solve* the problem. *Explain* how you can use guess and check to help you find factors of a polynomial.

5. *Write a Problem* that you can solve by using the guess-and-check method. Explain your answer.

6. *Look Ahead* Use guess and check to find the factors of $x^2 + 5x + 4$.

MIXED PROBLEM SOLVING

STRATEGIES

Look for a pattern.
Solve a simpler problem.
Act it out.
Guess and check.
Draw a diagram.
Make a chart.
Work backward.

Solve. Use any strategy.

7. *Framing* Forty-two inches of molding are needed to frame a picture. If molding sells for $5.80 a foot, how much will the molding cost?

8. *Critical Thinking* Edgar's mother is five times as old as Edgar. Five years from now she will be just three times as old as he is. How old is Edgar now?

9. *Business* Nami volunteers to get lunch for members of the Ecology Club. Their order of 13 burgers, 7 shakes, and 10 French fries is represented by the expression $13b + 7s + 10f$. Determine the cost of the order if burgers are $1.99 each, shakes are $1.49 each, and fries are $0.99.

10. *Number Theory* The product of a number and its next two consecutive whole numbers is 60. Use the guess-and-check strategy to find the number.

11. *Money Matters* Stamps for postcards cost $0.20, and stamps for first-class letters cost $0.32. Louise wants to send postcards and letters to 11 friends. If she has $2.80 for stamps, how many postcards and how many letters can she send?

12. *Retail Sales* In 1970, a desktop calculator sold for about $100. Today the same type of calculator sells for as little as $25. What is the percent of decrease of the cost of a desktop calculator?

13. **Standardized Test Practice** Teenagers were asked which they spent more time using, their computer, their video game system, or both equally. The graph shows the results of the survey.

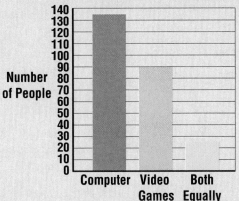

Favorite Electronic Entertainment

Source: Consumer Electronics Manufacturers Association

What was the total number of people who were surveyed?

A 135 **B** 140 **C** 200

D 250 **E** 420

14. **Standardized Test Practice** Which word, written as shown, does not have a vertical line of symmetry?

A M O T H **B** V O T E **C** W H Y **D** H A M

Factoring Polynomials

What you'll learn

You'll learn to factor polynomials by using algebra tiles.

When am I ever going to use this?

Knowing how to factor polynomials will help you find possible dimensions of a rectangle if you know its area.

Word Wise

factoring
trinomial

Sometimes you know a product and want to find its factors. This is called **factoring**. You can use algebra tiles to factor polynomials. You know the area of a rectangle and need to find the length and width.

HANDS-ON MINI-LAB

Work with a partner.   algebra tiles

Try This
Factor $2x + 6$.

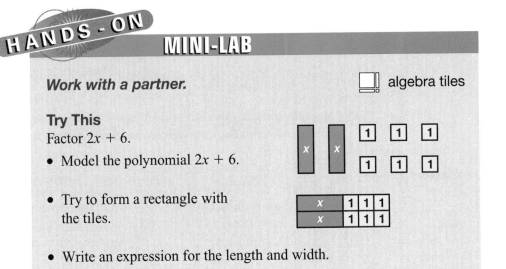

- Model the polynomial $2x + 6$.

- Try to form a rectangle with the tiles.

- Write an expression for the length and width.

Talk About It
1. What are the factors of $2x + 6$? How are these factors related to the length and width?
2. Explain how to factor $x^2 + 4x$.
3. Model the polynomial $x^2 + 3x + 5$. Can you form a rectangle?
4. What are the factors of $3x + 5$?

You can also use algebra tiles to factor **trinomials**. A trinomial is a polynomial with three terms.

Examples

1 Factor $x^2 + 3x + 2$.

Model the polynomial.   Try to form a rectangle with the tiles.

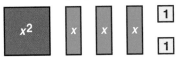

The rectangle has a width of $x + 1$ and a length of $x + 2$. Therefore, $x^2 + 3x + 2 = (x + 1)(x + 2)$.

2 Factor $x^2 + 2x + 2$.

Model the polynomial.

Try to form a rectangle with the tiles. You cannot form a rectangle with the tiles. Therefore, $x^2 + 2x + 2$ cannot be factored.

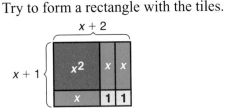

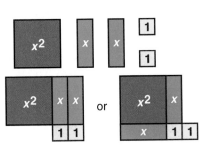

You can also use the guess-and-check strategy to find the factors of a trinomial. Look at $(x + 3)(x + 5) = x^2 + 8x + 15$. Notice that the product of the last terms of each polynomial, 3 and 5, equals the last term in the trinomial, 15. You can use this pattern to factor polynomials.

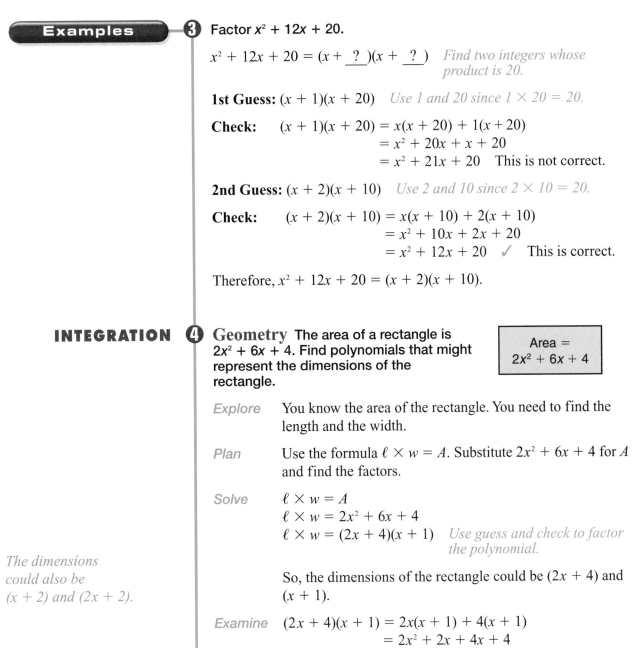

Examples

3 Factor $x^2 + 12x + 20$.

$x^2 + 12x + 20 = (x + \underline{\ ?\ })(x + \underline{\ ?\ })$ *Find two integers whose product is 20.*

1st Guess: $(x + 1)(x + 20)$ *Use 1 and 20 since $1 \times 20 = 20$.*

Check: $(x + 1)(x + 20) = x(x + 20) + 1(x + 20)$
$= x^2 + 20x + x + 20$
$= x^2 + 21x + 20$ This is not correct.

2nd Guess: $(x + 2)(x + 10)$ *Use 2 and 10 since $2 \times 10 = 20$.*

Check: $(x + 2)(x + 10) = x(x + 10) + 2(x + 10)$
$= x^2 + 10x + 2x + 20$
$= x^2 + 12x + 20$ ✓   This is correct.

Therefore, $x^2 + 12x + 20 = (x + 2)(x + 10)$.

INTEGRATION

4 **Geometry** The area of a rectangle is $2x^2 + 6x + 4$. Find polynomials that might represent the dimensions of the rectangle.

> Area = $2x^2 + 6x + 4$

Explore   You know the area of the rectangle. You need to find the length and the width.

Plan   Use the formula $\ell \times w = A$. Substitute $2x^2 + 6x + 4$ for A and find the factors.

Solve $\ell \times w = A$
$\ell \times w = 2x^2 + 6x + 4$
$\ell \times w = (2x + 4)(x + 1)$ *Use guess and check to factor the polynomial.*

The dimensions could also be $(x + 2)$ and $(2x + 2)$.

So, the dimensions of the rectangle could be $(2x + 4)$ and $(x + 1)$.

Examine $(2x + 4)(x + 1) = 2x(x + 1) + 4(x + 1)$
$= 2x^2 + 2x + 4x + 4$
$= 2x^2 + 6x + 4$ ✓

CHECK FOR UNDERSTANDING

Communicating Mathematics

Read and study the lesson to answer each question.

1. *Tell* how you know when a polynomial can be factored.

2. *You Decide* Tamika said that the factors of $x^2 + 3x + 9$ are $x + 3$ and $x + 3$. Is she correct? Explain.

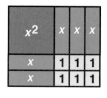

3a. *Model* the polynomial $2x^2 + 5x + 2$ and try to form a rectangle.

b. What are the factors? Explain how you know.

Guided Practice

4. Factor $x^2 + 5x + 6$.

If possible, factor each polynomial. Use drawings or algebra tiles if necessary.

5. $x^2 + 4x$

6. $x^2 + 6x + 8$

7. $x^2 + x + 3$

8. *Interior Design* When placing a rug under a dining room table, the rug should be 6 feet longer and 6 feet wider than the table to allow for pulling out chairs. A square table measures x feet on each side. Will a rug with an area of $x^2 + 12x + 36$ be suitable for underneath the table? Explain. (*Hint*: Factor the polynomial.)

EXERCISES

Practice

Factor each polynomial.

9. $3x + 6$

10. $x^2 + 6x + 5$

11. $2x^2 + 5x + 2$

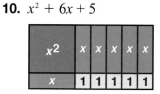

If possible, factor each polynomial. Use drawings or algebra tiles if necessary.

12. $5x + 5$

13. $18x + 6$

14. $3x^2 + 6x$

15. $x^2 + 5x + 4$

16. $x^2 + 2x + 3$

17. $x^2 + 8x + 16$

18. $2x^2 + 2x + 3$

19. $2x^2 + 7x + 3$

20. $2x^2 + 7x + 6$

21. Find the factors of $3x^2 + 4x$.

22. Factor the polynomial $x^2 + 7x + 12$.

Applications and Problem Solving

23. *Algebra* Simplify $\frac{2x^2 + 11x + 5}{x^2 + 7x + 10}$ by factoring the numerator and the denominator.

24. *Geometry* Find the length and width for a rectangle of area $3x^2 + 8x + 5$.

25. **Working on the** 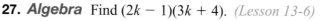**Project** Refer to Exercise 21 on page 585.

   a. Factor the polynomial. What must be true of $h + m$?

   b. A player scores 2 points if 2 shots are made, 1 point if 1 is made, and 0 points if both are missed. Find the number of expected points for a player. To do this, multiply the term for 2 shots made by 2, the term for 1 made by 1, and the term for both missed by 0. Write a polynomial for the number of points a player can be expected to score.

   c. Evaluate the polynomial from part b for each of your players.

26. **Critical Thinking** For the polynomial $2x^2 + 7x + \underline{}$, write a positive integer in the blank that makes the polynomial factorable.

Mixed Review

27. **Algebra** Find $(2k - 1)(3k + 4)$. *(Lesson 13-6)*

28. **Standardized Test Practice** Which would be the most precise measurement for a box of cereal? *(Lesson 11-7)*

   A 1 pound

   B 53 pounds

   C 510 grams

   D 538.5 grams

For **Extra Practice**, see page 644.

29. **Food** Three-fourths of a pan of peach cobbler is to be divided equally among 6 people. What part of the cobbler will each person receive? *(Lesson 7-8)*

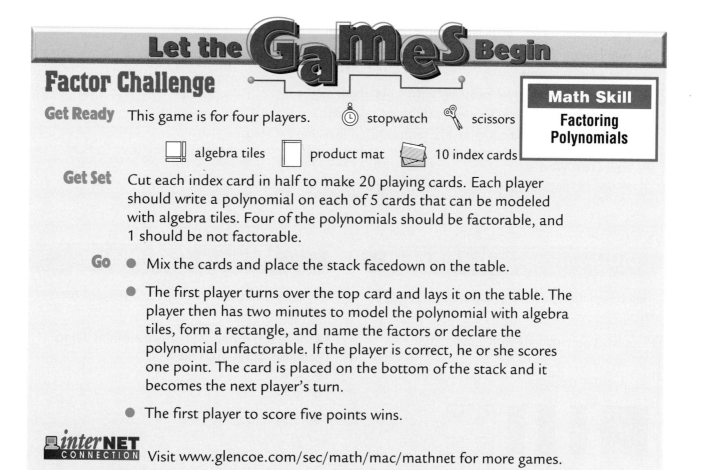

Let the Games Begin

Factor Challenge

Math Skill
Factoring Polynomials

Get Ready This game is for four players. stopwatch scissors algebra tiles product mat 10 index cards

Get Set Cut each index card in half to make 20 playing cards. Each player should write a polynomial on each of 5 cards that can be modeled with algebra tiles. Four of the polynomials should be factorable, and 1 should be not factorable.

Go
- Mix the cards and place the stack facedown on the table.

- The first player turns over the top card and lays it on the table. The player then has two minutes to model the polynomial with algebra tiles, form a rectangle, and name the factors or declare the polynomial unfactorable. If the player is correct, he or she scores one point. The card is placed on the bottom of the stack and it becomes the next player's turn.

- The first player to score five points wins.

inter NET CONNECTION Visit www.glencoe.com/sec/math/mac/mathnet for more games.

inter**NET**
CONNECTION Chapter Review **For additional lesson-by-lesson review, visit:**
www.glencoe.com/sec/math/mac/mathnet

Vocabulary

After completing this chapter, you should be able to define each term, concept, or phrase and give an example or two of each.

Algebra
binomial (p. 583)
factoring (p. 588)
like terms (p. 565)
monomial (p. 561)
polynomial (p. 562)
simplest form (p. 565)

term (p. 565)
trinomial (p. 588)
zero pair (p. 566)

Problem Solving
guess and check (p. 586)

Understanding and Using the Vocabulary

State whether each sentence is _true_ or _false_. If false, replace the underlined word to make a true sentence.

1. The expression $b^2 - 3b$ is an example of a <u>monomial</u>.

2. A polynomial is the sum or <u>difference</u> of two or more monomials.

3. The <u>additive</u> inverse of $9y^2 - 5y + 2$ is $-9y^2 + 5y - 2$.

4. A polynomial with two terms is called a <u>binomial</u>.

5. The product of $2m$ and $m^2 + 8m$ will have <u>three</u> terms.

6. When <u>factoring</u>, you combine like terms to find the simplest form.

In Your Own Words

7. _Write_ the definition of a monomial in your own words.

Objectives & Examples

Upon completing this chapter, you should be able to:

● represent polynomials with algebra tiles
(Lesson 13-1)

Model $x^2 + 3x + 4$.

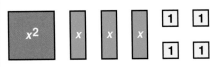

Review Exercises

Use these exercises to review and prepare for the chapter test.

Model each monomial or polynomial using drawings or algebra tiles.

8. $2x$

9. $3x^2 - 4$

10. $-2x^2 + 2x - 6$

11. $x^2 - 7$

Objectives & Examples

simplify polynomials by using algebra tiles
(Lesson 13-2)

Simplify $2x^2 + 6 + x^2$.

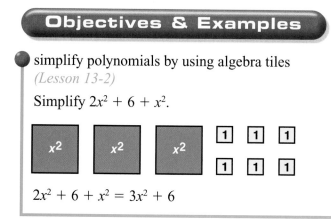

$2x^2 + 6 + x^2 = 3x^2 + 6$

add polynomials by using algebra tiles
(Lesson 13-3)

Find $(2x^2 + 6x) + (3x^2 - 2x)$.

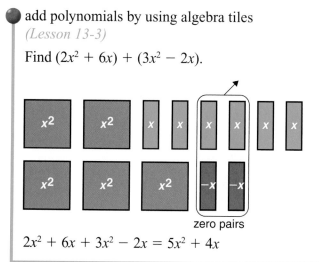

zero pairs

$2x^2 + 6x + 3x^2 - 2x = 5x^2 + 4x$

subtract polynomials by using algebra tiles
(Lesson 13-4)

Find $(6x^2 + 3) - (2x^2 - 1)$.

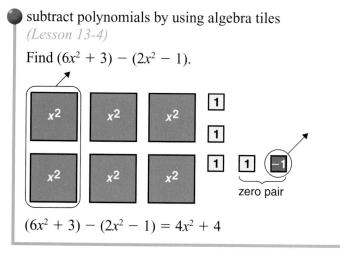

zero pair

$(6x^2 + 3) - (2x^2 - 1) = 4x^2 + 4$

Review Exercises

Simplify each polynomial. Use drawings or algebra tiles if necessary.

12. $3m^2 + 3m + 10m^2 + 2m$

13. $8p - 3p - 12$

14. $12x - 2x^2 - 9x + 6x^2$

15. $5a + 7b - a - 8b$

Find each sum. Use drawings or algebra tiles if necessary.

16. $(10m^2 - 3m) + (3m^2 + 2m)$

17. $(4d + 1) + (8d + 6)$

18. $(3a^2 + 6a) + (2a^2 - 5a)$

19. $(b^2 - 2b + 4) + (2b^2 + b - 8)$

20. $(2x^2 - 7x) + (x^2 + 5x)$

Find each difference. Use drawings or algebra tiles if necessary.

21. $(7g + 2) - (5g + 1)$

22. $(3c - 7) - (-3c + 4)$

23. $(2s^2 + 8) - (s^2 + 3)$

24. $(6k^2 - 3) - (k^2 - 5k - 2)$

25. $(7p^2 + 2p - 5) - (4p^2 + 6p - 2)$

<table>
<tr>
<td>

Objectives & Examples

multiply monomials and polynomials
(Lesson 13-5)

Find $b^4 \cdot b^7$.

$b^4 \cdot b^7 = b^{4+7}$ or b^{11}

Find $2x(2x + 3)$.

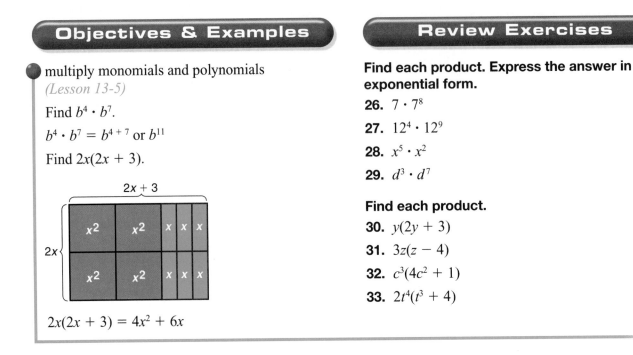

$2x(2x + 3) = 4x^2 + 6x$

</td>
<td>

Review Exercises

Find each product. Express the answer in exponential form.

26. $7 \cdot 7^8$

27. $12^4 \cdot 12^9$

28. $x^5 \cdot x^2$

29. $d^3 \cdot d^7$

Find each product.

30. $y(2y + 3)$

31. $3z(z - 4)$

32. $c^3(4c^2 + 1)$

33. $2t^4(t^3 + 4)$

</td>
</tr>
<tr>
<td>

multiply binomials by using algebra tiles
(Lesson 13-6)

Find $(x + 3)(x + 3)$.

$(x + 3)(x + 3) = x^2 + 6x + 9$

</td>
<td>

Find each product. Use drawings or algebra tiles if necessary.

34. $(x + 2)(x + 3)$

35. $(x - 5)(2x + 1)$

36. $(2x + 3)(3x + 1)$

37. $(x + 3)(x - 4)$

38. $(4x + 1)(x + 1)$

</td>
</tr>
<tr>
<td>

factor polynomials by using algebra tiles
(Lesson 13-7)

Factor $x^2 + 4x + 4$.

Try to form a rectangle with algebra tiles.

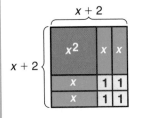

Therefore, $x^2 + 4x + 4 = (x + 2)(x + 2)$.

</td>
<td>

If possible, factor each polynomial. Use drawings or algebra tiles if necessary.

39. $14x + 7$

40. $x^2 + 6x$

41. $x^2 + 7x + 10$

42. $x^2 + 5x + 2$

43. $2x^2 + 3x + 1$

</td>
</tr>
</table>

Applications & Problem Solving

44. *Gardening* A diagram of Mr. Keith's flowerbed is shown below. Find its perimeter. *(Lesson 13-3)*

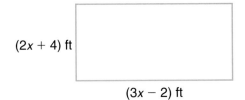

$(2x + 4)$ ft

$(3x - 2)$ ft

45. *Interior Design* Each side of a living room is $2x + 1$ yards long. Find the number of square yards of carpeting needed to cover the floor of the living room. *(Lesson 13-5)*

46. *Guess and Check* Lou's house number consists of three digits. The digits in the number are consecutive and have a sum of 21. The first digit is smaller than the last digit. What is Lou's house number? *(Lesson 13-7A)*

47. *Money Matters* Andrea borrowed $500 each year for technical school expenses. The amount she owes for the second year is $1{,}000 + 500r$, where r is the interest rate. The amount she owes for the third year is $1{,}500 + 1{,}500r + 500r^2$. How much did her debt increase between the second and third year? *(Lesson 13-4)*

Alternative Assessment

● **Open Ended**

You are remodeling a room in your house. The original room was square. The remodeled room is a rectangle with an area of $x^2 + 8x + 15$ square feet. By how much did each dimension of the room increase?

Suppose the area of the remodeled room is 143 square feet. Can you find the actual dimensions of the remodeled room? If so, state the dimensions. Can you find the area of the original room? If so, state the area.

● **Completing the** CHAPTER Project

Use the following checklist to make sure your poster is complete.

☑ The data for each player is clear and well organized.

☑ The drawings of your models are accurately drawn and labeled correctly.

☑ The geometric and algebraic models are accurate and correct.

●   PORTFOLIO Select one of the problems you solved in this chapter and place it in your portfolio. Attach a note explaining how your problem illustrates one or more of the important concepts covered in this chapter.

A practice test for Chapter 13 is provided on page 659.

Section One: Multiple Choice

There are nine multiple choice questions in this section. Choose the best answer. If a correct answer is *not here,* choose the letter for Not Here.

1. An electronic store advertised 20% off the price of a $14 CD. To find the amount saved on the CD, multiply $14 by —

   A 0.20.

   B 0.80.

   C 0.14.

   D 2.0.

2. A bagel shop offers 3 different cheeses and 3 different meats on 5 different types of bagels. If John chooses one bagel, one meat, and one cheese, how many different combinations can he order?

   F 11

   G 35

   H 45

   J 55

3. Find the value of x for the given triangle.

   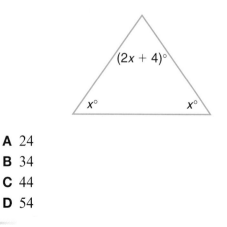

   A 24

   B 34

   C 44

   D 54

Please note that Questions 4–9 have five answer choices.

4. Jeremy was given an extra dollar on his paper route every time he sold 2 more subscriptions. Lori also earned an extra dollar every time she sold 2 subscriptions. The results of their sales are shown. How many dollars did Lori receive Monday through Wednesday?

   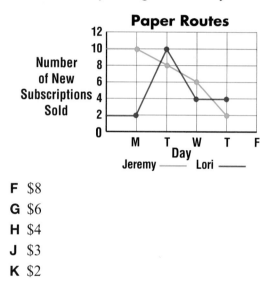

   F $8

   G $6

   H $4

   J $3

   K $2

5. Courtney owns a tent in the shape of a pyramid. The base of the tent is 8 feet by 10 feet. The tent is 7 feet tall at the center. How can you find the volume of the air inside the tent?

   A Find $8 \times 10 \times 7$ and divide by 3.

   B Add $2\left(\frac{1}{2} \cdot 7 \cdot 8\right)$ and $2\left(\frac{1}{2} \cdot 7 \cdot 10\right)$.

   C Find $7 \cdot 8 \cdot 10$.

   D Find $\frac{1}{2} \cdot 7 \cdot 8 \cdot 10$.

   E Find $8 \times 10 \times 7$ and divide by 2.

6. During an overnight camping trip, the temperature dropped 4°F every hour. If the temperature was 56°F at midnight, what was the temperature at 5 A.M.?

F 46°

G 36°

H −36°

J −46°

K Not Here

7. Kiar bought $12.48 worth of oranges at a fruit stand. If the oranges were $2.08 per pound, how many pounds were purchased?

A 4 lb   **B** 5 lb

C 6 lb   **D** 7 lb

E Not Here

8. On Fernando's last paycheck, 28% was deducted for income tax, and 7.5% was deducted for social security tax. What is the total percent deducted from his pay for these taxes?

F 20.5%   **G** 25.5%

H 100%   **J** 35.5%

K Not Here

9. Sandra bought $2\frac{1}{2}$ pounds of chocolate candy for $6.75. She also bought 2 pounds of popcorn for $3.24. What was the cost per pound of the chocolate candy?

A $1.62   **B** $1.89

C $3.37   **D** $2.70

E Not Here

Test Practice For additional test practice questions, visit:

www.glencoe.com/sec/math/mac/mathnet

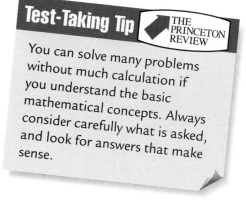

Section Two: Free Response

This section contains three questions for which you will provide short answers. Write your answers on your paper.

10. Which point is in the third quadrant?

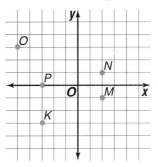

11. What is the probability of landing on an odd number?

12. Identify the solution of the system of equations graphed below.

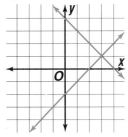

Student Handbook
Table of Contents

Basic Skills

Adding and Subtracting Decimals

1. 0.346
− 0.034

2. 4.26
+ 0.72

3. 23.4
− 0.526

4. 7.49
− 2.35

5. 133.06
− 48.114

6. 21.8
+ 434.64

7. 38.4
+ 0.77

8. 34.7
+ 31.3

9. 471.8
− 0.27

10. 365.57
− 222.4

11. 90
− 0.26

12. 31.666
− 5.885

13. 51.05
− 43.68

14. 70.02
− 59.08

15. 206.78
− 0.61

16. 208.888
+ 2.2

17. 61.42
+ 0.77

18. 549.558
− 0.65

19. 21
− 0.154

20. 0.67
+ 0.63

21. 7.2 − 4.563

22. 0.67 + 3.804

23. 6.7 + 4.53

24. 41.68 + 41.68

25. 7 − 3.26

26. 26.483 − 0.3

27. 0.421 + 0.39

28. 7.1 − 3.4

29. 2.365 + 4.453

30. 4.1 + 3.35

31. 0.774 − 0.5

32. 42.35 + 7.118

33. 10 − 0.45

34. 52.26 − 0.09

35. 5.361 + 0.8

36. 55.741 − 15.902

37. 8 − 0.002

38. 543.26 + 0.055

39. 200.431 + 5.006

40. 73.182 + 54.591

41. 7.8 + 2.2

42. 42 − 0.71

43. 943.007 + 0.16

44. 73.68 + 0.327

45. 5.2 − 3.8

46. 38.005 + 4.2

47. 700.421 + 0.9

48. 136.024 − 5.878

49. 0.7 + 5.3

50. 0.5 + 0.4

Basic Skills

Multiplying and Dividing Decimals

1. $\begin{array}{r} 4.02 \\ \times\, 0.16 \\ \hline \end{array}$

2. $0.4\overline{)10.32}$

3. $\begin{array}{r} 6.3 \\ \times\, 6.3 \\ \hline \end{array}$

4. $4.7\overline{)13.63}$

5. $4.32\overline{)2.592}$

6. $\begin{array}{r} 0.097 \\ \times\quad 8 \\ \hline \end{array}$

7. $\begin{array}{r} 22 \\ \times\, 8.1 \\ \hline \end{array}$

8. $53.6\overline{)804}$

9. $\begin{array}{r} 2.16 \\ \times\, 9.3 \\ \hline \end{array}$

10. $\begin{array}{r} 6.2 \\ \times\, 9.4 \\ \hline \end{array}$

11. $\begin{array}{r} 400 \\ \times\, 0.05 \\ \hline \end{array}$

12. $37.15\overline{)549.82}$

13. $10.72\overline{)269.072}$

14. $0.7\overline{)49}$

15. $\begin{array}{r} 23.6 \\ \times\, 0.74 \\ \hline \end{array}$

16. $200\overline{)8.464}$

17. $\begin{array}{r} 2.19 \\ \times\, 3.55 \\ \hline \end{array}$

18. $16.03\overline{)32.06}$

19. $6.43\overline{)0.02572}$

20. $\begin{array}{r} 6.73 \\ \times\, 8.91 \\ \hline \end{array}$

21. 4×0.4

22. $618 \div 0.4$

23. 0.6×8

24. $2{,}276.485 \div 77.3$

25. $60.125 \div 1.25$

26. 2.44×2.22

27. 0.1×17

28. $44.4 \div 222$

29. 1.68×7

30. $525 \div 0.10$

31. $0.01495 \div 2.99$

32. 0.9×1

33. 0.7×23

34. 14.5×14.7

35. $2.209 \div 0.047$

36. $18.8 \div 0.2$

37. $1.375 \div 0.05$

38. $65 \div 0.008$

39. 900×0.9

40. $360 \div 1.8$

41. 12×0.2

42. 0.39×0.5

43. $36 \div 0.036$

44. 0.008×9

45. $324.52 \div 7.6$

46. 29.1×82.7

47. $768.57 \div 41.1$

48. 0.64×0.3

Basic Skills

Adding and Subtracting Fractions

1. $\frac{13}{6} - \frac{1}{6}$

2. $\frac{1}{8} + \frac{1}{10}$

3. $\frac{3}{16} + \frac{1}{3}$

4. $\frac{3}{5} - \frac{1}{2}$

5. $\frac{3}{8} + \frac{1}{3}$

6. $\frac{4}{25} + \frac{7}{20}$

7. $\frac{29}{30} - \frac{7}{10}$

8. $\frac{6}{7} + \frac{2}{3}$

9. $\frac{9}{14} - \frac{3}{14}$

10. $\frac{7}{8} - \frac{3}{4}$

11. $\frac{6}{7} - \frac{2}{3}$

12. $\frac{1}{3} - \frac{1}{4}$

13. $\frac{7}{12} + \frac{1}{3}$

14. $\frac{5}{7} - \frac{1}{7}$

15. $\frac{13}{18} - \frac{5}{9}$

16. $\frac{8}{9} - \frac{1}{9}$

17. $\frac{9}{100} + \frac{11}{100}$

18. $\frac{9}{10} - \frac{2}{5}$

19. $\frac{5}{12} + \frac{1}{3}$

20. $\frac{2}{5} + \frac{5}{14}$

21. $\frac{7}{12} - \frac{1}{3}$

22. $\frac{4}{5} - \frac{3}{5}$

23. $\frac{31}{36} - \frac{3}{4}$

24. $\frac{8}{21} + \frac{10}{21}$

25. $\frac{14}{15} - \frac{1}{15}$

26. $\frac{1}{7} + \frac{9}{28}$

27. $\frac{2}{5} + \frac{1}{10}$

28. $\frac{4}{11} + \frac{6}{11}$

29. $\frac{11}{60} - \frac{2}{15}$

30. $\frac{5}{21} + \frac{2}{21}$

31. $\frac{5}{6} - \frac{1}{2}$

32. $\frac{3}{8} + \frac{3}{8}$

33. $\frac{3}{8} - \frac{1}{4}$

34. $\frac{8}{9} + \frac{1}{9}$

35. $\frac{5}{12} + \frac{1}{4}$

36. $\frac{5}{16} + \frac{1}{4}$

37. $\frac{7}{12} + \frac{1}{4}$

38. $\frac{3}{25} - \frac{1}{50}$

39. $\frac{1}{4} + \frac{1}{6}$

40. $\frac{89}{96} - \frac{11}{12}$

Basic Skills

Multiplying Fractions

1. $\frac{1}{4} \times \frac{1}{5}$

2. $\frac{2}{3} \times \frac{1}{2}$

3. $\frac{6}{7} \times \frac{7}{8}$

4. $\frac{1}{3} \times \frac{9}{10}$

5. $\frac{4}{5} \times \frac{3}{4}$

6. $80 \times \frac{7}{10}$

7. $36 \times \frac{2}{9}$

8. $2\frac{1}{3} \times \frac{3}{5}$

9. $\frac{1}{8} \times 20$

10. $\frac{1}{6} \times \frac{1}{4}$

11. $\frac{4}{5} \times \frac{12}{28}$

12. $\frac{16}{3} \times \frac{1}{8}$

13. $\frac{18}{8} \times \frac{4}{3}$

14. $1\frac{3}{5} \times 15$

15. $\frac{3}{50} \times \frac{10}{12}$

16. $\frac{9}{11} \times \frac{1}{4}$

17. $\frac{14}{15} \times \frac{45}{7}$

18. $\frac{12}{30} \times \frac{10}{18}$

19. $\frac{6}{17} \times \frac{17}{29}$

20. $16 \times \frac{1}{10}$

21. $3\frac{2}{5} \times \frac{10}{34}$

22. $9 \times 3\frac{1}{2}$

23. $\frac{5}{2} \times \frac{4}{10}$

24. $\frac{110}{121} \times \frac{11}{55}$

25. $\frac{1}{7} \times 49$

26. $\frac{81}{2} \times \frac{5}{6}$

27. $\frac{7}{4} \times 1\frac{3}{8}$

28. $8\frac{1}{2} \times 6\frac{2}{3}$

29. $16 \times \frac{9}{10}$

30. $\frac{52}{100} \times \frac{3}{4}$

31. $\frac{22}{50} \times \frac{18}{42}$

32. $\frac{5}{8} \times 36$

33. $\frac{70}{76} \times \frac{14}{28}$

34. $\frac{60}{3} \times \frac{3}{24}$

35. $\frac{5}{9} \times \frac{90}{100}$

36. $3\frac{6}{7} \times 2\frac{1}{3}$

37. $\frac{17}{60} \times \frac{100}{12}$

38. $\frac{42}{45} \times \frac{1}{6}$

39. $7\frac{1}{4} \times \frac{2}{3}$

40. $\frac{32}{38} \times \frac{78}{16}$

Basic Skills

Dividing Fractions

1. $\frac{1}{2} \div \frac{2}{3}$

2. $\frac{3}{4} \div \frac{1}{2}$

3. $3 \div \frac{6}{7}$

4. $\frac{1}{8} \div \frac{1}{3}$

5. $\frac{2}{3} \div \frac{1}{2}$

6. $\frac{3}{5} \div \frac{1}{4}$

7. $\frac{5}{6} \div \frac{2}{3}$

8. $\frac{1}{6} \div \frac{1}{4}$

9. $8 \div \frac{1}{2}$

10. $\frac{3}{8} \div \frac{6}{7}$

11. $\frac{4}{9} \div 2$

12. $\frac{5}{9} \div \frac{5}{6}$

13. $\frac{3}{4} \div \frac{3}{8}$

14. $\frac{1}{6} \div 15$

15. $\frac{10}{5} \div \frac{6}{5}$

16. $\frac{8}{9} \div \frac{4}{81}$

17. $\frac{1}{8} \div \frac{1}{16}$

18. $9 \div \frac{3}{4}$

19. $\frac{1}{20} \div 4$

20. $5 \div \frac{4}{3}$

21. $\frac{3}{4} \div 4\frac{1}{2}$

22. $2\frac{2}{3} \div 4$

23. $1\frac{1}{4} \div 3\frac{1}{2}$

24. $4\frac{2}{3} \div \frac{7}{8}$

25. $\frac{9}{10} \div 2\frac{1}{4}$

26. $\frac{2}{3} \div 2\frac{1}{2}$

27. $5 \div 1\frac{1}{3}$

28. $2\frac{1}{4} \div \frac{2}{3}$

29. $2\frac{2}{3} \div 5\frac{1}{3}$

30. $1\frac{1}{9} \div 1\frac{2}{3}$

31. $5\frac{1}{4} \div 3$

32. $4\frac{1}{2} \div 6\frac{3}{4}$

33. $3\frac{3}{4} \div \frac{5}{6}$

34. $2\frac{3}{4} \div 1\frac{3}{8}$

35. $6\frac{1}{2} \div 4\frac{1}{6}$

36. $2\frac{1}{7} \div 10$

37. $2\frac{4}{5} \div 5\frac{3}{5}$

38. $1\frac{5}{9} \div 2\frac{1}{3}$

39. $5\frac{3}{4} \div 2\frac{1}{2}$

40. $3\frac{1}{2} \div 1\frac{1}{2}$

Extra Practice

Lesson 1-1 *(Pages 4–7)*
Use the four-step plan to solve each problem.

1. Joseph is planting bushes around the perimeter of his lawn. If the bushes must be planted 4 feet apart and Joseph's lawn is 64 feet wide and 124 feet long, how many bushes will Joseph need to purchase?

2. The cost of a long distance phone call is $1.50 for the first two minutes and $0.60 for each additional minute. How much will Maria pay for a 24 minute phone call?

3. Find the next three numbers in the pattern.   1, 3, 7, 15, 31, ___, ___, ___

Lesson 1-2 *(Pages 8–10)*
Write each expression using exponents.

1. $4 \cdot 4 \cdot 4 \cdot 4$
2. $3 \cdot 3$
3. $7 \cdot 7 \cdot 7 \cdot 7 \cdot 7 \cdot 7$

Evaluate each expression.

4. 4^3
5. 6^2
6. 2^6

7. $5^2 \times 6^2$
8. 3×2^4
9. $10^4 \times 3^2$

10. $5^3 \times 1^9$
11. $2^2 \times 2^4$
12. $2 \times 3^2 \times 4^2$

13. 7^3
14. $9^2 + 3^2$
15. 0.5^2

Lesson 1-3 *(Pages 11–15)*
Evaluate each expression.

1. $15 - 5 + 9 - 2$
2. $6 \times 6 + 3.6$
3. $12 + 20 \div 4 - 5$

4. $6 \times 3 \div 9 - 1$
5. $(4^2 + 2^3) \times 5$
6. $24 \div 8 - 2$

7. $3 \times (4 + 5) - 7$
8. $4.3 + 24 \div 6$
9. $(5^2 + 2) \div 3$

Evaluate each expression if $a = 3$, $b = 6$, and $c = 5$.

10. $2a + bc$
11. ba^3
12. $\dfrac{bc}{a}$

13. $3a - 2b + c$
14. $(2c + b) \cdot a$
15. $\dfrac{2(ac)^2}{b}$

Find the solution for each equation from the given replacement set.

16. $g + 16 = 47$ {29, 31, 33, 35}
17. $y - 14 = 26$ {50, 45, 40, 35}

18. $7m = 161$ {21, 23, 25, 27}
19. $\dfrac{k}{16} = 19$ {300, 302, 304, 306}

Lesson 1-4 *(Pages 17–20)*

Solve each equation. Check your solution.

1. $g - 3 = 10$
2. $b + 7 = 12$
3. $a + 3 = 15$
4. $r - 3 = 4$
5. $t + 3 = 21$
6. $s + 10 = 23$
7. $9 + n = 13$
8. $13 + v = 31$
9. $s - 0.4 = 6$
10. $x - 1.3 = 12$
11. $18 = y + 3.4$
12. $7 + g = 91$
13. $63 + f = 71$
14. $0.32 = w - 0.1$
15. $c - 18 = 13$
16. $23 = n - 5$
17. $j - 3 = 7$
18. $18 = p + 3$
19. $12 + p = 16$
20. $25 = y - 50$
21. $x + 2 = 4$

Lesson 1-5 *(Pages 21–25)*

Solve each equation. Check your solution.

1. $4x = 36$
2. $39 = 3y$
3. $4z = 16$
4. $t \div 5 = 6$
5. $100 = 20b$
6. $8 = w \div 8$
7. $10a = 40$
8. $s \div 9 = 8$
9. $420 = 5s$
10. $8k = 72$
11. $2m = 18$
12. $\frac{m}{8} = 5$
13. $0.12 = 3h$
14. $\frac{w}{7} = 8$
15. $18q = 36$
16. $9w = 54$
17. $4 = p \div 4$
18. $14 = 2p$
19. $12 = 3t$
20. $\frac{m}{4} = 12$
21. $6h = 12$

Lesson 1-6 *(Pages 27–29)*

Write each phrase or sentence as an algebraic expression or equation.

1. 12 more than a number
2. 3 less than a number
3. a number divided by 4
4. a number increased by 7
5. a number decreased by 12
6. 8 times a number
7. 28 multiplied by m
8. 15 divided by a number
9. 54 divided by n
10. 18 increased by y
11. q decreased by 20
12. n times 41

13. The difference between a number and 12 is 37.

14. The product of a number and 7 is 42.

15. 6 less than the product of q and 4 is 18.

Lesson 1-7A *(Pages 30–31)*
Solve.

1. Dwayne's weight is twice Beth's weight minus 24 pounds. Dwayne weighs 120 pounds. How much does Beth weigh?

2. A store doubled its cost to determine the price for a sweater. Two months later the sweater was marked down $10.00. A week later the price was cut in half and the sweater sold for $12.52. How much did the store pay for the sweater? Did the store make or lose money on the sweater?

3. Jonathon Michaels just bought a new car. He received a $1,500 credit for his trade-in. He gave the car dealer a $2,500 down payment and is making 24 monthly payments of $279.00 each. What is the total cost of the car?

4. If a number is divided by 6 and then 14 is added to it, the result is 21. What is the number?

Lesson 1-7 *(Pages 32–36)*
Solve each equation. Check your solution.

1. $2x + 4 = 14$ **2.** $5p - 10 = 0$ **3.** $5 + 6a = 41$

4. $\frac{x}{3} - 7 = 2$ **5.** $18 = 6(q - 4)$ **6.** $18 = 4m - 6$

7. $3(r - 1) = 9$ **8.** $2x + 3 = 5$ **9.** $0 = 4x - 28$

10. $3x - 1 = 5$ **11.** $3z + 5 = 14$ **12** $3(x - 5) = 12$

13. $9a - 8 = 73$ **14.** $2x - 3 = 7$ **15.** $3t + 6 = 9$

16. $2y + 10 = 22$ **17.** $15 = 2y - 5$ **18.** $3c - 4 = 2$

19. $6 + 2p = 16$ **20.** $8 = 2 + 3x$ **21.** $4(b + 6) = 24$

Lesson 1-8 *(Pages 38–42)*
Find the perimeter and area of each figure.

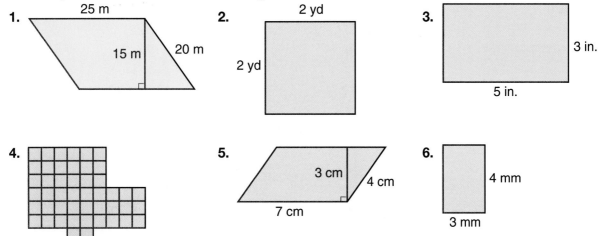

1. 25 m, 15 m, 20 m
2. 2 yd, 2 yd
3. 3 in., 5 in.
4.
5. 3 cm, 4 cm, 7 cm
6. 4 mm, 3 mm

Lesson 1-9 *(Pages 43–47)*
Solve each inequality. Show the solution on a number line.

1. $y + 3 > 7$

2. $c - 9 < 5$

3. $x + 4 \geq 9$

4. $y - 3 < 15$

5. $t - 13 \geq 5$

6. $5p \leq 25$

7. $4x < 12$

8. $15 \leq 3m$

9. $\frac{d}{3} > 15$

10. $8 < r \div 7$

11. $2y + 5 \geq 15$

12. $16 \leq 5d + 6$

13. $3x + 2 \geq 11$

14. $\frac{a}{3} - 2 > 1$

15. $9g < 27$

16. $14 \leq 2x + 4$

17. $\frac{x}{2} - 6 > 0$

18. $k + 5 \leq 6$

19. $15 > c - 2$

20. $4p \geq 24$

21. $24 < 14 + k$

Lesson 2-1 *(Pages 56–58)*
Graph each set of points on a number line.

1. $\{-8, -9, -6, -10\}$

2. $\{-3, 2, 0, -1\}$

3. $\{5, 6, 8, 7, 9\}$

Find each absolute value.

4. $|-1|$

5. $|-92|$

6. $|3|$

7. $|160 + 32|$

8. $|80 + 100|$

9. $|0|$

10. $|7 - 3|$

11. $|3| + |-7|$

12. $|-161|$

13. $|150|$

14. $|102| - |-2|$

15. $|-116|$

Lesson 2-2 *(Pages 59–61)*
Replace each ● with >, <, or = to make a true sentence.

1. $-3 \; ● \; 0$

2. $-1 \; ● \; -2$

3. $-5 \; ● \; -4$

4. $6 \; ● \; -7$

5. $8 \; ● \; 10$

6. $-6 \; ● \; 6$

7. $-11 \; ● \; -20$

8. $-8 \; ● \; 2$

9. $-13 \; ● \; -12$

10. $5 \; ● \; 2$

11. $9 \; ● \; -8$

12. $19 \; ● \; -19$

13. $|-2| \; ● \; |5|$

14. $|13| \; ● \; |-19|$

15. $|-6| \; ● \; |2|$

16. $|14| \; ● \; |-14|$

17. $|0| \; ● \; |-4|$

18. $|23| \; ● \; |-20|$

19. $|-75| \; ● \; |75|$

20. $-71 \; ● \; 72$

21. $-15 \; ● \; -35$

Lesson 2-3 *(Pages 62–65)*
Solve each equation.

1. $-7 + (-7) = h$

2. $k = -36 + 40$

3. $m = 18 + (-32)$

4. $47 + 12 = y$

5. $y = -69 + (-32)$

6. $-120 + (-2) = c$

7. $x = -56 + (-4)$

8. $14 + 16 = k$

9. $-18 + 11 = d$

10. $-42 + 29 = r$

11. $h = -13 + (-11)$

12. $x = 95 + (-5)$

13. $-120 + 2 = b$

14. $w = 25 + (-25)$

15. $a = -4 + 8$

16. $g = -9 + (-6)$

17. $42 + (-18) = f$

18. $-33 + (-12) = w$

19. $-96 + (-18) = g$

20. $-100 + 98 = a$

21. $5 + (-7) = y$

Lesson 2-4 *(Pages 66–68)*
Solve each equation. Check by solving another way.

1. $a = 7 + (-13) + 6 + (-7)$

2. $x = -6 + 12 + (-20)$

3. $4 + 9 + (-14) = k$

4. $c = -20 + 0 + (-9) + 25$

5. $b = 5 + 9 + 3 + (-17)$

6. $-36 + 40 + (-10) = y$

7. $(-2) + 2 + (-2) + 2 = m$

8. $6 + (-4) + 9 + (-2) = d$

9. $9 + (-7) + 2 = n$

10. $b = 100 + (-75) + (-20)$

11. $x = -12 + 24 + (-12) + 2$

12. $9 + (-18) + 6 + (-3) = c$

13. $(-10) + 4 + 6 = k$

14. $c = 4 + (-8) + 12$

Lesson 2-5 *(Pages 69–72)*
Solve each equation.

1. $3 - 7 = y$

2. $-5 - 4 = w$

3. $a = -6 - 2$

4. $12 - 9 = x$

5. $a = 0 - (-14)$

6. $a = 58 - (-10)$

7. $n = -41 - 15$

8. $c = -81 - 21$

9. $26 - (-14) = y$

10. $6 - (-4) = b$

11. $z = 63 - 78$

12. $-5 - (-9) = h$

13. $m = 72 - (-19)$

14. $-51 - 47 = x$

15. $-99 - 1 = p$

16. $r = 8 - 13$

17. $-2 - 23 = c$

18. $-20 - 0 = d$

19. $55 - 33 = k$

20. $84 - (-61) = a$

21. $z = -4 - (-4)$

Lesson 2-6 *(Pages 73–76)*

Find each sum or difference. If there is no sum or difference, write *impossible*.

1. $\begin{bmatrix} 2 & 0 \\ -6 & 4 \end{bmatrix} + \begin{bmatrix} 8 & 6 \\ 1 & -3 \end{bmatrix}$

2. $\begin{bmatrix} 8 & 5 & 2 \\ -1 & 3 & 11 \end{bmatrix} + \begin{bmatrix} -4 & 12 \\ 1 & 8 \\ 6 & -3 \end{bmatrix}$

3. $\begin{bmatrix} 1 & 2 \\ 6 & -3 \\ -5 & 7 \end{bmatrix} - \begin{bmatrix} -2 & 7 \\ 5 & 4 \\ -3 & 9 \end{bmatrix}$

4. $\begin{bmatrix} 0 & 8 & 6 \\ 7 & -2 & 4 \\ 0 & 0 & 9 \end{bmatrix} - \begin{bmatrix} -6 & 0 & -1 \\ 2 & 4 & 0 \\ -1 & 5 & 8 \end{bmatrix}$

5. $\begin{bmatrix} 13 & 2 \\ 7 & -1 \end{bmatrix} + \begin{bmatrix} 6 & -5 & 0 \\ 3 & 1 & -2 \\ 2 & 7 & 8 \end{bmatrix}$

6. $\begin{bmatrix} 13 & 5 & 7 \\ -6 & 12 & -1 \end{bmatrix} - \begin{bmatrix} 8 & -6 & 0 \\ -2 & 15 & 2 \end{bmatrix}$

Lesson 2-7 *(Pages 78–80)*

Solve each equation.

1. $5(-2) = d$

2. $-11(-5) = c$

3. $-5(-5) = z$

4. $x = -12(6)$

5. $b = 2(-2)$

6. $-3(2)(-4) = j$

7. $a = (-4)(-4)$

8. $4(21) = y$

9. $a = -50(0)$

10. $b = 3(-13)$

11. $a = 2(2)$

12. $d = -2(-2)$

13. $x = 5(-12)$

14. $2(2)(-2) = b$

15. $a = 6(-4)$

16. $x = -6(5)$

17. $-4(8) = a$

18. $3(-16) = y$

19. $c = -2(2)$

20. $6(3)(-2) = k$

21. $y = -3(12)$

Lesson 2-8 *(Pages 81–83)*

Solve each equation.

1. $a = 4 \div (-2)$

2. $16 \div (-8) = x$

3. $-14 \div (-2) = c$

4. $h = -18 \div 3$

5. $-25 \div 5 = k$

6. $n = -56 \div (-8)$

7. $x = 81 \div 9$

8. $-55 \div 11 = c$

9. $-42 \div (-7) = y$

10. $g = 18 \div (-3)$

11. $t = 0 \div (-1)$

12. $-32 \div 8 = m$

13. $81 \div (-9) = w$

14. $18 \div (-2) = a$

15. $x = -21 \div 3$

16. $d = 32 \div 8$

17. $8 \div (-8) = y$

18. $c = -14 \div (-7)$

19. $-81 \div 9 = y$

20. $q = -81 \div (-9)$

21. $-49 \div (-7) = y$

Lesson 2-9 *(Pages 86–89)*
Solve each equation. Check your work.

1. $-4 + b = 12$

2. $z - 10 = -8$

3. $-7 = x + 12$

4. $a + 6 = -9$

5. $r \div 7 = -8$

6. $-2a = -8$

7. $r - (-8) = 14$

8. $0 = 6r$

9. $\frac{y}{12} = -6$

10. $m + (-2) = 6$

11. $3m = -15$

12. $c \div (-4) = 10$

13. $5 + q = 12$

14. $\frac{16}{x} = -4$

15. $-6f = -36$

16. $81 = -9w$

17. $t + 12 = 6$

18. $8 + p = 0$

19. $0.12 = -3h$

20. $12 - x = 8$

21. $14 + t = 10$

Lesson 2-9B *(Pages 90–91)*
Solve.

1. At a rock concert, 2,143 sweatshirts were sold at $18.75 each. The receipts for the sale of the sweatshirts were:
   a. $4,108
   b. $40,181.25
   c. $401,812.50

2. Daisuke takes about 12 breaths per minute. In a week he will take about how many breaths?

3. Minnesota has at least 10,000 lakes. Its total area is 84,402 square miles. On the average, Minnesota has 1 lake for how many square miles of area?
   a. 840
   b. 84
   c. 8.4

4. If Daniel has 400 quarters, how many dollars worth of quarters does he have?
   a. $10
   b. $100
   c. $1,000

5. 600 raffle tickets are sold. If there are 3 prizes and you buy 2 tickets, what is your chance of winning a prize?
   a. 1 in 10
   b. 1 in 100
   c. 1 in 1,000

Lesson 2-10 *(Pages 92–95)*
Name the ordered pair for the coordinates of each point graphed on the coordinate plane.

1. A

2. B

3. C

4. D

5. E

6. F

7. G

8. H

9. I

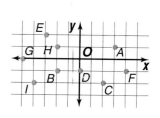

Graph each point on the same coordinate plane.

10. $J(3, -2)$

11. $K(2, 4)$

12. $L(-1, 6)$

13. $M(0, 5)$

14. $N(-2, -3)$

15. $P(-4, 0)$

16. $Q(4, -4)$

17. $R(0, 0)$

18. $S(3, 1)$

19. $T(-4, -1)$

Lesson 3-1 *(Pages 104–106)*
Express each ratio or rate in simplest form.

1. 27 to 9

2. 4 inches per foot

3. 16 out of 48

4. 10:50

5. 40 min. per hour

6. 35 is to 15

7. 16 wins, 16 losses

8. 7 out of 13

9. 5 out of 50

Express each rate as a unit rate.

10. $24 per dozen

11. 600 students to 30 teachers

12. 6 pounds gained in 12 weeks

13. $800 for 40 tickets

14. $6.50 for 5 pounds

15. 6 inches of rain in 3 weeks

Lesson 3-2 *(Pages 107–110)*
Express each ratio or fraction as a percent.

1. 3 out of 5

2. $\frac{1}{4}$

3. $\frac{7}{10}$

4. 39:100

5. 11 out of 25

6. 72.5:100

7. 3 out of 4

8. $\frac{1}{2}$

9. $\frac{7}{20}$

10. 93:100

Express each percent as a fraction in simplest form.

11. 30%

12. 4%

13. 20%

14. 85%

15. 3%

16. 80%

17. 17%

18. 55%

19. 82%

20. 48%

Lesson 3-3 *(Pages 111–113)*
Determine whether each pair of ratios forms a proportion.

1. $\frac{3}{5}, \frac{5}{10}$

2. $\frac{8}{4}, \frac{6}{3}$

3. $\frac{10}{15}, \frac{5}{3}$

4. $\frac{2}{8}, \frac{1}{4}$

5. $\frac{6}{18}, \frac{3}{9}$

6. $\frac{14}{21}, \frac{12}{18}$

7. $\frac{4}{20}, \frac{5}{25}$

8. $\frac{9}{27}, \frac{1}{3}$

Solve each proportion.

9. $\frac{2}{3} = \frac{a}{12}$

10. $\frac{7}{8} = \frac{c}{16}$

11. $\frac{3}{7} = \frac{21}{d}$

12. $\frac{2}{5} = \frac{18}{x}$

13. $\frac{3}{5} = \frac{n}{21}$

14. $\frac{5}{12} = \frac{b}{5}$

15. $\frac{4}{36} = \frac{2}{y}$

16. $\frac{3}{10} = \frac{z}{36}$

17. $\frac{2}{3} = \frac{t}{4}$

18. $\frac{9}{10} = \frac{r}{25}$

19. $\frac{16}{8} = \frac{y}{12}$

20. $\frac{7}{8} = \frac{a}{12}$

Lesson 3-4 *(Pages 114–117)*

Express each percent as a decimal.

1. 2%   **2.** 25%   **3.** 29%   **4.** 6.2%   **5.** 16.8%

6. 14%   **7.** 23.7%   **8.** 42%   **9.** 25.4%   **10.** 98%

Express each decimal as a percent.

11. 0.35   **12.** 14.23   **13.** 0.9   **14.** 0.13   **15.** 6.21

16. 0.23   **17.** 0.08   **18.** 0.036   **19.** 2.34   **20.** 0.39

Express each fraction as a percent.

21. $\frac{2}{5}$ **22.** $\frac{49}{50}$ **23.** $\frac{21}{50}$ **24.** $\frac{1}{3}$ **25.** $\frac{81}{100}$

26. $\frac{2}{25}$ **27.** $\frac{11}{20}$ **28.** $\frac{9}{75}$ **29.** $\frac{33}{40}$ **30.** $\frac{1}{50}$

Lesson 3-5 *(Pages 120–123)*

Compute mentally.

1. 10% of $206   **2.** 1% of 19.3   **3.** 20% of 15   **4.** 87.5% of 80

5. 50% of 46   **6.** 12.5% of 56   **7.** $33\frac{1}{3}$% of 93   **8.** 90% of 2,000

9. 30% of 70   **10.** 40% of 95   **11.** $66\frac{2}{3}$% of 48   **12.** 80% of 25

13. 25% of 400   **14.** 75% of 72   **15.** 37.5% of 96   **16.** 40% of 35

17. 60% of 85   **18.** 62.5% of 160   **19.** 90% of 205   **20.** 1% of 2,364

Lesson 3-6A *(Pages 124–125)*

Solve.

1. Paige wants to leave a 15% tip for the server at a restaurant. About how much should Paige leave if the meal cost $29.63?

2. In a survey of 1,138 college students, 4% said they would be willing to pay more for improved parking conditions. Is 4.5, 45 or 455 a reasonable estimate for the number of students willing to pay more?

3. You spend $15.03 plus $0.90 tax at the store for a new backpack and pay with a $20 bill. Would it be more reasonable to expect $3.00 or $4.00 in change?

4. Paddy wants to buy a pair of in-line skates that cost $89.95. The sales tax is 6%. What is a reasonable amount for the total cost of the skates plus tax?

5. An apple grower harvested 2,637 pounds of apples from one orchard and 1,826 pounds from another. What is a reasonable number of crates to have on hand if each crate holds 15 pounds of apples?

Lesson 3-6 (Pages 126–129)

Estimate the percent of the area shaded.

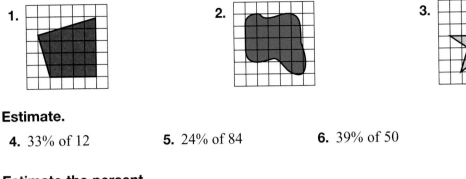

1. **2.** **3.**

Estimate.

4. 33% of 12   **5.** 24% of 84   **6.** 39% of 50   **7.** 1.5% of 135

Estimate the percent.

8. 11 out of 99   **9.** 28 out of 89   **10.** 9 out of 20   **11.** 25 out of 270

Lesson 4-1A (Pages 140–141)

1. The list shows prices of popular video games.

$24.99 $16.99 $44.99 $50.50 $35.99 $32.99 $10.99 $29.99 $29.99

$43.99 $45.99 $37.99 $14.99 $18.00 $34.89 $55.80 $37.90 $40.44

   a. Make a frequency table of the prices using $10 intervals.
   b. What is the most common interval of prices?

2. Use the frequency table. Which grade level has the greatest percent of students who play at least one sport?

Grade	Percent of Students Who Play at Least One Sport
9	46%
10	53%
11	71%
12	63%

3. Make a frequency table for the data.

12	21	4	15	22	5	9	6	19	11
6	0	2	7	6	17	13	1	24	12

Lesson 4-1 (Pages 142–146)

Use the histogram to answer each question.

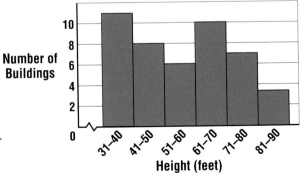

Heights of 45 Buildings

1. How large is each interval?

2. Which interval has the most buildings?

3. Which interval has the least buildings?

4. Compared to the total, how would you describe the number of buildings over 70 feet tall?

5. How does the number of buildings between 61 and 80 feet tall compare to the number of buildings between 31 and 50 feet tall?

Lesson 4-2 (Pages 148–151)

Make a circle graph for each set of data.

1.

Sporting Goods Sales	
Shoes	44%
Apparel	30%
Equipment	26%

2.

Energy Use in Home	
Heating/cooling	51%
Appliances	28%
Lights	21%

3.

Household income	
Primary job	82%
Secondary job	9%
Investments	5%
Other	4%

4.

Students in North High School	
White	30%
Black	28%
Hispanic	24%
Asian	18%

Lesson 4-3 (Pages 153–155)

Make a line plot for each set of data.

1. 8, 12, 10, 15, 11, 9, 12, 7, 14, 13, 8, 15, 17, 14, 11, 9, 8, 12, 15

2. 32, 41, 46, 38, 34, 51, 55, 49, 37, 42, 55, 46, 39, 58, 40, 35, 34, 52, 46

3. 161, 158, 163, 162, 165, 157, 159, 160, 163, 162, 158, 164, 161, 157, 166, 164

4. 78, 82, 83, 90, 58, 67, 95, 87, 97, 88, 75, 82, 78, 89, 86, 88, 79, 80, 91, 77

5. 49¢, 55¢, 77¢, 65¢, 51¢, 74¢, 68¢, 56¢, 49¢, 73¢, 62¢, 71¢, 54¢, 50¢, 70¢, 65¢

6. 2, 4, 12, 10, 2, 5, 7, 11, 7, 6, 3, 9, 12, 7, 5, 3, 7, 11, 2, 8, 7, 10, 8, 3, 7

7. 303, 298, 289, 309, 300, 294, 299, 301, 296, 308, 302, 289, 306, 308, 298, 299

8. 67, 73, 78, 61, 63, 77, 66, 75, 79, 66, 72, 69, 70, 74, 61, 63, 76, 64, 65, 78, 66

Lesson 4-4 (Pages 158–161)

Find the mean, median, and mode for each set of data. When necessary, round to the nearest tenth.

1. 2, 7, 9, 12, 5, 14, 4, 8, 3, 10

2. 58, 52, 49, 60, 61, 56, 50, 61

3. 122, 134, 129, 140, 125, 134, 137

4. 25.5, 26.7, 20.9, 23.4, 26.8, 24.0, 25.7

5. 36, 41, 43, 45, 48, 52, 54, 56, 56, 57, 60, 64, 65

6.

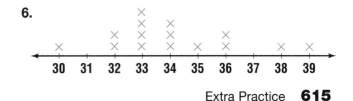

Lesson 4-5 *(Pages 163–166)*

Find the range, median, upper and lower quartiles, interquartile range and any outliers for each set of data.

1. 15, 12, 21, 18, 25, 11, 17, 19, 20

2. 2, 24, 6, 13, 8, 6, 11, 4

3. 189, 149, 155, 290, 141, 152

4. 451, 501, 388, 428, 510, 480, 390

5. 22, 18, 9, 26, 14, 15, 6, 19, 28

6. 245, 218, 251, 255, 248, 241, 250

7. 46, 45, 50, 40, 49, 42, 64

8. 128, 148, 130, 142, 164, 120, 152, 202

9. 2, 3, 2, 6, 4, 14, 13, 2, 6, 3

10. 88, 84, 92, 93, 90, 96, 87, 97

11. 378, 480, 370, 236, 361, 394, 345, 328, 388, 339

Lesson 4-6 *(Pages 168–170)*

Determine whether a scatter plot of the data below would show a *positive*, *negative*, or *no* relationship.

1. height and hair color

2. hours spent studying and test scores

3. income and month of birth

4. price of oranges and number available

5. size of roof and number of shingles

6. number of clouds and number of stars seen

7.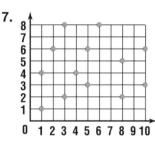

8.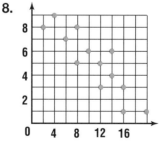

9.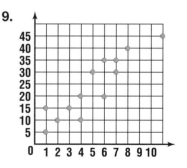

Lesson 4-7 *(Pages 171–173)*

Choose the most appropriate type of display for each data set and situation. Explain your choice.

1. number of fax machines in surveyed homes in newspaper article

2. the amount of sales by different people in a company for promotion decisions

3. ages of amusement park attendees in marketing information for the park

4. proficiency test scores for five consecutive years for comparing teaching techniques

5. numbers of Americans who own motorcycles, boats, and recreational vehicles in newspaper article

6. percent of people who own a certain type of car compared to all car owners in advertising

Lesson 4-8 *(Pages 174–177)*

Decide whether each location is a good place to find a representative sample for the selected survey. Justify your answer.

1. favorite kind of music at a jazz club

2. the distance students travel to school on a school bus

3. career choice in the student union of a large university

4. favorite fast food restaurant at a fast food restaurant

5. whether people own cats in a telephone poll

6. teenagers' favorite compact disc at three different middle schools

7. The two graphs at the right show the number of students in each grade who completed an obstacle course within the required time limit.
   a. Do both graphs contain the same information? Explain.
   b. One of the graphs is misleading. Explain why it is misleading.

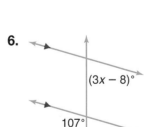

Lesson 5-1 *(Pages 188–192)*

Use the figure at the right for Exercises 1–4.

1. Find $m\angle 6$, if $m\angle 3 = 42°$.

2. Find $m\angle 4$, if $m\angle 3 = 71°$.

3. Find $m\angle 1$, if $m\angle 5 = 128°$.

4. Find $m\angle 7$, if $m\angle 2 = 83°$.

Find the value of x in each figure.

5.

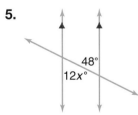

6.

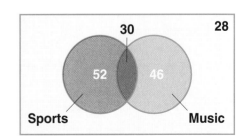

Lesson 5-2A *(Pages 194–195)*

1. The Venn diagram shows the number of eighth grade students at Franklin Middle School who participate in sports and music activities.
   a. How many eighth graders participate in sports?
   b. How many participate in musical events?
   c. How many do not participate in sports or music?

2. A tour bus is stopping at Calera City and Watertown. Fifteen people on the bus are only going to Calera City and seventeen are going only to Watertown. Twenty-two people are visiting both cities.
   a. Draw a Venn diagram that will represent this information.
   b. How many people are on the tour bus?

Lesson 5-2 *(Pages 196–199)*

Classify each triangle by its angles and by its sides.

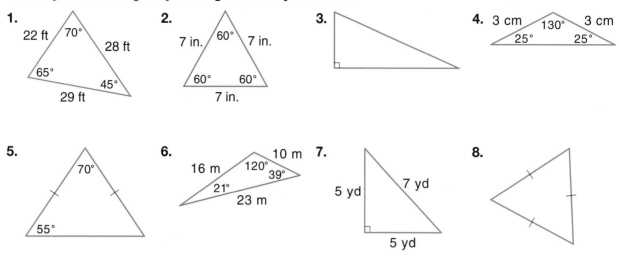

1.
22 ft / 70° / 28 ft / 65° / 45° / 29 ft

2.
7 in. / 60° / 7 in. / 60° / 60° / 7 in.

3.

4.
3 cm / 130° / 3 cm / 25° / 25°

5.
70° / 55°

6.
10 m / 16 m / 120° / 39° / 21° / 23 m

7.
5 yd / 7 yd / 5 yd

8.

Lesson 5-3 *(Pages 201–204)*

Sketch each figure. Let Q = quadrilateral, P = parallelogram, R = rectangle, S = square, RH = rhombus, and T = trapezoid. Write all of the letters that describe it inside the figure.

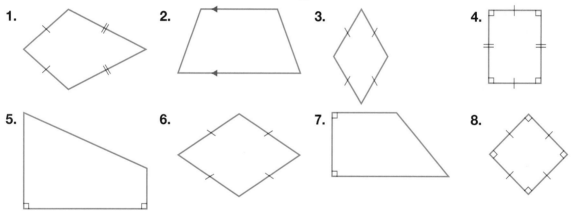

1.   2.   3.   4.

5.   6.   7.   8.

Lesson 5-4 *(Pages 206–209)*

Trace each figure. Determine if the figure has line symmetry. If so, draw the lines of reflection .

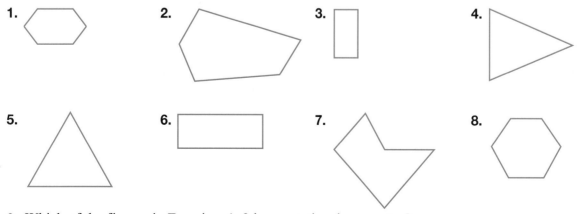

1.   2.   3.   4.

5.   6.   7.   8.

9. Which of the figures in Exercises 1–8 have rotational symmetry?

Lesson 5-5 *(Pages 210–212)*

Determine whether each pair of triangles is congruent. If so, write a congruence statement and tell why the triangles are congruent.

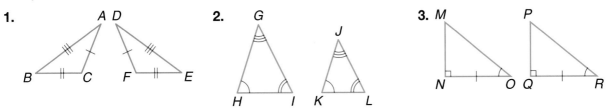

1.   A D
2.   G J
3.   M P

Find the value of *x* in each pair of congruent triangles.

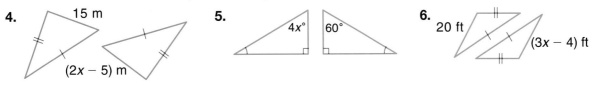

4. 15 m   (2*x* − 5) m
5. 4*x*°   60°
6. 20 ft   (3*x* − 4) ft

Lesson 5-6 *(Pages 215–218)*

Tell whether each pair of triangles is *congruent*, *similar*, or *neither*. Justify your answer.

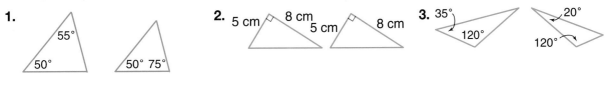

1.   55°   50°   50° 75°
2. 5 cm   8 cm   5 cm   8 cm
3. 35°   120°   20°   120°

Find the value of *x* in each pair of similar triangles.

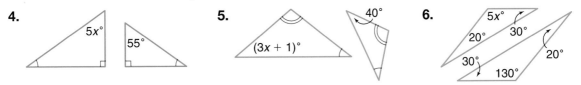

4. 5*x*°   55°
5. (3*x* + 1)°   40°
6. 5*x*°   30°   20°   30°   130°   20°

Lesson 5-7 *(Pages 220–223)*

Make an Escher-like drawing for each pattern described. For squares, use a tessellation of two rows of three squares as the base. For triangles, use a tessellation of two rows of five equilateral triangles as your base.

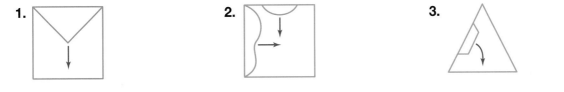

1.
2.
3.

Tell whether each pattern involves *translations* or *rotations*.

4. Exercise 1
5. Exercise 2
6. Exercise 3

Lesson 6-1 *(Pages 232-234)*
Determine whether each number is divisible by 2, 3, 4, 5, 6, 8, 9, or 10.

1. 210   **2.** 614   **3.** 985   **4.** 756

5. 432   **6.** 96   **7.** 87   **8.** 113

9. Is 6 a factor of 936?   **10.** Is 9 a factor of 216?

11. Is 1,346 divisible by 2?   **12.** Is 1,249 divisible by 8?

13. Is 752 divisible by 6?   **14.** Is 3 a factor of 1,687?

15. Is 4 a factor of 208?   **16.** Is 448 divisible by 8?

Lesson 6-2 *(Pages 235-238)*
Determine whether each number is *prime*, *composite*, or *neither*.

1. 17   **2.** 1,258   **3.** 37   **4.** 483   **5.** 97

6. 0   **7.** 25   **8.** 61   **9.** 45   **10.** 419

Find the prime factorization of each number.

11. 20   **12.** 65   **13.** 52   **14.** 30

15. 28   **16.** 72   **17.** 155   **18.** 50

19. 96   **20.** 201   **21.** 1,250   **22.** 2,648

Lesson 6-3A *(Pages 240-241)*
Solve.

1. Joshua and Martin are helping to make and sell submarine sandwiches to raise money for the marching band. They can add cheese, mushrooms, and/or green peppers to the basic submarine sandwich. How many different sandwiches can Joshua and Martin advertise?

2. Katrina wants to buy a soda from a vending machine. The soda costs $0.65. If Katrina uses exact change, how many different combinations of nickels, dimes, and quarters can she use?

3. How many three digit numbers can be formed from the digits 6, 7, 8, and 9 if no digit is repeated?

4. Shaney has a basket containing 21 muffins to share with friends. She takes the first one and each friend takes one as they continue around the table. There is one muffin left when the basket returns to Shaney. How many people could have shared the muffins if the basket can go around the table more than once?

Lesson 6-3 *(Pages 242-244)*
Find the GCF for each set of numbers.

1. 8, 18   **2.** 6, 9   **3.** 4, 12   **4.** 18, 24

5. 8, 24   **6.** 17, 51   **7.** 65, 95   **8.** 42, 48

9. 64, 32   **10.** 72, 144   **11.** 54, 72   **12.** 60, 75

13. 16, 24   **14.** 12, 27   **15.** 25, 30   **16.** 48, 60

17. 16, 20, 36   **18.** 12, 18, 42   **19.** 30, 45, 15

20. 20, 30, 40   **21.** 81, 27, 108   **22.** 9, 18, 12

Lesson 6-4 *(Pages 245-248)*
Name all sets of numbers to which each number belongs.

1. -7 **2.** $-6\frac{5}{9}$ **3.** 7.4   **4.** 86   **5.** -0.23

Write each fraction in simplest form.

6. $\frac{12}{16}$ **7.** $\frac{28}{32}$ **8.** $\frac{75}{100}$ **9.** $\frac{8}{16}$ **10.** $\frac{6}{18}$

11. $\frac{27}{36}$ **12.** $\frac{16}{64}$ **13.** $\frac{8}{32}$ **14.** $\frac{50}{100}$ **15.** $\frac{24}{40}$

16. $\frac{32}{80}$ **17.** $\frac{8}{24}$ **18.** $\frac{20}{25}$ **19.** $\frac{4}{10}$ **20.** $\frac{3}{5}$

21. $\frac{14}{19}$ **22.** $\frac{9}{12}$ **23.** $\frac{6}{8}$ **24.** $\frac{15}{18}$ **25.** $\frac{9}{20}$

Lesson 6-5 *(Pages 249–252)*
Express each decimal using bar notation.

1. 0.161616...   **2.** 0.12351235...   **3.** 0.6666...   **4.** 0.15151...

Write the first ten decimal places of each decimal.

5. $0.\overline{09}$ **6.** $0.\overline{076923}$ **7.** $0.8\overline{4563}$ **8.** $0.98\overline{745}$

Express each fraction or mixed number as a decimal.

9. $\frac{2}{5}$ **10.** $2\frac{3}{11}$ **11.** $-\frac{3}{4}$ **12.** $\frac{5}{7}$ **13.** $\frac{3}{4}$

14. $-\frac{2}{3}$ **15.** $\frac{17}{20}$ **16.** $\frac{14}{25}$ **17.** $-6\frac{7}{10}$ **18.** $\frac{7}{11}$

Express each decimal as a fraction or mixed number in simplest form.

19. 0.5   **20.** $0.\overline{8}$ **21.** 0.32   **22.** -0.75 **23.** $2.\overline{2}$

24. $0.\overline{38}$ **25.** -0.486 **26.** 20.08   **27.** -9.36 **28.** $10.1\overline{8}$

Lesson 6-6 *(Pages 253–256)*

A date is chosen at random from the calendar below. Find the probability of choosing each date.

1. The date is the thirteenth.

2. The date is Friday.

3. It is after the twenty-fifth.

4. It is before the seventh.

5. It is an odd-numbered date.

6. The date is divisible by 3.

		November				
S	**M**	**T**	**W**	**T**	**F**	**S**
		1	2	3	4	5
6	7	8	9	10	11	12
13	14	15	16	17	18	19
20	21	22	23	24	25	26
27	28	29	30			

Lesson 6-7 *(Pages 257–259)*

List the first six multiples of each number or algebraic expression.

1. 6
2. 9
3. 15
4. $3n$

Find the LCM of each set of numbers.

5. 5, 15
6. 13, 39
7. 16, 24

8. 18, 20
9. 21, 14
10. 25, 30

11. 28, 42
12. 7, 13
13. 6, 30

14. 12, 42
15. 8, 10
16. 30, 10

17. 12, 18, 6
18. 15, 75, 25
19. 6, 10, 15

20. 3, 6, 9
21. 21, 14, 6
22. 12, 35, 10

Lesson 6-8 *(Pages 261–264)*

Replace each ● with a <, >, or = to make a true sentence.

1. -5.6 ● 4.2
2. 4.256 ● 4.25
3. 0.233 ● $0.\overline{23}$

4. $\frac{5}{7}$ ● $\frac{2}{5}$
5. $\frac{6}{7}$ ● $\frac{7}{9}$
6. $\frac{2}{3}$ ● $\frac{2}{5}$

7. $\frac{3}{8}$ ● 0.375
8. $-\frac{1}{2}$ ● 0.5
9. 12.56 ● $12\frac{3}{8}$

Order each set of rational numbers from least to greatest.

10. $0.24, 0.2, 0.245, 2.24, 0.25$

11. $0.\overline{3}, 0.3, 0.3\overline{4}, 0.\overline{34}, 0.33$

12. $\frac{2}{5}, \frac{2}{3}, \frac{2}{7}, \frac{2}{9}, \frac{2}{1}$

13. $\frac{1}{2}, \frac{5}{7}, \frac{2}{9}, \frac{8}{9}, \frac{6}{6}$

Lesson 6-9 *(Pages 265–267)*

Express each number in standard form.

1. 4.5×10^3

2. 2×10^4

3. 1.725896×10^6

4. 9.61×10^2

5. 1×10^7

6. 8.256×10^8

7. 5.26×10^4

8. 3.25×10^2

9. 6.79×10^5

Express each number in scientific notation.

10. 720

11. 7,560

12. 892

13. 1,400

14. 91,256

15. 51,000

16. 145,600

17. 90,100

18. 123,568,000,000

Lesson 7-1 *(Pages 278–280)*

Solve each equation. Write the solution in simplest form.

1. $\frac{17}{21} + \left(-\frac{13}{21}\right) = m$

2. $t = \frac{5}{11} + \frac{6}{11}$

3. $k = -\frac{8}{13} + \left(-\frac{11}{13}\right)$

4. $-\frac{7}{12} + \frac{5}{12} = a$

5. $\frac{13}{28} - \frac{9}{28} = g$

6. $b = -1\frac{2}{9} - \frac{7}{9}$

7. $r = \frac{15}{16} + \frac{13}{16}$

8. $2\frac{1}{3} - \frac{2}{3} = n$

9. $-\frac{4}{35} - \left(-\frac{17}{35}\right) = c$

10. $\frac{3}{8} + \left(-\frac{5}{8}\right) = w$

11. $s = \frac{8}{15} - \frac{2}{15}$

12. $d = -2\frac{4}{7} - \frac{3}{7}$

13. $-\frac{29}{9} - \left(-\frac{26}{9}\right) = y$

14. $2\frac{3}{5} + 7\frac{3}{5} = z$

15. $x = \frac{5}{18} - \frac{13}{18}$

16. $j = -2\frac{2}{7} + \left(-1\frac{6}{7}\right)$

17. $p = -\frac{3}{10} + \frac{7}{10}$

18. $\frac{4}{11} + \frac{9}{11} = f$

Lesson 7-2 *(Pages 281–284)*

Solve each equation. Write the solution in simplest form.

1. $r = \frac{7}{12} + \frac{7}{24}$

2. $-\frac{3}{4} + \frac{7}{8} = z$

3. $\frac{2}{5} + \left(-\frac{2}{7}\right) = q$

4. $d = -\frac{3}{5} - \left(-\frac{5}{6}\right)$

5. $\frac{5}{24} - \frac{3}{8} = j$

6. $g = -\frac{7}{12} - \frac{3}{4}$

7. $-\frac{3}{8} + \left(-\frac{4}{5}\right) = x$

8. $t = \frac{2}{15} + \left(-\frac{3}{10}\right)$

9. $r = -\frac{2}{9} - \left(-\frac{2}{3}\right)$

10. $a = -\frac{7}{15} - \frac{5}{12}$

11. $\frac{3}{8} + \frac{7}{12} = s$

12. $-2\frac{1}{4} + \left(-1\frac{1}{3}\right) = m$

13. $3\frac{2}{5} - 3\frac{1}{4} = v$

14. $b = \frac{3}{4} + \left(-\frac{4}{15}\right)$

15. $f = -1\frac{2}{3} + 4\frac{3}{4}$

16. $-\frac{1}{8} - 2\frac{1}{2} = n$

17. $p = 3\frac{2}{5} - 1\frac{1}{3}$

18. $y = 5\frac{1}{3} + \left(-8\frac{3}{7}\right)$

Lesson 7-3 *(Pages 286–289)*

Solve each equation. Write the solution in simplest form.

1. $\frac{2}{11} \cdot \frac{3}{4} = m$

2. $4\left(-\frac{7}{8}\right) = r$

3. $d = -\frac{4}{7} \cdot \frac{3}{5}$

4. $g = \frac{6}{7}\left(-\frac{7}{12}\right)$

5. $b = \frac{7}{8} \cdot \frac{1}{3}$

6. $\frac{3}{4} \cdot \frac{4}{5} = t$

7. $-1\frac{1}{2} \cdot \frac{2}{3} = k$

8. $x = \frac{5}{6} \cdot \frac{6}{7}$

9. $c = 8\left(-2\frac{1}{4}\right)$

10. $-3\frac{3}{4} \cdot \frac{8}{9} = q$

11. $\frac{10}{21} \cdot -\frac{7}{8} = n$

12. $w = -1\frac{4}{5}\left(-\frac{5}{6}\right)$

13. $a = 5\frac{1}{4} \cdot 6\frac{2}{3}$

14. $-8\frac{3}{4} \cdot 4\frac{2}{5} = p$

15. $y = 6 \cdot 8\frac{2}{3}$

16. $n = \left(\frac{3}{5}\right)^2$

17. $-4\frac{1}{5}\left(-3\frac{1}{3}\right) = h$

18. $-8 \cdot \left(\frac{3}{4}\right)^2 = v$

Lesson 7-4 *(Pages 290–292)*

Name the multiplicative inverse of each of the following.

1. 3

2. -5

3. $\frac{2}{3}$

4. $2\frac{1}{8}$

5. $\frac{a}{b}$

6. -8

7. $\frac{1}{15}$

8. 0.75

9. c

10. $-\frac{3}{5}$

Solve each equation using properties of rational numbers.

11. $g = 9 \cdot 2\frac{2}{3}$

12. $-\frac{4}{5}\left(\frac{5}{6} \cdot \frac{3}{7}\right) = m$

13. $k = 12\frac{1}{2} \cdot \frac{3}{4}$

14. $w = -5 \cdot 4\frac{1}{4}$

15. $\frac{3}{8} \cdot 16\frac{1}{3} = d$

16. $q = \left(\frac{3}{7} \cdot 2\frac{1}{3}\right) \cdot \frac{7}{9}$

Evaluate each expression if $a = \frac{1}{2}$, $b = -\frac{3}{4}$, and $c = -2\frac{1}{6}$.

17. a^2

18. abc

19. $4a + 4b$

20. ac

Lesson 7-5A *(Pages 294–295)*

Solve.

1. The sum of two consecutive prime numbers is 966. What are the two numbers?

2. Ms. Flores started a job making $2,000 per month. At the same time, Mr. Williams started a job making $2,400 per month. If Ms. Flores gets a $400 per month raise each year and Mr. Williams gets a $300 per month raise each year, in how many years will they both be earning the same amount of money per month?

3. Daisuke started a savings account in January. He puts $100 in his savings account each month. In February, Liza started a savings account. She puts $120 in her savings account each month. In what month will Daisuke and Liza have the same amount of money in their savings accounts?

4. Sergio deposits $1 into his savings account the first week, $3 the second week, $5 the third week, $7 the fourth week, and so on. How much money will he have in his savings account on the 20th week?

Extra Practice

Lesson 7-5 *(Pages 296–299)*

Identify each sequence as *arithmetic*, *geometric* or *neither*. Then find the next three terms.

1. 1, 5, 9, 13, ...

2. 2, 6, 18, 54, ...

3. 1, 4, 9, 16, 25, ...

4. 729, 243, 81, ...

5. 2, −3, −8, −13, ...

6. 5, −5, 5, −5, ...

7. 810, −270, 90, −30, ...

8. 11, 14, 17, 20, 23, ...

9. 33, 27, 21, ...

10. 21, 15, 9, 3, ...

11. $\frac{1}{8}, -\frac{1}{4}, \frac{1}{2}, -1, ...$

12. $\frac{1}{81}, \frac{1}{27}, \frac{1}{9}, \frac{1}{3}, ...$

13. $\frac{3}{4}, 1\frac{1}{2}, 3, ...$

14. 2, 5, 9, 14, ...

15. $-1\frac{1}{4}, -1\frac{3}{4}, -2\frac{1}{4}, -2\frac{3}{4}, ...$

16. 9.9, 13.7, 17.5, ...

17. $\frac{1}{2}, 1\frac{1}{2}, 2\frac{1}{2}, 3\frac{1}{2}, ...$

18. 2, 12, 32, 62, ...

19. 3, −6, 12, −24, ...

20. 5, 7, 9, 11, 13, ...

21. −0.06, 2.24, 4.54, ...

22. 7, 14, 28, ...

23. −5.4, −1.4, 2.6, ...

24. −96, 48, −24, 12, ...

25. 4, 12, 36, ...

26. 20, 19, 18, 17, ...

27. 768, 192, 48, ...

Lesson 7-6 *(Pages 301–304)*

State the measures of the base(s) and the height of each triangle or trapezoid. Then find the area.

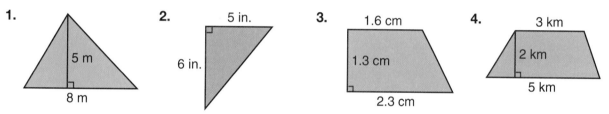

Find the area of each figure.

5. triangle: base, $2\frac{1}{2}$ in.; height, 7 in.

6. triangle: base, 12 cm; height, 3.2 cm

7. trapezoid: bases, 5 ft and 7 ft; height, 11 ft

8. trapezoid: bases, $4\frac{1}{4}$ yd and $3\frac{1}{2}$ yd; height, 5 yd

Lesson 7-7 *(Pages 309–311)*

Find the circumference of each circle to the nearest tenth. Use $\frac{22}{7}$ or 3.14 for π.

1. 14 mm, diameter

2. 18 cm, diameter

3. 24 in., radius

4. 42 m, diameter

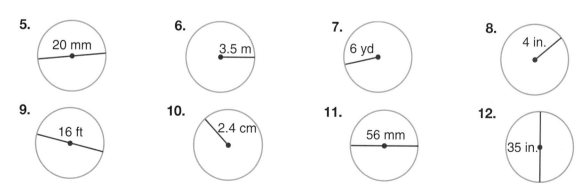

Lesson 7-8 *(Pages 312–314)*

Solve each equation. Write the solution in simplest form.

1. $x = \frac{2}{3} \div \frac{3}{4}$

2. $-\frac{4}{9} \div \frac{5}{6} = c$

3. $\frac{7}{12} \div \frac{3}{8} = q$

4. $m = \frac{5}{18} \div \frac{2}{9}$

5. $a = \frac{1}{3} \div 4$

6. $5\frac{1}{4} \div \left(-2\frac{1}{2}\right) = g$

7. $-6 \div \left(-\frac{4}{7}\right) = d$

8. $n = -6\frac{3}{8} \div \frac{1}{4}$

9. $p = \frac{6}{7} \div \frac{3}{5}$

10. $g = 3\frac{1}{3} \div (-4)$

11. $2\frac{5}{12} \div 7\frac{1}{3} = r$

12. $v = \frac{5}{6} \div 1\frac{1}{9}$

13. $\frac{3}{8} \div (-6) = b$

14. $k = \frac{5}{8} \div \frac{1}{6}$

15. $4\frac{1}{4} \div 6\frac{3}{4} = w$

16. $f = 4\frac{1}{6} \div 3\frac{1}{8}$

17. $t = 8 \div \left(-1\frac{4}{5}\right)$

18. $j = -5 \div \frac{2}{7}$

19. $\frac{3}{5} \div \frac{6}{7} = y$

20. $4\frac{8}{9} \div \left(-2\frac{2}{3}\right) = h$

21. $f = 8\frac{1}{6} \div 3$

22. $k = -\frac{3}{4} \div 9$

23. $s = 1\frac{11}{14} \div 2\frac{1}{2}$

24. $-2\frac{1}{4} \div \frac{4}{5} = z$

Lesson 7-9 *(Pages 315–317)*

Solve each equation. Check your solution.

1. $434 = -31y$

2. $6x = -4.2$

3. $\frac{3}{4}a = -12$

4. $-10 = \frac{b}{-7}$

5. $7.2 = \frac{3}{4}c$

6. $2r + 4 = 14$

7. $-2.4n = 7.2$

8. $7 = \frac{1}{2}d - 3$

9. $3.2n - 0.64 = -5.44$

10. $\frac{t}{3} - 7 = 2$

11. $\frac{3}{8} = \frac{1}{2}x$

12. $\frac{1}{2}h - 3 = -14$

13. $-0.46k - 1.18 = 1.58$

14. $4\frac{1}{2}s = -30$

15. $\frac{2}{3}f = \frac{8}{15}$

16. $\frac{2}{3}m + 10 = 22$

17. $\frac{2}{3}g + 4 = 4\frac{5}{6}$

18. $7 = \frac{1}{2}v + 3$

19. $\frac{g}{1.2} = -6$

20. $\frac{4}{7}z - 4\frac{5}{8} = 15\frac{3}{8}$

21. $-12 = \frac{1}{5}j$

Lesson 7-10 *(Pages 318–321)*

Solve each inequality. Graph the solution on a number line.

1. $5p < 25$

2. $c - 9 < 4$

3. $y + 4 \geq 6$

4. $-4 > -\frac{1}{3}k$

5. $2.5 \leq \frac{h}{0.4}$

6. $-\frac{4}{5}z > \frac{2}{5}$

7. $f + \frac{1}{5} \geq 4\frac{3}{5}$

8. $6 + 5x > 16$

9. $14 \leq 2x + 4$

10. $-2t + 28 < 5\frac{3}{5}$

11. $\frac{7}{8}n - \frac{1}{6} > \frac{3}{4}$

12. $\frac{4}{2.6}h - 11 \leq 6.8$

13. $\frac{2x + 4}{5} \geq 3$

14. $\frac{1}{3}j + 7 > -2\frac{1}{3}$

15. $2.4 < -1.8q + 5$

Lesson 8-1 *(Pages 330–333)*
Write a proportion to solve each problem. Then solve.

1. On a radar screen the distance between two planes is $3\frac{1}{2}$ inches. If the scale is 1 inch on the screen to 2 miles in the air, what is the actual distance between the two planes.

2. A car travels 144 miles on 4 gallons of gasoline. At this rate, how many gallons are needed to drive 450 miles?

3. A park ranger stocks a pond with 4 sunfish for every three perch. Suppose 296 sunfish are put in the pond, how many perch should be stocked?

4. A furniture store bought 8 identical sofas for $4,000. How much did each sofa cost?

Lesson 8-2 *(Pages 335–338)*
Express each fraction as a percent.

1. $\frac{2}{100}$ 2. $\frac{3}{25}$ 3. $\frac{20}{25}$ 4. $\frac{10}{16}$ 5. $\frac{4}{6}$ 6. $\frac{1}{4}$

7. $\frac{26}{100}$ 8. $\frac{3}{10}$ 9. $\frac{21}{50}$ 10. $\frac{7}{8}$ 11. $\frac{1}{3}$ 12. $\frac{2}{3}$

13. $\frac{2}{5}$ 14. $\frac{2}{50}$ 15. $\frac{8}{10}$ 16. $\frac{5}{12}$ 17. $\frac{7}{10}$ 18. $\frac{9}{20}$

19. $\frac{1}{2}$ 20. $\frac{3}{20}$ 21. $\frac{10}{25}$ 22. $\frac{3}{8}$ 23. $\frac{4}{20}$ 24. $\frac{19}{25}$

Write a percent proportion to solve each problem. Then solve.
Round answers to the nearest tenth.

25. 39 is 5% of what number?

26. What is 19% of 200?

27. 28 is what percent of 7?

28. 24 is what percent of 72?

29. 9 is $33\frac{1}{3}$% of what number?

30. Find 55% of 134.

Lesson 8-3 *(Pages 339–341)*
Write an equation in the form P = RB for each problem. Then solve.

1. Find 5% of $73.

2. What is 15% of 15?

3. Find 80% of $12.

4. What is 7.3% of 500?

5. Find 21% of $720.

6. What is 12% of $62.50?

7. Find 0.3% of 155.

8. What is 75% of $450?

9. Find 7.2% of 10.

10. What is 10.1% of $60?

11. Find 23% of 47.

12. What is 89% of 654?

13. $20 is what percent of $64?

14. Sixty-nine is what percent of 200?

15. Seventy is what percent of 150?

16. 26 is 30% of what number?

17. 7 is 14% of what number?

18. $35.50 is what percent of $150?

19. $17 is what percent of $25?

20. 152 is 2% of what number?

Lesson 8-3B *(Pages 342–343)*
Solve.

1. Out of 25 eighth-graders surveyed, 16 said that science was their favorite subject. If there are 173 eighth-graders at Monroe Middle School, about how many will choose science as their favorite subject?

2. What is 40% of 725?

3. What is the sum of the whole numbers from 1 to 100?

4. Thirty percent of the 320 people who visited the Museum of Natural History today were students. How many of the visitors were students?

5. How many cuts are needed to separate a long board into 25 smaller boards?

Lesson 8-4 *(Pages 344–347)*
Express each percent as a fraction or mixed number in simplest form.

1. 540%
2. $\frac{25}{50}$%
3. 0.02%
4. 620%
5. 0.7%

6. 111.5%
7. $\frac{7}{35}$%
8. 0.72%
9. 0.004%
10. 364%

11. 0.15%
12. 1,250%
13. $\frac{9}{10}$%
14. 730%
15. 100.01%

Express each percent as a decimal. Round to the nearest ten-thousandth.

16. 0.07%
17. $\frac{2}{3}$%
18. 310%
19. 6.05%
20. 7,652%

21. $\frac{12}{50}$%
22. 0.93%
23. 200%
24. 197.6%
25. 10.75%

26. 0.66%
27. 417%
28. 7.76%
29. 390%
30. $\frac{7}{10}$%

Lesson 8-5 *(Pages 348–351)*
Find each percent of change. Round to the nearest percent.

1. original: $35
   new: $29
2. original: $550
   new: $425
3. original: $72
   new: $88
4. original: $25
   new: $35

5. original: $28
   new: $19
6. original: $46
   new: $55
7. original: $78
   new: $44
8. original: $120
   new: $75

Find the sale price of each item to the nearest cent.

9. $4,220 piano, 35% off
10. $14 scissors, 10% off
11. $29 book, 40% off

12. $38 sweater, 25% off
13. $45 pants, 50% off
14. $280 VCR, 25% off

15. $3,540 motorcycle, 30% off
16. $15.95 compact disc, 20% off

Find the selling price for each item given the amount paid by the store and markup. Round to the nearest cent.

17. $250.00 golf clubs, 30% markup
18. $17.00 compact disc, 15% markup

19. $57.00 shoes, 45% markup
20. $26.00 book, 20% markup

Lesson 8-6 *(Pages 353–356)*

Find the simple interest to the nearest cent.

1. $500 at 7% for 2 years

2. $2,500 at 6.5% for 36 months

3. $8,000 at 6% for 1 year

4. $1,890 at 9% for 42 months

5. $760 at 4.5% for $2\frac{1}{2}$ years

6. $12,340 at 5% for 6 months

Find the total amount in each account to the nearest cent.

7. $300 at 10% for 3 years

8. $3,200 at 8% for 6 months

9. $20,000 at 14% for 20 years

10. $4,000 at 12.5% for 4 years

11. $450 at 11% for 5 years

12. $17,000 at 15% for $9\frac{1}{2}$ years

Lesson 8-7 *(Pages 357–360)*

Tell whether each pair of polygons is similar. Explain your reasoning.

1. 4 cm, 2 cm, 5 cm, 10 cm

2. 5.1 m, 4.6 m, 2.2 m, 4 m, 5 m, 3 m

Each pair of polygons is similar. Write a proportion to find the missing measure *x*. Then find the value of *x*.

3. 6 in., *x* in., 12 in., 8 in.

4. 12 cm, 16 cm, *x* cm, 8 cm, 10 cm

Lesson 8-8 *(Pages 361–364)*

Write a proportion for each problem and then solve it. Assume the triangles are similar.

1. A road sign casts a shadow 14 meters long, while a tree nearby casts a shadow 27.8 meters long. If the road sign is 3.5 meters high, how tall is the tree?

2. Find the distance across Catfish Lake.

3. A 7-foot tall flag stick on a golf course casts a shadow 21 feet long. A golfer standing nearby casts a shadow 16.5 feet long. How tall is the golfer?

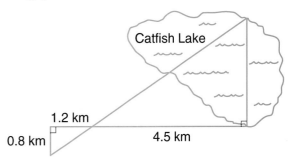

Catfish Lake

1.2 km

0.8 km

4.5 km

Lesson 8-9 *(Pages 366–369)*
Solve.

1. The distance between two cities on a map is 3.2 centimeters. If the scale on the map is 1 centimeter = 50 miles, find the actual distance between the two cities.

2. A scale model of the Empire State Building is 10 inches tall. If the Empire State Building is 1,250 feet tall, find the scale of this model.

3. On a scale drawing of a house, the dimensions of the living room are 4 inches by 3 inches. If the scale of the drawing is 1 inch = 6 feet, find the actual dimensions of the living room.

4. The trip from Columbus to Dayton is approximately 70 miles. If a scale on an Ohio map is 1 inch = 11 miles, about how far apart are the cities on the map?

Lesson 8-10 *(Pages 370–373)*
Find the coordinates of the image of each point for a dilation with a scale factor $\frac{5}{4}$.

1. $X(6, 8)$
2. $Y(4, 12)$
3. $Z(10, 2)$

Triangle *ABC* has vertices *A*(2, 2), *B*(−1, 4), and *C*(−3, −5). Find the coordinates of its image for a dilation with each given scale factor. Graph △*ABC* and its dilation.

4. 1
5. 0.5
6. 2
7. 3
8. $\frac{1}{4}$

In each figure, the green figure is a dilation of the blue figure. Find each scale factor.

9.

10.

Lesson 9-1 *(Pages 382–384)*
Find each square root.

1. $\sqrt{9}$
2. $\sqrt{0.16}$
3. $\sqrt{81}$
4. $\sqrt{0.04}$
5. $-\sqrt{625}$
6. $\sqrt{36}$
7. $-\sqrt{169}$
8. $\sqrt{144}$
9. $\sqrt{2.25}$
10. $\sqrt{961}$
11. $\sqrt{324}$
12. $-\sqrt{225}$
13. $\sqrt{0.01}$
14. $-\sqrt{4}$
15. $-\sqrt{0.09}$
16. $\sqrt{529}$
17. $-\sqrt{484}$
18. $\sqrt{196}$
19. $\sqrt{0.49}$
20. $\sqrt{1.69}$
21. $\sqrt{729}$
22. $\sqrt{0.36}$
23. $\sqrt{289}$
24. $-\sqrt{16}$
25. $\sqrt{1,024}$
26. $\sqrt{\dfrac{289}{10,000}}$
27. $\sqrt{\dfrac{169}{121}}$
28. $-\sqrt{\dfrac{4}{9}}$
29. $-\sqrt{\dfrac{81}{64}}$
30. $\sqrt{\dfrac{25}{81}}$

Lesson 9-2 (Pages 386–389)

Estimate to the nearest whole number.

1. $\sqrt{229}$
2. $\sqrt{63}$
3. $\sqrt{290}$
4. $\sqrt{27}$
5. $\sqrt{1.30}$
6. $\sqrt{8.4}$
7. $\sqrt{96}$
8. $\sqrt{19}$
9. $\sqrt{200}$
10. $\sqrt{76}$
11. $\sqrt{17}$
12. $\sqrt{34}$
13. $\sqrt{137}$
14. $\sqrt{540}$
15. $\sqrt{165}$
16. $\sqrt{326}$
17. $\sqrt{52}$
18. $\sqrt{37}$
19. $\sqrt{79}$
20. $\sqrt{18.35}$
21. $\sqrt{71}$
22. $\sqrt{117}$
23. $\sqrt{410}$
24. $\sqrt{25.70}$
25. $\sqrt{333}$
26. $\sqrt{23}$
27. $\sqrt{89}$
28. $\sqrt{47}$

Lesson 9-3 (Pages 390–394)

Name the set or sets of numbers to which each real number belongs. Let R = real numbers, Q = rational numbers, Z = integers, W = whole numbers, and I = irrational numbers.

1. 6.5
2. $\sqrt{25}$
3. $\sqrt{3}$
4. -7.2
5. $-0.\overline{61}$

Find an estimate for each square root. Then graph the square root on a number line.

6. $-\sqrt{12}$
7. $\sqrt{23}$
8. $\sqrt{2}$
9. $\sqrt{10}$
10. $-\sqrt{30}$

Solve each equation. Round solutions to the nearest tenth.

11. $y^2 = 49$
12. $x^2 = 225$
13. $x^2 = 64$
14. $y^2 = 79$
15. $x^2 = 16$
16. $y^2 = 24$
17. $y^2 = 625$
18. $x^2 = 81$

Lesson 9-4 (Pages 398–401)

Write an equation you could use to find the length of the missing side of each right triangle. Then find the missing length. Round to the nearest tenth.

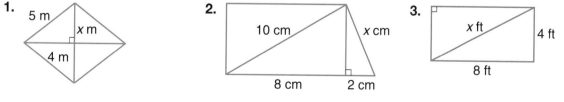

4. a, 6 cm; b, 5 cm
5. a, 12 ft; b, 12 ft
6. a, 8 in.; b, 6 in.
7. a, 20 m; c, 25 m
8. a, 9 mm; c, 14 mm
9. b, 15 m; c, 20 m
10. a, 5 ft; b, 50 ft
11. a, 4.5 yd; c, 8.5 yd

Determine whether each triangle with sides of given length is a right triangle.

12. 15 m, 8 m, 17 m
13. 7 yd, 5 yd, 9 yd
14. 5 in., 12 in., 13 in.

Lesson 9-5A *(Pages 402–403)*
Solve.

1. The streets in Sachi's city are arranged in square blocks. Sachi left her house and walked 5 blocks east and 2 blocks north to Jim's house. She and Jim walked 1 block north, 3 blocks west and 1 block north to school. After school Sachi walked 2 blocks west and 2 blocks south to Bella's house. How far from Bella's house is Sachi's house?

2. If there are 3 different ticket prices and 5 different days for attending a performance of a play, how many possible choices are there for date and ticket price?

3. Alice has 3 logs. The first is twice as long as the second. The third is 3 times as long as the second. Together they are 24 feet long. How long is each log?

4. The streets in a town are arranged in square blocks. Brad leaves his house and rollerblades 4 blocks south and 5 blocks west to the athletic field. He then rollerblades 2 blocks south and 5 blocks east to the library. How far is the library from Brad's home?

Lesson 9-5 *(Pages 404–407)*
State an equation that can be used to answer each question. Then solve. Round to the nearest tenth.

1. How far apart are the boats?

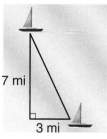

7 mi
3 mi

2. How high does the ladder reach?

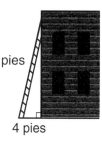

pies
4 pies

3. How long is each rafter?

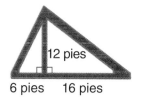

12 pies
6 pies   16 pies

4. Name the primitive Pythagorean triple to which 24-45-51 belongs.

Lesson 9-6 *(Pages 410–413)*
Find the distance between each pair of points whose coordinates are given. Round to the nearest tenth.

1.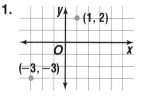

(1, 2)

(−3, −3)

2.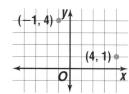

(−1, 4)

(4, 1)

3.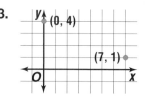

(0, 4)

(7, 1)

Graph each pair of ordered pairs. Then find the distance between the points. Round to the nearest tenth.

4. (−4, 2); (4, 17)   5. (5, −1); (11, 7)   6. (−3, 5); (2, 7)   7. (7, −9); (4, 3)

8. (5, 4); (−3, 8)   9. (−8, −4); (−3, 8)   10. (2, 7); (10, −4)   11. (9, −2); (3, 6)

Lesson 9-7 (Pages 414–417)

Find the missing lengths. Round to the nearest tenth.

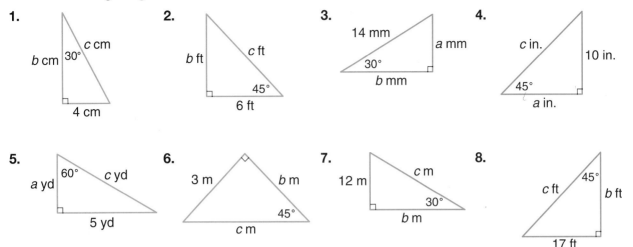

1. b cm, $30°$, c cm, 4 cm

2. b ft, c ft, $45°$, 6 ft

3. 14 mm, a mm, $30°$, b mm

4. c in., 10 in., $45°$, a in.

5. $60°$, a yd, c yd, 5 yd

6. 3 m, b m, $45°$, c m

7. 12 m, c m, $30°$, b m

8. $45°$, c ft, b ft, 17 ft

Lesson 10-1 (Pages 428–431)

Copy and complete each function table.

1. $f(n) = -4n$

n	−4n	f(n)
−2		
−1		
0		
1		
2		

2. $f(n) = n + 6$

n	n + 6	f(n)
−6		
−4		
−2		
0		
2		

3. $f(n) = 3n + 2$

n	3n + 2	f(n)
−3.5		
−2.5		
−1.5		
0		
1.5		

Find each function value.

4. $f\left(\frac{1}{2}\right)$ if $f(n) = 2n - 6$

5. $f(-4)$ if $f(n) = -\frac{1}{2}n + 4$

6. $f(1)$ if $f(n) = -5n + 1$

7. $f(6)$ if $f(n) = \frac{2}{3}n - 5$

8. $f(0)$ if $f(n) = 1.6n + 4$

9. $f(2)$ if $f(n) = 2n^2 - 8$

Lesson 10-2 (Pages 433–435)

Copy and complete each function table. Then graph the function.

1. $f(n) = 6n + 2$

n	f(n)	(n, f(n))
−3		
−1		
1		
$\frac{7}{3}$		

2. $f(n) = -2n + 3$

n	f(n)	(n, f(n))
−2		
−1		
0		
1		
2		

3. $f(n) = 4.5n$

n	f(n)	(n, f(n))
−4		
−2		
0		
1		
6		

Choose values for n and graph each function.

4. $f(n) = \frac{8}{n}$

5. $f(n) = \frac{2}{3}n + 1$

6. $f(n) = n^2 - 1$

7. $f(n) = 3.5n$

8. $f(n) = 4n - 1$

9. $f(n) = \frac{3}{5}n + \left(-\frac{1}{5}\right)$

Lesson 10-3 *(Pages 437–440)*

Copy and complete the table for each equation.

1. $y = 3x - 1$

x	y
−5	
−3	
−1	
0	
1	

2. $y = \frac{x}{4} + 2$

x	y
−8	
−4	
0	
4	

3. $y = -1.5x - 3$

x	y
−4	
−2	
2	
6	
10	

4. $y = 4x - 3$

x	y
−2	
$\frac{1}{2}$	
0	
$2\frac{1}{4}$	

Find four solutions of each equation.

5. $y = -3x + 5$

6. $y = 2x - 1$

7. $y = \frac{2}{3}x + 4$

8. $y = -0.4x$

9. $y = 12x - 8$

10. $y = \frac{3}{4}x + 2$

11. $y = -2.4x - 3$

12. $y = 5x + 7$

Lesson 10-4 *(Pages 442–444)*

Graph each function.

1. $y = -5x$

2. $y = 10x - 2$

3. $y = -2.5x - 1.5$

4. $y = 7x + 3$

5. $y = \frac{x}{4} - 8$

6. $y = 3x + 1$

7. $y = 25 - 2x$

8. $y = \frac{x}{6}$

9. $y = -2x + 11$

10. $y = 7x - 3$

11. $y = \frac{x}{2} + 5$

12. $y = 4 - 6x$

13. $y = -3.5x - 1$

14. $y = 4x + 10$

15. $y = 8x$

16. $y = -5x + \frac{1}{2}$

17. $y = \frac{x}{3} + 9$

18. $y = -7x + 15$

19. $y = 10x - 2$

20. $y = 1.5x - 7.5$

Lesson 10-5 *(Pages 446–449)*

Solve each system of equations by graphing.

1. $y = x$
$y = -x + 4$

2. $y = -x + 8$
$y = x - 2$

3. $y = -3x$
$y = -4x + 2$

4. $y = x - 1$
$y = -x + 11$

5. $y = -x$
$y = 2x$

6. $y = -x + 3$
$y = x + 3$

7. $y = x - 3$
$y = 2x + 8$

8. $y = -x + 6$
$y = x + 2$

9. $y = -x + 1$
$y = x - 4$

10. $y = -3x + 6$
$y = x - 2$

11. $y = 3x - 4$
$y = -3x - 4$

12. $y = 2x + 4$
$y = 3x - 9$

13. $y = -x + 4$
$y = x - 10$

14. $y = -x + 6$
$y = 2x$

15. $y = x - 4$
$y = -2x + 5$

Lesson 10-6A *(Pages 450–451)*

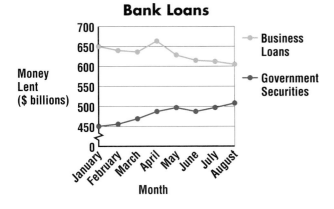

Bank Loans

Money Lent ($ billions)

Month

— Business Loans
— Government Securities

Use the graph above for Exercises 1–4.

1. Estimate the amount of money banks lent businesses in April.

2. In which month did business loans see its biggest increase?

3. By about how much did Government Securities increase from March to May?

4. Describe the overall trend in bank lending.

Lesson 10-6 *(Pages 452–455)*
Graph each quadratic function.

1. $y = x^2 - 1$
2. $y = 1.5x^2 + 3$
3. $f(n) = n^2 - n$
4. $y = 2x^2$

5. $y = x^2 + 3$
6. $y = -3x^2 + 4$
7. $y = -x^2 + 7$
8. $f(n) = 3n^2$

9. $f(n) = 3n^2 + 9n$
10. $y = -x^2$
11. $y = \frac{1}{2}x^2 + 1$
12. $y = 5x^2 - 4$

13. $y = -x^2 + 3x$
14. $f(n) = 2.5n^2$
15. $y = -2x^2$
16. $y = 8x^2 + 3$

17. $y = -x^2 + \frac{1}{2}x$
18. $y = -4x^2 + 4$
19. $f(n) = 4n^2 + 3$
20. $y = -4x^2 + 1$

21. $y = 2x^2 + 1$
22. $y = x^2 - 4x$
23. $y = 3x^2 + 5$
24. $f(n) = 0.5n^2$

25. $f(n) = 2n^2 - 5n$
26. $y = \frac{3}{2}x^2 - 2$
27. $y = 6x^2 + 2$
28. $f(n) = 5n^2 + 6n$

Lesson 10-7 *(Pages 456–459)*
Find the coordinates of the vertices of each figure after the translation described. Then graph the figure and its translation.

1. $\triangle ABC$ with vertices $A(-6, -2)$, $B(-1, 1)$, and $C(2, -2)$, translated by $(4, 3)$

2. $\triangle XYZ$ with vertices $X(-4, 3)$, $Y(0, 3)$, and $Z(-2, -1)$, translated by $(5, -2)$

3. rectangle $HIJK$ with vertices $H(1, 3)$, $I(4, 0)$, $J(2, -2)$, and $K(-1, 1)$, translated by $(-4, -6)$

4. rectangle $PQRS$ with vertices $P(-7, 6)$, $Q(-5, 6)$, $R(-5, 2)$, and $S(-7, 2)$, translated by $(9, -1)$

5. pentagon $DGLMR$ with vertices $D(1, 3)$, $G(2, 4)$, $L(4, 4)$, $M(5, 3)$, and $R(3, 1)$, translated by $(-5, -7)$

Lesson 10-8 *(Pages 460–463)*

Name the line of symmetry for each pair of figures.

1.

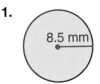

2.

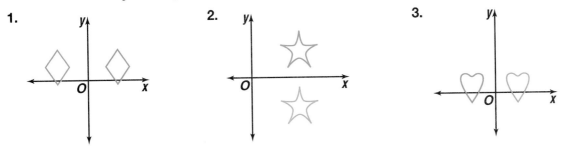

3.

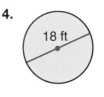

Graph each figure. Then draw its reflections over the *x*-axis and over the *y*-axis.

4. △*CAT* with vertices *C*(2, 3), *A*(8, 2), and *T*(4, −3)
5. trapezoid *TRAP* with vertices *T*(−2, 5), *R*(1, 5), *A*(4, 2), and *P*(−5, 2)
6. rectangle *ABCD* with vertices *A*(4, −1), *B*(7, −4), *C*(4, −7), and *D*(1, −4)

Lesson 10-9 *(Pages 464–467)*

Triangle *ABC* has vertices *A*(−2, −1), *B*(0, 1), and *C*(1, −1).

1. Graph △*ABC*.
2. Find the coordinates of the vertices after a 90° counterclockwise rotation.
3. Graph △*A*′*B*′*C*′.

Rectangle *WXYZ* has vertices *W*(1, 1), *X*(1, 3), *Y*(6, 3), and *Z*(6, 1).

4. Graph rectangle *WXYZ*.
5. Find the coordinates of the vertices after a rotation of 180°.
6. Graph rectangle *W*′*X*′*Y*′*Z*′.

Lesson 11-1 *(Pages 476–479)*

Find the area of each circle to the nearest tenth.

1.

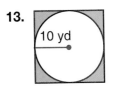

2.

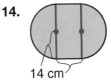

3.

4.

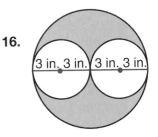

5. radius, 4 m
6. diameter, 6 in.
7. radius, 12 in.
8. diameter, 16 yd
9. diameter, 11 ft
10. radius, 5 in.
11. radius, 19 cm
12. diameter, 29 mm

Find the area of each shaded region to the nearest tenth.

13.

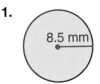

10 yd

14.

14 cm

15.

22 cm

16.

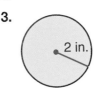

3 in. 3 in. 3 in. 3 in.

Lesson 11-2A *(Pages 480–481)*
Solve.

1. How many cubes are needed to make this display?

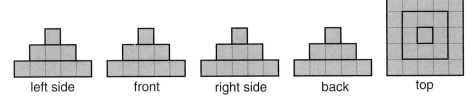

| left side | front | right side | back | top |

2. How many mirrored tiles would be needed to cover the display in Exercise 1 if each tile is the same size as the face of a cube? (The bottom of the display will not be covered with the tiles.)

3. How many cubes 3 cm on a side can be packaged in a 12 cm by 9 cm by 6 cm box?

4. How many cubes are needed to make the display shown in the plan below?

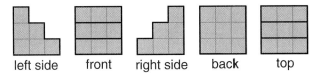

| left side | front | right side | back | top |

Lesson 11-2 *(Pages 482–485)*
Use isometric dot paper to draw each solid.

1. A rectangular prism that is 4 units high, 5 units long, and 6 units deep.

2. A triangular prism that is 4 units high.

3. A pyramid with a pentagonal base.

4. On grid paper, draw the front view, back view, two side views, and top view of the figure.

Lesson 11-3 *(Pages 486–489)*
Find the volume of each solid to the nearest tenth.

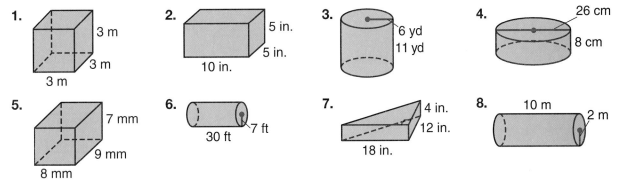

Lesson 11-4 (*Pages 490–493*)
Find the volume of each solid to the nearest tenth.

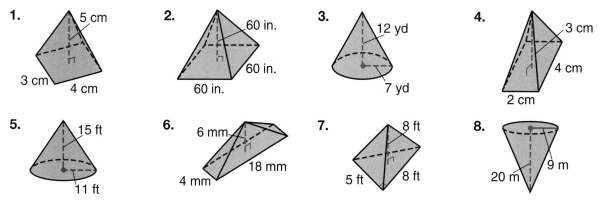

1. 5 cm, 3 cm, 4 cm
2. 60 in., 60 in., 60 in.
3. 12 yd, 7 yd
4. 3 cm, 4 cm, 2 cm
5. 15 ft, 11 ft
6. 6 mm, 18 mm, 4 mm
7. 8 ft, 5 ft, 8 ft
8. 9 m, 20 m

Lesson 11-5 (*Pages 495–498*)
Find the surface area of each prism to the nearest tenth.

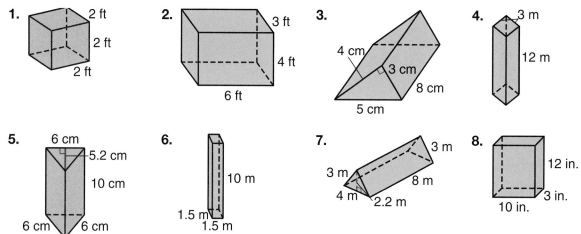

1. 2 ft, 2 ft, 2 ft
2. 3 ft, 4 ft, 6 ft
3. 4 cm, 3 cm, 8 cm, 5 cm
4. 3 m, 12 m
5. 6 cm, 5.2 cm, 10 cm, 6 cm, 6 cm
6. 10 m, 1.5 m, 1.5 m
7. 3 m, 3 m, 8 m, 4 m, 2.2 m
8. 12 in., 3 in., 10 in.

Lesson 11-6 (*Pages 499–502*)
Find the surface area of each cylinder to the nearest tenth.

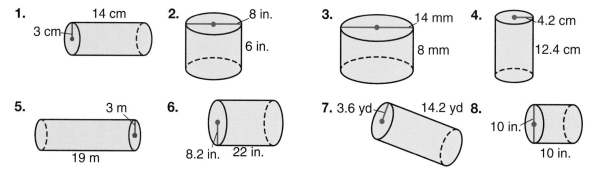

1. 14 cm, 3 cm
2. 8 in., 6 in.
3. 14 mm, 8 mm
4. 4.2 cm, 12.4 cm
5. 3 m, 19 m
6. 8.2 in., 22 in.
7. 3.6 yd, 14.2 yd
8. 10 in., 10 in.

Lesson 11-7 *(Pages 504–507)*

Analyze each measurement. Give the precision, significant digits if appropriate, greatest possible error, and relative error to two significant digits.

1. 18 min
2. $7\frac{1}{2}$ lb
3. 92.46 m
4. 7 ft
5. 0.067 kg
6. 61.7 cm
7. 8 mm
8. $4\frac{1}{4}$ in.

9. Which is a more precise measurement for the length of a pencil, 18 centimeters or 18.0 centimeters? Explain your answer.

10. A wire's length is measured as 0.02 meters.

   a. How many significant digits are there in 0.02?

   b. Suppose the measurement is actually 2.2 centimeters. How many significant digits are there in 2.2?

   c. Compare the preciseness of these two measurements.

Lesson 12-1 *(Pages 518–520)*

Draw a tree diagram to find the number of possible outcomes for each situation.

1. A bakery makes yellow, white, chocolate, or marble cake with a choice of chocolate or vanilla frosting.

2. A car comes in white, black, or red with standard or automatic transmission and with a 4-cylinder or 6-cylinder engine.

3. A customer can buy roses or carnations in red, yellow, pink, or white.

4. A bed comes in queen or king size with a firm or super firm mattress.

5. A pizza can be ordered with a regular or deep dish crust and with a choice of one topping, two toppings, or three toppings.

6. A woman's shoe comes in red, white, blue, or black with a choice of high, medium, or low heels.

Lesson 12-2 *(Pages 521–523)*

Find each value.

1. 8!
2. 10!
3. 0!
4. 7!
5. 6!
6. 5!
7. 2!
8. 11!
9. 9!
10. 4!
11. $P(5, 4)$
12. $P(3, 3)$
13. $P(12, 5)$
14. $P(8, 6)$
15. $P(10, 2)$
16. $P(6, 4)$
17. $P(7, 6)$
18. $P(9, 9)$

19. How many different ways can a family of four be seated in a row?

20. In how many different ways can you arrange the letters in the word *orange* if you take the letters five at a time?

21. How many ways can you arrange five different colored marbles in a row if the blue one is always in the center?

22. In how many different ways can Kevin listen to each of his ten CDs once?

Lesson 12-3 *(Pages 524–527)*
Find each value.

1. $C(8, 4)$
2. $C(30, 8)$
3. $C(10, 9)$
4. $C(7, 3)$
5. $C(12, 5)$
6. $C(17, 16)$
7. $C(24, 17)$
8. $C(9, 7)$

9. How many ways can you choose five compact discs from a collection of 17?

10. How many combinations of three flavors of ice cream can you choose from 25 different flavors of ice cream?

11. How many ways can you choose three books out of a selection of ten books?

12. How many ways can you choose seven apples out of a bag of two dozen apples?

13. How many ways can you choose two movies to rent out of ten possible movies?

Lesson 12-4 *(Pages 528–531)*
Use Pascal's Triangle to find each value.

1. $C(6, 4)$
2. $C(6, 2)$
3. $C(5, 1)$
4. $C(9, 3)$
5. $C(3, 2)$
6. $C(7, 6)$

Use Pascal's Triangle to answer each question.

7. Four coins are tossed. What is the probability that 2 will show heads and 2 will show tails?

8. How many combinations of 7 things taken 3 at a time are possible?

9. How many different 5-question quizzes can be formed from a test bank of 10 questions?

10. How many branches will there be in the last column of a tree diagram showing the outcomes of tossing 4 coins?

Lesson 12-5 *(Pages 534–537)*
Two socks are drawn from a drawer which contains one red sock, three blue socks, two black socks, and two green socks. Once a sock is selected, it is not replaced. Find the probability of each outcome.

1. a black sock and then a green sock
2. a red sock and then a green sock
3. a blue sock two times in a row
4. a green sock two times in a row

There are three quarters, five dimes, and twelve pennies in a bag. Once a coin is drawn from the bag it is not replaced. If two coins are drawn at random, find the probability of each outcome.

5. a quarter and then a penny
6. a nickel and then a dime
7. a dime and then a penny
8. a dime two times in a row

Lesson 12-6A *(Pages 538–539)*
Solve.

1. Estimate the probability that in a group of 7 people, exactly 5 people will be male.

2. In a survey of 130 people, 95 said they exercised regularly and 62 said they watched their diets. If 45 people said they both exercised regularly and watched their diets, how many do neither?

3. Act it out to find the probability that 4 coins tossed in the air all will land showing tails.

4. Ryuji has 8 compact discs in his glove compartment. Three are classical, four are rock, and one is country. If he chooses one at random each day, what is the probability that he will choose a rock tape 3 days in a row?

Lesson 12-6 *(Pages 540–543)*
Roll a number cube 50 times.

1. Record the results.

2. Based on your results, what is the probability of a 5?

3. Based on your results, what is the probability of an odd number?

4. What is the theoretical probability of an odd number?

Lesson 12-7 *(Pages 546–548)*
Use the survey on favorite type of music to answer each question.

1. What is the size of the sample?

2. What is the mode?

3. What fraction prefers country music?

4. What fraction prefers rap music?

Use the survey on favorite fruit to answer each question.

5. What is the size of the sample?

6. If 7,950 people were to choose one of these three fruits, how many would you expect to choose oranges?

Favorite Type of Music	
Country	72
Heavy Metal	41
Rap	45
Light Rock	92

Favorite Fruit	
Apple	155
Orange	300
Banana	145

Lesson 13-1 *(Pages 561–564)*

Write a monomial or polynomial for each model.

1. x^2 $-x$ $-x$ 1

2. -1 -1 -1 -1 -1

3. $-x^2$ $-x^2$ 1

4. x^2 $-x$ $-x$ $-x$

Model each monomial or polynomial using drawings or algebra tiles.

5. 7

6. $3x$

7. $-2x$

8. $-5x + 1$

9. $x + 2$

10. $-2x^2 + 3x$

11. $-x - 4$

12. $x^2 + 2x + 3$

13. $2x^2 - 2$

14. $2x - 5$

15. $-x^2 + 7x$

16. $2x^2 - 3x$

Lesson 13-2 *(Pages 565–569)*

Name the like terms in each list of terms.

1. $10x, 6y, y^2, 3x$

2. $-a, 2b^2, -3a, b^2$

3. $2m, n, -m, n$

4. $2, 7m, -m^2$

5. $3y, 6, 5, y^2$

6. $4x, 6, 2x, x^2$

Simplify each polynomial using the model.

7. $2x - x^2 + x - 1$

x x $-x^2$ x -1

8. $2a^2 + 1 + a^2$

a^2 a^2 1 a^2

Simplify each polynomial. Use drawings or algebra tiles if necessary.

9. $-2y + 3 + x^2 + 5y$

10. $m + m^2 + n + 3m^2$

11. $a^2 + b^2 + 3 + 2b^2$

12. $1 + a + b + 6$

13. $x + x^2 + 5x - 3x^2$

14. $-2y + 3 + y - 2$

Lesson 13-3 *(Pages 570–572)*

Find each sum. Use drawings or algebra tiles if necessary.

1. $\begin{aligned} & 2x^2 - 5x + 7 \\ +\ & x^2 - x + 11 \end{aligned}$

2. $\begin{aligned} & 2m^2 + m + 1 \\ +\ & (-m^2) + 2m + 3 \end{aligned}$

3. $\begin{aligned} & 2a - b + 6c \\ +\ & 3a - 7b + 2c \end{aligned}$

4. $\begin{aligned} & 5a + 3a^2 - 2 \\ +\ & 2a + 8a^2 + 4 \end{aligned}$

5. $\begin{aligned} & 3c + b + a \\ +\ & (-c) + b - a \end{aligned}$

6. $\begin{aligned} & -z^2 + x^2 + 2y^2 \\ +\ & 3z^2 + x^2 + y^2 \end{aligned}$

7. $(5x + 6y) + (2x + 8y)$

8. $(4a + 6b) + (2a + 3b)$

9. $(7r + 11m) + (4m + 2r)$

10. $(-z + z^2) + (-2z + z^2)$

11. $(3x - 7y) + (3y + 4x + 1)$

12. $(5m + 3n - 3) + (8m + 6)$

13. $(a + a^2) + (3a - 2a^2)$

14. $(3s - 5t) + (8t + 2s)$

Lesson 13-4 *(Pages 573–576)*

Find each difference. Use drawings or algebra tiles if necessary.

1. $\quad 5a - 6m$
$\quad - (2a + 5m)$

2. $\quad 2a - 7$
$\quad - (8a - 11)$

3. $\quad 9r^2 + r + 3$
$\quad - (11r^2 - r + 12)$

4. $(9x + 3y) - (9y + x)$

5. $(3x^2 + 2x - 1) - (2x + 2)$

6. $(a^2 + 6a + 3) - (5a^2 + 5)$

7. $(5a + 2) - (3a^2 + a + 8)$

8. $(3x^2 - 7x) - (8x - 6)$

9. $(3m + 3n) - (m + 2n)$

10. $(3m - 2) - (2m + 1)$

11. $(x^2 - 2) - (x + 3)$

12. $(5x^2 - 4) - (3x^2 + 8x + 4)$

13. $(7z^2 + 1) - (3z^2 + 2z - 6)$

Lesson 13-5 *(Pages 578–581)*

Find each product. Express the answer in exponential form.

1. $2^3 \cdot 2^4$

2. $5^6 \cdot 5$

3. $t^2 \cdot t^2$

4. $y^5 \cdot y^3$

Find each product.

5. $a(a + 2)$

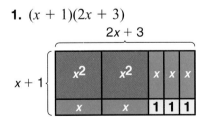

6. $x(2x - 3)$

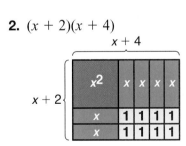

7. $a(a + 4)$

8. $m(m - 7)$

9. $z^2(z + 3)$

10. $6x(x + 10)$

11. $3y(5 + y)$

12. $2d(d^3 + 1)$

13. $m^3(m^2 - 2)$

14. $p^5(3p - 1)$

15. $b(9 + 4b)$

16. $4t^3(t + 3)$

17. $2r^4(5r + 9)$

18. $3n^2(6 - 7n^3)$

Lesson 13-6 *(Pages 583–585)*

Find each product.

1. $(x + 1)(2x + 3)$

2. $(x + 2)(x + 4)$

Find each product. Use drawings or algebra tiles if necessary.

3. $(r + 3)(r + 4)$

4. $(z + 5)(z + 2)$

5. $(3x + 7)(x + 1)$

6. $(x + 5)(2x + 3)$

7. $(c + 1)(c + 1)$

8. $(a + 3)(a + 7)$

9. $(b + 3)(b + 1)$

10. $(2y + 1)(y + 3)$

11. $(z + 8)(2z + 1)$

12. $(2m + 4)(m + 5)$

13. $(x + 3)(x + 2)$

14. $(c + 2)(c + 8)$

15. $(r + 4)(r + 4)$

16. $(2x + 4)(x + 4)$

Lesson 13-7A *(Pages 586–587)*
Solve.

1. Leslie has 6 times as much money as Inazo. If Inazo had $1.20 more, Leslie would have 2 times as much as Inazo. How much money does each person have?

2. Mei-yu bought some pens for $0.89 each and some pencils for $0.19 each. She spent $5.02. How many pens and how many pencils did she buy?

3. 70 more than a number is twice the number. What is the number?

4. The difference between a number and its double is 28. What is the number?

5. The product of two consecutive numbers is 2,450. What are the two numbers?

Lesson 13-7 *(Pages 588–591)*
Factor each polynomial.

1. $2x^2 + 4x + 2$

2. $x^2 + 5x + 4$

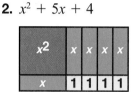

If possible, factor each polynomial. Use drawings or algebra tiles if necessary.

3. $5x + 10$

4. $x^2 + 8x$

5. $x^2 + 6x + 1$

6. $2x^2 + 8x + 6$

7. $2x^2 + 3x + 5$

8. $3x^2 + 16x + 5$

9. $x^2 + x + 4$

10. $x^2 + 10x + 21$

11. $3x^2 + 22x + 24$

Mixed Problem Solving

Solve using any strategy.

1. *Money Matters* Out-of-town newspapers cost 60¢, and local papers cost 35¢ each. Bill buys 2 out-of-town papers and 3 local papers. He hands the cashier $3. Should he expect more than 50¢ change?

2. The odometer in Sarah's car registered 68,364.7 miles last week. The odometer registers 71,429.2 miles today. Last week, the number of miles the car traveled was about—

   a. 900.

   b. 3,000.

   c. 10,000.

   d. 71,000.

3. *School* A section of the school auditorium is set up so that each row has the same number of seats. Cheri is seated in the seventh row from the back and the eighth row from the front of this section. Her seat is the fourth from the right and the seventh from the left. How many seats are in this section?

   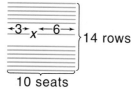

4. David pours 5 gallons of water into a tank of water. Then Carlos adds 8 gallons more to the tank. The tank now contains 32 gallons of water. How many gallons were in the tank before David added water?

5. Mia arranges her marbles into 2 even columns. Leah puts her marbles into 3 even columns, and Ryan puts his marbles into 5 even columns. Each person has the same number of marbles. What is the least number of marbles that Ryan can have?

6. *School* When Ruiz walks along the school corridor, he passes through three doorways. The first door is open half the time, the second door is open half the time, and the third door is open two times out of three. What is the probability that all three doors will be open?

7. *Design* A memorial waterfall is built of granite cubes using the plans below.

   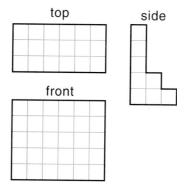

   a. How many cubes of granite are needed?

   b. If each cube measures 2 feet on an edge, how high is the waterfall?

8. A number plus half the number is 33. Find the number.

Mixed Problem Solving

Solve using any strategy.

1. *Sports* In the Sports Card Collectors' Club, 9 members collect basketball cards, 18 members collect baseball cards, and 15 members collect football cards. Four collect baseball and football cards, 3 collect baseball and basketball cards, 3 collect basketball and football cards, and 1 collects all three types of cards.
   a. Make a Venn diagram of this information.
   b. How many members are in the club?
   c. How many members collect baseball or football cards?
   d. How many collect only baseball cards?

2. Pencils are packed in boxes of 10 and 12. Tori buys 15 boxes of pencils containing a total of 160 pencils. How many of each type does Tori buy?

3. *Money* Juan has a mixture of pennies and dimes worth $2.28. He has between 39 and 56 pennies. How many dimes does Juan have?

4. Hannah buys 6 boxes of tissues containing 75 tissues each. Mick buys 2 boxes of tissues containing 175 tissues each. Hannah guesses that she has about twice as many tissues as Mick. Is her guess reasonable?

5. *Shopping* The graph shows the average amount of money spent per trip by buyers ages 8 through 17. These shoppers average 12.7 shopping trips a month.
   a. What is the average amount of money a 14-year old would spend shopping in one month?
   b. Find the average amount an 11-year old would spend in 6 months.

Spend, Spend, Spend

$31.20

$18.50

Ages 8-12 Ages 13-17

Source: International Mass Retail Association

6. *Money Matters* A model PL-3 shortwave radio costs $12 more than twice the cost of the model PL-1 radio. The PL-3 radio costs $191.98. What is the cost of the PL-1 radio?

7. Shelby is thinking of a number. Austin is thinking of a number that is 23 more than Shelby's number. Skylar's number is 18 less than Austin's number. Skylar's number is about—
   a. 5 less than Shelby's number.
   b. 41 more than Shelby's number.
   c. 5 more than Shelby's number.
   d. 41 less than Shelby's number.

8. Find the sum of whole numbers from 1 to 49.

Test

1. Write $4 \cdot 4 \cdot 6 \cdot 6 \cdot 6$ using exponents.

2. Evaluate $3^3 \cdot 2^2$.

3. Solve $8x = 88$ if the replacement set is $\{704, 80, 12, 11\}$.

Evaluate each expression if $a = 3$, $b = 2$, and $c = 5$.

4. $3(c - a)^2 + b - 12$

5. $(2c + b) \div a - 3$

Solve each equation. Check your solution.

6. $k - 10 = 65$

7. $0.3m = 4.8$

8. $x + 33 = 43$

9. $\$2.40 = \$1.85 + b$

10. $\dfrac{n}{1.5} = 8$

11. $10 = 0.8x$

12. $3(n - 5) = 30$

13. $\dfrac{a}{7} + 12 = 19$

14. $100 = 10 + 3x$

Write each phrase or sentence as an algebraic expression or equation.

15. 7 divided by x

16. the sum of 5 and y

17. Three times a number is 30.

Find the perimeter and area of each figure.

18.
19.
20.

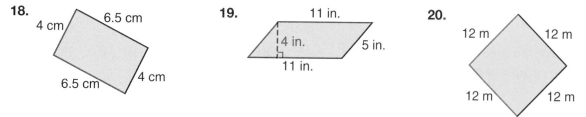

Solve each inequality. Graph the solution on a number line.

21. $x + 3 > 15$

22. $11 > 2c - 3$

23. $\dfrac{z}{3} + 4 \le 6$

24. Ray has $\$2,183$ in his savings account. He deposits $\$75$. How much will he have in his account after the deposit?

25. Taryn has $\$8.50$ left. She spent $\$17$ and $\$14.50$ at two stores. How much did she start with?

1. Find the absolute value of -8.

2. Graph the set $\{5, -3, -5, 0, 2\}$ on a number line.

Replace each ● with >, <, or = to make a true sentence.

3. 2 ● -19

4. $|-21|$ ● 21

5. -816 ● -125

6. $|-51|$ ● $|-18|$

Solve each equation.

7. $r = -572 + 58$

8. $p = -5(-13)$

9. $z = 60 + 12$

10. $m = -105 \div 15$

11. $-211 - 127 = d$

12. $-789 - (-54) = s$

13. $(7)(-10)(4) = h$

14. $2q - 12 = -40$

15. $g = -109 - (-34)$

16. $2x = 14$

17. $x + 6 = 2$

18. $y = 84 + (-13)$

19. $6t = 38 - (-22)$

20. $-7 + 22 + (-10) + 16 = s$

21. $-5(80)(-2) = m$

22. $k = \dfrac{700}{-100}$

23. $y = (-4) - 3$

24. $\dfrac{-160}{-5} = x$

Evaluate each expression if $c = -4$, $m = 5$, and $t = 10$.

25. $-5c + m$

26. $3ct \div m$

Find each sum or difference. If there is no sum or difference, write *impossible*.

27. $\begin{bmatrix} -3 & 0 \\ 5 & -2 \\ -4 & 6 \end{bmatrix} - \begin{bmatrix} 6 & -3 \\ -4 & 6 \end{bmatrix}$

28. $\begin{bmatrix} 3 & -2 \\ 4 & -1 \\ 5 & -3 \end{bmatrix} + \begin{bmatrix} -3 & 8 \\ 0 & 6 \\ -4 & -2 \end{bmatrix}$

Name the ordered pair for the coordinates of each point graphed on the coordinate plane.

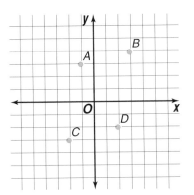

29. A

30. B

31. C

32. D

33. *School* Schools were closed in the district due to a winter storm. The temperature dropped 21°F over a three-hour period. If the temperature dropped at an even rate, how many degrees did the temperature fall each hour?

Express each ratio or rate in simplest form.

1. 6 misses in 10 tries

2. 8 out of 15

3. 9 wins:6 losses

Express each rate as a unit rate.

4. 250 miles in 5 hours

5. 180 pages in 6 hours

6. $15 in 4 days

Express each ratio or fraction as a percent. Round to the nearest hundredth of a percent.

7. 7 out of 10

8. $\frac{3}{4}$

9. $\frac{3}{8}$

10. $\frac{1}{6}$

11. 2 out of 6

12. 2:32

13. *Travel* According to the Travel Association of America, 4 out of 5 vacationers travel to their destination by car, truck, or recreational vehicle. Express the ratio as a percent.

Express each percent as a fraction in simplest form.

14. 30%

15. 8%

16. 45%

Express each percent as a decimal.

17. 25%

18. 8%

19. 13.5%

Express each decimal as a percent.

20. 0.55

21. 0.3

22. 0.065

Solve each proportion.

23. $\frac{3}{7} = \frac{x}{35}$

24. $\frac{10}{8.4} = \frac{5}{y}$

25. $\frac{3}{2} = \frac{z}{3.4}$

26. $\frac{18}{12} = \frac{24}{y}$

Compute mentally.

27. 10% of 99

28. 30% of 60

29. $33\frac{1}{3}$% of 90

Estimate.

30. 9% of 81

31. 97% of 16

32. 46% of 20

33. *Medicine* There are about 249 million people in the United States. Of these, 37% have type O+ blood. Estimate how many people have O+ blood.

Test

Use the histogram for Exercises 1–3.

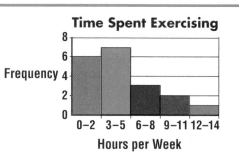

1. For which interval is the frequency the greatest?

2. How many people were surveyed?

3. How many people spend more than 8 hours per week exercising?

Use the table for Exercises 4–6.

On What Day Do Fatal Accidents Occur?							
Day	Sun.	Mon.	Tues.	Wed.	Thurs.	Fri.	Sat.
Accidents	6,590	4,920	4,620	4,920	5,160	6,630	7,560

Source: Carnegie Library of Pittsburgh

4. Make a bar graph of the data.   5. Make a circle graph of the data.

6. What type of display is most appropriate for showing this data in an article proposing more police patrols on the weekends? Explain your choice.

The scores on a 30-point math quiz are 25, 16, 25, 30, 29, 28, 19, 23, 22, 17, 19, 20, 25, 30, and 29.

7. Make a line plot for the data.

8. What is the frequency of the score that occurred most often?

9. How many students received a score of 19?

10. How many students received a score of 25 or above?

11. Find the mean, median, and mode of the data.

The ages of the customers at Hannah's Bagel Shop are 45, 36, 27, 16, 19, 46, 40, 38, 22, 23, 25, 40, and 17.

12. Find the range.   13. What is the median?

14. What is the upper quartile?   15. Find the lower quartile.

16. What is the interquartile range?   17. Are there any outliers? If so, name them.

18. *Travel* Would a scatter plot of data describing the gallons of gas used and the miles driven show a *positive, negative,* or *no* relationship?

19. *Finances* In order to make enough money to meet his budgeted monthly expenses, Hakeem needs to earn an average of $500 a month. Last year, he earned $540, $450, $560, $800, $350, $400, $350, $380, $500, $450, $600, and $200. Did he earn enough last year? If not, how much more would Hakeem have to earn to have an average income of $500?

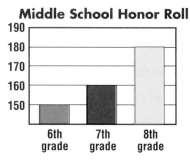

20. *School* The graph shows the number of students in each grade that had a B average or better. Is the graph misleading? Explain.

Find the measure of each angle if *a* ∥ *b* and *m* ∠ 4 = 50°.

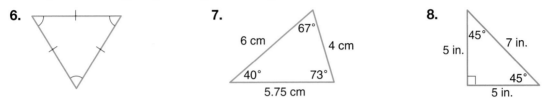

1. $m\angle 1$

2. $m\angle 2$

3. $m\angle 3$

4. $m\angle 5$

5. Can a triangle be scalene and obtuse? Draw a figure to justify your answer.

Classify each triangle by its angles and by its sides.

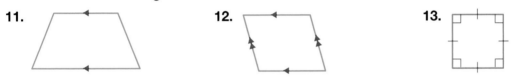

6.

7.

8.

9. Draw an isosceles obtuse triangle.

10. In $\triangle ABC$, $m\angle A = 35°$ and $m\angle B = 60°$. What is $m\angle C$?

Sketch each figure. Let Q = quadrilateral, P = parallelogram, R = rectangle, S = square, RH = rhombus, and T =trapezoid. Write all of the letters that describe it inside the figure.

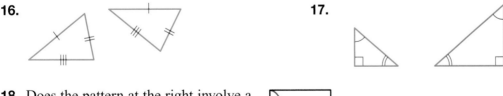

11.

12.

13.

14. Does a rectangle have line symmetry? If so, draw the lines of symmetry.

15. Does a rhombus have rotational symmetry?

Tell whether each pair of triangles is *congruent*, *similar*, or *neither*. Justify your answer.

16.

17.

18. Does the pattern at the right involve a translation or a rotation?

19. *Music* A survey of your class shows that 20 students have a portable cassette player and 20 students have a portable CD player. Ten students have both. Make a Venn diagram to help you find out how many students have a portable cassette or CD player.

20. *Business* A corporation is developing a new corporate logo. They can use either the letter X or the letter K. If they want to use the letter with the most lines of symmetry, which will they choose?

Test

Determine whether each number is divisible by 2, 3, 4, 5, 6, 8, 9 or 10.

1. 822   **2.** 285   **3.** 2,088

Find the prime factorization of each number.

4. 54   **5.** 100   **6.** 180

Find the GCF for each set of numbers.

7. 27, 54   **8.** 44, 66   **9.** 105, 60, 35

Write each fraction in simplest form.

10. $\frac{12}{16}$ **11.** $-\frac{36}{54}$ **12.** $\frac{36}{81}$

Express each fraction or mixed number as a decimal.

13. $\frac{13}{16}$ **14.** $-5\frac{3}{8}$ **15.** $\frac{9}{11}$

Express each decimal as a fraction or mixed number in simplest form.

16. -1.83 **17.** $0.\overline{67}$ **18.** $5.4\overline{5}$

The letters of the word "telephone" are written one each on 9 identical slips of paper and shuffled in a bag. A blindfolded student draws one slip of paper. Find each probability.

19. $P(e)$ **20.** $P(\text{o or p})$ **21.** $P(\text{vowel})$

Find the LCM for each set of numbers.

22. 7, 14   **23.** 6, 15, 18   **24.** 10, 20, 30

Replace each ⬤ with <, >, or = to make a true sentence.

25. $\frac{2}{9}$ ⬤ $\frac{5}{27}$ **26.** $5\frac{3}{5}$ ⬤ 5.61 **27.** -5.89 ⬤ -5.9

Express each number in standard form.

28. 4.3×10^{-4} **29.** 7.93×10^{5}

Express each number in scientific notation.

30. 65,460,000   **31.** 0.00000057

32. *Remodeling* Amanda wants to lay ceramic tiles on her countertop. The countertop is 60 inches long and 48 inches wide. What are the dimensions of the largest square tile that she can use without having to use any partial squares?

33. *Football* Oscar led the league by completing 50 out of 75 passes he threw. Write the fraction of his throws that were completed as a decimal.

Solve each equation. Write the solution in simplest form.

1. $-7\frac{3}{7} - 2\frac{5}{7} = m$ **2.** $\frac{8}{13} + \frac{8}{13} = a$ **3.** $r = \frac{9}{14} - \frac{5}{14}$ **4.** $\frac{3}{8} - \frac{4}{5} = t$

5. $y = -1\frac{5}{6} + (-8\frac{3}{8})$ **6.** $-8\frac{5}{9} - 2\frac{1}{6} = b$ **7.** $-\frac{3}{4} \cdot -\frac{8}{9} = g$ **8.** $x = -5\frac{1}{3} \cdot 2\frac{2}{3}$

9. $h = \left(\frac{4}{9}\right)^2$ **10.** $j = -4\frac{1}{2}\left(-\frac{2}{3}\right)$ **11.** $2\frac{4}{5}(10) = d$ **12.** $f = 12 \times 3\frac{3}{4}$

13. $n = \left(-\frac{4}{5} \cdot \frac{2}{3}\right) \cdot \frac{1}{2}$ **14.** $c = -\frac{7}{8} \div 1\frac{3}{4}$ **15.** $6\frac{1}{6} \div 1\frac{2}{3} = w$ **16.** $\dfrac{3\frac{3}{5}}{10} = k$

17. *Publishing* The width of a page of a newspaper is $13\frac{3}{4}$ inches. The left margin is $\frac{7}{16}$ inch, and the right margin is $\frac{1}{2}$ inch. What is the width of the page inside the margins?

18. *Food* Marcie wants to make enough pudding to serve 16 friends. The recipe that serves 4 requires $1\frac{3}{4}$ cups of milk. How much milk will she need?

Identify each sequence as *arithmetic, geometric,* or *neither.* Then find the next three terms.

19. $-10, -6, -2, 2, ...$ **20.** $88, 44, 22, 11, ...$ **21.** $50, 40, 31, 23, ...$

Find the area of each triangle.

	base	height
22.	11 m	18 m
23.	$5\frac{1}{2}$ ft	$6\frac{1}{4}$ ft

Find the area of each trapezoid.

	base (*a*)	base (*b*)	height
24.	10 in.	25 in.	$8\frac{1}{2}$ in.
25.	3.8 mm	5.3 mm	8.4 mm

Find the circumference of each circle described below. Use $\frac{22}{7}$ or 3.14 for π. Round decimal answers to the nearest tenth.

26. The diameter is $6\frac{2}{3}$ yards. **27.** The radius is 3.6 meters.

Solve each equation or inequality. Check your solution.

28. $2\frac{1}{2}w = 4\frac{3}{8}$ **29.** $\frac{3}{4} = -\frac{3}{4}x + \frac{9}{16}$ **30.** $\frac{a}{1.8} - 7.8 = 11$

31. $2.2 < \frac{b}{-10} - 2.4$ **32.** $\frac{1}{3}m + 7 \geq 11$ **33.** $-\frac{3}{4}j - 1 < -5\frac{1}{2}$

CHAPTER TEST

Write a proportion to solve each problem. Then solve.

1. A car uses 8 gallons to travel 120 miles. How many gallons would be used to travel 80 miles?

2. A muffin recipe calls for 3 cups of flour for 24 muffins. How many cups of flour are needed to make 36 muffins?

Write a percent proportion to solve each problem. Then solve. Round to the nearest tenth.

3. 18 is 25% of what number?   4. What is 2% of 3,600?   5. 62 is 90% of what number?

Write an equation in the form $RB = P$ for each problem. Then solve.

6. 30 is what percent of 50?   7. Find 45% of 600.   8. 1 is what percent of 30?

Express each percent as a fraction or mixed number in simplest form.

9. 135%   10. $\frac{1}{6}$%

Express each percent as a decimal.

11. 0.25%   12. 230%

Find each percent of change. Round to the nearest percent.

13. original: 10
    new: 16

14. original: 28
    new: 23

15. original: 350
    new: 240

Find the simple interest to the nearest cent.

16. $300 at 8% for 3 years   17. $1,800 at 6.5% for 6 months

18. Tell whether the pair of polygons are similar. Explain your reasoning.

19. A house casts a shadow that is 18 feet long. At the same time, the shadow of a tree in the backyard is 27 feet long. If the house is 24 feet tall, how tall is the tree?

20. A distance on a map is $1\frac{1}{2}$ inches. If the scale on the map is 1 inch = 50 miles, find the actual distance.

Find the coordinates of the image of each point for a dilation with a scale factor of 3.

21. A(2, 3)   22. B(4, 1)   23. C(6, 0)

24. **Budget** Stratman has budgeted $90 a month for entertainment expenses. If this is 15% of his total monthly budget, what is his total monthly budget?

25. **Travel** It took Tori 8 hours to travel 520 miles. At that rate, how far can she travel in $16\frac{1}{2}$ hours?

Find each square root.

1. $\sqrt{144}$

2. $\sqrt{\dfrac{36}{49}}$

3. $-\sqrt{0.16}$

Estimate to the nearest whole number.

4. $\sqrt{67}$

5. $\sqrt{510}$

6. $\sqrt{108.8}$

Let R = real numbers, Q = rational numbers, Z = integers, W = whole numbers, and I = irrational numbers. Name the set or sets of numbers to which each real number belongs.

7. $\sqrt{14}$

8. $-\sqrt{64}$

9. 6.1313...

10. $\sqrt{18.743}$

11. State the Pythagorean Theorem.

Find the missing measure for each right triangle. Round to the nearest tenth.

12. $a = 6$ km; $b = 8$ km

13. $b = 20$ in.; $c = 35$ in.

14. $a = 1.5$ cm; $c = 2.5$ cm

15. $a = 2.5$ ft; $b = 4.5$ ft

Determine whether each triangle with sides of given length is a right triangle.

16. 16 m, 34 m, 30 m

17. 12 ft, 20 ft, 24 ft

18. A ladder is leaning against a house. The top of the ladder is 16 feet from the ground, and the base of the ladder is 12 feet from the side of the house. How long is the ladder?

19. *Geometry* Find the perimeter of a right triangle with legs of 10 inches and 8 inches.

20. *Geometry* Find the distance between $R(-2, -2)$ and $S(5, 6)$. Round to the nearest tenth.

Find the missing lengths. Round decimals to the nearest tenth.

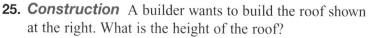

21.
32 in. 60°
30°
a in.
b in.

22.
b mm
30°
a mm 60°
55 mm

23.
45°
6 yd *c* yd
45°
b yd

24.
30 cm
45°
a cm
c cm
45°

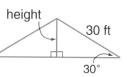

25. *Construction* A builder wants to build the roof shown at the right. What is the height of the roof?

height
30 ft
30°

CHAPTER TEST

Test

Make a function table for each function. Use −1, 0, and 2 for *n*. Then graph the function.

1. $f(n) = -3n$

2. $f(n) = 2n - 4$

Find four solutions of each equation.

3. $y = \frac{x}{4} + 2$

4. $y = -4 - 2.5x$

5. $y = 6 - x$

6. $y = 3x - 12$

Graph each function.

7. $y = \frac{x}{3} - 1$

8. $y = x^2 + 6$

9. $y = 2x^2 - 1$

10. $y = -4x + 10$

11. Find the solution of the system $y = 5x + 2$ and $y = 2x + 8$ by graphing.

12. Without graphing, determine whether $(3, -2)$ is the solution of the system of equations $y = 2x - 8$ and $y = -3x - 7$. Why or why not?

Rectangle *ABCD* has vertices *A*(−5, −2), *B*(−2, −2), *C*(−2, −1), and *D*(−5, −1). After a translation, the coordinates of *A′* are (2, 4).

13. Describe the translation using an ordered pair.

14. Find the coordinates of B', C', and D'.

Triangle *EFG* has vertices *E*(−5, 2), *F*(−2, 3), and *G*(−4, 5).

15. Find the coordinates of the vertices after a reflection over the *y*-axis.

16. Find the coordinates of the vertices after a reflection over the *x*-axis.

Triangle *HIJ* has vertices *H*(4, 4), *I*(1, 2), and *J*(2, 5).

17. Find the coordinates of $\triangle H'I'J'$ after a $180°$ rotation.

18. Find the coordinates of $\triangle H'I'J'$ after a rotation of $90°$ counterclockwise.

19. *Weather* By 6:00 P.M., 3 inches of rain had fallen. For the next three hours, 0.5 inch of rain fell each hour. How many inches of rain fell by 9:00 P.M.?

20. *Remodeling* One plumber charges a flat fee of $35 plus $25 per hour. Another plumber charges a flat fee of $20 plus $30 per hour.

   a. When are their charges equal?

   b. If it takes one hour to do a job, who will charge less, the first plumber or the second plumber?

Test

Find the area of each figure to the nearest tenth.

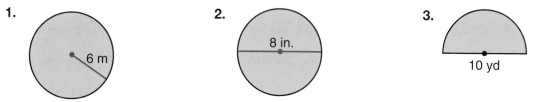

1. 6 m

2. 8 in.

3. 10 yd

4. Draw two views of a rectangular prism whose dimensions are 2 units by 3 units by 4 units.

Find the volume of each solid to the nearest tenth.

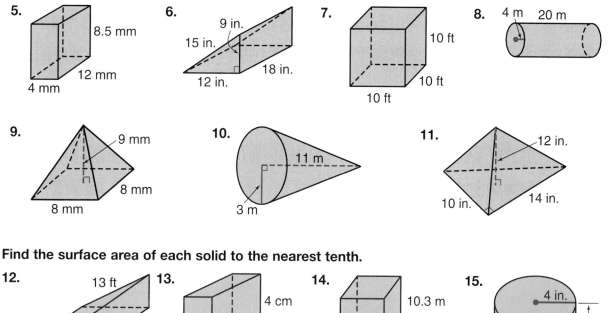

5. 8.5 mm, 12 mm, 4 mm

6. 9 in., 15 in., 18 in., 12 in.

7. 10 ft, 10 ft, 10 ft

8. 4 m, 20 m

9. 9 mm, 8 mm, 8 mm

10. 11 m, 3 m

11. 12 in., 10 in., 14 in.

Find the surface area of each solid to the nearest tenth.

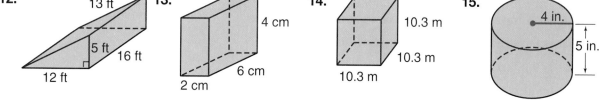

12. 13 ft, 5 ft, 16 ft, 12 ft

13. 4 cm, 6 cm, 2 cm

14. 10.3 m, 10.3 m, 10.3 m

15. 4 in., 5 in.

16. How many significant digits are in 10.4 centimeters?

17. What is the greatest possible error of 10.4 centimeters?

18. What is the relative error of 10.4 centimeters?

19. A grain silo has a diameter of 20 feet and is 50 feet tall. Another grain silo has a diameter of 30 feet and is 40 feet tall. Which silo has a greater volume? Explain.

20. Some cheeses are sealed in wax or a thin plastic film to protect their moisture. Find the surface area of the plastic film on the small cheese wheel.

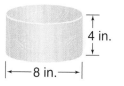

4 in.

8 in.

Video cameras come in three formats, VHS, VHS-C, or 8 mm, and with stereo or mono sound. They come with or without remote control.

1. Draw a tree diagram to find the number of possible choices of video cameras.

2. How many of the possible choices are mono video cameras?

Find each value.

3. $P(6, 3)$

4. 4!

5. $C(10, 5)$

6. Doug has 8 bowling trophies. How many ways can he arrange 4 of them in a row?

7. How many teams of 5 players can be chosen from 15 players?

8. In how many ways can 6 students stand in line?

Use Pascal's Triangle to answer each question.

9. How many combinations of 4 baseball cards can be chosen from 6 baseball cards?

10. What does the 6 in row 4 mean?

There are 4 blue marbles, 3 red marbles, and 2 white marbles in a bag. Once a marble is selected, it is not replaced. Find the probability of each outcome.

11. a red and then a white marble

12. two blue marbles

13. Does the probability in Exercise 12 represent dependent or independent events?

Two coins are tossed 20 times. No tails were tossed four times, one tail was tossed eleven times, and two tails were tossed five times.

14. What is the experimental probability of tossing no tails?

15. What is the theoretical probability of tossing no tails?

16. Why are the experimental probability and theoretical probability of tossing no tails different?

Use the survey on favorite flavors of gum to answer each question.

Favorite Flavors of Gum	
spearmint	12
bubble gum	20
fruit	8
peppermint	10

17. What is the size of this sample?

18. What percent of the sample choose bubble gum?

19. The school band plans to sell packages of gum to raise money for new uniforms. How many packages of bubble gum should they order if they plan to sell a total of 800 packages of gum?

20. Donna reaches into a bag containing 5 McIntosh apples, 3 Golden Delicious apples, and 2 Granny Smith apples. Explain how you could act it out to find the probability that she picks 2 Granny Smith apples.

CHAPTER TEST

Model each polynomial using drawings or algebra tiles.

1. $x^2 + 5x$

2. $-4x^2 - x + 3$

Simplify each expression. Then evaluate if $r = -1$ and $s = 2$.

3. $3r + 2s + 4r - s$

4. $r - 4s - 5r + 2s$

5. Identify the polynomial represented by the model at the right.

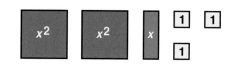

Simplify each polynomial. Use drawings or algebra tiles if necessary.

6. $2x^2 + 4x + 3x^2 + 5x$

7. $4c^2 + 3c - 3c^2 + c$

8. $3x^2 - y - 5x^2 + y$

9. $3a^2 + 7a - 4a^2 + 3a - 1$

Find each sum or difference. Use drawings or algebra tiles if necessary.

10. $(9z^2 - 3z) - (5z^2 + 8z)$

11. $(4c^2 + 2c) + (-4c^2 + c)$

12. $(-x^2 + 2x - 5) + (x^2 - 6x)$

13. $(5n^2 - 4n + 1) - (4n - 5)$

14. $\begin{array}{r} 5r^2 - 3r + 5 \\ + \ \ r^2 - 5r - 7 \\ \hline \end{array}$

15. $\begin{array}{r} 7y^2 + 4y - 5 \\ - (5y^2 + 7y - 10) \\ \hline \end{array}$

Find each product. Use algebra tiles or drawings if necessary.

16. $7^3 \cdot 7^5$

17. $c^6 \cdot c^8$

18. $8x(x + 3)$

19. $x^3(6x + 5)$

20. $(2x + 1)(x + 3)$

21. $(x + 4)(2x + 3)$

Factor each polynomial. Use drawings or algebra tiles if necessary.

22. $x^2 + 13x$

23. $x^2 + 7x + 10$

24. $x^2 + 6x + 5$

25. Randi rode her bicycle around a square city block. The length of each side of the block is $(4x + 3)$ feet. What is the total distance she rode?

Getting Acquainted with the Graphing Calculator

GRAPHING CALCULATORS

When some students first see a graphing calculator, they think, "Oh, no! Do we *have* to use one?", while others may think, "All right! We get to use these neat calculators!" There are as many thoughts and feelings about graphing calculators as there are students, but one thing is for sure: a graphing calculator *can* help you learn mathematics. Keep reading for answers to some frequently asked questions.

What is it?

So what is a graphing calculator? Very simply, it is a calculator that draws graphs. This means that it will do all of the things that a "regular" calculator will do, *plus* it will draw graphs of equations.

What does it do?

A graphing calculator can do more than just calculate and draw graphs. For example, you can program it and work with data to make statistical graphs and computations. If you need to generate random numbers, you can do that on the graphing calculator. If you need to find the absolute value of a number, you can do that too. It's really a very powerful tool, so powerful that it is often called a pocket computer.

What do all those different keys do?

As you may have noticed, graphing calculators have some keys that other calculators do not.

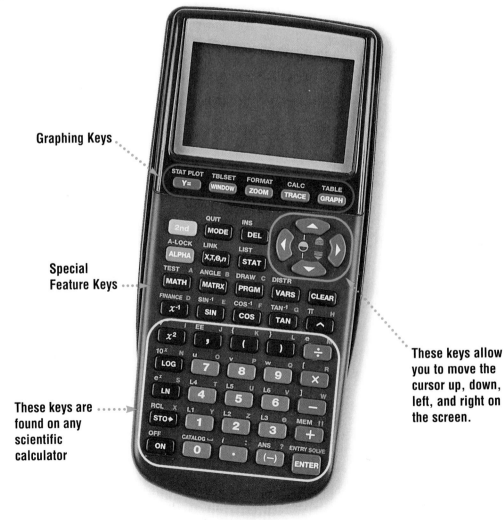

Graphing Keys

Special Feature Keys

These keys are found on any scientific calculator

These keys allow you to move the cursor up, down, left, and right on the screen.

Basic Keystrokes

- The yellow commands written above the calculator keys are accessed with the [2nd] key, which is also yellow. Similarly, the green characters above the keys are accessed with the [ALPHA] key, which is also green. In this text, commands that are accessed by the [2nd] and [ALPHA] keys are shown in brackets. For example, [2nd] [QUIT] means to press the [2nd] key followed by the key below the yellow [QUIT] command.

- [2nd] [ENTRY] copies the previous calculation so you can edit and use it again.

- [2nd] [QUIT] will return you to the home (or text) screen.

- Negative numbers are entered using the [(-)] key, not the minus sign, [−] .

- [2nd] [OFF] turns the calculator off.

Order of Operations

As with any scientific calculator, the graphing calculator observes the order of operations.

Example	Keystrokes	Display
4 + 13	4 [+] 13 [ENTER]	4 + 13 17
5^3	5 [∧] 3 [ENTER]	5 ∧ 3 125
4 (9 + 18)	4 [(] 9 [+] 18 [)] [ENTER]	4(9 + 18) 108
$\sqrt{24}$	[2nd] [√] 24 [ENTER]	√ (24 4.8989 79486

Programming

Programming features allow you to write and execute a series of commands for tasks that may be too complex or cumbersome to perform otherwise. Each program is given a name. Commands begin with a colon (:), which the calculator enters automatically, followed by an expression or an instruction.

When you press [PRGM] , you see three menus: EXEC, EDIT, and NEW. EXEC allows you to execute a stored program, EDIT allows you to edit or change a program, and NEW allows you to create a program.

- To begin entering a new program, press [PRGM] [▶] [▶] [ENTER] .

- You do not need to type each letter using the [ALPHA] key. Any command that contains lowercase letters should be entered by choosing it from a menu. Check your user's guide to find any commands that are unfamiliar.

- After a program is entered, press [2nd] [QUIT] to exit the program mode and return to the home screen.

- To execute a program, press [PRGM] . Then use the down arrow key to locate the program name and press [ENTER] twice, or press the number or letter next to the program name followed by [ENTER] .

- If you wish to edit a program, press [PRGM] [▶] and choose the program from the menu.

- To immediately re-execute a program after it is run, press [ENTER] when Done appears on the screen.

- To stop a program during execution, press [ON] or [2nd] [QUIT].

While a graphing calculator cannot do everything, it can make some things easier and help your understanding of math. To prepare for whatever lies ahead, you should try to learn as much as you can. Who knows? Maybe one day you will be designing the next satellite or building the next skyscraper with the help of a graphing calculator!

Getting Acquainted with Spreadsheets

What do you think of when people talk about computers? Maybe you think of computer games or using a word processor to write a school paper. But a computer is a powerful tool that can be used for many things.

One of the most common computer applications is a spreadsheet program. Here are answers to some of the questions you may have if you're new to using spreadsheets.

What is it?

You have probably seen tables of numbers in newspapers and magazines. Similar to those tables, a spreadsheet is a table that you can use to organize information. But a spreadsheet is more than just a table. You can also use a spreadsheet to perform calculations or make graphs.

Why use a spreadsheet?

The advantage a spreadsheet has over a simple calculator is that when a number is changed, the entire spreadsheet is recalculated and the new results are displayed. So with a spreadsheet, you can see patterns in data and investigate what happens if one or more of the numbers is changed.

How do I use a spreadsheet?

A spreadsheet is organized into boxes called *cells*. The cells are named by a letter, that identifies the column, and a number, that identifies the row. In the spreadsheet below, cell C4 is highlighted.

	A	B	C
1	Width	Length	Area
2	3	4	12
3	2	10	20
4	5	12	60
5	8	14	112

To enter information in a spreadsheet, simply move the cursor to the cell you want to access and click the mouse. Then type in the information and press Enter.

How do I enter formulas?

If you want to use the spreadsheet as a calculator, begin by choosing the cell where you want the result to appear.

- For a simple calculation, type = followed by the formula. For example, in the spreadsheet above, the formula in cell C2 is entered as "=A2*B2." *Notice that * is the symbol for multiplication in a spreadsheet.*

- Sometimes you will want similar formulas in more than one cell. First type the formula in one cell. Then select the cell and click the copy button. Finally select the cells where you want to copy the formula and click the paste button.

- Often it is useful to find the sum or average of a row or column of numbers. A spreadsheet allows you to choose from several functions like this instead of entering the formula manually. To enter a function, click the cell where you want the result to appear. Then click on the = button above the cells. A list of formulas will appear to the left. Click the down arrow button and choose your function. The spreadsheet will enter a range, which you may alter. For example, to find the average of row 2 of the spreadsheet below, the function chooses to find the average of cells B2, C2, and D2.

	A	B	C	D	E
1	Student	Test 1	Test 2	Test 3	Average
2	Kathy	88	85	91	88
3	Ben	86	89	92	89
4	Carmen	92	86	92	90
5	Anthony	80	88	87	85

The formula for cell E2 is =(B2+C2+D2)/3.

Spreadsheet software is one of the most common tools used in business today. You should try to learn as much as you can to prepare for your future. Who knows? Maybe you'll use what you're learning today as a company president tomorrow!

The Tangent Ratio

What you'll learn

You'll learn to find the tangent of an angle and find missing measures using the tangent.

When am I ever going to use this?

Knowing how to use the tangent ratio can help you find unknown measures in triangles.

Word Wise

tangent

The industrial technology class plans to add a wheelchair ramp to the emergency exit of the auditorium as a class project. They know that the landing is 3 feet high and that the angle the ramp makes with the ground cannot be greater than 6°. What is the minimum distance from the landing that the ramp should start? *This problem will be solved in Example 1.*

Problems like the one above involve a right triangle and ratios. One ratio, called the **tangent,** compares the measure of the leg opposite an angle with the measure of the leg adjacent to that angle. The symbol for the tangent of angle A is tan A.

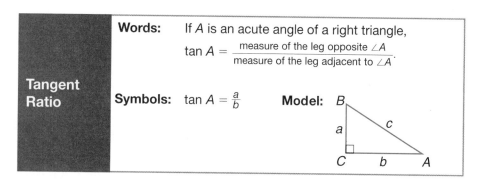

Tangent Ratio	**Words:** If A is an acute angle of a right triangle, $\tan A = \dfrac{\text{measure of the leg opposite } \angle A}{\text{measure of the leg adjacent to } \angle A}$.
	Symbols: $\tan A = \dfrac{a}{b}$ **Model:**

You can also use the symbol for tangent to write the tangent of an angle measure. The tangent of a 60° angle is written as tan 60°. If you know the measures of one leg and an acute angle of a right triangle, you can use the tangent ratio to solve for the measure of the other leg.

Example 1

Real World APPLICATION

Construction Solve the problem about the wheelchair ramp.

First, draw a diagram.

$m\angle A = 6°$

adjacent leg = x feet

opposite leg = 3 feet

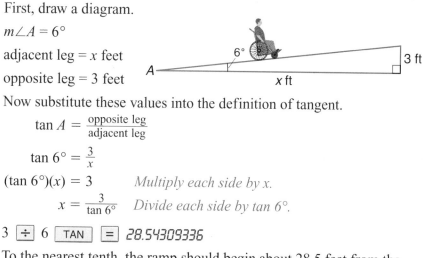

Now substitute these values into the definition of tangent.

$$\tan A = \frac{\text{opposite leg}}{\text{adjacent leg}}$$

$$\tan 6° = \frac{3}{x}$$

$(\tan 6°)(x) = 3$ *Multiply each side by x.*

$x = \dfrac{3}{\tan 6°}$ *Divide each side by tan 6°.*

3 ÷ 6 TAN = 28.54309336

To the nearest tenth, the ramp should begin about 28.5 feet from the landing.

If your calculator does not have a TAN *key, you can use the table on the back cover of this booklet to estimate answers.*

You can use the TAN^{-1} function on your calculator to find the measure of an acute angle of a right triangle when you know the measures of the two legs.

2 Find the measure of $\angle A$ to the nearest degree.

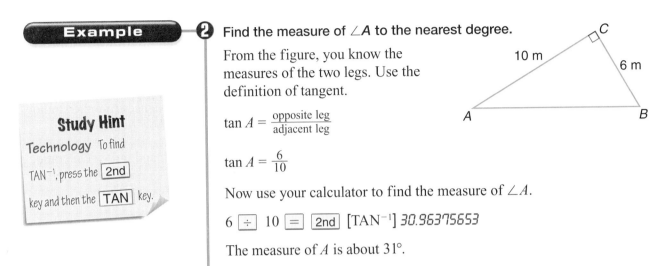

From the figure, you know the measures of the two legs. Use the definition of tangent.

$$\tan A = \frac{\text{opposite leg}}{\text{adjacent leg}}$$

$$\tan A = \frac{6}{10}$$

Now use your calculator to find the measure of $\angle A$.

6 ÷ 10 = 2nd [TAN^{-1}] *30.963755653*

The measure of A is about 31°.

Study Hint

Technology To find TAN^{-1}, press the 2nd key and then the TAN key.

CHECK FOR UNDERSTANDING

Communicating Mathematics

Read and study the lesson to answer each question.

1. *Write* a definition of the tangent ratio.

2. *Tell* how to use the tangent ratio to find the measure of a leg of a right triangle.

3. *Tell* how to find the measure of an angle in a right triangle when you know the measures of the two legs.

Guided Practice

Find each tangent to the nearest tenth. Find the measure of each angle to the nearest degree.

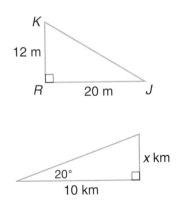

4. $\tan J$

5. $\tan K$

6. $m\angle J$

7. $m\angle K$

8. Find the value of x to the nearest tenth.

9. *Measurement* A guyline is fastened to a TV tower 50 feet above the ground and forms an angle of 65° with the tower. How far is it from the base of the tower to the point where the guyline is anchored into the ground? Round to the nearest foot.

Practice

Complete each exercise using the information in the figures. Find each tangent to the nearest tenth. Find the measure of each angle to the nearest degree.

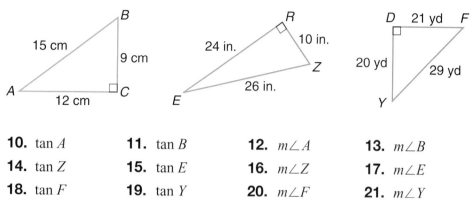

10. tan A

11. tan B

12. $m\angle A$

13. $m\angle B$

14. tan Z

15. tan E

16. $m\angle Z$

17. $m\angle E$

18. tan F

19. tan Y

20. $m\angle F$

21. $m\angle Y$

Find the value of x to the nearest tenth.

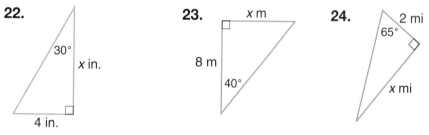

22.

23.

24.

25. If the leg opposite the 53° angle in a right triangle is 4 inches long, how long is the other leg to the nearest tenth?

26. If the leg adjacent to a 29° angle in a right triangle is 9 feet long, what is the measure of the other leg to the nearest tenth?

Applications and Problem Solving

27. *Measurement* A flagpole casts a shadow 25 meters long when the angle of elevation of the Sun is 40°. How tall is the flagpole to the nearest meter?

28. *Surveying* A surveyor is finding the width of a river for a proposed bridge. A theodolite is used by the surveyor to measure angles. The distance from the surveyor to the proposed bridge site is 40 feet. The surveyor measures a 50° angle to the bridge site across the river. Find the length of the bridge to the nearest foot.

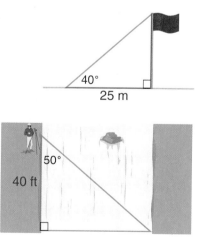

29. *Critical Thinking* In a right triangle, the tangent of one of the acute angles is 1. Describe how the measures of the two legs are related.

The Sine and Cosine Ratios

What you'll learn

You'll learn to find the sine and cosine of an angle and find missing measures using sine and cosine.

When am I ever going to use this?

Knowing how to use trigonometric ratios can help you solve problems using angles of elevation.

Word Wise

trigonometry
sine
cosine
angle of elevation

Toni decided to make a scale drawing of the Leaning Tower of Pisa for her project in art class. She knows the tower is 177 feet tall and tilts 16.5 feet off the perpendicular. First, she wants to draw the angle representing the tilt of the tower. What should be the measure of this angle? *This problem will be solved in Example 3.*

You know the measures of one leg and the hypotenuse of a right triangle. These are *not* the measures you need to use the tangent ratio. The tangent ratio is only one of several ratios used in the study of **trigonometry.**

Two other ratios are the **sine** ratio and the **cosine** ratio. These can be written as sin A and cos A. They are defined as follows.

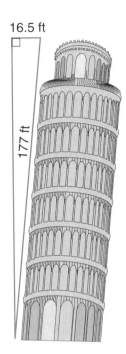

16.5 ft

177 ft

Sine and Cosine Ratios	**Words:**	If A is an acute angle of a right triangle,
		$\sin A = \dfrac{\text{measure of the leg opposite } \angle A}{\text{measure of the hypotenuse}}$ and
		$\cos A = \dfrac{\text{measure of the leg adjacent to } \angle A}{\text{measure of the hypotenuse}}$.
	Symbols: $\sin A = \dfrac{a}{c}$ $\cos A = \dfrac{b}{c}$	**Model:**

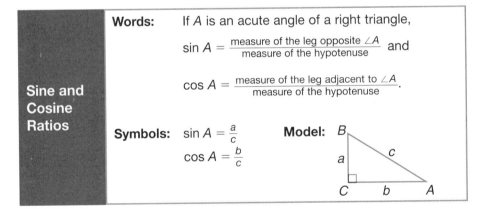

Example

1 Use △ ABC to find sin A, cos A, sin B, and cos B.

$$\sin A = \frac{BC}{AB}$$
$$= \frac{4}{5} \text{ or } 0.8$$

$$\sin B = \frac{AC}{AB}$$
$$= \frac{3}{5} \text{ or } 0.6$$

$$\cos A = \frac{AC}{AB}$$
$$= \frac{3}{5} \text{ or } 0.6$$

$$\cos B = \frac{BC}{AB}$$
$$= \frac{4}{5} \text{ or } 0.8$$

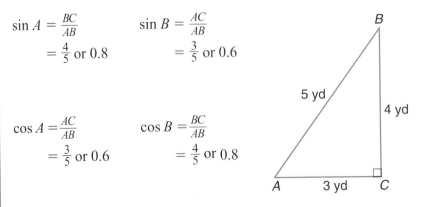

5 yd
4 yd
3 yd

You can find the sine and cosine of an angle by using a calculator.

sin 63° → 63 [SIN] *0.891006524*

cos 63° → 63 [COS] *0.4539905*

You can use the sine and cosine ratios to find missing lengths of sides or angle measures in a right triangle.

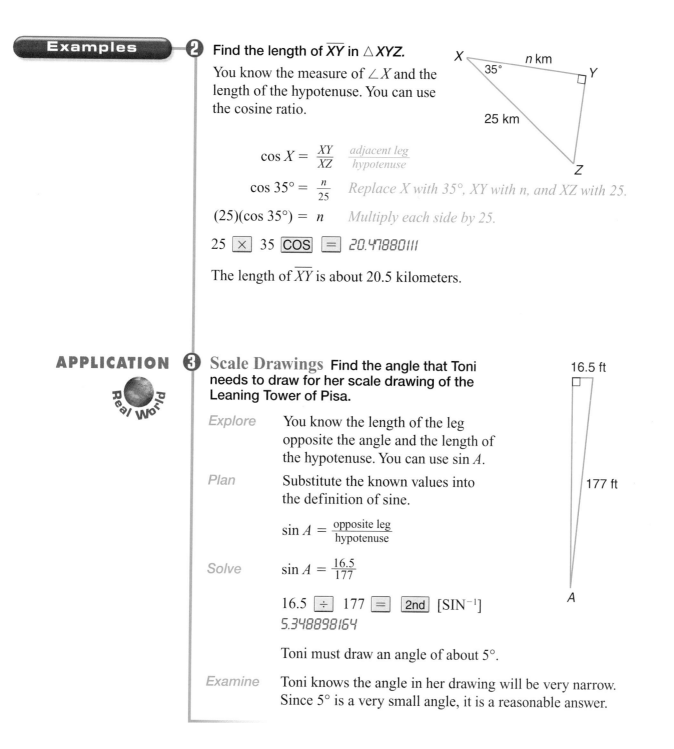

Examples

2 **Find the length of $\overline{XY}$ in △XYZ.**

You know the measure of ∠X and the length of the hypotenuse. You can use the cosine ratio.

$$\cos X = \frac{XY}{XZ} \quad \text{\textit{adjacent leg}}{\text{\textit{hypotenuse}}}$$

$$\cos 35° = \frac{n}{25} \quad \textit{Replace X with 35°, XY with n, and XZ with 25.}$$

$$(25)(\cos 35°) = n \quad \textit{Multiply each side by 25.}$$

25 [×] 35 [COS] [=] *20.47880111*

The length of $\overline{XY}$ is about 20.5 kilometers.

APPLICATION

3 **Scale Drawings** Find the angle that Toni needs to draw for her scale drawing of the Leaning Tower of Pisa.

Explore   You know the length of the leg opposite the angle and the length of the hypotenuse. You can use sin A.

Plan   Substitute the known values into the definition of sine.

$$\sin A = \frac{\text{opposite leg}}{\text{hypotenuse}}$$

Solve $\sin A = \frac{16.5}{177}$

16.5 [÷] 177 [=] [2nd] [SIN⁻¹]
5.348898164

Toni must draw an angle of about 5°.

Examine   Toni knows the angle in her drawing will be very narrow. Since 5° is a very small angle, it is a reasonable answer.

Many problems that can be solved using trigonometric ratios deal with angles of elevation. An **angle of elevation** is formed by a horizontal line and a line of sight above it.

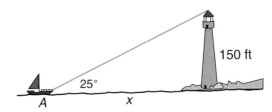

Angle of elevation
horizontal line

Example

INTEGRATION

4

Measurement The angle of elevation from a small boat to the top of a lighthouse is 25°. If the top of the lighthouse is 150 feet above sea level, find the distance from the boat to the base of the lighthouse.

150 ft

25°

A

x

Let x = the distance from the boat to the base of the lighthouse.

$\tan 25° = \dfrac{150}{x}$ *opposite leg*
                   adjacent leg

$(\tan 25°)\, x = 150$ *Multiply each side by x.*

$x = \dfrac{150}{\tan 25°}$ *Divide each side by tan 25°.*

150 ÷ 25 TAN = *321.8760381*

The boat is about 322 feet from the base of the lighthouse.

CHECK FOR UNDERSTANDING

Communicating Mathematics

Read and study the lesson to answer each question.

1. *Write* a definition for the sine and cosine ratios.

2. *Show* how you could use the sine or cosine to find the missing measure of one of the legs if you know the hypotenuse and an acute angle.

Guided Practice

Find each sine or cosine to the nearest tenth. Find the measure of each angle to the nearest degree.

3. $\cos A$

4. $\sin A$

5. $m\angle A$

6. $\sin B$

7. $\cos B$

8. $m\angle B$

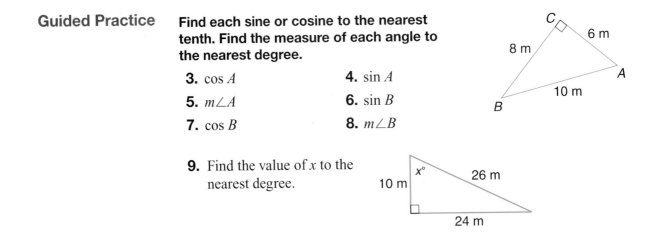

9. Find the value of x to the nearest degree.

10 m $x°$ 26 m

24 m

10. *Transportation* The end of an exit ramp from an interstate highway is 22 feet higher than the highway. If the ramp is 630 feet long, what angle does it make with the highway? Round to the nearest degree.

EXERCISES

Practice

Complete each exercise using the information in the figures. Find each sine or cosine to the nearest tenth. Find the measure of each angle to the nearest degree.

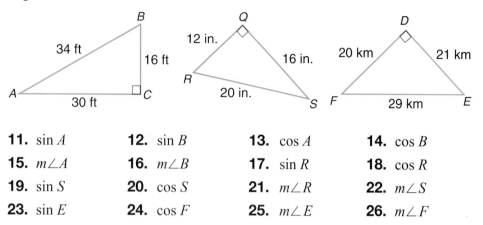

11. sin *A*	**12.** sin *B*	**13.** cos *A*	**14.** cos *B*
15. *m∠A*	**16.** *m∠B*	**17.** sin *R*	**18.** cos *R*
19. sin *S*	**20.** cos *S*	**21.** *m∠R*	**22.** *m∠S*
23. sin *E*	**24.** cos *F*	**25.** *m∠E*	**26.** *m∠F*

Find the value of *x* to the nearest tenth or nearest degree.

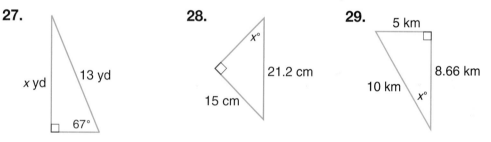

27.

28.

29.

Applications and Problem Solving

30. *Home Maintenance* A painter props a 20-foot ladder against a house. The angle it forms with the ground is 65°. To the nearest foot, how far up the side of the house does the ladder reach?

31. *Surveying* A surveyor is 85 meters from the base of a building. The angle of elevation to the top of the building is 20°. If her eye level is 1.6 meters above the ground, find the height of the building to the nearest meter.

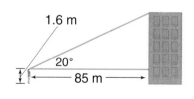

32. *Fire Fighting* A fire is sighted from a fire tower at an angle of depression of 2°. If the fire tower has a height of 125 feet, how far is the fire from the base of the tower? Round to the nearest foot.

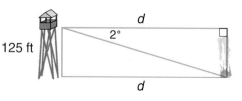

33. *Critical Thinking* Study your answers to Exercises 11–26. Make a conjecture about the relationship between the sine and cosine of complementary angles.

The Sine and Cosine Ratios **669**

Selected Answers

CHAPTER 1
Problem Solving and Algebra

Pages 6–7   Lesson 1-1
1. Explore – Identify what information is given and what you need to find. Plan – Estimate the answer and then select a strategy for solving. Solve – Carry out the plan and solve. Examine – Compare the answer to the estimate and determine if it is reasonable. If not, make a new plan.   **5.** Sample answer: Oxford Circus – Euston – Old Street   **7.** Sample answer: *Explore:* Locate and examine a website and investigate the software used to design homepages. *Plan:* Make a blueprint of how you would like the final page to look. *Solve:* Turn your blueprint into a home page using any available software. *Examine:* Test the code and see if the home page really looks like your blueprint. Make the necessary adjustments.   **9.** 4 by 4 by 4   **11.** Sample answer: clockwise from top – 17, 10, 45

Pages 9–10   Lesson 1-2
1. The 3 means to use 2 as a factor 3 times.
5. $16^2 \cdot 20^2$ **7.** 64   **9.** 288
11a. $10 \cdot 10 \cdot 10 \cdot 10 \cdot 10 \cdot 10 \cdot 10 \cdot 10$
11b. 484,000,000   **13.** 10^2 **15.** $4^3 \cdot 8^2$
17. $12 \cdot 14^2 \cdot 5^3$ **19.** $18^2 \cdot 5^3$ **21.** 1,296   **23.** 1
25. 2,000   **27.** 1,000,000   **29.** 22   **31.** 768
33. $10^3 \cdot 24^2 \cdot 50$ **35a.** 12^3 **35b.** 12.167 cubic meters   **37.** 16   **39.** 1 pen and 2 tablets

Pages 14–15   Lesson 1-3
1. Numerical expressions contain only numbers and algebraic expressions contain numbers and variables.   **3.** 33   **5.** 6   **7.** 6   **9.** 44   **11.** 15
13. 43,000   **15.** 40   **17.** 8   **19.** 2   **21.** 2
23. 37   **25.** 48   **27.** 96   **29.** 13.5   **31.** 9
33. 89   **35.** 31   **37.** 8   **39.** 4   **41.** 7   **43a.** y represents the number of years between Ms. Chisholm's and Ms. Braun's elections.   **43b.** 24
45. 64   **47.** 6 pounds

Page 16   Lesson 1-3B
1. 0; 27; 216   **3.** 48; 192; 1,200   **5.** 4 STO►
X,T,θ,n ALPHA [:] 7 STO► ALPHA [Y]

Pages 19–20   Lesson 1-4
1. Add when a number is subtracted from the variable and subtract when a number is added to the variable.   **3.** Tess is correct. You only need to use an inverse operation to solve if there is an operation performed on the side with the variable.   **5.** 7
7. 9   **9.** 43   **11.** \$247,500   **13.** 4   **15.** 35
17. 1   **19.** 27.4   **21.** 83   **23.** 52   **25.** 27
27. 14.46   **29.** 86   **31a.** 11,750 pounds
31b. 4,000 pounds   **33.** 4   **35.** 6   **37.** $5^2 \cdot 8^3$

Pages 23–25   Lesson 1-5
1. Multiply; To undo division, you multiply.   **3.** $3x = 12; x = 4$ **5.** 5   **7.** 78   **9.** 3.4   **11.** 225
13. 6   **15.** 448   **17.** 54   **19.** 0.53   **21.** 200
23. 7   **25.** 42   **27.** 183   **29.** 0.55   **31a.** \$161
31b. \$224   **33.** 435 representatives   **35.** 1.8 million years   **37.** $4^3 \cdot 8^2$

Page 25   Mid-Chapter Self Test
1. \$164,700   **3.** 6,912   **5.** 15   **7.** 20.1   **9.** 126

Pages 28–29   Lesson 1-6
1. Sample answer: three more than a number; a number increased by 3   **3.** Chapa is correct. *7 less than y* is translated as $y - 7$ and *7 less y* is translated $7 - y$.   **5.** $7 \cdot n$ or $7n$ **7.** $w - 9 = 15$ **9.** $d \div 4 = 7$ **11.** $k + 9$ **13.** $n + 5$ **15.** $16r = 80$
17. $6 + b$ **19.** $j + 10$ **21.** $48 - v = 15$ **23.** $2c$
25. $n - 8 = 25$ **27.** $n \div 3 = 25$ **29a.** $n + 2,249$
29b. $n + 2,249 = 2,425$ **31a.** $s - 9$ **31b.** $2.3a = 103.5$ **33.** 15   **35.** 97

Pages 30–31   Lesson 1-7A
1. They started with the money they had left and added on the amounts they spent in reverse order.
3. It is easier to work backward than forward when you know a result and need to know the start.
5. Work the problem forward and make sure the ending number is correct.   **7.** Undoing the subtraction first then undoing the multiplication is backward from the order of operations.   **9.** \$6.72
11. \$8,100   **13a.** \$393   **13b.** \$4,882

Pages 35–36   Lesson 1-7
1. Subtract 5 from each side.   **3.** Remove 3 counters from each side. Then divide the remaining counters into four groups. The solution is $x = 3$.

Check by substituting 3 counters for each cup and verifying that the result is a true equation.   **5.** 12
7. 64   **9.** 4   **11.** 32 clues   **13.** 3   **15.** 4
17. 152   **19.** 192   **21.** 0.5   **23.** 0.3   **25.** 0.25
27. 12   **29.** 18.9   **31.** 4   **33.** $\frac{n}{4} - 1 = 7$; 32
35. 355 miles   **39.** $c + 3 = 15$ **41.** 92   **43.** 196

Page 37   Lesson 1-7B

1. Multiply the input by 7. Then subtract 2 from the result.   **3.** Work backward to solve the two-step equation.   **5.** 12

Pages 40–42   Lesson 1-8

1. 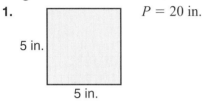 $P = 20$ in.

5 in.

5 in.

3. Sample answer: Area is the amount of space a figure takes up, the perimeter is the distance around the figure.   **5.** $P = 26$ yd; $A = 40$ yd^2 **7.** $P = 48$ cm; $A = 128$ cm^2 **9.** $P = 16$ units; 15 units^2
11. $P = 22.6$ m; $A = 29.2$ m^2 **13.** $P = 20$ m; $A = 23.5$ m^2 **15.** $P = 16$ units; $A = 12$ units^2 **17.** $P = 60$ in.; $A = 185$ in^2 **19.** $144 = 8b$; $b = 18$ yards
23a. 1 in^2 **23b.** 10,800 boxes   **25.** a $-\frac{1}{4}$ unit^2;
b $-\frac{1}{4}$ unit^2; c $-\frac{1}{16}$ unit^2; d $-\frac{1}{8}$ unit^2; e $-\frac{1}{16}$ unit^2;
f $-\frac{1}{8}$ unit^2; g $-\frac{1}{8}$ unit^2 **27.** A

Pages 46–47   Lesson 1-9

1. "In excess of" indicates that the speeds are more than 60 mph.
3. 0 1 2 3 4 5 6 7 8 9 10
5. $d < 9$ 4 6 8 10 12
7. $x < 9$ 4 6 8 10 12
9. $c > 5$ 0 2 4 6 8
11a. $3(x + 12) \le 90$ **11b.** $x \le 18$
13. $y \ge 25$ 23 24 25 26 27
15. $d \le 4$ 0 1 2 3 4 5
17. $m > 10$ 0 10 20 30
19. $9 > h$ 4 6 8 10 12

21. $r \le 7$ 0 1 2 3 4 5 6 7
23. $k \le 16$ 0 4 8 12 16 20
25. $y \le 9$ 0 3 6 9 12
27. $r < 50$ 0 20 40 60
29. $m \le 18$ 0 6 12 18 24
31. $4 + x > 15$; $x > 11$ **33.** $a > 16$
35. $2 < x \le 7$ **37.** $n = 28$ **39.** 35

Pages 48–51   Study Guide and Assessment

1. false; exponent   **3.** false; inequality   **5.** false; second   **7.** true   **9.** 8 minutes   **11.** $5^2 \cdot 6^3$
13. 625   **15.** 322   **17.** 10   **19.** 160   **21.** 50
23. 75   **25.** 2.9   **27.** 4.1   **29.** 294   **31.** 72
33. 12   **35.** $6x$ **37.** $4x = 48$ **39.** 66   **41.** 8
43. $P = 28$ m; $A = 36$ m^2
45. $a > 12$ 0 4 8 12 16 20 24
47. $n + 5 > 16$; $n > 11$ **49.** 4 pounds   **51.** $6 for each cup; $10 for each bowl

Pages 52–53   Standardized Test Practice

1. C   **3.** D   **5.** D   **7.** B   **9.** B   **11.** E   **13.** 62 pages   **15.** $t < 4$ **17.** 20   **19.** $A = 10$ in^2; $P = 14$ in.

····················

CHAPTER 2
Algebra: Using Integers

Pages 57–58   Lesson 2-1

1. Sample answer: Draw a number line. Locate the number and draw a dot at that point on the line. Label the point with a letter.
5.   −8 −6 −4 −2 0 2 4 6 8
7.   −8 −6 −4 −2 0 2 4 6 8
9. 23   **11.** 0   **13.** −92
15.   −8 −6 −4 −2 0 2 4 6 8
17.   −8 −6 −4 −2 0 2 4 6 8
   −7 −5   9 12
19.   −8 −6 −4 −2 0 2 4 6 8 10 12 14

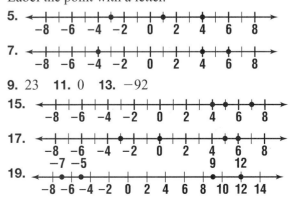

21.

A number line from -4 to 10 with points marked at -2, 4, 8.

23.

A number line from -8 to 8 with points marked at -2, 0, 2, 4.

25. 5 **27.** 12 **29.** 45 **31.** 88 **33.** 4 **35.** 5
37. -3 **39.** 12 **41.** False. For example, $|-4| > |2|$ but $-4 < 2$; and $|-8| > |1|$, but $-8 < 1$. **43.** C
45. 171

Pages 60–61   Lesson 2-2

1. $-3 < 4; 4 > -3$ **3.** Mariano is correct. **5.** $>$
7. $>$ **9.** $40 > -22; -22 < 40$ **11.** $=$ **13.** $>$
15. $<$ **17.** $=$ **19.** $>$ **21.** $=$ **23.** 564, 254,
$-100, -356, -450$ **25.** $1 > -1; -1 < 1$
27a. No; $|3| = 3$ and $3 \not< 3$
27b.

A number line from -6 to 6 with points marked at 0, 1, 2, 3.

29. 6

Pages 64–65   Lesson 2-3

1. Start at the first addend. Move to the left to add a
negative number. **3.** -1; no, addition is
commutative. **5.** -7 **7.** -5 **9.** 0 **11.** -10
13. -50 **15.** 0 **17.** 11 **19.** -26 **21.** 20
23. 108 **25.** 22 **27.** -130 **29.** -30 **31.** 31
33. 0 **35.** 8 **37a.** $v = -14,608 + 5,202$
37b. $-9,406$ million **39.** no; $|-3| + |5| = 8$,
$|-3 + 5| = 2; -3 + 5 = 2$ **41.** D **43.** 32

Pages 67–68   Lesson 2-4

1. Sample answer: Add $-13 + 13 = 0$;
$-45 + (-55) = -100$; and $15 + 25 = 40$. Then
$0 + (-100) + 40 = -60$. **3.** 13 **5.** 2 **7.** 4
9. -7 **11.** 14 **13.** -3 **15.** 12 **17.** 12
19. 214 **21.** 2 **23.** -1 **27a.** $389,100,000
loss **27b.** $-\$95,250,000$ **29.** B **31.** 64 m^2

Pages 71–72   Lesson 2-5

1. Sample answer: The additive inverse of an integer
is called its opposite. The opposite of an integer is
the same distance from 0, but has the opposite sign.
For example, 4 and -4 are additive inverses.

3.

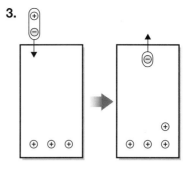

5. -1 **7.** 9 **9.** 0 **11.** 75 **13.** -7 **15.** 0
17. -21 **19.** -25 **21.** -131 **23.** 55 **25.** 78
27. -29 **29.** -412 **31.** 842 **33.** 6 **35.** -18
37. -11 **41a.** Answers will vary. For 1998, it was
6,710 years ago. **41b.** 7,979 years **43.** 47
45. B **47.** 18

Page 72   Mid-Chapter Self Test

1.

A number line from -10 to 10 with points marked at -6, -2, 0, 2, 4, 8.

3. -2 **5.** 0

Pages 74–75   Lesson 2-6

1. 3 rows, 2 columns **3.** $\begin{bmatrix} -1 & 7 \\ 5 & -2 \end{bmatrix}$

5. $\begin{bmatrix} 0 & 4 & 1 \\ -2 & -3 & -2 \\ 3 & 0 & 3 \end{bmatrix}$ **7.** $\begin{bmatrix} 8 & 4 & 7 & 6 \\ 13 & 2 & 0 & 0 \\ 7 & 5 & 47 & 24 \\ 13 & 12 & 47 & 36 \end{bmatrix}$

9. impossible **11.** $\begin{bmatrix} 11 & -12 \\ 1 & 26 \\ 3 & 28 \end{bmatrix}$ **13.** $\begin{bmatrix} 5 & 5 & 3 \\ 1 & -3 & 10 \end{bmatrix}$

15. $\begin{bmatrix} 16 & 4 & -7 \\ 5 & -1 & 5 \\ 1 & 9 & -12 \end{bmatrix}$ **17.** $\begin{bmatrix} 0 & 8 & -4 \\ 0 & -9 & 10 \\ 1 & 21 & -3 \end{bmatrix}$

19. $\begin{bmatrix} 0 & 10 & 20 \\ 9 & -1 & 48 \\ 3 & -11 & -8 \\ -6 & 34 & 9 \end{bmatrix}$ **21a.** $\begin{bmatrix} 4,362 & 5,917 \\ 4,904 & 4,838 \\ 3,353 & 3,344 \end{bmatrix}$

21b. $\begin{bmatrix} 399 & 554 \\ 157 & 299 \\ 165 & 139 \end{bmatrix}$ **23.** A **25.** more than 700,000

Page 77   Lesson 2-6B

1. $\begin{bmatrix} 5 & 5 & 5 \\ -1 & 11 & 7 \\ 9 & -8 & 2 \end{bmatrix}$ **3.** $\begin{bmatrix} 3 & -5 & -1 \\ 5 & -9 & -5 \\ -3 & 6 & 6 \end{bmatrix}$

5. Yes; $D + E = E + D$.

Page 80 Lesson 2-7

1. Sample answer: $[(-35)(45)](-2) = -1,575(-2) = 3,150$ or $(-35)[(-2)(45)] = (-35)(-90) = 3,150$
3. $(-2)(-4) = 8$

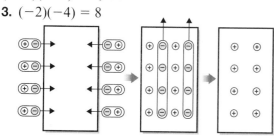

5. 6 **7.** -54 **9.** 16 **11.** -840 **13.** 48
15. 24 **17.** 78 **19.** 70 **21.** 504 **23.** 805
25. 441 **27.** $-2,592$ **29.** 400 **31.** 200
33. -60 or 60 heating degree-days **35.** y is
non-negative. If $y = 0$, then $y^2 = 0$. If y is any other
number, y^2 is positive because any two numbers with
the same sign have a positive product.
37. -13

Pages 82–83 Lesson 2-8

1. negative **3.** Rachel is correct. The divisor is 5
in all of the problems. **5.** -40 **7.** -7 **9.** -73
11. -3 **13.** about 2 million tons **15.** 2 **17.** 7
19. -7 **21.** 28 **23.** -49 **25.** -6 **27.** -12
29. -13 **31.** -98 **33.** -8 **35.** 13 **37.** 53
39. 8 **41a.** about $-15,533$ **41b.** about
6,472,000 **45a.** Division is not commutative. For
example, $9 \div 3 \neq 3 \div 9$. **45b.** Division is not
associative. For example, $(16 \div 4) \div 2 \neq 16 \div (4 \div 2)$. **47.** -37 **49.** B

Pages 84–85 Lesson 2-9A

1. 9 **3.** 2 **5.** $x - 5 = 8$ **7.** 6 **9.** 5 **11.** -8
13. 2 **15.** 6

Pages 88–89 Lesson 2-9

1.

$3r + 5 = -10$	
$3r + 5 - 5 = -10 - 5$	*Subtract 5 from each side.*
$3r = -15$	*Simplify.*
$\frac{3r}{3} = \frac{-15}{3}$	*Divide each side by 3.*
$r = -5$	*Simplify.*
$3r - (-5) = -10$	*Rewrite using the additive inverse.*
$3r + 5 = -10$	*Then follow the same steps as above.*

3. Sample answer: Your mistake could be in the
solution of the equation or could be in your check.
Check both. **5.** -3 **7.** -275 **9.** -20 **11.** -9
13. -72 **15.** -318 **17.** 58 **19.** 35 **21.** -15
23. 36 **25.** 17 **27.** -48 **29.** $-2 - x = 8$; -10
31. $2x + 7 = -21$; -14 **33.** 40 yards **35.** -13
37. 4

Pages 90–91 Lesson 2-9B

1. Crystal and Jackie eliminated the days that they
could not play to see which day or days they could.
5. To plan the solution, you need to identify the
possibilities for the solution. Then when you solve
you can eliminate the ones that are not the solution.
9a. They could take the following buses to
Baltimore and home: 8:00 A.M. bus and 3:00 P.M.;
9:00 A.M. and 4:00 P.M.; 10:00 A.M. and 5:00 P.M.;
11:00 A.M. and 6:00 P.M.; 12:00 P.M. and 7:00 P.M.;
1:00 P.M. and 8:00 P.M. **9b.** yes **11.** A
13. 111.485 mph

Pages 93–95 Lesson 2-10

1. For the x-coordinate, left is negative and right is
positive. For the y-coordinate, up is positive and
down is negative. **3.** Sample answer: x comes
before y in the alphabet. **5.** $(0, 3)$ **7.** $(-3, 2)$
9, 11, 13.

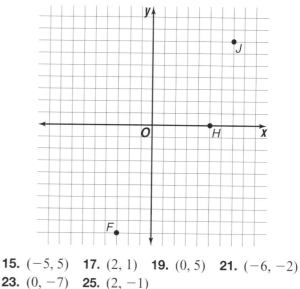

15. $(-5, 5)$ **17.** $(2, 1)$ **19.** $(0, 5)$ **21.** $(-6, -2)$
23. $(0, -7)$ **25.** $(2, -1)$

27, 29, 31, 33, 35, 37, 39, 41, 43.

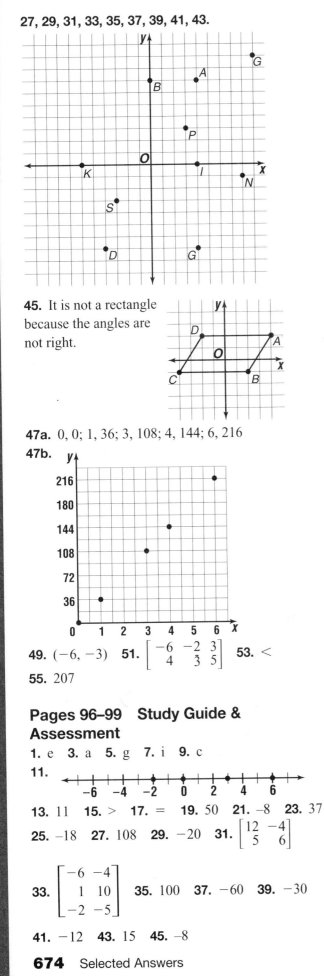

45. It is not a rectangle because the angles are not right.

47a. 0, 0; 1, 36; 3, 108; 4, 144; 6, 216
47b.

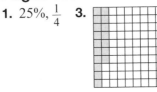

49. $(-6, -3)$ **51.** $\begin{bmatrix} -6 & -2 & 3 \\ 4 & \frac{2}{3} & 5 \end{bmatrix}$ **53.** $<$
55. 207

Pages 96–99 Study Guide & Assessment

1. e **3.** a **5.** g **7.** i **9.** c
11.

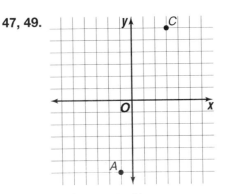

13. 11 **15.** $>$ **17.** $=$ **19.** 50 **21.** -8 **23.** 37
25. -18 **27.** 108 **29.** -20 **31.** $\begin{bmatrix} 12 & -4 \\ 5 & 6 \end{bmatrix}$

33. $\begin{bmatrix} -6 & -4 \\ 1 & 10 \\ -2 & -5 \end{bmatrix}$ **35.** 100 **37.** -60 **39.** -30

41. -12 **43.** 15 **45.** -8

47, 49.

51. -10 points **53.** blue

Pages 100–101 Standardized Test Practice

1. C **3.** A **5.** B **7.** B **9.** D **11.** C **13.** -8
15. -18 **17.** $12x - 8$ **19.** $41\frac{7}{16}$ ft²

CHAPTER 3
Using Proportion and Percent

Pages 105–106 Lesson 3-1

1. Ratios and rates are both comparisons by division. In a rate the units are different; in a ratio the units are the same. **5.** 4 to 5
7. 3 pounds/week **9.** 3:1 **11.** 3:2 **13.** 1:3
15. 1:3 **17.** 3 to 2 **19.** $0.35/minute
21. 15 students/teacher **23.** $0.80/pound **25.** 50
27. 3.8 dandelion plants, 17.7 grass plants
29. Rachel **31.** -20 **33.** C

Pages 109–110 Lesson 3-2

1. 25%, $\frac{1}{4}$ **3.**

5. 10% **7.** $\frac{3}{4}$ **9.** 1 **11.** 40% **13.** 90%
15. 84% **17.** 99% **19.** 46% **21.** $\frac{1}{20}$ **23.** $\frac{1}{4}$
25. $\frac{1}{100}$ **27.** $\frac{4}{5}$ **29.** $\frac{77}{100}$ **31.** 70%
33. 23 people **35.** 30% **37.** -8
39. $P = 36$ in., $A = 60$ in²

Pages 112–113 Lesson 3-3

1. reducing fractions and finding cross products
3. Sample answer: assigning a variable; solving equations **5.** yes **7.** 85 **9.** 0.2 **11.** no
13. yes **15.** no **17.** yes **19.** 300 **21.** 15
23. 10.5 **25.** 2.1 **27.** 0.5 cup **31.** $\frac{7}{20}$ **33.** 217

Pages 116–117   Lesson 3-4

1. $\frac{7}{20}$, 0.35, 35%

3. $15\% > \frac{1}{8}$

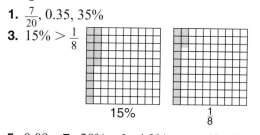

15% $\frac{1}{8}$

5. 0.09   **7.** 29%   **9.** 4.2%   **11.** 62.5%
13. 15%   **15.** 0.74   **17.** 0.01   **19.** 0.155
21. 0.038   **23.** 2%   **25.** 2.6%   **27.** 66.7%
29. 6.25%   **31.** 87.5%   **33.** 62.5%   **35.** 31.25%
37. 37.5%   **39.** >   **41.** =   **43.** <   **45.** 40%
47. 40%   **49.** 12   **51.** 2 out of 5   **53.** 81

Page 117   Mid-Chapter Self Test

1. $28/ticket   **3.** 22.5   **5.** 6%

Page 119   Lesson 3-4B

1. Except for 2, all are near 1.6.   **3.** Sample
answer: A golden rectangle is a rectangle in which
the ratio of the length to the width is 1.6.
5. Sample answer: United Nations building in New
York City   **7a.** 13, 21, 34, 55, 89, 144   **7b.** After
the first few terms in the sequence, the ratio of
successive numbers in the Fibonacci sequence is the
golden ratio.

Pages 121–122   Lesson 3-5

1. Move the decimal point in the number two places
to the left.   **3.** Elisa; you move the decimal point
one place to the left.   **5.** $13.55   **7.** 6
9. 450 Calories   **11.** 0.172   **13.** 140   **15.** 6
17. 900   **19.** 30   **21.** $3.84   **23.** <   **25.** =
27. $1.30   **29.** about 16,683,700 people
31. 0.395, 0.178   **33.** −17

Pages 124–125   Lesson 3-6A

1. Sample answer: $0.01 \times 90 = 0.9$ **3.** about
$3.75   **7.** $4   **9.** 4 miles   **11.** 50 cents
13. 2.7 million people

Pages 128–129   Lesson 3-6

1. $\frac{1}{4}$ of $100 is $25.   **3.** Darnell; 46% is less than
50%, so the answer will be less than half of 80.
5–27. Sample answers given.   **5.** about 20%
7. $\frac{7}{8}$ of 64 is 56   **9.** $\frac{22}{60} \approx \frac{20}{60}$ or 33.3%
11. $\frac{15}{49} \approx \frac{15}{50}$ or 30%   **13.** $\frac{64}{100}$ or 64%
15. about 12%   **17.** $\frac{1}{3}$ of 90 or 30

19. $\frac{1}{5}$ of 70 or 14   **21.** $\frac{2}{3}$ of 9 or 6   **23.** $\frac{8}{13} \approx \frac{8}{12}$
or 66.7%   **25.** $\frac{12}{60} = \frac{1}{5}$ or 20%   **27.** $\frac{9.2}{11} \approx \frac{9}{10}$
or 90%   **29.** $\frac{1}{3}$ and 150   **33.** $\frac{60}{206} \approx \frac{60}{200}$ or
about 30%   **37.** 135.8   **39.** 28

Pages 130–133   Study Guide and Assessment

1. ratio   **3.** proportion   **5.** 60%   **7.** 50%   **9.** $\frac{1}{4}$
11. 2:3   **13.** 25 to 9   **15.** $2.50 per minute
17. 22 miles per gallon   **19.** 16.5%   **21.** 56%
23. $\frac{7}{100}$ **25.** $\frac{9}{10}$ **27.** 4   **29.** 15   **31.** 0.90 or 0.9
33. 0.043   **35.** 70%   **37.** 65.5%   **39.** 35%
41. 16.67%   **43.** 12   **45.** $21   **47–51.** Sample
answers given.   **47.** $\frac{8}{100}$ of 100 or 8   **49.** $\frac{1}{1}$ of 35
or 35   **51.** $\frac{20}{52} \approx \frac{20}{50}$ or 40%   **53.** 24 to 65
55. 0.43, 0.16, 0.14, 0.11, 0.06, 0.04, 0.03

Pages 134–135   Standardized Test Practice

1. C   **3.** A   **5.** B   **7.** C   **9.** A   **11.** Sample
answer: $m + 4 = 3(14)$ **13.** 9 grams

CHAPTER 4
Statistics: Analyzing Data

Pages 140–141   Lesson 4-1A

1. Sample answer: Tally marks can be made quickly
and each person can add his or her mark without
having to erase and add to the total.   **5.** You would
need the lowest and highest numbers in the data set.
Knowing those numbers allow you to choose the
range of the set.   **7.** 15
9a. Sample answer:

Allowance	Number of 13- and 14-year-olds
$2.01-3.00	2
$3.01-4.00	1
$4.01-5.00	9
$5.01-6.00	7
$6.01-7.00	3
$7.01-8.00	0
$8.01-9.00	1
$9.01-10.00	2
$10.01-11.00	0
$11.01-12.00	1
$12.01-13.00	0
$13.01-14.00	0
$14.01-15.00	1

9b. $4.01-5.00   **11.** 144   **13.** 18-24 years old
15. B

Pages 144–146   Lesson 4-1

1. Sample answer: Because it is more visual, a histogram is more useful than a table when you are trying to show a general trend. Because the individual numbers are shown, a table is more useful when you need to know exact numbers.
5a. Sample answer: incubation times for eggs of different birds
5b.

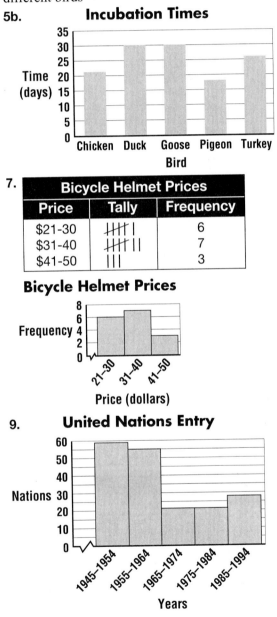

7.

Bicycle Helmet Prices		
Price	**Tally**	**Frequency**
$21-30	⦀⦀	6
$31-40	⦀⦀	7
$41-50	|||	3

Bicycle Helmet Prices

9. United Nations Entry

13. Sample answer: $60 ÷ 4 = $15

Page 147   Lesson 4-1B

1. 8   **3.** The scale for the x-axis is the interval.

Pages 150–151   Lesson 4-2

1. Tokyo Funland   **3a.** 100%   **3b.** 360°
3c.

Who pays when you date?			
Category	**Number**	**Ratio**	**Degrees in graph**
Boy	1200	0.48	172.8°
Split costs	800	0.32	115.2°
Girl	75	0.03	10.8°
Girl's parents	25	0.01	3.6°
Boy's parents	25	0.01	3.6°
Don't date	375	0.15	54°
Total	2,500	1	360°

3d. Who pays when you date?

5. Medals of Honor

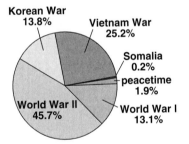

7. Giving to Charities

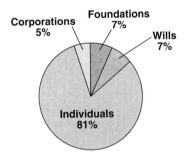

9. No; the percents are not parts of a whole because the categories overlap.   **11.** A

Pages 154–155   Lesson 4-3

1. Sample answer: Make a number line with appropriate numbers for the data points. Then mark an x for each data point above the appropriate number.

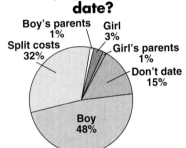

5.

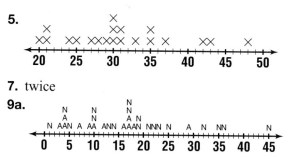

7. twice

9a.

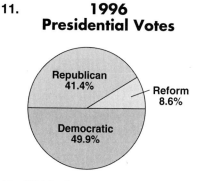

9b. Sample answer: The NFC has won more Super Bowls. The margins tend to be high, with NFC teams usually winning by larger margins than AFC teams.

11.

1996 Presidential Votes

Republican 41.4%
Reform 8.6%
Democratic 49.9%

13. 324 inches

Page 157   Lesson 4-3B

1. Sample answer: The data are clustered in these intervals.   **3.** The north-east quarter of the country, plus California and Florida. Sample answer: There are more birds to be viewed in these areas, there are more people in California than in other states, and there are more retirees with time for leisure activities in Florida.   **5.** Maps are more visual than tables.   **7.** Sample answer: Both use ranges of numbers to display data. However, a map need not use the same size intervals and a histogram does. Answers may vary. Generally a map is easier to change because there is more flexibility in the data intervals.

Pages 160–161   Lesson 4-4

1. Find the sum of the data and divide by the number of data.

$$\frac{1 + 3 + 4 + 6 + 6 + 6 + 8 + 9 + 10 + 10 + 14}{11} = 7$$

The data are written in order in the line plot. Since there are eleven pieces of data, the median is the sixth number. The median of the set is 6. In a line plot, you can find the mode by choosing the number that has the most ×s. In this data set, the mode is 6.   **3.** Carol is correct. The mean and the median may not be members of the set of data.   **5.** 34; 34; 34

7a. $71; $60; no mode   **7b.** Sample answer: There is a lot of variation in the prices. The median is a good representative price because half of the prices are above it and half are below.   **7c.** Because the value of the new data piece changes the total of the data drastically, the mean is affected most.   **9.** 6; 4.5, 3   **11.** 79.3; 79.5; 84   **13.** 9.0; 8.9; 8.3   **15.** 32.1; 33; 33 and 35   **17.** Sample answer: {1, 2, 3, 4, 5, 6, 7}   **19.** Sample answer: The mean, median, and mode of {1, 2, 3, 4, 4, 5, 6, 7} are all 4.   **21.** Sample answer: This is probably the mean, found by dividing the sum of the attendance figures for each showing of the movie by the number of people in the United States.   **23.** Because the value of the new data piece changes the total of the data drastically, the mean is affected most. Adding one very large or very small piece of data will only move the median slightly toward the new number. The mode will not be affected by adding a very large or very small number.   **25.** 15

Page 162   Lesson 4-4B

1. 90   **3.** 89

Pages 165–166   Lesson 4-5

1. The measures of variation describe the dispersal of data. The measures of central tendency describe the set as a whole.   **3.** 7; 16; 17, 13; 4; no outliers   **5.** 38; 52; 57, 48; 9; 22   **7.** 9; 58; 61, 56; 5; no outliers   **9.** 22; 37; 40.5, 34; 6.5; 51   **11.** 3.8; 2.9; 3.7, 2.3; 1.4; 6.1   **13.** 52; 62.5; 74.5, 56; 18.5; 104   **15.** 0.7; 0.55; 0.65, 0.25; 0.4; no outliers

17a. Sample answer: {1, 1, 2, 2, 2, 5, 9, 9, 9, 10, 10} and {1, 4, 4, 4, 4, 5, 5, 5, 9, 10, 10}
17b. Sample answer: {1, 2, 5, 7, 9, 10, 12, 14, 15, 17, 22} and {0, 2, 5, 7, 9, 10, 12, 14, 15, 17, 27}
19. 14.9; 17.5; 22

Page 166   Mid-Chapter Self Test

1.

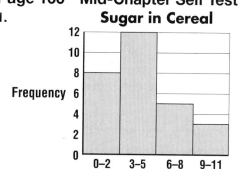

Sugar in Cereal

3.

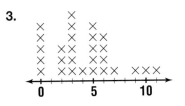

5. 11; 6, 2; 4; none

Page 167   Lesson 4-5B

1. minimum = 29, maximum = 76, upper quartile = 52.5, lower quartile = 36.5, median = 42
3. TI-83: Yes; the box-and-whisker plot for the actresses shows the two outliers using boxes. TI-82: No; the calculator does not evaluate for outliers. The ends of the whiskers are the extreme values.

Pages 169–170   Lesson 4-6

1. Write the data as ordered pairs and graph each ordered pair on a coordinate grid.   **5.** negative
7. no relationship   **9.** no relationship
11. positive   **13.** negative   **15.** negative
17. no relationship   **19.** positive

21a.

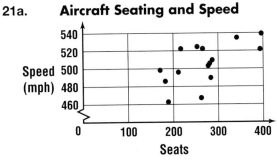

21b. positive   **23a.** Sample answer: Both are summer activities, so sales would increase at the same time.   **23b.** No. Warm weather could be causing the increase in both skateboard and swim suit sales.   **25.** A

Pages 172–173   Lesson 4-7

1. Sample answer: data on the number of immigrants who came to the United States in each decade   **5-17.** Sample answers given.
5. histogram or line plot; These are individual data. Which of these is the best choice depends on what the teacher is trying to show.   **7.** scatter plot; These are two sets of similar data and you want to show their relationship visually. A scatter plot would allow visual comparison.   **9.** circle graph; allows comparison of contribution of each division to the total amount of sales   **11.** scatter plot; A trend can be shown and accurate information is needed.
13. line plot; Shows outliers visually and shows

concentration of other prices.   **15.** histogram or scatter plot; Shows increase visually.   **17.** table or bar graph; Data is in categories.   **19.** No; a histogram requires that data be divided into intervals. This data is not well suited to intervals. A scatter plot is a better choice.

23.

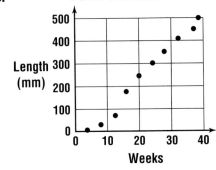

25. $5^2 \cdot 8^3$

Pages 175–177   Lesson 4-8

1. Sample answers: good places – schools, telephone surveys; bad places – computer stores, libraries   **3.** No; people at a Gloria Estefan concert will be more likely to choose her than the general public.   **5.** No; only people who are concerned enough about the situation will call.   **7.** No; people at a movie theater will be more likely to say watching movies than the general public.   **9.** Yes; a telephone poll will allow a selection of a number of people with different viewpoints.   **11.** No; because Texas has a higher concentration of immigrants than other states and immigrants are more likely to speak more than one language, a sample taken there would not be representative of all Americans.   **13.** Yes; if the schools are chosen in different types of neighborhoods, the sample will be representative of teenagers.   **15.** No; people who attend a community college will be more likely to have chosen careers that can be pursued with an associate's or bachelor's degree.   **17a.** The space between the tick marks is not consistent with the time between each date.   **19a.** The fewer the dentists surveyed, the less reliable the results.
19b. The commercial makes it sound as if the dentists recommended their brand of toothpaste over all others.   **21.** 0.048

Pages 178–181   Study Guide and Assessment

1. scatter plot   **3.** range   **5.** sample   **7.** circle graph   **9.** The mean is the average of the set and the median is the middle number when the set is arranged in order. **11.** 7-8 and 9-10 each have 7 data.

13.

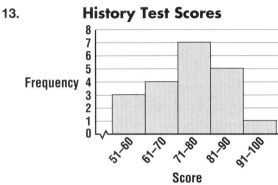

History Test Scores

17. 27 **19.** 7.7; 6.8; no mode **21.** 20; 20; 20 and 21 **23.** 74; 138; 150.5, 123.5; 27; 195
25–27. Sample answers are given. **25.** positive **27.** histogram **29.** Yes; a good cross section of the population visits a movie theater
33.

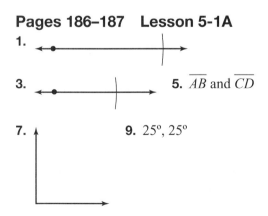

35. The graph on the right is misleading because the intervals are unequal on the vertical scale.

Pages 182–183 Standardized Test Practice

1. D **3.** C **5.** D **7.** B **9.** D **11.** A **13.** 3
15. $2n + 3 = 14$

CHAPTER 5
Geometry: Investigating Patterns

Pages 186–187 Lesson 5-1A

1.

3.

5. $\overline{AB}$ and $\overline{CD}$

7.

9. 25°, 25°

Pages 191–192 Lesson 5-1

1. Parallel lines are lines that are the same distance apart and never meet.

3.

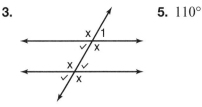

5. 110°

7. 120° **9.** 100; 80 **11.** 135° **13.** 138°
15. 68° **17.** 75° **19.** 105° **21.** 75° **23.** 60
25. 90°, 90°, 90° **27.** 115° **29.** C
31.

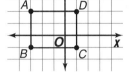

14 units

Page 193 Lesson 5-1B

1.

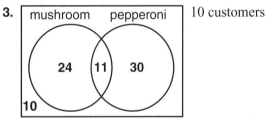

3. corresponding angles; yes

Pages 194–195 Lesson 5-2A

1. the students who like all three types of music; 8 students
3.

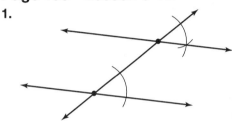

mushroom pepperoni 10 customers

24 11 30

10

5. The numbers in the Venn diagram must have a sum of 75.
7.

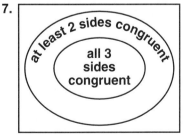

at least 2 sides congruent

all 3 sides congruent

9a. 4 times **9b.** $21 million **11.** D

Pages 198–199 Lesson 5-2

1. Both isosceles triangles and equilateral triangles have congruent sides. An isosceles triangle has at least two sides congruent, and an equilateral triangle

Selected Answers **679**

has three sides congruent.   **3.** Jaali is correct. A triangle has three angles whose sum is 180°. If a triangle had two obtuse angles, then the sum of two of the three angles would already be greater than 180°.   **5.** acute, scalene   **7.** false   **9.** obtuse, isosceles   **11.** acute, isosceles   **13.** acute, isosceles   **15.** 33   **17.** $m\angle A = 65°$, $m\angle B = 75°$, $m\angle C = 40°$

19. true

21. 60°, 60°, 60°   **25.** 40   **27.** B

Page 200   Lesson 5-3A

1. rectangle, 2, traceable   **3.** hexagon, 4, not traceable   **5.** octagon, 6, not traceable   **7.** Yes; any 4-sided figure with one diagonal will have 2 odd vertices.

Pages 203–204   Lesson 5-3

1. They each have four sides and four angles.
3a. yes   **3b.** The angles at each end of either of the parallel sides are congruent to each other.

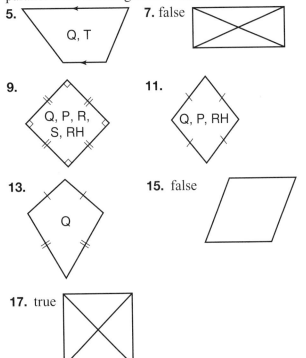

5. Q, T
7. false
9. Q, P, R, S, RH
11. Q, P, RH
13. Q
15. false
17. true

19. rectangle, square   **21a.** 80   **21b.** $m\angle A = 70°$, $m\angle B = 110°$, $m\angle C = 70°$, $m\angle D = 110°$
23. 12   **25.** A

Page 204   Mid-Chapter Self Test

1. 64   **3.** acute, scalene

5. false

Page 205   Lesson 5-4A

1.
3.

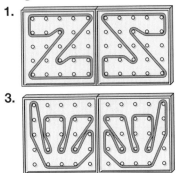

5. Yes, the corresponding sides are congruent. The figures are exactly the same except they are facing opposite directions.

Pages 208–209   Lesson 5-4

1. In line symmetry, half of the figure is the reflection of the other half.

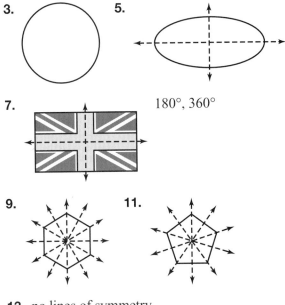

3.
5.
7.   180°, 360°
9.
11.

13. no lines of symmetry
15. isosceles triangles, equilateral triangles; equilateral triangles
17a.

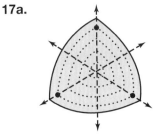

17b. 120°, 240°, 360°

19.

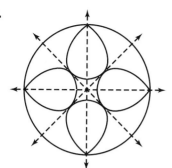

yes; 90°, 180°, 270°, 360°

21. (Numbers on the sides indicate the page numbers on the back of the sheet of paper.)

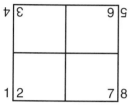

23. A

Pages 211–212   Lesson 5-5

1. Corresponding angles and corresponding sides must be congruent.   **3.** If each of the three sides of one triangle are congruent to each of the sides of another triangle, the triangles are congruent. SSS

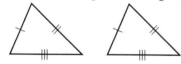

If 2 angles and the included side of one triangle are congruent to 2 angles and the included side of another triangle, the triangles are congruent. ASA

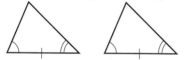

If 2 sides and the included angle of one triangle are congruent to 2 sides and the included angle of another triangle, the triangles are congruent. SAS

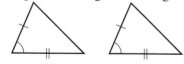

5. yes; $\triangle EFG \cong \triangle HIJ$; SAS   **7.** yes; the triangles formed by the sections for urban contemporary music and country music; SAS   **9.** no   **11.** yes; $\triangle ABD \cong \triangle CBD$; SAS   **13.** 25   **15.** The 4 triangles that form the large triangle in the center appear to be congruent. Three of these triangles are divided into 3 smaller triangles. These smaller triangles appear to be congruent.   **17.** no

Page 213   Lesson 5-5B

1.

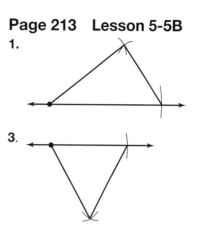

3.

Page 214   Lesson 5-6A

1.   **3.**

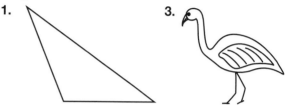

Pages 216–218   Lesson 5-6

1.

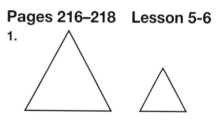

3. Helki; since the sum of 3 angles of a triangle is 180°, the unmarked angle in each of the triangles must be 75°. Since corresponding angles are congruent, the triangles are similar.   **5.** Similar; corresponding angles are congruent.   **7.** All of the triangles that form the Sierpinski Triangle are similar. The 4 triangles that form the large triangle are congruent. The 4 triangles that form three of these triangles are congruent, and so on.
9. Similar; corresponding angles are congruent.
11. congruent; ASA   **13.** 29   **15.** Similar; corresponding angles are congruent.   **17.** yes; $\triangle ABC \cong \triangle EFG$; SSS

Pages 222–223   Lesson 5-7

1. a tiling of a plane made with copies of the same shape or shapes fit together without gaps or overlaps.   **3a.** 60°, 60°, 60°   **3b.** 6
3c.

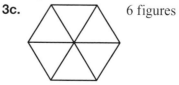   6 figures

3d. The number of figures needed to fill the space equals 360 divided by the measure of the angle.
5. translations   **7.**

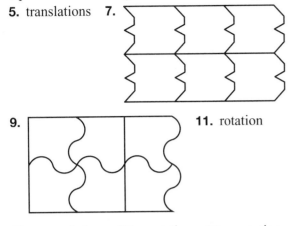

9.   **11.** rotation

15a. translation   **15b.** rotation   **15c.** rotation
17a. square
17b. translation   **19.** C

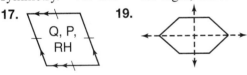

Pages 224–227   Study Guide and Assessment

1. i   **3.** g   **5.** e   **7.** a   **9.** If a figure can be turned less than 360° about its center and it looks like the original, then the figure has rotational symmetry.   **11.** 127°   **13.** right, scalene   **15.** 75°
17.   **19.**

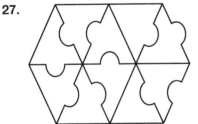

21. yes   **23.** yes; $\triangle WXY \cong \triangle PQR$; SSS
25. congruent; SAS
27.

29. 45 students   **31.** trapezoids, rectangles

Pages 228–229   Standardized Test Practice

1. A   **3.** C   **5.** D   **7.** B   **9.** B   **11.** $5a + 5b$
13. 28   **15.** 8

Pages 233–234   Lesson 6-1

1. b is a factor of a.   **3.** Alonzo; if the sum of the digits is divisible by 9, it is also divisible by 3.
5. 3, 9   **7.** yes   **9.** Sample answer: 1,110   **11.** 1 by 40, 2 by 20, 4 by 10, 5 by 8   **13.** 3, 5   **15.** 2, 3, 6, 9   **17.** 3, 5   **19.** 2, 4   **21.** no   **23.** yes
25. yes   **27.** Sample answer: 1,004   **29.** Sample answer: $2 \times 42, 3 \times 28, 4 \times 21$ **31.** Sample answer: $2 \times 8{,}592, 3 \times 5{,}728, 4 \times 4{,}296, 6 \times 2{,}864, 8 \times 2{,}148$ **33.** Yes; $3 \times 17 = 51$.   **35.** A
37. $\frac{\$0.32}{1 \text{ ounce}}$

Pages 237–238   Lesson 6-2

1. Both prime and composite numbers are whole numbers. 0 and 1 are neither prime nor composite. Prime numbers have exactly 2 factors, 1 and the number itself. Composite numbers have more than 2 factors.
3.

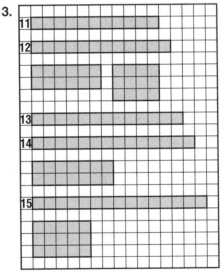

3a. 11, 13   **3b.** 12, 14, 15   **5.** prime   **7.** 7^2
9. $3^2 \cdot 5^2$ **11.** prime   **13.** composite   **15.** neither
17. prime   **19.** 5^2 **21.** $3 \cdot 5^2$ **23.** $3^2 \cdot 13$
25. $2^2 \cdot 5^2 \cdot 7^2$ **27.** 3   **29.** 61   **31a.** 128, 256, 512   **31b.** 2^n **33.** $9 = 3^2, 36 = 2^2 \cdot 3^2, 100 = 2^2 \cdot 5^2, 144 = 2^4 \cdot 3^2$ **33a.** All exponents are even.
33b. Yes; in order for a number to be a perfect square, all prime factors must be able to be arranged in pairs.   **35.** Similar; the figures are the same shape, but are different sizes.   **37.** $4x < 20; x < 5$

Page 239   Lesson 6-2B

1. 1, 3, 7, 9   **3.** There is a factor other than 1 that is the same for 10 and each number.

Pages 240–241 Lesson 6-3A

1. Atepa and Curtis have included all the types of pizzas that they can make. **3.** 8 combinations **5.** Sample answer: A sandwich can be made with ham, turkey, American cheese and/or Swiss cheese on white or wheat bread. How many different sandwiches can be made. **7.** 5 packages of 40 and 2 packages of 75 **9.** 10, 5, 4, or 2 people **11a.** bottom right **11b.** bottom left **13.** C

Pages 243–244 Lesson 6-3

1. 1, 2, 4, 5, 10, 20; 20

3. 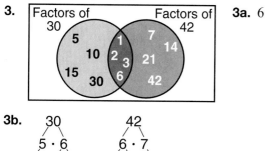 **3a.** 6

3b.

$$30 \quad\quad 42$$
$$5 \cdot 6 \quad\quad 6 \cdot 7$$
$$5 \cdot 2 \cdot 3 \quad\quad 2 \cdot 3 \cdot 7$$

5. 27 **7.** 15 **9.** 7 **11.** 21 **13.** 6 **15.** 7 **17.** 12 **19.** 42 **21.** 150 **23.** Sample answer: 7, 14, 21 **25a.** 18 in. **25b.** 7 shelves **27.** $2^3 \cdot 7$ **29.** D

Pages 247–248 Lesson 6-4

1. Sample answer: $\frac{1}{2}$ **3.** rationals **5.** $\frac{1}{9}$ **7.** $-\frac{4}{5}$ **9.** $1\frac{5}{8}$ lb **11.** whole numbers, integers, rationals **13.** whole numbers, integers, rationals **15.** rationals **17.** $\frac{3}{5}$ **19.** $-\frac{1}{3}$ **21.** $\frac{4}{9}$ **23.** $\frac{2}{3}$ **25.** $-\frac{3}{4}$ **27.** $-\frac{9}{20}$ **29.** $\frac{2}{5}$ **31.** $\frac{35}{73}$ **35.** 14 **37.** 3

Page 248 Mid-Chapter Self Test

1. 2, 4, 5, 8, 10 **3.** $2^2 \cdot 17$ **5.** $3^2 \cdot 7$ **7.** 6 **9.** $-\frac{9}{11}$

Pages 251–252 Lesson 6-5

1. 0.012; $\frac{3}{250}$ **3.** Nadia; $\frac{1}{4} = 0.25$, but $0.\overline{25} =$ 0.252525 **5.** -0.6171717171 **7.** $0.\overline{63}$ **9.** $\frac{33}{50}$ **11.** $-1\frac{5}{33}$ **13.** $0.2\overline{5}$ **15.** $7.0\overline{74}$ **17.** -0.3053053053 **19.** 0.75 **21.** -0.28 **23.** $-5.\overline{6}$ **25.** 12.625 **27.** $\frac{22}{25}$ **29.** $\frac{5}{6}$ **31.** $-1\frac{5}{9}$ **33.** $-5\frac{67}{99}$ **35.** terminating **37.** $\frac{19}{50}; \frac{191,919}{500,000}$

39. $\frac{13}{1,000}$ oz **43.** integers, rationals **45.** 12 **47.** C

Pages 254–256 Lesson 6-6

1. Sample answer: An even number picked at random is divisible by 2. **5.** 0 **7.** $\frac{1}{2}$ **9.** 1 **11.** $\frac{5}{6}$; $0.8\overline{3}$ **13.** 0 **15.** 1 **17.** $\frac{1}{8}$ **19.** $\frac{1}{2}$ **21.** $\frac{7}{8}$ **23.** $\frac{2}{11}$ **25.** 1 **27.** $\frac{5}{11}$ **31.** No; the number of ways something can occur cannot be negative. **33a.** $\frac{25}{108}$ **33b.** $\frac{2}{27}$ **33c.** $\frac{2}{27}$ **33d.** $\frac{1}{27}$ **35.** $\frac{4}{7}$ **37.** 80° **39.** 4

Pages 258–259 Lesson 6-7

1. (1) List the multiples of 10 and the multiples of 16. Find the least nonzero number that is a multiple of both numbers. (2) Find the prime factorization of 10 and 16. Multiply all the factors, using the common factors only once. **3a.** 36, 72 **3b.** 36 **5.** 0, t, $2t$, $3t$, $4t$, $5t$ **7.** 90 **9.** 840 **11.** 0, 7, 14, 21, 28, 35 **13.** 0, 14, 28, 42, 56, 70 **15.** 0, 150, 300, 450, 600, 750 **17.** 48 **19.** 100 **21.** 30 **23.** 105 **25.** 1,225 **27.** 340 **29.** yes **31.** when the GCF is 1 **33.** 8:00 A.M. **35.** C **37.** 358

Page 260 Lesson 6-8A

1. The midpoint is the mean of the endpoints. **3.** Add them together and divide by 2.

Pages 263–264 Lesson 6-8

1. 0.3 is represented by 30 small squares. 0.08 is represented by 8 small squares.

3. Sample answer: $\frac{1}{2}$, 0.53; Since $\frac{2}{5} = 0.4$, $\frac{4}{7} = 0.\overline{571428}$, and $\frac{1}{2} = 0.5$, $\frac{1}{2}$ and 0.53 are between the given numbers. **5.** < **7.** = **9.** $\frac{2}{5}$, 0.376, $\frac{3}{8}$, 0.367 **11.** 16 **13.** 90 **15.** < **17.** < **19.** < **21.** < **23.** $-\frac{1}{3}$, $-\frac{1}{4}$, $\frac{1}{10}$, $\frac{1}{9}$ **25.** -4.75, $-4\frac{2}{3}$, -4.5, $-4\frac{2}{5}$, $-4.1\overline{9}$ **27.** 0.6 **29a.** Q **29b.** S **29c.** P **29d.** R **31.** $\frac{1}{125}$ **33.** No; $0.\overline{4} = \frac{4}{9}$. **35.** Sample answer: 18 **37.** 35

Pages 266–267 Lesson 6-9

1. 36.2 and 0.362 are not greater than or equal to

1 and less than 10 **3.** 488,200 **5.** 234,000,000
7. 5.4×10^{-2} **9.** 14,000 lb **11.** 8,080,000,000
13. 0.000000075 **15.** 0.0000252 **17.** 202.1
19. 7.67×10^{-3} **21.** 4.004×10^{5}
23. 3.3×10^{-5} **25.** 7.6×10^{3} **27.** 2.5×10^{-8}
29. 3.2×10^{-8} **31.** 7.2×10^{6} dollars
33. $7\frac{2}{7}, 7.35, \frac{37}{5}$ **35.** -221

Pages 268–271 Study Guide and Assessment

1. multiple **3.** simplest form
5. 1 **7.** 3.2×10^{4} **9.** Write the digits of the decimal as the numerator. Use the appropriate power of 10 (10, 100, 1000, and so on) as the denominator. Then simplify. **11.** 5 **13.** 2 **15.** 3 **17.** $3^{2} \cdot 7$
19. $2 \cdot 101$ **21.** 3^{4} **23.** 14 **25.** 11 **27.** $\frac{5}{7}$
29. $\frac{6}{7}$ **31.** 0.625 **33.** 2.6 **35.** $\frac{13}{20}$ **37.** $\frac{16}{99}$
39. $8\frac{3}{8}$ **41.** $\frac{1}{2}$ **43.** $\frac{2}{3}$ **45.** 30 **47.** 168
49. 350 **51.** < **53.** < **55.** 6.48×10^{-5}
57. 5.5×10^{-4} **59.** 4.76×10^{-3} **61.** 0.250
63. $\frac{3}{5}$

Pages 272–273 Standardized Test Practice

1. C **3.** C **5.** B **7.** C **9.** B **11.** D **13.** C
15. I and II

CHAPTER 7
Algebra: Using Rational Numbers

Pages 279–280 Lesson 7-1

1. $\frac{2}{3} - \frac{1}{3} = \frac{1}{3}$
3. $1\frac{1}{2}$
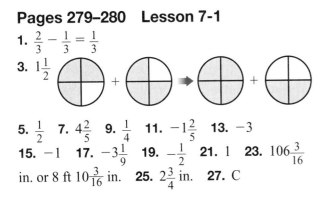
5. $\frac{1}{2}$ **7.** $4\frac{2}{5}$ **9.** $\frac{1}{4}$ **11.** $-1\frac{2}{5}$ **13.** -3
15. -1 **17.** $-3\frac{1}{9}$ **19.** $-\frac{1}{2}$ **21.** 1 **23.** $106\frac{3}{16}$
in. or 8 ft $10\frac{3}{16}$ in. **25.** $2\frac{3}{4}$ in. **27.** C

Pages 283–284 Lesson 7-2

1. $\frac{2}{3} + \frac{1}{4} = \frac{11}{12}$
3. $\frac{1}{4}$
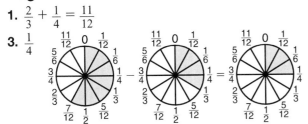

5. 7 **7.** $1\frac{1}{4}$ **9.** $\frac{7}{8}$ **11.** $9\frac{1}{12}$ **13.** $12\frac{1}{8}$
15. $1\frac{1}{24}$ **17.** $-2\frac{1}{10}$ **19.** $6\frac{5}{6}$ **21.** $1\frac{1}{4}$
23. $-9\frac{7}{8}$ **25.** $8\frac{7}{8}$ **27.** $15\frac{1}{4}$ **29.** $-10\frac{23}{40}$
31. $-3\frac{11}{24}$ **33.** $2\frac{5}{24}$ **35.** $66\frac{1}{4}$ in. **37a.** $a = \frac{4}{5}$,
$b = \frac{1}{2}$ **39.** $-2\frac{1}{3}$ **41.** positive

Pages 288–289 Lesson 7-3

1. Sample answer: Three out of four rows are shaded to represent $\frac{3}{4}$. Three out of eight columns are shaded to represent $\frac{3}{8}$. **3.** $\frac{3}{5}$
5. $2\frac{2}{5}$ **7.** $5\frac{19}{25}$
9. $4\frac{1}{4}$ in.; $\frac{5}{8}$ in^2
11. $-\frac{35}{48}$ **13.** $2\frac{1}{4}$ **15.** $-\frac{3}{5}$ **17.** $7\frac{1}{3}$ **19.** $\frac{9}{16}$
21. $-\frac{8}{27}$ **23.** $-\frac{1}{4}$ **25.** $7\frac{1}{2}$ **27.** $\frac{15}{32}$ **29a.** 27 in.
29b. 22 in. **31.** C **33.** 12

Pages 291–292 Lesson 7-4

1. $\frac{9}{5}$ **3.** Change $-3\frac{7}{8}$ to an improper fraction and then find the reciprocal.; $-\frac{8}{31}$ **5.** $\frac{8}{7}$ or $1\frac{1}{7}$ **7.** $\frac{1}{4}$
9. $-1\frac{2}{5}$ **11.** $-\frac{3}{2}$ **13.** -5 **15.** $\frac{4}{9}$ **17.** $-\frac{1}{x}$
19. $-\frac{1}{8}$ **21.** $-59\frac{1}{2}$ **23.** $-\frac{5}{11}$ **25.** No; the
reciprocal of $-2\frac{1}{2}$ is $-\frac{2}{5}$. **27.** -1 **29.** $\frac{7}{12}$
31. $36\frac{7}{8}$ in^2 **33.** $19\frac{1}{32}$ **35.** $\frac{1}{2} \times 180 = 90$

Pages 294–295 Lesson 7-5A

1. Santos; He made a rule after seeing several examples. **3.** $\frac{1}{64}$, 34,200; $\frac{1}{128}$, 39,900 **5.** Sample answer: Looking for a pattern involves determining what you know. **7a.** 20, 25 **7b.** 21, 15
9a. 25 units **9b.** 2.25 units **9c.**
11. 21 **13a.** 10
13b. Change cell B2 to 0.5.

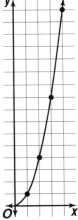

Pages 298–299 Lesson 7-5

1. When the difference between any two consecutive terms is the same, the sequence is arithmetic. **3.** Joanne is correct. The common difference is -4. **5.** N; $-11, -14, -16$ **7.** G;

48, −96, 196 **9.** A; 14, $16\frac{1}{2}$, 19 **11.** G; 81, 243, 729 **13.** G; $-\frac{1}{64}, -\frac{1}{256}, -\frac{1}{1,024}$ **15.** A; 36, 40, 44 **17.** G; 162, −486, 1,458 **19.** G; $\frac{7}{9}, \frac{7}{27}, \frac{7}{81}$ **21.** A; $3\frac{1}{6}, 2\frac{5}{6}, 2\frac{1}{2}$ **23.** N; 111, 136, 124 **25.** 4, $7\frac{1}{3}, 10\frac{2}{3}$, 14 **27.** 74 **29a.** 34, 55, 89; Neither; there is no common difference or common ratio. **29b.** Sample answer: The numbers are all Fibonacci numbers. **31.** $33.75 **33.** 24 **35.** 27

Page 299   Mid-Chapter Self Test

1. $\frac{1}{4}$ **3.** $6\frac{2}{5}$ **5.** $-96\frac{1}{4}$ **7.** $\frac{1}{2}\left(12\frac{4}{5}\right) = \frac{1}{2}\cdot 12 + \frac{1}{2}\cdot\frac{4}{5} = 6 + \frac{2}{5}$ or $6\frac{2}{5}$ **9.** G; 12, 3, $\frac{3}{4}$

Page 300   Lesson 7-5B

3. Each number is the sum of the previous two numbers. **5.** Sample answer: All the decimals are about 1.62.

Pages 303–304   Lesson 7-6

1. Sample answer:

3. The height of the parallelogram is half the height of the trapezoid and the length of the base of the parallelogram is the sum of the length of the bases of the trapezoid. **5.** $b = 2\frac{2}{3}$ ft, $h = 3\frac{3}{4}$ ft, $A = 5$ ft² **7.** $3\frac{63}{64}$ in² **9.** 234 ft² **11.** $b = 3$ ft, $h = 4$ ft, $A = 6$ ft² **13.** $a = 12$ yd, $b = 18$ yd, $h = 10$ yd, $A = 150$ yd² **15.** $b = 12$ cm, $h = 5$ cm, $A = 30$ cm² **17.** $22\frac{1}{2}$ in² **19.** 33 cm² **21.** $6\frac{1}{2}$ yd² **23.** $37\frac{1}{2}$ yd² **25.** 29.49 m² **27.** 9 ft² **29.** Sample answer:

3 in.
6 in.

31. $a = 12$ yd, $b = 18$ yd **33.** The area is quadrupled. **35.** 8.4×10^{-5} **37.** 27

Page 306   Lesson 7-6B

1. They are the same. **3.** yes **5a.** Count the squares. **5b.** Pick's Theorem works for all of the figures.

Page 308   Lesson 7-7A

1a. diameter **1b.** circumference **1c.** Answers will vary. The ratios should be close to $\frac{22}{7}$ or 3.14. **1d.** Answers will vary.; The ratios should be closer to $\frac{22}{7}$ or 3.14.

Pages 310–311   Lesson 7-7

1. $\frac{22}{7}$ and 3.14 **3.** Sample answer: Divide 22 by 3.14. Then round to the nearest tenth. **5.** 56.5 in. **7.** 106.8 cm **9.** 95.8 m **11.** 66 mm **13.** 185.3 cm **15.** 56.5 ft **17.** 33 in. **19.** 60.7 m **21.** $42\frac{3}{7}$ in. **23.** 0.325 m **25.** $9\frac{45}{56}$ or about 9.8 inches **27a.** 110^0 **27b.** 70^0 **27c.** 250^0 **27d.** 110^0 **27e.** 110^0 **27f.** 70^0 **27g.** 70^0 **27h.** 40^0 **29.** B

Pages 313–314   Lesson 7-8

1. There are 3 sets of $\frac{2}{3}$ to equal 2. **3.** $\frac{8}{9}$ **5.** $-\frac{5}{12}$ **7.** $\frac{2}{3}$ **9.** $-3\frac{3}{5}$ **11.** $-\frac{3}{20}$ **13.** 6 **15.** 6 **17.** $-2\frac{5}{8}$ **19.** $\frac{9}{25}$ **21.** $29\frac{3}{4}$ feet **23.** $\frac{m}{n}$ is greater; Dividing by a proper fraction is actually multiplying by an improper fraction. **25.** B

Pages 316–317   Lesson 7-9

1. Sample answer: $-1\frac{1}{4}a + 4 = 8$ **3.** Sample answer: Esohe's method uses the order of operations in reverse and is probably easier than Priya's method because it requires dividing two fractions. **5.** 3 **7.** −4.375 **9.** 1.8875 **11.** 60 **13.** $2\frac{3}{10}$ **15.** −3.2 **17.** −14.4 **19.** $-\frac{13}{30}$ **21.** 22.5 **23.** −0.5 **25.** −3.2 **27.** $\frac{1}{21}$ **29.** about 12.9 ft **31.** $-40°F = -40°C$ **33.** $2^4 \times 3$ **35.** 8

Pages 320–321   Lesson 7-10

1. The process for solving inequalities is the same. If the process includes dividing or multiplying by a negative number, the solution will have the opposite inequality sign.

3. $x < -3$ −5 −4 −3 −2 −1

5. $y \geq -16.5$ −18 −17 −16 −15

7. $h > 4\frac{1}{2}$ 3 4 5 6

9. at least 4 hours

11. $p < 3.17$ 3.17 0 1 2 3 4 5

13. $y \leq 84$

81 83 85

15. $k < -1\frac{1}{9}$

$-1\frac{1}{9}$

-3 -2 -1 0

17. $t < -4\frac{3}{8}$

-5 $-4\frac{1}{2}$ -4

19. $d > 2\frac{3}{10}$

2 $2\frac{1}{2}$ 3

21. $h \leq \frac{5}{22}$

0 $\frac{1}{2}$

23. $b \geq 11.4$

11 11.5 12

25. $g > -9\frac{1}{5}$

-10 -9 -8

27. $n > -8.97$

-8.97

-10 -9 -8 -7

29. b **31.** on days 12–15 **33.** -15 **35.** $\frac{1}{6}$

Pages 322–325 Study Guide and Assessment

1. g **3.** h **5.** j **7.** f **9.** Sample answer: Rewrite the mixed numbers as fractions. Multiply the numerators and multiply the denominators. Then simplify. **11.** $-\frac{1}{2}$ **13.** $\frac{1}{15}$ **15.** $-11\frac{1}{6}$
17. $-8\frac{1}{3}$ **19.** $2\frac{8}{9}$ **21.** $-\frac{3}{16}$ **23.** $-1\frac{3}{10}$
25. G; 2, 1, $\frac{1}{2}$ **27.** A; 32, 20, 8 **29.** 56 m^2
31. 16.3 m **33.** 20.4 in. **35.** $2\frac{1}{5}$ **37.** $\frac{1}{8}$
39. -11.52 **41.** 3.2
43. $d \geq -21$ **45.** $v < 7.6$

-21 -20 -19 7   8

47. 40 miles

Pages 326–327 Standardized Test Practice

1. D **3.** D **5.** A **7.** C **9.** B **11.** D
13. $a = 8$ ft, $b = 14$ ft, $h = 8$ ft, $A = 88$ ft^2
15. -484 **17.** 0.00349 **19.** yes;

CHAPTER 8
Applying Proportional Reasoning

Pages 331–333 Lesson 8-1
1a. Yes; each ratio represents the time to the cost.
1b. No; one ratio represents the time to the cost and the other represents the cost to the time. **1c.** Yes; each ratio represents the cost to the time.
3. Sample answer: $\frac{16}{1} = \frac{56}{p}$; $3\frac{1}{2}$ lb **5.** 48 min
7. 43,750 red blood cells **9.** Sample answer: $\frac{100}{1} = \frac{170}{m}$; 1.7 m **11.** Sample answer: $\frac{12}{96} = \frac{4}{x}$; 32¢ **13.** Sample answer: $\frac{35}{2.5} = \frac{420}{m}$; 30 min
15. 14,520 ft **17.** 335 mi **19a.** 21 lb
21. 5.25 pounds **23.** Sample answer: $\frac{15}{120} = \frac{25}{p}$, $\frac{120}{15} = \frac{p}{25}$, $\frac{15}{25} = \frac{120}{p}$, $\frac{25}{15} = \frac{p}{120}$ **25.** $2\frac{4}{9}$

Page 334 Lesson 8-1B
1. 9 cups cottage cheese, $2\frac{1}{4}$ cups chili sauce, $2\frac{1}{4}$ teaspoons onion powder, $2\frac{1}{4}$ cup skim milk, 27 tablespoons parmesan cheese
3.

9	hot sauce	B2/8	t.

Pages 336–338 Lesson 8-2
1. Percent means per one hundred and r is the number out of one hundred. **3.** Sample answer: Percentage: Find 20% of 40. $\frac{P}{40} = \frac{20}{100}$; Base: 3 is 40% of what number. $\frac{3}{B} = \frac{40}{100}$; Percent: 5 is what percent of 25? $\frac{5}{25} = \frac{r}{100}$ **5.** 5% **7.** 38%
9. $\frac{P}{60} = \frac{15}{100}$; 9 **11.** $\frac{15}{45} = \frac{r}{100}$; 33.3% **13.** 14%
15. 37% **17.** 20% **19.** 52% **21.** 82%
23. 50% **25.** 18.75% **27.** 25% **29.** $\frac{P}{66} = \frac{25}{100}$; 16.5 **31.** $\frac{16}{48} = \frac{r}{100}$; 33.3% **33.** $\frac{5}{B} = \frac{26}{100}$; 19.2 **35.** $\frac{7}{20} = \frac{r}{100}$; 35% **37.** $\frac{60}{B} = \frac{15}{100}$; 400
39. $\frac{45}{B} = \frac{35}{100}$; 128.6 **41.** 10% **43.** $48,700
47. 6 cups **49.** $\frac{2}{3}$ **51.** 40 shrimp/1 pound

Pages 340–341 Lesson 8-3
1. $R \cdot 34 = 28$ **3.** $0.47 \cdot 52 = P$; 24.44
5. $R \cdot 90 = 36$; 40% **7.** $0.30 \cdot B = 48$; $160
9. 4.5% **11.** $R \cdot 66 = 55$; $83\frac{1}{3}$%
13. $0.15 \cdot B = 30$; 200 **15.** $0.24 \cdot 72 = P$; 17.28
17. $\frac{2}{3} \cdot B = 16$; 24 **19.** $R \cdot 300 = 6$; 2%
21. $R \cdot 80 = 25$; 31.25% **23.** $R \cdot 50 = 6$; 12%
25. $R \cdot 80 = 70$; 87.5% **27.** $0.08 \cdot B = 54$; $675
29. $25,000 **31.** about 11 people
33. 1,062,000 tons **35.** D

Pages 342–344 Lesson 8-3B

1. Both 100 and 300 are divisible by 20, but they are not divisible by 18 or 22. Therefore, the computations are easier if 20 is used instead of 18 or 22. Sample answer: 25 **3.** vanilla, 15 cups; chocolate 75 cups; strawberry, 45 cups; chocolate chip, 135 cups; peanut butter, 30 cups **5.** Sample answer: Predict who will win a presidential election. Instead of surveying all voters, you would survey a smaller group that is representative of all voters. **7.** about 26 times **9.** 9 clothespins **11.** 50 Calories **13.** B

Pages 346–347 Lesson 8-4

1. 3 out of every 1,000 farmers in the world live in the U.S.

3a.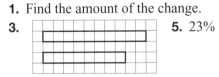

3b.   **5.** $1\frac{3}{4}$

7. 1.55 **9.** 1.235 **11a.** $1.17\% = 0.0117$; $1.06\% = 0.0106$; $0.92\% = 0.0092$; $1.08\% = 0.0108$; $2.21\% = 0.0221$ **11b.** 1994 **11c.** 1992 **13.** $1\frac{3}{20}$ **15.** $\frac{3}{5,000}$ **17.** $2\frac{9}{20}$ **19.** $134\frac{47}{100}$ **21.** $\frac{1}{3}$ **23.** 0.0003 **25.** 0.1025 **27.** 0.00079 **29.** 0.004 **31.** 1.104 **33.** 104%, 1, 1.04%, 0.4%, 0.04% **35.** 1.81 **37.** $200\% = \frac{200}{100} = \frac{2}{1} = 2$; therefore 200% means to double. **39.** about 51.5 cm **41.** C

Pages 350–351 Lesson 8-5

1. Find the amount of the change.

3.   **5.** 23%

7. $200.00 **9.** $16.80 **11.** 40% **13.** 67% **15.** 6% **17.** $13.05 **19.** $2.69 **21.** $10.40 **23.** $1,440.00 **25.** $19.60 **27.** $73.75 **29.** 8% **31.** $225 **33.** $19.50 **35a.** about 1.1% **35b.** decrease **39.** $\frac{1}{250}$ **41.** $6 + 7 + 2 = 15$, $15 \div 3 = 5$; yes **43.** 73

Page 352 Lesson 8-5B

1. sweaters, $22.49; jackets, $27.22; sweatshirts, $18.67; socks, $4.94; T-shirts, $5.99 **3.** $23.59

5.

8	suede jackets	$99.59	(100-B1)/ 100*B8

Pages 355–356 Lesson 8-6

1. $I = prt$; I represents the amount of interest, p represents the amount of money that is initially deposited in a savings account or the amount of money borrowed, r represents the percent of interest, and t represents the time in years. **3.** Janay; 8 months equals $\frac{8}{12}$ or $\frac{2}{3}$ year. Sonia forgot to change the time to years. **5.** $18.40 **7.** $760.52 **9.** $112.50 **11.** $77.24 **13.** $133.20 **15.** $840.00 **17.** $294.93 **19.** $421.38 **21.** $112.50 **23.** 14 years **25.** $70,200,000,000 **27.** about 31.6% **29.** -241

Page 356 Mid-Chapter Self Test

1. 1,176 bushels **3.** $\frac{63}{84} = \frac{r}{100}$; 75% **5.** $0.55 \cdot 86 = R$; 47.3 **7.** $2\frac{17}{20}$ **9.** $140.63

Pages 359–360 Lesson 8-7

1. If 2 polygons have corresponding congruent angles and have corresponding sides that are in proportion, then the polygons are similar. **5.** Sample answer: $\frac{10}{5} = \frac{6}{x}$; 3 **7.** $2\frac{1}{2}$ in. by $3\frac{1}{8}$ in. **9.** Yes; the corresponding angles are congruent and $\frac{3}{2} = \frac{3}{2} = \frac{3}{2} = \frac{3}{2}$. **11.** Sample answer: $\frac{5}{3} = \frac{x}{4}$; $6\frac{2}{3}$. **13.** Sample answer: $\frac{10}{6} = \frac{5}{x}$; 3 **15.** 8.75 m **17a.** yes **17b.** Sample answer: yes

17c. Yes; all angles are right angles, so all angles of one square are congruent to all angles of any other square. Since all 4 sides of a square are the same length, corresponding sides are always in proportion. **19.** $192 **21.** 9.6

Pages 362–364 Lesson 8-8

1. Indirect measurement means you did not actually measure the distance. Instead you used proportions and other measurements to find the distance.

3. Sample answer: The shadow of a flagpole is 12 feet at the same time the shadow of a nearby sign is 4 feet. If the sign is 8 feet tall, how high is the flagpole? To solve the problem, write the proportion and solve for h.

$$\textit{flagpole's shadow} \rightarrow \frac{12}{4} = \frac{h}{8} \leftarrow \textit{flagpole's height} \\ \textit{sign's shadow} \rightarrow \quad\quad\quad \leftarrow \textit{sign's height}$$

5. Sample answer: $\frac{h}{5} = \frac{9}{2\frac{1}{4}}$; 20 ft   **7.** Sample answer: $\frac{20}{8} = \frac{x}{5}$; 12.5 km   **9.** Sample answer: $\frac{90}{x} = \frac{60}{30}$; 45 mi   **11.** Sample answer: $\frac{s}{7} = \frac{26}{10}$; $18\frac{1}{5}$ ft   **13.** 630 ft   **15.** 2,000 mi   **17.** $\frac{451}{1,152}$

Page 365   Lesson 8-8B

1. 30° angle: ratio 1 ≈ 0.5774, ratio 2 ≈ 0.5000, ratio 3 ≈ 0.8660; 60° angle: ratio 1 ≈ 1.7321, ratio 2 ≈ 0.8660, ratio 3 ≈ 0.5000

3.
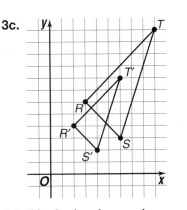
ratio 1 ≈ 1.0000, ratio 2 ≈ 0.7071, ratio 3 ≈ 0.7071

Pages 367–369   Lesson 8-9

1. Some objects are too small or too large to show the actual item in the space available.   **3.** Sierra; the scale is 1 inch = 3 feet or 1 inch = 36 inches which is 1:36.   **5.** 1 cm = 125 km or 1:12,500,000   **7a.** 16 ft by 18 ft   **7b.** 8 ft by 12 ft   **7c.** 8 ft by 8 ft   **7d.** 8 ft by 10 ft   **9.** 10 in. = 305 ft or 1:366   **11.** $2\frac{4}{5}$ in.   **13.** 2.6 cm   **15.** 25 in. by 35 in.   **17.** 60.8475 m

Pages 372–373   Lesson 8-10

1. A dilation is an enlargement or a reduction of an image.

3a.
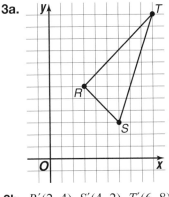

3b. $R'(2, 4)$, $S'(4, 2)$, $T'(6, 8)$

3c.
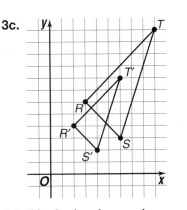

3d. The 2 triangles are the same shape. The dilated figure is smaller than the original.   **5.** 1.5
7a. 200%; 50% = $\frac{1}{2}$ so the image would be half the size, but 200% = 2.00 so the image would be twice as large.   **7b.** $\frac{3}{4}$ **9.** $B'\left(7\frac{1}{2}, 12\right)$
11. $D'(-8, 24)$, $R'(-4, -8)$, $T'(16, 12)$

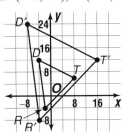

13. $D'(-5, 15)$, $R'\left(-2\frac{1}{2}, -5\right)$, $T'\left(10, 7\frac{1}{2}\right)$

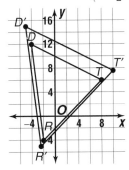

15. 2
17. Sample answer:

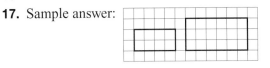

19.

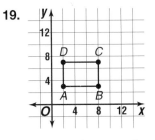

19a. 20 units; 24 units²

688   Selected Answers

19b.

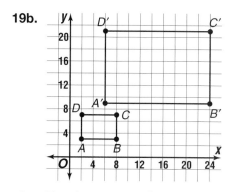

19c. 60 units; 216 units² **19d.** 3:1 **19e.** 9:1
19f. The perimeter of the dilation of a polygon is equal to the perimeter of the original polygon times the scale factor. The area of the dilation of a polygon is equal to the area of the original polygon times the square of the scale factor. **21.** 56 ft **23.** true

Pages 374–377 Study Guide and Assessment

1. d **3.** f **5.** a **7.** g **9.** Sample answer: $\frac{80}{10} = \frac{x}{15}$; $120 **11.** $\frac{P}{18} = \frac{45}{100}$; 8.1 **13.** $\frac{P}{80} = \frac{86}{100}$; 68.8 **15.** $0.66 \cdot 7,000 = P$; 4,620 **17.** $0.30 \cdot B = 15$; 50 **19.** $2\frac{3}{20}$ **21.** 0.006 **23.** 25% **25.** 33%
27. 40% **29.** $31.20 **31.** $68.25 **33.** $5\frac{1}{3}$ in.
35. 36 ft **37.** $13\frac{3}{4}$ mi **39.** $22\frac{1}{2}$ mi **41.** (60, 12)
43. (4, 16) **45.** 18 cuts **47.** 107 mi

Pages 378–379 Standardized Test Practice

1. A **3.** A **5.** D **7.** B **9.** A **11.** B **13.** 2
15. 5

CHAPTER 9
Algebra: Exploring Real Numbers

Pages 383–384 Lesson 9-1

1. $-\sqrt{100}$ **3.** When each square root is squared, it equals the perfect square. $5^2 = 25$ • • • • •
• • • • •
• • • • •
• • • • •
• • • • •
5. 9 **7.** -8 **9.** 776 feet **11.** -3 **13.** -6
15. $\frac{2}{3}$ **17.** $-\frac{4}{5}$ **19.** -14 **21.** -1.7
23. $0.5, -0.5$ **25.** $13, -13$; 13 is the principal square root because it is the positive square root.
27. 17 rows of 17 trees in each row **29.** Yes, because the factorization of the product is simply

the product of the two perfect square factorizations, and therefore it is also a perfect square.
31. 115.5 cm² **33.** 8.5; 8.5; 6, 12

Page 385 Lesson 9-2A

1. 4 **3.** 12 **5.** You can square numbers by guess and check until you find a number when squared is less than the square root you are finding and then the next greater number when squared is just greater than the square root.

Pages 387–389 Lesson 9-2

1. If three more tiles are shaded, then 12 tiles are shaded with four unshaded. Therefore, $\sqrt{12}$ is closer to $\sqrt{9}$ than $\sqrt{16}$; $\sqrt{12} \approx 3$. **3.** 8 **5.** 12
7. about 2.7 miles **9.** 4 **11.** 4 **13.** 14 **15.** 6
17. 2 **19.** 12 **21.** 9 **23.** about 526 square feet
25a. yes, about 3 mph
25b. Sample answer:

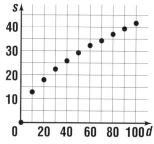

27. 30 **29.** E

Pages 392–394 Lesson 9-3

1.
$$\sqrt{15}$$
+—+—+—+—+
0 1 2 3 4

3. Their teacher is correct because it depends on what number of which you are taking the square root. Examples, $\sqrt{9} = 3$ and $3 < 9$, but $\sqrt{0.09} = 0.3$ and $0.3 > 0.09$. **5.** Q, R **7.** $\sqrt{20} \approx 4.5$;
$$\sqrt{20}$$
+—+—+—+
2 3 4 5

9. $-7.7, 7.7$ **11.** W, Z, Q, R **13.** I, R
15. Z, Q, R **17.** 2.4;
$$\sqrt{6}$$
+—+—+—+
0 1 2 3

19. 7.1;
$$\sqrt{50}$$
+—+—+—+
5 6 7 8

21. 10.4;
$$\sqrt{108}$$
+—+—+—+—+
8 9 10 11

23. $-8.3, 8.3$ **25.** $-8.7, 8.7$ **27.** $-30.4, 30.4$
29. $\frac{8}{3}$ **31.**

Distance (ft)	Time (s)
128	$\sqrt{8} \approx 2.8$
64	$\sqrt{4} = 2$
32	$\sqrt{2} \approx 1.4$
16	$\sqrt{1} = 1$
8	$\sqrt{\frac{1}{2}} \approx 0.7$

33a. $\sqrt{10} \approx 3.2$ units **33b.** $4\sqrt{10} \approx 12.6$ units
35. about 700 **37.** E

Page 394 Mid-Chapter Self Test
1. 1 **3.** 1.5 m **5.** 5 **7.** Q, R **9.** I, R

Page 397 Lesson 9-4A

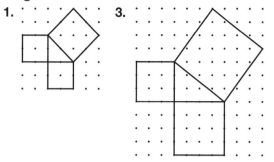

5. 32 units **7.** (one small side)^2 + (another small side)^2 = (longest side)^2

Pages 399–401 Lesson 9-4
1. 100 square units **3.** $6^2 + 12^2 = 36 + 144$ or 180 and $14^2 = 196$. Since $180 \neq 196$, the triangle is not a right triangle. **5.** $9^2 + b^2 = 41^2$; $b = 40$
7. $5^2 + 5^2 = c^2$; $c \approx 7.1$ **9.** yes **11.** $12^2 + 9^2 = c^2$; $c = 15$ **13.** $a^2 + 12^2 = 30^2$; $a \approx 27.5$
15. $1^2 + 1^2 = h^2$; $h^2 + 1^2 = x^2$; $x \approx 1.7$
17. $3^2 + b^2 = 8^2$; $b \approx 7.4$ **19.** $40^2 + b^2 = 41^2$; $b = 9$ **21.** $48^2 + 55^2 = c^2$; $c = 73$ **23.** yes
25. no **27.** yes **29.** 24 inches **31.** between 18 and 20 inches
33.   **35.** 44 feet **37.** C

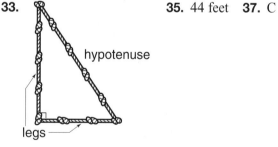

Page 403 Lesson 9-5A
1. Yes, each path is along the faces of the cube, so they are feasible. **3.** There are 49 seats in this section. **7.** 20 people **9.** 9 meters **11.** 11 quarters and 19 nickels **13.** C

Pages 406–407 Lesson 9-5
1. Sample answer: 10-24-26 and 15-36-39 **3.** $x^2 + 21^2 = 30^2$; $x = \sqrt{459} \approx 21.4$ miles **5.** $5^2 + 8^2 = x^2$; $x = \sqrt{89} \approx 9.4$ miles **7.** $x^2 = 9^2 + 18^2$; $x = \sqrt{405} \approx 20.1$ feet; $x^2 = 9^2 + 12^2$; $x = \sqrt{225} =$

15 feet **9.** $\sqrt{512} \approx 22.6$ feet **11.** about 3.7 feet longer **13.** about 13.2 km **15.** B

Page 409 Lesson 9-5B

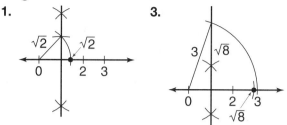

5. Draw a number line. At 1, construct a perpendicular line segment 1 unit in length. Draw a line from O to the top of the perpendicular line segment. Label it c. Open the compass to the length of c. With the tip of the compass at O, draw an arc that intersects the number line to the right of O at A. The distance from O to A is $\sqrt{2}$ units. **7.** Draw a number line. At 1, construct a perpendicular line segment 2 units in length. Draw a line from O to the top of the perpendicular line segment. Label it c. Open the compass to the length of c. With the tip of the compass at O, draw an arc that intersects the number line to the right of O at B. The distance from O to B is $\sqrt{5}$ units. **9.** $(\sqrt{11})^2 + 5^2 = 6^2$; none for 12; $(\sqrt{13})^2 = 3^2 + 2^2$; none for 14; $(\sqrt{15})^2 + 7^2 = 8^2$; $(\sqrt{17})^2 + 8^2 = 9^2$; none for 18; $(\sqrt{19})^2 + 9^2 = 10^2$; $(\sqrt{20})^2 + 4^2 = 6^2$

Pages 411–413 Lesson 9-6
1.   **3.** Both distances are 5 units or about 0.25 miles. Because the Madison Building is at $(4, -3)$ and the Sewall-Belmont House is at $(4, 3)$,
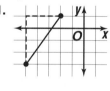
the distances are the same. **5.** 7.1 units **7.** about 9.4 miles **9.** 4.1 units **11.** 7.1 units
13. 4.5 units **15.** 7 units **17.** 4.2 units
19. 4.2 units
21a. about 8 feet;

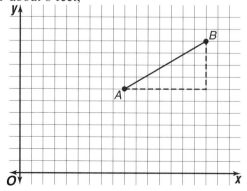

23. B **25.** -384

Pages 416–417 Lesson 9-7

1. $c = 2a$ where a is leg opposite 30° angle
5. $a = 12$ mm; $b \approx 20.8$ mm **7.** $b = 3.2$ m; $c \approx$ 4.5 m **9.** $a = 6$ in.; $b \approx 10.4$ in. **11.** $b \approx 34.6$ ft; $c = 40$ ft **13.** 3.75 inches **15a.** $\sqrt{2}, \sqrt{3}, \sqrt{4}$ or $2, \sqrt{5}$ **15b.** yes, the smallest triangle; yes, the fourth triangle **15c.** $\sqrt{6}$ **17.** $24\sqrt{3}$ square units; 24 units **19.** D

Pages 418–421 Study Guide and Assessment

1. b **3.** c **5.** j **7.** i **9.** Answers will vary. Sample answer: Rational numbers can be expressed in the form $\frac{a}{b}$, where a and b are integers and $b > 0$, but irrational numbers cannot be expressed in that form; rational: 0.75, irrational: $\sqrt{3}$ **11.** 2.5
13. $\frac{2}{3}$ **15.** 1.8 **17.** 7 **19.** 18 **21.** 15
23. Z, Q, R **25.** Q, R **27.** $4^2 + b^2 = 9.5^2$; $b \approx 8.6$ m **29.** 20.6 km
31. 7.8 units; **33.** about 3.6 units

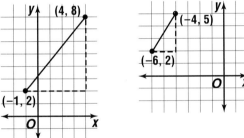

35. $b \approx 10.4$ km; $c = 12$ km **37.** $b = 7$ m; $c \approx 9.9$ m **39.** 9.4 feet **41.** 10 feet

Pages 422–423 Standardized Test Practice

1. B **3.** B **5.** D **7.** C **9.** A **11.** C **13.** -7
15. isosceles triangle **17.** 4.5 units

CHAPTER 10
Algebra: Graphing Functions

Pages 429–431 Lesson 10-1

1. domain, range **3.** Raul is correct. Candace's solution did not follow the order of operations.
5. 4 **7.** 42

9.

n	$-5n$	$f(n)$
-4	$-5(-4)$	20
-2	$-5(-2)$	10
0	$-5(0)$	0
$2\frac{1}{4}$	$-5\left(2\frac{1}{4}\right)$	$-11\frac{1}{4}$
5.7	$-5(5.7)$	-28.5

11. -2 **13.** 0 **15.** 5.84 **17.** -8

19a.

Time (seconds) n	Distance (feet) $f(n)$
2	2,200
5	5,500
11	12,100
18	19,800

19b. 9.6 seconds **23.** 9.1 feet **25.** B

Page 432 Lesson 10-1B

1.

X	$Y = 15X$
-3	-45
-2	-30
0	0
1.5	22.5
4.8	72

3.

X	$Y = 3 - 2X$
-14	31
-3.3	9.6
1.3	0.4
4.5	-6
9.4	-15.8

Pages 434–435 Lesson 10-2

1. Use the n values for x and the $f(n)$ values for y and graph the coordinates (x, y).

3.

n	$f(n)$	$(n, f(n))$
-2	3	$(-2, 3)$
-1	4	$(-1, 4)$
0	5	$(0, 5)$
1	6	$(1, 6)$
2	7	$(2, 7)$

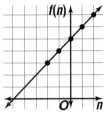

5a.

n	$f(n)$	$(n, f(n))$
10	14.8	$(10, 14.8)$
20	49.6	$(20, 49.6)$
30	84.4	$(30, 84.4)$
40	119.2	$(40, 119.2)$
50	154	$(50, 154)$

5b.

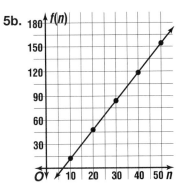

7.

n	f(n)	(n, f(n))
−2	160	(−2, 16)
0	0	(0, 0)
0.5	−4	(0.5, −4)
1	−8	(1, −8)
2	−16	(2, −16)

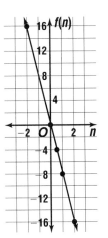

9.

n	f(n)	(n, f(n))
−4	12	(−4, 12)
−1	9	(−1, 9)
0	8	(0, 8)
6	2	(6, 2)
10	−2	(10, −2)

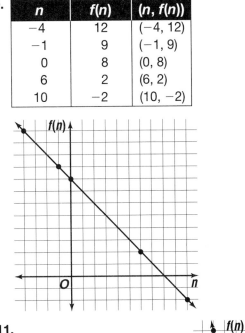

11.

n	f(n)	(n, f(n))
−3	12	(−3, 12)
−2	7	(−2, 7)
−1	4	(−1, 4)
0	3	(0, 3)
1	4	(1, 4)
2	7	(2, 7)
3	12	(3, 12)

13.

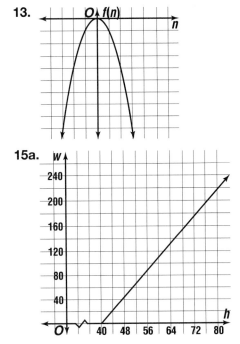

15a.

15b. 132 pounds   **17.** 3,809.48 m

Pages 439–440   Lesson 10-3

1. Sample answers: (−2, −1), (−1, 1), (2, 7), (4, 11)   **5.**

x	y
−4	−2
0	0
5	2.5
16	8

7. Sample answers: (3, 180), (4, 360), (5, 540), (6, 720)   **9.**

x	y
−6	9
−1	1.5
3	−4.5
16	−24

11.

x	y
−5	6
0	7
5	8
10	9

13.

x	y
−3	3
3	5
6	6
8	$6\frac{2}{3}$

15–17. Sample answers given.
15. (−1, 2), (0, 1), (1, 0), (2, −1)   **17.** (0, −57), (2, −27), (4, 3), (6, 33)   **19.** Sample answers: (−459.67, 0), (32, 273.15), (98.6, 310.15)

21.

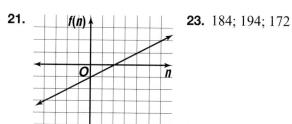

23. 184; 194; 172

Page 441 Lesson 10-4A

1. Sample answer: yes; each additional washer stretches the rubber band about the same additional distance

3. Sample answer:

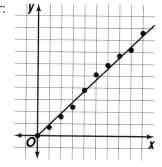

Pages 443–444 Lesson 10-4

1. The third point is for a check. **3.** 5 hours

5.

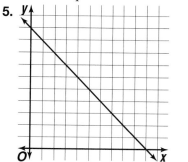

7.

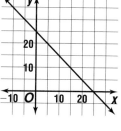

9.

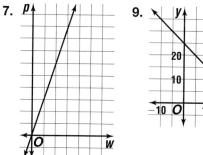

11.

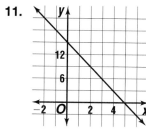

13.

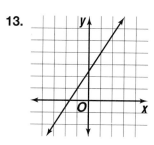

15.

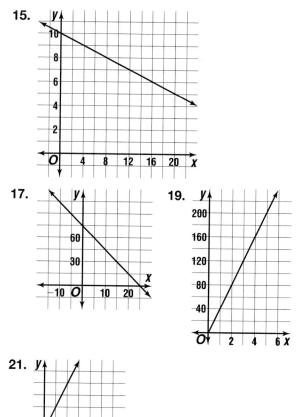

17.

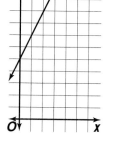

19.

21.

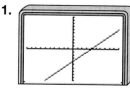

23. Sample answer: {(1, 13), (2, 16), (3, 19), (4, 22)} **25.** D

Page 445 Lesson 10-4B

1.

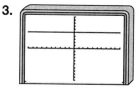

3.

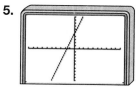

5.

7. perpendicular lines

Pages 448–449 Lesson 10-5

1. two or more equations

3. Sample answer:

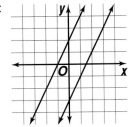

5. $(2, 2)$ **7.** No; The lines meet at a point where x is negative, so the time when imports and exports has passed.   **9.** $(1, 2)$ **11.** no solution   **13.** $(6, 9)$
15. $(4, 3)$ **17.** $(2, 7)$

21a.

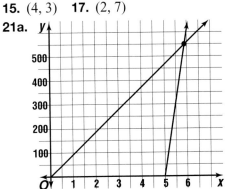

21b. about 6 hours after the Fokker leaves

23.

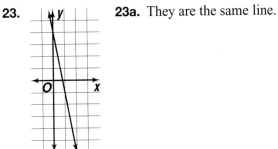

23a. They are the same line.

23b. Yes; all of the points on the graph represent solutions.   **25.** $2\frac{6}{7}$

Page 449   Mid-Chapter Self Test

1.

n	3n + 5	f(n)
−2	3(−2) + 5	−1
−1	3(−1) + 5	2
0	3(0) + 5	5
3	3(3) + 5	14
5	3(5) + 5	20

3. Sample answers: $(0, 1)$, $(1, 3.5)$, $(2, 6)$, $(4, 11)$
5. $(2, 5)$

Pages 450–451   Lesson 10-6A

1. The change in the lumber production from one year to the next is not constant.   **5.** y-values
7. $58,451   **9.** 15   **11.** Eric - amethyst; Julian - onyx; Louisa - sapphire   **13.** A

Pages 453–455   Lesson 10-6

1. 2   **3.** Nicole is correct. The function can be written as $y = \frac{1}{3}x^2$. The highest power in the equation is 2, so the function is quadratic.

5.

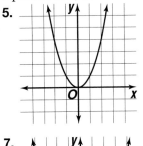

7.

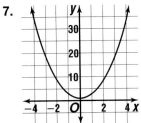

9.

x	y	(x, y)
−3	−9	(−3, −9)
−1	−1	(−1, −1)
0	0	(0, 0)
1.5	−2.25	(1.5, −2.25)
2	−4	(2, −4)

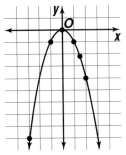

11.

n	f(n)	(n, f(n))
−2	4	(−2, 4)
−1.5	0.5	(−1.5, 0.5)
0	−4	(0, −4)
3	14	(3, 14)
4	28	(4, 28)

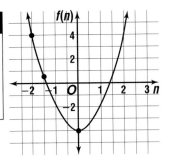

13.

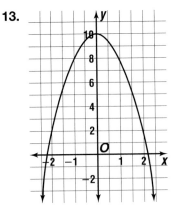

694   Selected Answers

15.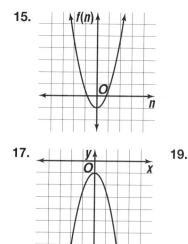

17. 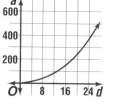 **19.**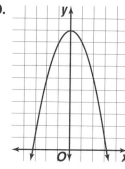

21. $(-2, 8), (0, -4), (1, -1)$ **22.** $(-1.5, -2.75)$, $(3.5, -7.75), (0, 1)$ **23a.**

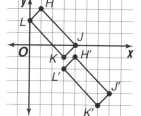

23b. about 22 miles

25a.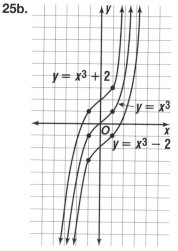

The graphs are all the same shape. The graph of $y = x^2$ has a y-intercept of 0; the graph of $y = x^2 + 2$ has a y-intercept of 2, and the graph of $y = x^2 + (-2)$ has a y-intercept of -2. The graph of $y = x^2$ has a y-intercept of 0 and the graph of $y = x^2 + c$ has a y-intercept of c.

25b.

The graphs are all the same shape. The graph of $y = x^3$ has a y-intercept of 0; the graph of $y = x^3 + 2$ has a y-intercept of 2, and the graph of $y = x^3 + (-2)$ has a y-intercept of -2.

27. 3.5%

Pages 458–459   Lesson 10-7

1. slide

3.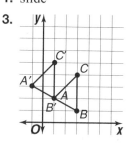

5. $H'(4, -1), J'(7, -4), K'(6, -5), L'(3, -2)$

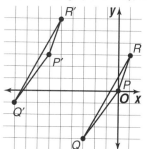

7. yes

9. $P'(-6, 3), Q'(-9, -1), R'(-5, 6)$

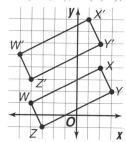

11. $W'(-5, 5), X'(1, 8), Y'(2, 6), Z'(-4, 3)$

13. $A'(-4, 0), B'(-2, 0), C'(-1, 2), D'(-3, 4), E'(-5, 2)$

15a. $(2, -5)$ **15b.** $Q'(-8, -3), R'(-2, -5), T'(2, -1)$ **17.** Sample answer: In early years the center moved west, then it moved south and west. **19.** The final position of the figure is the same as the original position of the figure. **21.** D

1. Answers may vary. Sample answer: In a mirror, the object seen is the flip of the real object. In a geometric reflection, what is on one side of a line is flipped from what is on the other side of the line.

3.

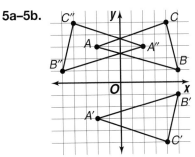

5a–5b.

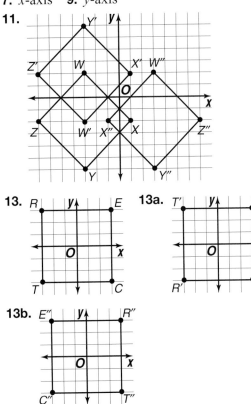

5c. yes; The corresponding sides are all congruent.

7. *x*-axis   **9.** *y*-axis

11.

13.   **13a.**

13b.

13c. They are all the same figure.

17. $(0, -1), (-2, 0), (-2, 5)$

19. Infomercial viewing

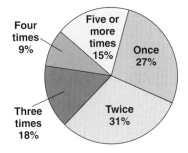

1. second quadrant

5.

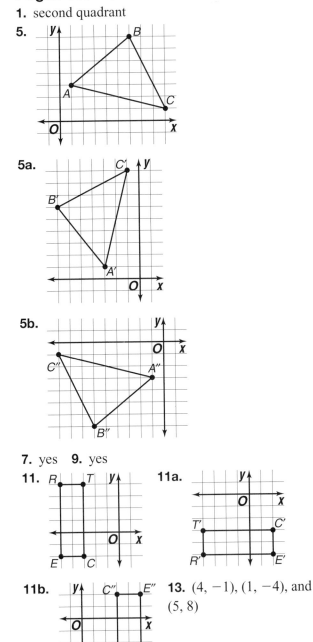

5a.

5b.

7. yes   **9.** yes

11.   **11a.**

11b.   **13.** $(4, -1), (1, -4)$, and $(5, 8)$

15a. Second and third   **15b.** Second: 120°, 240°;

Third: 180°   **17.** 2, 4, 10, J, Q, K of hearts; 2, 4, 10, J, Q, K of clubs; 2, 3, 4, 5, 6, 8, 9, 10, J, Q, K, A of diamonds; 2, 4, 10, J, Q, K of spades
19. $(y, -x)$ **21.** D

Pages 468–471   Study Guide and Assessment

1. domain   **3.** system of equations   **5.** two
7. reflection   **9.** Substitute the values into the original equation and check to see that a true sentence results.

11.

n	f(n)	(n, f(n))
−3	−8	(−3, −8)
0	1	(0, 1)
1	4	(1, 4)

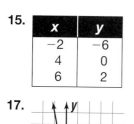

13. Sample table given.

n	f(n)	(n, f(n))
−2	4	(−2, 4)
0	0	(0, 0)
2	−4	(2, −4)

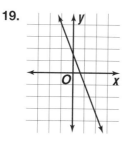

15.

x	y
−2	−6
4	0
6	2

17.

19.

21. (1, 5)   **23.** (1, 6)

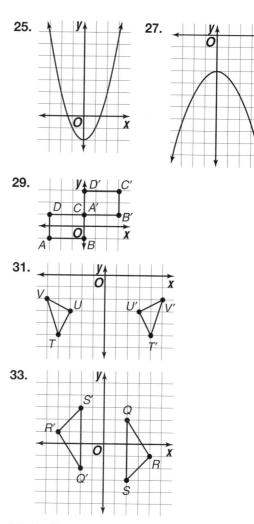

25.

27.

29.

31.

33.

35. A: line symmetry to a horizontal line; E: line symmetry to a vertical line; N: line symmetry to a diagonal line and rotational symmetry for 180°; T: line symmetry to a horizontal line; Z: no symmetry   **37a.** Sample answer: about 3.3 million   **37b.** from 1950 to 1960

Pages 472–473   Standardized Test Practice

1. B   **3.** B   **5.** D   **7.** A   **9.** B   **11.** E   **13.** 2
15. (−1, 0), (4, 5), (8, 1), (3, −4)   **17.** 10

CHAPTER 11
Geometry: Using Area and Volume

Pages 478–479   Lesson 11-1

1. Sample answer: Multiply 3 times 10^2. The area is about 300 m^2.   **3.** Alisa; since the diameter is 10 m, the radius is 5 m. Therefore the area of the whole circle equals $\pi \cdot 5^2$ and the area of the semicircle is $\frac{1}{2}\pi \cdot 5^2$.

5. 530.9 ft² **7.** about 251.93 mm² **9.** 380.1 ft²
11. 1,134.1 m² **13.** 254.5 cm² **15.** 514.2 cm²
17. 8 in. **19.** $\frac{8}{9}$ **21.** 9:1 **23.** C

Pages 480–481 Lesson 11-2A

1. a pre-fabricated unit **3.** Sample answer: Start
with a tower of 3 blocks and 1 block in front of it.
Then add a tower of 2 blocks to the right of the
original tower, and place 1 block in front of it.
Finally, make a tower of 3 blocks to the right of the
tower of 2, and place 1 block in front of it.
5. Check to make sure the model satisfies all of the
information. **9.** A

11.

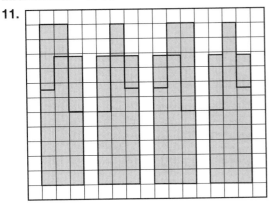

13. C

Pages 484–485 Lesson 11-2

1. Solid lines are used for the edges that you can see
and dashed lines are used for the edges you cannot
see. **3.** Both the pentagonal prism and the
pentagonal pyramid have a base in the shape of a
pentagon. The pentagonal prism has 2 pentagonal
bases and the other faces are quadrilaterals. The
pentagonal pyramid has 1 pentagonal base and the
other faces are triangles.

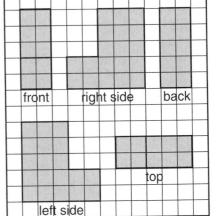

5a. rectangular prism **5b.** 2 units by 5 units by
3 units **5c.** 6 faces **5d.** 12 edges **5e.** 8 vertices
5f. Sample answer:

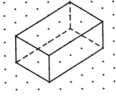

7. 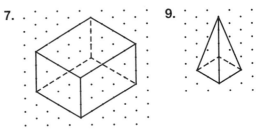 **9.**

11. Lightly draw an oval for
the base, draw two edges, and
draw another oval for the top.
Use dashed lines to show the
part you cannot see.

13.

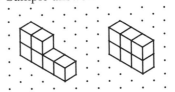

15. Sample answer:

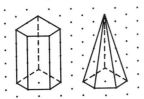

17a. True; a revolving door made from
3 intersecting planes **17b.** True; a corner of a
room where two walls meet the floor **17c.** True;
three parallel shelves on a book shelf

19. A

Pages 488–489 Lesson 11-3

1. The formula $V = Bh$ can be used to find the
volume of a rectangular prism, a triangular prism,
or a cylinder. Since the area of the base of a
rectangular prism equals the length times the width,
another formula for the volume of a rectangular
prism is $V = \ell wh$. Since the area of the base of a
cylinder equals pi times the radius squared, another
formula for the volume of a cylinder is $V = \pi r^2 h$.
3c. 18; they are the same. **5.** 1,125.7 cm³
7. 1,331 in³ **9.** 2,880 cm³ **11.** 5.5 cm³
13. Sample answer: 1,500 ft³ **15a.** 24 units³

15b. 192 units³ **15c.** 8:1 **15d.** It is multiplied by 8. **17a.** No, the inside dimensions of the refrigerator are $1\frac{5}{12}$ feet by $1\frac{1}{2}$ feet by $3\frac{1}{2}$ feet. Therefore the volume is $1\frac{5}{12} \times 1\frac{1}{2} \times 3\frac{1}{2}$ or $7\frac{7}{16}$ ft³ which is less than 8 ft³. **17b.** Double one of the dimensions. **19a.** $V = a^3$ **19b.** $V = 5s^2$ **19c.** $V = b^{2\sqrt{3}}$ **21.** B

Pages 492–493 Lesson 11-4

1. Since the area of the base of a cone is πr^2, $V = \frac{1}{3}Bh$ becomes $V = \frac{1}{3}(\pi r^2)h$ or $V = \frac{1}{3}\pi r^2 h$.
3c. They are the same. **3d.** 3 times **3e.** 1:3
5. 121.8 m³ **7.** 256 in³ **9.** 56 yd³ **11.** 85 yd³
13. Sample answer: 300 m³ **15.** It is doubled.
17. about 31,672,000 ft³ **19.** The new height would be $\frac{1}{4}$ of the original height. **21.** about 9%

Page 493 Mid-Chapter Self Test

1. about 38.5 ft² **3.** 512 cm³ **5.** 20.9 m³

Page 494 Lesson 11-5A

1. Nets of boxes that are rectangular prisms will be made up of rectangles. **3.** Nets of boxes that are rectangular prisms will have 3 sets of congruent shapes.

Pages 496–498 Lesson 11-5

1.

Side	Area
front	ℓh
back	ℓh
top	ℓw
bottom	ℓw
side	wh
side	wh

The sum of the 6 faces is $\ell h + \ell h + \ell w + \ell w + wh + wh$ or $2\ell h + 2\ell w + 2wh$.

3.

3a. 52 units²; 208 units² **3b.** 1:2 **3c.** 1:4
3d. No; $\frac{1}{2} \neq \frac{1}{4}$. **5.** 630 in² **7.** 17.04 m²
9. 98 m² **11.** 1,330 cm² **13.** 527.4 cm²
15. Sample answer: 600 m² **17.** Sample answer: Mary is redecorating her room. She wants to cover the 5 sides of an open storage cube with contact paper. If each edge of the cube is 2 ft long, what is the area that needs to be covered with the contact paper? **19.** Block #4; its surface area is 96 in² which is less than the surface areas of the other blocks (196 in², 112 in², and 168 in²).
21. about 50.3 m³ **23.** D

Pages 501–502 Lesson 11-6

1.

area of top	πr^2
area of bottom	πr^2
area of curved surface	πdh or $\pi(2r)h$ or $2\pi rh$

The surface area of the cylinder is $\pi r^2 + \pi r^2 + 2\pi rh$ or $2\pi r^2 + 2\pi rh$.

3. Geraldo

Surface Area of Rectangular Prism

top	10 cm²
bottom	10 cm²
front	25 cm²
back	25 cm²
side	10 cm²
side	10 cm²
Total	90 cm²

Surface Area of Cylinder

top	about 10.2 cm²
bottom	about 10.2 cm²
curved surface	about 56.5 cm²
Total	about 76.9 cm²

5. 766.9 cm² **7.** about 2,136.3 ft² **9.** 326.7 cm²
11. about 169.6 ft² **13.** Sample answer: 450 in²
15. about 227.4 in² **17.** Double the radius; consider the expression for the surface area of a cylinder, $2\pi r^2 + 2\pi rh$. If you double the height, you will double the second addend. If you double the radius, you will quadruple the first addend and double the second addend. **19.** D

Pages 505–507 Lesson 11-7

1a. true **1b.** false **1c.** false **1d.** true
1e. true **1f.** false **1g.** true **3.** Sample answer: An odometer measures length. It measures to the nearest 0.1 mile. It could be used to measure the distance to school or the distance between 2 cities.

5. The measurement is to the nearest 0.1 km. There are 3 significant digits. The greatest possible error is 0.05 km, and the relative error is $\frac{0.05}{76.4}$ or about 0.00065. **7.** The measurement is to the nearest millimeter. There is 1 significant digit. The greatest possible error is 0.5 mm, and the relative error is $\frac{0.5}{7}$ or about 0.071. **9.** The measurement is to the nearest $\frac{1}{2}$ lb. The greatest possible error is $\frac{1}{4}$ lb, and the relative error is $\frac{1}{38}$ or about 0.026. **11.** The measurement is to the nearest minute. There are 2 significant digits. The greatest possible error is $\frac{1}{2}$ min, and the relative error is $\frac{1}{94}$ or about 0.011. **13.** 34.3 oz; 2 lb is measured to the nearest pound, 34 oz is measured to the nearest ounce, and 34.3 oz is measured to the nearest 0.1oz. Therefore 34.3 is most precise. **15.** 78 in.; the relative error of 17 in. is $\frac{1}{34}$ and the relative error of 78 in. is $\frac{1}{156}$. Since $\frac{1}{34} > \frac{1}{156}$, the relative error of 78 in. is less than the relative error of 17 in. **17.** 10 significant digits **19.** 55 cm² **21.** A

Pages 508–511 Study Guide and Assessment

1. f **3.** g **5.** l **7.** a **9.** j **11.** 153.9 m²

13.

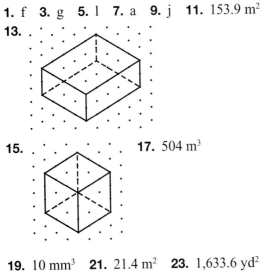

15.

17. 504 m³

19. 10 mm³ **21.** 21.4 m² **23.** 1,633.6 yd²
25. The measurement is to the nearest foot. There are 2 significant digits. The greatest possible error is $\frac{1}{2}$ ft, and the relative error is $\frac{1}{24}$ or about 0.042.
27. The measurement is to the nearest 0.1 m. There are 3 significant digits. The greatest possible error is 0.05 m, and the relative error is $\frac{0.05}{24.2}$ or about 0.0021. **29.** 15 balls **31a.** about 30.5 ft²
31b. about $6\frac{1}{2}$ trash cans

Pages 512–513 Standardized Test Practice

1. D **3.** D **5.** A **7.** B **9.** B **11.** 132 ft²
13. $A'(2, 0)$, $B'(2, -5)$, $C'(-4, -5)$, $D'(-4, 0)$

CHAPTER 12
Investigating Discrete Math and Probability

Pages 516–517 Lesson 12-1A

1.

Player A	Player B
scissors	scissors
scissors	paper
scissors	stone
paper	scissors
paper	paper
paper	stone
stone	scissors
stone	paper
stone	stone

3. 3 ways

5. 3 outcomes **7.** Yes; each player's chance of winning is $\frac{1}{3}$. **9.** 36 different outcomes
11. 9 ways **13.** no **15.** Unfair; player X's chance of winning is $\frac{4}{9}$, and players Y's chance of winning is $\frac{5}{9}$.

Pages 519–520 Lesson 12-1

1. Sample answer: Grandma's Diner has two flavors of ice cream, vanilla and chocolate. They serve the ice cream with one of four toppings; hot fudge, strawberry, butterscotch, and marshmallow. How many different ice cream desserts does Grandma's Diner serve? **3.** Both a tree diagram and the Fundamental Principle of Counting can be used to find the number of outcomes. A tree diagram actually shows the different outcomes. The Fundamental Principle of Counting does not show the actual outcomes, but it requires less time to solve the problem. **5.** 40 different T-shirts

7. First Spin / Second Spin / Third Spin / Outcome

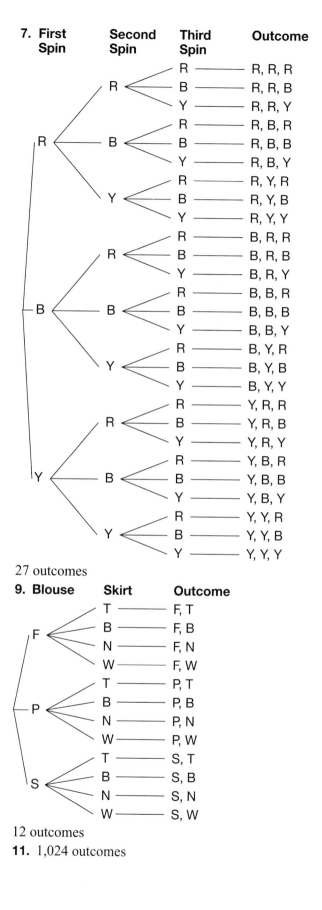

27 outcomes

9. Blouse / Skirt / Outcome

Blouse	Skirt	Outcome
F	T	F, T
	B	F, B
	N	F, N
	W	F, W
P	T	P, T
	B	P, B
	N	P, N
	W	P, W
S	T	S, T
	B	S, B
	N	S, N
	W	S, W

12 outcomes

11. 1,024 outcomes

13a. First Die / Second Die / Third Die / Outcome

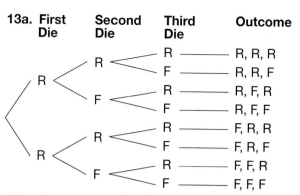

13b. 8 outcomes   **15.** 35,152 call letters
17. 3 significant digits

Pages 522–523   Lesson 12-2

1. Sample answer: How many 2-digit whole numbers can you write using the digits 1, 2, 3, 4, and 5 if no digit can be used twice.   **3a.** 6 ways
3b. They are the same.   **5.** 72   **7.** 2
9. 504 ways   **11.** 336   **13.** 720   **15.** 5,040
17. 3,628,800   **19.** 24,024   **21.** 40,320
23. 720 ways   **25.** 240 ways   **27.** They are the same. $P(n, n) = n \cdot (n - 1) \cdot (n - 2) \cdot \ldots \cdot 2 \cdot 1$ and $P(n, n - 1) = n \cdot (n - 1) \cdot (n - 2) \cdot \ldots \cdot 2$ which are equal.   **29.** 15 yd, 20 yd; 10 yd; 175 yd²

Pages 526–527   Lesson 12-3

1. $C(20, 6)$ **3.** Montega; the three positions are different, so order is important.   **5.** 56   **7.** 10
9. combination   **11.** 495 ways   **13.** 21   **15.** 210
17. 126   **19.** 286   **21.** permutation
23. combination   **25.** permutation   **27.** 2,598,960 hands   **29.** 253 combinations   **31.** 16 points
33. 840 numbers   **35.** $712\frac{1}{2}$ miles

Pages 530–531   Lesson 12-4

1. Add pairs of number in row 8.

row 8   1   8   28   56   70   56   28   8   1
row 9   1   9   36   84   126   126   84   36   9   1

3a. 1 way; 6 ways; 15 ways; 20 ways; 15 ways; 6 ways; 1 way   **3b.** row 6   **5.** 4   **7.** $\frac{1}{32}$ **9.** 56
11. 10   **13.** 126   **15.** 20 combinations
17. 64 branches   **19.** combination of 4 things taken 2 at a time   **21a.** 4   **21b.** The number of ways is equal to the number in the corresponding position in Pascal's Triangle.   **23.** They are the same. Pascal's Triangle is symmetric. Therefore, $C(9, 0) = C(9, 9)$, $C(9, 1) = C(9, 8)$, $C(9, 2) = C(9, 7)$, and so on.

25.

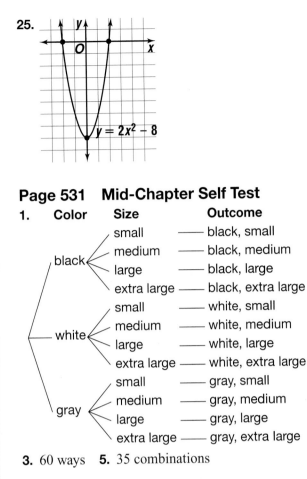

$y = 2x^2 - 8$

Page 531   Mid-Chapter Self Test

1.

Color	Size	Outcome
black	small	black, small
	medium	black, medium
	large	black, large
	extra large	black, extra large
white	small	white, small
	medium	white, medium
	large	white, large
	extra large	white, extra large
gray	small	gray, small
	medium	gray, medium
	large	gray, large
	extra large	gray, extra large

3. 60 ways   **5.** 35 combinations

Pages 532–533   Lesson 12-4B

1. Fibonacci sequence   **3.** 30   **5.** Inverted triangles; yes, because when two numbers next to each other have the same factor, the number below them will also have the same factor.   **7.** The right and left sides of Pascal's Triangle are ones. The inside numbers are the sum of the two numbers above it. If the rows are numbered from the top starting with 0, each row represents the number of combinations of items corresponding to the number of the row. The sums of the diagonals form the Fibonacci sequence. The product of the 6 numbers surrounding any number is always a perfect square. The multiples of a number form inverted triangles.

Pages 536–537   Lesson 12-5

1. The person is not likely to be 65 years old or older and living in an urban area. Only about 8 out of 100 people would satisfy these conditions.
3. Tonya; the first spin does not affect the second spin. Therefore the probability is $\frac{2}{5} \cdot \frac{2}{5}$ or $\frac{4}{25}$.
5. $\frac{1}{12}$ **7.** $\frac{1}{5}$ **9.** $\frac{1}{18}$ **11.** $\frac{3}{20}$ **13.** $\frac{3}{10}$ **15.** $\frac{2}{5}$
17. $\frac{2}{9}$ **19.** $\frac{1}{18}$ **21.** $\frac{1}{6}$ **23.** $\frac{1}{27}$ **25.** $\frac{1}{240}$

27. 15   **29.** 120

Pages 538–539   Lesson 12-6A

1. Since there are 6 possible balloons and 6 possible outcomes when a number cube is rolled, a roll of a 1 could represent the balloon Rosita wants. Since there is a 1 in 4 chance of the right wind and 4 possible outcomes for the spinner, a spin of one section could represent the wind Kelsey wanted. By rolling the number cube and spinning the spinner, you can see how many times both favorable outcomes occur out of a total number of tries.
3. No; the simulation will probably be similar to the computed results. However, simulations will vary.
5. You must plan a simulation that will correctly represent the situation.   **7.** The act it out strategy can find the probability of an event by experimenting.   **9.** jacket, $50.00; shirt, $12.50
11. 30 students   **13b.** No number would guarantee all 3 toys. You could be very unlucky and never get one of the toys.

Pages 541–543   Lesson 12-6

1. If the outcomes cannot be broken down into equally likely events, theoretical probability cannot be calculated.   **3.** No; if you draw more marbles, the experimental probability will come closer to the theoretical probability.   **5a.** $\frac{1}{5}$ **5b.** $\frac{1}{4}$

7d.

First Coin	Second Coin	Outcome
H	H	H H
	T	H T
T	H	T H
	T	T T

7e. $\frac{1}{4}$ **7f.** $\frac{1}{2}$ **9a.** Experimental probability; it is based on what happened in the past.   **9b.** 12,000 seeds   **13.** D

Page 545   Lesson 12-7A

1. 10, 20, 10   **3.** $\frac{1}{4}, \frac{1}{2}, \frac{1}{4}$

5a.

	F	F
F	FF	FF
f	Ff	Ff

$P(FF) = \frac{1}{2}, P(Ff) = \frac{1}{2}$

5b.

	F	F
F	FF	FF
F	FF	FF

$P(FF) = 1$

5c.

	f	f
f	ff	ff
f	ff	ff

$P(ff) = 1$

5d.

	F	f
f	Ff	ff
f	Ff	ff

$P(\mathbf{Ff}) = \frac{1}{2}, P(\mathbf{ff}) = \frac{1}{2}$

Pages 547–548 Lesson 12-7

1. by setting up and solving a proportion
3. Sample answer: If the sample is not random, any predictions will be incorrect. Taking every tenth student from an alphabetical list of students would be an appropriate sampling. Taking a survey of the football players before practice would be an inappropriate sampling.   **5a.** about 760 people
5b. Sample answer: how many T-shirts to stock at future concerts   **7a.** 25 students   **7b.** 48%
7c. $\frac{6}{25}$ **7d.** about 50 votes   **9a.** No; 327 prefer Sunshine Orange Juice and 273 prefer the competitor's juice which is not 2 to 1.   **9b.** Yes; over 50% of the people in one survey did prefer Sunshine Orange Juice.   **9c.** No; another survey may show a different result.   **13.** probability determined by conducting an experiment or simulation

Pages 550–553 Study Guide and Assessment

1. f   **3.** h   **5.** d   **7.** b   **9.** j   **11.** 8 outcomes
13. 12 outcomes   **15.** 120   **17.** 120 numbers
19. 66   **21.** 1,140 groups   **23.** 70 combinations
25. $\frac{1}{4}$ **27c.** $\frac{1}{4}; \frac{1}{2}$ **29.** $\frac{9}{50}$
31.

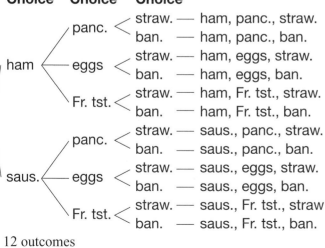

First Choice	Second Choice	Third Choice	Outcome

12 outcomes

33. 1,140 combinations

Pages 554–555 Standardized Test Practice

1. A   **3.** C   **5.** B   **7.** E   **9.** C   **11.** C   **13.** 12
15. 1, 2, 3, 4, 6, 8, 9, 12, 18, 24, 36, 72

CHAPTER 13
Algebra: Exploring Polynomials

Page 560 Lesson 13-1A

1. $1, x, x^2$ **3.** Sample answer: Each set of tiles will have 3 tiles with areas 1, x, and x^2.
5.

Pages 562–564 Lesson 13-1

1a. polynomial   **1b.** monomial   **1c.** monomial
1d. polynomial   **5.** $-x^2 + 2x - 4$
7.

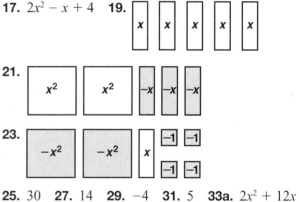

9. -3
11. 70 ft   **13.** $x^2 + 3x$ **15.** $-4x + 2$
17. $2x^2 - x + 4$ **19.**

21.

23.

25. 30   **27.** 14   **29.** -4 **31.** 5   **33a.** $2x^2 + 12x$
33b. 32 sq. units   **33c.** rectangular prism if all sides do not have the same measure; cube if $x = 3$
35.

	b
a	ab

37. 10 units   **39.** 60%

Pages 567–569 Lesson 13-2

1. $2x^2, -3x^2; -y, 5y$
3. $-x^2 - x - 3;$

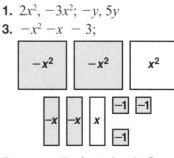

5. none   **7.** $4x + 1$ **9.** $8x - y; -14$

11a. $4q + 8d + 5n$ **11b.** \$2.05   **13.** $3x^2, -2x^2$
15. $7a, 10a; 6b, 14b$ **17.** $4y, -2y, -3y; 8, 9$
19. $-2x + 2y$ **21.** $5a + 5$ **23.** $-7y$
25. $3x^2 + 6$ **27.** $17a + 11b; 46$ **29.** $-a + 9b;$
-22 **31.** 52 **33a.** $50r + 100, \$103$
33b. $50r^2 + 150r + 150$ **35a.** $114O + 96G$
35b. \$2,232   **37.** 15 **39.** B

Pages 571–572   Lesson 13-3

1. Add like terms.
3. $(x^2 - 2x - 3) + (-2x^2 + 2x + 4) = -x^2 + 1$

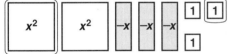

5. $9a + 5b$ **7.** $-x + 6$ **9a.** $(x + 4) +$
$(2x - 3) + (3x + 1) = (6x + 2)$ cm   **9b.** 26 cm
11. $6y^2 + 2y + 2$ **13.** $x^2 + 5$ **15.** $16q - r + 1$
17. $12a + 12b$ **19.** $10a^2 + 2a - 3$ **21.** $2x^2 -$
$3x - 10$ **23.** $7g + 3h + 4; 51$ **25.** $15g + h +$
$11; 90$ **27.** 24 **29.** $3x^2 + 21x + 6$
31. $-10a^2 + 15a$ **33.** $3\frac{7}{20}$

Pages 574–576   Lesson 13-4

1. $-2x$
3. $(2x^2 - 3x + 3) - (x^2 + 1) = x^2 - 3x + 2$

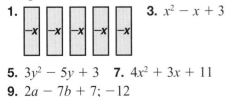

5. $2p^2 - p + 1$ **7.** $2x^2 - 3x + 5$ **9.** \$1.03
11. y^2 **13.** $r^2 - 8rt + 2t^2$ **15.** $-5a + 6$
17. $2a^2 - 5a - 9$ **9.** $3x^2 - 3x$ **21.** $2c - 11d -$
$4; -63$ **23.** $9c + 6d - 10; 12$ **25.** $2p^2 + 2pq +$
q^2 **27a.** $-2A + 4B + 3C$ **27b.** 5 **29.** $6y^2 +$
$15y + 3$ **31.** $-\frac{1}{3}$ **32.** B

Page 576   Mid-Chapter Self Test

1.   **3.** $x^2 - x + 3$

5. $3y^2 - 5y + 3$ **7.** $4x^2 + 3x + 11$
9. $2a - 7b + 7; -12$

Page 577   Lesson 13-5A

1. $2x + 4$

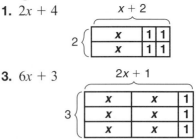

3. $6x + 3$

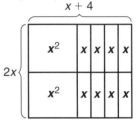

Pages 580–581   Lesson 13-5

1. Multiply $3x$ by $5x$ and multiply $3x$ by 2. Add the results.   **3.** $2x(x + 4) = 2x^2 + 8x$

5. n^8 **7.** $b^2 - 2b$ **9a.**

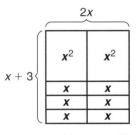

9b. $2x(x + 3); 2x^2 + 6x$ **9c.** 260 ft²   **11.** 2^{10}
13. p^{14} **15.** $z^2 - z$ **17.** $b^2 + 3b$ **19.** $2m^3 + 4m^2$
21. $d^8 + 15d^3$ **23.** a^{11} **25.** $15d^2 + 28d$
27. $(3x + 6)$ dollars   **29.** $5c + 3d; 21$ **31.** D

Pages 584–585   Lesson 13-6

1. b, c   **3.** $4x^2 + 10x + 6$

5. $x^2 + 2x - 3$ **7.** $x^2 + 7x + 12$
9. $2x^2 + 7x + 3$ **11.** $b^2 + 10b + 24$
13. $4m^2 + 6m + 2$ **15.** $d^2 + 6d + 9$
17. $3a^2 - 8a - 3$ **19a.** $x^2 + 14x + 40$
19b. $9x^2 + 2x - 48$ **21.** $h^2 + 2mh + m^2$; yes;
only h and m will vary from player to player.
23. $t^2 - 4t$ **25.** 306

Pages 586–587   Lesson 13-7A

1. Sample answer: Ramón guessed an equal number of bottles and six-packs; Sheila guessed more than twice as many six-packs than bottles.   **3.** 29, 31
7. $20.30   **9.** $46.20   **11.** 6 postcards, 5 letters
13. D

Pages 589–591   Lesson 13-7

1. Sample answer: You can form a rectangle with the algebra tiles.   **3a.**

x^2	x^2	x
x	x	1
x	x	1

3b. $2x + 1$ and $x + 2$; these are the length and the width of the rectangle.   **5.** $x(x + 4)$ **7.** not factorable   **9.** $3(x + 2)$ **11.** $(2x + 1)(x + 2)$
13. $6(3x + 1)$ **15.** $(x + 4)(x + 1)$
17. $(x + 4)(x + 4)$ **19.** $(2x + 1)(x + 3)$
21. $x(3x + 4)$ **23.** $\frac{2x + 1}{x + 2}$ **25a.** $h^2 + 2mh + m^2 = (h + m)(h + m)$; $h + m = 1$
25b. $2 \cdot h^2 + 1 \cdot 2mh + 0 \cdot m^2 = 2h^2 + 2mh$
27. $6k^2 + 5k - 4$ **29.** $\frac{1}{8}$

Pages 592–595   Study Guide and Assessment

1. false; binomial   **3.** true   **5.** false; two
7. Sample answer: A monomial is a number, a variable, or a product of a number and one or more variables.   **9.**

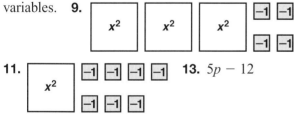

11.   **13.** $5p - 12$

15. $4a - b$ **17.** $12d + 7$ **19.** $3b^2 - b - 4$
21. $2g + 1$ **23.** $s^2 + 5$ **25.** $3p^2 - 4p - 3$
27. 12^{13} **29.** d^{10} **31.** $3z^2 - 12z$ **33.** $2t^7 - 8t^4$
35. $2x^2 - 9x - 5$ **37.** $x^2 - x - 12$
39. $7(2x + 1)$ **41.** $(x + 5)(x + 2)$
43. $(2x + 1)(x + 1)$ **45.** $(4x^2 + 4x + 1)$ yd^2
47. $500 + 1,000r + 500r^2$

Pages 596–597   Standardized Test Practice

1. A   **3.** C   **5.** A   **7.** C   **9.** D   **11.** 0.5

Photo Credits

PHOTO CREDITS

Glossary

absolute value (57)   The number of units a number is from zero on the number line.

acute (196)   An angle with a measure greater than 0° and less than 90°.

addition property of equality (18)   If you add the same number to each side of an equation, the two sides remain equal. If $a = b$, then $a + c = b + c$.

additive inverse (70)   Two integers that are opposites of each other are called additive inverses. The sum of any number and its additive inverse is zero, $a + (-a) = 0$.

algebraic expression (12)   A combination of variables, numbers, and at least one operation.

alternate exterior angles (189)   In the figure, transversal t intersects lines ℓ and m. $\angle 1$ and $\angle 7$, $\angle 2$ and $\angle 8$ are alternate exterior angles. If lines ℓ and m are parallel, then these angles are congruent.

alternate interior angles (189)   In the figure, transversal t intersects lines ℓ and m. $\angle 3$ and $\angle 5$, $\angle 4$ and $\angle 6$ are alternate interior angles. If lines ℓ and m are parallel, then these angles are congruent.

altitude (40, 301)   A segment in a quadrilateral that is perpendicular to both bases, with endpoints on the base lines.

altitude (490)   The segment that goes from the vertex of a cone to its base and is perpendicular to the base.

area (39)   The number of square units needed to cover a surface enclosed by a geometric figure.

arithmetic sequence (296)   A sequence of numbers in which you can find the next term by adding the same number to the previous term.

associative property (8)   For any numbers a, b, and c, $(a + b) + c = a + (b + c)$. Also for any numbers a, b, and c, $(ab)c = a(bc)$.

B

bar graph (142)   A graphic form using bars to make comparisons of statistics.

bar notation (249)   In repeating decimals the line or bar placed over the digits that repeat. Another way to write 2.6363636. . . is $2.\overline{63}$.

base (8)   In a power, the number used as a factor. In 10^3, the base is 10. That is, $10^3 = 10 \times 10 \times 10$.

base (40, 301, 302)   The base of a parallelogram or a triangle is any side of the figure. The bases of a trapezoid are the parallel sides.

base (335)   In a percent proportion, the number to which the percentage is compared. In $\frac{3}{4} = \frac{x}{100}$, the total number of parts, 4, is called the base.

base (483)   The bases of a prism are the two parallel congruent faces.

binomial (583)   A polynomial with exactly two terms.

box-and-whisker plot (167)   A diagram that summarizes data using the median, the upper and lower quartiles, and the extreme values. A box is drawn around the quartile value and whiskers extend from each quartile to the extreme data points.

cell (162)   The basic unit of a spreadsheet. A cell can contain data, labels, or formulas.

center (309)   The given point from which all points on a circle or a sphere are the same distance.

circle (309)   The set of all points in a plane that are the same distance from a given point called the center.

circle graph (148)   A type of statistical graph used to compare parts of a whole.

circular cone (490)   A shape in space that has a circular base and one vertex.

circular cylinder (487)   A cylinder with two bases that are parallel, congruent circular regions.

circumference (309)   The distance around a circle.

column (73)   Numbers stacked on top of each other in a vertical arrangement.

combination (524)   An arrangement or listing of objects in which order is not important.

common difference (296)   The difference between any two consecutive terms in an arithmetic sequence.

common ratio (297)   The constant factor used to multiply consecutive terms in a geometric sequence.

commutative property (8)   For any real numbers a and b, $a + b = b + a$ and $ab = ba$.

compatible numbers (126)   Two numbers that are easy to divide mentally. They are often members of fact families and can be used as an estimation.

complementary angles (198)   Two angles are complementary if the sum of their measures is 90°.

composite number (235)   Any whole number greater than 1 that has more than two factors.

compound event (534)   A compound event consists of two or more simple events.

congruent triangles (210)   Triangles that have the same size and shape.

converse (399)   The converse of the Pythagorean Theorem can be used to test whether a triangle is a right triangle. If the sides of the triangle have lengths a, b, and c, such that $c^2 = a^2 + b^2$, then the triangle is a right triangle.

coordinate (56)   A number associated with a point on the number line.

coordinate system (92)   A plane in which a horizontal number line and a vertical number line intersect at their zero points.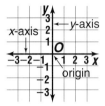

corresponding angles (189)   Angles that have the same position on two different parallel lines cut by a transversal. $\angle 1$ and $\angle 5$, $\angle 2$ and $\angle 6$, $\angle 3$ and $\angle 7$, and $\angle 4$ and $\angle 8$ are corresponding angles.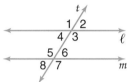

corresponding parts (210)   Parts on congruent or similar figures that match.

Counting Principle (518)   This principle is used to find the number of outcomes by multiplying. If event M can occur in m ways and is followed by event N that can occur in n ways, then the event M followed by the event N can occur in $m \cdot n$ ways.

cross products (111)   The products of the terms on the diagonals when two ratios are compared. If the cross products are equal, then the ratios form a proportion. In the proportion $\frac{2}{3} = \frac{8}{12}$, the cross products are 2×12 and 3×8.

D

data analysis (154)   To study data and draw conclusions from the numbers observed.

density property (260)   Between every pair of rational numbers there are infinitely many rational numbers.

dependent events (535)   Two or more events in which the outcome of one event does affect the outcome of the other event or events.

diameter (309)   The distance across a circle through its center.

dilation (214, 370)   The process of reducing or enlarging an image in mathematics.

discount (349)   The amount by which the regular price is reduced.

distributive property (34)   The sum of two addends multiplied by a number is the sum of the product of each addend and the number. For any numbers a, b, and c, $a(b + c) = ab + ac$ and $(b + c)a = ba + ca$.

divisible (232)   A number is divisible by another if the quotient is a whole number and the remainder is zero.

division property of equality (22)   If each side of an equation is divided by the same nonzero number, then the two sides remain equal. If $a = b$, then $\frac{a}{c} = \frac{b}{c}$, $c \neq 0$.

domain (428)   The set of input values in a function.

edge (483)   The intersection of faces of a three-dimensional figure.

element (73)   Each number in a matrix is called an element.

equation (13)   A mathematical sentence that contains the equal sign, $=$.

equilateral (196)   All sides of a figure are congruent.

evaluate (9)   To find the value of an expression by replacing the variables with numerals.

event (253)   A specific outcome or type of outcome.

experimental probability (540)   An estimated probability based on the relative frequency of positive outcomes occurring during an experiment.

exponent (8)   In a power, the number of times the base is used as a factor. In 10^3, the exponent is 3.

F

face (483)   Any surface that forms a side or a base of a prism.

factor (8)   When two or more numbers are multiplied, each number is a factor of the product. In $4 \times 5 = 20$, 4 and 5 are factors.

factorial (522)   The expression $n!$ is the product of all counting numbers beginning with n and counting backward to 1.

factoring (588)   Finding the factors of a product.

factor tree (236)   A diagram showing the prime factorization of a number.

Fibonacci sequence (300)   A list of numbers in which the first two numbers are both 1 and each number that follows is the sum of the previous two numbers, 1, 1, 2, 3, 5, 8, . . .

frequency table (142)   A table for organizing a set of data that shows the number of times each item or number appears.

function (428)   A relation in which each element of the input is paired with exactly one element of the output according to a specified rule.

function table (428)   A table organizing the input, rule, and output of a function.

Fundamental Theorem of Arithmetic (236)   A property of numbers that states every number has a unique set of prime factors.

GLOSSARY

G

geometric sequence (297)   When consecutive terms of a sequence are formed by multiplying by a constant factor, the sequence is called a geometric sequence.

graph (56)   To draw or plot the points named by those numbers on a number line or coordinate plane.

greatest common factor (GCF) (242)   The greatest of the common factors of two or more numbers. The GCF of 18 and 24 is 6.

greatest possible error (504)   Half the smallest unit used to make a measurement.

H

height (40, 301)   The length of the altitude of a triangle or a quadrilateral.

histogram (142)   A special kind of bar graph that displays the frequency of data that has been organized into equal intervals. The intervals cover all possible values of data, therefore there are no spaces between the bars of the graph.

hypotenuse (398)   The side opposite the right angle in a right triangle.

I

independent events (534)   Two or more events in which the outcome of one event does not affect the outcome of the other event(s).

indirect measurement (361)   A technique using proportions to find a measurement.

inequality (43)   A mathematical sentence that contains $<$, $>$, $\neq$, $\leq$, or $\geq$.

integer (56)   The whole numbers and their opposites.

$$\ldots, -3, -2, -1, 0, 1, 2, 3, \ldots$$

interest (353)   The amount charged or paid for the use of money.

interquartile range (163)   The range of the middle half of data.

inverse operation (18)   Pairs of operations that undo each other. Addition and subtraction are inverse operations. Multiplication and division are inverse operations.

inverse property of multiplication (290)   For every nonzero number $\frac{a}{b}$, where a and b do not equal 0, there is exactly one number $\frac{b}{a}$ such that $\frac{a}{b} \times \frac{b}{a} = 1$.

irrational number (391)   A number that cannot be expressed as $\frac{a}{b}$, where a and b are integers and $b \neq 0$.

isosceles (196)   A triangle that has at least two congruent sides.

L

least common denominator (LCD) (261)   The least common multiple of the denominators of two or more fractions.

least common multiple (LCM) (257)   The least of the nonzero common multiples of two or more numbers. The least common multiple of 2 and 3 is 6.

legs (398)   The two sides of a right triangle that form the right angle.

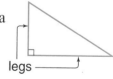

like terms (565)   Expressions that contain the same variables, such as $3ab$ and $7ab$.

linear function (442)   An equation in which the graphs of the solutions form a line.

line of symmetry (460)   A line that divides a figure into two halves that are reflections of each other.

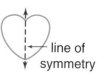

GLOSSARY

line plot (153)   A graph that uses an × above a number on a number line each time that number occurs in a set of data.

line symmetry (206)   Figures that match exactly when folded in half have line symmetry.

lower quartile (164)   The median of the lower half of data in an interquartile range.

markup (349)   The difference between the price paid by the merchant and the increased selling price.

matrix (73)   A rectangular arrangement of elements in rows and columns.

mean (158)   The sum of the numbers in a set of data divided by the number of pieces of data.

measures of central tendency (158)   Numbers or pieces of data that can represent the whole set of data.

median (158, 159)   The middle number in a set of data when the data are arranged in numerical order. If the data has an even number, the median is the mean of the two middle numbers.

mixed number (278)   The sum of a whole number and a fraction. $6\frac{2}{3}$ is a mixed number.

mode (158)   The number(s) or item(s) that appear most often in a set of data.

monomial (561)   A number, a variable, or a product of a number and one or more variables.

multiple (257)   The product of a number and any whole number. 28 is a multiple of 4 and 7.

multiplication property of equality (22)   If each side of an equation is multiplied by the same number, then the two sides remain equal. If $a = b$, then $ac = bc$.

multiplicative inverse (290)   A number times its multiplicative inverse is equal to 1. The multiplicative inverse of $\frac{2}{3}$ is $\frac{3}{2}$.

N

numerical expression (11)   A mathematical expression that has a combination of numbers and at least one operation. $4 + 2$ is a numerical expression.

O

obtuse (196)   An obtuse angle is any angle that measures greater than 90° but less than 180°.

open sentence (13)   An equation that contains a variable.

opposite (70)   Two integers are opposites if they are represented on the number line by points that are the same distance from zero, but on opposite sides of zero. The sum of opposites is zero.

order of operations (11)   The rules to follow when more than one operation is used.

1. Simplify the expressions inside grouping symbols; start with the innermost grouping symbols first.
2. Evaluate all powers.
3. Then do all multiplications and divisions from left to right.
4. Then do all additions and subtractions from left to right.

ordered pair (92)   A pair of numbers used to locate a point in the coordinate system. The ordered pair is written in this form: (x-coordinate, y-coordinate).

origin (92)   The point of intersection of the x-axis and y-axis in a coordinate system.

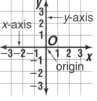

outcome (253, 518)   One possible result of a probability event. For example, 4 is an outcome when a number cube is rolled.

GLOSSARY

outlier (164)   Data that is more than 1.5 times the interquartile range from the upper or lower quartiles.

P

parallel lines (188)   Lines in the same plane that do not intersect. The symbol ‖ means parallel.

parallelogram (39, 201)   A quadrilateral with two pairs of parallel sides.

Pascal's Triangle (528)   A triangular arrangement of numbers in which each number is the sum of the two numbers to the right and to the left of it in the row above.

pentagon (358)   A polygon having five sides.

percent (107)   A ratio that compares a number to 100.

percentage (335)   In a percent proportion, a number (P) that is compared to another number called the base (B).

percent of change (348)   The ratio of the amount of increase or decrease to the original amount.

percent of decrease (349)   The ratio of an amount of decrease to the previous amount, expressed as a percent.

percent of increase (349)   The ratio of an amount of increase to the previous amount, expressed as a percent.

percent proportion (335)   Is $\frac{P}{B} = \frac{r}{100}$ where P represents the percentage, B represents the base, and r represents the number per hundred. Also written as $\frac{\text{Percentage}}{\text{Base}}$ = Rate.

perfect square (382)   A rational number whose square root is a whole number. 25 is a perfect square because $\sqrt{25} = 5$.

perimeter (38)   The distance around a geometric figure.

permutation (521)   An arrangement or listing in which order is important.

perpendicular (197)   Two lines or line segments that intersect to form right angles.

polygon (196)   A simple closed figure in a plane formed by three or more line segments.

polynomial (562)   The sum or difference of two or more monomials.

population (546)   The entire group of items or individuals from which the samples under consideration are taken.

power (8)   A number that can be written using an exponent. The power 7^3 is read *seven to the third power,* or *seven cubed.*

precision (504)   The precision of a measurement depends on the unit of measure. The smaller the unit the more precise the measurement is.

prime factorization (235)   Expressing a composite number as the product of prime numbers. The prime factorization of 63 is $3 \times 3 \times 7$.

prime number (235)   A whole number greater than 1 that has exactly two factors, 1 and itself.

principal (353)   The amount of an investment or a debt.

principal square root (382)   A nonnegative square root. $\sqrt{49}$ indicates the *principal* square root of 49.

prism (483)   A three-dimensional figure that has two parallel and congruent bases in the shape of polygons.

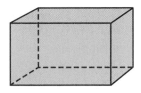

probability (253)   The chance that some event will happen. It is the ratio of the number of ways a certain event can occur to the number of possible outcomes.

proportion (111)   A statement of equality of two or more ratios, $\frac{a}{b} = \frac{c}{d}, b \neq 0, d \neq 0.$

pyramid (483)   A solid figure that has a polygon for a base and triangles for sides.

Pythagorean Theorem (396–398)   In a right triangle, the square of the length of the hypotenuse is equal to the sum of the squares of the lengths of the legs. $c^2 = a^2 + b^2$

Pythagorean triple (405)   A set of three integers that satisfy the Pythagorean Theorem.

quadrant (92)   One of the four regions into which two perpendicular number lines separate the plane.

quadratic function (452)   A function in which the greatest power is 2.

quadrilateral (201)   A polygon having four sides.

quartiles (163)   Values that divide data into four equal parts.

radical sign (382)   The symbol used to indicate a nonnegative square root. $\sqrt{}$

radius (309)   The distance from the center of a circle to any point on the circle.

random (253)   Outcomes occur at random if each outcome is equally likely to occur.

range (428)   The set of output values in a function.

range (163)   The difference between the greatest number and the least number in a set of data.

rate (105)   A ratio of two measurements having different units.

rate (335)   In a percent proportion, the ratio of a number to 100.

ratio (104)   A comparison of two numbers by division. The ratio of 2 to 3 can be stated as 2 out of 3, 2 to 3, 2:3, or $\frac{2}{3}$.

rational numbers (245)   Numbers of the form $\frac{a}{b}$, where a and b are integers and $b \neq 0$.

real numbers (391)   The set of rational numbers together with the set of irrational numbers.

reciprocal (290)   The multiplicative inverse of a number.

rectangle (38, 201)   A quadrilateral with four congruent angles.

reflection (206, 460)   A type of transformation where a figure is flipped over a line of symmetry.

relative error (504)   The comparison of the greatest possible error with the measurement itself.

$$relative\ error = \frac{greatest\ possible\ error}{measurement}$$

repeating decimal (249)   A decimal whose digits repeat in groups of one or more. Examples are 0.181818... and 0.8333.... Using bar notation, these numbers are written as $0.\overline{18}$ and $0.8\overline{3}$

rhombus (201)   A parallelogram with four congruent sides.

right (196)   An angle that measures 90°.

rotation (206, 464)   When a figure is turned around a central point.

rotational symmetry (207)   A figure has rotational symmetry if it can be turned less than 360° about its center and still looks like the original.

row (73)   In a matrix the numbers side-by-side horizontally form a row.

sample (174, 546)   A randomly-selected group chosen for the purpose of collecting data.

sample space (253)   The set of all possible outcomes.

scale (366)   The ratio of a given length on a drawing or model to its corresponding length in reality.

scale drawing (366)   A drawing that is similar but either larger or smaller than the actual object.

scale factor (370)   The ratio of a dilated image to the original image.

scale model (366)   A replica of an original object that is too large or too small to be built at actual size.

scalene (196)   A triangle with no congruent sides.

scatter plot (168)   A graph that shows the general relationship between two sets of data.

scientific notation (265)   A way of expressing numbers as the product of a number that is at least 1 but less than 10 and a power of 10. In scientific notation, 5,500 is 5.5×10^3.

selling price (349)   The amount a customer pays for an item.

sequence (296)   A list of numbers in a certain order, such as, 0 ,1, 2, 3, or 2, 4, 6, 8.

significant digits (504)   All of the digits of a measurement that are known to be accurate plus one estimated digit.

similar polygons (357)   Two polygons are similar if their corresponding angles are congruent and their corresponding sides are in proportion. They have the same shape but may not have the same size.

similar triangles (215)   Triangles that have the same shape but may not have the same size.

simplest form (246)   The form of a fraction where the GCF of the numerator and denominator is 1.

simplest form (565)   An expression that has no like terms in it. In simplest form $6x + 8x$ is $14x$.

simulation (540)   The process of acting out a problem.

solid (482)   A three-dimensional figure.

solution (13)   A value for the variable that makes an equation true. The solution for $10 + y = 25$ is 15.

spreadsheet (162)   A tool used for organizing and analyzing data.

square (38, 201)   A parallelogram with all sides congruent and all angles congruent.

square root (382)   One of the two equal factors of a number. If $a^2 = b$, then a is the square root of b. A square root of 144 is 12 since $12^2 = 144$.

statistics (142)   The branch of mathematics that deals with collecting, organizing, and analyzing data.

substitute (11)   To replace a variable in an algebraic expression with a number, creating a numerical expression.

subtraction property of equality (18)   If you subtract the same number from each side of an equation, the two sides remain equal. For any numbers a, b, and c, if $a = b$, then $a - c = b - c$.

supplementary angles (190)   Two angles are supplementary if the sum of their measures is 180°.

surface area (495)   The sum of the areas of all the faces of a three-dimensional figure.

symmetric (460)   Figures that can be folded into two identical parts.

system of equations (446)   A set of equations with the same variables.

term (565)   A number, a variable, or a product of numbers and variables.

term (296)   A number in a sequence.

terminating decimal (249)   A decimal whose digits end. Every terminating decimal can be written as a fraction with a denominator of 10, 100, 1,000, and so on.

tessellation (220)   A repetitive pattern of polygons that fit together with no holes or gaps.

theoretical probability (540)   The long-term probability of an outcome based on mathematical principles.

transformation (220)   Movements of geometric figures.

translation (220, 456)   One type of transformation where a figure is slid horizontally, vertically, or both.

transversal (188)   A line that intersects two parallel lines to form eight angles.

trapezoid (201, 302)   A quadrilateral with exactly one pair of parallel sides.

tree diagram (518)   A diagram used to show the total number of possible outcomes in a probability experiment.

triangle (196)   A polygon having three angles and three sides.

trigonometry (365)   The study of triangle measurement.

trinomial (588)   A polynomial with three terms.

unit rate (105)   A rate with denominator of 1.

upper quartile (163)   The median of the upper half of a set of numbers.

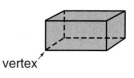

variable (12)   A symbol, usually a letter, used to represent a number in mathematical expressions or sentences.

variation (163)   The spread in the values in a set of data.

vertex (483)   The vertex of a prism is the point where all the faces intersect.

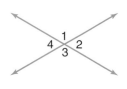

vertical angles (190)   Congruent angles formed by the intersection of two lines. In the figure, the vertical angles are $\angle 1$ and $\angle 3$, and $\angle 2$ and $\angle 4$.

volume (486)   The number of cubic units needed to fill the space occupied by a solid.

x-axis (92)   The horizontal number line which helps to form the coordinate system.

x-coordinate (92)   The first number of an ordered pair.

y-axis (92)   The vertical line of the two perpendicular number lines in a coordinate plane.

y-coordinate (92)   The second number of an ordered pair.

Z

zero pair (63)   The result of pairing one positive counter with one negative counter.

Spanish Glossary

A

absolute value / valor absoluto (57)   Número de unidades en la recta numérica que un número dista de cero.

acute / agudo (196)   Ángulo que mide más de 0° y menos de 90°.

addition property of equality / propiedad de adición de la igualdad (18)   Si sumas el mismo número a ambos lados de una ecuación, los dos lados permanecen iguales. Si $a = b$, entonces $a + c = b + c$.

additive inverse / inverso aditivo (70)   Dos enteros que son opuestos mutuos reciben el nombre de inversos aditivos. La suma de cualquier número y su inverso aditivo es cero, $a + (-a) = 0$.

algebraic expression / expresión algebraica (12)   Combinación de variables, números y al menos una operación.

alternate exterior angles / ángulos alternos externos (189)   En la figura, la transversal t interseca las rectas ℓ y m. El $\angle 1$ y $\angle 7$, así como el $\angle 2$ y el $\angle 8$, reciben el nombre de ángulos alternos externos. Si las rectas ℓ y m son paralelas, estos ángulos son todos congruentes.

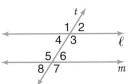

alternate interior angles / ángulos alternos internos (189)   En la figura, la transversal t interseca las rectas ℓ y m. El $\angle 3$ y el $\angle 5$, así como el $\angle 4$ y el $\angle 6$ reciben el nombre de ángulos alternos internos. Si las rectas ℓ y m son paralelas, estos ángulos son congruentes.

altitude / altitud (40, 301)   Segmento en un cuadrilátero que es perpendicular a ambas bases y cuyos extremos yacen en las bases del cuadrilátero.

altitude / altitud (490)   Segmento trazado desde el vértice de un cono hasta su base y que es perpendicular a la base.

area / área (39)   Número de unidades cuadradas que se requieren para cubrir la superficie encerrada por una figura geométrica.

arithmetic sequence / sucesión aritmética (296)   Sucesión de números en que se puede calcular el próximo término sumando siempre el mismo número al término anterior.

associative property / propiedad asociativa (8)   Para números a, b y c cualesquiera, $(a + b) + c = a + (b + c)$ y $(ab)c = a(bc)$.

B

bar graph / gráfica de barras (142)   Tipo de gráfica que usa barras para comparar estadísticas.

bar notation / notación de barra (249)   En los decimales periódicos, la línea o barra que se escribe encima de los dígitos que se repiten. Otra forma de escribir 2.6363636... es $2.\overline{63}$.

base / base (8)   Número que se usa como factor en una potencia. En 10^3, la base es 10, es decir, $10^3 = 10 \times 10 \times 10$.

base / base (40, 301, 302)   La base de un paralelogramo o de un triángulo es cualquier lado de la figura. Las bases de un trapecio son sus lados paralelos.

base / base (335)   Número con el cual se compara el porcentaje, en una proporción porcentual. En $\frac{3}{4} = \frac{x}{100}$, el número total de partes, 4, recibe el nombre de base.

base / base (483)   Las bases de un prisma son sus caras congruentes y paralelas.

binomial / binomio (583)   Polinomio que tiene exactamente dos términos.

box-and-whisker plot / diagrama de caja y patillas (167)   Diagrama que resume información usando la mediana, los cuartiles superior e inferior y los valores extremos. Se dibuja una caja alrededor de los cuartiles y se trazan patillas que los unan a los valores extremos respectivos.

cell / celda (162) Unidad básica de una hoja de cálculos. Las celdas pueden contener datos, rótulos o fórmulas.

center / centro (309) Punto en el plano del cual equidistan todos los puntos de un círculo o de una esfera.

circle / círculo (309) Conjunto de todos los puntos en un plano que equidistan de un punto dado llamado centro.

circle graph / gráfica circular (148) Tipo de gráfica estadística que se usa para comparar las partes de un todo.

circular cone / cono circular (490) Forma espacial que posee una base circular y un vértice.

circular cylinder / cilindro circular (487) Cilindro que posee dos bases circulares paralelas y congruentes.

circumference / circunferencia (309) La distancia alrededor de un círculo.

column / columna (73) Números colocados uno encima de otro en un arreglo vertical.

combination / combinación (524) Arreglo o lista de objetos en que el orden no es importante.

common difference / diferencia común (296) Diferencia entre dos términos consecutivos cualesquiera de una sucesión aritmética.

common ratio / razón común (297) Factor constante que se usa para calcular cualquier término, a partir del segundo, de una sucesión geométrica multiplicando el término anterior por este factor.

commutative property / propiedad conmutativa (8) Para números reales a y b cualesquiera, $a + b = b + a$ y $ab = ba$.

compatible numbers / números compatibles (126) Dos números que son fáciles de dividir mentalmente. Son a menudo miembros de la misma familia de factores y se pueden usar en estimaciones.

complementary angles / ángulos complementarios (198) Dos ángulos son complementarios si la suma de sus medidas es 90°.

composite number / número compuesto (235) Cualquier número entero mayor que 1 que posee más de dos factores.

compound event / evento compuesto (534) Un evento compuesto consiste en dos o más eventos simples.

congruent triangles / triángulos congruentes (210) Triángulos que tienen la misma forma y tamaño.

converse / recíproco (399) El recíproco del Teorema de Pitágoras puede usarse para averiguar si un triángulo es un triángulo rectángulo. Si las longitudes de los lados de un triángulo son a, b y c y si $c^2 = a^2 + b^2$, entonces el triángulo es un triángulo rectángulo.

coordinate / coordenada (56) Número asociado con un punto en una recta numérica.

coordinate system / sistema de coordenadas (92) Plano en el cual se han trazado dos rectas numéricas, una horizontal y una vertical, que se intersecan en sus puntos cero.

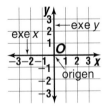

corresponding angles / ángulos correspondientes (189) Ángulos que ocupan la misma posición en dos rectas paralelas distintas atravesadas por una transversal. $\angle 1$ y $\angle 5$, $\angle 2$ y $\angle 6$, $\angle 3$ y $\angle 7$, $\angle 4$ y $\angle 8$ son ángulos correspondientes.

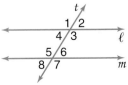

corresponding parts / partes correspondientes (210)   Partes de figuras congruentes o semejantes que coinciden.

Counting Principle / Principio Fundamental de Contar (518)   Este principio se usa para calcular el número de resultados mediante multiplicación. Si el evento *M* puede ocurrir de *m* maneras y es seguido por el evento *N*, que puede ocurrir de *n* maneras, entonces el evento *M* seguido del evento *N* puede ocurrir de *m* · *n* maneras.

cross products / productos cruzados (111)   Los productos que resultan de la comparación de los términos de las diagonales de dos razones. Las razones forman una proporción si y sólo si los productos son iguales. En la proporción $\frac{2}{3} = \frac{8}{12}$, los productos cruzados son 2 × 12 y 3 × 8.

data analysis / análisis de datos (154)   El estudio de datos y la extracción de conclusiones de los números observados.

density property / propiedad de densidad (260)   Entre cada par de números racionales hay números racionales infinitos.

dependent events / eventos dependientes (535)   Dos o más eventos en que el resultado de uno de ellos afecta el resultado de los otros eventos.

diameter / diámetro (309)   Longitud de cualquier segmento de recta cuyos extremos yacen en un círculo y que pasa por su centro.

diámetro

dilation / dilatación (214, 370)   En matemáticas, proceso de reducción o ampliación de una imagen.

discount / descuento (349)   La cantidad de reducción del precio normal.

distributive property / propiedad distributiva (34)   La adición de dos sumandos multiplicada por un número es igual a la adición del producto de cada sumando por el número. Para números *a*, *b* y *c* cualesquiera, $a(b + c) = ab + ac$ y $(b + c)a = ba + ca$.

divisible / divisible (232)   Un número es divisible entre otro si el cociente es un número entero y el residuo es cero.

division property of equality / propiedad de división de la igualdad (22)   Si cada lado de una ecuación se divide entre el mismo número no nulo, los dos lados de la ecuación permanecen iguales. Si $a = b$, entonces $\frac{a}{c} = \frac{b}{c}, c \neq 0$.

domain / dominio (428)   Conjunto de valores de entrada de una función.

edge / arista (483)   Intersección de las caras de una figura tridimensional.

element / elemento (73)   Cada número en una matriz recibe el nombre de elemento.

equation / ecuación (13)   Enunciado matemático que contiene el signo de igualdad, =.

equilateral / equilátero (196)   Figura con todos los lados congruentes entre sí.

evaluate / evaluar (9)   Calcular el valor de una expresión sustituyendo las variables con números.

event / evento (253)   Resultado específico o tipo de resultado de un experimento probabilístico.

experimental probability / probabilidad experimental (540)   Probabilidad de un evento que se calcula o estima basándose en la frecuencia relativa de los resultados favorables al evento en cuestión, que ocurren durante un experimento probabilístico.

exponent / exponente (8)   Número de veces que la base de una potencia se usa como factor. En 10^3, el exponente es 3.

face / cara (483)   Cualquier superficie que forma un lado o una base de un prisma.

factor / factor (8)   Cuando se multiplican dos o más números enteros, cada número es un factor del producto. En 4 × 5 = 20, 4 y 5 son factores de 20.

factorial / factorial (522)   La expresión *n*! es el producto de los *n* primeros números de contar, contando al revés.

factoring / factorización (588)   Encontrar los factores de un producto.

factor tree / árbol de factores (236)   Diagrama que muestra la factorización prima de un número.

Fibonacci sequence / sucesión de Fibonacci (300)   Sucesión de números en que los dos primeros términos son iguales a 1 y cada término que sigue es igual a la suma de los dos términos anteriores, 1, 1, 2, 3, 5, 8,... .

frequency table / tabla de frecuencia (142)   Tabla que se usa para organizar un conjunto de datos y que muestra cuántas veces aparece cada dato.

function / función (428)   Relación en que cada elemento de entrada es apareado con un único elemento de salida, según una regla específica.

function table / tabla de funciones (428)   Tabla que organiza las entradas, la regla y las salidas de una función.

Fundamental Theorem of Arithmetic / Teorema Fundamental de la Aritmética (236)   Teorema que afirma que todo número entero tiene un conjunto único de factores primos.

geometric sequence / sucesión geométrica (297)   Sucesión de números en que se puede calcular cualquier término, después del segundo, multiplicando el término anterior por el mismo número.

graph / graficar (56)   Dibujar o trazar, en una recta numérica o en un plano de coordenadas, los puntos indicados.

greatest common factor (GCF) / máximo común divisor (MCD) (242)   El mayor factor común de dos o más números. El MCD de 18 y 24 es 6.

greatest possible error / error máximo posible (504)   Mitad de la unidad más pequeña que se usó para tomar una medida.

height / altura (40, 301)   La longitud de la altitud de un triángulo o de un cuadrilátero.

histogram / histograma (142)   Tipo especial de gráfica de barras que exhibe la frecuencia de los datos una vez que los datos han sido organizados en intervalos iguales. Los intervalos cubren todos los valores posibles de los datos, de modo que no hay espacios entre las barras de la gráfica.

hypotenuse / hipotenusa (398)   Lado de un triángulo rectángulo opuesto a su ángulo recto.

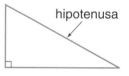

independent events / eventos independientes (534)   Dos o más eventos en los cuales el resultado de uno de ellos no afecta el resultado de los otros eventos.

indirect measurement / medida indirecta (361)   Técnica que usa proporciones para calcular medidas.

inequality / desigualdad (43)   Enunciado matemático que contiene < , >, ≠, ≤ o ≥.

integer / entero (56)   Los números enteros no negativos y sus opuestos.
$$\ldots, -3, -2, -1, 0, 1, 2, 3, \ldots$$

SPANISH GLOSSARY

interest / interés (353)   Cantidad que se cobra o se paga por el uso del dinero.

interquartile range / amplitud intercuartílica (163)   El rango de la mitad central de un conjunto de datos.

inverse operation / operaciones inversas (18)   Pares de operaciones que se anulan mutuamente. La adición y la sustracción son operaciones inversas. La multiplicación y la división son operaciones inversas.

inverse property of multiplication / propiedad de inverso multiplicativo (290)   Para cada número no nulo $\frac{a}{b}$, donde a y b no son cero, existe un único número $\frac{b}{a}$ tal que $\frac{a}{b} \times \frac{b}{a} = 1$.

irrational number / número irracional (391)   Un número que no puede escribirse como $\frac{a}{b}$, donde a y b son enteros y $b \neq 0$.

isosceles / isósceles (196)   Triángulo que tiene por lo menos dos lados congruentes.

L

least common denominator (LCD) / mínimo común denominador (mcd) (261)   El menor múltiplo común de los denominadores de dos o más fracciones.

least common multiple (LCM) / mínimo común múltiplo (mcm) (257)   El menor múltiplo común no nulo de dos o más números. El mínimo común múltiplo de 2 y 3 es 6.

legs / catetos (398)   Los lados que forman el ángulo recto de un triángulo rectángulo.

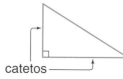
catetos

like terms / términos semejantes (565)   Expresiones que contienen las mismas variables, como por ejemplo $3ab$ y $7ab$.

linear function / función lineal (442)   Ecuación en que la gráfica de las soluciones es una recta.

line of symmetry / eje de simetría (460)   Recta que divide una figura en dos mitades que son reflexiones mutuas.

eje de simetría

line plot / esquema lineal (153)   Gráfica que usa una recta numérica y un $\times$, sobre un número en la recta numérica, cada vez que el número aparece en un conjunto de datos.

line symmetry / simetría lineal (206)   Exhiben simetría lineal las figuras que coinciden exactamente cuando se doblan.

lower quartile / cuartil inferior (164)   Mediana de la mitad inferior de un conjunto de datos, en la amplitud intercuartílica.

M

markup / margen de utilidad (349)   Diferencia entre el precio que paga el comerciante y el precio de venta al consumidor.

matrix / matriz (73)   Un arreglo rectangular de elementos en filas y columnas.

mean / media (158)   Suma de los números de un conjunto de datos dividida entre el número total de datos.

measures of central tendency / medidas de tendencia central (158)   Números que pueden representar el conjunto total de datos.

median / mediana (158, 159)   Número central de un conjunto de datos, una vez que los datos han sido ordenados numéricamente. Si hay un número par de datos, la mediana es el promedio de los dos datos centrales.

mixed number / número mixto (278)   Suma de un entero y una fracción. $6\frac{2}{3}$ es un número mixto.

mode / modal (158)   Número(s) de un conjunto de datos que aparece(n) más frecuentemente.

monomial / monomio (561)   Un número, una variable o el producto de un número por una o más variables.

multiple / múltiplo (257)   El producto de un número entero por cualquier otro número entero. 28 es múltiplo de 4 y de 7.

multiplication property of equality / propiedad de multiplicación de la igualdad
(22) Si cada lado de una ecuación se multiplica por el mismo número, los lados permanecen iguales. Si $a = b$, entonces $ac = bc$.

multiplicative inverse / inverso multiplicativo
(290) El producto de un número por su inverso multiplicativo es igual a 1. El inverso multiplicativo de $\frac{2}{3}$ es $\frac{3}{2}$.

numerical expression / expresión numérica
(11) Expresión matemática que tiene una combinación de números y por lo menos una operación. $4 + 2$ es una expresión numérica.

O

obtuse / obtuso (196) Ángulo que mide más de 90°, pero menos de 180°.

open sentence / enunciado abierto (13) Ecuación que contiene una variable.

opposite / opuestos (70) Dos enteros son opuestos si, en la recta numérica, están representados por puntos que equidistan de cero, pero en direcciones opuestas. La suma de opuestos es cero.

order of operations / orden de las operaciones
(11) Reglas a seguir cuando hay más de una operación involucrada.

1. Primero ejecuta todas las operaciones dentro de los símbolos de agrupamiento, comenzando con los más interiores.

2. Calcula todas las potencias.

3. Luego multiplica y divide, ordenadamente, de izquierda a derecha.

4. Finalmente, suma y resta, ordenadamente, de izquierda a derecha.

ordered pair / par ordenado (92) Par de números que se usa para ubicar un punto en un plano de coordenadas. Se escribe de la siguiente forma: (coordenada x, coordenada y).

origin / origen (92)
Punto de intersección axial en un plano de coordenadas.

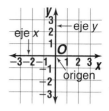

outcome / resultado
(253, 518) Uno de los resultados posibles de un experimento probabilístico. Por ejemplo, 4 es un resultado posible cuando se lanza un dado.

outlier / valor atípico (164) Dato o datos que dista(n) de los cuartiles respectivos más de 1.5 veces la amplitud intercuartílica.

P

parallel lines / rectas paralelas (188) Rectas que yacen en un mismo plano y que no se intersecan. El símbolo ∥ significa paralela a.

parallelogram / paralelogramo (39, 201) Cuadrilátero que posee dos pares de lados paralelos.

Pascal's Triangle / Triángulo de Pascal (528) Arreglo triangular de números en el cual cada número se obtiene sumando los dos números a su izquierda y a su derecha, en la fila superior.

pentagon / pentágono (358) Polígono de cinco lados.

percent / tanto por ciento (107) Razón que compara un número con 100.

percentage / porcentaje (335) Número (P) de una proporción porcentual que se compara con otro número llamado base (B).

percent of change / tanto por ciento de cambio (348) Razón de la cantidad de aumento o disminución comparada con la cantidad original.

percent of decrease / tanto por ciento de disminución (349) Razón de la cantidad de disminución comparada con la cantidad original, escrita como tanto por ciento.

percent of increase / tanto por ciento de aumento (349)   Razón de la cantidad de aumento comparada con la cantidad original, escrita como tanto por ciento.

percent proportion / proporción porcentual (335)   La proporción $\frac{P}{B} = \frac{r}{100}$ en que P representa el porcentaje, B representa la base y r representa el número por cada 100. También se escribe como $\frac{\text{Porcentaje}}{\text{Base}} = \text{Tasa}$.

perfect square / cuadrado perfecto (382)   Número racional cuya raíz cuadrada es un número entero. 25 es un cuadrado perfecto porque $\sqrt{25} = 5$.

perimeter / perímetro (38)   La medida del contorno de una figura geométrica cerrada.

permutation / permutación (521)   Arreglo o lista en que el orden es importante.

perpendicular / perpendiculares (197)   Dos rectas o segmentos de recta que se intersecan formando un ángulo recto.

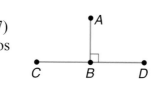

polygon / polígono (196)   Figura simple cerrada en un plano, formada por tres o más segmentos de recta.

polynomial / polinomio (562)   Suma o diferencia de uno o más monomios.

population / población (546)   El grupo total de individuos del cual se toman las muestras que están bajo estudio.

power / potencia (8)   Número que puede escribirse usando un exponente. La potencia 7^3 se lee *siete a la tercera potencia* o *siete al cubo*.

precision / precisión (504)   La precisión de una medición depende de la unidad de medida. Mientras más pequeña sea la unidad de medida, más precisa es la medición.

prime factorization / factorización prima (235)   Manera de expresar un número compuesto como producto de números primos. La factorización prima de 63, por ejemplo, es $3 \times 3 \times 7$.

prime number / número primo (235)   Número entero mayor que 1 que sólo tiene dos factores, 1 y sí mismo.

principal / capital (353)   Cantidad de dinero invertido o adeudado.

principal square root / raíz cuadrada principal (382)   La raíz cuadrada no negativa. $\sqrt{49}$ indica la raíz cuadrada *principal* de 49.

prism / prisma (483)   Figura tridimensional que tiene dos bases poligonales paralelas y congruentes.

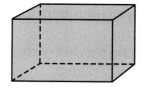

probability / probabilidad (253)   La posibilidad de que suceda un evento. Es la razón del número de maneras en que puede ocurrir un evento al número total de resultados posibles.

proportion / proporción (111)   Ecuación que afirma la igualdad de dos o más razones, $\frac{a}{b} = \frac{c}{d}, b \neq 0, d \neq 0$.

pyramid / pirámide (483)   Figura sólida que tiene una base poligonal y caras triangulares.

Pythagorean Theorem / Teorema de Pitágoras (396–398)   En un triángulo rectángulo, el cuadrado de la longitud de la hipotenusa es igual a la suma de los cuadrados de las longitudes de los catetos. $c^2 = a^2 + b^2$

Pythagorean triple / triplete pitagórico (405)   Conjunto de tres enteros que satisfacen el Teorema de Pitágoras.

Q

quadrant / cuadrante (92)   Una de las cuatro regiones en que dos rectas perpendiculares dividen un plano.

quadratic function / función cuadrática (452)   Función polinómica en que la potencia mayor de la variable es 2.

quadrilateral / cuadrilátero (201)   Polígono de cuatro lados.

quartiles / cuartiles (163)   Valores que dividen un conjunto de datos en cuatro partes iguales.

radical sign / signo radical (382)   Símbolo con que se indica la raíz cuadrada no negativa. $\sqrt{}$

radius / radio (309)   Distancia desde el centro de un círculo hasta cualquier punto del círculo.

random / al azar (253)   Los resultados ocurren aleatoriamente o al azar si cada resultado tiene la misma posibilidad de ocurrir y es imposible predecirlos.

range / rango (428)   Conjunto de los valores de salida de una función.

range / rango (163)   Diferencia entre los valores máximo y mínimo de un conjunto de datos.

rate / tasa (105)   Razón de dos mediciones que tienen distintas unidades de medida.

rate / tasa (335)   Razón de un número a 100 en una proporción porcentual.

ratio / razón (104)   Comparación de dos números mediante división. La razón 2 a 3 se puede escribir como 2 de cada 3, 2 a 3, 2:3 ó $\frac{2}{3}$.

rational numbers / números racionales (245)   Números de la forma $\frac{a}{b}$, donde a y b son enteros y $b \neq 0$.

real numbers / números reales (391)   Conjunto de los números racionales junto con el conjunto de los números irracionales.

reciprocal / recíproco (290)   El inverso multiplicativo de un número.

rectangle / rectángulo (38, 201)   Cuadrilátero cuyos cuatro ángulos son congruentes entre sí.

reflection / reflexión (206, 460)   Transformación en que a una figura se le da vuelta de campana por encima de un eje de simetría.

relative error / error relativo (504)   Comparación del error máximo posible con la medida misma.

$$error\ relativo = \frac{error\ máximo\ posible}{medida}$$

repeating decimal / decimal periódico (249)   Decimal en que los dígitos, en algún momento, comienzan a repetirse en bloques de uno o más números. Por ejemplo, 0.181818... y 0.8333... . Usando notación de barra, estos decimales se escriben $0.\overline{18}$ y $0.8\overline{3}$.

rhombus / rombo (201)   Paralelogramo cuyos lados son todos congruentes entre sí.

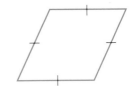

right / recto (196)   Ángulo que mide 90°.

rotation / rotación (206, 464)   Cuando se gira una figura en torno a un punto central.

rotational symmetry / simetría rotacional (207)   Una figura posee simetría rotacional si se puede girar menos de 360° en torno a su centro, sin que esto cambie su apariencia con respecto de la figura original.

row / fila (73)   Los números que están horizontalmente uno al lado del otro en una matriz.

sample / muestra (174, 546)   Grupo escogido aleatoriamente con el objeto de recoger información.

sample space / espacio muestral (253)   Conjunto de todos los resultados posibles de un experimento probabilístico.

scale / escala (366)   Razón de una longitud dada en un dibujo o modelo a su longitud real.

scale drawing / dibujo a escala (366)   Dibujo que es semejante, pero más grande o más pequeño que el objeto real.

scale factor/ factor de escala (370)   Razón de la imagen dilatada a la imagen original.

scale model / modelo a escala (366)   Réplica de un objeto real el cual es demasiado grande o demasiado pequeño como para construirlo de tamaño natural.

scalene / escaleno (196)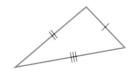
Triángulo sin ningún par de lados congruentes.

scatter plot / gráfica de dispersión (168)
Gráfica que muestra la relación general entre dos conjuntos de datos.

scientific notation / notación científica (265)
Manera de expresar números como el producto de un número que es al menos igual a 1, pero menor que 10, por una potencia de diez. En notación científica, $5,500 = 5.5 \times 10^3$.

selling price / precio de venta (349)   Dinero que paga un consumidor por un artículo.

sequence / sucesión (296) Lista de números en cierto orden, como, por ejemplo, 0, 1, 2, 3 ó 2, 4, 6, 8.

significant digits / dígitos significativos (504)
Todos los dígitos de una medición que se sabe que son exactos, más un dígito aproximado.

similar polygons / polígonos semejantes (357)
Dos polígonos son semejantes si sus ángulos correspondientes son iguales y sus lados correspondientes son igualmente proporcionales. Tienen la misma forma, pero pueden no tener el mismo tamaño.

similar triangles / triángulos semejantes (215)
Triángulos que tienen la misma forma, pero no necesariamente el mismo tamaño.

simplest form / forma reducida (246)
La forma de una fracción en la cual el MCD de su numerador y denominador es 1.

simplest form / forma reducida (565)   Una expresión que carece de términos semejantes. La forma reducida de $6x + 8x$ es $14x$.

simulation / simulación (540)   Proceso de representar un problema.

solid / sólido (482)   Figura tridimensional.

solution / solución (13)   Valor de la variable de una ecuación que hace verdadera la ecuación. La solución de $10 + y = 25$ es 15.

spreadsheet / hoja de cálculos (162)
Herramienta que se usa para organizar y analizar datos.

square / cuadrado (38, 201)   Paralelogramo con todos los lados, y también los ángulos, congruentes entre sí.

square root / raíz cuadrada (382)   Uno de dos factores iguales de un número. Si $a^2 = b$, entonces a es una raíz cuadrada de b. La raíz cuadrada no negativa de 144 es 12 porque 12 es un número no negativo y $12^2 = 144$.

statistics / estadística (142)   Rama de las matemáticas que trata de la recopilación, organización y análisis de datos.

substitute / sustituir (11)   Reemplazar una variable en una expresión algebraica con un número, obteniendo así una expresión numérica.

subtraction property of equality / propiedad de sustracción de la igualdad (18)   Si sustraes el mismo número de ambos lados de una ecuación, los lados permanecen iguales. Para números a, b y c cualesquiera, si $a = b$, entonces $a - c = b - c$.

supplementary angles / ángulos suplementarios (190)   Dos ángulos son suplementarios si la suma de sus medidas es 180°.

surface area / área de superficie (495)   Suma de las áreas de todas las superficies de una figura tridimensional.

symmetric / simétricas (460)   Figuras que se pueden doblar en dos partes exactamente correspondientes.

system of equations / sistema de ecuaciones (446)   Conjunto de ecuaciones que tienen las mismas variables.

T

term / término (565)   Número, variable o producto de números y variables.

term / término (296)   Número de una sucesión.

terminating decimal / decimal terminal (249)   Decimal cuyos dígitos terminan. Todo decimal terminal puede escribirse como una fracción con un denominador de 10, 100, 1,000, etc.

tessellation / teselado (220)   Un patrón repetitivo de polígonos que coinciden perfectamente, sin dejar huecos o espacios.

theoretical probability / probabilidad teórica (540)   La probabilidad a largo plazo de un resultado o evento, que se basa en principios matemáticos.

transformation / transformación (220)   Movimientos de figuras geométricas.

translation / traslación (220, 456)   Tipo de transformación en que una figura se desliza horizontalmente, verticalmente o de ambas maneras.

transversal / transversal (188)   Recta que interseca dos rectas paralelas formando así ocho ángulos.

trapezoid / trapecio (201, 302)   Cuadrilátero con un único par de lados paralelos.

tree diagram / diagrama de árbol (518)   Diagrama que se usa para encontrar y mostrar el número total de resultados posibles de un experimento probabilístico.

triangle / triángulo (196)   Polígono de tres lados y con tres ángulos.

trigonometry / trigonometría (365)   Estudio de los triángulos.

trinomial / trinomio (588)   Polinomio de tres términos.

U

unit rate / tasa unitaria (105)   Tasa que tiene un denominador de 1.

upper quartile / cuartil superior (163)   La mediana de la mitad superior de un conjunto de números.

V

variable / variable (12)   Un símbolo, por lo general, una letra, que se usa para representar números en expresiones o enunciados matemáticos.

variation / variación (163)   Dispersión de los valores de un conjunto de datos.

vertex / vértice (483)   El vértice de un prisma es el punto en que se intersecan todas las caras del prisma.

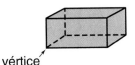
vértice

vertical angles / ángulos opuestos por el vértice (190)   Ángulos congruentes que se forman de la intersección de dos rectas. En la figura, los ángulos opuestos por el vértice son $\angle 1$ y $\angle 3$, así como $\angle 2$ y $\angle 4$.

volume / volumen (486)   Número de unidades cúbicas que se requieren para llenar el espacio que ocupa un sólido.

X

x-axis / eje x (92)   La recta numérica horizontal que ayuda a formar el sistema de coordenadas.

x-coordinate / coordenada x (92)   Primer número de un par ordenado.

y-axis / eje y (92)   La recta numérica vertical que ayuda a formar el sistema de coordenadas.

y-coordinate / coordenada y (92)   Segundo número de un par ordenado.

zero pair / par nulo (63)   Resultado de aparear una ficha positiva con una negativa.

Index

Edges, 200, 483
Element, 73
Eliminate possibilities, 90–91
Ellipses, 56
Equally-likely, 253
Equations, 13
  addition, 13–15, 17–20
  cubic, 455, 489
  division, 13–15, 21–25
  with fractions, 315–317
  functions written as, 428
  with integer solutions, 84–89
  with irrational solutions, 392
  modeling, 17, 21, 23, 84–85
  multiplication, 13–15, 21–25
  nonlinear, 455, 489
  open sentence, 13
  percent, 339–341
  replacement set for, 13
  solutions of, 13
  solving, 13–15, 17–25, 32–36,
    84–89, 315–317, 383
  square root, 383
  subtraction, 13–15, 17–20
  two-step, 32–36
  with two variables, 437–440
  using guess-and-check to solve, 45
  writing, 28–29
Equilateral triangles, 196
Error
  greatest possible, 504
  relative, 504
Escher, M. C., 220
Estimation
  area of irregular figures, 126
  with percents, 126–129
  products of fractions, 317
  of square roots, 385–389
  study hints, 149, 358, 399,
    443, 465
  using compatible numbers, 126
  using to sketch a circle graph,
    126–127
Evaluate, 11–12
Events, 253
  compound, 534–537
  dependent, 535–536
  equally–likely, 253
  independent, 534–535
Experimental probability, 540–543
  simulation, 540
Exponents, 8–10
  evaluating expressions
    containing, 9
  negative, 267, 581
  and rational numbers, 287
  writing expression using, 8
Expressions
  algebraic, 12, 27
  evaluating, 9, 11–16
  exponents in, 8–10

grouping symbols in, 11
numerical, 11
substitution in, 11
using graphing calculators to
  evaluate, 16
variables in, 12
writing, 27–29

Faces, 483
Factorial, 522
  on a calculator, 522
Factoring, 588
  trinomials, 588–589
Factors, 8
  greatest common, 242–244
  scale, 370
  of trinomials, 588–589
Factor trees, 236
Fair games, 516
Family Activities, 29, 94, 117, 145,
  208, 266, 304, 340, 407, 467, 497,
  542, 585
Fibonacci sequence, 300
Figurate numbers, 382
Formulas
  for area of circles, 476
  for area of parallelograms, 40
  for area of rectangles, 40
  for area of squares, 40
  for area of trapezoids, 302
  for area of triangles, 302
  for circumference of circles, 309
  for distance, 21
  for perimeter of parallelograms, 39
  for perimeter of rectangles, 38
  for perimeter of squares, 38
  for simple interest, 353
  in spreadsheets, 71, 162, 295, 305,
    334, 352
  for surface area of cylinders, 500
  for volume of cones, 491
  for volume of cylinders, 487
  for volume of prisms, 486
  for volume of pyramids, 491
Four-step plan, 4–7, 34, 79, 87, 127,
  197, 243, 262, 282, 300–331, 340,
  354, 404, 467–468, 477, 496, 525,
  529, 574, 579, 589
Fractions, 245
  adding like, 278–280
  adding unlike, 281–284
  comparing, 261
  decimals as, 250–251
  as decimals, 249, 261
  dividing, 312–314
  least common denominator, 261
  like, 278
  modeling, 278

multiplicative inverse, 312
multiplying, 286–289
ordering, 261
as percents, 107–108, 115, 121
percents as, 107–108, 114, 121
renaming, 281–282
simplest form, 246
subtracting like, 278–280
subtracting unlike, 281–284
Frequency tables, 140–141
Functions, 37, 428–471
  domain, 428
  graphing, 433–435, 442–455
  input, 37, 428
  linear, 351, 442, 445
  machines, 37
  nonlinear, 435, 455, 489
  notation, 433
  output, 37, 428
  quadratic, 452
  range, 428
  rule, 428
  tables, 428, 432
  written as equations, 428
Function tables, 428, 432
  using to graph, 433–435
**Fundamental Theorem of
  Arithmetic,** 236

Games
  fair, 516–517
  unfair, 516–517
  See also Let the Games Begin
Geometric sequence, 296–297
  common ratio, 297
Geometry
  30°-60° right triangle, 365,
    414–415
  45°-45° right triangle, 365,
    415–416
  acute triangles, 196
  alternate exterior angles, 189
  alternate interior angles, 189
  altitude, 40, 301, 490
  angles, 186–187
  area, 39–40, 301–302, 476–479,
    588–589
  ASA congruence, 210
  base, 40, 301–302, 335, 483, 487,
    490
  complementary angles, 198
  congruent triangles, 210–212
  construction of congruent line
    segments, 186
  construction of congruent
    triangles, 213
  converse of Pythagorean
    Theorem, 399

dividing, 81–83
in equations, 86–87
graphing, 56–57
in matrices, 74–75
modeling addition of, 62–63
modeling multiplication of, 78
modeling subtraction of, 69–70
multiplying, 78–80
negative, 56
on number line, 56–60
opposite of, 70
ordering, 60–61
positive, 56
in real number system, 390
subtracting, 69–72
Integration, *See* Applications, Connections, and Integration Index on pages xxii–1
Interdisciplinary Investigations
Extra! Extra! Newspapers May Take Over the Planet!, 556–557
Math at the Mall, 424–425
What in the World is "WYSIWYG?", 274–275
What's for Dinner?, 136–137
Interest, 353
compound, 356
simple, 353–356
Internet Connections, 3, 15, 26, 47, 55, 61, 76, 103, 122, 123, 137, 139, 146, 152, 170, 185, 212, 218, 219, 231, 238, 247, 275, 277, 285, 293, 316, 329, 333, 356, 369, 381, 389, 394, 395, 425, 427, 430, 431, 436, 455, 475, 498, 506, 515, 537, 543, 549, 557, 559, 561, 582, 591
Interquartile range, 164
Inverse
additive, 70
multiplicative, 290
operation, 18, 21
Inverse property of multiplication, 290
Investigations, *See* Interdisciplinary Investigations
Irrational numbers, 391
approximating, 391
graphing, 408–409
on number line, 391
in real number system, 391
Isometric crystal, 482
Isosceles triangles, 196, 417

Labs, *See* Hands-On Labs, Problem Solving Labs, Technology Labs, and Thinking Labs
Least common denominator, 261
using to rename fractions, 283

Least common multiple, 257–259
using prime factorization to find, 258
Legs, 398
Leonardo, 300
Let the Games Begin
Absolutely, 61
Architest, 498
Collecting Factors, 238
Estimate and Eliminate, 389
Factor Challenge, 591
Fraction Track, 285
Guess My Rule, 431
Map Sense, 369
Matrix Madness, 76
Parabola Hit or Miss, 455
Per-fraction, 123
Shape Relations, 218
What's Missing?, 333
Win the Lottery, 543
You're the Greatest, 47
You're the Winner...Bar None!, 146
Like fractions,
adding, 278–280
subtracting, 278–280
Like terms, 565
Linear functions, 442, 445
graphing, 442–444
on graphing calculator, 445
Line graphs
interpreting, 570
Line plots, 153–155
data analysis using, 154
Lines
parallel, 188–192
of symmetry, 206
Line segments, 186
constructing congruent, 186
measuring, 186
Line symmetry, 206
Line of symmetry, 206, 460
Look for a pattern, 294–295
Lower quartile, 164

M

Major arc, 311
Make a List, 240–241
Make a Model, 480–481
Make a Table, 140–141
Markup, 349
percent of increase, 349
Mathematical Techniques
estimation, 126–129, 149, 358, 399, 443, 465
mental math, 108, 111, 115, 120, 249, 262, 330, 478, 525
number sense, 311, 317
See also Problem Solving

Mathematical Tools
paper/pencil, 9, 120, 477
real objects
algebra tiles, 385, 561–577, 579, 580, 583–585, 588, 590, 591
base-ten blocks, 385
compass, 127, 149, 186, 193, 213, 408, 409, 414, 490
counters, 17, 21, 32, 63, 69, 78, 84, 535, 544, 545
cubes, 486, 488
dot paper, 305, 396, 397, 482–485
equation mat, 17, 21, 32, 84, 85
geoboard and geobands, 205, 305, 396, 397
geomirror, 460
grid paper, 40, 107, 118, 168, 235, 238, 274, 278, 300, 301, 307, 308, 344, 348, 371, 410, 424, 442, 447, 455, 456, 460, 495
hexagonal grid, 532–533
integer mat, 63, 69, 78
number cubes, 255, 280, 516, 517
product mat, 577, 583
protractor, 186, 188, 190, 202, 358, 365, 414, 415, 464
ruler, 168, 186, 202, 214, 286, 358, 365, 410, 414, 415, 441, 442, 490, 494, 499, 556
spinner, 253–255, 455, 520, 536, 537
straightedge, 188, 190, 193, 200, 213, 371, 408, 409, 447, 456, 460
tape measure, 118, 153, 168, 424
technology, *See* Calculators, Computers, Technology Labs, Technology Mini-Labs, and Technology Tips
See also Problem Solving
Math Journal, 6, 7, 9, 40, 57, 88, 93, 105, 109, 112, 172, 211, 244, 263, 291, 310, 337, 362, 399, 439, 462, 466, 484, 506, 519, 547, 562
Math in the Media, 20, 89, 151, 267, 289, 364, 440, 507
Matrix, 73
adding, 74
column, 73
element, 73
on a graphing calculator, 77
rows, 73
subtracting, 74
Mean, 158–166
using spreadsheets to find, 162
Measurement
customary system, 106, 186, 505
greatest possible error, 504

indirect, 361–364
metric system, 186, 505
precision, 504–507
relative error, 504
significant digits, 504–507
Measures of central tendency,
158–161
average, 158–160
mean, 158–160
median, 159–160
mode, 158–160
range, 163
Measures of variation, 163–166
interquartile range, 163
outlier, 164–165
quartiles, 163–164
range, 163
Median, 158–160
Mental math, 120
study hints, 108, 111, 115, 249,
262, 330, 478, 525
Metric system, 186, 505
Mid-Chapter Self Test, 25, 72, 117,
166, 204, 248, 299, 356, 394, 449,
493, 531, 576
Minor arc, 311
Misleading
graphs, 174–177
statistics, 174–177
Mixed numbers, 245, 278
as improper fractions, 245
Mode, 158–160
Monoclinic crystal, 482
Monomials, 561
adding, 570–572
multiplying, 578–581
subtracting, 573–576
Motion geometry, 456
Multiples, 257
common, 258
least common, 257
Multiplication
equations, 13–15, 21–25
of fractions, 286–289
of integers, 78–80
of polynomials, 578–581, 583–585
Multiplication property of equality,
21
Multiplicative inverse, 290

Nets, 494
Networks, 200
edges, 200
nodes, 200
traceable, 200
Nodes, 200
Nonexamples, 44, 111–113,
171–172, 174, 175, 186, 200, 204,

206, 208, 211, 212, 215, 216, 217,
226, 232, 233, 234, 235, 237, 249,
266, 331, 359, 360, 364, 390–391,
447, 481, 505, 517, 540, 548, 588
Nonlinear equations, 455, 489
Number line
graphing points on, 56–57
inequalities on, 44–45, 318–319
integers on, 56–60
irrational numbers on, 408–409
percents on, 121
pi on, 307–308
real numbers on, 391
square roots on, 386
using to add integers, 62
using to compare integers, 59
using to draw a box-and-whisker
plot, 167
using for a line plot, 153–155
using to order integers, 60
whole numbers on, 44
Numbers
compatible, 126
composite, 235
figurate, 382
integers, 390
irrational, 391
prime, 235
rational, 390
real, 390–394
whole, 390
Numerical expression, 11

Obtuse triangle, 196
of 10, 265
on number line, 386
Open sentence, 13
Operation
inverse, 18, 21
Opposite, 70
Ordered pairs, 92–93
graphing, 92–93
x-coordinate, 92
y-coordinate, 92
Ordering
decimals, 261
fractions, 261
integers, 60–61
rational numbers, 261–264
Order of operations, 11–15
grouping symbols in, 11
Origin, 92
Orthorhombic crystal, 482
Outcomes, 253, 518
counting, 518–520
Counting Principle, 518
equally-likely, 253
random, 253

Outliers, 164
Output, 428

Parabola, 453
Parallel lines, 188–192
alternate exterior angles, 189
alternate interior angles, 189
constructing, 193
corresponding angles, 189
transversal, 188
Parallelogram, 39, 201
altitude of, 40
area of, 40
base of, 40
height of, 40
perimeter of, 39
rectangle, 201
rhombus, 201
square, 201
Parents, *See* Family Activities
Pascal, Blaise, 528
Pascal's Triangle, 528–531
patterns in, 532–533
Patterns, 42, 99, 141
divisibility, 232–234
look for a pattern, 284–295
number, 237, 304, 405, 451
Pentagon, 358
similar, 358
Percentage, 335
Percent of change, 348–351
Percent of decrease, 349
Percent equation, 339–341
base, 339
rate, 339
Percent of increase, 349
Percent proportion, 335–338
base, 335
rate, 335
Percents, 107, 114, 335
base, 335
on a calculator, 114
of change, 348–351
and circle graphs, 108–109,
126–127
comparing and ordering, 121
compatible numbers, 126
decimals as, 115, 121
as decimals, 114, 121
of decrease, 349
discount, 349
equation, 339–341
estimation with, 126–129
finding, 120–123
fractions as, 107–108, 115, 121
as fractions, 107–108, 121
of increase, 349
large, 344–347

base of, 301
classifying, 196–199
congruent, 210–212
equilateral, 196
formula for area of, 302
height of, 301
isosceles, 196, 417
obtuse, 196
Pythagorean Theorem, 396–410
right, 196, 417
scalene, 196
similar, 215–218
using Venn diagram to show relationships, 196
Triangular prism, 482, 489
Triclinic crystal, 482
Trigonometry, 365
Trinomial, 588
using algebra tiles to factor, 588–589
Two-dimensional figures
congruent, 210–212
nets, 494
parallelogram, 39–40, 201
polygons, 196, 357
quadrilateral, 201–204
rectangle, 38–40, 201
similar, 357–360
square, 38–40, 201
triangle, 196–199

Unfair games, 516
Unit rate, 105–106
Upper quartile, 163
Use a Graph, 450–451
Use a Venn Diagram, 194–195

Vanishing point, 371
Variables, 12
Variation, 163
Venn diagram, 194, 242, 243, 245
of real number system, 390, 391
Verbal phrase, 27
Vertex
of prisms, 483
Vertical angles, 190
Volume
of circular cones, 490–491
of circular cylinders, 487–488
modeling, 486, 490
of prisms, 486–487, 489
of pyramids, 491, 498
and surface area, 498, 503

Whole numbers
on a number line, 44
in real number system, 390
Work backward, 30–31
Working on the Chapter Project
Be True to Your School, 209, 223, 227
Consider the Probabilities, 543, 548, 553
Games People Play, 444, 449, 467, 471
Hit or Miss, 564, 585, 591, 595
Home Page Bound, 7, 25, 41, 51
Oldies But Goodies!, 146, 161, 173, 181
Pack Your Bags, 106, 113, 129, 133
Patterns in Nature, 248, 252, 264, 271

Proceed with Caution, 388, 393, 413, 421
Stay Tuned!, 338, 351, 377
Take Me Out to the Ball Game, 248, 252, 264, 271
Up, Up, and Away, 485, 502, 507, 511
Where the Wild Things Are, 65, 83, 95, 99
Write a problem, 6, 31, 36, 57, 68, 72, 80, 83, 91, 125, 129, 141, 145, 195, 241, 289, 295, 331, 343, 362, 401, 403, 417, 431, 448, 451, 481, 497, 501, 519, 522, 539, 587

x-axis, 92
reflection over, 461
x-coordinate, 92

y-axis, 92
reflection over, 461–462
y-coordinate, 92

Zero pair, 63

INDEX

Number and Operations

$+$	plus or positive
$-$	minus or negative
$a \cdot b$	
$a \times b$	a times b
ab or $a(b)$	
$\div$	divided by
$\pm$	positive or negative
$=$	is equal to
$\neq$	is not equal to
$<$	is less than
$>$	is greater than
$\leq$	is less than or equal to
$\geq$	is greater than or equal to
$\approx$	is approximately equal to
$\%$	percent
$a{:}b$	the ratio of a to b, or $\frac{a}{b}$

Geometry and Measurement

$\cong$	is congruent to
$\sim$	is similar to
$^\circ$	degree(s)
$\overleftrightarrow{AB}$	line AB
$\overline{AB}$	segment AB
$\overrightarrow{AB}$	ray AB
$\llcorner$	right angle
$\perp$	is perpendicular to
$\parallel$	is parallel to
AB	length of $\overline{AB}$, distance between A and B
$\triangle ABC$	triangle ABC
$\angle ABC$	angle ABC
$\angle B$	angle B
$m\angle ABC$	measure of angle ABC
$\odot C$	circle C
$\overset{\frown}{AB}$	arc AB
π	pi $\left(\text{approximately } 3.14159 \text{ or } \frac{22}{7}\right)$
(a, b)	ordered pair with x-coordinate a and y-coordinate b
$\sin A$	sine of angle A
$\cos A$	cosine of angle A
$\tan A$	tangent of angle A

Algebra and Functions

a'	a prime
a^n	a to the nth power
a^{-n}	$\frac{1}{a^n}$ (one over a to the nth power)
$\lvert x \rvert$	absolute value of x
$\sqrt{x}$	principal (positive) square root of x
$f(n)$	function, f of n

Probability and Statistics

$P(A)$	the probability of event A
$n!$	n factorial
$P(n, r)$	permutation of n things taken r at a time
$C(n, r)$	combination of n things taken r at a time